Lecture Notes in Computer Science 9797

Commenced Publication in 1973
Founding and Former Series Editors:
Gerhard Goos, Juris Hartmanis, and Jan van Leeuwen

Editorial Board

More information about this series at http://www.springer.com/series/7407

Thang N. Dinh · My T. Thai (Eds.)

Computing and Combinatorics

22nd International Conference, COCOON 2016
Ho Chi Minh City, Vietnam, August 2–4, 2016
Proceedings

Springer

Editors
Thang N. Dinh
Virginia Commonwealth University
Richmond, VA
USA

My T. Thai
University of Florida
Gainesville, FL
USA

ISSN 0302-9743   ISSN 1611-3349   (electronic)
Lecture Notes in Computer Science
ISBN 978-3-319-42633-4   ISBN 978-3-319-42634-1   (eBook)
DOI 10.1007/978-3-319-42634-1

Library of Congress Control Number: 2016944821

LNCS Sublibrary: SL1 – Theoretical Computer Science and General Issues

Printed on acid-free paper

This Springer imprint is published by Springer Nature
The registered company is Springer International Publishing AG Switzerland

Preface

The 22nd International Computing and Combinatorics Conference (COCOON 2016) was held during August 2–4, 2016, in Ho Chi Minh City, Vietnam. COCOON 2016 provided a forum for researchers working in the area of theoretical computer science and combinatorics.

The technical program of the conference included 50 regular papers selected by the Program Committee from 113 full submissions received in response to the call for papers. All the papers were peer reviewed by at least three Program Committee members or external reviewers. The papers cover various topics, including algorithms and data structures, algorithmic game theory, approximation algorithms and online algorithms, automata, languages, logic, and computability, complexity theory, computational learning theory, cryptography, reliability and security, database theory, computational biology and bioinformatics, computational algebra, geometry, number theory, graph drawing and information visualization, graph theory, communication networks, optimization, and parallel and distributed computing. Some of the papers will be selected for publication in special issues of *Theoretical Computer Science* (TCS) and *Journal of Combinatorial Optimization* (JOCO). It is expected that the journal version of the papers will be in a more complete form.

We would like to thank the Program Committee members and external reviewers for volunteering their time to review conference papers. We would like to extend special thanks to the publication, publicity, and local organization chairs for their hard work in making COCOON 2016 a successful event. Last but not least, we would like to thank all the authors for presenting their works at the conference.

August 2016

Thang N. Dinh
My T. Thai

Organization

Program Chairs

Thang N. Dinh   Virginia Commonwealth University, USA
My T. Thai   University of Florida, USA

Publicity Chairs

Chunyu Ai   University of South Carolina Upstate, USA
Subhankar Mishra   University of Florida, USA

Local Organization Chair

Hien T. Nguyen   Ton Duc Thang University, Vietnam

Program Committee

Eric Allender	Rutgers University, USA
Yossi Azar	Tel Aviv University, Israel
Yixin Cao	Hong Kong Polytechnic University, Hong Kong, SAR China
Xi Chen	Columbia University, USA
Francis Chin	Hang Seng Management College, Hong Kong, SAR China
Bhaskar DasGupta	University of Illinois at Chicago, USA
David Eppstein	University of California, Irvine, USA
Uriel Feige	Weizmann Institute of Science, Israel
Zachary Friggstad	University of Alberta, Canada
Raffaele Giancarlo	University of Palermo, Italy
Mohammadtaghi Hajiaghayi	University of Maryland, USA
Lenwood Heath	Virginia Tech, USA
Pinar Heggernes	University of Bergen, Norway
Xiaodong Hu	Chinese Academy of Sciences, China
Hiro Ito	The University of Electro-Communications, Japan
Valentine Kabanets	Simon Fraser University, Canada
Ming-Yang Kao	Northwestern University, USA
Donghyun Kim	North Carolina Central University, USA
Stavros Kolliopoulos	National and Kapodistrian University of Athens, Greece
Nam Nguyen	Towson University, USA
Huy Nguyen	Toyota Technological Institute at Chicago, USA

Kunsoo Park	Seoul National University, South Korea
Desh Ranjan	Old Dominion, USA
Marc Uetz	University of Twente, The Netherlands
Dorothea Wagner	Karlsruhe Institute of Technology (KIT), Germany
Gerhard Woeginger	Eindhoven University of Technology, The Netherlands
Shengyu Zhang	The Chinese University of Hong Kong, Hong Kong, SAR China
Hong-Sheng Zhou	Virginia Commonwealth University, USA

Additional Reviewers

Antonios Antoniadis	Ragesh Jaiswal	Simona E. Rombo
Daniel Apon	Shaofeng Jiang	Anamitra Roy Choudhury
Yuichi Asahiro	Jun Kawahara	Alan Roytman
Moritz Baum	Walter Kern	Ignaz Rutter
Liu Bei	Thomas Kesselheim	Toshiki Saitoh
Rémy Belmonte	Konstantinos Kollias	Kanthi Sarpatwar
Manuel Bodirsky	Dieter Kratsch	Saeed Seddighin
Niv Buchbinder	Fleszar Krzysztof	Shinnosuke Seki
Valentin Buchhold	Nirman Kumar	Igor Shparlinski
Kevin Buchin	Bundit Laekhanukit	Junggab Son
Dimitris Chatzidimitriou	Elmar Langetepe	Manuel Sorge
Rajesh Chitnis	Mun-Kyu Lee	Ben Strasser
Janka Chlebikova	Zengpeng Li	Takeyuki Tamura
Sherman S.M. Chow	Vahid Liaghat	Qiang Tang
Ilan Cohen	Wei-Kai Lin	Junichi Teruyama
Radu Curticapean	Tian Liu	RN Uma
Konrad Dabrowski	Yao Lu	Filippo Utro
Bireswar Das	Spyridon Maniatis	Adi Vardi
Sina Dehghani	Arnaud Mary	Thomas Veale
Tuyet Duong	Tamara Mchedlidze	Wei Wang
Soheil Ehsani	Nicole Megow	Franziska Wegner
Michael Elberfeld	Matthias Mnich	Daniel Wichs
Hossein Esfandiari	Jérôme Monnot	Marcin Wrochna
Ophir Friedler	Benjamin Moseley	Hadi Yami
Takuro Fukunaga	Wolfgang Mulzer	Jie You
Loukas Georgiadis	Atsuki Nagao	Victor Zamaraev
Konstantinos Georgiou	Benjamin Niedermann	Chihao Zhang
Daniel Goncalves	Kenta Ozeki	Stanislav Zivny
Kasper Green Larsen	Ulrich Pferschy	Uri Zwick
Alexander Grigoriev	Roman Prutkin	Tobias Zündorf
Michael Hamann	Zhenzhong Qi	Erik Jan van Leeuwen
Johann Hurink	Marcel Radermacher	Suzanne van der Ster
Falk Hüffner	Felix Reidl	
Sungjin Im	Mohsen Rezapour	

Contents

Game Theory and Algorithms

Parameterized Complexity and Algorithms

Database and Data Structures

Network and Algorithms

Graph Theory and Algorithms

Logic, Algebra and Automata

Game Theory and Algorithms

Clairvoyant Mechanisms for Online Auctions

Philipp Brandes[1]([✉]), Zengfeng Huang[2], Hsin-Hao Su[3], and Roger Wattenhofer[1]

[1] ETH Zurich, Zürich, Switzerland
{pbrandes,wattenhofer}@ethz.ch
[2] UNSW, Kensington, Australia
zengfeng.huang@unsw.edu.au
[3] MIT, Cambridge, USA
hsinhao@csail.mit.edu

Abstract. In this paper we consider online auctions with buyback; a form of auctions where bidders arrive sequentially and the bidders have to be accepted or rejected immediately. Each bidder has a valuation for being allocated the good and a preemption price. Sold goods can be bought back from the bidders for a preemption price. We allow unbounded valuations and preemption prices independent from each other. We study the clairvoyant model, a model sitting between the traditional offline and online models. In the clairvoyant model, a sequence of all potential customers (their bids and compensations) is known in advance to the seller, but the seller does not know when the sequence stops. In the case of a single good, we present an algorithm for computing the difficulty Δ, the optimal ratio between the clairvoyant mechanism and the pure offline mechanism (which knows when the sequence stops, and can simply sell the good to the customer with the highest bid, without having to pay any compensations). We also present an optimal clairvoyant mechanism if there are multiple goods to be sold. If the number of goods is unbounded, however, we show that the problem in the clairvoyant model becomes $\mathcal{NP}$-hard. Based on our results in the clairvoyant model, we study the Δ-online problem (where the sequence is unknown to the mechanism, but the difficulty Δ of the input sequence is known). We show that there is a tight gap of $\Theta(\Delta^5)$ between the offline and the online model.

1 Introduction

Traditional auctions have a rich theory but only make sense in the presence of at least two bidders. In reality, however, many auctions have a rather low demand, and bidders do not compete concurrently. Instead, bidders appear *online*, one after the other.

A familiar example is booking a seat in an airplane. Prices for a flight fluctuate over time, a known pattern is that seats become more expensive as a flight fills up, because the airline starts to learn that there is demand for the flight. Selling seats in an airplane is not a traditional auction since customers are not bidding against each other. Rather, potential customers check the price well in

© Springer International Publishing Switzerland 2016
T.N. Dinh and M.T. Thai (Eds.): COCOON 2016, LNCS 9797, pp. 3–14, 2016.
DOI: 10.1007/978-3-319-42634-1_1

advance of a flight. If the price is right, they book a seat, sealing the deal with the airline. Airlines generally try to marginally overbook flights, i.e., they sell more tickets than available, assuming that not all customers will actually show up at the gate. Sometimes there are more customers than seats, and the airline must get some customers off the plane. This is usually achieved by having them fly later and giving them some cash as compensation. We believe that such compensations are easily covered by the high premium of late customers.[1]

In this paper we analyze these online auctions. Our bidders come in an online fashion and name their price for a good. The seller can choose to sell the good for that price, or not sell the good (and hope for a better bid to come in later). Bidder and seller also establish a compensation, in case the good is sold to the customer but the deal is later canceled (in the case of a better bidder showing up, worth paying the compensation). These online auctions need two ingredients: First, a good with a price that may fluctuate over time. Second, customers which want to receive the good (or a reservation for the good) quickly. In particular, the time between the arrivals of two customers should generally be larger than the time a customer is willing to wait for the outcome of her bid. In this case online auctions seem to be a better suitable model than traditional auctions. We believe that such online auctions happen often in practice. Booking flights is the running example in this paper, but there are plenty of other examples. Selling ad slots on web pages is a popular one. Since the number of page views is not known beforehand, some sold slots might not be served and thus those slots need to be bought back. More examples are real estate sales, selling network services with quality of service guarantees, or concert tickets.

A simple example will show that online auctions become academically interesting for a worst case analysis only if reasonable compensations are present. Let us assume that a first customer offers a low price but a prohibitively high compensation. If the seller accepts the deal, a next customer offering a much higher price will show up. On the other hand, if the seller does not accept the deal, no other customer will show up. No matter how the seller decides regarding the first customer, the mistake could be devastating.

The starting point for our analysis is what we call the *clairvoyant model*, a hybrid online/offline model. In the clairvoyant model, a sequence of all potential customers (their bids and compensations) is known in advance to the seller, but the seller does not know when the sequence stops, i.e., who the last customer of the sequence is. No matter who the last customer is, the seller wants to do a good job, i.e., the seller wants to sell the good to a customer with a high bid *and* keep compensations that accumulated so far low. It turns out that the clairvoyant model is a stepping stone for a deeper understanding of online auctions, sitting nicely between the pure online and offline models. It introduces a novel technique for analyzing online auctions from a theoretical point of view.

---

[1] In reality, airlines do not implement online auctions in the clean form described in this paper. Airlines do not seem to maximize their profits with this mechanism, probably for psychological reasons. As such, on web pages, flights still can be sold out, instead of just asking for a higher and higher premium for an unexpectedly popular flight.

Our contributions are as follows: After introducing the clairvoyant model, we present an optimal mechanism for it in the case of a single good. The result of that mechanism is a factor Δ worse than a pure offline mechanism (that knows when the sequence stops, and can simply sell the good to the customer with the highest bid, without having to pay any compensations). In other words, the parameter Δ tells us how nasty the compensations are. It directly tells us the difficulty of an input sequence. If compensations are minimal (just return the money to canceled customers), then we have by definition $\Delta = 1$. We also show an optimal clairvoyant mechanism if there are multiple goods to be sold. If the number of goods is unbounded, however, we prove that the clairvoyant model becomes $\mathcal{NP}$-hard. Based on the results in the clairvoyant model, we study the pure online problem (where the sequence is unknown to the mechanism) in a deterministic setting. If Δ is known, we show that there is a tight gap of $\Theta(\Delta^5)$ between the online and the offline model.

2   Related Work

There has been a lot of research of traditional ("offline") auctions, inspired by the seminal papers of Vickrey, Clarke, and Groves ("VCG") [5,11,26]. They introduce the notion of truthfulness, which means that no bidder has an advantage if she is not telling the truth about her valuation. There is a large amount of work on traditional auctions, for an overview see, e.g., Nisan et al. [22].

Online mechanisms have been introduced in [8,18]. In those online mechanisms, the bidders have an arrival and departure time and a valuation for the good. It is assumed that the good expires after a certain period of time, and that a replacement becomes available. In this setting, it was shown that something similar to VCG style second price auctions is still a viable allocation strategy. The initial motivation behind these kind of online auctions is the WiFi at Starbucks [8]. Customers arrive and then depart some time later with each customer having a valuation for the WiFi. Many papers on online mechanisms mainly focus on truthfulness or other incentive compatible solution concepts, e.g., [12,19,23,24]. An overview of online auctions can be found in [22].

Somewhat related to our online auctions are not even auctions, but the secretary problem [20]. In the classic setting one employer interviews n secretaries, with the goal to hire the best secretary. The employer has to decide right after an interview whether to hire or discard a secretary. Unlike our model, previous decisions cannot be recalled. If secretaries are interviewed in random order, it has been shown that the optimal strategy is to first interview n/e secretaries, and then simply hire the first secretary that is better than all previously interviewed secretaries [20]. It has also been shown that, if the input is adversarial (as in our work), the situation is hopeless; the best strategy is to just hire a random secretary, without any interview process [10]. This setting has been adapted to the online auctions in [13]. Instead of secretaries, there are buyers and instead of a job there is a single indivisible good. They present a mechanism that is, if the buyers appear in random order – as in the original problem – $e + o(1)$

competitive for efficiency and $e^2 + o(1)$ competitive for revenue. Since we have the possibility to cancel previous decisions with financial compensations, our model allows more freedom.

The work closest to ours considers online auctions with buyback, introduced independently by Babaioff et al. and Constantin et al. [3,6]. Both limit the preemption price (paid to reacquire the good) to a constant fraction of the valuation v of a bidder and this fraction is independent of the individual bidder. Lower and upper bounds for deterministic and randomized algorithms depending on the fraction of the preemption price are presented in their work. Our work allows arbitrary values for the preemption price (that can depend on the specific customer) and we analyze how to deal with this very heterogeneous set of customers. This kind of auction is not truthful since a buyer can overstate her preemption price and thus gain if her good is bought back [6]. In [2] the goods cannot be allocated to any subset of bidders, but bidders form a matroid, This is extended to an intersection of matroids in [1], while still limiting the buyback factor. The concept of buyback has also been applied to the knapsack problem [3,14,17] where the goods appear in an online fashion and can be removed later on from the knapsack. Buyback is also used in scheduling with eviction [9].

Online algorithms often face two different types of problems: First, they do not know the future, and second, they have to deal with past mistakes. Hartline and Sharp [15,16] formalized the two types of problems. When problems are analyzed in this framework, they are called incremental problems. This approach has been applied to various problems, e.g., to maximum flow, online median, facility location, and clustering [4,7,21,25]. Our setting is different as we can potentially fix past mistakes with compensations. Nevertheless, our clairvoyant analysis is a relative of incremental problems.

3   Model

We consider an online auction. There are r indivisible and identical *goods*. Each bidder b_i is willing to buy exactly one good, and has a *valuation* v_i for being allocated a good. The bidders arrive one after another; whether to allocate a good to a bidder must be decided immediately. Bidders that are not allocated a good cannot be recalled, but bidders that are allocated a good can be recalled. A recalled bidder b_i is willing to return her good if she receives adequate compensation. We call the value *preemption price*, which is paid if the good is *bought back*. The preemption price of bidder b_i is denoted by π_i. In summary, bidder b_i is fully specified by $b_i = (v_i, \pi_i)$. Neither v_i nor π_i are bounded, any value in $\mathbb{R}^+$ is allowed. We assume that the input sequence of bidders $b_1, \ldots, b_n$ is created in advance by an adversary who knows the mechanism that is used to allocate the goods. As described above, if the good of a bidder b_i is bought back, the mechanism has to pay the preemption price. For now, we assume that the mechanism retains the initial valuation v_i of the bidder. We denote this the *retaining* model. In this model we assume that $v_i \leq \pi_i$ for every bidder b_i. We will show later that this is not necessary and in fact use the model when the value is not retained, which is called the *non-retaining* model.

Let us concentrate on the case of a single good ($r = 1$). Let offline(ℓ) denote the highest valuation of the first ℓ bidders, i.e., offline(ℓ) = $\max_{1 \leq i \leq \ell} v_i$. Since the pure offline mechanism knows the whole input sequence and when it stops, it can sell the good just to one single bidder, the bidder with the highest valuation.

As discussed in the introduction, the online mechanism cannot be competitive with the offline model. Essentially, an online mechanism has to deal with two different issues: First, it does not know the future, and second, it needs to offer a solution at all times. We will now introduce the *clairvoyant* model, a model between pure online and offline. The clairvoyant model knows the whole sequence $b_1, \ldots, b_n$ of future potential bidders, but does not know when the sequence stops, i.e., who the last bidder of the sequence is. Because of this, a clairvoyant mechanism must offer a solution at all times.

Both pure online and clairvoyant mechanisms may need to accept more than one bidder (and hence buy the good back). Let S be the set of all bidders that have been accepted during the course of a mechanism and let $[\ell]$ denote the set of the first ℓ bidders, i.e., $\{b_1, \ldots, b_\ell\}$. We define gain(S, ℓ) = $\sum_{b_i \in S \cap [\ell]} (v_i - \pi_i) +$ $\max_{b_i \in S \cap [\ell]} \pi_i$. It is the sum of valuations of bidders in S up to bidder b_ℓ, minus the preemption prices for the bidders whose good were bought back. Since we retain the value of a bidder, we have $v_i \leq \pi_i$ for every bidder b_i and thus the bidder with the highest preemption price is also the last accepted bidder.

Since the mechanism does not know when the input sequence stops and it thus can stop anytime, we evaluate any mechanism in its worst round. Specifically, given S, the *gain competitiveness* is defined to be $\max_{1 \leq \ell \leq n} \frac{\text{offline}(\ell)}{\text{gain}(S,\ell)}$. If we now minimize this over the best mechanism (the set S of accepted bidders), we get the optimal gain competitiveness

$$\Delta = \min_S \max_{1 \leq \ell \leq n} \frac{\text{offline}(\ell)}{\text{gain}(S, \ell)}.$$

This can be interpreted as the difficulty of the input sequence. In other words, our mechanisms are evaluated in their worst round, i.e., the round in which it has the highest competitive ratio compared to the pure offline mechanism. This forces our mechanisms into accepting bidders early, and possibly repeatedly, thus paying preemption prices repeatedly. The task is to design mechanisms that choose a set S and thereby allocate the goods to the bidders minimizing gain competitiveness.

We will clarify the terms defined above by presenting a simple example. Let the input sequence be $(1, 2), (4, 100), (50, 60)$. A pure offline mechanism will accept $b_3 = (50, 60)$ since this is the bidder with the highest valuation. A clairvoyant mechanism must always accept the first bidder since it could also be the last one. Assume that it also accepts the third bidder. We now calculate the gain competitiveness for this set as

$$\max \left\{ \frac{\text{offline}(1)}{v_1} = \frac{1}{1}, \frac{\text{offline}(2)}{v_1} = \frac{4}{1}, \frac{\text{offline}(3)}{v_1 + v_3 - \pi_1} = \frac{50}{1 + 50 - 2} \right\} = 4.$$

Note that this is also optimal since accepting bidder b_2 prevents the mechanism from choosing b_3, hence $\Delta = 4$. This gives us a theoretical insight on the input sequence. No online mechanism could have done better. As explained, the clairvoyant model sits between pure offline and online models. It turns out that it is comparable to both pure models, even though the pure models are not comparable to each other.

4   Auctioning Off a Single Good

We start our analysis by considering the special case of just a single good being sold, i.e., $r = 1$.

4.1   Clairvoyant Mechanism

We now present a mechanism that optimally solves the clairvoyant model, giving us insights into what is possible for an online mechanism.

Theorem 1. *There exists a clairvoyant mechanism that calculates the set of bidders that should be accepted to solve the online auction for one good optimally, i.e., it calculates Δ. If the inputs are integers, its runtime is polynomial; otherwise it is a FPTAS.*

We now formalize and extend the impossibility result from the introduction. Due to space limitations, the proofs have been moved to the full version.

Lemma 1.

(1) The value of Δ depends on the input sequence and is unbounded.
(2) The gain competitiveness of the pure online mechanism is unbounded and independent of Δ.
(3) No randomized online mechanism can achieve bounded gain competitiveness if the number r of items is in $o(n)$, i.e., $r \in o(n)$.

4.2   Bounded Preemption Prices

The impossibility results from the introduction and the previous section exploited that the preemption price could be arbitrarily large. Thus, in the following we restrict the previously arbitrarily large preemption prices to be at most ρ times as large as the valuation, i.e., $\rho \geq \frac{\pi_i}{v_i}$ for all $1 \leq i \leq n$. Intuitively, this can either be seen as a simple, reasonable constraint for the customers. If someone values a seat on an airplane with some value v, then losing this seat should not be arbitrarily larger than v. One could also model this scenario in such a way that every customer also has to buy an insurance whose compensation depends on the premium. If she loses her seat, then the insurance will pay her the preemption price. Now the price of the insurance is closely related to the preemption price. This interpretation also guarantees us that at most a factor of ρ between v_i and π_i for every bidder b_i. The following results resemble closely those in [3,6]. The factor ρ allows us to design a mechanism that is 4ρ gain competitive. It accepts a bidder if her valuation is at least by a factor 2 larger than the preemption price of the bidder that is currently allocated the good.

Theorem 2. *There exists a mechanism that has 4ρ factor gain competitiveness.*

Corollary 1. *If $\rho \geq \frac{\pi_i}{v_i}$ for every bidder b_i, then $\Delta \leq 4\rho$.*

4.3   Online Mechanism with Δ

This raises the question whether restricting the preemption price is the only way to go. We already know that Δ contains valuable information about the input sequence. But does it contain all the necessary information for an online mechanism to be competitive? We now provide the mechanisms with this information and denote them Δ-online mechanisms. These more powerful online mechanisms can achieve a $\mathcal{O}(\Delta^5)$ factor approximation of the clairvoyant mechanisms. Note that this information is not as strong as knowing that the preemption price of every bidder is at most a factor of ρ larger. The clairvoyant mechanism might accept someone whose preemption price is much larger than its valuation. We briefly describe the mechanism. Simply put, this mechanism accepts bidders with a sufficiently small preemption price (and a high enough valuation to pay back the last bidder). Furthermore, it also accepts bidders that have such a high valuation that the clairvoyant mechanism also had to accept it.

We denote the current bidder with $b = (v, \pi)$. We call the last accepted bidder $b^* = (v^*, \pi^*)$. The online mechanism accepts the first bidder for sure, so initially $b^* = (v_1, \pi_1)$. After the first bidder, the current bidder b is accepted for two different reasons: We call bidders *good* if $\pi \leq 2\Delta^2 v$; if a bidder is not good, it is *bad*. The mechanism will accept a good bidder if its valuation $v > 2\pi^*$. We call bidders *crucial* if $v > 2\Delta v^{**}$, where $v^{**} \geq v^*$ is the largest valuation seen so far. The mechanism will accept a crucial bidder if its valuation $v > \pi^*/(1 - \frac{1}{\Delta^2})$. The pseudocode is shown in Algorithm 1.

In this section a Δ-online mechanism is presented that is $\mathcal{O}(\Delta^5)$ competitive. But first, we need some additional notation.

Theorem 3. *Given the value of Δ, there exists a mechanism that has gain competitiveness $O(\Delta^5)$ compared to the offline solution.*

Algorithm 1. A Δ-online mechanism

---

accept the first bidder and set $(v^*, \pi^*) = (v_1, \pi_1)$ and $v^{**} = v_1$;
while *there is a new bidder b_i* **do**
  if $\pi_i \leq 2\Delta^2 v_i$ *and* $v_i > 2\pi^*$ **then**
     buy good back and give it to bidder b_i;
     $\pi^* \leftarrow \pi_i$ and $v^* \leftarrow v_i$;
  end
  else if $v_i \geq 2\Delta v^{**}$ *and* $v_i > \pi^*/(1 - \frac{1}{\Delta^2})$ **then**
     buy good back and give it to bidder b_i;
     $\pi^* \leftarrow \pi_i$ and $v^* \leftarrow v_i$;
  end
  $v^{**} = \max\{v^{**}, v_i\}$;
end

Proof. Notice that the clairvoyant mechanism will accept every crucial bidder. Let $\bar{b}_1 = (\bar{v}_1, \bar{\pi}_1), \bar{b}_2 = (\bar{v}_2, \bar{\pi}_2), \ldots$ be the subsequence of bidders who are crucial, and let $\bar{b}_0 = b_1$ be the very first bidder, who will also be accepted by the clairvoyant mechanism. We will prove the theorem by induction over the crucial bidders. Our induction hypothesis is that before $\bar{b}_i$ came, the gain competitiveness of the mechanism is at most $8\Delta^5$, we then prove that before $\bar{b}_{i+1}$ came, the gain competitiveness remains $8\Delta^5$. Before we can continue our proof, we need two helper lemmas.

As before, let $b^* = (v^*, \pi^*)$ be the last bidder our mechanism has accepted.

Lemma 2. *If the clairvoyant mechanism accepts a bad bidder $\hat{b} = (\hat{v}, \hat{\pi})$, then the next bidder it will accept must be the first crucial bidder that comes afterward.*

Proof. Let $\bar{b} = (\bar{v}, \bar{\pi})$ be the next bidder clairvoyant mechanism accepts after $\hat{b} = (\hat{v}, \hat{\pi})$, and v^{**} be the maximum valuation of all bidders before $\bar{b}$. Then we must have $\bar{v} > \hat{\pi} > 2\Delta^2\hat{v}$. Note that $v^{**} \leq \Delta\hat{v}$, since otherwise the gain competitiveness of the clairvoyant mechanism will be larger than Δ, and thus we have $\bar{v} > 2\Delta v^{**}$, and therefore $\bar{b}$ must be crucial. As clairvoyant mechanism needs to accept all crucial bidders, $\bar{b}$ must be the first crucial bidder after $(\hat{v}, \hat{\pi})$.

Lemma 3. *If b^* is bad, then the next bidder our mechanism accepts must be the first crucial bidder $\bar{b} = (\bar{v}, \bar{\pi})$ that comes afterward. Furthermore, the gain after accepting $\bar{b}$ is at least $\frac{1}{\Delta^2}\bar{v}$.*

Proof. Let $\bar{b}$ be the next crucial bidder after b^*. If b^* is bad, then b^* must be crucial since our mechanism only accepts bad bidders that are crucial. So the clairvoyant mechanism will also accept b^* since it accepts every crucial bidder. By Lemma 2, the next bidder after b^* the clairvoyant mechanism will accept is $\bar{b}$. So $\bar{v} - \pi^* \geq \frac{1}{\Delta}\bar{v}$, since otherwise the gain of clairvoyant mechanism will be less than $\frac{1}{\Delta}\bar{v}$. This implies that $\bar{v} \geq \pi^* + \frac{1}{\Delta}\bar{v} > \pi^* + \frac{1}{\Delta^2}\bar{v}$ and therefore $\bar{v} > \pi^*/(1 - \frac{1}{\Delta^2})$. Thus, our mechanism will also accept $\bar{b}$. Let v^{**} be the maximum valuation before $\bar{b}$, then $v^{**} \leq \Delta v^*$. So between b^* and $\bar{b}$, our mechanism will not accept any bidder.

By our assumption, b^* is last bidder our mechanism accepts before $\bar{b}_i$, so if b^* is bad, $\bar{b}_i$ must be the first crucial bidder after b^*, and our mechanism will accept $\bar{b}_i$. The gain after accepting $\bar{b}_i$ is at least $\frac{1}{\Delta^2}\bar{v}_i$, and the gain competitiveness is at most Δ^2.

If our mechanism does not accept $\bar{b}_i$, then $\bar{v}_i < \pi^* + \frac{1}{\Delta^2}\bar{v}_i$. Moreover, by Lemma 3, if our mechanism does not accept $\bar{b}_i$, then b^* is good, and thus $\bar{v}_i < \pi^* + \frac{1}{\Delta^2}\bar{v}_i \leq 2\Delta^2 v^* + \frac{1}{\Delta^2}\bar{v}_i$. Thus, we have $\bar{v}_i - \frac{1}{\Delta^2}\bar{v}_i < 2\Delta^2 v^*$ or equivalently $\bar{v}_i < 2\Delta^2 v^*/(1 - \frac{1}{\Delta^2}) \leq 3\Delta^2 v^*$ (wlog assuming $\Delta > 2$, otherwise we can achieve constant factor competitiveness by treating Δ as two in the mechanism). This implies that the current gain competitiveness is at most $6\Delta^2$ using that b^* is a good bidder.

Based on the above analysis and a simple induction we conclude that if our mechanism accepts a bad bidder $b^* = (v^*, \pi^*)$, the gain is at least $\frac{1}{\Delta^2}v^*$ at this moment. It is also easy to see, if b^* is good, then the gain is at least $v^*/2$ (analogue to the proof of Theorem 2).

We now combine the previous observations. Let $c = \{\bar{b}_i, c_1, c_2, \cdots c_t\}$ be the sequence of bidders that arrive between $\bar{b}_i$ and $\bar{b}_{i+1}$ (excluding $\bar{b}_{i+1}$). Let $b' = (v', \pi')$ be the last bidder the clairvoyant mechanism accepts. If the clairvoyant mechanism only accepts good bidders in c, then the gain competitiveness between our online mechanism and the clairvoyant mechanism is at most $4\Delta^4$, because $v' \leq 2\pi^* \leq 4\Delta^2 v^*$ holds at all time (otherwise, our online mechanism will accept (v', π')) and the gain of our online mechanism is at least v^*/Δ^2, which implies the gain competitiveness is at most $4\Delta^4$.

Thus, we only need to consider the case when clairvoyant mechanism accepts at least one bad bidder in c (possibly $\bar{b}_i$). By the above analysis, we know that if the clairvoyant mechanism accepts some bad bidder $\hat{c} = (\hat{v}, \hat{\pi})$, then the next bidder it accepts is $\bar{b}_{i+1}$. Furthermore, $v^{**} \leq \Delta\hat{v}$, where v^{**} is maximum valuation before $\bar{b}_{i+1}$.

Before accepting $\hat{c} = (\hat{v}, \hat{\pi})$ the clairvoyant mechanism only accepts good bidders. Now suppose we are at the time right before $\hat{c}$ comes. Suppose, at this time, our online mechanism accepts $b^* = (v^*, \pi^*)$ and clairvoyant mechanism accepts (v', π'). We first consider the case when b^* is good. Then we have $v' \leq 2\pi^* \leq 4\Delta^2 v^*$. Let m be the maximum valuation before $\hat{c}$. We have $m \leq \Delta v'$, and $\hat{b}v \leq 2\Delta m$ (otherwise $\bar{b}_{i+1} = \hat{c}$). Remember that v^{**} is the maximum valuation before $\bar{b}_{i+1}$. Hence, $v^{**} \leq \Delta\hat{v} \leq 2\Delta^2 m \leq 2\Delta^3 v' \leq 8\Delta^5 v^*$.

We now conclude this proof with a simple case distinction. If $b^* = (v^*, \pi^*)$ is good, then the gain competitiveness of our mechanism will never be worse than $8\Delta^5$ after it accepts $b^* = (v^*, \pi^*)$, as the gain is at least $v^*/2$. Moreover, before $\hat{b}$ (with $\hat{v}$) came, both our mechanism and clairvoyant mechanism only accept good bidders, so the gain competitiveness of our mechanism is at most $8\Delta^5$ before this time. So the gain competitiveness of our mechanism is at most $8\Delta^5$ before $\bar{b}_{i+1}$ comes.

On the other hand, if $b^* = (v^*, \pi^*)$ is bad, which implies that $\hat{c} = b^* = \bar{b}_i$, and that the clairvoyant mechanism does not accept any bidder before $\bar{b}_{i+1}$. This implies that $v^{**} \leq \Delta v^*$, and the gain competitiveness of our mechanism in this period is at most Δ^3, since the gain is at least $\frac{1}{\Delta^2}v^*$.

The bound from Theorem 3 is tight. We proceed by showing the matching lower bound for any deterministic mechanism.

Theorem 4. *Any deterministic Δ-online mechanism has gain competitiveness of $\Omega(\Delta^5)$ compared to the offline solution.*

Proof. For any $d > 0$, we will present a sequence of bidders, for which the gain competitiveness between the offline mechanism and the clairvoyant mechanism is at most $2d$, but for any online mechanism, the gain competitiveness is at least $4d^5$. Given Δ, we can set $d = \Delta/2$. Thus, any online mechanism is at least $\Omega(\Delta^5)$ worse than the offline mechanism. The input sequence is depicted in Fig. 1.

The input sequence starts with bidder b_1 with $(v_1, \pi_1) = (1, 1)$, then the adversary inserts a sequence of bidders $b_{i+1} = (v_{i+1}, \pi_{i+1})$, for $i = 1, 2 \ldots,$

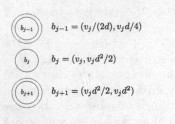

$$b_{j-1} \quad b_{j-1} = (v_j/(2d), v_j d/4)$$
$$b_j \quad b_j = (v_j, v_j d^2/2)$$
$$b_{j+1} \quad b_{j+1} = (v_j d^2/2, v_j d^2)$$

$$b_{j-1} \quad b_{j-1} = (v_j/(2d), v_j d/4)$$
$$b_j \quad b_j = (v_j, v_j d^2/2)$$
$$b_{j+1} \quad b_{j+1} = (v_j d^2/2, v_j d^2)$$
$$b_{j+2} \quad b_{j+2} = (2v_j d^3, v_j d^{1000})$$
$$b_{j+3} \quad b_{j+3} = (v_j d^{999}, v_j d^{1337})$$

(a) The bidders b_{j-1} and b_{j+1} are accepted by the clairvoyant mechanism. The bidders b_j and b_{j+1} are accepted by the online mechanism resulting in negative gain.

(b) The bidders b_{j-1} and b_{j+1} are accepted by the clairvoyant mechanism. The bidders b_j and b_{j+2} are accepted by the online mechanism. Thus, a bidder $b_{j+3} = (v_j d^{999}, v_j d^{1337})$ would inevitably lead to a gain competitiveness of $\omega(\Delta^5)$.

$$b_{j-1} \quad b_{j-1} = (v_j/(2d), v_j d/4)$$
$$b_j \quad b_j = (v_j, v_j d^2/2)$$
$$b_{j+1} \quad b_{j+1} = (v_j d^2/2, v_j d^2)$$
$$b_{j+2} \quad b_{j+2} = (2v_j d^3, v_j d^{1000})$$
$$b_{j+3} \quad b_{j+3} = (2v_j d^4, v_j d^{1337})$$
$$b_{j+4} \quad b_{j+4} = (v_j d^{1336}, v_j d^{2000})$$

$$b_{j-1} \quad b_{j-1} = (v_j/(2d), v_j d/4)$$
$$b_j \quad b_j = (v_j, v_j d^2/2)$$
$$b_{j+1} \quad b_{j+1} = (v_j d^2/2, v_j d^2)$$
$$b_{j+2} \quad b_{j+2} = (2v_j d^3, v_j d^{1000})$$
$$b_{j+3} \quad b_{j+3} = (2v_j d^4, v_j d^{1337})$$
$$b_{j+4} \quad b_{j+4} = (4v_j d^5, v_j d^{2000})$$
$$b_{j+5} \quad b_{j+5} = (4v_j d^{1999}, v_j d^{2000})$$

(c) The bidders b_{j-1}, b_{j+1}, and b_{j+2} are accepted by the clairvoyant mechanism. The bidders b_j and b_{j+3} are accepted by the online mechanism. Thus, a bidder $b_{j+4} = (v_j d^{1336}, v_j d^{2000})$ would inevitably lead to a gain competitiveness of $\omega(\Delta^5)$.

(d) The bidders b_{j-1}, b_{j+1}, and b_{j+3} are accepted by the clairvoyant mechanism. The bidders b_j and b_{j+4} are accepted by the online mechanism. Thus, a bidder $b_{j+5} = (v_j d^{1999}, v_j d^{2000})$ would inevitably lead to a gain competitiveness of $\omega(\Delta^5)$.

Fig. 1. The bidders accepted by the clairvoyant mechanism are marked with (thinly) dashed lines. The online mechanism accepts by definition b_j. If the online mechanism accepts the bottom left bidder, the bidder on the bottom right appears; resulting in a $\omega(\Delta^5)$ gain competitiveness.

where $(v_{i+1}, \pi_{i+1}) = (2d^i, d^{i+2})$. Let b_j be the first bidder in this sequence that the online mechanism accepts. Notice that the online mechanism has to accept one, since otherwise the gain competitiveness is infinity. The clairvoyant mechanism accepts bidder b_{j-1}, but not b_j. The adversary then sets $(v_{j+1}, \pi_{j+1}) = ((d^2/2)v_j, d^2 v_j)$, so that the online mechanism cannot accept this bidder because the new gain would be at most $v_j d^2/2 - v_j d^2/2 - \pi_1 < 0$. The clairvoyant mechanism accepts bidder b_{j+1} to maintain gain competitiveness $\mathcal{O}(\Delta)$.

The next bidder b_{j+2} that comes has $(v_{j+2}, \pi_{j+2}) = (d^3 v_j, d^{1000} v_j)$, so the online mechanism cannot accept this bidder either, since otherwise the adversary can make the next bidder have a valuation of $d^{999} v_j$, which makes the gain competitiveness much larger than $4d^5$. The clairvoyant mechanism does not accept bidder b_{j+2} and still maintains gain competitiveness $\mathcal{O}(\Delta)$.

Bidder b_{j+3} is then $(v_{j+3}, \pi_{j+3}) = (2d^4 v_j, d^{1337} v_j)$. For the same reason, the online mechanism cannot accept this one. The clairvoyant mechanism accepts bidder b_{j+3} to maintain gain competitiveness $\mathcal{O}(\Delta)$. If the online mechanism accepts this bidder, then the clairvoyant mechanism accepts bidder b_{j+2}, but not bidder b_{j+3} (see Fig. 1).

Bidder b_{j+4} is $(v_{j+4}, \pi_{j+4}) = (4d^5 v_j, d^{2000} v_j)$, and again the online mechanism cannot accept this one. The clairvoyant mechanism does not accepts bidder b_{j+4} and still maintains gain competitiveness $\mathcal{O}(\Delta)$.

At this point, the online mechanism accepted (v_j, π_j), and the gain competitiveness is at most $\frac{4d^5 v_j}{v_j} = 4d^5$. Thus, the claim follows.

5   Auctions with Several Goods

In this section we consider auctions with r goods. The pure offline mechanism chooses the best r bidders and never has to pay a preemption price. If r is constant, then we show the following constructive result.

Theorem 5. *Checking whether there is a solution with gain competitiveness of δ in an online auction is r goods can be computed in $\mathcal{O}(n^{r+1})$.*

Similar to the problem of checking whether there is a k-clique in a graph, the general version of this problem is $\mathcal{NP}$-hard.

Theorem 6. *Checking whether there is a solution with gain competitiveness of δ in an online auction is $\mathcal{NP}$-hard.*

References

1. Varadaraja, A.B.: Buyback problem - approximate matroid intersection with cancellation costs. In: Aceto, L., Henzinger, M., Sgall, J. (eds.) ICALP 2011, Part I. LNCS, vol. 6755, pp. 379–390. Springer, Heidelberg (2011)

2. Varadaraja, A.B., Kleinberg, R.: Randomized online algorithms for the buyback problem. In: Leonardi, S. (ed.) WINE 2009. LNCS, vol. 5929, pp. 529–536. Springer, Heidelberg (2009)
3. Babaioff, M., Hartline, J.D., Kleinberg, R.: Selling ad campaigns: online algorithms with cancellations. In: EC, pp. 61–70 (2009)
4. Chrobak, M., Kenyon, C., Noga, J., Young, N.E.: Incremental medians via online bidding. Algorithmica 50, 455–478 (2008)
5. Clarke, E.H.: Multipart pricing of public goods. Public Choice 11(1), 17–33 (1971)
6. Constantin, F., Feldman, J., Muthukrishnan, S., Pál, M.: An online mechanism for ad slot reservations with cancellations. In: SODA, pp. 1265–1274 (2009)
7. Dasgupta, S., Long, P.M.: Performance guarantees for hierarchical clustering. J. Comput. Syst. Sci. 70(4), 555–569 (2005)
8. Friedman, E.J., Parkes, D.C.: Pricing WiFi at Starbucks: issues in online mechanism design. In: EC, pp. 240–241 (2003)
9. Fung, S.P.Y.: Online scheduling of unit length jobs with commitment and penalties. In: Diaz, J., Lanese, I., Sangiorgi, D. (eds.) TCS 2014. LNCS, vol. 8705, pp. 54–65. Springer, Heidelberg (2014)
10. Gilbert, J., Mosteller, F.: Recognizing the maximum of a sequence. J. Am. Stat. Assoc. 61(313), 35–73 (1966)
11. Groves, T.: Incentives in teams. Econom.: J. Econom. Soc. 41, 617–631 (1973)
12. Hajiaghayi, M.T.: Online auctions with re-usable goods. In: EC (2005)
13. Hajiaghayi, M.T., Kleinberg, R.D., Parkes, D.C.: Adaptive limited-supply online auctions. In: EC, pp. 71–80 (2004)
14. Han, X., Kawase, Y., Makino, K.: Online knapsack problem with removal cost. In: Gudmundsson, J., Mestre, J., Viglas, T. (eds.) COCOON 2012. LNCS, vol. 7434, pp. 61–73. Springer, Heidelberg (2012)
15. Hartline, J., Sharp, A.: An incremental model for combinatorial maximization problems. In: Àlvarez, C., Serna, M. (eds.) WEA 2006. LNCS, vol. 4007, pp. 36–48. Springer, Heidelberg (2006)
16. Hartline, J., Sharp, A.: Incremental flow. Networks 50(1), 77–85 (2007)
17. Kawase, Y., Han, X., Makino, K.: Unit cost buyback problem. In: Cai, L., Cheng, S.-W., Lam, T.-W. (eds.) Algorithms and Computation. LNCS, vol. 8283, pp. 435–445. Springer, Heidelberg (2013)
18. Lavi, R., Nisan, N.: Competitive analysis of incentive compatible on-line auctions. In: EC, pp. 233–241 (2000)
19. Lavi, R., Nisan, N.: Online ascending auctions for gradually expiring items. In: SODA, pp. 1146–1155 (2005)
20. Lindley, D.V.: Dynamic programming and decision theory. j-APPL-STAT 10(1), 39–51 (1961)
21. Mettu, R.R., Plaxton, C.G.: The online median problem. In: FOCS (2000)
22. Nisan, N., Roughgarden, T., Tardos, E., Vazirani, V.V.: Algorithmic Game Theory. Cambridge University Press, New York (2007)
23. Parkes, D.C., Singh, S.P.: An MDP-based approach to online mechanism design. In: NIPS (2003)
24. Parkes, D.C., Singh, S.P., Yanovsky, D.: Approximately efficient online mechanism design. In: NIPS (2004)
25. Plaxton, C.G.: Approximation algorithms for hierarchical location problems. In: STOC, pp. 40–49 (2003)
26. Vickrey, W.: Counterspeculation, auctions, and competitive sealed tenders. J. Financ. 16(1), 8–37 (1961)

Truthfulness for the Sum of Weighted Completion Times

Eric Angel[1], Evripidis Bampis[2(✉)], Fanny Pascual[2], and Nicolas Thibault[3]

[1] IBISC, Université d'Évry Val d'Essonne, Evry, France
[2] Sorbonne Universités, UPMC Univ Paris 06, CNRS,
LIP6 UMR 7606, Paris, France
evripidis.bampis@lip6.fr
[3] CRED, Université Panthéon-Assas, Paris 2, Paris, France

Abstract. We consider the problem of designing truthful mechanisms for scheduling selfish tasks on a single machine or on a set of m parallel machines. The objective of every selfish task is the minimization of its completion time while the aim of the mechanism is the minimization of the sum of weighted completion times. For the model *without payments*, we prove that there is no $(2 - \epsilon)$-approximate deterministic truthful algorithm and no $(\frac{3}{2} - \epsilon)$-approximate randomized truthful algorithm when the tasks' lengths are private data. When both the lengths and the weights are private data, we show that it is not possible to get an α-approximate deterministic truthful algorithm for any $\alpha > 1$. In order to overcome these negative results we introduce a new concept that we call *preventive preemption*. Using this concept, we are able to propose a simple optimal truthful algorithm with no payments for the single-machine problem when the lengths of the tasks are private. For multiple machines, we present an optimal truthful algorithm for the unweighted case. For the weighted-multiple-machines case, we propose a truthful randomized algorithm which is $\frac{3}{2}$-approximate in expectation based on preventive preemption. For the model *with payments*, we prove that there is no optimal truthful algorithm even when only the lengths of the tasks are private data. Then, we propose an optimal truthful mechanism using preventive preemption and appropriately chosen payments.

1 Introduction

A lot of attention has been devoted to scheduling problems in the literature of algorithmic game theory starting from the seminal paper of Koutsoupias and Papadimitriou [18]. Most of these papers consider that the social welfare is expressed as the makespan of the obtained schedule [2–7,9,18,19]. However, in environments where jobs are owned by independent and competing agents for the same resource(s), it is more natural to measure the social welfare using another classical measure of performance, the *average* (weighted) completion time of the tasks [21]. A few papers consider this objective [1,11,12,15], but not in the context of truthfulness (they focus on coordination mechanisms and the price of anarchy). Given the interest of the algorithmic-game-theory community

© Springer International Publishing Switzerland 2016
T.N. Dinh and M.T. Thai (Eds.): COCOON 2016, LNCS 9797, pp. 15–26, 2016.
DOI: 10.1007/978-3-319-42634-1_2

to mechanism design aspects of scheduling problems, it is a natural question to know what is the difficulty of conceiving a truthful mechanism when the social welfare is the weighted completion time of the tasks. In some applications, for ethical or practical reasons, pricing is undesirable and so it is important to conceive mechanisms without payments [8,16]. In other applications however this is not the case. Hence we consider both cases in the sequel. We focus on the following problem: we are given a set of tasks where each task is owned by a selfish agent who is the only one to know the length and/or the weight of his task. The tasks have to be executed on a single-machine or on a set of identical machines. The valuation of each agent/task is the opposite of his completion time. The weight of a task models the importance of the task for the system (and not the agent) and in that case it is more natural to consider that the valuation of the agent is just the completion time of his task[1]. We study this problem both with payments and without payments. When we use payments, the objective of each agent is the maximization of his utility which is defined as the difference between his valuation and his payment. When payments are not allowed, the objective of each agent is the minimization of his (weighted) completion time. Agents may lie concerning their length and/or weight if by doing so, they are able to increase their utility. Our aim is to find a truthful mechanism that minimizes the weighted sum of completion times.

Our contribution. In the first part of the paper, we study the model *without payments*. When the lengths of the tasks are private data, we prove that there is no $(2 - \epsilon)$-approximate deterministic truthful algorithm even in the case of a single machine where the weights of all the tasks are unitary. We also show that there is no $(\frac{3}{2} - \epsilon)$-approximate randomized truthful algorithm for the same environment. When both the lengths and the weights are private data, then we show that it is not possible to get an α-approximate deterministic truthful algorithm for any $\alpha > 1$. In order to overcome these negative results we introduce a new concept that we call *preventive preemption*. The intuitive idea behind preventive preemption is simple: whenever a task bids a length smaller than its real length, the scheduler will preempt it at the end of the declared processing time and he will resume it later. Think for instance a planning of a meeting room. Once the schedule of meetings is done, then every meeting has to finish or be interrupted at the planned time. An interrupted meeting could continue only after all other meetings are finished. Notice that as our mechanism is proved to be truthful no task will be interrupted during the constructed schedule. This is in the same vein as the approach used recently by Fotakis et al. [13] where *selective verification* is used as a threat in order to construct a truthful mechanism. Using preventive preemption as a threat, we are able to propose a simple optimal truthful algorithm with no payments for the single-machine problem where the lengths of the tasks are private and the weights are public. For multiple machines, we are able to prove that this approach gives an optimal truthful algorithm for

---

[1] Notice however that our results can be generalized to the case where the valuation of the tasks is their weighted completion time.

the unweighted case. For the case of multiple machines with weights, given that the problem is NP-hard even if all data are public, we turn our attention to the development of approximate truthful mechanisms. We propose a truthful randomized algorithm which is $\frac{3}{2}$-approximate in expectation based on preventive preemption. We also show that the natural WSPT algorithm of Smith [21] is not truthful. In the second part of the paper, we consider the model *with payments*. For the single-machine case, given that the optimal solution can be computed in polynomial time and the social welfare is utilitarian, one may think that it is sufficient to apply the well known Vickrey-Clarke-Groves (VCG) mechanism [10,14,22]. However, in what follows we prove that this is not true even when only the lengths of the tasks are private data. Then, we propose an optimal truthful mechanism for the single-machine case using preventive preemption and appropriately chosen payments. Our results are summarized in Table 1.

Table 1. Summary of the results presented in this paper. *TA* means "truthful algorithm", *det* means "deterministic" and *rand* means "randomized". The number before TA is the approximation ratio. For example, the sentence "$\nexists$ det $(2 - \varepsilon)$ TA (thm 1)" in the first cell means that Theorem 1 shows that there does not exist any deterministic truthful algorithm which has an approximation ratio of $2 - \varepsilon$ (when payment and preemption are not allowed, and when the lengths of the tasks are private). Unless otherwise specified, the results hold for any number of machines.

	Without preemption	With preventive preemption
Without payment	*Private lengths:* • $\nexists$ det $(2 - \varepsilon)$ TA (thm 1) • $\nexists$ rand $(1.5 - \varepsilon)$ TA (thm 2)	*Private lengths:* • $m = 1$: $\exists$ optimal det TA (thm 4) • $m \geq 2$, identical w: $\exists$ optimal det TA (thm 5) • $m \geq 2$: $\exists$ rand 1.5 TA (thm 6)
	Private lengths and weights: • $\nexists$ det α TA, for all α (thm 3)	*Private lengths and weights:* • $\nexists$ det $(2 - \varepsilon)$ TA (thm 8)
With payment	*Private lengths:* • $\nexists$ optimal TA (thm 7)	*Private lengths and weights:* • $m = 1$: $\exists$ optimal det TA (thm 9) • $m \geq 2$: $\exists$ rand 1.5 TA (cor 2)

1.1   Formal Definition of the Problem

We consider n agents, $N = \{1, 2, \cdots, n\}$, and a single machine or a set of m parallel identical machines. Each agent i is the owner of a single task and he is the only one to know the private data of his task. The private data of a task can be either its length $t_i > 0$ or both its length $t_i > 0$ and its weight $w_i > 0$. When both the length and the weight of a task are private, we call these data (t_i, w_i), the *agent's true data* or the *agent's type* (if only the length of the task is private, then the agent's type is just t_i). Everything else is public knowledge. From now on in this section, we assume for simplicity that both the length and the weight of the tasks are private data. Each agent will report a pair (b_i, w_i^b) to the mechanism that we call the *agent's bid*. By B, we denote the set of all bids, i.e. $B = \{(b_1, w_1^b), \ldots, (b_n, w_n^b)\}$. We adopt an extension of the *strong model of execution* [4] where, once task i starts to be executed, it is executed during t_i units of time, independently of the value of his bid b_i (i.e. even if $b_i \neq t_i$). In the model of [4], the bid value b_i should always be larger than or equal to t_i while here, b_i may get any positive value ($b_i < t_i$ or $b_i \geq t_i$). By C_i, we denote the completion time of task i.

For the model *with payments*, a mechanism is a pair $\mathcal{M} = (A, P)$, where A is an algorithm that finds an output $o(B)$ and P is a payment function: $P(o(B), B) = (p_1, p_2, \ldots, p_n)$. The output $o(B)$ computed by A is a function of the bids, B, of the agents, while the payment is a function of the output $o(B)$ and of the agents' bids B. This means that, contrary to the framework with *verification* introduced by Nisan and Ronen for scheduling problems [19], the payments have to be computed without knowing the true types of the tasks. Let us now define the output of A. Since the true types of the tasks are not known by the mechanism, A is not able to produce a feasible schedule in which the completion time of every task is known in advance. In the case where the preemption of the tasks is not allowed, $o(B)$ is defined as the order in which the tasks will be executed on each machine along with the lengths of the idle-periods that precede the tasks, if such idle periods exist. More formally, in the single-machine case when the preemption of the tasks (the possibility of interrupting and resuming the execution of the task later) is not allowed, we define the output $o(B)$ of algorithm A as a sequence of n pairs (I_i, i) where i is a task and I_i is the length of the idle-period just before task i. Notice that when no idle-periods exist between the tasks, all I_i's will be equal to 0 and we will simply denote the output by a sequence of n tasks. In the case where the preemption of the tasks is allowed, the output $o(B)$ will be defined in a similar way, the only difference being that more than one time-intervals may represent a task, one time-interval for each piece of the preempted task. For multiple machines, the above definitions generalize in the natural way. The objective of the mechanism is to determine a schedule of the tasks minimizing the sum of weighted completion times, or equivalently maximizing the social welfare which is defined as $-\sum_{1 \leq i \leq n} w_i C_i$. For every task i, we define S_i as the set of tasks scheduled before i on the same machine in the output $o(B)$, and T_i as the set of real lengths of the tasks of S_i (i.e. $T_i = \{t_j : j \in S_i\}$). The completion time of task i is $C_i = \sum_{j \in S_i} (I_j + t_j) + I_i + t_i$ and the utility of task i is $u_i(t_i, o(B), B, T_i) = -C_i(t_i, o(B), B, T_i) - p_i(o(B), B)$, where $p_i(o(B), B)$ is the payment, or in other words the amount that i must pay. It is important here to notice that the payments are computed before the real execution of the tasks.

For the model *without payments*, a mechanism for this problem is an algorithm A that determines an output $o(B)$.

In both models, every task/agent i is considered as selfish: the strategy of agent i is to declare a bid (b_i, w_i^b) in order to maximize his utility u_i. Our aim is to propose a *truthful mechanism*, i.e. a mechanism that gives incentive to the agents/tasks to declare their true types. We say that a mechanism is *truthful* if and only if for every i, $1 \leq i \leq n$, and for every bid (b_j, w_j^b), $j \neq i$, the utility u_i of task i reaches its maximum when i bids its true data, i.e. $(b_i, w_i^b) = (t_i, w_i)$.

In other words, a mechanism is truthful if truth-telling is the best strategy for a player i regardless of the strategies adopted by the other players.

2   No Payments

In this section, we consider the problem of designing a truthful mechanism without payments. We start by proving some negative results for truthful deterministic or randomized algorithms. Then, we introduce the notion of preventive preemption, and we show that by using it we are able to design optimal or approximate truthful mechanisms.

2.1   Negative Results: Private Lengths

We first consider deterministic algorithms.

Theorem 1. Let $\varepsilon > 0$. There is no truthful deterministic $(2 - \varepsilon)$-approximate algorithm, even if all the tasks have the same weights.

Proof. Let $\mathcal{A}$ be a deterministic algorithm which is α-approximate, with $\alpha < 2$. Let us show that $\mathcal{A}$ is not a truthful algorithm.

Let us consider a first instance I_1: a single machine and two tasks T_1 and T_2 of lengths M and M^2 respectively (with $M > 1$). Both tasks have the same weight (in the sequel we will thus consider the criteria $\sum C_i$, which is equivalent to $\sum w_i C_i$ in this case). In an optimal schedule, T_1 is executed at time 0 and T_2 starts when T_1 has been executed, at time M. The cost of such a schedule is $\sum_{i \in \{1,2\}} C_i = M + (M + M^2) = M^2 + 2M$. In such a schedule task T_2 starts at time M.

Let S be a schedule of I_1 in which task T_2 starts *before time M*. In such a schedule task T_1 cannot be completed before the start of T_2. The cost of S is thus larger than or equal to $M^2 + (M^2 + M) = 2M^2 + M$ (in the best case there is no idle time: task T_2 is scheduled at time 0 and task T_1 starts as soon as T_2 is completed, i.e. at time M^2). The ratio between the cost of S and the optimal cost is larger than or equal to $\frac{2M^2+M}{M^2+2M} = \frac{2M+1}{M+2}$, which tends towards 2 when M tends towards the infinity. Since $\mathcal{A}$ is an α-approximate algorithm, with $\alpha < 2$, $\mathcal{A}$ cannot return schedule S. Therefore, in the schedule returned by $\mathcal{A}$ on instance I_1, T_2 starts at the soonest at time M.

Consider now a second instance, I_2: a single machine and two tasks T_1 and T_3 of lengths M and 1 respectively. Both tasks have the same weight. In an optimal schedule T_3 is executed at time 0 and T_1 starts when T_3 has been executed, at time 1. The cost of such a schedule is $1 + (1 + M) = M + 2$.

Let S be a schedule of I_2 in which task T_3 *does not start before time M*. The cost of S is thus larger than or equal to $M + (M + 1) = 2M + 1$ (in the best case task T_1 is scheduled at time 0 and task T_3 starts as soon as T_1 is completed, i.e. at time M). The ratio between the cost of S and the optimal cost is larger than or equal to $\frac{2M+1}{M+2}$, which tends towards 2 when M tends towards the infinity. Since $\mathcal{A}$ is an α-approximate algorithm, with $\alpha < 2$, $\mathcal{A}$ cannot return schedule S. Therefore, in the schedule returned by $\mathcal{A}$ on instance I_2, T_3 starts before time M.

Let us now consider the following situation: task T_1 bids a length M and task T_2 has a true length of M^2. Given the values bid by T_1, if T_2 bid its true value, then the instance corresponds to instance I_1. As seen above, in the schedule returned by $\mathcal{A}$ on instance I_1, T_2 starts *at the soonest at time* M.

Assume that task T_2 lies and bids a length of 1 instead of M^2. The input of the algorithm is now two tasks of length M and 1: it is instance I_2 (the algorithm cannot know that T_2 lies). As seen above, since $\mathcal{A}$ is an α-approximate algorithm, with $\alpha < 2$, in the schedule returned by $\mathcal{A}$ on instance I_2, T_2 starts *before time* M. Task T_2 decreases its starting time (and thus its completion time) by bidding a false value. Therefore $\mathcal{A}$ is not a truthful algorithm.

If we consider the case of randomized algorithms, we are able to prove the following result (the proof is omitted).

Theorem 2. *Let $\mathcal{A}$ be a (randomized) truthful algorithm which does not introduce idle times between the tasks. Then $\mathcal{A}$ is not α-approximate, with $\alpha < \frac{3}{2}$.*

2.2   Negative Results: Private Lengths and Weights

If both the lengths and the weights of the tasks are private data then it is not possible to obtain a truthful deterministic approximation algorithm.

Theorem 3. *Let $\alpha > 1$. There is no truthful deterministic α-approximate algorithm if both the lengths and the weights of the tasks are private values.*

Proof. Let $\mathcal{A}$ be a deterministic algorithm which is α-approximate. Let us show that $\mathcal{A}$ is not a truthful algorithm. Let $M = 3\alpha$.

Let us consider a first instance I_1: a single machine and two tasks T_1 and T_2. Task T_1 has a length of M^2 and a weight of 1. Task T_2 has a length of M and a weight of M. In an optimal schedule, T_2 is executed at time 0 and T_1 starts when T_2 has been executed, at time M. The cost of such a schedule is $M^2 + (M + M^2) = 2M^2 + M$.

Let S be a schedule of I_1 in which task T_1 starts *before time* M. In such a schedule, task T_2 cannot be completed before the start of T_1: since no preemption is allowed, T_1 is executed before T_2. The cost of S is thus larger than or equal to $M^2 + (M^2 + M)M = M^3 + 2M^2$ (in the best case there is no idle time: task T_1 is scheduled at time 0 and task T_2 starts as soon as T_1 is completed, i.e. at time M^2). The ratio between the cost of S and the optimal cost is thus larger than or equal to $\frac{M^3+2M^2}{2M^2+M} = \frac{M^2+2M}{2M+1} > \frac{M}{3} = \alpha$. Since $\mathcal{A}$ is an α-approximate algorithm, $\mathcal{A}$ cannot return schedule S. Therefore, in the schedule returned by $\mathcal{A}$ on instance I_1, T_1 starts at the soonest at time M.

Let us now consider a second instance, I_2: a single machine and two tasks T_1 and T_2. Task T_1 has a length of 1 and a weight of M^2. Task T_2 has a length of M and a weight of M. In an optimal schedule T_1 is executed at time 0 and

T_2 starts when T_1 has been executed, at time 1. The cost of such a schedule is $M^2 + (1 + M)M = 2M^2 + M$.

Let S be a schedule of I_2 in which task T_1 *does not start before time M*. The cost of S is thus larger than or equal to $M^2 + (M + 1)M^2 = M^3 + 2M^2$ (in the best case task T_2 is scheduled at time 0 and task T_1 starts as soon as T_2 is completed, i.e. at time M). The ratio beween the cost of S and the optimal cost is larger than or equal to $\frac{M^3+2M^2}{2M^2+M} = \frac{M^2+2M}{2M+1} > \frac{M}{3} = \alpha$. Since $\mathcal{A}$ is an α-approximate algorithm, $\mathcal{A}$ cannot return schedule S. Therefore, in the schedule returned by $\mathcal{A}$ on instance I_2, T_1 starts before time M.

Let us now consider the following situation: task T_1 has a length M^2 and weight 1 and task T_2 bids a length M and a weight M. Given the values bid by T_2, if T_1 bids its true values, then the instance corresponds to instance I_1. As seen above, in the schedule returned by $\mathcal{A}$ on instance I_1, T_1 starts *at the soonest at time M*.

Let us now consider that task T_1 lies and bids a length of 1 and a weight of M^2. The input of the algorithm is now identical to instance I_2 (the algorithm cannot know that T_1 lies). As seen above, since $\mathcal{A}$ is an α-approximate algorithm, in the schedule returned by $\mathcal{A}$ on instance I_2, T_1 starts *before time M*. Task T_1 decreases its starting time (and thus its completion time) by bidding false values. Therefore $\mathcal{A}$ is not a truthful algorithm.

2.3   Positive Results: Single Machine with Preventive Preemption

In the remaining of this section, we show that if preventive preemption is used, then it becomes possible to design a truthful mechanism without payments which is optimal with respect to the social welfare. A preemptive schedule on a single machine can be defined as a vector $\sigma = (\rho_1, \ldots, \rho_n)$ where for every task i, $1 \leq i \leq n$, ρ_i corresponds to the set of time-intervals during which task i is executed, i.e. $\rho_i = [l_i^1, r_i^1) \cup \cdots \cup [l_i^k, r_i^k)$ with $l_i^1 < r_i^1 \leq l_i^2 < r_i^2 \leq \cdots \leq l_i^k < r_i^k$ and $\sum_{j=1}^k \left(r_i^j - l_i^j \right) = t_i$, where t_i is the true length of task i. In addition, for every pair of tasks i, j, we have $\rho_i \cap \rho_j = \emptyset$. Hence, in schedule σ, task i starts at time l_i^1, it is preempted at time r_i^1, then its execution continues at time l_i^2, it is again preempted at time r_i^2 and so on until its completion. Clearly, for the considered objective function, i.e. the sum of weighted completion times, any schedule where at least one task is preempted is strictly worse than the optimal non-preemptive schedule. Hence, given that we are interested in obtaining a truthful algorithm which outputs an optimal outcome, we need to design an algorithm which preempts the execution of a task only when the task bids a false value of its length. However, there is no possibility for the mechanism to know *a priori* if a task lies, and the mechanism has to define a (perhaps preliminary) schedule based only on the values that the tasks bid, i.e. *before* their real execution. Our algorithm is the following one: it schedules the tasks following the increasing order of the ratio of the declared length to weight, i.e. following Smith's rule, and it executes each task i during b_i units of time in the time interval $[l_i^1, l_i^1 + b_i)$. Whenever the real length of a task is greater than its declared one, then the task will be preempted at $l_i^1 + b_i$

and restarted after the completion of all the b_i's, $1 \leq i \leq n$, following a round robin policy if more than one tasks are preempted. We now introduce what we will call *preventive preemption*.

Definition 1. *An algorithm uses* preventive preemption *if it constructs a schedule in which a task i is preempted (and resumed later), if and only if, $b_i < t_i$.*

Our algorithm, that we call Weighted Shortest Processing Time with Preventive Preemption (WSPT-PP), uses the concept of preventive preemption. Our algorithm is based on the classical Smith's rule WSPT (Weighted Shortest Processing Time) which is optimal for the sum of the weighted processing times for the single-machine case. As we prove below an important property of WSPT-PP is that it is *truthful* and consequently no task is finally preempted, since for every task i, we have $b_i = t_i$. Let us now define more formally this algorithm[2].

Algorithm WSPT-PP

1 Sort all tasks in the WSPT order (i.e. such that $\frac{b_1}{w_1^b} \leq \frac{b_2}{w_2^b} \leq \cdots \leq \frac{b_n}{w_n^b}$).

2 Schedule the first interval $[l_i^1, r_i^1)$ of every task i such that $l_i^1 = \sum_{j=1}^{i-1} b_j$ and $r_i^1 = l_i^1 + b_i$.

3. After time $t = \sum_{j=1}^{n} b_j$, schedule the tasks which are not already completed using the round robin policy: For each $x \geq 2$, if Task i is not completed at time $\left(\sum_{j=1}^{n} b_j\right) + n(x-2) + i - 1$, schedule this task in the time interval $[l_i^x, r_i^x)$, with $l_i^x = \left(\sum_{j=1}^{n} b_j\right) + n(x-2) + i - 1$ and $r_i^x = \left(\sum_{j=1}^{n} b_j\right) + n(x-2) + i$.

Theorem 4. WSPT-PP *is a polynomial-time, optimal and truthful algorithm for the single machine case where the private data of every task is its length and the social welfare is the weighted sum of completion times.*

Proof. Assume that task i bids $b_i > t_i$. By the definition of WSPT-PP, task i will not start earlier than if it bids $b_i = t_i$ (and thus it will not decrease its completion time by lying). On the other hand, if task i bids $b_i < t_i$, again by the definition of WSPT-PP, it will be preempted b_i units of time after its starting time and it will be continued after date $\sum_{j=1}^{n} b_j$. Thus, its completion time will be at least $t_i - b_i + \sum_{j=1}^{n} b_j = t_i + \sum_{\substack{j=1 \\ j \neq i}}^{n} b_j$. If it bids $b_i = t_i$, it will not be preempted and its completion time will be at most $\sum_{j=1}^{n} b_j = t_i + \sum_{\substack{j=1 \\ j \neq i}}^{n} b_j$. In both cases task i has no incentive to lie, and so WSPT-PP is truthful. Thus the obtained schedule is without preemption, i.e. identical to the one obtained by the classical WSPT algorithm. Given the optimality of WSPT, we obtain that WSPT-PP is also optimal.   □

Remark. Notice that the previous results hold also if the valuation of each task is defined as its weighted completion time.

---

[2] Recall that in this section $w_i^b = w_i$.

2.4   Positive Results: Parallel Machines with Preventive Preemption

It is well known that the Shortest Processing Time (SPT) algorithm computes an optimal solution for the problem of minimizing the sum of completion times on identical parallel machines [21]. Based on that, we can apply SPT with preventive preemption (SPT-PP) on identical parallel machines and obtain a polynomial-time optimal and truthful algorithm for the parallel machines case where the social welfare is the minimization of the sum of completion times.

The proof of the truthfulness of SPT-PP is similar than the one of WSPT-PP for the single-machine case and it is omitted here. Given the truthfulness of SPT-PP, it is easy to see that no task will be preempted by SPT-PP and the produced schedule will be the same as the one of SPT.

Theorem 5. SPT-PP *is an optimal and truthful algorithm for the parallel machine case where the private data of every task is its length and the social welfare is the sum of completion times.*

For the multiple machines case with weights, given that the problem is NP-hard even if all data are public, we turn our attention to the development of approximate truthful mechanisms. We propose the following simple algorithm that we call RAND-WSPT-PP: Assign tasks independently and uniformly at random to the machines, and on each machine schedule the tasks using the WSPT rule by applying preventive preemption if necessary. It is easy to see that a task i has no influence on the choice of the machine on which it will be scheduled by lying on its length. In addition, according to the proof of Theorem 4 whatever the machine it is scheduled on, its best strategy is to declare $b_i = t_i$. This means that all the tasks will declare their true lengths and the algorithm will produce a non-preemptive schedule. It has been proved in [20] that this algorithm is 3/2-approximate in expectation. Consequently, we get the following result.

Theorem 6. RAND-WSPT-PP *is a truthful randomized 3/2-approximate in expectation algorithm for the parallel machine case where the private data of every task is its length and the social welfare is the weighted sum of completion times.*

Remark. The derandomization of this algorithm is WSPT-PP: the tasks are sorted according to the non decreasing ratio of b_i/w_i's, and they are scheduled following this order as soon as a machine becomes available [21]. If we impose large penalties on liars, e.g. by starting the exceeding part of a task at a time equal to the sum of all the declared processing times of the tasks, then it is easy to see that preventive preemption guarantees that no agent will lie when we apply WSPT-PP. This gives a $(1 + \sqrt{2})/2$-approximation [17]. If however, we impose that the exceeding part is started after the completion of the last task on the same machine or on any machine, then the tasks have incentive to lie. To see this consider the following example.

Example. Consider the following instance: two machines and three tasks: $w_1 = t_1 = 1$, $w_2 = t_2 = 1$, $w_3 = 2$ and $t_3 = 2 + \varepsilon$ (where ϵ is a small positive value, e.g. $\varepsilon = 0.1$). The schedule returned by WSPT-PP is the following one: each task of length 1 is scheduled at time 0 on a machine. Task 3 is scheduled at time 1, after a task of length 0. Its completion time is thus $3 + \varepsilon$. Task 3 has incentive to bid $2 - \varepsilon$. In this case, WSPT-PP schedules task 3 at time 0, and tasks 1 and 2 are scheduled on the other machine. Since task 3 is alone on its machine, it will be completed at time $2 + \epsilon$ even with preventive preemption. Even if we consider a stronger version of preventive preemption, that we may call preventive preemption with migration, where we execute the remaining part of the preempted task on the machine of maximum load, then task 3 will finish at time $2 + 2\epsilon$ instead of $3 - \varepsilon$: task 3 has still incentive to bid a false value.

3   Introducing Payments

3.1   Private Lengths

Let us first prove that the VCG method cannot be applied for the single-machine case without preventive preemption.

Theorem 7. *There is no optimal truthful mechanism with payment for the single machine case even in the unweighted case.*

Proof. By contradiction, assume that there is an optimal truthful mechanism minimizing the sum of completion times of the tasks on a single machine. It is well known that the Shortest Processing Time first (SPT) algorithm, which schedules the tasks in non-decreasing order of their lengths, is the only algorithm that maximizes the social welfare $-\sum_{1 \leq i \leq n} C_i$. Given that SPT does not insert any idle time, a schedule can be defined as an ordering of the tasks. Let 1 and 2 be the two tasks to schedule (i.e. $N = \{1, 2\}$) and consider the following scenario: when task 2 tells the truth, we have $t_2 = b_2 > b_1$. In this case, SPT constructs a schedule σ where task 1 is scheduled before task 2 ($\sigma = (1, 2)$). Then the utility of task 2 is $u_2 = -C_2 - p_2 = -t_1 - t_2 - p_2$. On the other hand, when task 2 lies and bids $b'_2 < b_1$, SPT constructs σ' where task 2 is scheduled before task 1 ($\sigma' = (2, 1)$) and the utility of task 2 becomes $u'_2 = -C'_2 - p'_2 = -t_2 - p'_2$. Given that the mechanism is assumed to be truthful, we must have $u_2 \geq u'_2$ (i.e. task 2 should not have incentive to lie) and thus $-t_1 - t_2 - p_2 \geq -t_2 - p'_2 \Rightarrow p'_2 - p_2 \geq t_1$. However, since t_1 is not known to the mechanism when the payments are computed, it is clear that there is no any payment function satisfying this property.   □

Corollary 1. *The VCG method cannot be applied for the single-machine case.*

3.2   Private Lengths and Weights

In this section, we show that preventive preemption associated with payments helps even when both the length and the weight of the tasks are private data. Since now each agent can lie on his weight, algorithm WSPT-PP is not truthful anymore. Indeed any task i has incentive to bid $b_i = t_i$ and $w_i^b > w_i$ in order to get a smaller ratio $\frac{b_i}{w_i^b}$, and then to decrease its completion time C_i. Moreover, as shown by Theorem 8 below, when both weights and lengths are private values, there is no optimal algorithm even if preemptive preemption is allowed (the proof is omitted due to lack of space). We then propose an optimal truthful algorithm which uses payment and preventive preemption.

Theorem 8. *Let $\varepsilon > 0$. There is no truthful deterministic $(2 - \varepsilon)$-approximate algorithm which does not use payment when the weights of the tasks is a private value, even when preventive preemption is allowed.*

Theorem 9. *For every task i, let s_i be the starting time of task i in the schedule obtained by WSPT-PP. The mechanism using algorithm WSPT-PP and the following payment function $p_i = -s_i + \sum_{j \neq i} b_j$ is polynomial-time computable, optimal and truthful for the single machine case.*

Proof. By the definition of algorithm WSPT-PP, $-s_i + \sum_{j \neq i} b_j$ is a positive value and it can be computed by the scheduler using only the values $(b_1, w_1^b), \ldots, (b_n, w_n^b)$. Thus, $p_i = -s_i + \sum_{j \neq i} b_j$ is a valid payment function. Moreover, for every task i, if i tells the truth, we have $u_i = -C_i - p_i = -(s_i + t_i) - (-s_i + \sum_{j \neq i} b_j) = -t_i - \sum_{j \neq i} b_j$ whereas if i lies, by the definition of algorithm WSPT-PP, it cannot be completed before time $s_i + t_i$ and thus we have $u_i \leq -t_i - \sum_{j \neq i} b_j$. Hence, task i takes no advantage of not telling the truth and so the mechanism is truthful. Moreover, given the truthfulness of the mechanism, WSPT-PP constructs the same schedule as WSPT without preemption. Thus, as WSPT constructs an optimal solution minimizing the sum of the weighted completion times, so does WSPT-PP. $\square$

For applications where the valuation of a task is its weighted completion time, it is also possible to obtain payments that ensure that WSPT-PP is truthful (the details will be given in the full version of the paper).

Multiple machines. Notice that for multiple machines we can use the algorithm RAND-WSPT-PP (see Sect. 2.4) with appropriate payments in order to obtain a randomized truthful approximation algorithm.

Corollary 2. *There exists a truthful $\frac{3}{2}$-approximate in expectation algorithm for the parallel machine case with payments when the private data of every task are its length and its weight.*

Acknowledgments. The work of Evripidis Bampis and Fanny Pascual was partly supported by the French ANR grant ANR-14-CE24-0007-01 "CoCoRICo-CoDec".

References

1. Abed, F., Correa, J.R., Huang, C.-C.: Optimal coordination mechanisms for multi-job scheduling games. In: Schulz, A.S., Wagner, D. (eds.) ESA 2014. LNCS, vol. 8737, pp. 13–24. Springer, Heidelberg (2014)

2. Ambrosio, P., Auletta, V.: Deterministic monotone algorithms for scheduling on related machines. In: Persiano, G., Solis-Oba, R. (eds.) WAOA 2004. LNCS, vol. 3351, pp. 267–280. Springer, Heidelberg (2005)
3. Andelman, N., Azar, Y., Sorani, M.: Truthful approximation mechanisms for scheduling selfish related machines. In: Diekert, V., Durand, B. (eds.) STACS 2005. LNCS, vol. 3404, pp. 69–82. Springer, Heidelberg (2005)
4. Angel, E., Bampis, E., Pascual, F.: Truthful algorithms for scheduling selfish tasks on parallel machines. Theoret. Comput. Sci. **369**, 157–168 (2006)
5. Angel, E., Bampis, E., Pascual, F., Tchetgnia, A.: On truthfulness and approximation for scheduling selfish tasks. J. Sched. **12**, 437–445 (2009)
6. Angel, E., Bampis, E., Thibault, N.: Randomized truthful algorithms for scheduling selfish tasks on parallel machines. Theor. Comput. Sci. **414**(1), 1–8 (2012)
7. Archer, A., Tardos, E.: Truthful mechanisms for one-parameter agents. In: FOCS, pp. 482–491 (2001)
8. Braverman, M., Chen, J., Kannan, S.: Optimal provision-after-wait in healthcare. In: ITCS 2014, Princeton, NJ, pp. 541–542 (2014)
9. Christodoulou, G., Gourvès, L., Pascual, F.: Scheduling selfish tasks: about the performance of truthful algorithms. In: Lin, G. (ed.) COCOON 2007. LNCS, vol. 4598, pp. 187–197. Springer, Heidelberg (2007)
10. Clarke, E.: Multipart pricing of public goods. Public Choice **11**(1), 17–33 (1971)
11. Cohen, J., Pascual, F.: Scheduling tasks from selfish multi-tasks agents. In: Träff, J.L., Hunold, S., Versaci, F. (eds.) Euro-Par 2015. LNCS, vol. 9233, pp. 183–195. Springer, Heidelberg (2015)
12. Cole, R., Correa, J.R., Gkatzelis, V., Mirrokni, V.S., Olver, N.: Inner product spaces for minsum coordination mechanisms. In: ACM STOC 2011, pp. 539–548 (2011)
13. Fotakis, D., Tzamos, C., Zampetakis, E.: Who to trust for truthfully maximizing welfare? CoRR abs/1507.02301 (2015)
14. Groves, T.: Incentive in teams. Econometrica **41**(4), 617–631 (1973)
15. Hoeksma, R., Uetz, M.: The price of anarchy for minsum related machine scheduling. In: Solis-Oba, R., Persiano, G. (eds.) WAOA 2011. LNCS, vol. 7164, pp. 261–273. Springer, Heidelberg (2012)
16. Hurst, J., Siciliani, L.: Tackling excessive waiting times for elective surgery: a comparison of policies in 12 OECD countries. Health Policy **72**(2), 201–215 (2005)
17. Kawaguchi, T., Kyan, S.: Worst case bound of an LRF schedule for the mean weighted flow-time problem. SIAM J. Comput. **15**(4), 1119–1129 (1986)
18. Koutsoupias, E., Papadimitriou, C.: Worst-case equilibria. In: Meinel, C., Tison, S. (eds.) STACS 1999. LNCS, vol. 1563, p. 404. Springer, Heidelberg (1999)
19. Nisan, N., Ronen, A.: Algorithmic mechanism design. In: STOC, pp. 129–140 (1999)
20. Schulz, A.S., Skutella, M.: Scheduling unrelated machines by randomized rounding. SIAM J. Discret. Math. **15**(4), 450–469 (2002)
21. Smith, W.E.: Various optimizers for single stage production. Naval Res. Logist. Q. **3**, 59–66 (1956)
22. Vickrey, W.: Counterspeculation, auctions and competitive sealed tenders. J. Financ. **16**, 8–37 (1961)

Network Topologies for Weakly Pareto Optimal Nonatomic Selfish Routing

Xujin Chen and Zhuo Diao[✉]

Institute of Applied Mathematics, AMSS, Chinese Academy of Sciences,
Beijing 100190, China
{xchen,diaozhuo}@amss.ac.cn

Abstract. In this paper we study the model of nonatomic selfish routing and characterize the topologies of undirected/directed networks in which every Nash equilibrium is weakly Pareto optimal, meaning that no deviation of all players could make everybody better off. In particular, we first obtain the characterizations for single-commodity case by applying relatively standard graphical arguments, and then the counterpart for two-commodity undirected case by introducing some new algorithmic ideas and reduction techniques.

Keywords: Nonatomic selfish routing · Weakly Pareto optimal · Single-commodity networks · Multi-commodity networks · Extension-parallel networks

1 Introduction

A basic task of network management is routing traffic to achieve the highest possible network efficiency. However, it is usually difficult or even impossible to implement centralized optimal routing in many large systems, as modeled by selfish routing games [9]. In these games, a number of players (network users) selfishly choose routes in the network for traveling from their origins to their destinations, aiming to minimize their own latencies. The selfish behaviors often lead to Braess's paradox [2], which exposes the seemingly counterintuitive phenomenon that less route options lead to shorter travel time at the Nash Equilibrium (NE). The paradox in particular reflects the fact that there is a feasible routing which is better for all players than the NE. This stands on the contrary to the spirit of *weak Pareto optimality – no alternative solution could make every individual strictly gain*. The absence of weak Pareto optimality exhibits not only the inefficiency, but also a kind of unstable state where players might have incentive to form a grand coalition to deviate. A natural question on network design arises as to in which network topologies the NE of any routing

Research supported in part by NNSF of China under Grant No. 11531014 and 11222109, and by CAS Program for Cross & Cooperative Team of Science & Technology Innovation.

© Springer International Publishing Switzerland 2016
T.N. Dinh and M.T. Thai (Eds.): COCOON 2016, LNCS 9797, pp. 27–38, 2016.
DOI: 10.1007/978-3-319-42634-1_3

instance is always Weakly Pareto Optimal (WPO). Once such a WPO network is established, regardless of the latency functions and the locations of origins and destinations, the strategic interactions among players would lead to equilibrium outcomes that enjoy sort of efficiency and stability, and the occurrence of Braess's paradox is particularly prevented. The purpose of this paper is to identify the network structures that inherently guarantee weak Pareto optimality for not only any single origin-destination pair (i.e., the single-commodity case) and but also any multiple origin-destination pairs (i.e., the multi-commodity case).

Related Work. Milchtaich [8] studied under the model of nonatomic selfish routing the weak Pareto optimality of NE in undirected networks w.r.t. a fixed origin-destination pair (s, t). In the nonatomic model, there are an infinite number of players each controlling a negligible portion of the total traffic from s to t. It was shown that all NE are WPO for any nonnegative, continuous and nondecreasing latency functions if and only if the network has *linearly independent routes*, meaning that every s-t path has at least an edge which does not belong to any other s-t path. Milchtaich's result [8] parallels an earlier necessary and sufficient condition of Holzam and yone (Lev-tov) [5] for atomic selfish routing games played by a finite number of players each controlling a nonsplittable unit traffic from origin s to destination t in a directed network. The condition ensures that all (pure strategy) NE are WPO by excluding from the network the so-called bad configuration. Holzam and yone (Lev-tov) [6] then related the forbidden structure with a recursive extension-parallel construction for *irredundant* networks, i.e., networks that are unions of their s-t paths. The authors proved that an irredundant directed network does not contain any bad configuration if and only if it is *extension-parallel*. Later, Milchtaich [8] established the equivalence between the extension-parallel structure of an irredundant network and the linearly independent route property of its underlying undirected network.

Strengthening the stability of weak Pareto optimality, which in some sense only excludes the coalition of all players, a Strong Equilibrium (SE) prevents any subset of players from deviating. In particular, every SE (if exists) is an NE that is WPO. For atomic routing restricted to irredundant single-commodity directed networks, Holzman and yone (Lev-tov) [6] proved that extension-parallel networks are exactly the ones that guarantee the existence of SE. Regarding the multi-commodity counterpart, the network characterization was given in terms of forbidden bad configurations [5]. Recently, Holzman and Monderer [7] studied the atomic routing game on a special directed network consisting of paths from a specific source to a specific sink, and proved for the multi-commodity case that the sets of NE and SE are identical if and only if the network is extension-parallel.

The network structures that guarantee NE of selfish routing to possess other kinds of properties stronger or weaker (in some sense) than weak Pareto optimality were also discussed in literatures, such as Pareto optimality [5,8], social optimality [4] and Braess's paradox freeness [3,8].

Our Contributions. We focus on nonatomic selfish routing model. First, we extend the Milchtaich's linearly independent route characterization [8] for the single-commodity undirected networks to directed ones; we prove that

- Extension-parallel networks are *essentially* the networks that guarantee the NE of every single-commodity routing instance is always WPO (Theorem 1).

The proof relies on applications and extension of previous results from [6–8]. In particular, for the undirected case, we further transfer the relatively local picture, expressed in terms of forbidden minors (Theorems 2(iii) and 4(ii)) or every two-terminal subnetwork (Theorem 1), to a global one that gives the explicit structure of the whole graph (Theorem 3). Then, by utilizing algorithmic ideas on flow, graph theory tools, and double-counting method, we show that

- A connected undirected network with the NE of every 2-commodity routing instance being WPO is either a tree, or contains only one non-edge block (a maximal subgraph without cut-vertices); in the latter case, the non-edge block is a cycle or consists of a number of parallel edges or is obtained from a triangle by duplicating an edge for a number of times (Theorem 5).

The theoretical result and technical methods constitute our main contribution. The restrictive topologies indicate more or less the scarcity of WPO NE in multi-commodity routing practice. The ideas and approaches might be useful for future research on selfish routing. Furthermore, for k-commodity case with $k \geq 3$, we show that undirected WPO networks are extremely limited (Theorem 6).

2 Routing Model

We consider both undirected and directed networks, and model them by graphs or digraphs $G = (V, E)$ with *vertex* set V and *link* set E, respectively. Loops are not allowed, while more than one link can join the same pair of vertices. Each link $e \in E$ is associated with a *nonnegative, continuous, nondecreasing* latency function $\ell_e(\cdot)$ which specifies the time needed to traverse e as a function of the link congestion on e. Undirected links are called *edges* while directed ones are called *arcs*. Let $u, v \in V$, a path in G from u to v is called a u-v path. We use the standard definition of a path that does not allow any vertex repetition. We will often abbreviate "undirected graphs" as "graphs", and collectively refer to graphs and digraphs as (di)graphs.

Let k be a positive integer. Given k origin-destination pairs of vertices (s_i, t_i), $i \in [k] = \{1, \ldots, k\}$, in G, we call $(G, (s_i, t_i)_{i=1}^{k})$ a *k-commodity network embedded in G* if for each $i \in [k]$, $s_i \neq t_i$ and G contains at least an s_i-t_i path.

We focus on nonatomic selfish routing for traffic flow. Given a positive *demand* $\mathbf{r} = (r_i)_{i=1}^{k}$, the traffic in $(G, (s_i, t_i)_{i=1}^{k})$ comprises k flows, each for one commodity. The flow of commodity $i \in [k]$ with a total amount of r_i is formed by an infinite number of players traveling from s_i and t_i. Each player (who is associated to a unique origin-destination pair) selects a single path from his origin to his destination that has a minimum latency, given the congestion imposed by the rest of players. The nonatomic routing model assumes that the choice of each individual player has a negligible impact on the experiences of others.

Formally, let $(G, (s_i, t_i)_{i=1}^{k}, \mathbf{r}, \ell)$ denote a k-commodity selfish routing instance, where latency functions $\ell_e(\cdot)$, $e \in E$, are collectively represented by ℓ.

For each $i \in [k]$, let $\mathcal{P}_i$ be the set of s_i-t_i paths in G; a *flow of commodity* i is a nonnegative vector $\mathbf{f}_i = (f_i(P))_{P \in \mathcal{P}_i}$ with $\sum_{P \in \mathcal{P}_i} f_i(P) = r_i$. The combination of $\mathbf{f}_1, \ldots, \mathbf{f}_k$ gives rise to a *(k-commodity) flow* $\mathbf{f} = (\mathbf{f}_i)_{i=1}^k$ for $(G, (s_i, t_i)_{i=1}^k, \mathbf{r})$. Under $\mathbf{f}$, each link e that is contained by some path in $\cup_{i=1}^k \mathcal{P}_i$ experiences a *congestion* $f(e) = \sum_{i=1}^k \sum_{P \in \mathcal{P}_i : e \in P} f_i(P)$, and thus a *link latency* $\ell_e(f(e))$. Accordingly, each path P contained by $\cup_{Q \in \cup_{i=1}^k \mathcal{P}_i} Q$ and any player traveling through P suffer from a *path latency* $\ell_P(\mathbf{f}) = \sum_{e \in P} \ell_e(f(e))$.

In nonatomic routing games, Nash equilibria are characterized by Wardrop's principle in a way that all players travel only on the minimum latency paths from their own origins to their own destinations. A flow $\boldsymbol{\pi}$ of $(G, (s_i, t_i)_{i=1}^k, \mathbf{r}, \ell)$ is called an *NE (flow)* of the instance if it satisfies the following *NE property*:

$$\forall\, i \in [k] \text{ and } \forall\, P \in \mathcal{P}_i \text{ with } \pi_i(P) > 0, \text{ there holds } \ell_P(\boldsymbol{\pi}) = \min_{Q \in \mathcal{P}_i} \ell_Q(\boldsymbol{\pi}).$$

By the classical result of Beckmann et al. [1], the NE of $(G, (s_i, t_i)_{i=1}^k, \mathbf{r}, \ell)$ exist, and are essentially unique in the sense that the link latencies are invariant under any NE of $(G, (s_i, t_i)_{i=1}^k, \mathbf{r}, \ell)$. Thus, for each $i \in [k]$, the common latency experienced by all players traveling from s_i to t_i in any NE of $(G, (s_i, t_i)_{i=1}^k, \mathbf{r}, \ell)$ is also an invariant, which we denote by $\ell_i(G, (s_j, t_j)_{j=1}^k, \mathbf{r})$.

Given a k-commodity routing instance $(G, (s_i, t_i)_{i=1}^k, \mathbf{r}, \ell)$, its "unique" NE flow is *weakly Pareto optimal* (WPO) if for every feasible flow $\mathbf{f}$, there exist $h \in [k]$ and $P \in \mathcal{P}_i$ such that $f_i(P) > 0$ and $\ell_P(\mathbf{f}) \geq \ell_h(G, (s_i, t_i)_{i=1}^k, \mathbf{r})$, i.e., some players travelling from s_h to t_h experience a lentency under $\mathbf{f}$ not smaller than that they experience under a NE flow. We say that a k-commodity network $(G, (s_i, t_i)_{i=1}^k)$ embedded in G is *WPO* if for any positive traffic demand $\mathbf{r} \in \mathbb{R}_{>0}^k$, and any nonnegative, continuous, nondecreasing latency functions ℓ on E, the NE of $(G, (s_i, t_i)_{i=1}^k, \mathbf{r}, \ell)$ is WPO. A (di)graph G is called *WPO w.r.t. k commodities* if every k-commodity network embedded in G is WPO.

The concept of minors in graph theory is useful in characterizing WPO (di)graphs. Given (di)graphs G and H, we call H a *minor* of G if it could be obtained from a sub(di)graph of G by contracting links (possibly none); we call H a *topological minor* of G if G contains a subdivision of H as a sub(di)graph. If H is not a (topological) minor of G, then we say that G does not have a (topological) minor isomorphic to H, or simply G has no H-(topological) minor.

Lemma 1. *Let G be a (di)graph and G' a topological minor of G. If G is WPO w.r.t. k commodities, then so is G'.*   □

Due to the limitation on pages, we omit some proofs in this extended abstract, and postpone them the full version of the paper.

3   Single-Commodity Networks

We start this section by introducing some definitions for both undirected and directed networks, which are followed by a unified characteristic description for

WPO (di)graphs w.r.t. single commodity. Then we discuss the undirected and directed cases in Sects. 3.1 and 3.2, respectively.

Considering a network G with origin-destination pair (s, t), we say that G *has linearly independent routes*, or *linearly independent s-t routes* to be more specific, if in G every s-t path has at least a link that does not belong to any other s-t path. We call G a *two-terminal network* (with terminals s, t), or an (s, t)-*terminal network*, if each link and each vertex of G are contained in at least an s-t path. Note that if a two-terminal network is directed, then its terminals must be its unique source and unique sink.

Definition 1 [6,8]. *The* terminal extension *of an (s, t)-terminal network G is the operation that adds a new vertex together with a new link from it to s (or from t to it), where the new vertex becomes the origin (or destination) of the resulting network.*

Definition 2 [6]. *A single-commodity network G with origin-destination pair (s, t) is (s, t)-extension parallel or extension-parallel for short if*

– G *has a single link with ends s and t; or*
– G *is a terminal extension of a smaller extension-parallel network; or*
– G *is obtained by connecting two smaller extension-parallel networks in parallel – identifying their origins (resp. destinations) to form s (resp. t).*

It has been known that an (s, t)-terminal undirected network is (s, t)-extension-parallel if and only if it has linearly independent s-t routes (see Proposition 5 of [8]). In view of the 1–1 correspondence between the sets of s-t paths of an (s, t)-extension-parallel directed network and its underlying undirected network, we have the following equivalent definition for extension-parallel (un)directed networks, where a graph is considered as the underlying graph of itself.

Definition 3 [6,8]. *A two-terminal network is extension-parallel if its underlying undirected network has linearly independent routes.*

The following unified characterization combines Theorems 2(ii) and 4(iii) to be discussed in the next two subsections.

Theorem 1. *A (di)graph G is WPO w.r.t. single commodity if and only if every maximal two-terminal network embedded in G is extension-parallel.*   □

3.1 Undirected Networks

In this subsection, we formally state Milchtaich's result [8] on single-commodity networks where the origin and destination are fixed. The result implies a forbidden minor characterization for WPO graphs straightforwardly. Our efforts are devoted to transferring the forbidden minor description to a constructive one (Theorem 3) which gives the explicit global graphical structures.

Fig. 1. The forbidden terminal-reduced topological minors for WPO single-commodity networks with origin-destination pair (s,t).

Let G be an undirected network with origin-destination pair (s,t). A *sub-network* of G is a network with the same origin and destination obtained from G by removing vertices and edges (possibly none). A single-commodity network G' is an *terminal-reduced topological minor* of G if some subnetwork of G can be obtained from G' by applying edge subdivisions and terminal extensions (cf. Definition 1) for any number of times (possibly none) in any order.[1] An (s,t)-*minor* of G is a network obtained from some sub(di)graph of G that contains s,t by contracting links (possibly none), where when contracting any link incident with s (or t), the resulting vertex is named as s (or t). Milchtaich [8] characterized WPO graphs G w.r.t. a fixed single origin-destination pair via two (equivalent) necessary and sufficient conditions: (i) none of F_i, $i = 1,2,3$, depicted in Fig. 1 is a terminal-reduced topological minor of G, and (ii) the maximal two-terminal network embedded in G has linearly independent routes.

We observe that in any graph with origin-destination pair (s,t), the presence of any F_i, $i \in [3]$ as a terminal-reduced topological minor implies the presence of F_1 as an (s,t)-minor, and vice versa. It follows that a graph is WPO w.r.t. single commodity if and only if it does not contain F_1 as a minor.

Theorem 2 [8]. *Let G be a graph and $(G,(s,t))$ be a single commodity network embedded in G. Then:*

(i) *$(G,(s,t))$ is WPO if and only if G has linearly independent s-t routes.*

(ii) *$(G,(s,t))$ is WPO if and only if the maximal (s,t)-subnetwork of G is (s,t)-extension parallel.*

(iii) *G is WPO w.r.t. single commodity if and only if G does not contain any minor isomorphic to F_1.*

To interpret the above forbidden minor characterization (iii) in a constructive way, we next study the structures of graphs without F_1-minor. Given a graph G, a maximal connected subgraph of G without cut-vertices is called a *block* of G. A block is *trivial* if it consists of an edge or a vertex. Let $\mathcal{S}$ denote the set of graphs each of which is obtained from a cycle by adding duplications of one of its edges (possibly none). See Fig. 2(a) for an illustration.

Lemma 2. *A connected graph G does not contain a minor isomorphic to F_1 if and only if G has at most one nontrivial block and this nontrivial block (if exists) belongs to $\mathcal{S}$.*   □

---

[1] In terminologies of [8], G' is said to be embedded in G.

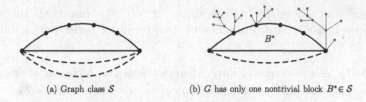

(a) Graph class $\mathcal{S}$ (b) G has only one nontrivial block $B^* \in \mathcal{S}$

Fig. 2. Graphs without F_1-minor.

The combination of Theorem 2(iii) and Lemma 2 provides the following constructive characterization for WPO graphs w.r.t single commodity.

Theorem 3. *A graph G is WPO w.r.t. single commodity if and only if every component of G has at most one nontrivial block, and all nontrivial blocks of G (if any) belong to $\mathcal{S}$.* □

3.2 Directed Networks

In this subsection, we characterize WPO digraphs w.r.t. single commodity by excluding the orientations $\vec{F}_i$ of F_i, $i = 1, 2, 3$ (see Fig. 3) as topological minors. The result, on one hand, parallels to the exclusion of terminal-reduced topological minors F_1, F_2, F_3 to assure a WPO undirected network w.r.t. a fixed origin-destination pair [8]. On the other hand, it stands in contrast to the single forbidden minor (i.e., F_1) characterization for WPO graphs (see Theorem 2(iii)).

Our proof involves series-parallel networks, for which the following recursive definition (see e.g., [10]) turns out to be helpful.

Definition 4. *A single-commodity directed network G with origin-destination pair (s, t) is* two-terminal series-parallel *or* (s, t)-series-parallel *if*

(i) *G has a single arc from s to t; or*
(ii) *G is obtained by connecting two smaller (o_i, d_i)-series-parallel directed networks H_i, $i = 1, 2$, in series – identifying d_1 and o_2, and naming o_1 as s, and d_2 as t; or*
(iii) *G is obtained by connecting two smaller (o_i, d_i)-series-parallel directed networks H_i, $i = 1, 2$, in parallel – identifying o_1 and o_2 to form s and identifying d_1 and d_2 to form t.*

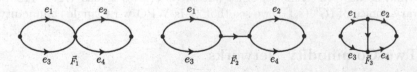

Fig. 3. The forbidden topological minors for WPO digraphs w.r.t. single commodity.

Holzman and Monderer (Proposition 2 of [7]) obtained the following forbidden-minor characterization of two-terminal series-parallel directed networks.

Lemma 3 [7]. *A two-terminal directed network is two-terminal series-parallel if and only if it does not have any topological minor isomorphic to $\vec{F}_3$.*   □

Theorem 4. *Let G be a digraph. The following are equivalent:*

(i) G is WPO w.r.t. single commodity;

(ii) G does not have any topological minor isomorphic to $\vec{F}_1$, $\vec{F}_2$ or $\vec{F}_3$;

(iii) Every maximal two-terminal network embedded in G is extension-parallel.

Proof. $(i) \Rightarrow (ii)$: By Lemma 1, it suffices to show that $\vec{F}_i = (V_i, E_i)$, $i \in \{1, 2, 3\}$, is not WPO. Let s and t denote the unique source and sink of $\vec{F}_i$, respectively. Let the four links $e_1, e_2, e_3, e_4 \in E_i$ be as depicted in Fig. 3, and let $\mathcal{P}$ denote the set of s-t paths in $\vec{F}_i$. Let $P_1, P_2, P_3 \in \mathcal{P}$ be the unique s-t paths in $\vec{F}_i$ that contain $\{e_1, e_2\}$, $\{e_3, e_4\}$ and $\{e_1, e_4\}$, respectively. Consider the single-commodity routing instance $(\vec{F}_i, (s.t), \mathbf{r}, \ell)$, where $\mathbf{r} = (2)$, $\ell_{e_1}(x) = x = \ell_{e_4}(x)$, $\ell_{e_2}(x) = 2 = \ell_{e_3}(x)$, and $\ell_e(x) = 0$ for any $e \in E_i \setminus \{e_1, e_2, e_3, e_4\}$. It is easy to see that the unique NE flow $\boldsymbol{\pi}$ of this instance is given by $\pi(P_3) = 2$ and $\pi(P) = 0$ for each $P \in \mathcal{P} \setminus \{P_3\}$, and incurs a latency $\ell(\vec{F}_i, (s, t), \mathbf{r}) = 4$. On the other hand, $(\vec{F}_i, (s.t), \mathbf{r}, \ell)$ admits a feasible flow $\mathbf{f}$ given by $f(P_1) = 1 = f(P_2)$ and $f(P) = 0$ for each $P \in \mathcal{P} \setminus \{P_1, P_2\}$. Now $\max_{P \in \mathcal{P}, f(P) > 0} \ell(P) = \ell(P_1) = \ell(P_2) = 3 < \ell(\vec{F}_i, (s, t), \mathbf{r})$ implies that $\boldsymbol{\pi}$ is not WPO.

$(ii) \Rightarrow (iii)$: By contradiction, take $G = (V, E)$ to be a counterexample with a minimum number m of arcs. Clearly $m \geq 2$. The minimality of G says that G is a two-terminal network, which is therefore not extension-parallel. By Lemma 3, condition (ii) implies that G is two-terminal series-parallel. Since $m \geq 2$, by Definition 4, there exist two smaller two-terminal series-parallel digraphs H_1 and H_2 whose connection in series or in parallel gives G. It follows from the minimality of G that both H_1 and H_2 are extension-parallel. Because G is not extension-parallel, it must be the case that H_1 and H_2 are connected in series and both H_1 and H_2 have more than one arc, which shows that G contains a subdivision of $\vec{F}_1$ or $\vec{F}_2$, a contradiction to (ii).

$(iii) \Rightarrow (i)$: Consider an arbitrary maximal two-terminal network $(G', (s, t))$ embedded in G. Since G' is extension-parallel, its underlying graph, written as $\underline{G}'$, has linearly independent s-t routes. Thus $(\underline{G}', (s, t))$ is WPO by Theorem 2(i). As $\underline{G}'$ and G' have the same set of s-t paths, $(G', (s, t))$ is also WPO. From the arbitrary choice of $(G', (s, t))$, we see that G is WPO w.r.t single commodity. □

4   Two-Commodity Networks

In this section, we characterize WPO undirected graph w.r.t. two commodities, and leave the characterization of the directed counterpart as an open question.

Let $\mathcal{R}$ be the set of graphs each of which is formed by a number (at least two) of parallel edges between two vertices. Let $\mathcal{T}$ be the set of graphs each of which is obtained from a graph in $\mathcal{R}$ by dividing exactly one edge with a new vertex. Let $\mathcal{C}$ be the set of undirected cycles of lengths at least four. Note that $\mathcal{R} \cup \mathcal{T} \cup \mathcal{C}$ is a proper subset of the graph class $\mathcal{S}$ defined in Sect. 3.1.

The goal of this section is to establish the following theorem. Since $\mathcal{R} \cup \mathcal{T} \cup \mathcal{C} \subsetneqq \mathcal{S}$, the result can be viewed as a natural evolvement of Theorem 3. Let F_4 be the graph as depicted in Fig. 4.

Theorem 5. *Let G be an undirected graph. The following are equivalent:*

(i) G is WPO w.r.t. two commodities;
(ii) Every component of G has at most one nontrivial block, and all nontrivial blocks of G (if any) belong to $\mathcal{R} \cup \mathcal{T} \cup \mathcal{C}$.
(iii) G does not contain any minor isomorphic to F_1 or F_4.

The equivalence between (ii) and (iii) is straightforward. We concentrate on the proof of (i) being equivalent to (ii).

The proof for implication from (i) to (ii) is relatively easy: any WPO graph must be F_1-minor free; the freeness enforces that each component of the graph can contain at most one nontrivial block and this block can only belong to $\mathcal{S}$ (Lemma 2 in Sect. 3.1); in turn the weak Pareto optimality w.r.t. two commodities guarantees the membership of $\mathcal{R} \cup \mathcal{T} \cup \mathcal{C}$ for this block, as otherwise routing instances with non-WPO NE flows could be constructed (see Lemma 4).

The proof of the reverse implication reduces to proving that every graph in $\mathcal{R} \cup \mathcal{T} \cup \mathcal{C}$ is WPO. By Milchtaich's linearly independent route characterization (Theorem 2(i)), we only need to consider two cases: (1) the graph belongs to $\mathcal{C}$ and (2) the graph belongs to $\mathcal{T}$ (see Lemmas 6 and 7).

Non-WPO Graphs. We focus on 2-connected graphs without F_1-minor which do not belong to $\mathcal{R} \cup \mathcal{T} \cup \mathcal{C}$, i.e., graphs in $\mathcal{S} \setminus (\mathcal{R} \cup \mathcal{T} \cup \mathcal{C})$.

Lemma 4. *If $G \in \mathcal{S} \setminus (\mathcal{R} \cup \mathcal{T} \cup \mathcal{C})$, then G is not WPO w.r.t. two commodities.*

Proof. It is easy to see that G contains F_4 as a topological minor. By Lemma 1, it suffices to show that F_4 is not WPO w.r.t. two commodities. Indeed, a 2-commodity routing instance $(F_4, (s_i, t_i)_{i=1}^2, \mathbf{r}, \ell)$ whose NE is not WPO can be constructed as follows: Let the vertices and edges of F_4 be labeled as in Fig. 4. Let the demand $\mathbf{r}$ be defined by $r_1 = r_2 = 10$. The nonnegative, continuous nondecreasing latency functions ℓ satisfy $\ell_{e_1}(5) = 0 < 2 = \ell_{e_1}(6)$, $\ell_{e_2}(3) = 2 = \ell_{e_2}(5)$, $\ell_{e_3}(12) = 0 < 2 = \ell_{e_3}(13)$, $\ell_{e_4}(10) = 0 < 2 = \ell_{e_4}(11)$ and $\ell_{e_5}(7) = 2 = \ell_{e_5}(8)$.

Observe that the set of s_1-t_1 paths in F_4 is $\{e_1 \cup e_5, e_2 \cup e_5, e_3 \cup e_4\}$ and the set of s_2-t_2 paths in F_4 is $\{e_1 \cup e_3, e_2 \cup e_3, e_5 \cup e_4\}$. Clearly $(F_4, (s_i, t_i)_{i=1}^2, \mathbf{r}, \ell)$ admits 2-commodity flows $\boldsymbol{\pi}$ and $\mathbf{f}$ defined respectively by $\pi_1(e_1 \cup e_5) = 3$, $\pi_1(e_3 \cup e_4) = 7$, $\pi_1(e_2 \cup e_5) = 0$, $\pi_2(e_1 \cup e_3) = 3 = \pi_2(e_2 \cup e_3) = 3$, $\pi_2(e_5 \cup e_4) = 4$ and $f_1(e_1 \cup e_5) = 4$, $f_1(e_3 \cup e_4) = 6$, $f_1(e_2 \cup e_5) = 0$, $f_2(e_1 \cup e_3) = 1$, $f_2(e_2 \cup e_3) = 5$, $f_2(e_5 \cup e_4) = 4$. A routine check gives the congestions and latencies in Table 1.

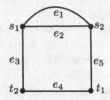

Fig. 4. Graph F_4.

Table 1. Edge congestions and latencies in F_4.

j	1	2	3	4	5
$\pi(e_j)$	6	3	13	11	7
$\ell_{e_j}(\pi(e_j))$	2	2	2	2	2
$f(e_j)$	5	5	12	10	8
$\ell_{e_j}(f(e_j))$	0	2	0	0	2

Since each s_i-t_i path $(i = 1, 2)$ has exactly two edges, each of which suffers from a link latency 2 under π, we deduce that π is a NE for $(F_4, (s_i, t_i)_{i=1}^2, \mathbf{r}, \ell)$, with $\ell_1(F_4, (s_i, t_i)_{i=1}^2, \mathbf{r}) = \ell_2(F_4, (s_i, t_i)_{i=1}^2, \mathbf{r}) = 4$. On the other hand, from $f_1(e_2 \cup e_5) = 0$ we see that for any $i \in [2]$ and any path $P \in \mathcal{P}_i$ with $f_i(P) > 0$, there holds $\ell_P(\mathbf{f}) \leq 2$. So F_4 is not WPO, proving the lemma. □

Building Blocks of WPO Graphs. Our goal is to prove weak Pareto optimality for building blocks of WPO graphs w.r.t. two commodities. As each graph in $\mathcal{R}$ has linearly independent routes, in view of Theorem 2(i), we focus on building blocks in $\mathcal{C} \cup \mathcal{T}$. The following lemma is crucial to our proofs; it exhibits a basic property of flow allocations on cycles.

Lemma 5. *If $\mathbf{f}$ and $\mathbf{g}$ are flows for 2-commodity routing instance $(G, (s_i, t_i)_{i=1}^2, \mathbf{r})$ on undirected cycle G, then there exist $i \in \{1, 2\}$ and $P \in \mathcal{P}_i$ with $f_i(P) > 0$ such that $f(e) \geq g(e)$ for all edges $e \in P$.* □

Lemma 6. *All undirected cycles are WPO w.r.t. two commodities.*

Proof. Given any 2-commodity routing instance $(G, (s_i, t_i)_{i=1}^2, \mathbf{r}, \ell)$ on cycle G, for any flow $\mathbf{f}$ and the NE flow π of the instance, by Lemma 5, there exist $i \in \{1, 2\}$ and $P \in \mathcal{P}_i$ with $f_i(P) > 0$ such that $f(e) \geq \pi(e)$ for all links $e \in P$. It follows from the nondecreasing property of ℓ and the NE property of π that $\ell_P(\mathbf{f}) \geq \ell_P(\pi) \geq \ell_i(G, (s_i, t_i)_{i=1}^2, \mathbf{r})$, implying that π is WPO. □

When studying any flow $\mathbf{f}$ and the NE flow π of a 2-commodity routing instance on $G \in \mathcal{T}$, an important step is to "sum up" flows on the parallel edges of G to obtain an imaginary flow on a triangle (which is a cycle), and then apply Lemma 5 to the imaginary flows on the triangle. This provides four inequalities comparing the flow allocations of $\mathbf{f}$ and π such that one of them must be true. From the valid inequality, we elaborate on the detailed flow allocations on the parallel edges and reach contradictions assuming $\mathbf{f}$ shows that π is not WPO.

Lemma 7. *All graphs in $\mathcal{T}$ are WPO w.r.t. two commodities.* □

The proof of Theorem 5. We are now ready to give a wrap-up proof of our main result on 2-commodity networks.

Proof (of Theorem 5). Assume without of generality that G is connected.

$(i) \Rightarrow (ii)$: It suffices to consider the case where G contains a nontrivial block B. Since G is WPO w.r.t. two commodities, it is WPO w.r.t. single commodity (considering two origins and two destinations as one, respectively). By Theorem 2(ii), F_1 is not a minor of G. It follows from Lemma 2 that B is the only one nontrivial block of G and $B \in \mathcal{S}$. If $B \notin \mathcal{R} \cup \mathcal{T} \cup \mathcal{C}$, then B is not WPO w.r.t. two commodities by Lemma 4. On the other hand, as B is a subgraph, and hence a topological minor of G, it follows from Lemma 1 that B is WPO w.r.t. two commodities, a contradiction. So we have $B \in \mathcal{R} \cup \mathcal{T} \cup \mathcal{C}$ as desired.

$(ii) \Rightarrow (i)$: If G has no nontrivial block, then G is a tree. Any routing instance on G has a unique flow because, given any pair of vertices of G, there is only one path between them in G. It is instant that G is WPO w.r.t. two commodities.

So we consider the case where G contains a unique nontrivial block B and $B \in \mathcal{R} \cup \mathcal{C} \cup \mathcal{T}$. Lemmas 6 and 7 say that B is WPO w.r.t. two commodities in case of $B \in \mathcal{C} \cup \mathcal{T}$. When $B \in \mathcal{R}$, it reduces to a single-commodity network, and Theorem 2(i) implies B is WPO w.r.t. two commodities.

Suppose for a contradiction that the NE flow π of some 2-commodity instance $(G, (s_i, t_i)_{i=1}^2, \mathbf{r}, \ell)$ is not WPO. So the instance admits a flow $\mathbf{f}$ such that

$$\text{for each } i \in [2], \ell_P(\mathbf{f}) < \min_{Q \in \mathcal{P}_i} \ell_Q(\pi) \text{ holds for all } P \in \mathcal{P}_i \text{ with } f_i(P) > 0. \quad (4.1)$$

Observe that for any trivial block (which must be a single edge) of G and any $i \in [2]$, either all s_i-t_i paths pass through this edge or none of them passes through it. Thus $f(e) = \pi(e)$ and $\ell_e(f(e)) = \ell_e(\pi(e))$ for any trivial block e of G. Therefore (4.1) implies that for each $i \in [2]$, B contains at least one edge from some s_i-t_i path, and therefore at least one edge from all s_i-t_i paths. Since B is a block, it follows that for each $i \in [2]$, there exist two vertices s_i' and t_i' in B such that $\mathcal{P}_i' = \{P \cap B : P \in \mathcal{P}_i\}$ is the set of s_i'-t_i' paths in B. We now construct a 2-commodity routing instance $(B, (s_i', t_i')_{i=1}^2, \mathbf{r}, \ell')$, where ℓ' is the restriction of ℓ to B. For each $\mathbf{g} \in \{\mathbf{f}, \pi\}$, define $\mathbf{g}_i' : \mathcal{P}_i' \to \mathbb{R}_+$ $(i = 1, 2)$ by $g_i'(P \cap B) = g(P)$, $P \in \mathcal{P}_i$. It is easy to see that $\mathbf{f}'$ and π' are, respectively, a flow and a NE flow of $(B, (s_i', t_i')_{i=1}^2, \mathbf{r}, \ell')$. Notice from (4.1) that for each $i \in [2]$, $\ell_{P'}(\mathbf{f}') < \min_{Q' \in \mathcal{P}_i'} \ell_{Q'}(\pi')$ holds for all $P' \in \mathcal{P}_i'$ with $f_i'(P') > 0$, which shows that B is not WPO w.r.t. two commodities, a contradiction.

$(ii) \Leftrightarrow (iii)$: The implication $(ii) \Rightarrow (iii)$ is trivial. Conversely, given graph G without F_1-minor or F_4-minor, no nontrivial block of G belongs to $\mathcal{S} \setminus (\mathcal{R} \cup \mathcal{T} \cup \mathcal{C})$ as noted in the proof of Lemma 4. Then (ii) is instant from Lemma 2.   □

WPO Digraphs. The task of characterizing WPO digraphs w.r.t. two commodities would be very challenging if not intractable. A good starting point might be investigating the relations between WPO digraphs w.r.t. two commodities and those w.r.t. single commodity. Clearly, the former digraph class $\mathcal{D}_2$ is a subset of the latter digraph class $\mathcal{D}_1$. On the other hand, all digraphs we have found in $\mathcal{D}_1$ belong to $\mathcal{D}_2$. It would be interesting to discover a digraph to show the nonemptyness of $\mathcal{D}_1 - \mathcal{D}_2$, or prove the surprising relation that $\mathcal{D}_1 = \mathcal{D}_2$.

5 Concluding Remarks

In the paper, we have obtained network characterizations for weak Pareto optimality of nonatomic selfish routing in the cases of single commodity and undirected two commodities.

As far as directed networks are concerned, Theorem 4 is a natural extension of Milchtaich's characterization [8] for single-commodity undirected networks to the directed case. On the other hand, the study for two or more commodities might be inherently difficult, due to directional structures which usually do not admit concise descriptions, and are not so algorithmically friendly.

Regarding the undirected case, we have an almost complete solution for characterizing weak Pareto optimality. Complementary to Theorems 2 and 5, the investigation of 3-commodity case provides the following negative result.

Theorem 6. *If G is a graph that contains a cycle of length at least 6, then G is not WPO w.r.t. k commodities for any $k \geq 3$.* $\square$

Theorem 6 implies that the class of WPO graphs G w.r.t more than two commodities would be extremely limited. Assuming G is connected, we deduce from Theorems 5 and 6 that G is a tree, or G has a unique nontrivial block and it belongs to $\mathcal{R}$, or G has a unique nontrivial block and it belongs to $\mathcal{T}$ or is a cycle of length at most 5. In the first two cases, G is WPO w.r.t. any number of commodities. It remains to investigate the weak Pareto optimality w.r.t. k (≥ 3) commodities for graphs in $\mathcal{T}$ and cycles of length at most 5. To fulfill the task, we need develop new tools that help us to avoid tedious case analysis.

References

1. Beckmann, M.J., McGuire, C.B., Winsten, C.B.: Studies in the Economics of Transportation. Yale University Press, New Haven (1956)
2. Braess, D.: Über ein paradoxon aus der verkehrsplanung. Unternehmensforschung **12**(1), 258–268 (1968)
3. Chen, X., Chen, Z., Hu, X.: Excluding braess's paradox in nonatomic selfish routing. In: Hoefer, M., et al. (eds.) SAGT 2015. LNCS, vol. 9347, pp. 219–230. Springer, Heidelberg (2015). doi:10.1007/978-3-662-48433-3_17
4. Epstein, A., Feldman, M., Mansour, Y.: Efficient graph topologies in network routing games. Games Econ. Behav. **66**(1), 115–125 (2009)
5. Holzman, R., yone (Lev-tov), N.L.: Strong equilibrium in congestion games. Games Econ. Behav **21**(1–2), 85–101 (1997)
6. Holzman, R., yone (Lev-tov), N.L.: Network structure and strong equilibrium in route selection games. Math. Soc. Sci. **46**(2), 193–205 (2003)
7. Holzman, R., Monderer, D.: Strong equilibrium in network congestion games: increasing versus decreasing costs. Int. J. Game Theory **44**, 647–666 (2014)
8. Milchtaich, I.: Network topology and the efficiency of equilibrium. Games and Econ. Behav. **57**(2), 321–346 (2006)
9. Roughgarden, T., Tardos, É.: How bad is selfish routing? J. ACM **49**(2), 236–259 (2002)
10. Valdes, J., Tarjan, R.E., Lawler, E.L.: The recognition of series parallel digraphs. SIAM J. Comput. **11**(2), 298 (1982)

New Results for Network Pollution Games

Eleftherios Anastasiadis[1], Xiaotie Deng[2], Piotr Krysta[1(✉)],
Minming Li[3], Han Qiao[4], and Jinshan Zhang[1]

[1] Department of Computer Science, University of Liverpool, Liverpool, UK
{e.anastasiadis,p.krysta,jinshan.zhang}@liverpool.ac.uk
[2] Department of Computer Science and Engineering, Shanghai Jiao Tong University,
Shanghai, China
deng-xt@cs.sjtu.edu.cn
[3] Department of Computer Science, City University of Hong Kong, Hong Kong, China
minming.li@cityu.edu.hk
[4] School of Management, University of Chinese Academy of Sciences, Beijing, China
qiaohan@ucas.ac.cn

Abstract. We study a newly introduced network model of the pollution control
and design approximation algorithms and truthful mechanisms with objective to
maximize the social welfare. On a high level, we are given a graph whose nodes
represent the agents (sources of pollution), and edges between agents represent
the effect of pollution spread. The government is responsible to maximize the
social welfare while setting bounds on the levels of emitted pollution both locally
and globally. We obtain a truthful in expectation FPTAS when the network is
a tree (modelling water pollution) and a deterministic truthful 3-approximation
mechanism. On planar networks (modelling air pollution) the previous result was
a huge constant approximation algorithm. We design a PTAS with a small violation of local pollution constraints. We also design approximation algorithms for
general networks with bounded degree. Our approximations are near best possible under appropriate complexity assumptions.

Keywords: Algorithmic mechanism design · Approximation algorithms · Planar
and tree networks

1 Introduction

Environmental degradation accompanies the advance in technology, resulting in global
water and air pollution. As an example, in 2012, China discharged 68.5 billion tons
of industrial wastewater, and the SO_2 emissions reached 21.2 million tons (National
Bureau of Statistics of China, 2013). The recent annual State of the Air report of the
American Lung Association finds 47 % of Americans live in counties with frequently
unhealthy levels of either ozone or particulate pollution [2]. The latest assessment of

X. Deng was supported by the National Science Foundation of China (Grant No. 61173011)
and a Project 985 grant of Shanghai Jiaotong University. P. Krysta and J. Zhang were supported
by the Engineering and Physical Sciences Research Council under grant EP/K01000X/1. M. Li
was partly supported by a grant from the Research Grants Council of the Hong Kong Special
Administrative Region, China (Project No. CityU 117913). H. Qiao was supported by the
National Science Foundation of China (Grant No. 71373262).

© Springer International Publishing Switzerland 2016
T.N. Dinh and M.T. Thai (Eds.): COCOON 2016, LNCS 9797, pp. 39–51, 2016.
DOI: 10.1007/978-3-319-42634-1_4

air quality, by the European Environment Agency, finds that around 90% of city inhabitants in the European Union are exposed to one of the most damaging air pollutants at harmful levels [1]. Environmental research suggests that water pollution is on of the very significant factor affecting water security worldwide [19]. It is the role of regulatory authorities to make efficient environmental pollution control policies in balancing economic growth and environment protection.

We give new algorithmic results on the pollution control model called a Pollution Game (PG), introduced in [3], and inspired by [6, 15]. We briefly describe applications of PG to air pollution control presented in [3]; for precise definition of PG see Sect. 2. In the first application, the graph's vertices represent pollution sources (agents) and edges are routes of pollution transition from one source to another. The government as the regulator can decide to either shut down or keep open a pollution source (by selling licences to agents) taking into account the diffusion nature of pollution (emission at one source affects the neighbors at diminishing level). It sets bounds on global and local levels of pollution (called global and local constraint(s), resp.), aiming to optimize the social welfare. The emissions exceeding licences, if any, must be cleaned-up (hence, agent's clean-up cost). In the second application [3], vertices represent mayors of cities and edges the roads between cities. The percentage of cars moving from one city to another is represented by the weight of the corresponding edge. The model allows the regulator to auction pollution licences for cars to mayors. The pollution level of an agent (mayor), i.e., the number of allocated licences and their prices, is set by the regulator.

Here we also consider an application of PG to water pollution in rivers, modelled by tree networks. In water pollution the government decides which pollution sources should be shut down so that the effluent level in water is as low as possible. Water pollution cost sharing was introduced in [17] and the network is a path (single river). This model was extended to tree networks (a system of rivers) in [10]. We also model a system of rivers as a tree, but study a different pollution control model, i.e., [3].

Our Results. We present best possible algorithmic results for trees and planar graphs when we allow a small violation of the constraints on local pollution of every agent (called a local constraint). Suppose first that the objective function is linear. Then, for PG on trees we obtain an FPTAS and this is the best we can achieve as PG is weakly NP-hard [3] on stars. For planar graphs the best known result was a big constant approximation algorithm [3]. We design a PTAS with $(1 + \delta)$-violation of the local pollution constraints for any $\delta > 0$, and this is tight as we prove that the problem is strongly NP-hard on planar graphs even with $(1 + \delta)$-violations. By using a Lavi-Swamy technique [16] we prove that our FPTAS for trees leads to a randomized truthful in expectation mechanism. In addition, we also design a deterministic truthful mechanism on trees with an approximation ratio $3 + \epsilon$. Suppose now that the objective function is 2-piecewise linear or general and monotone. Then for graphs with degree at most Δ we obtain $O(\Delta)$-approximation algorithms and a Unique Games-hardness within $\Delta/\log^2 \Delta$.

Technical Contributions/Approaches. Suppose that the objective functions are linear. When the network is a directed tree, a somehow non-standard two level dynamic programming approach is designed to obtain an FPTAS for PG with binary variables. This approach is crucial to deal with the global constraint. For that we design an FPTAS for a

Table 1. Our results. TiE/DT: truthful in expectation/deterministic truthful mechanism. PG(poly) is PG with poly-size integer variables, PG(general) without this assumption.

	General objective function	Linear objective function	
	Bounded degree Δ	Trees	Planar
Lower bound	$\Omega(\frac{\Delta}{\log \Delta^2})$	NP-hard	Strongly NP-hard (δ violation)
PG(poly)	$O(\Delta)^a$	FPTAS TiE | $O(1)$ DT	PTAS (δ violation)
PG(general)	$O(\Delta)$ TiE^b	FPTAS TiE^c	$O(1)$ TiE [3]

^{a}Monotone increasing obj. function. ^{b}Piece-wise linear obj. function with one shift and an additional mild assumption. ^{c}Running time is polynomial in q.

special multiple choice, multi-dimensional knapsack problem where coefficients of all constraints except one are bounded by a polynomial of the input size; this generalizes the results in [7]. A similar idea is applied to design deterministic truthful mechanisms on trees and a PTAS for PG on planar graphs with $(1 + \delta)$-violations.

To obtain our PTAS for planar PG with $(1 + \delta)$-violations, we first use known rounding techniques (e.g., [8, 14]) to make all the coefficients polynomially bounded. Then, we design a dynamic programming approach to solve PG on graphs with bounded tree-width tree decomposition. Finally, we combine a special (called nice) tree decomposition of k-outerplanar graphs, Baker's shifting technique and our two-level dynamic programming approach for dealing with the global constraint, obtaining our PTAS.

Even when polluters' cost functions are linear with a single parameter, simple monotonicity is not sufficient to turn our algorithms into truthful mechanisms. This is because polluters' utility functions have externalities – they are affected by their neighbours. Thus, we need to use general techniques to obtain truthful mechanisms: maximal in range mechanisms (for deterministic truthfulness) and maximal in distributional range mechanisms (for truthfulness in expectation). The deterministic truthful mechanism for trees uses a maximum in range technique (Chaps. 11 and 12 in [18]).

For piece-wise linear objective functions on bounded degree graphs we prove that PG is Δ column sparse so a randomized algorithm of [5] is applicable. For general monotone objective functions on bounded degree graphs we prove that the objective function is submodular and use randomized rounding with alterations.

Organization. Our results are summarized in (Table 1). Section 2 contains definitions and preliminaries, and our results on trees are in Sect. 3. Section 4 presents our results on planar graphs, and, finally, Sect. 5 discusses general objective functions. All missing details and proofs will appear in the full version.

2 Preliminaries

Model and Applications. We describe the model and mention two applications following [3] to gain an intuition. Consider an area of pollution sources (e.g. factories) each owned by an agent. The government's goal as a regulator is to optimize the social welfare, restricting levels of emitted pollution. Thus, given a weighted digraph $G = (V, E)$, where V is the set of n pollution sources (players, agents) and edge $(u, v) \in E$ means u and v are geographic neighbours, i.e., $(u, v) \in E$ if the pollution emitted by u affects

v. For each $(u, v) \in E$ weight $w_{(u,v)} = w_{uv}$ is a discount factor of the pollution discharged by player u affecting its neighbour v. W.l.o.g., $w_{uv} \in (0, 1]$, $\forall (u, v) \in E$.

The government sets the total pollution quota discharged to the environment (by the number of pollution sources that remain open) to be $p \geq \sum_{v \in V} x_v$, where $x_v \in \{0, 1\}$ denotes if pollution source $v \in V$ will be shut down or not. Each agent v has a non-decreasing benefit function $b_v : \mathbb{R}_{\geq 0} \longrightarrow \mathbb{R}_{\geq 0}$, where $b_v(x_v)$ is a concave increasing function with $b_v(0) = 0$, representing v's benefit. Each v has a non-decreasing damage function $d_v : \mathbb{R}_{\geq 0} \longrightarrow \mathbb{R}_{\geq 0}$, and b_v is concave increasing, $b_v(0) = 0$ and d_v is convex increasing[1]. Player v's total welfare r_v is v's benefit minus damage cost: $b_v(x_v) - d_v\left(x_v + \sum_{u \in \delta_G^-(v)} w_{uv} x_u\right)$, where, $\delta_G^-(v) = \{u \in V : (u, v) \in E\}$, $\delta_G^+(v) = \{u \in V : (v, u) \in E\}$. Thus, v is affected via the damage function by his own pollution if $x_v \neq 0$ and by the total discounted pollution neighbours. This models that pollution spreads along the edges of G. The government decides on the allowable local level of pollution p_v, for every $v \in V$, which imposes the following constraints for every $v \in V$: $x_v \leq q_v$, $x_v + \sum_{u \in \delta_G^-(v)} w_{uv} x_u \leq p_v$. The first application assumes $x_v \in \{0, 1\}$ and $q_v = 1$, $\forall v \in V$ and the second $x_v \in \{0, 1, \ldots, q_v\}$ and $q_v \in \mathbb{N}$.

The problem of social welfare maximization is the following convex integer program (1)–(4), called a pollution game (PG) on G, where (2) is called global constraint, (3) are local constraints,

$$\max \; R(x) = \sum_{v \in V} (b_v(x_v) - d_v(x_v + \sum_{u \in \delta_G^-(v)} w_{uv} x_u)) \quad (1)$$

$$\text{s.t.} \; \sum_{v \in V} x_v \leq p \quad (2)$$

$$x_v + \sum_{u \in \delta_G^-(v)} w_{uv} x_u \leq p_v, \; \forall v \in V \quad (3)$$

$$x_v \in \{0, 1, \ldots, q_v\}, \; \forall v \in V \quad (4)$$

and $x_v + \sum_{u \in \delta_G^-(v)} w_{uv} x_u$ is the local level of pollution of v. Value q_v is decided by the government and for this application $q_v = 1$. We call (1)–(4), PG with integer variables (if $x_v \in \mathbb{Z}$) or with binary variables (if $x_v \in \{0, 1\}$). For an instance I of PG, $|I|$ is the number of bits to encode I, and if $q \in poly(|I|)$, $q = \max_{v \in V}\{q_v\} + 1$, we call (1)–(4), PG with polynomial size integer variables.

Basic Definitions. Let $I = (G, \mathbf{b}, \mathbf{d}, \mathbf{p}, \mathbf{q})$ be an instance of PG, $\mathbf{b} = (b_v)_{v \in V}$, $\mathbf{d} = (d_v)_{v \in V}$, $\mathbf{p} = (p_v)_{v \in V}$ and $\mathbf{q} = (q_v)_{v \in V}$ (b_v is private information of v and other parameters are public). Let $\mathcal{I}$ be the set of all instances, and $\mathcal{X}$ the set of feasible allocations. Given a digraph $G = (V, E)$, $G^{un} = (V, E^{un})$, where $E^{un} = \{(u, v) : (u, v) \in E \text{ or } (v, u) \in E\}$. A mechanism $\phi = (X, P)$ consists of an allocation $X : \mathcal{I} \to \mathcal{X}$ and payment function $P : \mathcal{I} \to \mathbb{R}_{\geq 0}^{|V|}$ ($X(I)$ satisfies (2)–(4)). For any vector x, x_{-u} denotes vector x without its u-th component. Note, $r_v(X(I)) = b_v(X_v(I)) - d_v(X_v(I) + \sum_{u \in \delta_G^-(v)} w_{uv} X_u(I))$ is the welfare of player v under $X(I)$. A mechanism $\phi = (X, P)$ is truthful, if for any b_{-v}, b_v and b_v', $r_v(X(b_v, b_{-v})) - P_v(b_v, b_{-v}) \geq r_v(X(b_v', b_{-v})) - P_v(b_v', b_{-v})$. A randomized mechanism is truthful in expectation if for any b_{-v}, b_v and b_v', $\mathbb{E}(r_v(X(b_v, b_{-v}))$ –

---

[1] [15] uses cost function rather than benefit function, viewed as $M_v - b_v(x_v)$, with M_v a large constant for any $v \in V$. The cost function is convex decreasing and it is equivalent to $b_v(x_v)$ being a concave increasing function. We use benefit function rather than cost function.

$P_v(b_v, b_{-v})) \geq \mathbb{E}(r_v(X(b'_v, b_{-v})) - P_v(b'_v, b_{-v}))$, where $\mathbb{E}(\cdot)$ is over the algorithm's random bits. $OPT_G^{fr}(PG)$ $(OPT_G^{in}(PG)$, resp.) denotes the value of the optimal fractional (integral, resp.) solution of PG on G. A mechanism is individually rational if each agent v has non-negative utility when he declares b_v, regardless of the other agents' declarations. The approximation ratio of an algorithm $\mathcal{A}$ w.r.t. $OPT_G^{in}(PG)$ (resp. $OPT_G^{fr}(PG)$) is $\eta^{in}(\mathcal{A}) = \frac{OPT_G^{in}(PG)}{R(\mathcal{A})}$ $(\eta^{fr}(\mathcal{A}) = \frac{OPT_G^{fr}(PG)}{R(\mathcal{A})})$, where $R(\mathcal{A})$ is the objective value of the $\mathcal{A}$'s solution. If unspecified, the approximation ratio refers to η^{in}. An FPTAS (PTAS, resp.) for a problem $\mathcal{P}$ is an algorithm $\mathcal{A}$ that for any $\epsilon > 0$ and any instance I of $\mathcal{P}$, outputs a solution with the objective value at least $(1-\epsilon)OPT_I^{in}(\mathcal{P})$ and terminates in time $poly(\frac{1}{\epsilon}, |I|)$ $((\frac{1}{\epsilon}|I|)^{g(\frac{1}{\epsilon})}$, resp.), where g is a function independent from I. Let $\gamma_k = \min\{2k^2 + 2, 8k, \frac{k}{(1-\frac{1}{k}(1+(\frac{2}{k})^{\frac{1}{3}}))^k}\} = (e+o(1))k = O(k)$, and $[n] = \{1, \ldots, n\}$. We use 'vertex' to denote the vertex in a graph and 'node' to denote a vertex of the tree obtained from a tree decomposition of a graph. An undirected graph is an outerplanar if it can be drawn in the plane without crossings in such a way that all of the vertices belong to the unbounded face of the drawing. An undirected graph G is k-outerplanar if for $k = 1$, G is outerplanar and for $k > 1$, G has a planar embedding such that if all vertices on the exterior face are deleted, the connected components of the remaining graph are all $(k-1)$-outerplanar. An planar graph is k outerplanar where k can be equal to $+\infty$. A digraph is called a planar graph if its undirected version is planar. We consider some standard embedding of a planar graph and define level k vertices in a planar embedding E of a planar graph G. A vertex is at level 1 if it is on the exterior face. Call a cycle of level i vertices a level i face if it is an interior face in the subgraph induced by the level i vertices. For each level i face f, let G_f be the subgraph induced by all vertices placed inside f in this embedding. Then the vertices on the exterior face of G_t are at level $i + 1$.

In Sects. 3 and 4 we assume that b_v and d_v are both linear with slopes s_v^0 and s_v^1 respectively, i.e., $b_v(x) = s_v^0 x$ and $d_v(y) = s_v^1 y$, for any $v \in V$. The social welfare function is $R(x) = \sum_{v \in V} \omega_v x_v$, where $\omega_v = s_v^0 - s_v^1 - \sum_{u \in \delta_G^+(v)} s_u^1 w_{vu}$ $(R(x) = \sum_{v \in V} b_v(x_v) - d_v(x_v + \sum_{u \in \delta_G^-(v)} w_{uv} x_u) = \sum_{v \in V} s_v^0 x_v - s_v^1(x_v + \sum_{u \in \delta_G^-(v)} w_{uv} x_u) = \sum_{v \in V} \omega_v x_v)$.

3 Directed Trees

Truthful in Expectation Mechanisms. A digraph G is called a *directed tree* if the undirected graph G^{un} is a tree. We consider trees where arcs are directed towards the leaves. We obtain our truthful in expectation FPTAS for PG with binary variables on any directed trees by a two-level dynamic programming (DP) approach (used also in Sect. 4). The first bottom-up level is based on a careful application of the standard single-dimensional knapsack FPTAS. The second level is by an interesting generalization of an FPTAS of [7] for a special multi-dimensional knapsack problem, see (IP_2) below, with a constant number of constraints most of which have $poly(|I|)$ size of coefficients. This FPTAS generalizes the results in [7], where the authors consider the one dimensional knapsack problem with cardinality constraint; it will appear in our paper's full version.

We will also need the following tool from mechanism design for packing problems. An integer linear packing problem with binary variables is a problem of maximising a linear objective function over a set of linear packing constraints, i.e., constraints of form $a \cdot x \le b$ where $x \in \{0,1\}^n$ is a vector of binary variables, and $a, b \in \mathbb{R}^n_{\ge 0}$.

Proposition 1 *[11]. Given an FPTAS for an integer linear packing problem with binary variables, there is a truthful in expectation mechanism that is an FPTAS.*

We first present an FPTAS without constraint (2) which captures our main technique.

Warmup (Without Global Constraint). The algorithm uses a DP and FPTAS for knapsack as a subroutine. Note, on a star, any instance of knapsack can be reduced

$$\max \sum_{i \in [n_v]} (M_{u_i\text{in}}^{v\text{in}} x_{u_i} + M_{u_i\text{out}}^{v\text{in}} (1 - x_{u_i})) + \omega_v$$
$$\text{s.t. } 1 + w_{v'v} + \sum_{i \in [n_v]} w_{u_iv} x_{u_i} \le p_v, \quad (IP_1)$$
$$x_{u_i} \in \{0,1\}, \ \forall i \in [n_v]$$

to PG without global constraint. Thus FPTAS is the best possible for such PG unless $P = NP$.

We keep four values for each $v \in V$. Suppose v's father is v', let $M_{v\text{in}}^{v'\text{in}}$ denote the optimal value of PG on subtree rooted at v when both v' and v are selected in the solution. Similarly, we have $M_{v\text{out}}^{v'\text{in}}$, $M_{v\text{in}}^{v'\text{out}}$ and $M_{v\text{out}}^{v'\text{out}}$. Let u_i, $i = 1, 2, ..., n_v$ denote children of v. Suppose $M_{u_i\text{in}}^{v\text{in}}$, $M_{u_i\text{out}}^{v\text{in}}$, $M_{u_i\text{in}}^{v\text{out}}$ and $M_{u_i\text{out}}^{v\text{out}}$ have been calculated, for any $i = 1, ..., n_v$. Some of them are undefined due to infeasibility. Now, calculate $M_{v\text{in}}^{v'\text{in}}$. Observe, $M_{v\text{in}}^{v'\text{in}}$ is equal to the optimal value of the knapsack (IP_1), where $M_{u_i\text{in}}^{v\text{in}}$ and $M_{u_i\text{out}}^{v\text{in}}$ have finite values (otherwise remove them). If this knapsack problem has a feasible solution, we get value $M_{v\text{in}}^{v'\text{in}}$, otherwise set $M_{v\text{in}}^{v'\text{in}}$ undefined. Similarly, calculate $M_{v\text{out}}^{v'\text{in}}$, $M_{v\text{in}}^{v'\text{out}}$ and $M_{v\text{out}}^{v'\text{out}}$. Thus, at each step if we calculate an optimal solution, it will be obtained by above DP approach. For knapsack with n_v variables, there is an FPTAS. Hence, at each step we get approximate value $\bar{M}_{v\text{in}}^{v'\text{in}} \ge (1 - \epsilon) M_{v\text{in}}^{v'\text{in}}$ in poly-time in n_v and $\frac{1}{\epsilon}$ by knapsack's FPTAS; similarly for other three values. Thus, in the final solution, $\bar{M}_{root} \ge (1 - \epsilon)^k M_{root}$, where k is the number of levels of the tree and M_{root} is PG's optimal value without global constraint, terminating in $poly(|I|, \frac{1}{\epsilon})$ time; $|I|$ is the input size. Set $1 - \epsilon' = (1 - \epsilon)^k$, then $\epsilon = \Theta(\frac{\epsilon'}{k})$. The run time is $poly(|I|, \frac{k}{\epsilon'}) = poly(|I|, \frac{1}{\epsilon'})$ due to $k \le |I|$, giving FPTAS for PG without global constraint.

W.l.o.g., suppose $p \le n$, otherwise let $p = n$. For each vertex v, we keep $4p$ values. Let v's father be v', and let $M_{v\text{in}}^{v'\text{in}}(s)$ be the optimal value of PG on the subtree rooted at v when

$$\max \sum_{i \in [n_v]} \sum_{s \in [p]} (M_{u_i\text{in}}^{v\text{in}}(s) x_{is} + M_{u_i\text{out}}^{v\text{in}}(s) y_{is})$$
$$\text{s.t. } \sum_{i \in [n_v]} \sum_{s \in [p]} s(x_{is} + y_{is}) \le \ell - 1,$$
$$\sum_{s=0}^{p} (x_{is} + y_{is}) = 1, \ \forall i \in [n_v] \quad (IP_2)$$
$$1 + w_{v'v} + \sum_{i \in [n_v]} [w_{u_iv} (\sum_{s=0}^{p} x_{is})] \le p_v,$$
$$x_{is}, y_{is} \in \{0,1\}, \ \forall i \in [n_v], s \in [p]$$

both v' and v are selected in the solution, and the total pollution level allocated to the subtree rooted at v is $\le s$, $s = 0, 1, ..., p$. Similarly, we have $M_{v\text{out}}^{v'\text{in}}(s)$, $M_{v\text{in}}^{v'\text{out}}(s)$ and $M_{v\text{out}}^{v'\text{out}}(s)$. Let u_i, $i \in [n_v]$ denote the children of v. Suppose $M_{u_i\text{in}}^{v\text{in}}(s)$, $M_{u_i\text{out}}^{v\text{in}}(s)$, $M_{u_i\text{in}}^{v\text{out}}(s)$ and $M_{u_i\text{out}}^{v\text{out}}(s)$ have been calculated, for any $i \in [n_v]$ and $s = 0, 1, ..., p$. Some of them are undefined due to infeasibility. Note, $M_{u_i\text{in}}^{v\text{in}}(0)$, $M_{u_i\text{in}}^{v\text{out}}(0)$ are undefined and $M_{u_i\text{out}}^{v\text{in}}(0) = M_{u_i\text{out}}^{v\text{out}}(0) = 0$. Now, calculate $M_{v\text{in}}^{v'\text{in}}(\ell)$. Observe, $M_{v\text{in}}^{v'\text{in}}(\ell)$ is equal to the optimal value of the knapsack problem (IP_2) (called $\text{KNAPSACK}_v(\ell)$) plus ω_v. If $M_{u_i\text{in}}^{v\text{out}}(s)$ and $M_{u_i\text{out}}^{v\text{out}}(s)$ are undefined, they are removed from KNAPSACK_v

(ℓ). Note, $x_{i0} \equiv 0$, for any $i \in [d]$. If KNAPSACK_v (ℓ) has a feasible solution, we get the value $M_{vin}^{v'in}(\ell)$, otherwise set $M_{vin}^{v'in}(\ell)$ undefined. Similarly, calculate $M_{vout}^{v'in}(\ell)$, $M_{vin}^{v'out}(\ell)$, $M_{vout}^{v'out}(\ell)$, $\ell = 0, 1, ..., p$. From the analysis of DP without global constraint, if there is an FPTAS for KNAPSACK_v (ℓ), then there is one for KNAPSACK_{root} (p), and so an FPTAS for PG with binary variables on directed trees. Note, the second constraint in (IP_2) can be replaced by $\sum_{s=1}^{p}(x_{is} + y_{is}) \leq 1, \forall i \in [n_v]$. Then, by Proposition 1:

Theorem 1. *There is a truthful in expectation mechanism for PG with binary variables on directed trees, which is an FPTAS.*

For $x_v \in \mathbb{Z}$, we can replace each x_v by q_v duplicated variables $x_{vj}, j = 1, \cdots, q_v$, i.e., $\{x_v \in \{0, 1, ..., q_v\}\} = \{\sum_{j \in [q_v]} jx_{vj} \mid \sum_{j \in [q_v]} x_{vj} \leq 1, x_{vj} \in \{0, 1\}\}$. This transforms a poly-size integer constraint into a multiple choice, one dimensional knapsack constraint. Hence, for directed trees, by a DP, we can construct a pseudo poly-time algorithm to compute the exact optimal value of PG with integer variables, in time $poly(|V|, q, OPT^{in}(PG))$. And, we can remove $OPT^{in}(PG)$ from the running time losing an ϵ by scaling techniques, implying a $(1 - \epsilon)$-approximation algorithm for PG with integer variables with time $poly(|V|, q, 1/\epsilon)$. By Proposition 1:

Theorem 2. *There is a truthful in expectation mechanism for PG with polynomial size integer variables on directed trees, which is an FPTAS.*

Deterministic Truthful Mechanisms. We use a maximal in range (MIR) mechanism for PG with polynomial size integer variables on directed trees. By transformation from integer constraint into multiple choice and one dimensional knapsack constraint, we know we only need to show such approximation algorithm for binary variables. Based on recent deterministic truthful PTAS for 2 dimensional knapsack[2] [8,9,14] we obtain:

Theorem 3. *There is a deterministic $(\eta^{in} = 3 + \epsilon)$-approximation truthful mechanism for PG with polynomial size integer variables on directed trees, which for binary variables terminates in $O(|V|^2 \Delta^{6+\frac{1}{\epsilon}})$ time.*

4 Planar Graphs

A PTAS with δ-violation: Our approach to obtain a PTAS has three main steps:

1. Round PG to an equivalent problem $\bar{PG}_2$ with polynomial size integer variables.
2. Using the nice tree decomposition, we present a dynamic programming approach to solve $\bar{PG}_2$ optimally on an k-outerplanar graph.
3. By a shifting technique similar to [4], we obtain a PTAS with $1 + \delta$ violation.

Step 1: Rounding Procedure. Recall that PG is equivalent to maximizing $\sum_{v \in V} \omega_v x_v$ subject to constraints (1)–(3) where $\omega_v = \max\{0, s_v^0 - s_v^1 - \sum_{u \in \delta_G^+(v)} s_u^1 w_{vu}\}$ and $w_{v,v} = 1 \ \forall v \in V$, and b_v and d_v are both linear with slopes s_v^0 and s_v^1. For each

---

[2] This PTAS also works for multiple choice and constant dimensional knapsack problem, which will be used for PG with polynomial size integer variables.

$v \in V$, suppose $q_v \in [2^{o_v-1} - 1, 2^{o_v} - 1)$. Let $o_v = \lfloor \log_2(q_v) \rfloor + 1$ if $q_v \neq 2^{o_v-1} - 1$ and $o_v = \lfloor \log_2(q_v) \rfloor + 2$ otherwise; $c_v^i = 2^{i-1}$, $i \in [o_v - 1]$ and $c_v^{o_v} = q_v - 2^{o_v-1} + 1$. Notice, $\{x_v \mid x_v \in \mathbb{Z}, 0 \leq x_v \leq q_v\} = \{\sum_{i=1}^{o_v} c_v^i y_v^i \mid y_v^i \in \{0,1\}, i \in [o_v]\}$, for any $v \in V$. Thus, PG is equivalent to the following integer program (denoted as PG'):

$$\begin{array}{ll}
\max \sum_{v \in V} \sum_{i=1}^{o_v} \omega_v c_v^i y_v^i & (PG') \\
\text{s.t.} \sum_{v \in V} \sum_{i=1}^{o_v} c_v^i y_v^i \leq p, \\
\forall v \in V: \ \sum_{i=1}^{o_v} w_{vv} c_v^i y_v^i + \\
\quad + \sum_{u \in \delta_G^-(v)} \sum_{i=1}^{s_u} w_{uv} c_u^i y_u^i \leq p_v, \\
\forall v \in V, i \in [o_v]: \ y_v^i \in \{0,1\},
\end{array}
\qquad
\begin{array}{ll}
\max \sum_{v \in V} \sum_{i=1}^{o_v} \omega_v b_v^i y_v^i & (\bar{PG}_1) \\
\text{s.t.} \sum_{v \in V} \sum_{i=1}^{o_v} c_v^i y_v^i \leq p \\
\forall v \in V: \ \sum_{i=1}^{o_v} \bar{w}_{vv}^i y_v^i + \\
\quad + \sum_{u \in \delta_G^-(v)} \sum_{i=1}^{s_u} \bar{w}_{uv}^i y_v^i \leq \bar{p}_v \\
\forall v \in V, i \in [o_v]: \ y_v^i \in \{0,1\}
\end{array}$$

Let $o^* = \max_{v \in V} o_v$ and $\rho = o^* |V|$. Recall that $q = \max_{v \in V}\{q_v\} + 1$. For any $\delta > 0$, let $\bar{w}_{uv}^i = \lfloor \frac{2 w_{uv} c_v^i \rho}{p_v \delta} \rfloor$ and $\bar{p}_v = \lceil \frac{2 p_v \rho}{p_v \delta} \rceil = \lceil \frac{2\rho}{\delta} \rceil$, for any $u, v \in V$. Then we have the following modified PG' (denoted as $\bar{PG}_1$ – see above).

Lemma 1. *Any feasible solution of PG' is feasible in $\bar{PG}_1$, and any feasible solution of $\bar{PG}_1$ is feasible for PG except violating each local constraint by a factor of $1 + \delta$.*

Proof. We only prove local constraints for each direction since the proof of the global constraint is similar. Let $\{y_v^i\}_{v \in V, i \in [o_v]}$ be a feasible solution of PG'. We know that $\sum_{i=1}^{o_v} w_{vv} c_v^i y_v^i + \sum_{u \in \delta_G^-(v)} \sum_{i=1}^{s_u} w_{uv} c_u^i y_v^i \leq p_v$, $\forall v \in V$. Then $\sum_{i=1}^{o_v} \bar{w}_{vv}^i y_v^i + \sum_{u \in \delta_G^-(v)} \sum_{i=1}^{s_u} \bar{w}_{uv}^i y_v^i \leq \frac{2\rho}{p_v}(\sum_{i=1}^{o_v} w_{vv} c_v^i y_v^i + \sum_{u \in \delta_G^-(v)} \sum_{i=1}^{s_u} w_{uv} c_u^i y_v^i) \leq \frac{2\rho}{p_v \delta} p_v \leq \bar{p}_v$ as desired. On the other hand, suppose $\{y_v^i\}_{v \in V, i \in [o_v]}$ is a feasible solution of $\bar{PG}_1$. We know $\sum_{i=1}^{o_v} \bar{w}_{vv}^i y_v^i + \sum_{u \in \delta_G^-(v)} \sum_{i=1}^{s_u} \bar{w}_{uv}^i y_v^i \leq \bar{p}_v$, $\forall v \in V$. Then $\sum_{i=1}^{o_v} w_{vv} c_v^i y_v^i + \sum_{u \in \delta_G^-(v)} \sum_{i=1}^{s_u} w_{uv} c_u^i y_v^i \leq \frac{p_v \delta}{2\rho} [\sum_{i=1}^{o_v}(\bar{w}_{vv}^i + 1) y_v^i + \sum_{u \in \delta_G^-(v)} \sum_{i=1}^{s_u}(\bar{w}_{uv}^i + 1) y_v^i] \leq \frac{p_v \delta}{2\rho} \bar{w}_{vv}^i y_v^i + \sum_{u \in \delta_G^-(v)} \sum_{i=1}^{s_u} \bar{w}_{uv}^i y_v^i + \frac{p_v \delta \rho}{2\rho} \leq \frac{p_v \delta \bar{p}_v}{2\rho} + \frac{p_v \delta}{2} \leq \frac{p_v \delta}{2\rho}(\frac{2\rho}{\delta} + 1) + \frac{p_v \delta}{2} \leq p_v(1 + \delta)$, $\forall v \in V$. $\qquad \square$

$$\begin{array}{ll}
\max & \sum_{v \in V} \omega_v x_v \qquad (\bar{PG}_2) \\
\text{s.t.} & \sum_{v \in V} x_v \leq p \\
\forall v \in V: & \bar{w}_{vv}(x_v) + \sum_{u \in \delta_G^-(v)} \bar{w}_{uv}(x_u) \leq \bar{p}_v \\
\forall v \in V: & x_v \in \Lambda_v
\end{array}$$

Note, for each $\ell \in [q_v]$, there is a solution $\{y_v^i\}_{i \in [o_v]}$ s.t. $\sum_{i=1}^{o_v} c_v^i y_v^i = \ell$. We use the following solution: If $\ell \leq 2^{o_v-1} - 1$, set $y_v^{o_v} = 0$ and there is a unique solution $\sum_{i=1}^{o_v} c_v^i y_v^i = \ell$; If $2^{o_v-1} - 1 < \ell \leq q_v$, set $y_v^{o_v} = q_v - 2^{o_v-1} + 1$ and there is also a unique solution s.t. $\sum_{i=1}^{o_v} c_v^i y_v^i = \ell$. Hence, there is one-to-one correspondence from x_v to $\{y_v^i\}_{i \in [o_v]}$. Notice that for a given x_v, the above defined solution $\{y_v^i\}_{i \in [o_v]}$ is the one such that $\sum_{i=1}^{o_v} \bar{w}_{vv}^i y_v^i + \sum_{u \in \delta_G^-(v)} \sum_{i=1}^{s_u} \bar{w}_{uv}^i y_v^i$ is minimized. Now let $\bar{w}_{vu}(x_v) = \sum_{i=1}^{o_v} \bar{w}_{vu}^i y_v^i$, for any $v, u \in V$, where $\{y_v^i\}_{i \in [o_v]}$ is according to the above solution corresponding to x_v. Let $\Lambda_v = [q_v] \cup \{0\}$. Thus, $\bar{PG}_1$ (also PG) is equivalent to the integer program (denoted as $\bar{PG}_2$, see above).

Step 2: Preliminaries of Tree Decompositions on k-Outerplanar Graphs.

Definition 1. *A tree decomposition of an undirected graph* $G = (V, E)$ *is a pair* $(\{X_i | i \in I\}, T = (I, F))$, *with* $\{X_i | i \in I\}$ *a family of subsets of* V, *one for each node of* T, *and* T *a tree such that: (1)* $\bigcup_{i \in I} X_i = V$, *(2) for all edges* $(v, w) \in E$, *there exists an* $i \in I$ *with* $v \in X_i$ *and* $w \in X_i$, *(3) for all* i, j, $k \in I$: *if* j *is on the path from* i *to* k *in* T, *then* $X_i \cap X_k \subseteq X_j$. *The width of a tree decomposition* $(\{X_i | i \in I\}, T = (I, F))$ *is* $\max_{i \in I} |X_i| - 1$. *The minimum width of all tree decompositions of* G *is called treewidth.*

Definition 2. *A tree decomposition* $(\{X_i | i \in I\}, T = (I, F))$ *of* $G = (V, E)$ *is called a nice tree decomposition if* T *is a rooted binary tree and (1) if a node* $i \in I$ *has two children* j *and* k, *then* $X_i = X_j = X_k$ *(joint node), (2) if a node* $i \in I$ *has one child* j, *then either* $X_i \subset X_j$, *and* $|X_i| = |X_j| - 1$ *(forget node), or* $X_j \subset X_i$ *and* $|X_j| = |X_i| - 1$ *(introduce node), (3) if node* $i \in I$ *is a leaf of* T, *then* $|X_i| = 1$ *(leaf node).*

Lemma 2 [12]. *For any* k-*outerplanar graph* $G = (V, E)$, *there is an algorithm to compute a tree decomposition* $(\{X_i | i \in I\}, T = (I, F))$ *of* G *with treewidth at most* $3k - 1 = O(k)$, *and* $I = O(|V|)$ *in* $O(k|V|)$ *time.*

Given a tree decomposition $(\{X_i | i \in I\}, T = (I, F))$ for $G = (V, E)$ with treewidth k and $I = O(|V|)$, we can obtain a nice tree decomposition with the same treewidth k and the number of nodes $O(k|V|)$ in $O(k^2|V|)$ time [13]. Thus, for any k-outerplanar graph $G = (V, E)$, we can compute a nice tree decomposition $(\{X_i | i \in I\}, T = (I, F))$ of G with treewidth at most $3k - 1 = O(k)$, and $I = O(k|V|)$ in $O(k^2|V|)$ time. In the following, we suppose there is a nice tree decomposition for any k-outerplanar graph.

Dynamic Programming (DP). A DP to solve $\bar{P}G_2$ on a k-outerplanar digraph is presented by using a nice tree decomposition of its undirected version. Note, a nice tree decomposition of an undirected version of digraph is also a nice tree decomposition of itself. Given nice tree decomposition $(\{X_i | i \in I\}, T = (I, F))$ of a k-outerplanar digraph $G = (V, E)$, using a bottom-up approach, DP for $\bar{P}G_2$ works as follows.

For any node $i \in I$, suppose $X_i = \{v_1^i, v_2^i, \cdots, v_t^i\}$, where $t \leq 3k$. We also say vertex v_1^i belongs to node X_i, similarly we can say a vertex belongs to a subtree of T, meaning this vertex belongs to some node of this subtree. Given any emission amount $\{x_v\}_{v \in V}$, recall $\bar{w}_{vv}(x_v) + \sum_{u \in \delta_G^-(v)} \bar{w}_{uv}(x_u)$ is the local level of pollution of vertex v. We use $\boldsymbol{a^i} = (a_1^i, a_2^i, \cdots, a_t^i)$ to denote the emission amount allocated to vertices in X_i, i.e., a_s^i denotes the emission amount allocated to the vertex v_s^i, $s \in [t]$. Similarly $\boldsymbol{\ell^i}$ denotes the local levels of pollution of vertices in X_i. Let G_i denote the subgraph generated by all the vertices belonging to the subtree (node X_i) rooted at X_i. We use Q^i to denote the total emission quota allocated to G_i. Let $\Omega_i(\boldsymbol{a^i}, \boldsymbol{\ell^i}, Q^i)$ denote the optimal objective value of $\bar{P}G_2$ restricted on the subgraph G_i, when the emission amount and local level of pollution of v_s^i are exactly a_s^i and ℓ_s^i, $s \in [t]$, and the total emission amount allocated to G_i is exactly Q^i. If there is no feasible solution for $\Omega_i(\boldsymbol{a^i}, \boldsymbol{\ell^i}, Q^i)$, we will see that our DP approach will automatically set $\Omega_i(\boldsymbol{a^i}, \boldsymbol{\ell^i}, Q^i)$ to be $-\infty$. Let $\bar{w}_{uv}(x_v) \equiv 0$ if (u, v) is not an edge in G. Note that the range of a_s^i we need to compute is in Λ_v, and ℓ_s^i is from 0 to $\bar{p}_{v_s^i}$, $s \in [t]$, Q^i is from 0 to p. We present the DP approach

- X_i is a leaf node or a start node, where $t = 1$. $\Omega_i(a_1^i, \ell_1^i, Q^i) = \omega_{v_1^i} a_1^i$ if the triple (a^i, ℓ^i, Q^i) is feasible, which can be verified easily e.g. $Q^i = a_1^i$ and $\ell_1^i = \bar{w}_{v_1^i v_1^i}(a_1^i)$. Let $\Omega_i(a_1^i, \ell_1^i, Q^i) = -\infty$ if the triple (a^i, ℓ^i, Q^i) is not feasible.
- X_i is a forget node, and suppose its child is $X_j = X_i \cup \{v_{t+1}^j\}$.
  $$\Omega_i(\boldsymbol{a^i}, \boldsymbol{\ell^i}, Q^i) = \max_{a_{t+1}^j, \ell_{t+1}^j} \Omega_j(\boldsymbol{a^i}, a_{t+1}^j, \boldsymbol{\ell^i}, \ell_{t+1}^j, Q^i)$$
- X_i is an introduce node, and suppose its child is $X_j = X_i \backslash \{v_t^i\}$. Let $a_s^j = a_s^i$ and $\ell_s^j = \ell_s^i - \bar{w}_{v_t^i v_s^i}(a_t^i), \forall s \in [t-1]$. $\Omega_i(\boldsymbol{a^i}, \boldsymbol{\ell^i}, Q^i) = \Omega_j(\boldsymbol{a^j}, \boldsymbol{\ell^j}, Q^i - a_t^i) + \omega_{v_t^i} a_t^i$ if $\sum_{s \in [t]} \bar{w}_{v_t^i v_t^i}(a_s^i) = \ell_t^i$, and $\Omega_i(\boldsymbol{a^i}, \boldsymbol{\ell^i}, Q^i) = -\infty$ otherwise.
- X_i is a joint node, and suppose its two children are $X_j = X_k = X_i$. $\Omega_i(\boldsymbol{a^i}, \boldsymbol{\ell^i}, Q^i) = \max_A \{\Omega_j(\boldsymbol{a^j}, \boldsymbol{\ell^j}, Q^j) + \Omega_k(\boldsymbol{a^k}, \boldsymbol{\ell^k}, Q^k)\}$, where the condition $A = \{(\boldsymbol{a^j}, \boldsymbol{\ell^j}, Q^j), (\boldsymbol{a^k}, \boldsymbol{\ell^k}, Q^k) \mid \boldsymbol{a^j} + \boldsymbol{a^k} = \boldsymbol{a^i}, \boldsymbol{\ell^j} + \boldsymbol{\ell^k} = \boldsymbol{\ell^i}, Q^j + Q^k = Q^i\}$.
- X_i is the root of T, $OPT(Q^i) = \max_{\boldsymbol{a^i}, \boldsymbol{\ell^i}} \{\Omega_i(\boldsymbol{a^i}, \boldsymbol{\ell^i}, Q^i)\}$ is the optimal value (social welfare) of $\bar{P}G_2$ when total scaled emission amount is exactly Q^i, i.e., the global constraint satisfies $\sum_{v \in V} b_v x_v = Q^i$.

Analysis of Running Time of DP. It is not difficult to see that the above DP approach gives the correct solution of $\bar{P}G_2$ on k-outerplanar graphs. For each node X_i, we need to keep $O(pq^{3k} \lceil \frac{2\rho}{\delta} \rceil^{3k}) = O(|V|q^{3k+1} \lceil \frac{2\rho}{\delta} \rceil^{3k})$ number of Ω_i values. Each Ω_i can be computed in $O(|V|q^{3k+1} \lceil \frac{2\rho}{\delta} \rceil^{3k})$ time (this is the worst case running time when X_i is a joint node). There are $O(k|V|)$ nodes in T. Therefore, the total running time of the DP approach (by multiplying above three numbers) is $O(k|V|^3 q^{6k+2} \lceil \frac{2\rho}{\delta} \rceil^{6k})$.

Based on the above DP approach, we can solve $\bar{P}G_2$ on any k-outerplanar graph optimally for any fixed k (which includes any directed tree whose treewidth is 2). Therefore, for any $\delta > 0$ and fixed k, we can use VCG (see, e.g., Chap. 9 in [18]) to get an optimal deterministic truthful mechanism for PG on any directed k-outerplanar graph that violates each local constraint by a factor of δ and runs in $O(k|V|^3 q^{6k+2} \lceil \frac{2\rho}{\delta} \rceil^{6k})$ time (note that Theorem 4 also works for bounded treewidth graphs).

Theorem 4. *For any $\delta > 0$ and fixed k, there is an optimal deterministic truthful mechanism for PG on any directed k-outerplanar graph $G = (V, E)$ that violates each local constraint by a factor of $1 + \delta$ and runs in $O(k|V|^3 q^{6k+2} \lceil \frac{2\rho}{\delta} \rceil^{6k})$ time, where $\rho = |V|(\lfloor \log_2(q) \rfloor + 2)$.*

Step 3: PTAS for Planar Graphs. Observe that when there are some boundary conditions on k-outerplanar, the above DP approach still works. For example, if the emission amount of any vertex in any first and last face (level 1 and level k face) of the k-outerplanar graph is zero, we just modify the dynamic programming approach in a bottom-up manner to set $\Omega_i = -\infty$ if any vertex v in any first and last face is a parameter of Ω_i and its emission amount $a_v^i > 0$. Then the modified DP approach is the desired algorithm for $\bar{P}G_2$ on the k-outerplanar graph under this boundary condition.

Proposition 2. *PG is strongly NP-hard on planar graphs with degree at most 3 when we allow a $(1 + \delta)$-violation of local constraints.*

Theorem 5. *For any fixed k and $\delta > 0$, there is an $O(k^2 |V|^3 q^{6k+2} \lceil \frac{2\rho}{\delta} \rceil^{6k})$ algorithm for PG with integer variables on directed planar graph $G = (V, E)$ that achieves*

$(\eta^{in} = \frac{k}{k-2})$-approximation and violates each local constraint by a factor of $1 + \delta$, where $\rho = |V|(\lfloor \log_2(q) \rfloor + 2)$.

Proof. We use $OPT(\bar{PG}_2)$ to denote $OPT^{in}_G(\bar{PG}_2)$ and omit the superscript and subscript. By Lemma 1, we know $OPT = OPT(PG) \leq OPT(\bar{PG}_2)$. Let $\bar{PG}_2(i)$ denote the $\bar{PG}_2$ restricted on G by setting $x_v = 0$ for each v who belongs to any face $f \equiv i$ or $i + 1 \pmod{k}$. Let $\{x^*_v\}_{v \in V}$ be an optimal solution for $\bar{PG}_2$. Then we know $\sum_{i \in [k]} \sum_{v \in f: f \equiv i \text{ or } i+1 (mod\,k)} x^*_v = 2OPT(\bar{PG}_2)$. As a consequence, there exists $i \in [k]$ such that $\sum_{v \in f: f \equiv i \text{ or } i+1(mod\,k)} x^*_v \leq \frac{2OPT(\bar{PG}_2)}{k}$. Observe that $\{x_v\}_{v \in V}$ is a feasible solution for $\bar{PG}_2(i)$, where $x_v = 0$ if v belongs to any face $f \equiv i$ or $i+1 \pmod{k}$ and $x_v = x^*_v$ otherwise. Thus, $OPT(\bar{PG}_2(i)) \geq (1 - \frac{2}{k})OPT(\bar{PG}_2) \geq (1-\frac{2}{k})OPT$. Solving each $\bar{PG}_2(i), i \in [k]$, then choosing $\max_{i \in [k]}\{OPT(\bar{PG}_2(i))\}$ (which is at least $(1 - \frac{2}{k})OPT$) gives the desired result. Now let us see how to solve $\bar{PG}_2(i)$. Note that for $\bar{PG}_2(i)$, $x_v = 0$ for any v who belongs to any face $f \equiv i$ or $i + 1 \pmod{k}$. $\bar{PG}_2(i)$ consists of independent k'-outerplanar graphs, each of which has some boundary condition i.e. the emission amount of any vertex in any first and last face is zero and $k' \leq k$. Suppose the number of these independent k'-outerplanar graphs is L^i. W.l.o.g. suppose these k'-outerplanar graphs are ordered from exterior to interior as $G_s = (V_s, E_s)$, $s \in [L^i]$ (e.g. G_s is the subgraph of G constructed by all the vertices of levels from $(s - 2)k + i + 1$ to $(s - 1)k + i$, $s = 2, \cdots, L^i - 1$, with boundary $x_v = 0$ if v is of level $(s - 2)k + i + 1$ or $(s - 1)k + i$).

Let $\Omega_s(Q^s)$ denote the optimal value if there is a solution such that the total allocated scaled emission amount to G_s is exactly Q^s with boundary condition and $\Omega_s(Q^s) = 0$ otherwise, which can be solved by the above DP approach on k'-outerplanar graphs with boundary conditions. Then, it is not difficult to see the optimal solution for $\bar{PG}_2(i)$ is the optimal solution of the following integer linear program (denoted SUB):

$$\max \sum_{s \in [L^i]} \sum_{Q^s=0}^{p} \Omega_s(Q^s) y_{sQ^s}$$
$$\text{s.t.} \sum_{s \in [L^i]} \sum_{Q^s=0}^{p} Q^s y_{sQ^s} \leq p$$
$$\sum_{Q^s=0}^{p} y_{sQ^s} = 1$$
$$y_{sQ^s} \in \{0,1\} \forall s \in [L^i], Q^s \in [p]$$

Let $g_t(Q)$ denote the optimal integer value of SUB when only G_s, $s \in [t]$ is considered and the total emission amount allocated to these graphs is exactly Q.

Then we have the following recursion function: $g_t(Q) = \max_{Q^t = 0,1,\cdots,Q}\{g_{t-1}(Q - Q^t) + \Omega_t(Q^t)\}$. The optimal value of SUB is $\max_{Q = 0,1,\cdots,p}\{g_{L^i}(Q)\}$, which gives the optimal solution of $\bar{PG}_2(i)$ by tracking the optimal value of this dynamic programming approach. The running time of this approach is $O(|L^i|p^2)$. Hence, the total running time for obtaining and solving $\bar{PG}_2(i)$ is $O(|L^i|p^2) + \sum_{s \in [L^i]} O(k|V_s|^3 q^{6k+2} \lceil \frac{2\rho}{\delta} \rceil^{6k}) = O(k|V|^3 q^{6k+2} \lceil \frac{2\rho}{\delta} \rceil^{6k})$. We need to solve $\bar{PG}_2(i)$, for each $i \in [k]$ and then get $\max_{i \in [k]}\{OPT(\bar{PG}_2(i))\}$. Therefore, the overall running time is $O(k^2|V|^3 q^{6k+2} \lceil \frac{2\rho}{\delta} \rceil^{6k})$, and Theorem 5 is proved. □

Let $\frac{2}{k} = \epsilon$ in Theorem 5. Also note that $\rho = |V|(\lfloor \log_2(q) \rfloor + 2)$. We have:

Theorem 6. *There is* $O\left(\frac{1}{\epsilon^2}|V|^{12/\epsilon+3}q^2 \lceil \frac{2(\lfloor \log_2 q \rfloor + 2)q}{\delta} \rceil^{12/\epsilon+1}\right) = \left(\frac{|V|q(\log_2 q+2)}{\delta}\right)^{O(\frac{1}{\epsilon})}$ *time algorithm for PG for fixed* $\delta, \epsilon > 0$ *on directed planar graph* $G = (V, E)$ *that* *achieves social welfare* $(1 - \epsilon)OPT^{in}(PG)$ *and violates each local constraint by a factor of* $1 + \delta$. *This is a PTAS for PG with polynomial size integer variables.*

5 General Objective Function for Bounded Degree Graphs

Full details of our results for general objective functions will appear in the full version of the paper. Our most general algorithmic result is given in Theorem 7.

Theorem 7. *Let* $x_v \in \{0, 1\}$ *for any* $v \in V$. *Assume that* $R(x)$ *is monotone increasing as set function on sets* $S \subseteq V$ *s.t.* $v \in S$ *iff* $x_v = 1$. *Then there is an* $(\eta^{fr} = \frac{e\gamma\Delta+2}{e-1} + 1)$-*approximation algorithm for PG with integer variables on graphs with degree* $\leq \Delta$.

Our hardness results for general objective functions are Theorems 8 and 9. By a reduction from independent set we get the following:

Theorem 8. *PG is Unique Games-hard to approximate within* $n^{1-\epsilon}$ *and within* $\frac{\Delta}{\log^2 \Delta}$ *for G with degree* Δ *when* p_v *is any constant number* ≥ 1, $b_v(x_v)$ *is linear and* $d_v(y)$ *is piecewise linear (with 2 pieces)* $\forall v \in V$ *and* w_{vu} *is positive constant* $\forall(v, u) \in E$.

Theorem 9. *It is strongly NP-hard to find an optimal solution to Pollution Game (PG) when* p_v *is any constant number* ≥ 1, $b_v(x_v)$ *is linear and* $d_v(y)$ *is piecewise linear (with two pieces)* $\forall v \in V$ *and* w_{vu} *is positive constant for any* $(v, u) \in E$.

References

1. Air quality in Europe - 2014 Report. European Environment Agency Report No. 5/2014. http://www.eea.europa.eu/publications/air-quality-in-europe-2014
2. State of the Air 2014 Report. American Lung Association, 30 April 2014. http://www.stateoftheair.org/2014/key-findings/
3. Anastasiadis, E., Deng, X., Krysta, P., Li, M., Qiao, H., Zhang, J.: Network pollution games. In: AAMAS (2016)
4. Baker, B.S.: Approximation algorithms for NP-complete problems on planar graphs. J. ACM **41**(1), 153–180 (1994)
5. Bansal, N., Korula, N., Nagarajan, V., Srinivasan, A.: Solving packing integer programs via randomized rounding with alterations. Theor. Comput. **8**(1), 533–565 (2012)
6. Belitskaya, A.V.: Network game of pollution cost reduction. Contrib. Game Theor. Manag. **6**, 24–34 (2013)
7. Caprara, A., Kellerer, H., Pferschy, U., Pisinger, D.: Approximation algorithms for knapsack problems with cardinality constraints. Eur. J. Oper. Res. **123**(2), 333–345 (2000)
8. Chau, C., Elbassioni, K., Khonji, M.: Truthful mechanisms for combinatorial AC electric power allocation. In: Proceedings of the 13th AAMAS, pp. 1005–1012 (2014)
9. Dobzinski, S., Nisan, N.: Mechanisms for multi-unit auctions. In: Proceedings of the 8th ACM Conference on Electronic Commerce, pp. 346–351. ACM (2007)
10. Dong, B., Ni, D., Wang, Y.: Sharing a polluted river network. Environ. Resour. Econ. **53**(3), 367–387 (2012)
11. Dughmi, S., Roughgarden, T.: Black-box randomized reductions in algorithmic mechanism design. SIAM J. Comput. **43**(1), 312–336 (2014)

12. Katsikarelis, I.: Computing bounded-width tree, branch decompositions of k-outerplanar graphs (2013). arXiv preprint arXiv: 1301.5896
13. Kloks, T.: Treewidth: Computations and Approximations. Lecture Notes in Computer Science, vol. 842. Springer, Heidelberg (1994)
14. Krysta, P., Telelis, O., Ventre, C.: Mechanisms for multi-unit combinatorial auctions with a few distinct goods. In: Proceedings of 12th AAMAS, pp. 691–698 (2013)
15. Kwerel, E.: To tell the truth: imperfect information and optimal pollution control. Rev. Econ. Stud. **44**, 595–601 (1977)
16. Lavi, R., Swamy, C.: Truthful and near-optimal mechanism design via linear programming. J. ACM (JACM) **58**(6), 25 (2011)
17. Ni, D., Wang, Y.: Sharing a polluted river. Games Econ. Behav. **60**(1), 176–186 (2007)
18. Nisan, N., Roughgarden, T., Tardos, E., Vazirani, V.V.: Algorithmic Game Theory, vol. 1. Cambridge University Press, Cambridge (2007)
19. Vorosmarty, C.J., McIntyre, P.B., Gessner, M.O., Dudgeon, D., Prusevich, A., Green, P., Glidden, S., Bunn, S.E., Sullivan, C.A., Reidy Liermann, C., Davies, P.M.: Global threats to human water security, river biodiversity. Nature **467**, 555–561 (2010)

Parameterized Complexity
and Algorithms

Polynomial-Time Algorithm for Isomorphism of Graphs with Clique-Width at Most Three

Bireswar Das, Murali Krishna Enduri, and I. Vinod Reddy[✉]

IIT Gandhinagar, Gujarat, India
{bireswar,endurimuralikrishna,reddy_vinod}@iitgn.ac.in

Abstract. The clique-width is a measure of complexity of decomposing graphs into certain tree-like structures. The class of graphs with bounded clique-width contains bounded tree-width graphs. We give a polynomial time graph isomorphism algorithm for graphs with clique-width at most three. Our work is independent of the work by Grohe and Schweitzer [17] showing that the isomorphism problem for graphs of bounded clique-width is polynomial time.

1 Introduction

Two graphs $G_1 = (V_1, E_1)$ and $G_2 = (V_2, E_2)$ are *isomorphic* if there is a bijection $f : V_1 \rightarrow V_2$ such that $\{u, v\} \in E_1$ if and only if $\{f(u), f(v)\} \in E_2$. Given a pair of graphs as input the problem of deciding if the two graphs are isomorphic is known as *graph isomorphism problem* (GI). Despite nearly five decades of research the complexity status of this problem still remains unknown. The graph isomorphism problem is not known to be in P. It is in NP but very unlikely to be NP-complete [5]. The problem is not even known to be hard for P. Recently Babai [2] designed a quasi-polynomial time algorithm to solve the GI problem improving the previously best known $2^{O(\sqrt{n \log n})}$ time algorithm [1,27]. Although the complexity of the general graph isomorphism problem remains elusive, many polynomial time algorithms are known for restricted classes of graphs e.g., bounded degree [21], bounded genus [23], bounded tree-width [3], etc.

The graph parameter *clique-width*, introduced by Courcelle et al. in [7], has been studied extensively. The class of bounded clique-width graphs is fairly large in the sense that it contains distance hereditary graphs, bounded tree-width graphs, bounded rank-width graphs [19], etc. Fellows et al. [15] shows that the computing the clique-width of a graph is NP-hard. Oum and Seymour [24] gave an elegant algorithm that computes a $(2^{3k+2} - 1)$-expression for a graph G of clique-width at most k or decides that the clique-width is more than k.

The parameters tree-width and clique-width share some similarities, for example many NP-complete problems admit polynomial time algorithms when the tree-width or the clique-width of the input graph is bounded. A polynomial

---

B. Das—Part of the research was done while the author was a DIMACS postdoctoral fellow.

M.K. Enduri—Supported by Tata Consultancy Services (TCS) research fellowship.

© Springer International Publishing Switzerland 2016
T.N. Dinh and M.T. Thai (Eds.): COCOON 2016, LNCS 9797, pp. 55–66, 2016.
DOI: 10.1007/978-3-319-42634-1_5

time isomorphism algorithm for bounded tree-width graphs has been known for a long time [3]. Recently Lokhstanov et al. [20] gave an fpt algorithm for GI parameterized by tree-width. The scenario is different for bounded clique-width graphs. The complexity of GI for bounded clique-width graphs is not known. Polynomial time algorithm for GI for graphs with clique-width at most two, which coincides with the class of co-graphs, is known probably as a folklore. The complexity of recognizing graphs with clique-width at most three was unknown until Corneil et al. [6] came up with the first polynomial time algorithm. Their algorithm (henceforth called the CHLRR algorithm) works via an extensive study of the structure of such graphs using split and modular decompositions. Apart from recognition, the CHLRR algorithm also produces a 3-expression for graphs with clique-width at most three. For fixed $k > 3$, though algorithms to recognize graphs with clique-width at most k are known [25], computing a k-expression is still open. Recently in an independent work by Grohe and Schweitzer [17] designed an isomorphism algorithm for graphs of bounded clique-width subsuming our result. Their algorithm uses group theory techniques and has worse runtime. However our algorithm has better runtime and uses different simpler intuitive techniques.

In this paper we give isomorphism algorithm for graphs with clique-width at most three with runtime $O(n^3 m)$. Our algorithm works via first defining a notion of equivalent k-expression and designing $O(n^3)$ algorithm to test if two input k-expressions are equivalent under this notion. Next we modify the CHLRR algorithm slightly to output a linear sized set $parseG$ of 4-expressions for an input graph G of clique-width at most three which runs in $O(n^3 m)$ time. Note that modified CHLRR algorithm will not output a canonical expression. However we show that for two isomorphic graphs G and H of clique-width at most three, $parseG$ contains an equivalent k-expression for each k-expression in $parseH$ and vice versa. Moreover, if G and H are not isomorphic then no pair in $parseG \times parseH$ is equivalent.

2   Preliminaries

In this paper, the graphs we consider are without multiple edges and self loops. The complement of a graph G is denoted as $\overline{G}$. The *coconnected components* of G are the connected components of $\overline{G}$. We say that a vertex v is *universal* to a vertex set X if v is adjacent to all vertices in $X \setminus \{v\}$. A *biclique* is a bipartite graph (G, X, Y), such that every vertex in X is connected to every vertex of Y. A *labeled graph* is a graph with labels assigned to vertices such that each vertex has exactly one label. In a labeled graph G, $lab(v)$ is the label of a vertex v and $lab(G)$ is the set of all labels. We say that a graph is *bilabeled* (*trilabeled*) if it is labeled using exactly two (three) labels. The set of all edges between vertices of label a and label b is denoted E_{ab}. We say E_{ab} is complete if it corresponds to a biclique.

The subgraph of G induced by $X \subseteq V(G)$ is denoted by $G[X]$, the set of vertices adjacent to v is denoted $N_G(v)$. The closed neighborhood $N_G[v]$ of v is $N_G(v) \cup \{v\}$. We write $G \cong_f H$ if f is an isomorphism between graphs

G and H. For labeled graphs G and H, we write $G \cong_f^\pi H$ if $G \cong_f H$ and $\pi : lab(G) \to lab(H)$ is a bijection such that for all $x \in V(G)$ if $lab(x) = i$ then $lab(f(x)) = \pi(i)$. The set of all isomorphisms from G to H is denoted $\mathsf{ISO}(G, H)$.

Definition 1. *The **clique-width** of a graph G is defined as the minimum number of labels needed to construct G using the following four operations:*

   i. $v(i)$: Creates a new vertex v with label i
  ii. $G_1 \oplus G_2 \cdots \oplus G_l$: Disjoint union of labeled graphs $G_1, G_2, \cdots, G_l$
 iii. $\eta_{i,j}$: Joins each vertex with label i to each vertex with label j $(i \neq j)$
 iv. $\rho_{i \to j}$: Renames all vertices of label i with label j

Every graph can be constructed using the above four operations, which is represented by an algebraic expression known as k-*expression*, where k is the number of labels used in expression. The *clique-width* of a graph G, denoted by $cwd(G)$, is the minimum k for which there exists a k-expression that defines the graph G. From the k-expression of a graph we can construct a tree known as *parse tree* of G. The leaves of the parse tree are vertices of G with their initial labels, and the internal nodes correspond to the operations ($\eta_{i,j}$, $\rho_{i \to j}$ and $\oplus$) used to construct G. For example, C_5 (cycle of length 5) can be constructed by

$$\eta_{1,3}((\rho_{3 \to 2}(\eta_{2,3}((\eta_{1,2}(a(1) \oplus b(2))) \oplus (\eta_{1,3}(c(3) \oplus d(1)))))) \oplus e(3)).$$

The k-expression for a graph need not be unique. The clique-width of any induced subgraph is at most the clique-width of its graph [9].

Now we describe the notions of modular and split decompositions. A set $M \subseteq V(G)$ is called a *module* of G if all vertices of M have the same set of neighbors in $V(G) \setminus M$. The *trivial modules* are $V(G)$, and $\{v\}$ for all v. In a labeled graph, a module is said to be a l-*module* if all the vertices in the module have the same label. A *prime* (l-*prime*) graph is a graph (labeled graph) in which all modules (l-modules) are trivial. The modular decomposition of a graph is one of the decomposition techniques which was introduced by Gallai [16]. The *modular decomposition* of a graph G is a rooted tree T_M^G that has the following properties:

1. The leaves of T_M^G are the vertices of G.
2. For an internal node h of T_M^G, let $M(h)$ be the set of vertices of G that are leaves of the subtree of T_M^G rooted at h. ($M(h)$ forms a module in G).
3. For each internal node h of T_M^G there is a graph G_h (*representative graph*) with $V(G_h) = \{h_1, h_2, \cdots, h_r\}$, where $h_1, h_2, \cdots, h_r$ are the children of h in T_M^G and for $1 \leq i < j \leq r$, h_i and h_j are adjacent in G_h iff there are vertices $u \in M(h_i)$ and $v \in M(h_j)$ that are adjacent in G.
4. G_h is either a clique, an independent set, or a prime graph and h is labeled *Series* if G_h is clique, *Parallel* if G_h is an independent set, and *Prime* otherwise.

James et al. [18] gave first polynomial time algorithm for finding a modular decomposition which runs in $O(n^4)$ time. Linear time algorithms to find modular decompositions are proposed in [10, 26].

A vertex partition (A, B) of a graph G is a *split* if $\tilde{A} = A \cap N(B)$ and $\tilde{B} = B \cap N(A)$ forms a biclique. A split is trivial if $|A|$ or $|B|$ is one. Split decomposition was introduced by Cunningham [11]. Loosely it is the result of a recursive process of decomposing a graph into components based on the splits. Cunningham [11] showed that a graph can be decomposed uniquely into components that are stars, cliques, or prime (i.e., without proper splits). This decomposition is known as the *skeleton*. For details see [12]. A polynomial time algorithm for computing the skeleton of a graph is given in [22].

Theorem 1 [12] (See [6]). *Let G be a connected graph. Then the skeleton of G is unique, and the proper splits of G correspond to the special edges of its skeleton and to the proper splits of its complete and star components.*

Organization of the Paper: In Sect. 3 we discuss GI-completeness of prime graph isomorphism. In Sect. 4 we define a notion of equivalence of parse trees called *structural isomorphism*, and give an algorithm to test if two parse trees are structurally isomorphic. We give an overview of the CHLRR algorithm [6] in Sect. 5. In Sect. 6, we present the isomorphism algorithm for prime graphs of clique-width at most three. We modify the CHLRR algorithm suitably to output structurally isomorphic parse trees for isomorphic graphs, the proof of this can be found in full version of the paper [14].

3   Completeness of Prime Graph Isomorphism

It is known that isomorphism problem for prime graphs is GI-complete [4]. There is an easy polynomial time many-one reduction from GI to prime graph isomorphism[1] (see [14]). Unfortunately, this reduction does not preserve the clique-width. We also give a clique-width preserving Turing reduction from GI to prime graph isomorphism which we use in our main algorithm. The reduction hinges on the following lemma.

Lemma 1 [8]. *G is a graph of clique-width at most k iff each prime graph associated with the modular decomposition of G is of clique-width at most k.*

We next show that if we have an oracle for GI for colored prime graphs of clique-width at most k then there is a GI algorithm for graphs with clique-width at most k.

Theorem 2. *Let $\mathcal{A}'$ be an algorithm that given two colored prime graphs G' and H' of clique-width at most k, decides if $G' \cong H'$ via a color preserving isomorphism. Then there exists an algorithm $\mathcal{A}$ that on input any colored graphs G and H of clique-width at most k decides if $G \cong H$ via a color preserving isomorphism.*

Proof. Let G and H be two colored graphs of clique-width at most k. The algorithm is similar to [13], which proceeds in a bottom up approach in stages starting from the leaves to the root of the modular decomposition trees T_G and

---

[1] In fact, it is an AC^0 reduction.

T_H of G and H respectively. Each stage corresponds to a level in the modular decomposition. In every level, the algorithm $\mathcal{A}$ maintains a table that stores whether for each pair of nodes x and y in T_G and T_H the subgraphs $G[x]$ and $H[y]$ induced by leaves of subtrees of T_G and T_H rooted at x and y are isomorphic. For the leaves it is trivial to store such information. Let u and v be two internal nodes in the modular decomposition trees of T_G and T_H in the same level. To decide if $G[u]$ and $H[v]$ are isomorphic $\mathcal{A}$ does the following.

If u and v are both *series* nodes then it just checks if the children of u and v can be isomorphically matched. The case for *parallel* node is similar. If u and v are *prime* nodes then the vertices of representative graphs G_u and H_v are colored by their isomorphism type i.e., two internal vertices u_1 and u_2 of the representative graphs will get the same color iff subgraphs induced by leaves of subtrees of T_G (or T_H) rooted at u_1 and u_2 are isomorphic. To test $G[u] \cong H[v]$, $\mathcal{A}$ calls $\mathcal{A}'(\widehat{G}_u, \widehat{H}_v)$, where $\widehat{G}_u$ and $\widehat{H}_v$ are the colored copies of G_u and H_v respectively. At any level if we can not find a pairwise isomorphism matching between the internal nodes in that level of T_G and T_H then $G \cong H$. In this manner we make $O(n^2)$ calls to algorithm $\mathcal{A}'$ at each level. The total runtime of the algorithm is $O(n^3)T(n)$, where $T(n)$ is run time of $\mathcal{A}'$. Note that by Lemma 1 clique-width of G_u and H_v are at most k.   □

4   Testing Isomorphism Between Parse Trees

In this section we define a notion of equivalence of parse trees called *structural isomorphism*, and we give an algorithm to test if two given parse trees are equivalent under this notion. As we will see, the graphs generated by equivalent parse trees are always isomorphic. Thus, if we have two equivalent parse trees for the two input graphs, the isomorphism problem indeed admits a polynomial time algorithm. In Sect. 6, we prove that the CHLRR algorithm can be tweaked slightly to produce structurally isomorphic parse trees for isomorphic graphs with clique-width at most three and thus giving a polynomial-time algorithm for such graphs.

Let G and H be two colored graphs. A bijective map $\pi : V(G) \to V(H)$ is *color consistent* if for all vertices u and v of G, $color(u) = color(v)$ iff $color(\pi(u)) = color(\pi(v))$. Let $\pi : V(G) \to V(H)$ be a color consistent mapping, define $\pi/color : color(G) \to color(H)$ as follows: for all c in $color(G)$, $\pi/color(c) = color(\pi(v))$ where $color(v) = c$. It is not hard to see that the map $\pi/color$ is well defined. Recall that the internal nodes of a parse tree are $\eta_{i,j}$, $\rho_{i \to j}$ and $\oplus$ operations. The levels of a parse tree correspond to $\oplus$ nodes. Let T_g be a parse tree of G rooted at $\oplus$ node g. Let g_1 be descendant of g which is neither η nor ρ. We say that g_1 is an *immediate significant descendant* of g if there is no other $\oplus$ node in the path from g to g_1. For an immediate significant descendant g_1 of g, we construct a *colored quotient graph* Q_{g_1} that corresponds to graph operations appearing in the path from g to g_1 performed on graph G_{g_1}, where G_{g_1} is graph generated by parse tree T_{g_1}. The vertices of Q_{g_1} are labels of G_{g_1}. The colors and the edges of Q_{g_1} are determined by the operations on the path from g_1 to g. We start with coloring a vertex a by color a and no edges. If the operation performed is $\eta_{a,b}$ on G_{g_1} then add edges between vertices of color a and color b. If the operation is $\rho_{a \to b}$ on G_{g_1} then recolor the vertices of color a with color b. After taking care of an operation we move to the next operation on the path from g_1 to g until we reach $\oplus$ node g. Notice that if the total number of labels used in a parse tree is k then the size of any colored quotient graph is at most k.

Definition 2. *Let T_g and T_h be two parse trees of G and H rooted at $\oplus$ nodes g and h respectively. We say that T_g and T_h are structurally isomorphic via a label map π (denoted $T_g \cong^\pi T_h$)*

1. *If T_g and T_h are single nodes[2] or inductively,*
2. *If T_g and T_h are rooted at g and h having immediate significant descendants $g_1, \cdots, g_r$ and $h_1, \cdots, h_r$, and there is a bijection $\gamma : [r] \to [r]$ and for each i there is a $\pi_i \in \mathsf{ISO}(Q_{g_i}, Q_{h_{\gamma(i)}})$ such that $T_{g_i} \cong^{\pi_i} T_{h_{\gamma(i)}}$ and $\pi_i/color = \pi|_{color(Q_{g_i})}$, where $T_{g_1}, \cdots, T_{g_r}$ and $T_{h_1}, \cdots, T_{h_r}$ are the subtrees rooted at $g_1, \cdots, g_r$ and $h_1, \cdots, h_r$ respectively[3]*

We say that T_g and T_h are structurally isomorphic if there is a π such that $T_g \cong^\pi T_h$.

The structural isomorphism is an equivalence relation: reflexive and symmetric properties are immediate from the above definition. The following lemma shows that it is also transitive.

Lemma 2. *Let T_{g_1}, T_{g_2} and T_{g_3} be the parse trees of G_1, G_2 and G_3 respectively such that $T_{g_1} \cong^{\pi_1} T_{g_2}$ and $T_{g_2} \cong^{\pi_2} T_{g_3}$ then $T_{g_1} \cong^{\pi_2 \pi_1} T_{g_3}$.*

Proof. The proof is by induction on the height of the parse trees. The base case trivially satisfies the transitive property. Assume that g_1, g_2 and g_3 are nodes of height $d + 1$. Let g_{1i} be an immediate significant descendant of g_1. Since $T_{g_1} \cong^{\pi_1} T_{g_2}$, there is an immediate significant descendant g_{2j} of g_2 and $\pi_{1i} \in \mathsf{ISO}(Q_{g_{1i}}, Q_{g_{2j}})$ such that $\pi_{1i}/color = \pi|_{color(Q_{g_{1i}})}$ and $T_{g_{1i}} \cong^{\pi_{1i}} T_{g_{2j}}$. Similarly, g_{2j} will be matched to some immediate significant descendant g_{3k} of g_3 via $\pi_{2j} \in \mathsf{ISO}(Q_{g_{2j}}, Q_{g_{3k}})$ such that $\pi_{2j}/color = \pi|_{color(Q_{g_{2j}})}$ and $T_{g_{2j}} \cong^{\pi_{2j}} T_{g_{3k}}$. The nodes g_{1i}, g_{2j} and g_{3k} has height at most d. Therefore, by induction hypothesis $T_{g_{1i}} \cong^{\pi_{2j} \pi_{1i}} T_{g_{3k}}$. By transitivity of isomorphism we can say $\pi_{2j} \pi_{1i} \in \mathsf{ISO}(Q_{g_{1i}}, Q_{g_{3k}})$. To complete the proof we just need to show $\pi_{2j} \pi_{1i}/color = \pi_2 \pi_1|_{color(Q_{g_{1i}})}$. This can be inferred from the following two facts:

(1) $\pi_{2j} \pi_{1i}/color = \pi_{2j}/color \; \pi_{1i}/color$
(2) $\pi_2 \pi_1|_{color(Q_{g_{1i}})} = \pi_2|_{color(Q_{g_{2j}})} \; \pi_1|_{color(Q_{g_{1i}})}$. $\qquad \square$

---

[2] In this case they are trivially structurally isomorphic via π.
[3] Notice that this definition implies that G_{g_i} and $H_{h_{\gamma(i)}}$ are isomorphic via the label map π_i where G_{g_i} and $H_{h_{\gamma(i)}}$ are graphs generated by the parse trees T_{g_i} and $T_{h_{\gamma(i)}}$ respectively.

Algorithm to Test Structural Isomorphism: Next we describe an algorithm that given two parse trees T_G and T_H tests if they are structurally isomorphic. From the definition if $T_G \cong^\pi T_H$ then we can conclude that G and H are isomorphic. We design a dynamic programming algorithm that basically checks the local conditions 1 and 2 in Definition 2.

The algorithm starts from the leaves of parse trees and proceeds in levels where each level corresponds to $\oplus$ operations of parse trees. Let g and h denotes the $\oplus$ nodes at level l of T_G and T_H respectively. At each level l, for each pair of $\oplus$ nodes $(g, h) \in (T_G, T_H)$, the algorithm computes the set $R_l^{g,h}$ of all bijections $\pi : lab(G_g) \to lab(H_h)$ such that $G_g \cong_f^\pi H_h$ for some f, and stores in a table indexed by (l, g, h), where G_g and H_h are graphs generated by sub parse trees T_g and T_h rooted at g and h respectively. To compute $R_l^{g,h}$, the algorithm uses the already computed information $R_{l+1}^{g_i, h_j}$ where g_i and h_j are immediate significant descendants of g and h.

The base case correspond to finding $R_l^{g,h}$ for all pairs (g, h) such that g and h are leaves. Since in this case G_g and H_h are just single vertices, it is easy to find $R_l^{g,h}$. For the inductive step let $g_1, \cdots, g_r$ and $h_1, \cdots, h_{r'}$ be the immediate significant descendants of g and h respectively. If $r \neq r'$ then $R_l^{g,h} = \emptyset$. Otherwise we compute $R_l^{g,h}$ for each pair (g, h) at level l with help of the already computed information up to level $l + 1$ as follows.

For each $\pi : lab(G_g) \to lab(H_h)$ and pick g_1 and try to find a h_{i_1} such that $T_{g_1} \cong^{\pi_1} T_{h_{i_1}}$ for some $\pi_1 \in \mathsf{ISO}(Q_{g_1}, Q_{h_{i_1}}) \cap R_{l+1}^{g_1, h_{i_1}}$ such that $\pi_1/color = \pi|_{color(Q_{g_1})}$. We do this process to pair g_2 with some unmatched h_{i_2}. Continue in this way until all immediate significant descendants are matched. By Lemma 3, we know that this greedy matching satisfies the conditions of Definition 2. If all the immediate significant descendants are matched we add π to $R_l^{g,h}$. It is easy to see that if $R_l^{g,h} \neq \emptyset$ then the subgraphs $G_g \cong_f^\pi H_h$ for $\pi \in R_l^{g,h}$. From the definition of structurally isomorphic parse trees it is clear that if $R_0^{g,h} \neq \emptyset$ then $G \cong H$. The algorithm is polynomial time as the number of choices for π and π_1 is at most $k!$ which is a constant, where $|lab(G)| = k$.

Note that for colored graphs, by ensuring that we only match vertices of same color in the base case, the whole algorithm can be made to work for colored graphs. In Lemma 2 we prove that structural isomorphism satisfies transitivity. In fact, structural isomorphism satisfies a stronger notion of transitivity as stated in the following lemma.

Lemma 3. *Let T_g and T_h be two parse trees of graphs G and H. Let g_1 and g_2 be two immediate significant descendants of g, and h_1 and h_2 be two immediate significant descendants of h. Suppose for $i = 1, 2$, $T_{g_i} \cong^{\pi_i} T_{h_i}$ for some $\pi_i \in \mathsf{ISO}(Q_{g_i}, Q_{h_i})$ with $\pi_i/color = \pi|_{color(Q_{g_i})}$. Also assume that $T_{g_1} \cong^{\pi_3} T_{h_2}$ where $\pi_3 \in \mathsf{ISO}(Q_{g_1}, Q_{h_2})$ and $\pi_3/color = \pi|_{color(Q_{g_1})}$. Then, $T_{g_2} \cong^{\pi_1 \pi_3^{-1} \pi_2} T_{h_1}$ where $\pi_1 \pi_3^{-1} \pi_2 \in \mathsf{ISO}(Q_{g_2}, Q_{h_1})$ and $\pi_1 \pi_3^{-1} \pi_2/color = \pi|_{color(Q_{g_2})}$.*

Proof. By Lemma 2, $T_{g_2} \cong^{\pi_1 \pi_3^{-1} \pi_2} T_{h_1}$. The rest of the proof is similar to the proof of the inductive case of Lemma 2. $\square$

5   Overview of the CHLRR Algorithm

Corneil et al. [6] gave the first polynomial time algorithm (the *CHLRR algorithm*), to recognize graphs of clique-width at most three. We give a brief description of their algorithm in this section. We mention that our description of this fairly involved algorithm is far from being complete. The reader is encouraged to see [6] for details. By Lemma 1 we assume that the input graph G is prime.

To test whether clique-width of prime graph G is at most three the algorithm starts by constructing a set of bilabelings and trilabelings of G. In general the number of bilabelings and trilabelings are exponential, but it was shown (Lemmas 8 and 9 in [6]) that it is enough to consider the following linear size subset denoted by *LabG*.

1. For each vertex v in $V(G)$
   [B_1] Generate the bilabeling[4]$\{v\}$ and add it to *LabG*.
   [B_2] Generate the bilabeling $\{x \in N(v) \mid N[x] \subseteq N[v]\}$ and add it to *LabG*.
2. Compute the skeleton of G search this skeleton for the special edges, clique and star components.
   [T_1] For each special edge s (corresponds to a proper split), generate the trilabeling $\tilde{X}, \tilde{Y}, V(G) \setminus (\tilde{X} \cup \tilde{Y})$ where (X,Y) is the split defined by s and add it to *LabG*.
   [B_3] For all clique components C, generate the bilabeling C and add it to *LabG*.
   [B_4] For all star components S, generate the bilabeling $\{c\}$, where c is the special center of S, and add it to *LabG*.

Lemma 4 [6]. *Let G be a prime graph. Clique-width of G is at most three if and only if at least one of the bilabelings or trilabelings in LabG has clique-width at most three.*

By Lemma 4 the problem of testing whether G is of clique-width at most three is reduced to checking one of labeled graph in *LabG* is of clique-width at most three. To test if a labeled graph A taken from *LabG* is of clique-width at most three, the algorithm follows a top down approach by iterating over all possible last operations that arise in the parse tree representation of G. For example, for each vertex x in G the algorithm checks whether the last operation must have joined x with its neighborhood. In this case the problem of testing whether G can be constructed using at most three labels is reduced to test whether $G \setminus \{x\}$ can be constructed using at most three lables. Once the last operations are fixed the original graph decomposes into smaller components, which can be further decomposed recursively.

For each A in *LabG*, depending on whether it is bilabeled or trilabeled the algorithm makes different tests on A to determine whether A is of clique-width at most three. Based on the test results the algorithm either concludes clique-width of A is more than three or returns top operations of the parse tree for A along with some connected components of A which are further decomposed recursively.

---

[4]   *Bilabeling* of a set $X \subseteq V$ indicates that all the vertices in X are labeled with one label and $V \setminus X$ is labeled with another label.

If A in $LabG$ is connected, trilabeled (with labels l_1, l_2, l_3) and l-prime then by the construction of $LabG$, A corresponds to a split (possibly trivial). If A has a proper split then there exists $a \neq b$ in $\{l_1, l_2, l_3\}$ such that A will be disconnected with the removal of edges E_{ab}. This gives a decomposition with top operations $\eta_{a,b}$ followed by a $\oplus$ node whose children are connected components of $A \setminus E_{ab}$. If A has a universal vertex v (trivial split) labeled a in A then by removing edges E_{ab} and E_{ac} we get a decomposition with top operations $\eta_{a,b}$ and $\eta_{a,c}$ followed by a $\oplus$ operation with children connected components of $A \setminus (E_{ab} \cup E_{ac})$.

To describe the bilabeled case we use V_i to denote the set of vertices of A with label i. If A in $LabG$ is connected, bilabeled (with labels l_1, l_2) and l-prime, then the last operation is neither η_{l_1, l_2} (otherwise A will have a l-module) nor $\oplus$ (A is connected). So the last operation of the decomposition must be a relabeling followed by a join operation i.e., we have to introduce a third label set V_{l_3} such that all the edges are present between the two of three labeled sets.

After introducing third label if there is only one join to undo, then we have a unique way to decompose the graph into smaller components. If there are more than one possible join to be removed, then it is enough to consider one of them and proceed (see Sect. 5.2 in [6]). There are four ways to introduce the third label to decompose the graph, but they might correspond to overlapping cases. To overcome this the algorithm first checks whether A belongs to any of three simpler cases described below.

PC1: A has a universal vertex x of label $l \in \{l_1, l_2\}$. In this case relabel vertex x with l_3 and remove the edges $E_{l_3 l_2}$, and $E_{l_3 l_1}$ to decompose A. This gives a decomposition with $\rho_{l_3 \to l}, \eta_{l_3, l_2}, \eta_{l_3, l_1}$ followed by $\oplus$ operation with children x and $A \setminus \{x\}$.

PC2: A has a vertex x of label $l \in \{l_1, l_2\}$ that is universal to all vertices of label $l' \in \{l_1, l_2\}$, but is not adjacent to all vertices with the other label, say $\bar{l'}$. In this case relabel vertex x with l_3 and remove the edges $E_{l_3 l'}$. This gives a decomposition with $\rho_{l_3 \to l}, \eta_{l_3, l'}$ above a $\oplus$ operation with children x and $A \setminus \{x\}$.

PC3: A has two vertices x and y of label l, where y is universal to everything other than x, and x is universal to all vertices of label l other than y, and non-adjacent to all vertices with the other label $\bar{l}$. In this case the algorithm relabels vertices x and y with l_3, and by removing edges $E_{l_3 l}$ disconnects the graph A, with two connected components x and $A \setminus \{x\}$. Now in graph $A \setminus \{x\}$ again remove the edges $E_{l_3 \bar{l}}$ to decompose the graph into two parts y and $A \setminus \{x, y\}$.

If A does not belongs to any of above three simpler cases then there are four different ways to introduce the third label set to decompose the graph as described below.

Let $\mathcal{E}$ be the set of all connected, bilabeled, l-prime graphs with clique-width at most three and not belonging to above three simpler cases. For $l \in \{1, 2\}$ we define the following four subsets of $\mathcal{E}$.

1. $\mathcal{U}_l$: $V_l^a \neq \emptyset$ and removing the edges between the V_l^a and $V_{\bar{l}}$ disconnects the graph.
2. $\overline{\mathcal{D}}_l$: $\overline{V}_l$ is not connected and removing the edges between the coconnected components of $\overline{V}_l$ disconnects the graph.

In these four cases the algorithm introduces a new label l_3 and removes the edges E_{ll_3}, $l \in \{l_1, l_2\}$ to disconnect A. This gives a decomposition with $\rho_{l_3 \to l}$ and η_{l,l_3} followed by $\oplus$ operation with children that are the connected components of $A \setminus E_{ll_3}$. For more details about decomposition process when A is in $\mathcal{U}_l$ or $\overline{\mathcal{D}}_l$, $l \in \{1, 2\}$ the reader is encouraged to see Sect. 5.2 in [6].

The following Lemma shows that there is no other possible way of decomposing a clique-width at most three graphs apart from the cases described above.

Lemma 5 [6]. $\mathcal{E} = \mathcal{U}_1 \cup \mathcal{U}_2 \cup \overline{\mathcal{D}}_1 \cup \overline{\mathcal{D}}_2$, and this union is disjoint.

In summary, for any labeled graph A in $LabG$ the CHLRR algorithm tests whether A belongs to any of the above described cases, if it is then it outputs suitable top operations and connected components. The algorithm continues the above process repeatedly on each connected component of A until it either returns a parse tree or concludes clique-width of A is more than three.

6   Isomorphism Algorithm for Prime Graphs of Clique-Width at Most Three

In Sect. 4 we described algorithm to test structural isomorphism between two parse trees. In this Section we show that given two isomorphic prime graphs G and H of clique-width at most three, the CHLRR algorithm can be slightly modified to get structurally isomorphic parse trees. We have used four labels in order to preserve structural isomorphism in the modified algorithm [14]. Recall that the first step of the CHLRR algorithm is to construct a set $LabG$ of bilabelings and trilabelings of G as described in Sect. 5.

Definition 3. We say that $LabG$ is equivalent to $LabH$ denoted as $LabG \equiv LabH$ if there is a bijection $g : LabG \to LabH$ such that for all $A \in LabG$, there is an isomorphism $f : V(A) \to V(g(A))$ and a bijection $\pi : lab(A) \to lab(g(A))$ such that $A \cong_f^\pi g(A)$.

Lemma 6 [14]. $LabG \equiv LabH$ iff $G \cong_f H$.

Lemma 7. Let $A \in LabG$ and $B \in LabH$. If $A \cong_f^\pi B$ for some f and π then parse trees generated from Decompose function (Algorithm 2 [14]) for input graphs A and B are structurally isomorphic. i.e., $Decompose(A) \cong_f^\pi Decompose(B)$.

Proof. Follows from Lemma 11 and Lemma 12 described in [14]. The major modifications are done in PC2 case, where we have used four labels in order to preserve structural isomorphism between parse trees.   □

Isomorphism Algorithm
For two input prime graphs G and H the algorithm works as follows. Using modified CHLRR algorithm, first a parse tree T_G of clique-width at most three is computed for G. The parse tree T_G of G is not canonical but from Lemmas 6 and 7, we know that if $G \cong H$ then there exists parse tree T_H of H, structurally

isomorphic to T_G. Therefore we compute parse tree of clique-width at most three for each labeled graph in $LabH$. For each such parse tree T_H, the algorithm uses the structural isomorphic algorithm described in Sect. 4 to test the structural isomorphism between parse trees T_G and T_H. If $T_G \cong T_H$ for some T_H, then we conclude that $G \cong H$. If there is no parse tree of H which is structurally isomorphic to T_G then G and H can not be isomorphic.

Computing a parse tree T_G of G takes $O(n^2 m)$ time. As there are $O(n)$ many labeled graphs in $LabH$, computing all possible parse trees for labeled graphs in $LabH$ takes $O(n^3 m)$ time. Testing structural isomorphism between two parse trees need $O(n^3)$ time. Therefore the running time to check isomorphism between two prime graphs G and H of clique-width at most three is $O(n^3 m)$. $\qquad \square$

The correctness of the algorithm follows from Lemma 8 and Theorem 3. Lemma 8 shows that if $G \cong H$ then we can always find two structurally isomorphic parse trees T_G and T_H using the modified CHLRR algorithm.

Lemma 8. *Let G and H be prime graphs with clique-width at most three. If $G \cong_f H$ then for every T_G in $parseG$ there is a T_H in $parseH$ such that T_G is structurally isomorphic to T_H where $parseG$ and $parseH$ are the set of parse trees generated by Algorithm 1 [14] on input $LabG$ and $LabH$ respectively.*

Proof. If $G \cong_f H$ then from Lemma 6 we have $LabG \equiv LabH$ i.e., for every A in $LabG$ there is a $B = g(A)$ in $LabH$ such that $A \cong_f^\pi B$ for some f and π. On input such A and B to Lemma 7 we get two parse trees T_A and T_B which are structurally isomorphic. $\qquad \square$

Theorem 3. *Let G and H be graphs with clique-width at most three. Then there exists a polynomial time algorithm to check whether $G \cong H$.*

Proof. The proof follows from the prime graph isomorphism of graphs with clique-width at most three described in Lemma 8 and Theorem 2. $\qquad \square$

References

1. Babai, L.: Moderately exponential bound for graph isomorphism. In: Gécseg, F. (ed.) Fundamentals of Computation Theory. LNCS, vol. 117, pp. 34–50. Springer, Heidelberg (1981)
2. Babai, L.: Graph isomorphism in quasipolynomial time (2015). arXiv preprint arXiv:1512.03547
3. Bodlaender, H.L.: Polynomial algorithms for graph isomorphism and chromatic index on partial k-trees. J. Algorithms **11**(4), 631–643 (1990)
4. Bonamy, M.: A small report on graph and tree isomorphism (2010). http://bit.ly/1ySeNBn
5. Boppana, R.B., Hastad, J., Zachos, S.: Does co-NP have short interactive proofs? Inf. Process. Lett. **25**(2), 127–132 (1987)
6. Corneil, D.G., Habib, M., Lanlignel, J.M., Reed, B., Rotics, U.: Polynomial-time recognition of clique-width 3 graphs. Discrete Appl. Math. **160**(6), 834–865 (2012)
7. Courcelle, B., Engelfriet, J., Rozenberg, G.: Handle-rewriting hypergraph grammars. J. Comput. Syst. Sci. **46**(2), 218–270 (1993)

8. Courcelle, B., Makowsky, J.A., Rotics, U.: Linear time solvable optimization problems on graphs of bounded clique-width. Theor. Comput. Syst. **33**(2), 125–150 (2000)

9. Courcelle, B., Olariu, S.: Upper bounds to the clique width of graphs. Discrete Appl. Math. **101**(1), 77–114 (2000)

10. Cournier, A., Habib, M.: A new linear algorithm for modular decomposition. CAAP'94. LNCS, vol. 787, pp. 68–84. Springer, Heidelberg (1994)

11. Cunningham, W.H.: A combinatorial decomposition theory. Can. J. Math. **32**(3), 734–765 (1980)

12. Cunningham, W.H.: Decomposition of directed graphs. SIAM J. Algebraic Discrete Methods **3**(2), 214–228 (1982)

13. Das, B., Enduri, M.K., Reddy, I.V.: Logspace and FPT algorithms for graph isomorphism for subclasses of bounded tree-width graphs. In: Rahman, M.S., Tomita, E. (eds.) WALCOM 2015. LNCS, vol. 8973, pp. 329–334. Springer, Heidelberg (2015)

14. Das, B., Enduri, M.K., Reddy, I.V.: Polynomial-time algorithm for isomorphism of graphs with clique-width at most 3. arXiv preprint (2015). arXiv:1506.01695

15. Fellows, M.R., Rosamond, F.A., Rotics, U., Szeider, S.: Clique-width is NP-complete. SIAM J. Discrete Math. **23**(2), 909–939 (2009)

16. Gallai, T.: Transitiv orientierbare graphen. Acta Mathematica Hungarica **18**(1), 25–66 (1967)

17. Grohe, M., Schweitzer, P.: Isomorphism testing for graphs of bounded rank width. In: 2015 IEEE 56th Annual Symposium on Foundations of Computer Science (FOCS), pp. 1010–1029. IEEE (2015)

18. James, L.O., Stanton, R.G., Cowan, D.D.: Graph decomposition for undirected graphs. In: Proceedings of 3rd Southeastern Conference on Combinatorics, Graph Theory, and Computing, pp. 281–290 (1972)

19. Kamiński, M., Lozin, V.V., Milanič, M.: Recent developments on graphs of bounded clique-width. Discrete Appl. Math. **157**(12), 2747–2761 (2009)

20. Lokshtanov, D., Pilipczuk, M., Pilipczuk, M., Saurabh, S.: Fixed-parameter tractable canonization and isomorphism test for graphs of bounded treewidth. In: IEEE 55th Annual Symposium on (FOCS), pp. 186–195 (2014)

21. Luks, E.M.: Isomorphism of graphs of bounded valence can be tested in polynomial time. J. Comput. Syst. Sci. **25**(1), 42–65 (1982)

22. Ma, T.H., Spinrad, J.: An $O(n^2)$ algorithm for undirected split decomposition. J. Algorithms **16**(1), 145–160 (1994)

23. Miller, G.: Isomorphism testing for graphs of bounded genus. In: Proceedings of 12th Annual ACM Symposium on Theory of Computing, pp. 225–235. ACM (1980)

24. Oum, S., Seymour, P.: Approximating clique-width and branch-width. J. Comb. Theor. Ser. B **96**(4), 514–528 (2006)

25. Oum, S., Seymour, P.: Testing branch-width. J. Comb. Theor. Ser. B **97**(3), 385–393 (2007)

26. Tedder, M., Corneil, D.G., Habib, M., Paul, C.: Simpler linear-time modular decomposition via recursive factorizing permutations. In: Aceto, L., Damgård, I., Goldberg, L.A., Halldórsson, M.M., Ingólfsdóttir, A., Walukiewicz, I. (eds.) ICALP 2008, Part I. LNCS, vol. 5125, pp. 634–645. Springer, Heidelberg (2008)

27. Zemlyachenko, V., Konieko, N., Tyshkevich, R.: Graph isomorphism problem (Russian). In: The Theory of Computation I. Notes Sci. Sem. LOMI, vol. 118 (1982)

Fixed Parameter Complexity of Distance Constrained Labeling and Uniform Channel Assignment Problems

(Extended Abstract)

Jiří Fiala[✉], Tomáš Gavenčiak, Dušan Knop,
Martin Koutecký, and Jan Kratochvíl

Department of Applied Mathematics, Charles University,
Malostranské nám. 25, Prague, Czech Republic
{fiala,gavento,knop,koutecky,honza}@kam.mff.cuni.cz

Abstract. We study computational complexity of the class of distance-constrained graph labeling problems from the fixed parameter tractability point of view. The parameters studied are neighborhood diversity and clique width.

We rephrase the distance constrained graph labeling problem as a specific uniform variant of the CHANNEL ASSIGNMENT problem and show that this problem is fixed parameter tractable when parameterized by the neighborhood diversity together with the largest weight. Consequently, every $L(p_1, p_2, \ldots, p_k)$-LABELING problem is FPT when parameterized by the neighborhood diversity, the maximum p_i and k.

Finally, we show that the uniform variant of the CHANNEL ASSIGNMENT problem becomes NP-complete when generalized to graphs of bounded clique width.

1 Introduction

The frequency assignment problem in wireless networks yields an abundance of various mathematical models and related problems. We study a group of such discrete optimization problems in terms of parameterized computational complexity, which is one of the central paradigms of contemporary theoretical computer science. We study parameterization of the problems by *clique width* and particularly by *neighborhood diversity* (nd), a graph parameter lying between clique width and the size of a minimum vertex cover.

All these problems are NP-hard even for constant clique width, including the uniform variant, as we show in this paper. On the other hand, we prove that they are in FPT with respect to nd. Such fixed parameter tractability has so far only been known only for the special case of $L(p, 1)$ labeling when parameterized by vertex cover [7].

Paper supported by project Kontakt LH12095 and by GAUK project 1784214.
Second, third and fourth authors are supported by the project SVV-2016-260332.
First, third and fifth authors are supported by project CE-ITI P202/12/G061 of GAČR.

T.N. Dinh and M.T. Thai (Eds.): COCOON 2016, LNCS 9797, pp. 67–78, 2016.
DOI: 10.1007/978-3-319-42634-1_6

1.1 Distance Constrained Labelings

Given a k-tuple of positive integers $p_1, \ldots, p_k$, called *distance constraints*, an $L(p_1, \ldots, p_k)$-labeling of a graph is an assignment l of integer labels to the vertices of the graph satisfying the following condition: Whenever vertices u and v are at distance i, the assigned labels differ by at least p_i. Formally, $\text{dist}(u, v) = i \implies |l(u) - l(v)| \geq p_i$ for all $u, v : \text{dist}(u, v) \leq k$. Often only non-increasing sequences of distance constraints are considered.

Any $L(1)$-labeling is a graph coloring and vice-versa. Analogously, any coloring of the k-th distance power of a graph is an $L(1, \ldots, 1)$-labeling. The concept of $L(2, 1)$-labeling is attributed to Roberts by Griggs and Yeh [13]. It is not difficult to show that whenever l is an optimal $L(p_1, \ldots, p_k)$-labeling within a range $[0, \lambda]$, then the so called *span* λ is a linear combination of $p_1, \ldots, p_k$ [13,16]. In particular, a graph G allows an $L(p_1, \ldots, p_k)$-labeling of span λ if and only if it has an $L(cp_1, \ldots, cp_k)$-labeling of span $c\lambda$ for any positive integer c.

For computational complexity purposes, we define the following class of decision problems:

Problem 1. $L(p_1, \ldots, p_k)$-LABELING:

Parameters:	Positive integers $p_1, \ldots, p_k$
Input:	Graph G, positive integer λ
Query:	Is there an $L(p_1, \ldots, p_k)$ labeling of G using labels from the interval $[0, \lambda]$?

The $L(2, 1)$-LABELING problem was shown to be NP-complete by Griggs and Yeh [13] by a reduction from HAMILTONIAN CYCLE (with $\lambda = |V_G|$). Fiala et al. [8] showed that $L(2, 1)$-LABELING remains NP-complete also for all fixed $\lambda \geq 4$, while for $\lambda \leq 3$ it is solvable in linear time.

Despite a conjecture that $L(2, 1)$-LABELING remains NP-complete on trees [13], Chang and Kuo [2] showed a dynamic programming algorithm for this problem, as well as for all $L(p_1, p_2)$-labelings where p_2 divides p_1. All the remaining cases of the $L(p_1, p_2)$-LABELING problem on trees have been shown to be NP-complete by Fiala et al. [6]. The same authors showed that $L(2, 1)$-LABELING is already NP-complete on series-parallel graphs [5], which have of tree width at most 2. Note that these results imply NP-hardness of $L(3, 2)$-LABELING on graphs of clique width at most 3 and of $L(2, 1)$-LABELING for clique width at most 6 [3].

On the other hand, when λ is fixed, then the existence of an $L(p_1, \ldots, p_k)$-labeling of G can be expressed in MSO_1, hence it allows a linear time algorithm on any graph of bounded clique width [15].

1.2 Channel Assignment

Channel assignment is a concept closely related to distance constrained graph labeling. Here, every edge has a prescribed weight $w(e)$ and it is required that the labels of adjacent vertices differ at least by the weight of the corresponding edge. The associated decision problem is defined as follows:

Problem 2. CHANNEL ASSIGNMENT:

Input:	Graph G, a positive integer λ, edge weights $w : E_G \to \mathbb{N}$
Query:	Is there a labeling l of the vertices of G by integers from $[0, \lambda]$ such that $|l(u) - l(v)| \geq w(u, v)$ for all $(u, v) \in E_G$?

The maximal edge weight is an obvious necessary lower bound for the span of any labeling. Observe that for any bipartite graph, in particular also for all trees, it is also an upper bound — a labeling that assigns 0 to one class of the bipartition and $w_{\max} = \max\{w(e), e \in E_G\}$ to the other class satisfies all edge constraints. McDiarmid and Reed [19] showed that it is NP-complete to decide whether a graph of tree width 3 allows a channel assignment of given span λ. This NP-hardness hence applies on graphs of clique width at most 12 [3]. It is worth noting that for graphs of tree width 2, i.e. for subgraphs of series-parallel graphs, the complexity characterization of CHANNEL ASSIGNMENT is still open. Only a few partial results are known [20], among others that CHANNEL ASSIGNMENT is polynomially solvable on graphs of bounded tree width if the span λ is bounded by a constant.

Any instance G, λ of the $L(p_1, \ldots, p_k)$-LABELING problem can straightforwardly be reduced to an instance G^k, λ, w of the CHANNEL ASSIGNMENT problem. Here, G^k arises from G by connecting all pairs of vertices that are in G at distance at most k, and for the edges of G^k we let $w(u, v) = p_i$ whenever $\operatorname{dist}_G(u, v) = i$.

The resulting instances of CHANNEL ASSIGNMENT have by the construction some special properties. We explore and generalize these to obtain a uniform variant of the CHANNEL ASSIGNMENT problem.

1.3 Neighborhood Diversity

Lampis significantly reduced (from the tower function to double exponential) the hidden constants of the generic polynomial algorithms for MSO_2 model checking on graphs with bounded vertex cover [17]. To extend this approach to a broader class of graphs he introduced a new graph parameter called the neighborhood diversity of a graph as follows:

Definition 1 (Neighborhood Diversity). *A partition $V_1, \ldots, V_d$ is called a* neighborhood diversity decomposition *if it satisfies*

- *each V_i induces either an empty subgraph or a complete subgraph of G, and*
- *for each distinct V_i and V_j there are either no edges between V_i and V_j, or every vertex of V_i is adjacent to all vertices of V_j.*

We write $u \sim v$ to indicate that u and v belong to the same class of the decomposition.

The neighborhood diversity *of a graph G, denoted by $\mathrm{nd}(G)$, is the minimum τ such that G has a neighborhood diversity decomposition with τ classes.*

Observe that for the optimal neighborhood diversity decomposition it holds that $u \sim u'$ is equivalent with $N(u) \setminus v = N(v) \setminus u$. Therefore, the optimal neighborhood diversity decomposition can be computed in $O(n^3)$ time [17].

Classes of graphs of bounded neighborhood diversity reside between classes of bounded vertex cover and graphs of bounded clique width. Several non-MSO$_1$ problems, e.g. HAMILTONIAN CYCLE can be solved in polynomial time on graphs of bounded clique width [21]. On the other hand, Fomin et al. stated more precisely that the HAMILTONIAN CYCLE problem is $W[1]$-hard, when parameterized by clique width [9]. In sequel, Lampis showed that some of these problems, including HAMILTONIAN CYCLE, are indeed fixed parameter tractable on graphs of bounded neighborhood diversity [17].

Ganian and Obdržálek [12] further deepened Lampis' results and showed that also problems expressible in MSO$_1$ with cardinality constraints (cardMSO$_1$) are fixed parameter tractable when parameterized by nd(G).

Observe that a sufficiently large n-vertex graph of bounded neighborhood diversity can be described in significantly more effective way, namely by using only $O(\log n \, \mathrm{nd}(G)^2)$ space:

Definition 2 (Type Graph). *The* type graph $T(G)$ *for a neighborhood diversity decomposition* $V_1, \ldots, V_d$ *of a graph* G *is a vertex weighted graph on vertices* $\{t_1, \ldots, t_d\}$, *where each* t_i *is assigned weight* $s(t_i) = |V_i|$, *i.e. the size of the corresponding class of the decomposition. Distinct vertices* t_i *and* t_j *are adjacent in* $T(G)$ *if and only if the edges between the two corresponding classes* V_i *and* V_j *form a complete bipartite graph. Moreover,* $T(G)$ *contains a loop incident with vertex* t_i *if and only if the corresponding class* V_i *induces a clique.*

For our purposes, i.e. to decide existence of a suitable labeling of a graph G, it suffices to consider only its type graph, as G can be uniquely reconstructed from $T(G)$ (up to an isomorphism) and vice-versa.

Moreover, the reduction of $L(p_1, \ldots, p_k)$-LABELING to CHANNEL ASSIGNMENT preserves the property of bounded neighborhood diversity:

Observation 3. *For any graph* G *and any positive integer* k *it holds that* $\mathrm{nd}(G) \geq \mathrm{nd}(G^k)$.

Proof. The optimal neighborhood diversity decomposition of G is a neighborhood diversity decomposition of G^k. □

1.4 Our Contribution

Our goal is an extension of the FPT algorithm for $L(2,1)$-LABELING on graphs of bounded vertex cover to broader graph classes and for rich collections of distance constraints. In particular, we aim at $L(p_1, \ldots, p_k)$-LABELING on graphs of bounded neighborhood diversity.

For this purpose we utilize the aforementioned reduction to the CHANNEL ASSIGNMENT problem, taking into account that the neighborhood diversity remains bounded, even though the underlying graph changes.

It is worth to note that we must adopt additional assumptions for the CHAN-NEL ASSIGNMENT since otherwise it is NP-complete already on complete graphs, i.e. on graphs with $nd(G) = 1$. To see this, we recall the construction of Griggs and Yeh [13]. They show that a graph H on n vertices has a Hamiltonian path if and only if the complement of H extended by a single universal vertex allows an $L(2,1)$-labeling of span $n+1$. As the existence of a universal vertex yields diameter at most two, the underlying graph for the resulting instance of CHANNEL ASSIGNMENT is K_{n+1}.

On the other hand, the additional assumptions on the instances of CHANNEL ASSIGNMENT will still allow us to reduce any instance of the $L(p_1, \ldots, p_k)$-LABELING problem. By the reduction, all edges between classes of the neighborhood diversity decomposition are assigned the same weight. We formally adopt this as our additional constraint as follows:

Definition 3. *The edge weights w on a graph G are nd-uniform if $w(u,v) = w(u',v')$ whenever $u \sim u'$ and $v \sim v'$ with respect to the optimal neighborhood diversity decomposition. In a similar way we define uniform weights with respect to a particular decomposition.*

Our main contribution is an algorithm for the following scenario:

Theorem 4. *The CHANNEL ASSIGNMENT problem on nd-uniform instances is FPT when parameterized by nd and w_{max}, where $w_{max} = \max\{w(e), e \in E_G\}$.*

Immediately, we get the following consequence:

Theorem 5. *For $p_1, \ldots, p_k$, the $L(p_1, \ldots, p_k)$-LABELING problem is FPT when parameterized by nd, k and maximum p_i (or equivalently by nd and the k-tuple $(p_1, \ldots, p_k)$).*

One may ask whether the uniform version of CHANNEL ASSIGNMENT allows an FPT algorithm also for a broader class of graphs. Finally, we show that a natural generalization of this concept on graphs of bounded clique width yields an NP-complete problem on graphs of clique width at most 5.

2 Representing Labelings as Sequences and Walks

We now focus on the nd-uniform instances of the CHANNEL ASSIGNMENT problem. It has been already mentioned that the optimal neighborhood diversity decomposition can be computed in cubic time. The test, whether it is nd-uniform, could be computed in quadratic additional time. On the other hand, on nd-uniform instances it suffices to consider only the type graph, whose edges take weights from the edges of the underlying graph (see Fig. 1), since such a weighted type graph corresponds uniquely to the original weighted graph, up to an isomorphism.

Hence without loss of generalization assume that our algorithms are given the type graph whose edges are weighted by separation constraints w, however we express the time complexity bounds in terms of the size of the original graph.

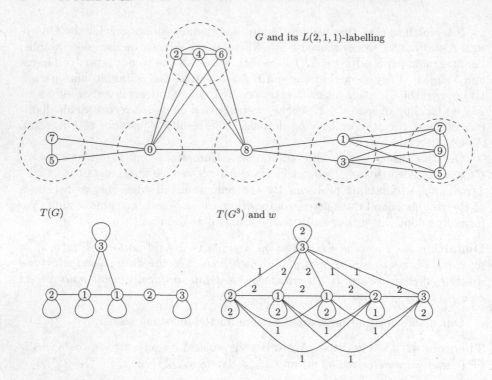

Fig. 1. An example of a graph with its neighborhood diversity decomposition. Vertex labels indicate one of its optimal $L(2,1,1)$-labelings. The corresponding type graph. The weighted type graph corresponding to the resulting instance of the CHANNEL ASSIGNMENT problem.

Without loss of generality we may assume that the given graph G and its type graph $T(G)$ are connected, since connected components can be treated independently.

If the type graph $T(G)$ contains a type t not incident with a loop, we may reduce the channel assignment problem to the graph G', obtained from G by deleting all but one vertices of the type t. Any channel assignment of G' yields a valid channel assignment of G by using the same label on all vertices of type t in G as was given to the single vertex of type t in G'. Observe that adding a loop to a type, which represents only a single vertex, does not affect the resulting graph G'. Hence we assume without loss of generality that all types are incident with a loop. We call such type graph *reflexive*.

Observation 6. *If the type graph $T(G)$ is reflexive, then vertices of G of the same type have distinct labels in every channel assignment.*

Up to an isomorphism of the graph G, any channel assignment l is uniquely characterized by a sequence of type sets as follows:

Lemma 1. *Any weighted graph G corresponding to a reflexive weighted type graph $T(G), w$ allows a channel assignment of span λ, if and only if there exists a sequence of sets $\mathcal{T} = T_0, \ldots, T_\lambda$ with the following properties:*

(i) $T_i \subseteq V_{T(G)}$ for each $i \in [0, \lambda]$,
(ii) for each $t \in V_{T(G)} : s(t) = |\{T_i : t \in T_i\}|$,
(iii) for all $(t, r) \in E_{T(G)} : (t \in T_i \wedge r \in T_j \wedge (t \neq r \vee i \neq j)) \Rightarrow |i - j| \geq w(t, r)$

Proof. Given a channel assignment $l : V_G \to [0, \lambda]$, we define the desired sequence $\mathcal{T}$, such that the i-th element is the set of types that contain a vertex labeled by i. Formally $T_i = \{t : \exists u \in V_t : l(u) = i\}$. Now

(i) each element of the sequence is a set of types, possibly empty,
(ii) as all vertices of V_i are labeled by distinct labels by Observation 6, any type t occurs in $s(t)$ many elements of the sequence
(iii) if u of type t is labeled by i, and it is adjacent to v of type r labeled by j, then $|i - j| = |l(u) - l(v)| \geq w(u, v) = w(t, r)$, i.e. adjacent types t and r may appear in sets that are in the sequence at least $w(t, r)$ apart.

In the opposite direction assume that the sequence $\mathcal{T}$ exists. Then for each set T_i and type $t_j \in T_i$ we choose a distinct vertex $u \in V_j$ and label it by i, i.e. $l(u) = i$.

Now the condition (ii) guarantees that all vertices are labeled, while condition (iii) guarantees that all distance constraints are fulfilled. □

Observe that Lemma 1 poses no constraints on pairs of sets T_i, T_j that are at distance at least $w_{\max}$. Hence, we build an auxiliary directed graph D on all possible sequences of sets of length at most $z = w_{\max} - 1$.

The edges of D connect those sequences that overlap on a fragment of length $z - 1$, i.e. when they could be consecutive in $\mathcal{T}$. This construction is well known from the so-called shift register graph.

Definition 4. *For a general graph F and weights $w : E_F \to [1, z]$ we define a directed graph D such that*

– *the vertices of V_D are all z-tuples $(T_1, \ldots, T_z)$ of subsets of V_F such that for all $(t, r) \in E_F : (t \in T_i \wedge r \in T_j) \Rightarrow |i - j| \geq w(t, r)$*
– *$((T_1, \ldots, T_z), (T_1', \ldots, T_z')) \in E_D \Leftrightarrow T_i' = T_{i+1}$ for all $i \in [1, z - 1]$.*

As the first condition of the above definition mimics (iii) of Lemma 1 with $F = T(G)$, any sequence $\mathcal{T}$ that justifies a solution for $(T(G), w, \lambda)$, can be transformed into a walk of length $\lambda - z + 1$ in D.

In the opposite direction, namely in order to construct a walk in D, that corresponds to a valid channel assignment, we need to guarantee also an analogue of the condition (ii) of Lemma 1. In other words, each type should occur sufficiently many times in the resulting walk. Indeed, the construction of D is independent on the function s, which specifies how many vertices of each type are present in G.

In this concern we consider only special walks that allow us to count the occurrences of sets within z-tuples. Observe that V_D also contains the z-tuple $\emptyset^z = (\emptyset, \dots, \emptyset)$. In addition, any walk of length $\lambda - z + 1$ can be converted into a closed walk from $\emptyset^z$ of length $\lambda + z + 1$, since the corresponding sequence $\mathcal{T}$ can be padded with z additional empty sets at the front, and another z empty sets at the end. From our reasoning, the following claim is immediate:

Lemma 2. *A closed walk $\mathcal{W} = W_1, \dots, W_{\lambda+z+1}$ on D where $W_1 = W_{\lambda+z+1} = \emptyset^z$, yields a solution of the* CHANNEL ASSIGNMENT *problem on a nd-uniform instance G, w, λ with reflexive $T(G)$, if and only if $s(t) = |\{W_i : t \in (W_i)_1\}|$ holds for each $t \in V_{T(G)}$.*

We found interesting that our representation of the solution resembles the NP-hardness reduction found by Griggs and Yeh [13] (it was briefly outlined in Sect. 1.4) and later generalized by Bodlaender et al. [1]. The key difference is that in their reduction, a Hamilton path is represented by a sequence of vertices of the constructed graph. In contrast, we consider walks in the type graph, which is assumed to be of limited size.

3   The Algorithm

In this section we prove the following statement, which directly implies our main result, Theorem 4:

Proposition 1. *Let G, w be a weighted graph, whose weights are uniform with respect to a neighborhood diversity partition with τ classes.*

Then the CHANNEL ASSIGNMENT *problem can be decided on G, w and any λ in time $2^{2^{O(\tau w_{\max})}} \log n$, where n is the number of vertices of G, provided that G, w are described by a weighted type graph $T(G)$ on τ nodes.*

A suitable labeling of G can be found in additional $2^{2^{O(\tau w_{\max})}} n$ time.

Proof. According to Lemma 2, it suffices to find a closed walk $\mathcal{W}$ (if it exists) corresponding to the desired labeling l. From the well-known Euler's theorem it follows that any directed closed walk $\mathcal{W}$ yields a multiset of edges in D that induces a connected subgraph and that satisfies Kirchhoff's law. In addition, any such suitable multiset of edges can be converted into a closed walk, though the result need not be unique.

For this purpose we introduce an integer variable $\alpha_{(W,U)}$ for every directed edge $(W, U) \in E_D$. The value of the variable $\alpha_{(W,U)}$ is the number of occurrences of (W, U) in the multiset of edges.

Kirchhoff's law is straightforwardly expressed as:

$$\forall W \in V_D : \sum_{U : (W,U) \in E_D} \alpha_{(W,U)} - \sum_{U : (U,W) \in E_D} \alpha_{(U,W)} = 0$$

In order to guarantee connectivity, observe first that an edge (W, U) and $\emptyset^z$ would be in distinct components of a subgraph of D, if the subgraph is formed by removing edges that include a cut C between (W, U) and $\emptyset^z$. Now, the chosen multiset of edges is disconnected from $\emptyset^z$, if there is such an edge (W, U) together with a cut set C such that $\alpha_{(W,U)}$ has a positive value, while all variables corresponding to elements of C are zeros. As all variable values are bounded above by λ, we express that C is not a cutset for the chosen multiset of edges by the following condition:

$$\alpha_{(W,U)} - \lambda \sum_{e \in C} \alpha_e \leq 0$$

To guarantee the overall connectivity, we apply the above condition for every edge $(W, U) \in E_D$, where $W, U \neq \emptyset^z$, and for each set of edges C that separates W or U from $\emptyset^z$.

The necessary condition expressed in Lemma 2 can be stated in terms of variables $\alpha_{(W,U)}$ as

$$\forall t \in V_{T(G)} : \sum_{W : t \in (W)_1} \sum_{U : (W,U) \in E_D} \alpha_{(W,U)} = s(t)$$

Finally, the size of the multiset is the length of the walk, i.e.

$$\sum_{(W,U) \in E_D} \alpha_{(W,U)} = \lambda + z + 1$$

Observe that these conditions for all (W, U) and all suitable C indeed imply that the $\emptyset^z$ belongs to the subgraph induced by edges with positively evaluated variables $\alpha_{(W,U)}$.

Frank and Tardos [10] (improving the former result due to Lenstra [18]) showed that the time needed to solve the system of inequalities with p integer variables is $O(p^{2.5p+o(p)} L)$, where L is the number of bits needed to encode the input. As we have $2^{O(\tau z)}$ variables and the conditions are encoded in space $2^{2^{O(\tau z)}} \log n$, the time needed to resolve the system of inequalities is $2^{2^{O(\tau z)}} \log n$. □

We are aware the the double exponential dependency on nd and $w_{\max}$ makes our algorithm interesting mostly from the theoretical perspective. Naturally, one may ask, whether the exponential tower height might be reduced or whether some nontrivial lower bounds on the computational complexity could be established (under usual assumptions on classes in the complexity hierarchy).

4   NLC-Uniform Channel Assignment

One may ask whether the concept of nd-uniform weights could be extended to broader graph classes. We show, that already its direct extension to graphs of bounded clique width makes the CHANNEL ASSIGNMENT problem NP-complete. Instead of clique width we express our results in terms of NLC-width [21]

(NLC stands for node label controlled). The parameter NLC-width is linearly dependent on clique width, but it is technically simpler.

We now briefly review the related terminology. A NLC-decomposition of a graph G is a rooted tree whose leaves are in one-to-one correspondence with the vertices of G. For the purpose of inserting edges, each vertex is given a label (the labels for channel assignment are now irrelevant), which may change during the construction of the graph G. Internal nodes of the tree are of two kinds: *relabel* nodes and *join* nodes.

Each relabel node has a single child and as a parameter takes a mapping ρ on the set of labels. The graph corresponding to a relabel node is isomorphic to the graph corresponding to its child, only ρ is applied on each vertex label.

Each join node has a two children and as a parameter takes a binary relation S on the set of labels. The graph corresponding to a join node is isomorphic to the disjoint union of the two graphs G_1 and G_2 corresponding to its children, where further edges are inserted as follows: $u \in V_{G_1}$ labeled by i is made adjacent to $v \in V_{G_2}$ labeled by j if and only if $(i, j) \in S$.

The minimum number of labels needed to construct at least one labeling of G in this way is the NLC width of G, denoted by $\mathrm{nlc}(G)$.

Observe that $\mathrm{nlc}(G) \leq \mathrm{nd}(G)$ as the vertex types could be used as labels for the corresponding vertices and the adjacency relation in the type graph could be used for S in all join nodes. In particular, in this construction the order of performing joins is irrelevant and no relabel nodes are needed.

Definition 5. *The edge weights w on a graph G are nlc-uniform with respect to a particular NLC-decomposition, if $w(u, v) = w(u', v')$ whenever edges (u, v) and (u', v') are inserted during the same join operation and at the moment of insertion u, u' have the same label in G_1 and v, v' have the same label in G_2.*

Observe that our comment before the last definition justifies that weights that are uniform with respect to a neighborhood diversity decomposition are uniform also with respect to the corresponding NLC-decomposition.

Gurski and Wanke showed that the NLC-width remains bounded when taking powers of trees [14]. It is well known that NLC-width of a tree is at most three. Fiala et al. proved that $L(3, 2)$-LABELING is NP-complete on trees [6]. To combine these facts together we show that the weights on the graph arising from a reduction of the $L(3, 2)$-labeling on a tree to CHANNEL ASSIGNMENT are nlc-uniform.

Theorem 7. *The CHANNEL ASSIGNMENT problem is NP-complete on graphs with edge weights that are nlc-uniform with respect to an NLC-decomposition of width at most four.*

5   Conclusion

We have shown an algorithm for the CHANNEL ASSIGNMENT problem on nd-uniform instances and several complexity consequences for the

$L(p_1, \ldots p_k)$-LABELING problem. In particular, Theorem 5 extends known results for the $L(p, 1)$-LABELING problem to labelings with arbitrarily many distance constraints, answering an open question of [7]. Simultaneously, we broaden the considered graph classes by restricting neighborhood diversity instead of vertex cover.

While the main technical tools of our algorithms are bounded-dimension ILP programs, ubiquitous in the FPT area, the paper shows an interesting insight on the nature of the labelings over the type graph and the necessary patterns of such labelings of very high span. Note that the span of a graph is generally not bounded by any of the considered parameters and may be even proportional to the order of the graph.

Solving a generalized problem on graphs of bounded neighborhood diversity is a viable method for designing FPT algorithms for a given problem on graphs of bounded vertex cover, as demonstrated by this and previous papers. This promotes neighborhood diversity as a parameter that naturally generalizes the widely studied parameter vertex cover.

We would like to point out that the parameter *modular width*, proposed by Gajarský et al. [11], offers further generalization of neighborhood diversity towards the clique width [4].

As an interesting open problem we ask whether it is possible to strengthen our results to graphs of bounded modular width or whether the problem might be already NP-complete for fixed modular width, as is the case with clique width. For example, the GRAPH COLORING problem ILP-based algorithm for bounded neighborhood diversity translates naturally to an algorithm for bounded modular width. On the other hand, there is no apparent way how our labeling results could be adapted to modular width in a similar way.

Acknowledgement. We thank Andrzej Proskurowski, Tomáš Masařík and anonymous referees for valuable comments.

References

1. Bodlaender, H.L., Kloks, T., Tan, R.B., van Leeuwen, J.: λ-Coloring of graphs. In: Reichel, H., Tison, S. (eds.) STACS 2000. LNCS, vol. 1770, pp. 395–406. Springer, Heidelberg (2000)
2. Chang, G.J., Kuo, D.: The $L(2, 1)$-labeling problem on graphs. SIAM J. Discret. Math. **9**(2), 309–316 (1996)
3. Corneil, D.C., Rotics, U.: On the relationship between clique-width and treewidth. SIAM J. Comput. **34**(4), 825–847 (2005)
4. Courcelle, B., Olariu, S.: Upper bounds to the clique width of graphs. Discret. Appl. Math. **101**(1–3), 77–114 (2000)
5. Fiala, J., Golovach, P.A., Kratochvíl, J.: Distance constrained labelings of graphs of bounded treewidth. In: Caires, L., Italiano, G.F., Monteiro, L., Palamidessi, C., Yung, M. (eds.) ICALP 2005. LNCS, vol. 3580, pp. 360–372. Springer, Heidelberg (2005)

6. Fiala, J., Golovach, P.A., Kratochvíl, J.: Computational complexity of the distance constrained labeling problem for trees (extended abstract). In: Aceto, L., Damgård, I., Goldberg, L.A., Halldórsson, M.M., Ingólfsdóttir, A., Walukiewicz, I. (eds.) ICALP 2008, Part I. LNCS, vol. 5125, pp. 294–305. Springer, Heidelberg (2008)
7. Fiala, J., Golovach, P.A., Kratochvíl, J.: Parameterized complexity of coloring problems: treewidth versus vertex cover. Theor. Comput. Sci. **412**(23), 2513–2523 (2011)
8. Fiala, J., Kratochvíl, J., Kloks, T.: Fixed-parameter complexity of λ-labelings. Discret. Appl. Math. **113**(1), 59–72 (2001)
9. Fomin, F.V., Golovach, P.A., Lokshtanov, D., Saurabh, S.: Intractability of clique-width parameterizations. SIAM J. Comput. **39**(5), 1941–1956 (2010)
10. Frank, A., Tardos, É.: An application of simultaneous diophantine approximation in combinatorial optimization. Combinatorica **7**(1), 49–65 (1987)
11. Gajarský, J., Lampis, M., Ordyniak, S.: Parameterized algorithms for modular-width. In: Gutin, G., Szeider, S. (eds.) IPEC 2013. LNCS, vol. 8246, pp. 163–176. Springer, Heidelberg (2013)
12. Ganian, R., Obdržálek, J.: Expanding the expressive power of monadic second-order logic on restricted graph classes. In: Lecroq, T., Mouchard, L. (eds.) IWOCA 2013. LNCS, vol. 8288, pp. 164–177. Springer, Heidelberg (2013)
13. Griggs, J.R., Yeh, R.K.: Labelling graphs with a condition at distance 2. SIAM J. Discret. Math. **5**(4), 586–595 (1992)
14. Gurski, F., Wanke, E.: The NLC-width and clique-width for powers of graphs of bounded tree-width. Discret. Appl. Math. **157**(4), 583–595 (2009)
15. Kobler, D., Rotics, U.: Polynomial algorithms for partitioning problems on graphs with fixed clique-width (extended abstract). In: 12th ACM-SIAM of the Symposium on Discrete Algorithms, SODA 2001, Washington, pp. 468–476 (2001)
16. Král, D.: The channel assignment problem with variable weights. SIAM J. Discret. Math. **20**(3), 690–704 (2006)
17. Lampis, M.: Algorithmic meta-theorems for restrictions of treewidth. Algorithmica **64**(1), 19–37 (2012)
18. Lenstra Jr., H.W.: Integer programming with a fixed number of variables. Math. Oper. Res. **8**(4), 538–548 (1983)
19. McDiarmid, C., Reed, B.: Channel assignment on graphs of bounded treewidth. Discret. Math. **273**(1–3), 183–192 (2003)
20. Škvarek, M.: The channel assignment problem for series-parallel graphs. Bachelor's thesis, Charles University, Prague (2010). (in Czech)
21. Wanke, E.: k-NLC graphs and polynomial algorithms. Discret. Appl. Math. **54**(2–3), 251–266 (1994)

A Parameterized Algorithm for Bounded-Degree Vertex Deletion

Mingyu Xiao[✉]

School of Computer Science and Engineering, University of Electronic Science
and Technology of China, Chengdu, China
myxiao@gmail.com

Abstract. The d-bounded-degree vertex deletion problem, to delete at
most k vertices in a given graph to make the maximum degree of the
remaining graph at most d, finds applications in computational biology,
social network analysis and some others. It can be regarded as a special case of the $(d + 2)$-hitting set problem and generates the famous
vertex cover problem. The d-bounded-degree vertex deletion problem
is NP-hard for each fixed $d \geq 0$. In terms of parameterized complexity, the problem parameterized by k is W[2]-hard for unbounded d and
fixed-parameter tractable for each fixed $d \geq 0$. Previously, (randomized)
parameterized algorithms for this problem with running time bound
$O^*((d + 1)^k)$ are only known for $d \leq 2$. In this paper, we give a uniform parameterized algorithm deterministically solving this problem in
$O^*((d+1)^k)$ time for each $d \geq 3$. Note that it is an open problem whether
the d'-hitting set problem can be solved in $O^*((d' - 1)^k)$ time for $d' \geq 3$.
Our result answers this challenging open problem affirmatively for a special case. Furthermore, our algorithm also gets a running time bound of
$O^*(3.0645^k)$ for the case that $d = 2$, improving the previous deterministic
bound of $O^*(3.24^k)$.

Keywords: Parameterized algorithms · Graph algorithms · Bounded-
degree vertex deletion · Hitting set

1 Introduction

The d-bounded-degree vertex deletion problem is a natural generation of the
famous vertex cover problem, which is one of the best studied problems in combinatorial optimization. An application of the d-bounded-degree vertex deletion
problem in computational biology is addressed by Fellows et al. [5]: A clique-
centric approach in the analysis of genetic networks based on micro-array data
can be modeled as the d-bounded-degree vertex deletion problem. The problem also plays an important role in the area of property testing [12]. Its "dual
problem" – the s-plex problem was introduced in 1978 by Seidman and Foster [14]
and it becomes an important problem in social network analysis now [1].

M. Xiao—Supported by NFSC of China under the Grant 61370071 and Fundamental
Research Funds for the Central Universities under the Grant ZYGX2015J057.

T.N. Dinh and M.T. Thai (Eds.): COCOON 2016, LNCS 9797, pp. 79–91, 2016.
DOI: 10.1007/978-3-319-42634-1_7

The d-bounded-degree vertex deletion problem is also extensively studied in theory, especially in parameterized complexity. It has been shown that the problem parameterized by the size k of the deletion set is W[2]-hard for unbounded d and fixed-parameter tractable for each fixed $d \geq 0$ [5]. Betzler et al. [2] also studied the parameterized complexity of the problem with respect to the treewidth tw of the graph. The problem is FPT with parameters k and tw and W[2]-hard with only parameter tw. Fellows et al. [5] generated the NT-theorem for the vertex cover problem to the d-bounded-degree vertex deletion problem, which can imply a linear vertex kernel for the problem with $d = 0, 1$ and a polynomial vertex kernel for each fixed $d \geq 2$. A linear vertex kernel for the case that $d = 2$ was developed in [4]. Recently, a refined generation of the NT-theorem was proved [17], which can get a linear vertex kernel for each fixed $d \geq 0$.

In terms of parameterized algorithms, the case that $d = 0$, i.e., the vertex cover problem, can be solved in $O^*(1.2738^k)$ time now [3]. When $d = 1$, the problem is known as the P_3 vertex cover problem. Tu [15] gave an $O^*(2^k)$-time algorithm and the running time bound was improved to $O^*(1.882^k)$ by Wu [16] and to $O^*(1.8172^k)$ by Katrenič [11]. When $d = 2$, the problem is known as the co-path/cycle problem. For this problem, there is an $O^*(3.24^k)$-time deterministic algorithm [4] and an $O^*(3^k)$-time randomized algorithm [6]. For $d \geq 3$, a simple branch-and-reduce algorithm that tries all $d + 2$ possibilities for a $(d + 1)$-star in the graph gets the running time bound of $O^*((d + 2)^k)$. In fact, the d-bounded-degree vertex deletion problem can be regarded as a special case of the $(d + 2)$-hitting set problem and the latter problem has been extensively studied in parameterized algorithms [7–9,13]. For a graph G, we regard each vertex in the graph as an element and each $(d + 1)$-star as a set of size $d + 2$ (a vertex of degree $d_0 > d$ will form $\binom{d_0}{d+1}$ sets). Then the d-bounded-degree vertex deletion problem in G becomes an instance of the $(d+2)$-hitting set problem. There are several parameterized algorithms for the d'-hitting set problem running in $O^*((d' - 1 + c)^k)$ time [9,13], where $0 < c < 1$ is a function of d'^{-1}. It leaves as an interesting open problem whether the d'-hitting set problem can be solved in $O^*((d' - 1)^k)$ time. Note that it is marked in [9] that "$(d' - 1)^k$ seems an unsurpassable lower bound". By using fastest algorithms for the $(d + 2)$-hitting set problem, we can get an algorithm with running time bound of $O^*((d + 1 + c_0)^k)$ with $0 < c_0 < 1$ for each fixed d.

In this paper, we design a uniform algorithm for the d-bounded-degree vertex deletion problem, which achieves the running time bound of $O^*((d+1)^k)$ for each $d \geq 3$. Although our problem is a special case of the $(d + 2)$-hitting set problem, the above bound is not easy to reach. We need a very careful analysis and some good graph structural properties. It is also worthy to mention that our algorithm also works on the case that $d = 2$ and runs in $O^*(3.0645^k)$ time, improving the previous deterministic bound of $O^*(3.24^k)$ [4] and comparable with the previous randomized bound of $O^*(3^k)$ [6].

2 Preliminaries

Let $G = (V, E)$ be a simple undirected graph, and $X \subseteq V$ be a subset of vertices. The subgraph induced by X is denoted by $G[X]$, and $G[V \setminus X]$ is written as $G \setminus X$. We may simply use v to denote the set $\{v\}$ of a single vertex v. Let $N(X)$ denote the set of *neighbors* of X, i.e., the vertices in $V \setminus X$ adjacent to a vertex $x \in X$, and denote $N(X) \cup X$ by $N[X]$. The *degree* $d(v)$ of a vertex v is defined to be $|N(v)|$. A graph of maximum degree p is also called a *degree-p graph*. For an integer $q \geq 1$, a star with $q + 1$ vertices is called a *q-star*. A set S of vertices is called a *d-deletion set* of a graph G, if $G \setminus S$ has maximum degree at most d. In our problem, we want to find a d-deletion set of size at most k in a graph. Formally, our problem is defined as following.

---

d-Bounded-Degree Vertex Deletion
Instance: A graph $G = (V, E)$ and two nonnegative integers d and k.
Question: To decide whether there is a subset $S \subseteq V$ of vertices such that $|S| \leq k$ and the induced graph $G[V \setminus S]$ has maximum degree at most d.

In the above definition, S is also called a *solution set*.

2.1 Some Basic Properties

The following lemmas are basic structural properties used to design branching rules in our algorithms.

Lemma 1. *Let v be a vertex of degree $\geq d + 1$ in a graph G. Any d-deletion set contains either v or $d(v) - d$ neighbors of v.*

A vertex v *dominates* a vertex u if all vertices of degree $\geq d + 1$ in $N[u]$ are also in $N[v]$. Note that in this definition, we do not require $N[u] \subseteq N[v]$.

Lemma 2. *If a vertex v of degree $d + 1$ dominates a neighbor u of it, then there is a minimum d-deletion set containing at least one vertex in $N[v] \setminus \{u\}$.*

Proof. Since v is of degree $d+1$, any d-deletion set S contains at least one vertex in $N[v]$. Assume that S contains only u in $N[v]$. We can see that $S' = S \cup \{v\} \setminus \{u\}$ is still a d-deletion set and $|S'| \leq |S|$. Thus, the lemma holds. $\square$

Lemma 3. *If a vertex u dominates a vertex v of degree $d + 1$, then there is a minimum d-deletion set containing at least one neighbor of v.*

Proof. Since u dominates v and v is of degree $d + 1$, we know that u is a neighbor of v. Any d-deletion set S contains at least one vertex in $N[v]$ since it is of degree $d+1$. Assume that $S \cap N[v] = \{v\}$. We can see that $S' = S \cup \{u\} \setminus \{v\}$ is a d-deletion set containing a neighbor of v and $|S'| \leq |S|$. Thus, the lemma holds. $\square$

If there is a vertex of degree $\geq d + 1$ dominating a neighbor of it or being dominated by another vertex, we say that the graph has a *proper domination*. Note that if a vertex u of degree $\geq d + 1$ has at most one neighbor v of degree $\geq d + 1$, then u is dominated by v and then there is a proper domination. In fact, we have:

Lemma 4. *If a graph has no proper domination, then each vertex of degree $\geq d + 1$ in it has at least two nonadjacent neighbors of degree $\geq d + 1$.*

2.2 Branch-and-Search Algorithms

Our algorithm is a typical branch-and-search algorithm. In our algorithm, we search a solution for an instance by recursively branching on the current instance into several smaller instances until the instances become trivial instances. Each simple branching operation creates a recurrence relation. Assume that the branching operation branches on an instance with parameter k into l branches such that in the i-th branch the parameter decreases by at least a_i. Let $C(k)$ denote the worst size of the search tree to search a solution to any instance with parameter k. We get a recurrence relation[1]

$$C(k) \leq C(k - a_1) + C(k - a_2) + \cdots + C(k - a_l) + 1.$$

The largest root of the function $f(x) = 1 - \sum_{i=1}^{l} x^{-a_i}$ is called the *branching factor* of the recurrence relation. Let α be the maximum branching factor among all branching factors in the algorithm. The size of the search tree that represents the branching process of the algorithm applied to an instance with parameter k is given by $O(\alpha^k)$. More details about the analysis and how to solve recurrences can be found in the monograph [10].

3   The Idea and Organization of the Algorithm

Our purpose is to design a branch-and-search algorithm for the d-bounded-degree vertex deletion problem such that the branching factor of each recurrence relation with respective to the parameter k is at most $d + 1$. Lemma 1 provides a simple branching rule: for a vertex v of degree $\geq d + 1$, branching by either including v or each set of $d(v) - d$ neighbors of v to the solution set. We will show that when $d(v) \geq d + 2$, this simple branching operation is good enough to get a branching factor $\leq d + 1$ for each $d \geq 2$ (See Step 1 in Sect. 4). Thus, we can use this operation to deal with vertices of degree $\geq d + 2$. Lemma 1 for a degree-$(d + 1)$ vertex v can be interpreted as: at least one vertex in $N[v]$ is in a d-deletion set. This branching operation will only get a branching factor of $d + 2$ for this case. But when there is a proper domination in a degree-$(d + 1)$ graph, we still can

---

[1] In fact, we may simply write a recurrence relation as $C(k) \leq C(k - a_1) + C(k - a_2) + \cdots + C(k - a_l)$. This difference will only affect a constant behind O in the finial running time.

branch with branching factor $d+1$, since we can ignore one branch by Lemmas 2 and 3. The detailed analysis is given in Step 2 in Sect. 4. When the graph is of maximum degree $d+1$ and has no proper domination, we need to use more structural properties.

To find a d-deletion set in a degree-$(d+1)$ graph is equivalent to find a vertex subset intersecting $N[v]$ for each degree-$(d+1)$ vertex v. If there are some vertices in $N[v_1] \cap N[v_2]$ for two degree-$(d+1)$ vertices v_1 and v_2, some information may be useful for us to design a good branching rule. Note that for two adjacent degree-$(d+1)$ vertices v_1 and v_2, there are at least two vertices in the intersection of $N[v_1]$ and $N[v_2]$. Lemma 4 guarantees that each degree-$(d+1)$ vertex has at least two nonadjacent degree-$(d+1)$ neighbors if a degree-$(d+1)$ graph has no proper domination. So we will focus on adjacent degree-$(d+1)$ vertices.

We define three relations between two degree-$(d+1)$ vertices. A pair of adjacent degree-$(d+1)$ vertices is a *good pair* if they have at least one and at most $d-2$ common neighbors. A pair of adjacent degree-$(d+1)$ vertices is a *close pair* if they have exactly $d-1$ common neighbors. A pair of nonadjacent degree-$(d+1)$ vertices is a *similar pair* if they have the same neighbor set. We have a good branching rule to deal with good pairs. See Step 3 in Sect. 4. After dealing with all good pairs, for any pair of adjacent degree-$(d+1)$ vertices, either it is a close pair or the two vertices have no common neighbor. We do not have a simple branching rule with branching factor $d+1$ for these two cases. Then we change to consider three adjacent degree-$(d+1)$ vertices.

Let v_1, v_2 and v_3 be three degree-$(d+1)$ vertices such that v_2 is adjacent to v_1 and v_3. We find that the hardest case is that exact one pair of vertices in $\{v_1, v_2, v_3\}$ is a close or similar pair, for which we still can not get a branching factor $\leq d+1$. We call this case a *bad case*. If no pair of vertices in $\{v_1, v_2, v_3\}$ is a close or similar pair, we call $\{v_1, v_2, v_3\}$ a *proper triple* of degree-$(d+1)$ vertices. Our idea is to avoid bad cases and only branch on proper triples.

Consider four degree-$(d+1)$ vertices v_1, v_2, v_3 and v_4 such that there is an edge between v_i and v_{i+1} for $i = 1, 2, 3$. If at most one pair of vertices in $\{v_1, v_2, v_3, v_4\}$ is a close or similar pair, then at least one of $\{v_1, v_2, v_3\}$ and $\{v_2, v_3, v_4\}$ will be a proper triple. Thus the only left cases are that at least two pairs of vertices in $\{v_1, v_2, v_3, v_4\}$ are close or similar pairs. Luckily, we find good branching rules to deal with them. When both of $\{v_1, v_2\}$ and $\{v_2, v_3\}$ are close pairs, $\{v_1, v_2, v_3\}$ is called a *close triple*. See Fig. 1(a) for an illustration of close triple. Our algorithm deals with close triples in Step 4 in Sect. 4. When both of $\{v_1, v_2\}$ and $\{v_3, v_4\}$ are close pairs, $\{v_1, v_2, v_3, v_4\}$ is called a *type-I close quadruple*. See Fig. 1(b) for an illustration of type-I close quadruple. Our algorithm deals with type-I close quadruples in Step 5 in Sect. 4. When both of $\{v_1, v_3\}$ and $\{v_2, v_4\}$ are similar pairs, $\{v_1, v_2, v_3, v_4\}$ is called a *type-II close quadruple*. See Fig. 1(c) for an illustration of type-II close quadruple. Our algorithm deals with type-II close quadruples in Step 6 in Sect. 4. When $\{v_1, v_2, v_3, v_4\}$ has one close pair and one similar pair, we can see that there is always a close triple in it. Therefore, we have considered all possible cases. The last step of our algorithm is then to deal with proper triples.

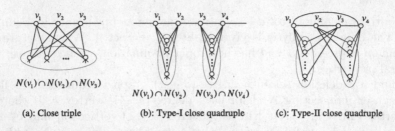

(a): Close triple   (b): Type-I close quadruple   (c): Type-II close quadruple

Fig. 1. Illustrations of some structures

4   The Algorithm and Its Analysis

We are ready to describe the whole algorithm. Our algorithm works for any $d \geq 0$ but can only achieve the running time bound of $O^*((d+1)^k)$ for each $d \geq 3$. Our algorithm is a recursive algorithm containing seven major steps, each of which will branch on the current instance into several sub-instances and invoke the algorithm itself on each sub-instance. Next, we describe these steps. When we introduce one step, we assume that all previous steps can not be applied anymore. For the purpose of presentation, we will analyze the correctness and running time of each step after describing it.

Step 1 (Vertices of degree $\geq d+2$)

If there is a vertex v of degree $\geq d+2$ in the graph, we branch on v into $1 + \binom{d(v)}{d(v)-d}$ branches according to Lemma 1 by either including v or each set of $d(v) - d$ neighbors of v to the solution set.

In the branch where v is included to the solution set, we delete v from the graph and decrease the parameter k by 1. In the branch where a set $N' \subseteq N(V)$ of $d(v) - d$ neighbors of v are included to the solution set, we delete N' from the graph and decrease the parameter k by $d(v) - d$. For this operation, we get a recurrence relation

$$C(k) \leq C(k-1) + \binom{d(v)}{d(v)-d} \cdot C(k - (d(v) - d)). \tag{1}$$

Let γ denote the branching factor of (1).

Lemma 5. *If $d(v) - d \geq 2$, the branching factor γ of (1) satisfies that*

$$\gamma \leq \frac{1 + \sqrt{2d^2 + 6d + 5}}{2}. \tag{2}$$

A proof of this lemma can be found the full version of this paper. It is easy to verify that $\gamma \leq d+1$ for $d \geq 2$. After Step 1, the graph has maximum degree $d+1$.

Step 2 (Proper dominations)

If a vertex v of degree $d + 1$ is dominated by a vertex u (or dominates a neighbor u of it), we branch on v into $d(v)$ branches by including each vertex in $N(v)$ (or $N[v] \setminus \{u\}$) to the solution set. The correctness of this step is based on Lemmas 2 and 3.

In each branch, a vertex is included to the solution set and k decreases by 1. Vertex v is of degree $d + 1$ since the graph has maximum degree at most $d + 1$ after Step 1. We get a recurrence relation

$$C(k) \leq d(v) \cdot C(k - 1) = (d + 1) \cdot C(k - 1),$$

the branching factor of which is $d + 1$.

Step 3 (Good pairs of degree-$(d + 1)$ vertices)

Recall that a pair of adjacent degree-$(d+1)$ vertices is a *good pair* if they have at least one and at most $d - 2$ common neighbors. we use the following branching rule to deal with a good pair $\{v_1, v_2\}$. Let $N^+ = (N(v_1) \cap N(v_2)) \cup \{v_1, v_2\}$, $N_1 = N(v_1) \setminus N^+$ and $N_2 = N(v_2) \setminus N^+$. Assume that v_1 and v_2 have x common neighbors. Note that for any d-degree deletion set S', if S' does not contain any vertex in N^+, then S' contains at least one vertex in N_1 and one vertex in N_2. We branch into $|N^+| + |N_1||N_2| = (x + 2) + (d - x)^2$ branches. In the first $|N^+|$ branches each vertex in N^+ is included to the solution set; and in the last $|N_1||N_2|$ branches each pair of vertices in N_1 and N_2 is included to the solution set. In each branch, if z vertices are included to the solution set, then the parameter k in this branch decreases by z. This branching operation gives a recurrence relation

$$C(k) \leq (x + 2) \cdot C(k - 1) + (d - x)^2 \cdot C(k - 2),$$

the branching factor of which is

$$\frac{1}{2} \left(2 + x + \sqrt{5x^2 - 8dx + 4d^2 + 4x + 4} \right).$$

It is easy to verify that when $1 \leq x \leq d - 2$, the branching factor is at most $d + 1$.

Step 4 (Close triples of degree-$(d + 1)$ vertices)

Recall that a pair of adjacent degree-$(d + 1)$ vertices is a *close pair* if they have exactly $d - 1$ common neighbors. The formal definition of close triple is that: the set of three degree-$(d + 1)$ vertices v_1, v_2 and v_3 is called a *close triple* if $\{v_1, v_2\}$ and $\{v_2, v_3\}$ are two close pairs and v_1 and v_3 are not adjacent. According to the definition of close triples, we can see that $N(v_1) \cap N(v_2) \cap N(v_3) = N(v_2) \setminus \{v_1, v_3\}$. For a close triple $\{v_1, v_2, v_3\}$, we observe the following. Vertex v_1 (resp., v_3) is adjacent to a degree-$(d+1)$ vertex $v_0 \notin N[v_2]$ (resp., $v_4 \notin N[v_2]$) by Lemma 4. Let $N_2^- = N[v_2] \setminus \{v_1, v_3\}$. For any d-degree deletion set S', if $S \cap N_2^- = \emptyset$, then S' contains either v_1 and a vertex in $\{v_3, v_4\}$ (since S' must contain a vertex in $N[v_2]$ and a vertex in $N[v_3]$) or v_3 and a vertex in $\{v_0, v_1\}$ (since S' must contain a vertex in $N[v_2]$ and a vertex in $N[v_1]$). Then we can branch by either including each vertex in N_2^- to the solution set or including each of $\{v_1, v_3\}$, $\{v_1, v_4\}$ and $\{v_0, v_3\}$ to the solution set. This branching operation gives a recurrence relation

$$C(k) \leq (d - 1) \cdot C(k - 1) + 3 \cdot C(k - 2),$$

the branching factor of which is

$$\frac{1}{2} \left(d - 1 + \sqrt{d^2 - 4d + 13} \right).$$

It is easy to verify that when $d \geq 2$, the branching factor is less than $d + 1$.

Step 5 (Type-I close quadruples of degree-$(d+1)$ vertices)

A set of four degree-$(d + 1)$ vertices $\{v_1, v_2, v_3, v_4\}$ is called a *type-I close quadruple* if $\{v_1, v_2, v_3, v_4\}$ induces a cycle or a path of 4 vertices, and $\{v_1, v_2\}$ and $\{v_3, v_4\}$ are two close pairs. Let $N_{12}^- = N(v_1) \cap N(v_2)$ and $N_{34}^- = N(v_3) \cap N(v_4)$. When the graph has no proper dominations, good pairs or close triples, it holds that $N_{12}^- \cap N_{34}^- = \emptyset$.

Let S' be an arbitrary d-degree deletion set. Our branching rule for type-I close quadruples is different for the cases whether $\{v_1, v_2, v_3, v_4\}$ induces a cycle or a path.

Case 1. $\{v_1, v_2, v_3, v_4\}$ induces a cycle of 4 vertices: We consider the following different subcases.

Case 1.1. $S' \cap \{v_1, v_2, v_3, v_4\} = \emptyset$: Then $S' \cap N_{12}^- \neq \emptyset$ and $S' \cap N_{34}^- \neq \emptyset$. For this case, we included each pair of vertices in N_{12}^- and N_{34}^- to the solution set to create $|N_{12}^-||N_{34}^-| = (d-1)^2$ branches, each of which decreases k by 2.

Case 1.2. $S' \cap \{v_1, v_2, v_3, v_4\} = \{v_1\}$ or $S' \cap \{v_1, v_2, v_3, v_4\} = \{v_2\}$: Then $S' \cap N_{34}^- \neq \emptyset$, otherwise no vertex in $N[v_3]$ or $N[v_4]$ would be in S' and then S' would not be a d-degree deletion set. Furthermore, if $S' \cap \{v_1, v_2, v_3, v_4\} = \{v_2\}$, then $S' \setminus \{v_2\} \cup \{v_1\}$ is still a d-degree deletion set of the same size, since $N[v_2] \setminus N[v_1] = \{v_3\}$, v_3 is adjacent to all vertices in N_{34}^- and $S' \cap N_{34}^- \neq \emptyset$. So for this case, we include $\{v_1, x\}$ to the solution set for each $x \in N_{34}^-$ to create $|N_{34}^-| = d - 1$ branches, each of which decreases k by 2.

Case 1.3. $S' \cap \{v_1, v_2, v_3, v_4\} = \{v_3\}$ or $S' \cap \{v_1, v_2, v_3, v_4\} = \{v_4\}$: Then $S' \cap N_{12}^- \neq \emptyset$. For the same reason, we include $\{v_3, x\}$ to the solution set for each $x \in N_{12}^-$ to create $|N_{12}^-| = d - 1$ branches, each of which decreases k by 2.

Case 1.4. $|S' \cap \{v_1, v_2, v_3, v_4\}| \geq 2$: Then $S' \setminus \{v_1, v_2, v_3, v_4\} \cup \{v_1, v_3\}$ is a d-degree deletion set of size not greater than that of S', since $N[\{v_1, v_2, v_3, v_4\}] \subseteq N[\{v_1, v_3\}]$. For this case, we can simply include $\{v_1, v_3\}$ to the solution set.

The branching operation gives a recurrence relation

$$C(k) \leq (d-1)^2 \cdot C(k-2) + (d-1) \cdot C(k-2) + (d-1) \cdot C(k-2) + C(k-2)$$
$$= d^2 \cdot C(k-2), \tag{3}$$

the branching factor of which is $d < d + 1$.

Case 2. $\{v_1, v_2, v_3, v_4\}$ induces a path of 4 vertices: Let $\{v_0\} = N(v_1) \setminus N[v_2]$ and $\{v_5\} = N(v_4) \setminus N[v_3]$, where it is possible that $v_0 = v_5$. We observe the following different cases.

Case 2.1. S' does not contain any vertex in $N_{12}^- \cup N_{34}^-$: Then S' contains at least one vertex in $\{v_0, v_1, v_2\}$ and at least one vertex in $\{v_3, v_4, v_5\}$, since S' must contain at least one vertex in $N[v_1]$ and at least one vertex in $N[v_4]$. If $|S' \cap \{v_1, v_2, v_3, v_4\}| \geq 2$, then $S'' = S' \setminus \{v_1, v_2, v_3, v_4\} \cup \{v_1, v_4\}$ is still a d-degree deletion set with $|S''| \leq |S'|$, since $N[\{v_1, v_2, v_3, v_4\}] \subseteq N[\{v_1, v_4\}]$.

Otherwise, it holds either $S' \cap \{v_0, v_1, v_2\} = \{v_0\}$ or $S' \cap \{v_3, v_4, v_5\} = \{v_5\}$. If $S' \cap \{v_0, v_1, v_2\} = \{v_0\}$, then $v_3 \in S'$ since S' must contain at least one vertex in $N[v_2]$. If $S' \cap \{v_3, v_4, v_5\} = \{v_5\}$, then $v_2 \in S'$ since S' must contain at least one vertex in $N[v_3]$. So for this case, we conclude that there is a solution contains one of $\{v_1, v_4\}$, $\{v_0, v_3\}$ and $\{v_2, v_5\}$. In our algorithm, we generate three branches by including each of $\{v_1, v_4\}$, $\{v_0, v_3\}$ and $\{v_2, v_5\}$ to the solution set. In each of the three branches, the parameter k decreases by 2.

Case 2.2. S' does not contain any vertex in N_{12}^- but contain some vertex in N_{34}^-: Since $S' \cap N[v_1] \neq \emptyset$, we know that S' contains at least one vertex in $\{v_0, v_1, v_2\}$. If $v_2 \in S'$, then $S'' = S' \setminus \{v_2\} \cup \{v_1\}$ is still a d-degree deletion set. The reason relies on that $N[v_2] \setminus N[v_1] = \{v_3\}$, v_3 is adjacent to each vertex in N_{34}^-, and S'' contains at least one vertex in N_{34}^-. So for this case, there is a solution contains one vertex in $\{v_0, v_1\}$. In our algorithm, we create $2|N_{34}^-| = 2(d-1)$ branches by including to the solution each pair of vertices x and y such that $x \in \{v_0, v_1\}$ and $y \in N_{34}^-$. In each of the $2(d-1)$ branches, the parameter k decreases by 2.

Case 2.3. S' does not contain any vertex in N_{34}^- but contain some vertex in N_{12}^-: For the same reason in Case 2.2, there is a solution contains one vertex in $\{v_4, v_5\}$. In our algorithm, we create $2|N_{12}^-| = 2(d-1)$ branches by including to the solution each pair of vertices x and y such that $x \in \{v_4, v_5\}$ and $y \in N_{12}^-$. In each of the $2(d-1)$ branches, the parameter k decreases by 2.

Case 2.4. S' contains some vertex in N_{12}^- and some vertex in N_{34}^-: For this case, Our algorithm simply generates $|N_{12}^-||N_{34}^-| = (d-1)^2$ branches by including to the solution each pair of vertices x and y such that $x \in N_{12}^-$ and $y \in N_{34}^-$. In each of the $(d-1)^2$ branches, the parameter k decreases by 2.

The above branching operation gives a recurrence relation

$$C(k) \leq 3C(k-2) + 2(d-1) \cdot C(k-2) + 2(d-1) \cdot C(k-2) + (d-1)^2 \cdot C(k-2)$$
$$= d(d+2) \cdot C(k-2),$$

the branching factor of which is $\sqrt{d(d+2)} < d+1$.

Step 6 (Type-II close quadruples of degree-$(d+1)$ vertices)

Two nonadjacent degree-$(d+1)$ vertices are *similar* if they have the same neighbor set. A set of four degree-$(d+1)$ vertices $\{v_1, v_2, v_3, v_4\}$ is called a *type-II close quadruple* if $\{v_1, v_3\}$ and $\{v_2, v_4\}$ are two similar pairs and there is an edge between v_i and v_{i+1} for $i = 1, 2, 3$. Note that there must be an edge between v_1 and v_4 since $\{v_1, v_3\}$ is a similar pair. So as a type-II close quadruple, $\{v_1, v_2, v_3, v_4\}$ always induces a cycle of 4 vertices.

Let $\{v_1, v_2, v_3, v_4\}$ be a type-II close quadruple. We use N_{13}^- to denote $N(v_1) \setminus \{v_2, v_4\}$ and N_{24}^- to denote $N(v_2) \setminus \{v_1, v_3\}$. Note that it holds $N_{13}^- \cap N_{24}^- = \emptyset$, if we assume that there is no good pairs or close triples. Let S' be a d-degree deletion set. We consider the following different subcases.

Case 1. $S' \cap \{v_1, v_2, v_3, v_4\} = \emptyset$: Then $S' \cap N_{13}^- \neq \emptyset$ and $S' \cap N_{24}^- \neq \emptyset$. For this case, we included each pair of vertices in N_{13}^- and N_{24}^- to the solution set to create $|N_{13}^-||N_{24}^-| = (d-1)^2$ branches, each of which decreases k by 2.

Case 2. $S' \cap \{v_1, v_2, v_3, v_4\} = \{v_1\}$ or $S' \cap \{v_1, v_2, v_3, v_4\} = \{v_3\}$: Then $S' \cap N_{13}^- \neq \emptyset$, otherwise S' would not be a d-degree deletion set since no vertex in $N[v_3]$ or $N[v_1]$ is in S'. Furthermore, if $S' \cap \{v_1, v_2, v_3, v_4\} = \{v_3\}$, then $S' \setminus \{v_3\} \cup \{v_1\}$ is still a d-degree deletion set of the same size. So for this case, we include $\{v_1, x\}$ to the solution set for each $x \in N_{13}^-$ to create $|N_{13}^-| = d - 1$ branches, each of which decreases k by 2.

Case 3. $S' \cap \{v_1, v_2, v_3, v_4\} = \{v_2\}$ or $S' \cap \{v_1, v_2, v_3, v_4\} = \{v_4\}$: Then $S' \cap N_{24}^- \neq \emptyset$. For the same reason, we include $\{v_2, x\}$ to the solution set for each $x \in N_{24}^-$ to create $|N_{24}^-| = d - 1$ branches, each of which decreases k by 2.

Case 4. $|S' \cap \{v_1, v_2, v_3, v_4\}| \geq 2$: Then $S' \setminus \{v_1, v_2, v_3, v_4\} \cup \{v_1, v_2\}$ is a d-degree deletion set of size not greater than S', since $N[\{v_1, v_2, v_3, v_4\}] \subseteq N[\{v_1, v_2\}]$. For this case, we can simply include $\{v_1, v_2\}$ to the solution set.

The branching operation gives a recurrence relation

$$C(k) \leq (d-1)^2 \cdot C(k-2) + (d-1) \cdot C(k-2) + (d-1) \cdot C(k-2) + C(k-2)$$
$$= d^2 \cdot C(k-2),$$

the branching factor of which is $d < d + 1$.

Step 7 (Proper triples of degree-$(d+1)$ vertices)
A set of three degree-$(d+1)$ vertices $\{v_1, v_2, v_3\}$ is called a *proper triple* if $\{v_1, v_2, v_3\}$ induces a path and no pair of vertices in $\{v_1, v_2, v_3\}$ is close or similar.

Lemma 6. *Let G be a graph of maximum degree $d+1$ for any integer $d > 0$. If G has no proper dominations, good pairs, close triples, type-I close quadruples or type-II close quadruples, then G has some proper triples.*

A proof of this lemma can be found in the full version.

For a proper triple $\{v_1, v_2, v_3\}$ in a graph having none of dominated vertices, good pairs, close triples, type-I close quadruples and type-II close quadruples, we have the following properties: $N(v_1) \cap N(v_2) = \emptyset$, $N(v_2) \cap N(v_3) = \emptyset$ and $1 \leq |N(v_1) \cap N(v_3)| \leq d$.

Let $N_{13}^- = N(v_1) \cap N(v_3) \setminus \{v_2\}$, $N_1^- = N(v_1) \setminus N(v_3)$, $N_3^- = N(v_3) \setminus N(v_1)$, $N_2^- = N(v_2) \setminus \{v_1, v_3\}$ and $x = |N_{13}^-|$. Since $\{v_1, v_3\}$ is not a similar pair, we know that $0 \leq x \leq d - 1$. Let S' be a d-deletion set. To design our branching rule, we consider the following different cases.

Case 1. $v_2 \in S'$: We simply include v_2 to the solution set and the parameter k decreases by 1. For all the remaining cases, we assume that $v_2 \notin S'$.

Case 2. $v_2 \notin S'$ and $v_1, v_3 \in S'$: We simply include v_1 and v_3 to the solution set and the parameter k decreases by 2.

Case 3. $v_1, v_2 \notin S'$ and $v_3 \in S'$: For the case, $S' \cap (N(v_1) \setminus \{v_2\}) \neq \emptyset$. We create $|N(v_1) \setminus \{v_2\}| = d$ branches by including v_3 and each vertex in $N(v_1) \setminus \{v_2\}$ to the solution set and the parameter k in each branch decreases by 2.

Case 4. $v_2, v_3 \notin S'$ and $v_1 \in S'$: For the case, $S' \cap (N(v_3) \setminus \{v_2\} \neq \emptyset$. We create $|N(v_3) \setminus \{v_2\}| = d$ branches by including v_1 and each vertex in $N(v_3) \setminus \{v_2\}$ to the solution set and the parameter k in each branch decreases by 2.

Case 5. $v_1, v_2, v_3 \notin S'$: Then S' must contains (i) a vertex in N_2^- and (ii) either a vertex in N_{13}^- or two vertices from N_1^- and N_3^- respectively. Our algorithm generates $|N_2^-||N_{13}^-| + |N_2^-|||N_1^-||N_3^-| = (d-1)x + (d-1)(d-x)^2$ branches. Each of the first $(d-1)x$ branches includes a vertex in N_2^- and a vertex in N_{13}^- to the solution set and the parameter k decreases by 2. The last $(d-1)(d-x)^2$ branches are generated by including each triple $\{w_1 \in N_2^-, w_2 \in N_1^-, w_3 \in N_3^-\}$ to the solution set, where the parameter k decreases by 3.

The above branching operation gives a recurrence relation

$$
\begin{aligned}
C(k) \leq \ &C(k-1) + C(k-2) + d \cdot C(k-2) + d \cdot C(k-2) + \\
&(d-1)x \cdot C(k-2) + (d-1)(d-x)^2 \cdot C(k-3) \\
= \ &C(k-1) + ((2d+1) + (d-1)x) \cdot C(k-2) + (d-1)(d-x)^2 \cdot C(k-3),
\end{aligned} \tag{4}
$$

where $0 \leq x \leq d - 1$.

Lemma 7. *When $d \geq 3$, the branching factor of (4) is at most $d + 1$ for each $0 \leq x \leq d - 1$.*

A proof of this lemma can be found in the full version.

4.1   The Results

Lemma 6 guarantees that when the graph has a vertex of degree $\geq d+1$, one of the above seven steps can be applied. When $d \geq 3$, the branching factor in each of the seven steps is at most $d + 1$. Thus,

Theorem 1. *The d-bounded-degree vertex deletion problem for each $d \geq 3$ can be solved in $O^*((d+1)^k)$ time.*

Note that all the seven steps of our algorithm work for $d = 2$. In the first six steps, we still can get branching factors at most $d + 1$ for $d = 2$. In Step 7, when $d = 2$ and $x = d - 1 = 1$, (4) becomes

$$
C(k) \leq C(k-1) + 6C(k-2) + C(k-3),
$$

which has a branching factor of 3.0645. This is the biggest branching factor in the algorithm. Then

Theorem 2. *The co-path/cycle problem can be solved in $O^*(3.0645^k)$ time.*

Note that previously the co-path/cycle problem could only be solved deterministically in $O^*(3.24^k)$ time [4].

5   Concluding Remarks

In this paper, by studying the structural properties of graphs, we show that the d-bounded-degree vertex deletion problem can be solved in $O^*((d+1)^k)$ time for each $d \geq 3$. Our algorithm is the first nontrivial parameterized algorithm for the d-bounded-degree vertex deletion problem with $d \geq 3$.

Our problem is a special case of the $(d+2)$-hitting set problem. It is still left as an open problem that whether the d'-hitting set problem can be solved in $O^*((d'-1)^k)$ time. Our result is a step toward to this interesting open problem. However, our method can not be extended to the d'-hitting set problem directly, since some good graph structural properties do not hold in the general d'-hitting set problem.

References

1. Balasundaram, B., Butenko, S., Hicks, I.V.: Clique relaxations in social network analysis: the maximum k-plex problem. Oper. Res. **59**(1), 133–142 (2011)
2. Betzler, N., Bredereck, R., Niedermeier, R., Uhlmann, J.: On bounded-degree vertex deletion parameterized by treewidth. Discrete Appl. Math. **160**(1–2), 53–60 (2012)
3. Chen, J., Kanj, I.A., Xia, G.: Improved upper bounds for vertex cover. Theoret. Comput. Sci. **411**, 3736–3756 (2010)
4. Chen, Z.-Z., Fellows, M., Fu, B., Jiang, H., Liu, Y., Wang, L., Zhu, B.: A linear kernel for co-path/cycle packing. In: Chen, B. (ed.) AAIM 2010. LNCS, vol. 6124, pp. 90–102. Springer, Heidelberg (2010)
5. Fellows, M.R., Guo, J., Moser, H., Niedermeier, R.: A generalization of Nemhauser and Trotter's local optimization theorem. J. Comput. Syst. Sci. **77**, 1141–1158 (2011)
6. Feng, Q., Wang, J., Li, S., Chen, J.: Randomized parameterized algorithms for P_2-packing and co-path packing problems. J. Comb. Optim. **29**(1), 125–140 (2015)
7. Fernau, H.: A top-down approach to search-trees: improved algorithmics for 3-hitting Set. Algorithmica **57**, 97–118 (2010)
8. Fernau, H.: Parameterized algorithms for d-hitting set: the weighted case. Theor. Comput. Sci. **411**(16–18), 1698–1713 (2010)
9. Fernau, H.: Parameterized algorithmics for d-hitting set. Int. J. Comput. Math. **87**(14), 3157–3174 (2010)
10. Fomin, F.V., Kratsch, D.: Exact Exponential Algorithms. Springer, Heidelberg (2010)
11. Katrenič, J.: A faster FPT algorithm for 3-path vertex cover. Inf. Process. Lett. **116**(4), 273–278 (2016)
12. Newnan, I., Sohler, C.: Every proerty of hyperfinite graphs is testable. SIAM J. Comput. **42**(3), 1095–1112 (2013)
13. Niedermeier, R., Rossmanith, P.: An efficient fixed-parameter algorithm for 3-hitting set. J. Discrete Algorithms **1**, 89–102 (2003)
14. Seidman, S.B., Foster, B.L.: A graph-theoretic generalization of the clique concept. J. Math. Soc. **6**, 139–154 (1978)
15. Tu, J.: A fixed-parameter algorithm for the vertex cover P3 problem. Inf. Process. Lett. **115**, 96–99 (2015)

16. Wu, B.Y.: A measure and conquer approach for the parameterized bounded degree-one vertex deletion. In: Xu, D., Du, D., Du, D. (eds.) COCOON 2015. LNCS, vol. 9198, pp. 469–480. Springer, Heidelberg (2015)
17. Xiao, M.: On a generalization of Nemhauser and Trotter's local optimization theorem. In: Elbassioni, K., Makino, K. (eds.) ISAAC 2015. LNCS, vol. 9472, pp. 442–452. Springer, Heidelberg (2015). doi:10.1007/978-3-662-48971-0_38

The Monotone Circuit Value Problem
with Bounded Genus Is in NC

Faisal N. Abu-Khzam[1,4], Shouwei Li[2(✉)], Christine Markarian[3],
Friedhelm Meyer auf der Heide[2], and Pavel Podlipyan[2]

[1] Department of Computer Science and Mathematics,
Lebanese American University, Beirut, Lebanon
[2] Heinz Nixdorf Institute and Department of Computer Science,
Paderborn University, Fürstenallee 11, 33102 Paderborn, Germany
sli@mail.uni-paderborn.de
[3] Department of Mathematical Sciences, Haigazian University, Beirut, Lebanon
[4] School of Engineering and Information Technology,
Charles Darwin University, Darwin, Australia

Abstract. We present an efficient parallel algorithm for the general
Monotone Circuit Value Problem (MCVP) with n gates and an underlying graph of bounded genus k. Our algorithm generalizes a recent
result by Limaye et al. who showed that MCVP with toroidal embedding (genus 1) is in NC when the input contains a toroidal embedding
of the circuit. In addition to extending this result from genus 1 to any
bounded genus k, and unlike the work reported by Limaye et al., we do
not require a precomputed embedding to be given. Most importantly,
our results imply that given a P-complete problem, it is possible to find
an algorithm that makes the problem fall into NC by fixing one or more
parameters. Hence, we deduce the interesting analogy: *Fixed Parameter
Parallelizable* (FPP) is with respect to P-complete what *Fixed Parameter Tractable* (FPT) is with respect to NP-complete. Similar work that
uses treewidth as parameter was also presented by Elberfeld et al. in [6].

1 Introduction

Parameterized complexity theory provides a refined classification of computationally intractable problems based on a multivariate complexity analysis of
(exact) algorithms. The class FPT occupies the bottom of parameterized complexity hierarchy just as the class P is in the classical (polynomial) hierarchy. In
short, a problem is *Fixed Parameter Tractable* (FPT) if it has an algorithm that
runs in $\mathcal{O}(f(k) \cdot n^{\mathcal{O}(1)})$, where n is the problem size and k is the input parameter
that is independent of n, for an arbitrary computable function f. A well-known
example is the parameterized version of the Vertex Cover Problem which can be

This work was partially supported by the German Research Foundation (DFG)
within the Collaborative Research Center "On-The-Fly Computing" (SFB 901) and
the International Graduate School "Dynamic Intelligent Systems".

T.N. Dinh and M.T. Thai (Eds.): COCOON 2016, LNCS 9797, pp. 92–102, 2016.
DOI: 10.1007/978-3-319-42634-1_8

solved in $\mathcal{O}(kn + 1.274^k)$ time, where n is the number of vertices of the input graph and k is an upper bound on the size of the sought vertex cover [4].

The study of Parameterized Complexity has been extended to parallel computing and this is broadly known as *Parameterized Parallel Complexity*. The first systematic work on *Parameterized Parallel Complexity* appeared in [3] where the authors introduced two classes of efficiently parallelizable parameterized problems known as PNC and FPP respectively, both based on the degree of efficiency required. The class of PNC (*parameterized analog of NC*) contains all parameterized problems that have a parallel deterministic algorithm with running time $\mathcal{O}(f(k) \cdot (\log n)^{h(k)})$ and $\mathcal{O}(g(k) \cdot n^\beta)$ processors, where n is the size of input, k is the parameter, f, g, h are arbitrary computable functions, and β is a constant independent of n and k. A noticeable drawback to the definition of PNC is the exponent in the logarithm bounding the running time. Since the latter depends on k, it grows at a rapid rate thus making the running time very close to a linear function, even for not too large values of the parameter. On the other hand, the class of FPP (*Fixed Parameter Parallelizable*) contains all parameterized problems that have a parallel deterministic algorithm with running time $\mathcal{O}(f(k) \cdot (\log n)^\alpha)$ and $\mathcal{O}(g(k) \cdot n^\beta)$ processors, where n is the size of input, k is the parameter, f and g are arbitrary computable functions, and α, β are constants independent of n and k. It was shown in [3,6] that some problems with bounded treewidth (i.e., treewidth is the parameter) belong to FPP. In fact, this parameter is somehow coarse since using it leads to numerous NP-complete problems ending up in FPP as well.

Motivated by the above, we restrict our attention to P-complete problems and consider a more natural parameter — the genus of a graph. Our result becomes more helpful in understanding the intrinsic difficulty of P-complete problems and whether P = NC. This is the primary motivation behind this paper. Consider, as an example, one of the most studied P-complete problems, the *Circuit Value Problem* (CVP). A *Boolean Circuit* is a directed acyclic graph consisting of NOT, AND and OR gates. CVP is the problem of evaluating a Boolean Circuit on a given input. It was shown to be P-complete with respect to logarithmic space reductions in [9]. In addition, some restricted variants of CVP have also been studied. The *Planar Circuit Value Problem* (PCVP), for example, is a variant of CVP in which the underlying graph of the circuit has a planar embedding. Another variant is the *Monotone Circuit Value Problem* (MCVP) in which the circuit has only AND and OR gates. Both PCVP and MCVP were also shown to be P-complete in [7]. Interestingly, if the circuit is simultaneously *planar* and *monotone*, then it can be evaluated in NC. This variant is known as PMCVP and its first NC algorithm was given in [13]. The latter employs the straight-line code parallel evaluation technique and has a running time in $\mathcal{O}(\log^3 n)$ with $\Omega(n^6)$ processors. Subsequently, a more sophisticated algorithm with the same running time but requiring only a linear number of processors was presented in [11].

Our Contribution: In this paper, we explore the class of Parameterized Parallel Complexity by presenting an efficient parallel algorithm for the general MCVP with n gates and an underlying graph of bounded genus k. Our algorithm improves and generalizes the result in [10], which shows that the MCVP with toroidal embedding (genus 1) is in NC. The work in [10] was non-constructive and assumed that the input contains a fixed toroidal embedding of the circuit. We extend the result from genus 1 to a bounded constant genus k and unlike in [10], we do not require such an embedding to be given. Moreover, our results imply that given a P-complete problem, it is possible to find an algorithm that makes the problem fall into NC by fixing one or more parameters. Hence, we deduce the interesting analogy: FPP is with respect to P-complete what FPT is with respect to NP-complete. Similar work appeared in [6] where they consider the treewidth as a parameter.

Structure of the Paper: The rest of this paper is structured as follows. In Sect. 2, we give some preliminaries, including the PQ-tree data structure along with some parallel operations on it. These are essential for the understanding of our algorithm, which we present in the following sections. In Sect. 3, we give a sketch of our algorithm. Then we decompose the algorithm and prove its correctness in Sects. 4 to 6. We conclude our paper with some remarks and future work in Sect. 7.

2   Preliminaries

We assume familiarity with basic graph theoretic terminology but we shall give a few general definitions and notations.

The (orientable) *genus* of a graph is the minimal integer g such that the graph can be drawn without edge-crossing on a sphere with g handles. In a connected graph, a *block* (or 2-connected component) is a maximal 2-connected subgraph. The *block decomposition* of a graph is the set of all the blocks of the graph. It is not hard to compute a block decomposition of a connected graph G in NC [12]. We may assume throughout the rest of this paper that G is a 2-connected graph. This is especially needed for the relation between genus and block decomposition described in Lemma 1. Otherwise, we can apply the algorithms mentioned above to obtain a block decomposition of the input graph and process each block independently.

Lemma 1 [1]. *The genus of a connected graph is the sum of the genus of its blocks.*

Given a universe $U = \{e_1, \ldots, e_m\}$, a *PQ-tree* is a tree-based data structure that represents a class of permissible permutations over the set U in which the leaves are elements of U and the internal nodes are distinguished as being labeled either P-*nodes* or Q-*nodes*. Let T be a PQ-tree over the universe U. We denote by $L(T)$ the set of linear orders represented by T, and say that T generates $L(T)$. One element of $L(T)$ is obtained by reading off the leaves from left to right in the order in which they appear in T. The other elements are those linear orders obtained in the same way according to the following conditions:

– Every element of U appears precisely once as a leaf node;
– The P-node has at least two children and they might be arbitrarily permuted;

– The Q-node has at least three children and they are allowed only to be placed in reverse order.

Since there is no way a PQ-tree over a non-empty ground set can represent the empty set of orderings, we use a special *null* tree T_{null} to represent the empty set. With each linear ordering λ we associate the cyclic ordering $co(\lambda)$ obtained from λ by letting the first element of λ follow the last. Then the PQ-tree T represents the set of cyclic orderings $CO(T) = co(L(T))$.

Let A be a subset of the universe U. We say a linear ordering $\lambda = e_1, \ldots, e_m$ of U *satisfies* the set A if all the elements of A are consecutive in λ. For a PQ-tree T, let

$$\Psi(T, A) = \{\lambda : \lambda \in L(T), \lambda \text{ satisfies } A\} \tag{1}$$

Given any T and $A \subseteq U$, there is a PQ-tree $\hat{T}$ such that

$$L(\hat{T}) = \Psi(T, A) \tag{2}$$

called the *reduction* of T with respect to A. In order to parallelize the planarity testing algorithm presented in [2], Klein and Reif in [8] introduced three new operations on PQ-trees: *multiple-disjoint-reduction, intersection* and *join*.

Given any T and $A_1, \ldots, A_k \subseteq U$, there is a PQ-tree $\hat{T}$ such that

$$L(\hat{T}) = \Psi(T, \{A_1, \ldots, A_k\}) \tag{3}$$

called the multiple-disjoint-reduce of T with respect to $\{A_1, \ldots, A_k\}$. They proposed Algorithm MREDUCE($T, \{A_1, \ldots, A_k\}$) which modifies T to obtain a PQ-tree $\hat{T}$ such that $L(\hat{T}) = \Psi(T, \{A_1, \ldots, A_k\})$ if all subsets A_i's are pairwise disjoint. Their algorithm works in $\mathcal{O}(\log m)$ time using a linear number of processors, where $m = |U|$. Note that if no ordering generated by T satisfies $\{A_1, \ldots, A_k\}$, then the result of multiple-disjoint-reduce $\hat{T}$ is just the null tree T_{null}.

A PQ-tree $\hat{T}$ is the intersection of two PQ-trees T and T' over the same ground set if $L(\hat{T}) = L(T) \cap L(T')$. Klein and Reif also proposed Algorithm INTERSECT(T, T') to reduce the given PQ-trees simultaneously with respect to multiple sets that are not necessarily disjoint, using the multiple-disjoint-reduce as a subroutine. The algorithm modifies T' to be the intersection of the two original trees. INTERSECT can be computed in $\mathcal{O}(\log^2 m)$ time using m processors, where m is the size of the ground set.

The last operation is join. Suppose $T_0, \ldots, T_k$ are PQ-trees over $U_0, \ldots, U_k$, respectively, and for some U_i, U_j may overlap. We say that T is the *join* of T_0 with $T_1, \ldots, T_k$ if $CO(T) = CO(T_0)$ join $(CO(T_1), \ldots, CO(T_k))$. To be more specific, we can compute a new PQ-tree T such that the cyclic ordering of T satisfies $U_0, \ldots, U_k$. The join of T_0 with $T_1, \ldots, T_k$ can be computed in $\mathcal{O}(\log^2 m)$ time using m processors, where m is the total number of ground elements, using the multiple-disjoint-reduce and the intersection as subroutines.

3   The Algorithm

Our main result can be summarized in the following theorem.

Theorem 1. *Given a general monotone boolean circuit with n gates and an underlying graph of bounded genus k, the circuit can be evaluated in $\mathcal{O}\left((k+1)\cdot\log^3 n\right)$ time using $\mathcal{O}(n^c)$ processors, where $\mathcal{O}(n^c)$ is the best processors boundary for parallel matrix multiplication. Therefore, the monotone circuit value problem with bounded genus k is in FPP.*

Throughout the rest of this paper, we will respectively use the terms "graph" and "circuit", "node" and "gate", "edge" and "wire" interchangeably. The basic strategy of our algorithm is as follows, first we transform the input graph into a layered graph by subdividing the edges. It is well known that subdivision will not increase the genus of a input graph. Then the algorithm greedily contracts planar subgraphs in parallel until it observes a subgraph that is not planar. Due to the particular contraction process, which is based on working on the layered version of the input graph, this situation allows to split the circuit into two parts of a smaller total genus. Then the algorithm proceeds on both parts in parallel. In fact, we cannot merely choose a single embedding for each subgraph since the embeddings of two subgraphs might be inconsistent thus preventing the embeddings from being combined. Instead, we use PQ-trees to represent the set of all embeddings of each planar subgraph.

It is obvious that any circuit can be considered as a directed acyclic graph (DAG). If the circuit has multiple sinks, then each of them can be evaluated independently. We therefore assume throughout the rest of this paper that there is only one single sink t in the circuit.

Algorithm 1. Parallel algorithm for MCVP

1: **procedure** ALG(G)
2:     Transform the input graph G into a layered graph G' with procedure SPLIT(G).
3:     Parallel construct PQ-tree for each node $v \in G'$ except the sink node t. Each PQ-tree represents a plan subgraph.
4:     Take subgraphs that consist of nodes in consequent layers and apply the join operation on the PQ-trees in even layer with odd layer in parallel.
5:     **if** the join operation is not T_{null} **then**
6:         Contract the subgraphs represented by PQ-trees and update layer numbers.
7:         Run $ALG(\cdot)$ on the new graph.
8:     **else**
9:         Delete all directed edges between layers i and $i+1$ and then add one new node t' as the sink node of the second part.
10:        Run $ALG(\cdot)$ on both subgraphs recursively.
11:    **end if**
12:    Evaluate the planar sub-circuits in $k+1$ steps.
13: **end procedure**

---

4   Layered Computation

The first step of our algorithm is to transform the input graph into a layered graph with the same genus. To do so, we first flip the directions of edges in G to obtain G^*, and let $d(t,v)$ be the length of the longest directed path from sink t to node v in G^*.

The layer number $\mathfrak{d}(v)$ of the node v defined as follows:

$$\mathfrak{d}(v) = d(t, v). \tag{4}$$

In this step, we assign to each node v of the given DAG G the layer number $\mathfrak{d}(v)$, and then add some dummy nodes to make the input graph layered. Algorithm 2 formally describes this procedure.

Algorithm 2. Split of DAG

1: **procedure** SPLIT(G)
2:     Calculate the layer number $\mathfrak{d}(v)$ for every node $v \in G$.
3:     **for all** directed edges (u, v) in G **do**
4:         Let $l = \mathfrak{d}(u) - \mathfrak{d}(v) - 1$.
5:         **if** $l > 0$ **then**
6:             Add dummy nodes $n_1, \ldots, n_l$ and directed edges $(u, n_1), (n_1, n_2), \ldots,$ (n_l, v) to the graph G.
7:         **end if**
8:     **end for**
9:     Together with added dummy nodes and edges we obtain DAG G', such that for any edge (u, v), $l = \mathfrak{d}(u) - \mathfrak{d}(v) - 1 = 0$.
10: **end procedure**

Note that for any directed edge $(u, v) \in G'$, it holds that $l = 0$. In other words, all edges are situated between adjacent layers, i.e., there are no edges crossing more than two layers and no edges in the same layer.

We still need to show that Algorithm 2 can be accomplished in NC time and the layered graph G' has the same genus as G. It is easy to see that the layer number for each node can be computed in NC time, and more precisely in NC^2 by using parallel topological sorting algorithms, such as that in [5]. So, we only need to observe that:

Lemma 2. *Graphs G and G' have the same genus.*

Proof. Since graph G' is obtained only by edge subdivision of graph G and the edge subdivision operation will not change the genus of a graph, then G and G' have the same genus. □

5   Parallel Contraction

The second step of our algorithm can be viewed as a greedy contraction process in parallel on the layers. First, we describe how to represent the set of embeddings of a subgraph with a PQ-tree in the following lemma.

Lemma 3. *For any gate in a planar circuit, all input wires and all output wires of the gate are placed consecutively in the cyclic ordering of the wires around the gate in the plane.*

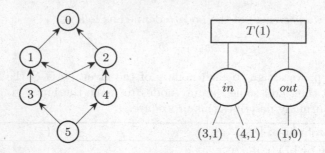

Fig. 1. An example of a DAG and a PQ-tree constructed for node 1 with P-nodes represented as circles and a Q-node represented as a rectangle

Proof. Let c be a gate in C'. Assume that i_1 and i_2 are the two input wires of c and o_1 and o_2 are the two output wires of c, such that o_1 and o_2 interlace with i_1 and i_2 in the cyclic ordering of the wires around c. Suppose s is the single source of C' and t is the single sink of C', then there are four directed paths P_1, P_2, P_3 and P_4 in C': $P_1 = (c, o_1, \ldots, t)$, $P_2 = (c, o_2, \ldots, t)$, $P_3 = (s, \ldots, i_1, c)$, and $P_4 = (s, \ldots, i_2, c)$. These four paths cannot be embedded in a plane without having crossing edges. This contradicts with the fact that C' is a plane graph and concludes the proof. $\square$

It has been shown in [8] that any planar graph can be represented by a PQ-tree. Suppose $v \in G$, then we can directly construct a valid PQ-tree $T(v)$ corresponding to node v, and any cyclic ordering of the edges incident to v will be an arrangement of v, provided that the incoming edges and the outgoing edges are consecutive, respectively. In this case, we let $T(v)$ be the tree whose root is a Q-node with two P-nodes children, *in* and *out*, where the children of *in* are the incoming edges of v and the children of *out* are the outgoing edges of v. Figure 1 illustrates an example of DAG and an initial PQ-tree $T(1)$ constructed from node 1.

It is easy to see that this parallel construction of PQ-trees for each node $v \in G'$, except for the sink node t, takes only constant time with a linear number of processors, since we only need to rearrange the input and output edges of a node.

Next we describe how to contract the subgraphs. We start with the original layered graph and let $G^{(0)} = G'$. In the ith stage, we choose a collection of subgraphs of the graph $G^{(i)}$ in accordance with the layer number, contract these subgraphs and update the layer number. This results in the graph $G^{(i+1)}$.

For each node $v \in G^{(i+1)}$ and each stage $j \leq i$, we denote by $H^{(j)}(v)$ the subgraph of $G^{(j)}$ that was contracted over steps $j+1, \ldots, i$ forming v. We use $H(v)$ for $H^{(0)}(v)$. If $u \in H^{(j)}(v)$ for $v \in G^{(i+1)}$, we use u^{i+1} to denote v.

We choose our subgraphs to contract at each stage i such that the following properties are always guaranteed:

- At most $\mathcal{O}(\log n)$ stages are needed.
- The sink node is never contracted with any other node;
- For each node $v \neq t$ in $G^{(i)}$, the subgraph $H(v)$ permits a PQ-tree representation of the set of its embeddings;
- The layer number is easy to update, following the contraction of the edges.

We first show that the algorithm terminates in $\mathcal{O}(\log n)$ stages. Then we show how the subgraphs are chosen and prove that our method of choosing the subgraphs satisfies the above properties.

Lemma 4. *Algorithm 1 terminates in $\mathcal{O}(\log n)$ stages.*

Proof. We already showed in Algorithm 2 how to transform the input DAG into a layered DAG such that all directed edges go from layer $i + 1$ to layer i in the layered DAG. To ensure that only $\mathcal{O}(\log n)$ stages are needed, we contract the nodes as follows. Suppose that there is one node u at layer $i + 1$ and its neighbor set in layer i is $\{v_1, \ldots, v_m\}$. Moreover, assume that the other neighbor of each v_i in the neighbor set of u in layer $i + 1$ is $\{u_1, \ldots, u_h\}$. Since the input is a monotone boolean circuit, there are only two input wires for each gate. If (u, v_i) is already an input wire for gate v_i, then there must exist another input wire coming from gate $\{u_1, \ldots, u_h\}$ to gate v_i. We will contract the edges incident to u, $\{v_1, \ldots, v_m\}$, and $\{u_1, \ldots, u_h\}$ together.

In each stage, either the number of layers are reduced by two (contract success) or the nodes in adjacent layers cannot be contracted to form a larger planar subgraph. Hence, for the latter, we cut all edges between layer $i + 1$ and i. After cutting the edges between layer $i + 1$ and i, the graph is split into two parts. One part is below layer i (including layer i) which has some hanging incoming edges, and the other part is above layer $i + 1$ (including layer $i + 1$) which has some hanging outgoing edges. It is easy to see that the first part is still a connected directed acyclic graph with sink node t. However, the second part could be a disconnected graph because the adjacent layers may not be a complete bipartite graph. So, we add a new sink node t' to the second part and draw all the hanging outgoing edges to t'. Clearly, this step guarantees that the second part is a connected directed acyclic graph (refer to Fig. 2 for an illustration). Then, we handle these two subgraphs in parallel. □

At every stage i, we compute these PQ-trees for each new node $v \in G^{(i+1)}$ in parallel from the PQ-trees for the nodes $H^{(i)}(v)$ identified to form v.

If a null tree T_{null} arises as $T(v)$ for some node v, then there are no arrangements of v. So, the candidate subgraphs cannot form a larger planar subgraph. Assume on the other hand that the contraction process continues until there is only one node left other than the sink node in G^i. Then every internal node is adjacent only to the sink node. Let $\{v_1, \ldots, v_m\}$ be these nodes. For $j = i, \ldots, m$, if $T(v_j)$ is not T_{null}, then there is a planar embedding for each internal node v_j. So both the input edges and the output edges form a consecutive subsequence, and the graph is planar.

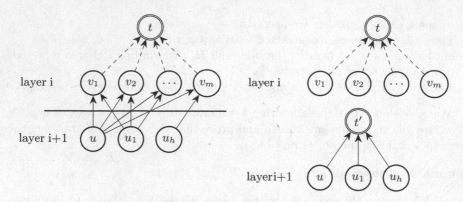

Fig. 2. Cut operation with a new sink node.

6   Split Process

In this section, we show why the cut operation reduces the genus of both parts by at least 1 and how the circuit will be split into $k + 1$ planar sub-circuits.

Lemma 5. *The cut operation reduces the genus of subgraphs by at least 1.*

Proof. Suppose that some nodes in layer $i + 1$ and i cannot be contracted, then either there exists at least one node u in layer $i + 1$ such that its incoming edges from layer $i + 2$ interlace with the outgoing edges to layer i, or there exists at least one node v in layer i such that its incoming edges from layer $i + 1$ interlace with the outgoing edges to layer $i - 1$. Irrespective of both of these scenarios, we do the following and obtain a new graph containing only two blocks. We delete all directed edges between layers i and $i + 1$ and then add one new node t' as the sink node of the second part and the source node of the first part. Suppose the genus of the first block is g_1 and the genus of the second block is g_2. Then the genus of the first block is equal to the genus of the first part of the graph after cutting, and the genus of the second block is equal to the genus of the second part of the graph after cutting. The cut operation will definitely reduce the genus of the layered graph by 1. Hence, we have $g_1 + g_2 + 1 = k$. This implies that $g_1 + g_2 < k$. Since $g_1 \geq 0$ and $g_2 \geq 0$, then $g_1 < k$ and $g_2 < k$.   □

Lemma 6. *Algorithm 1 splits the input circuit with bounded genus k into $k + 1$ planar circuits.*

Proof. We prove this lemma by induction. Without loss of generality, after the first cut, we observe that $g_1 \leq g_2$ and the genus of the layered graph reduces by 1. Now, we have two separate graphs. We consider different combinations of g_1 and g_2 as follows:

– if g_1 is 0, then it is a planar subgraph. This graph is represented by one PQ-tree. The other part will have genus $g_2 = k - 1$. In this case, the genus reduces by 1 and the number of subgraphs increases by 1.

– if g_1 is 1, then it is not a planar subgraph and it will be cut into two subgraphs later. The other part will have genus $g_2 = k - 2$.

Hence we conclude that every cut will reduce the genus, at least by 1, and hence there are at most $k + 1$ planar subgraphs.   □

Our main result follows from the lemmas above together with the parallel evaluation technique for PMCVP that has a running time in $\mathcal{O}(\log^3 n)$ with a linear number of processors. We note that the number of processors needed is bounded by $\mathcal{O}(n^c)$, which is the processors boundary for the parallel matrix multiplication algorithm. This is only because we need a parallel topological sorting algorithm to compute the layer number. Otherwise, our algorithm will only need a linear number of processors.

7 Concluding Remarks and Future Work

We presented an efficient parallel algorithm for the general MCVP problem with bounded genus. We deduce that MCVP with genus k is in FPP. This implies that given a P-complete problem, it is possible to find an algorithm that makes the problem fall into NC by fixing one or more parameters. Hence, with the results in this paper, we initiate the study of a new class of problems analogous to the class FPT. Subsequently, many questions remain unanswered. For example, can we construct a hierarchy for P-problems analogous to the one for NP-problems?

Acknowledgments. We wish to thank the anonymous referees for their valuable comments to improve the structure and presentation of this paper.

References

1. Battle, J., Harary, F., Kodama, Y.: Additivity of the genus of a graph. Bull. Am. Math. Soc. **68**(6), 565–568 (1962)
2. Booth, K.S., Lueker, G.S.: Testing for the consecutive ones property, interval graphs, and graph planarity using PQ-tree algorithms. J. Comput. Syst. Sci. **13**(3), 335–379 (1976)
3. Cesati, M., Di Ianni, M.: Parameterized parallel complexity. In: Pritchard, D., Reeve, J.S. (eds.) Euro-Par 1998. LNCS, vol. 1470, pp. 892–896. Springer, Heidelberg (1998)
4. Chen, J., Kanj, I.A., Xia, G.: Improved upper bounds for vertex cover. Theor. Comput. Sci. **411**(40), 3736–3756 (2010)
5. Cook, S.A.: A taxonomy of problems with fast parallel algorithms. Inf. Control **64**(1), 2–22 (1985)
6. Elberfeld, M., Jakoby, A., Tantau, T.: Logspace versions of the theorems of bodlaender and courcelle. In: 2010 51st Annual IEEE Symposium on Foundations of Computer Science (FOCS), pp. 143–152. IEEE (2010)
7. Goldschlager, L.M.: The monotone and planar circuit value problems are log space complete for P. SIGACT News **9**(2), 25–29 (1977)

8. Klein, P.N., Reif, J.H.: An efficient parallel algorithm for planarity. J. Comput. Syst. Sci. **37**(2), 190–246 (1988)
9. Ladner, R.E.: The circuit value problem is log space complete for P. SIGACT News **7**(1), 18–20 (1975)
10. Limaye, N., Mahajan, M., Sarma, J.M.: Upper bounds for monotone planar circuit value and variants. Comput. Complex. **18**(3), 377–412 (2009)
11. Ramachandran, V., Yang, H.: An efficient parallel algorithm for the general planar monotone circuit value problem. SIAM J. Comput. **25**(2), 312–339 (1996)
12. Tarjan, R.E., Vishkin, U.: An efficient parallel biconnectivity algorithm. SIAM J. Comput. **14**(4), 862–874 (1985)
13. Yang, H.: An NC algorithm for the general planar monotone circuit value problem. In: Proceedings of the Third IEEE Symposium on Parallel and Distributed Processing, pp. 196–203, December 1991

Database and Data Structures

Locality-Sensitive Hashing Without False Negatives for l_p

Andrzej Pacuk, Piotr Sankowski, Karol Wegrzycki, and Piotr Wygocki[✉]

Institute of Informatics, University of Warsaw, Warsaw, Poland
{apacuk,sank,k.wegrzycki,wygos}@mimuw.edu.pl

Abstract. In this paper, we show a construction of locality-sensitive hash functions without false negatives, i.e., which ensure collision for every pair of points within a given radius R in d dimensional space equipped with l_p norm when $p \in [1, \infty]$. Furthermore, we show how to use these hash functions to solve the c-approximate nearest neighbor search problem without false negatives. Namely, if there is a point at distance R, we will certainly report it and points at distance greater than cR will not be reported for $c = \Omega(\sqrt{d}, d^{1-\frac{1}{p}})$. The constructed algorithms work:
- with preprocessing time $\mathcal{O}(n \log(n))$ and sublinear expected query time,
- with preprocessing time $\mathcal{O}(\text{poly}(n))$ and expected query time $\mathcal{O}(\log(n))$.

Our paper reports progress on answering the open problem presented by Pagh [8], who considered the nearest neighbor search without false negatives for the Hamming distance.

1 Introduction

The *Nearest Neighbor* problem is of major importance to a variety of applications in machine learning and pattern recognition. Ordinarily, points are embedded in $\mathbb{R}^d$, and distance metrics usually measure similarity between points. Our task is the following: given a preprocessed set of points $S \subset \mathbb{R}^d$ and a query point $q \in \mathbb{R}^d$, find the point $v \in S$, with the minimal distance to q. Unfortunately, the existence of an efficient algorithm (i.e., whose query and preprocessing time would not depend exponentially on d), would disprove the strong exponential time hypothesis [8,10]. Due to this fact, we consider the *c-approximate nearest neighbor* problem: given a distance R, a query point q and a constant $c > 1$, we need to find a point within distance cR from point q [4]. This point is called a *cR-near neighbor* of q.

Definition 1. *Point v is an r-near neighbor of q in metric $\mathcal{M}$ iff $\mathcal{M}(q, v) \leq r$.*

One of the most interesting methods for solving the c-approximate nearest neighbor problem in high-dimensional space is *locality-sensitive hashing* (LSH). The algorithm offers a sub-linear query time and a sub-quadratic space complexity. The rudimentary component on which LSH method relies is *locality-sensitive*

© Springer International Publishing Switzerland 2016
T.N. Dinh and M.T. Thai (Eds.): COCOON 2016, LNCS 9797, pp. 105–118, 2016.
DOI: 10.1007/978-3-319-42634-1_9

hashing function. Intuitively, a hash function is *locality-sensitive* if the probability of collision is much higher for "nearby" points than for "far apart" ones. More formally:

Definition 2. *A family $H = \{h : S \to U\}$ is called (r, c, p_1, p_2)-sensitive for distance $\mathcal{D}$ and induced ball $\mathcal{B}(q, r) = \{v : \mathcal{D}(q, v) < r\}$, if for any $v, q \in S$:*

- *if $v \in \mathcal{B}(q, r)$ then $\mathbb{P}[h(q) = h(v)] \geq p_1$,*
- *if $v \notin \mathcal{B}(q, cr)$ then $\mathbb{P}[h(q) = h(v)] \leq p_2$.*

For $p_1 > p_2$ and $c > 1$.

Indyk and Motwani [7] considered randomized c-approximate R-near neighbor (Definition 3).

Definition 3 (The Randomized c-Approximate R-Near Neighbor or (R,c)-NN). *Given a set of points in a $P \subset \mathbb{R}^d$ and parameters $R > 0$, $\delta > 0$. Construct a data structure D such that for any query point q, if there exists a R-near neighbor of q in P, D reports some cR-near neighbor of q in P with probability $1 - \delta$.*

In this paper, we study guarantees for LSH based (R,c)-NN such that for each query point q, every close enough point $\|x - q\|_p < R$ will be certainly returned, i.e., there are no false negatives.[1] In other words, given a set S of size n and a query point q, the result is a set $P \subseteq S$ such that:

$$\{x : \|x - q\|_p < r\} \subseteq P \subseteq \{x : \|x - q\|_p \leq cr\}.$$

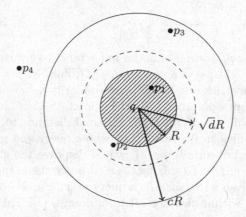

Fig. 1. The presented algorithms guarantee that points in the dashed area (p_1) will be reported as neighbors. Points within the dotted circle (p_2) will be reported as neighbor with high probability. Points (p_3) within a distance cR might be reported, but not necessarily. Points (p_4) outside circle cR cannot be reported. The schema picture presents an example for the euclidean distance $(p = 2)$.

---

[1] $\|\cdot\|_p$ denotes the standard l_p norm for fixed p.

Moreover, for each distant point ($\|x - q\|_p > cR$), the probability of being returned is bounded by p_{fp} – probability of false positives. In [8] this type of LSH is called *LSH without false negatives*. The fact that the probability of false negatives is 0 is our main improvement over Indyk and Motwani algorithm [7]. Furthermore, Indyk and Motwani showed that p-stable distributions (where $p \in (0, 2]$) are (r, c, p_1, p_2)-sensitive for l_p. We generalized their results on any distribution with mean 0, bounded second and fourth moment and any $p \in [1, \infty]$ (see Lemma 1, for rigorous definitions). Finally, certain distributions from this abundant class guarantee that points within given radius will always be returned (see Fig. 1). Unfortunately, our results come with a price, namely $c \geq \max\{\sqrt{d}, d^{1-1/p}\}$.

2 Related Work

2.1 Nearest Neighbor in High Dimensions

Most common techniques for solving the approximate nearest neighbor search, such as the spatial indexes or k-d trees [3] are designed to work well for the relatively small number of dimensions. The query time for k-d trees is $\mathcal{O}(n^{1-\frac{1}{d}})$ and when the number of dimensions increases the complexity basically converges to $\mathcal{O}(n)$. For interval trees, query time $\mathcal{O}(\log^d n)$ depends exponentially on the number of dimensions. The major breakthrough was the result of Indyk and Motwani [7]. Their algorithm has expected complexity of $\mathcal{O}(dn^{\frac{1}{c}})$ for any approximation constant $c > 1$ and the complexity is tight for any metric l_p (where $p \in (0, 2]$). Indyk and Motwani introduced the following LSH functions:

$$h(v) = \left\lfloor \frac{\langle a, v \rangle + b}{r} \right\rfloor,$$

where a is the d-dimensional vector of independent random variables from a p-stable distribution and b is a real number chosen uniformly from the range $[0, r]$.

Our algorithm is based on similar functions and we prove compelling results for more general family of distributions (we show bounds for any distribution with a bounded variance and an expected value equal to 0). Furthermore, our algorithm is correct for any $p \in [1, \infty]$. Indyk and Motwani's LSH algorithm was showed to be optimal for l_1 metric. Subsequently, Andoni et al. [1] showed near optimal results for l_2. Recently, data dependant techniques have been used to further improve LSH by Andoni and Razenshteyn [2]. However, the constant ρ in a query time $\mathcal{O}(n^\rho)$ remains:

$$\rho = \frac{\log p_1}{\log p_2}.$$

When a formal guarantee that $p_1 = 1$ is needed their algorithm does not apply.

2.2 LSH Without False Negatives

Recently, Pagh [8] presented a novel approach to nearest neighbor search in Hamming space. He showed the construction of an efficient locality-sensitive hash function family that guarantees collision for any close points. Moreover, Pagh showed that bounds of his algorithm for $cr = \log n/k$ (where $k \in \mathbb{N}$) essentially match bounds of Indyk and Motwani (differ by at most factor $\ln 4$ in the exponent). More precisely, he showed that the problem of false negatives can be avoided in the Hamming space at some cost in efficiency. He proved bounds for general values of c. This paper is an answer to his open problem: whether is it possible to get similar results for other distance measures (e.g., l_1 or l_2).

Pagh introduced the concept of an *r-covering* family of hash function:

Definition 4. *For $A \subseteq \{0,1\}^d$, the Hamming projection family $\mathcal{H}_A$ is r-covering if for every $x \in \{0,1\}^d$ with $\|x\|_H \leq r$, there exist $h \in \mathcal{H}_A$ such that $h(x) = \mathbf{0}$.*

Then, he presented a fast method of generating such an r-covering family. Finally, he showed that the expected number of false positives is bounded by $2^{r+1-\|x-y\|_H}$.

3 Basic Construction

We will consider the l_p metric for $p \in [1, \infty]$ and n fixed points in $\mathbb{R}^d$ space. Let v be a d-dimensional vector of independent random variables drawn from distribution $\mathcal{D}$. We define a function h_p as:

$$h_p(x) = \left\lfloor \frac{\langle x, v \rangle}{r\rho_p} \right\rfloor,$$

where $\langle, \rangle$ is a standard inner product and $\rho_p = d^{1-\frac{1}{p}}$. The scaling factor ρ_p is chosen so that: $\|z\|_1 \leq \rho_p \|z\|_p$. The rudimentary distinction between the hash function h_p and LSH is that we consider two hashes equal when they differ at most by one. In Indyk and Motwani [7] version of LSH, there were merely probabilistic guarantees, and close points (say 0.99 and 1.01) could be returned in different buckets with small probability. Since our motivation is to find all close points with absolute certainty, we need to check the adjacent buckets as well.

First, observe that for given points, the probability of choosing a hash function that will classify them as equal is bounded on both sides as given by the following observations. The proofs of these observations are in Appendices A and B.

Observation 1 (Upper Bound on the Probability of Point Equivalence)

$$\mathbb{P}\left[|h_p(x) - h_p(y)| \leq 1\right] \leq \mathbb{P}\left[|\langle x - y, v \rangle| < 2\rho_p r\right].$$

Observation 2 (Lower Bound on the Probability of Point Equivalence)

$$\mathbb{P}\left[|h_p(x) - h_p(y)| \leq 1\right] \geq \mathbb{P}\left[|\langle x - y, v \rangle| < \rho_p r\right].$$

Interestingly, using the aforementioned observations we can configure a distribution $\mathcal{D}$ so the close points must end up in the same or adjacent bucket.

Observation 3 (Close Points Have Close Hashes). For distribution $\mathcal{D}$ such that every $v_i \sim \mathcal{D}$: $-1 \leq v_i \leq 1$ and for $x, y \in \mathbb{R}^d$, if $\|x - y\|_p < r$ then $\forall_{h_p} |h_p(x) - h_p(y)| \leq 1$.

Proof. We know that $\|z\|_1 \leq \rho_p \|z\|_p$ and $|v_i| \leq 1$ (because v_i is drawn from bounded distribution $\mathcal{D}$), so:

$$\rho_p \|x - y\|_p \geq \|x - y\|_1 = \sum_i |x_i - y_i| \geq \sum_i |v_i(x_i - y_i)| \geq \left| \sum_i v_i(x_i - y_i) \right|$$
$$= |\langle x - y, v \rangle|.$$

Now, when points are close in l_p:

$$\|x - y\|_p < r \iff \rho_p \|x - y\|_p < \rho_p r \implies |\langle x - y, v \rangle| < \rho_p r.$$

Next, by Observation 2:

$$1 = \mathbb{P}\left[|\langle x - y, v \rangle| < \rho_p r\right] \leq \mathbb{P}\left[|h_p(x) - h_p(y)| \leq 1\right].$$

Hence, the points will inevitably hash into the same or adjacent buckets.   □

Now we will introduce the inequality that will help to bound the probability of false positives.

Observation 4 (Inequality of Norms in l_p). Recall that $\rho_p = d^{1 - \frac{1}{p}}$. For every $z \in \mathbb{R}^d$ and $p \in [1, \infty]$:

$$\|z\|_2 \geq \frac{\rho_p}{\max\{d^{\frac{1}{2}}, d^{1 - \frac{1}{p}}\}} \|z\|_p.$$

This technical observation is proven in Appendix C.

The major question arises: what is the probability of false positives? In contrast to the Indyk and Motwani [7], we cannot use p-stable distributions because these distributions are not bounded. We will present the proof for a different class of functions.

Lemma 1 (The Probability of False Positives for General Distribution).
Let $\mathcal{D}$ be a random variable such that $\mathbb{E}(\mathcal{D}) = 0$, $\mathbb{E}(\mathcal{D}^2) = \alpha^2$, $\mathbb{E}(\mathcal{D}^4) \leq 3\alpha^4$ (for any $\alpha \in \mathbb{R}^+$). Define constant $\tau_1 = \frac{2}{\alpha} \max\{d^{\frac{1}{2}}, d^{1 - \frac{1}{p}}\}$.
When $\|x - y\|_p > cr$, $x, y \in \mathbb{R}^d$ and $c > \tau_1$ then:

$$p_{fp_1} = \mathbb{P}\left[|h_p(x) - h_p(y)| \leq 1\right] < 1 - \frac{\left(1 - \frac{\tau_1^2}{c^2}\right)^2}{3},$$

for every metric l_p, where $p \in [1, \infty]$ (p_{fp_1} is the probability of false positive).

Proof. By Observation 4:

$$\|z\|_2 \geq \frac{2\|z\|_p}{\alpha\tau_1}\rho_p$$

Subsequently, let $z = x - y$ and define a random variable $X = \langle z, v \rangle$. Therefore:

$$\mathbb{E}(X^2) = \alpha^2\|z\|_2^2 \geq \left(\frac{2\|z\|_p}{\tau_1}\rho_p\right)^2 > \left(2r\rho_p\frac{c}{\tau_1}\right)^2.$$

Because $\frac{c}{\tau_1} > 1$ we have $\theta = \frac{(2r\rho_p)^2}{\mathbb{E}X^2} < 1$. Variable θ and a random variable $X^2 > 0$ satisfy Paley-Zygmunt inequality (analogously to [9]):

$$\mathbb{P}\left[|h_p(x) - h_p(y)| > 1\right] \geq \mathbb{P}\left[|\langle z, v \rangle| \geq 2r\rho_p\right] \geq \mathbb{P}\left[X^2 > (2r\rho_p)^2\right]$$

$$\geq \left(1 - \frac{(2r\rho_p)^2}{\mathbb{E}(X^2)}\right)^2 \frac{\mathbb{E}(X^2)^2}{\mathbb{E}(X^4)}.$$

Eventually, we assumed that $\mathbb{E}(X^4) \leq 3(\alpha\|z\|_2)^4$:

$$\mathbb{P}\left[|h_p(x) - h_p(y)| > 1\right] \geq \frac{\left(1 - \frac{(2r\rho_p)^2}{\mathbb{E}(X^2)}\right)^2}{3} > \frac{\left(1 - \frac{\tau_1^2}{c^2}\right)^2}{3}.$$

$\square$

Simple example of a distribution that satisfies both Observation 3 and Lemma 1 is a uniform distribution on $(-1, 1)$ with a standard deviation α equal to $\sqrt{\frac{1}{3}}$. Another example of such distribution is a discrete distribution with uniform values $\{-1, 1\}$. As it turns out, Lemma 2 shows that the discrete distribution leads to even better bounds.

Lemma 2 (Probability of False Positives for the Discrete Distribution). *Let $\mathcal{D}$ be a random variable such that $\mathbb{P}\left[\mathcal{D} = \pm 1\right] = \frac{1}{2}$. Define constant $\tau_2 = \sqrt{8}\max\{d^{\frac{1}{2}}, d^{1-\frac{1}{p}}\}$. Then for every $p \in [1, \infty]$, $x, y \in \mathbb{R}^d$ and $c > \tau_2$ such that $\|x - y\|_p > cr$, it holds:*

$$p_{fp_2} = \mathbb{P}\left[|h_p(x) - h_p(y)| \leq 1\right] < 1 - \frac{(1 - \frac{\tau_2}{c})^2}{2}.$$

Proof. Because of Observation 4 we have the inequality:

$$\|z\|_2 \geq \sqrt{8}\frac{\|z\|_p}{\tau_2}\rho_p.$$

Let $z = x - y$ and $X = \langle z, v \rangle$, be a random variable. Then:

$$\mathbb{P}\left[|h_p(x) - h_p(y)| > 1\right] \geq \mathbb{P}\left[|X| > 2r\rho_p\right].$$

Khintchine inequality [5] states $\mathbb{E}|X| \geq \frac{\|z\|_2}{\sqrt{2}}$, so:

$$\mathbb{E}(|X|) \geq \frac{\|z\|_2}{\sqrt{2}} \geq \frac{2\rho_p\|z\|_p}{\tau_2} > 2r\rho_p\frac{c}{\tau_2}.$$

Note that, a random variable $|X|$ and $\theta = \frac{2r\rho_p}{\mathbb{E}(|X|)} < 1$, satisfy the Paley-Zygmunt inequality (because $\frac{c}{\tau_2} > 1$), though:

$$\mathbb{P}\left[|h_p(x) - h_p(y)| > 1\right] \geq \left(1 - \frac{2r\rho_p}{\mathbb{E}(|X|)}\right)^2 \frac{\mathbb{E}(|X|)^2}{\mathbb{E}(|X|^2)}$$

$$> \left(1 - \frac{2r\rho_p}{2r\rho_p \frac{c}{\tau_2}}\right)^2 \frac{1}{2} = \frac{\left(1 - \frac{\tau_2}{c}\right)^2}{2}.$$

$\square$

Altogether, in this section we have introduced a family of hash functions h_p which:

– Guarantees that, with an absolute certainty, points within the distance R will be mapped to the same or adjacent buckets (see Observation 3),
– Maps "far away" points to the non-adjacent hashes with high probability (Lemmas 1 and 2).

These properties will enable us to construct an efficient algorithm for solving the c-approximate nearest neighbor search problem without false negatives.

3.1 Tightness of Bounds

We showed that for two distant points $x, y : \|x - y\|_p > cr$, the probability of a collision is small when $c = \max\{\rho_p, \sqrt{d}\}$. The natural question arises: Can we bound the probability of a collision for points $\|x - y\|_p > c'r$ for some $c' < c$?

We will show that such c' does not exist, i.e., there always exists $\tilde{x}$ such that $\|\tilde{x}\|_p$ will be arbitrarily close to cr, so $\tilde{x}$ and $\vec{0}$ will end up in the same or adjacent bucket with high probability. More formally, for any $p \in [1, \infty]$, for $h_p(x) = \left\lfloor \frac{\langle x, v \rangle}{r\rho_p} \right\rfloor$, where coordinates of d-dimensional vector v are random variables v_i, such that $-1 \leq v_i \leq 1$ with $\mathbb{E}(v_i) = 0$. We will show that there always exists $\tilde{x}$ such that $\|\tilde{x}\|_p \approx r \max\{\rho_p, \sqrt{d}\}$ and $|h_p(\tilde{x}) - h_p(\vec{0})| \leq 1$ with high probability.

For $p \geq 2$ denote $x_0 = (r\rho_p - \epsilon, 0, 0, \ldots, 0)$. We have $\|x_0 - \vec{0}\|_p = r\rho_p - \epsilon$ and:

$$|h_p(x_0) - h_p(\vec{0})| = \left| \left\lfloor \frac{r\rho_p - \epsilon}{r\rho_p} \cdot v_1 \right\rfloor - 0 \right| \leq 1.$$

For $p \in [1, 2)$, denote $x_1 = rd^{-\frac{1}{p} + \frac{1}{2} - \epsilon}\vec{1}$. We have $\|x_1\|_p = rd^{\frac{1}{2} - \epsilon}$ and by applying Observation 2 for complementary probabilities:

$$\mathbb{P}\left[|h_p(x_1) - h_p(\vec{0})| > 1\right] \leq \mathbb{P}\left[|\langle x_1, v \rangle| \geq \rho_p r\right] = \mathbb{P}\left[|\langle \vec{1}, v \rangle| \geq d^{\frac{1}{2} + \epsilon}\right]$$

$$= \mathbb{P}\left[\left|\frac{\sum_{i=1}^d v_i}{d}\right| \geq d^{-\frac{1}{2} + \epsilon}\right] \leq 2 \cdot \exp\left(\frac{-d^{2\epsilon}}{2}\right).$$

The last inequality follows from Hoeffding [6] (see Appendix D for technical details).

So the aforementioned probability for $p \in [1, 2)$ is bounded by an expression exponential in $d^{2\epsilon}$. Even if we would concatenate k random hash functions (see proof of Theorem 1 for more details), the chance of collision would be at least $(1 - 2e^{\frac{-d^{2\epsilon}}{2}})^k$. To bound this probability, the number k needs to be at least $\Theta(e^{\frac{d^{2\epsilon}}{2}})$. The probability bounds do not work for ϵ arbitrary close to 0: we proved that introduced hash functions for $c = d^{1/2-\epsilon}$ do not work (may give false positives).[2]

Hence, to obtain a significantly better approximation factor c, one must introduce a completely new family of hash functions.

4   The Algorithm

In this section, we apply the LSH family introduced in Sect. 3 to construct an c-approximate algorithm without false negatives. To begin with, we will define a general algorithm that will satisfy our conditions. Subsequently, we will show that complexity of the query is sublinear, and it depends linearly on the number of dimensions.

Theorem 1. *For any $c > \tau$ and the number of iterations $k \geq 0$, there exists a c-approximate nearest neighbor algorithm without false negatives for l_p, where $p \in [1, \infty]$:*

- *Preprocessing time: $\mathcal{O}(n(kd + 3^k))$,*
- *Memory usage: $\mathcal{O}(n3^k)$,*
- *Expected query time: $\mathcal{O}(d(|P| + k + np_{fp}^k))$.*

Where $|P|$ is the size of the result and p_{fp} is the upper bound of probability of false positives (note that p_{fp} depends on a choice of τ from Lemmas 1 or 2).

Proof. Let $g(x) := (h_p^1(x), h_p^2(x), \ldots, h_p^k(x))$ be a hash function defined as a concatenation of k random LSH functions presented in Sect. 3. We introduce the clustering $m : g(\mathbb{R}^d) \to 2^n$, where each cluster is assigned to the corresponding hash value. For each hash value α, the corresponding cluster $m(\alpha)$ is $\{x : g(x) = \alpha\}$.

Since we consider two hashes to be equal when they differ at most by one (see Observation 3), for hash α, we need to store the reference for every point that satisfies $\|\alpha - x\| \leq 1$. The number of such clusters is 3^k, because the result of each hash function can vary by one of $\{-1, 0, 1\}$ and the number of hash functions is k. Thus, the memory usage is $\mathcal{O}(n3^k)$ (see Fig. 2).

To preprocess the data, we need to compute the value of the function g for every point in the set and then put its reference into 3^k cells. Hence, the preprocessing time complexity equals $\mathcal{O}(n(kd + 3^k))$.

---

[2] However, one may try to obtain tighter bound (e.g., $c = d^{1/2}/\log(d)$) or show that for every $\epsilon > 0$, the approximation factor $c = d^{1/2} - \epsilon$ does not work.

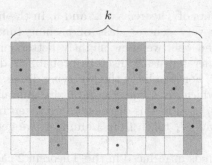

Fig. 2. Blue dots represent value of $g(q)$ for query. Green dots are always distant by 1, hence green and blue points are considered close. At least one red dot is distant from blue dot by more than 1, hence red dots will not be considered close to blue. Thus, algorithm needs to check 3^k various possibilities. (Color figure online)

Eventually, to answer a query, we need to compute $g(q)$ in time $\mathcal{O}(kd)$ and then for every point in $\|g(x) - g(q)\|_\infty \leq 1$ remove distant points $\|x - q\|_p > cR$. Hence, we need to look up every false-positive to check whether they are within distance cr from the query point. We do that in expected time $\mathcal{O}(d(|P| + k + np_{fp}{}^k))$, because $np_{fp}{}^k$ is the expected number of false positives.   □

The number of iterations k can be chosen arbitrarily, so we will choose the optimal value to minimize the query time. This gives the main result of this paper:

Theorem 2. *For any $c > \tau$ and for large enough n, there exists a c-approximate nearest neighbor algorithm without false negatives for l_p, where $p \in [1, \infty]$:*

- *Preprocessing time: $\mathcal{O}(n(\gamma d \log n + (\frac{n}{d})^\gamma)) = \mathrm{poly}(n)$,*
- *Memory usage: $\mathcal{O}(n(\frac{n}{d})^\gamma)$,*
- *Expected query time: $\mathcal{O}(d(|P| + \gamma \log(n) + \gamma d))$.*

Where $|P|$ is the size of the result, $\gamma = \frac{\ln 3}{-\ln p_{fp}}$ and p_{fp} and τ are chosen as in Theorem 1.

Proof. Denote $a = -\ln p_{fp}$, $b = \ln 3$ and set $k = \left\lceil \frac{\ln \frac{na}{d}}{a} \right\rceil$.

Let us assume that n is large enough so that $k \geq 1$. Then using the fact that $x^{1/x}$ is bounded for $x > 0$ we have:

$$3^k \leq 3 \cdot (3^{\ln \frac{na}{d}})^{1/a} = 3 \cdot (\frac{na}{d})^{b/a} = \mathcal{O}((\frac{n}{d})^{b/a}) = \mathcal{O}((\frac{n}{d})^\gamma),$$

$$np_{fp}{}^k = ne^{-ak} \leq ne^{-a\frac{\ln(\frac{na}{d})}{a}} = \frac{d}{a} = \mathcal{O}(d\gamma),$$

$$k = \mathcal{O}(\gamma \log(n)).$$

Substituting these values in the Theorem 1 gives needed complexity guaranties.   □

There are two variants of Theorems 1, 2 and 3. In the first variant, we show complexity bounds for very general class of hashing functions introduced in Lemma 1. In the second one, we show slightly better guaranties for hashing functions which are generated using discrete probability distribution on $\{0,1\}$ introduced in Lemma 2. For simplicity the following discussion is restricted only to the second variant which gives better complexity guaranties. The definitions of constants p_{fp_2} and τ_2 used in this discussion are taken from Lemma 2. For a general case, i.e., p_{fp_1} and τ_1 taken from Lemma 1, we get only slightly worse results.

The complexity bounds introduced in the Theorem 2 can be simplified using the fact that $\ln(x) < x - 1$. Namely, we have:

$$\gamma = \frac{\ln 3}{-\ln p_{fp_2}} = \frac{\ln 3}{-\ln(1 - \frac{(1-\frac{\tau_2}{c})^2}{2})} < \frac{2\ln 3}{(1-\frac{\tau_2}{c})^2}.$$

However, the preprocessing time is polynomial in n for any constant c, it strongly depends on the bound for probability p_{fp_2} and c. Particularly when c is getting close to τ_2, the exponent of the preprocessing time might be arbitrarily large.

To the best of our knowledge, this is the first algorithm that will ensure that no false negatives will be returned by the nearest neighbor approximated search and does not depend exponentially on the number of dimensions. Note that for given c, the parameter γ is fixed. By Lemma 2, we have: $p_{fp_2} = 1 - \frac{(1-\frac{\tau_2^2}{c^2})^2}{2}$, so:

$$\lim_{c\to\infty} \gamma = \lim_{c\to\infty} \frac{\ln 3}{-\ln p_{fp_2}} = \log_2 3 \approx 1.58.$$

If we omit terms polynomial in d, the preprocessing time of the algorithm from Theorem 2 converges to $\mathcal{O}(n^{2.58})$ ($\mathcal{O}(n^{3.71})$ for general case - see Appendix E).

4.1 Light Preprocessing

Although the preprocessing time $\mathcal{O}(n^{2.58})$ may be reasonable when there are multiple, distinct queries and the data set does not change (e.g., static databases, pre-trained classification, geographical map). Still, unless the number of points is small, this algorithm does not apply. Here, we will present an algorithm with a light preprocessing time $\mathcal{O}(dn \log n)$ and $\mathcal{O}(n \log n)$ memory usage where the expected query time is $o(n)$.

The algorithm with light preprocessing is very similar to the algorithm described in Theorem 1, but instead of storing references to the point in all 3^k buckets during preprocessing, this time searching for every point x that matches $\|x - q\|_\infty \le 1$ is done during the query time.

The expected query time with respect to k is $\mathcal{O}(d(|P| + k + n p_{fp}{}^k) + 3^k)$. During the preprocessing phase we only need to compute k hash values for each of n points and store them in memory. Hence, preprocessing requires $\mathcal{O}(kdn)$ time and uses $\mathcal{O}(nk)$ memory.

Theorem 3. *For any $c > \tau$ and for large enough n, there exists a c-approximate nearest neighbor algorithm without false negatives for l_p, where $p \in [1, \infty]$:*

- *Preprocessing time: $\mathcal{O}(nd \log n)$,*
- *Memory usage: $\mathcal{O}(n \log n)$,*
- *Expected query time: $\mathcal{O}(d(|P| + n^{\frac{b}{a+b}}(\frac{b}{a})^{\frac{a}{b+a}}))$.*

Where $|P|$ is the size of the result, $a = -\ln p_{fp}$, $b = \ln 3$, p_{fp} and τ are chosen as in Theorem 1.

Proof. We set the number of iterations $k = \left\lceil \frac{\ln \frac{na}{b}}{a+b} \right\rceil$. Assume n needs to be large enough so that $k \geq 1$. Since a is upper bounded for both choices of p_{fp}:

$$3^k \leq 3 \cdot 3^{\frac{\ln(\frac{na}{b})}{a+b}} = 3(\frac{na}{b})^{\frac{b}{a+b}} = \mathcal{O}(n^{\frac{b}{a+b}}).$$

Analogously:

$$n p_{fp}{}^k = n(e^{-a})^k \leq ne^{-a\frac{\ln(\frac{na}{b})}{a+b}} = n \cdot \left(\frac{b}{a}\right)^{\frac{a}{a+b}} \cdot \left(\frac{1}{n}\right)^{\frac{a}{a+b}} = n^{\frac{b}{a+b}}\left(\frac{b}{a}\right)^{\frac{a}{a+b}}.$$

Hence, for this choice of k we obtain the expected query time is equal to:

$$\mathcal{O}(d(|P| + k + n p_{fp}{}^k)) + 3^k = \mathcal{O}(d(|P| + \log n + n^{\frac{b}{a+b}}\left(\frac{b}{a}\right)^{\frac{a}{a+b}}) + n^{\frac{b}{a+b}})$$

$$= \mathcal{O}(d(|P| + n^{\frac{b}{a+b}}\left(\frac{b}{a}\right)^{\frac{a}{a+b}})).$$

Substituting k, we obtain formulas for preprocessing time and memory usage. $\square$

Eventually, exactly as previously for a general distribution from Lemma 1, when $c \to \infty$ we have: $a \to \ln \frac{3}{2}$ (see Theorem 3 for the definition of constant a). Hence, for a general distribution we have a bound for complexity equal to $\mathcal{O}(n^{\log_{4.5} 3}) \approx \mathcal{O}(n^{0.73})$. For the discrete distribution from Lemma 2, the constant a converges to $\ln 2$. Hence, the expected query time converges to $\mathcal{O}(n^{0.61})$.

5   Conclusion and Future Work

We have presented the c-approximate nearest neighbor algorithm without false negatives in l_p for all $p \in [1, \infty]$ and $c > \max\{\sqrt{d}, d^{1-1/p}\}$. Due to this inequality our algorithm can be used cognately to the original LSH [7] but with additional guarantees about very close points (one can set $R' = \sqrt{d}R$ and be certain that all points within distance R will be returned). In contrast to the original LSH, our algorithm does not require any additional parameter tunning.

The future work concerns relaxing restriction on the approximation factor c and reducing time complexity of the algorithm or proving that these restrictions are essential. We wish to match the time complexities given by [7] or show that achieved bounds are optimal. We show the tightness of our construction, hence to break the bound of $\sqrt{d}$, one would need to introduce a new technique.

Acknowledgments. This work was supported by ERC PoC project PAAl-POC 680912 and FET project MULTIPLEX 317532. We would also like to thank Rafał Latała for meaningful discussions.

A   Proof of Observation 1

Proof. We will use, the fact that for any $x, y \in \mathbb{R}$ we have $|\lfloor x \rfloor - \lfloor y \rfloor| \leq 1 \Rightarrow |x - y| < 2$. Then the following implications hold:

$$|h_p(x) - h_p(y)| \leq 1 \iff \left| \left\lfloor \frac{\langle x, v \rangle}{\rho_p r} \right\rfloor - \left\lfloor \frac{\langle y, v \rangle}{\rho_p r} \right\rfloor \right| \leq 1 \implies \left| \frac{\langle x, y \rangle}{\rho_p r} - \frac{\langle y, v \rangle}{\rho_p r} \right| < 2 \iff$$

$$\iff |\langle x - y, v \rangle| < 2\rho_p r.$$

So, based on the increasing property of the probability:

$$\text{if } A \subset B \text{ then } \mathbb{P}[A] \leq \mathbb{P}[B],$$

the inequality of the probabilities holds.   □

B   Proof of Observation 2

Proof. We will use the fact that for $x, y \in \mathbb{R} : |x - y| < 1 \Rightarrow |\lfloor x \rfloor - \lfloor y \rfloor| \leq 1$).

$$\left| \langle x - y, v \rangle \right| < \rho_p r \iff \left| \frac{\langle x, v \rangle}{\rho_p r} - \frac{\langle x, v \rangle}{\rho_p r} \right| < 1 \implies \left| \left\lfloor \frac{\langle x, v \rangle}{\rho_p r} \right\rfloor - \left\lfloor \frac{\langle x, v \rangle}{\rho_p r} \right\rfloor \right| \leq 1 \iff$$

$$\iff |h_p(x) - h_p(y)| \leq 1$$

                                                                                              □

C   Proof of Observation 4

Proof. For every $0 < b \leq a$ vectors in $\mathbb{R}^d$ satisfy the inequality:

$$\|z\|_a \leq \|z\|_b \leq d^{(\frac{1}{b} - \frac{1}{a})} \|z\|_a. \tag{1}$$

For $p > 2$ we have $\max\{d^{\frac{1}{2}}, d^{1-\frac{1}{p}}\} = d^{1-\frac{1}{p}}$. Then, using ineqaulity (1) for $a = p$ and $b = 2$ we have:

$$\|z\|_2 \geq \|z\|_p = \frac{\rho_p}{d^{1-\frac{1}{p}}} \|z\|_p = \frac{\rho_p}{\max\{d^{\frac{1}{2}}, d^{1-\frac{1}{p}}\}} \|z\|_p$$

For $1 \leq p \leq 2$ we have $\max\{d^{\frac{1}{2}}, d^{1-\frac{1}{p}}\} = d^{\frac{1}{2}}$. Analogously by using inequality (1) for $a = 2$ and $b = p$:

$$\|z\|_p \leq d^{\frac{1}{p} - \frac{1}{2}} \|z\|_2 = \|z\|_2 \frac{d^{\frac{1}{2}}}{\rho_p}$$

Hence, by dividing both sides we have:

$$\|z\|_p \frac{\rho_p}{\max\{d^{\frac{1}{2}}, d^{1-\frac{1}{p}}\}} \leq \|z\|_2$$

                                                                                              □

D   Hoeffding Bound

Here we are going to show all technical details used in the proof in the Sect. 3.1. Let us start with the Hoeffding inequality. Let $X_1, \ldots, X_d$ be bounded independent random variables: $a_i \leq X_i \leq b_i$ and $\overline{X}$ be the mean of these variables $\overline{X} = \sum_{i=1}^{d} X_i/d$. Theorem 2 of Hoeffding [6] states:

$$\mathbb{P}\left[|\overline{X} - \mathbb{E}\left[\overline{X}\right]| \geq t\right] \leq 2 \cdot \exp\left(-\frac{2d^2t^2}{\sum_{i=1}^{d}(b_i - a_i)^2}\right).$$

In our case, $D_1, \ldots, D_d$ are bounded by $a_i = -1 \leq D_i \leq 1 = b_i$ with $\mathbb{E}D_i = 0$. Hoeffding inequality implies:

$$\mathbb{P}\left[\left|\frac{\sum_{i=1}^{d} D_i}{d}\right| \geq t\right] \leq 2 \cdot \exp\left(-\frac{2d^2t^2}{\sum_{i=1}^{d}(b_i - a_i)^2}\right) = 2 \cdot \exp\left(-\frac{dt^2}{2}\right).$$

Taking $t = d^{-1/2+\epsilon}$ we get the claim:

$$\mathbb{P}\left[\left|\frac{\sum_{i=1}^{d} D_i}{d}\right| \geq d^{-1/2+\epsilon}\right] \leq 2 \cdot \exp\left(-\frac{d^{2\epsilon}}{2}\right).$$

E   Preprocessing Complexity Bounds for the Distributions Introduced in Lemma 1

By Lemma 1, we have: $\mathsf{p}_{\mathsf{fp}_1} = 1 - \frac{(1-\frac{\tau_1^2}{c^2})^2}{3}$, so:

$$\lim_{c \to \infty} \gamma = \lim_{c \to \infty} \frac{\ln 3}{-\ln \mathsf{p}_{\mathsf{fp}_1}} = \frac{\ln 3}{\ln 1.5} \approx 2.71.$$

If we omit terms polynomial in d, the preprocessing time of the algorithm from Theorem 2 converges to $\mathcal{O}(n^{3.71})$.

References

1. Andoni, A., Indyk, P.: Near-optimal hashing algorithms for approximate nearest neighbor in high dimensions. Commun. ACM **51**(1), 117–122 (2008)
2. Andoni, A., Razenshteyn, I.: Optimal data-dependent hashing for approximate near neighbors. In: Servedio, R.A., Rubinfeld, R. (eds.) Proceedings of the Forty-Seventh Annual ACM on Symposium on Theory of Computing, STOC 2015, Portland, OR, USA, 14–17 June 2015, pp. 793–801. ACM (2015)
3. Bentley, J.L.: K-d trees for semidynamic point sets. In: Proceedings of the Sixth Annual Symposium on Computational Geometry, SCG 1990, pp. 187–197. ACM, New York (1990)
4. Datar, M., Indyk, P.: Locality-sensitive hashing scheme based on p-stable distributions. In: Proceedings of the Twentieth Annual Symposium on Computational Geometry, SCG 2004, pp. 253–262. ACM Press (2004)

5. Haagerup, U.: The best constants in the Khintchine inequality. Stud. Math. **70**(3), 231–283 (1981)
6. Hoeffding, W.: Probability inequalities for sums of bounded random variables. J. Am. Stat. Assoc. **58**(301), 13–30 (1963)
7. Indyk, P., Motwani, R.: Approximate nearest neighbors: towards removing the curse of dimensionality. In: Proceedings of the Thirtieth Annual ACM Symposium on Theory of Computing, STOC 1998, pp. 604–613. ACM, New York (1998)
8. Pagh, R.: Locality-sensitive hashing without false negatives. In: Krauthgamer, R. (ed.) Proceedings of the Twenty-Seventh Annual ACM-SIAM Symposium on Discrete Algorithms, SODA 2016, Arlington, VA, USA, 10–12 January 2016, pp. 1–9. SIAM (2016)
9. Veraar, M.: On Khintchine inequalities with a weight. Proc. Am. Math. Soc. **138**, 4119–4121 (2010)
10. Williams, R.: A new algorithm for optimal 2-constraint satisfaction and its implications. Theor. Comput. Sci. **348**(2), 357–365 (2005)

Improved Space Efficient Algorithms for BFS, DFS and Applications

Niranka Banerjee[(✉)], Sankardeep Chakraborty, and Venkatesh Raman

The Institute of Mathematical Sciences, CIT Campus,
Taramani, Chennai 600 113, India
{nirankab,sankardeep,vraman}@imsc.res.in

Abstract. Recent work by Elmasry et al. (STACS 2015) and Asano et al. (ISAAC 2014), reconsidered classical fundamental graph algorithms focusing on improving the space complexity. Elmasry et al. gave, among others, implementations of breadth first search (BFS) and depth first search (DFS) in a graph on n vertices and m edges, taking $O(m + n)$ time using $O(n)$ and $O(n \lg \lg n)$ bits of space respectively improving the naive $O(n \lg n)$(We use lg to denote logarithm to the base 2.) bits implementation. We continue this line of work focusing on space.

Our first result is a simple data structure that can maintain any subset S of a universe of n elements using $n + o(n)$ bits and support in constant time, apart from the standard insert, delete and membership queries, the operation *findany* that finds and returns any element of the set (or outputs that the set is empty). Using this we give a BFS implementation that takes $O(m + n)$ time using at most $2n + o(n)$ bits. Later, we further improve the space requirement of BFS to at most $1.585n + o(n)$ bits albeit with a slight increase in running time to $O(m \lg n f(n))$ time where $f(n)$ is any extremely slow growing function of n. These improve the space by a constant factor from earlier representations.

We demonstrate the use of our data structure by developing another data structure using it that can represent a sequence of n non-negative integers $x_1, x_2, \ldots x_n$ using at most $\sum_{i=1}^{n} x_i + 2n + o(\sum_{i=1}^{n} x_i + n)$ bits and, in constant time, determine whether the i-th element is 0 or decrement it otherwise. We use this data structure to output the vertices of a

- directed acyclic graph in topological sorted order in $O(m + n)$ time and $O(m + n)$ bits, and
- graph with degeneracy d in degeneracy order in $O(nd)$ time using $O(nd)$ bits.

We also discuss an algorithm for finding a minimum weight spanning tree of a weighted undirected graph using at most $n + o(n)$ bits.

For DFS we give an $O(m + n)$ bits implementation for finding a chain decomposition of a connected undirected graph, and to find cut vertices, bridges and maximal two connected subgraphs of a connected graph. We also provide a $O(n)$ bits implementations for finding strongly connected components of a directed graph, to output the vertices of a directed acyclic graph in a topologically sorted manner, and to find a sparse biconnected subgraph of a biconnected graph. These improve the space required for earlier implementations from $\Omega(n \lg n)$ bits.

© Springer International Publishing Switzerland 2016
T.N. Dinh and M.T. Thai (Eds.): COCOON 2016, LNCS 9797, pp. 119–130, 2016.
DOI: 10.1007/978-3-319-42634-1_10

1 Introduction

Motivated by the rapid growth of huge data set ("big data"), algorithms that utilize space efficiently are becoming increasingly important than ever before. Another reason for the importance of space efficient algorithms is the proliferation of specialized handheld devices and embedded systems that have a limited supply of memory. Hence, there is a growing body of work that considers algorithms that do not modify the input and use only a limited amount of work space, and this paper continues this line of research for fundamental graph algorithms.

1.1 Our Results and Organization of the Paper

Asano et al. [2], in a recent paper, show that DFS of a directed or undirected graph on n vertices and m edges can be performed using $n + o(n)$ bits and (an unspecified) polynomial time. Using $2n + o(n)$ bits, they can bring down the running time to $O(mn)$ time, and using a larger $O(n)$ bits, their running time is $O(m \lg n)$. In a similar vein,

- we show in Sect. 3 that the vertices of a directed or undirected graph can be listed in BFS order using $1.585n + o(n)$ bits and $O(mf(n) \lg n)$ time where $f(n)$ is any (extremely slow-growing) function of n i.e. $\lg^* n$ (the o term in the space is a function of $f(n)$), while the runtime can be brought down to the optimal $O(m + n)$ time using $2n + o(n)$ bits.
  En route to this algorithm, we develop in Sect. 2,
- a data structure that maintains a set of elements from a universe of size n, say $[1..n]$ using $n + o(n)$ bits to support, apart from insert, search and delete operations, the operation *findany* of finding an arbitrary element of the set, and returning its value all in constant time. It can also output all elements of the set in no particular order in $O(k + 1)$ time where k is the number of elements currently belonging to the set.
  Our structure gives an explicit implementation, albeit for a weaker set of operations than that of Elmasry et al. [15] whose space requirement was $cn + o(n)$ bits for an unspecified constant $c > 2$; furthermore, our structure is simple and is sufficient to implement BFS space efficiently, improving by a constant factor of their BFS implementation keeping the running time same[1].

We could support the *findany* operation by keeping track of one of the elements, but once that element is deleted, we need to find another element to answer a subsequent *findany* query. This is easy to support in constant time if we have the elements stored in a linked list which takes $O(n \lg n)$ bits, or if we have a dynamic rank-select structure [20] where each operation takes $O(\frac{\lg n}{\lg \lg n})$ time.

In the same section we improve the space for BFS further at the cost of slightly increased runtime. We also provide a similar tradeoff for the minimum spanning tree problem. Our algorithm takes $n + O(n/f(n))$ bits and

---

[1] Since our initial submission to COCOON, Hagerup and Kammer [19] have reported a structure with $n + o(n)$ bits for the data structure and hence obtaining a similar bound as ours for BFS.

$O(m \lg n f(n))$ time, for any function $f(n)$ such that $1 \leq f(n) \leq n$. While this algorithm is similar in spirit to that of Elmasry et al. which works in $O(m \lg n)$ time using $O(n)$ bits or $O(m + n \lg n)$ time using $O(n \lg(2 + \frac{m}{n}))$ bits, we work out the constants in the higher order term for space, and improve them slightly though with a slight degration in time.

- Using our data structure, in Sect. 4 we develop another data structure to represent a sequence $x_1, x_2, \ldots x_n$ of n integers using $m + 2n + o(m + n)$ bits where $m = \sum_{i=1}^{n} x_i$. In this, we can determine whether the i-th element is 0 and if not, decrement it, all in constant time. In contrast, the data structure claimed (without proof) in [15] can even change (not just decrement) or access the elements, but in constant *amortized* time. However, their structure requires an $O(\lg n)$ limit on the x_i values while we pose no such restriction. Using this data structure in Sect. 4,
    - we determine whether a given directed graph is acyclic and give an implementation of topological sort of the graph if it is in $O(m + n)$ time and $O(m + n)$ bits of space. This improves an earlier bound of $O(m + n)$ time and $O(n \lg \lg n)$ space [15], and is more space efficient for sparse directed graphs (that includes those directed graphs whose underlying undirected graph is planar or has bounded treewidth or degeneracy).
    - A graph has a degeneracy d if every induced subgraph of the graph has a vertex with degree at most d (for example, planar graphs have degeneracy 5, and trees have degeneracy 1). An ordering $v_1, v_2, \ldots v_n$ of the vertices in such a graph is a degenerate order if for any i, the i-th vertex has degree at most d among vertices $v_{i+1}, v_{i+2}, \ldots v_n$. There are algorithms [8,16] that can find the degeneracy order in $O(m + n)$ time using $O(n)$ words. We show that, given a d, we can output the vertices of a d-degenerate graph in $O(m + n)$ time using $O(m + n)$ bits of space in the degeneracy order. We can even detect if the graph is d-degenerate in the process. As m is $O(nd)$, we have an $O(nd)$ bits algorithm which is more space efficient if d is $o(\lg n)$ (this is the case, for example, in planar graphs or trees).
- For DFS, we have two kinds of results improving on the result of Asano et al. [2] who showed that DFS in a directed or undirected graph can be performed in $O(m \lg n)$ time and $O(n)$ bits of space, and of Elmasry et al. [15] who improved the time to $O(m \lg \lg n)$ time still using $O(n)$ bits of space.
    - In Sect. 5, we first show that for sparse graphs (graphs where $m = O(n)$), we can perform DFS in linear time using $O(m + n)$ (i.e. $O(n)$ in sparse graphs) bits. Building on top of this encoding and other observations, we show how to efficiently compute the *chain decomposition* of a connected undirected graph. This lets us perform a variety of applications of DFS (including testing 2-vertex and 2-edge connectivity, finding cut vertices and edges, maximal 2-connected components and (open) ear decompositions) in the same time and space. Our algorithms for these applications improve the space requirement of all the previous algorithms from $\Theta(n \lg n)$ bits to $O(m + n)$ bits, preserving the same linear runtime.
    - Section 6 talks about applications of DFS using $O(n)$ bits. Using $O(n)$ bits of space, we show that

* we can compute the strongly connected components of a directed graph in $O(m \lg n \lg \lg n)$ time,
* we can output the vertices of a directed acyclic graph in a topologically sorted fashion in $O(m \lg \lg n)$ time, and
* we can find a sparse spanning biconnected subgraph of a biconnected undirected graph in $O(m \lg \lg n)$ time.

1.2   Model of Computation

We assume that the input graph is given in a read-only memory (and so cannot be modified). If an algorithm must do some outputting, this is done on a separate write-only memory. When something is written to this memory, the information can not be read or rewritten again. So the input is "read only" and the output is "write only". In addition to the input and the output media, a limited random-access workspace is available. The data on this workspace is manipulated wordwise as on the standard word RAM, where the machine consists of words of size w in $\Omega(\lg n)$ bits and any logical, arithmetic, and bitwise operations involving a constant number of words take a constant amount of time. We count space in terms of the number of bits used by the algorithms in workspace. This model is called the *register input model* and it was introduced by Frederickson [17] while studying some problems related to sorting and selection.

We assume that the input graphs are represented using the standard adjacency list throughout the paper. For the algorithms in Sect. 5 we require that the input graph must be represented using the standard adjacency list along with cross pointers, i.e. for undirected graphs given a vertex u and the position in its list of a neighbor v of u, there is a pointer to the position of u in the list of v. When we work with directed graphs, we assume that the graphs are represented as in and out adjacency lists i.e. given a vertex u, we have a list of out-neighbors and in-neighbors of u. We then augment these two lists for every vertex with cross pointers, i.e. for each $(u, v) \in E$, given u and the position of v in out-neighbors of u, there is a pointer to the position of u in in-neighbors of v. This representation was used by Elmasry et al. [15]. When discussing graph algorithms below, we always use n and m to denote the number of vertices and the number of edges respectively, in the input graph.

1.3   Related Work

In computational complexity theory, the constant work-space model is represented by the complexity class LOGSPACE [1]. There are several algorithmic results for this class, most celebrated being Reingold's method for checking reachability between two vertices in an undirected graph [24]. Barnes et al. gave a sub-linear space algorithm for directed graph reachability [7]. Recent work has focused on space requirement in special classes of graphs like planar and H-minor free graphs [3, 11]. In the algorithms literature, where the focus is also on improving time, a large amount of research has been devoted to memory constrained algorithms, even as early as in the 1980s [22]. Early work on this focused on the selection problem [17,22,23], but more recently on computational geometry problems [4,6,14] and graph algorithms [2,5,15]. Regarding the data structure we develop to support *findany* operation, Elmasry et al. [Lemma 2.1, [15]] state a data structure (without proof) that supports all the operations i.e. insert, search, delete and findany (they call it *some_id*) among others, in constant time. But their data structure takes $O(n)$ bits of space where the constant in the O term is not explicitly stated. Our data structure, on the other hand, is probably simpler and takes just $n + o(n)$ bits of space.[2]

1.4 Preliminaries

We will use the following well-known lemma:

Lemma 1. *A sequence of n integers in the range $\{1, \cdots, c\}$ where c is a constant, can be represented using $n \lg c + o(n)$ bits where the i-th integer can be accessed or modified in constant time.*

We also need the following theorem.

Theorem 1 [12,18,21]. *We can store a bitstring O of length n with additional $o(n)$ bits such that rank and select operations (defined below) can be supported in $O(1)$ time. Such a structure can also be constructed from the given bitstring in $O(n)$ time.*

Here the rank and select operations are defined as following:

- $rank_a(O, i) =$ number of occurrences of $a \in \{0, 1\}$ in $O[1, i]$, for $1 \leq i \leq n$;
- $select_a(O, i) =$ position in O of the ith occurrence of $a \in \{0, 1\}$.

2 Maintaining Dictionaries Under Findany Operation

We consider the data structure problem of maintaining a set S of elements from $\{1, 2, \ldots n\}$ to support the following operations in constant time.
insert (i): Insert element i into the set.
search (i): Determine whether the element i is in the set.
delete (i): Delete the element i from the set if it exists in the set.
findany: Find any element from the set and return its value. If the set is empty, return a NIL value.

It is trivial to support the first three operations in constant time using n bits. Our main result in this section is that the *findany* operation can also be supported in constant time using $o(n)$ additional bits.

Theorem 2. *A set of elements from a universe of size n can be maintained using $n + o(n)$ bits to support insert, delete, search and findany operations in constant time. We can also output all elements of the set (in no particular order) in $O(k + 1)$ time where k is the number of elements in the set.*

---

[2] Hagerup and Kammer [19] have recently reported a structure with $n + o(n)$ bits for the data structure supporting the same set of operations.

Proof. Let S be the characteristic bit vector of the set having n bits. We follow a two level blocking structure of S, as in the case of succinct structures supporting rank and select [12, 21]. However, as S is 'dynamic' (in that bit values can change due to insert and delete), we need more auxiliary information. In the discussion below, sometimes we omit floors and ceilings to keep the discussion simple, but they should be clear from the context.

We divide the bit vector S into $n/\lg^2 n$ blocks of consecutive $\lg^2 n$ bits each, and divide each such block into up to $2 \lg n$ small blocks of size $\lceil (\lg n)/2 \rceil$ bits each. We refer to the small blocks explicitly as *small* blocks, and by *blocks* we refer to the (big) blocks of size $\lg^2 n$ bits. We call a block (big or small) non-empty if it contains at least a 1. We maintain the non-empty (big) blocks, and the non-empty small blocks within each (big) block in linked lists (not necessarily in order). Within a small block, we find the first 1 or the next 1 by a table look up. We provide the specific details below.

First, we maintain an array *number* indicating the number of 1s in each block, i.e. *number*$[i]$ gives the number of 1s in the i-th block of S. It takes $O(n \lg \lg n/\lg^2 n)$ bits as each block can have at most $\lg^2 n$ elements of the given set. Then we maintain a queue (say implemented in a space efficient resizable array [10]) *block-queue* having the block numbers that have a 1 bit, and new block numbers are added to the list as and when new blocks get 1. It can have at most $n/\lg^2 n$ elements and so has $O(n/\lg^2 n)$ indices taking totally $O(n/\lg n)$ bits. In addition, every element in *block-queue* has a pointer to another queue of small block numbers of that block that have an element of S. Each such queue has at most $2 \lg n$ elements each of size at most $2 \lg \lg n$ bits each (for the small block index). Thus the queue *block-queue* along with the queues of small block indices takes $O(n \lg \lg n/\lg n)$ bits. We also maintain an array, *block-array*, of size $n/\lg^2 n$ where *block-array*$[i]$ points to the position of block i in *block-queue* if it exists, and is a NIL pointer otherwise and array, *small-block-array*, of size $2n/\lg n$ where *small-block-array*$[i]$ points to the position of the subblock i in its block's queue if its block was present in *block-queue*, and is a NIL pointer otherwise. So, *block-array* takes $n/\lg n$ bits and *small-block-array* takes $2n \lg \lg n/\lg n$ bits.

We also maintain a global table T precomputed that stores for every bitstring of size $\lceil (\lg n)/2 \rceil$, and a position i, the position of the first 1 bit after the i-th position. If there is no 'next 1', then the answer stored is -1 indicating a NIL value. The table takes $O(\sqrt{n}(\lg \lg n)^2)$ bits. This concludes the description of the data structure that takes $n + O(n \lg \lg n/\lg n)$ bits.

Now we explain how to support each of the required operations. Membership is the easiest, as it is a static operation, just look at the i-th bit of S and answer accordingly. In what follows, when we say the 'corresponding bit or pointer', we mean the bit or the pointer corresponding to the block or the small block corresponding to an element (inserted or deleted) which can be determined in constant time from the index of the element. To insert an element i, first determine from the table T, whether there is a 1 in the corresponding small block (before the element is inserted), set the i-th bit of S to 1, and increment the corresponding value in *number*. If the corresponding pointer of *block-array* was

NIL, then insert the block index to *block-queue* at the end of the queue, and add the small block corresponding to the i-th bit into the queue corresponding to the index of the block in *block-queue*, and update the corresponding pointers of *block-array* and *small-block-array*. If the corresponding bit of *block-array* was not NIL (the big block already had an element), and if the small block did not have an element before (as determined using T), then find the position of the block index in *block-queue* from *block-array*, and insert the small block index into the queue of that block at the end of the queue. Update the corresponding pointer of *small-block-array*.

To support the delete operation, set the i-th bit of S to 0 (if it was already 0, then there is nothing more to do) and decrement the corresponding number in *number*. Determine from the table T if the small block of i has a 1 (after the i-th bit has been set to 0). If not, then find the index of the small block from the arrays *block-array* and *small-block-array* and delete that index from the block's queue from *block-queue*. If the corresponding number in *number* remains more than 0, then there is nothing more to do. If the number becomes 0, then find the corresponding block index in *block-queue* from the array *block-array*, and delete that block (along with its queue that will have only one small block) from *block-queue*. Update the pointers in *block-array* and *small-block-array* respectively. As we don't maintain any order in the queues in *block-queue*, if we delete an intermediate element from the queue, we can always replace that element by the last element in the queue updating the pointers appropriately.

To support the findany operation, we go to the tail of the queue *block-queue*, if it is NIL, we report that there is no element in the set, and return the NIL value. Otherwise, go to the block at the tail of *block-queue*, and get the first (non-empty) small block number from the queue, and find the first element in the small block from the table T, and return the index of the element.

To output the elements of the set, we traverse the list *block-queue* and the queues of each element of *block-queue*, and for each small block in the queues, we find the next 1 in constant time using the table T and output the index. □

We can generalize to maintain a collection of more than one disjoint subsets of the given universe to support the insert, delete, membership and findany operations. In this case, insert, delete and findany operations should come with a set index (to be searched, inserted or deleted). Specifically, we show the following.

Theorem 3 (♠)[3]. *A collection of c disjoint sets that partition the universe of size n can be maintained using $n \lg c + o(n)$ bits to support insert, delete, search and findany operations in constant time, where c is a fixed constant. We can also output all elements of any given set (in no particular order) in $O(k + 1)$ time where k is the number of elements in the set.*

3 Breadth First Search

Following the observations of [15], the space efficient implementation follows using the data structure of Theorem 3. We explain the details for completeness.

---

[3] Proofs of results marked with (♠) will appear in full version.

Our goal is to output the vertices of the graph in the BFS order. We start as in the textbook BFS by coloring all vertices white. The algorithm grows the search starting at a vertex s, making it grey and adding it to a queue. Then the algorithm repeatedly removes the first element of the queue, and adds all its white neighbors at the end of the queue (coloring them grey), coloring the element black after removing it from the queue. As the queue can store up to $O(n)$ elements, the space for the queue can be $O(n \lg n)$ bits. To reduce the space to $O(n)$ bits, the two crucial observations on the properties of BFS are that: (i) elements in the queue are only from two consecutive levels of the BFS tree, and that the (ii) elements belonging to the same level can be processed in any order, but elements of the lower level must be processed before processing elements of the higher level.

The algorithm maintains four colors: white, grey0, grey1 and black, and represents the vertices with each of these colors as sets W, S_0, S_1 and B respectively using the data structure of Theorem 3. It starts with initializing S_0 (grey 0) to s, S_1 and B as empty sets and W to contain all other vertices. Then it processes the elements in each set S_0 and S_1 switching between the two until both sets are empty. As we process an element from S_i, we add its white neighbors to $S_{i+1 \bmod 2}$ and delete it from S_i and add it to B. When S_0 and S_1 become empty, we scan the W array to find the next white vertex and start a fresh BFS again from that vertex. As insert, delete, membership and findany operations take constant time, and we are maintaining four sets, we have from Theorem 3,

Theorem 4. *Given a directed or undirected graph, its vertices can be output in a BFS order starting at a vertex using $2n + o(n)$ bits in $O(m + n)$ time.*

Note that it is sufficient to build findany structures only on sets S_0 and S_1 to efficiently find grey vertices.

3.1 Improving the Space to $n \lg 3 + o(n)$ Bits

There are several ways to implement BFS using just two of the three colors used in the standard BFS [13], but the space restriction, hence our inability to maintain the standard queue, provides challenges.

We give a 3 color implementation overloading grey and black vertices, i.e. we use one color to represent grey and black vertices. Grey vertices remain grey even after processing. This poses the challenge of separating the grey vertices from the black ones correctly before exploring. We will have three colors, one (color 2) for the unexplored vertices and two colors (0 and 1) for those explored including those currently being explored. The two colors indicate the parity of the level (the distance from the starting vertex) of the explored vertices. Thus the starting vertex s is colored 0 to mark that its distance from s is of even length and every other vertex is colored 2 to mark them as unexplored (or white). We will have these values stored in the representation of Lemma 1 using $1.585n + o(n)$ bits and we call this as the color array. The algorithm repeatedly scans this array and in the i-th scan, it changes all the 2 neighbors of $i \bmod 2$ to $i + 1 \bmod 2$.

The exploration (of the connected component) stops when in two consecutive scans of the list, no 2 neighbor is found. Each scan list takes $O(m)$ time and at most $n+2$ scans of the list are performed resulting in an $O(mn)$ time algorithm.

The $O(m)$ time for each scan of the previous algorithm is because while looking for vertices labelled 0 that are supposed to be 'grey', we might cross over spurious vertices labelled 0 that are 'black' (in the normal BFS coloring). To improve the runtime further, we maintain two queues Q_0 and Q_1 each storing up to $n/\lg^2 n$ values of the grey 0 and grey 1 vertices We also store two boolean variables, *overflow-Q0, overflow-Q1*, initialized to 0 and to be set to 1 when more elements are to be added to these queues (but they don't have room). Now the algorithm proceeds in a similar fashion as the previous algorithm except that, along with marking corresponding vertices 0 or 1 in the color array, it also inserts them into the appropriate queues. i.e. when it expands vertices from Q_0 (Q_1), it inserts their (white) neighbors colored 2 to Q_1 (Q_0 respectively) apart from setting their color entries to 1 (0 respectively). When it runs out of space in any of these queues to insert the new elements (as we have limited only $n/\lg^2 n$ values in each of the queues), it continues to make the changes (i.e. 2 to 1 or 2 to 0) in the color array directly without adding those vertices to the queue, but set the corresponding overflow bit. Now instead of scanning the color array for vertices labelled 0 or 1, we traverse the appropriate queues spending time proportional to the sum of the degree of the vertices in the level. If the overflow bit in the corresponding queue is 0, then we simply move on to the next queue and continue. Otherwise, we switch to our previous algorithm and scan the array appropriately changing the colors of their white neighbors and adding them to the appropriate queue if possible. It is easy to see that this method correctly explores all the vertices of the graph using $1.585n + o(n)$ bits.

To analyse the runtime, notice that as long as the overflow bit of a queue is 0, we spend time proportional the number of neighbors of the vertices in that level, and we spend $O(m)$ time otherwise. When an overflow bit is 1, then the number of nodes in the level is at least $n/\lg^2 n$ and this can not happen for more than $\lg^2 n$ levels where we spend $O(m)$ time each. Hence, the total runtime is $O(m\lg^2 n)$ proving the following.

Theorem 5. *Given a directed or undirected graph, its vertices can be output in a BFS order starting at a vertex using $1.585n + o(n)$ bits and in $O(m\lg^2 n)$ time.*

By making the sizes of the two queues to $O(n/(f(n)\lg n))$ for any (slow growing) function $f(n)$, we obtain[4]

Theorem 6. *Given a directed or undirected graph, its vertices can be output in a BFS order starting at a vertex using $1.585n + O(n/f(n))$ bits and in $O(mf(n)\lg n)$ time where $f(n)$ is any slow-growing function of n.*

We do not know whether we can reduce the space to $n+o(n)$ bits while still maintaining the runtime to $O(m\lg^c n)$ for some constant c. However, we provide such an algorithm for the Minimum Spanning Tree problem to prove the following.

---

[4] Hagerup and Kammer [19] in their recent paper obtain a better time bound using the same space.

Theorem 7 (♠). *A minimum spanning forest of a given undirected weighted graph, where the weights of any edge can be represented in $O(\lg n)$ bits, can be found using $n + O(n/f(n))$ bits and in $O(m \lg n f(n))$ time, for any function $f(n)$ such that $1 \le f(n) \le n$.*

4   Applications of Findany Dictionary

In what follows we use our findany data structure of Sect. 2 to develop a data structure as below.

Theorem 8 (♠). *Let $x_1, x_2, \ldots x_n$ be a sequence of non-negative integers, and let $m = \sum_{i=1}^{n} x_i$. Then the sequence can be represented using at most $m + 2n + o(m+n)$ bits such that we can determine whether the i-th element of the sequence is 0 and decrement it otherwise, in constant time.*

Proof (sketch:). Encode each integer x_i in unary delimited by a separate bit. Treat the unary representation of x_i as a representation of the full subset of the universe of size x_i and apply our data structure of Theorem 3 for the decrement operation.   □

Using the data structure we just developed, we show the following theorems,

Theorem 9 (♠). *Given a directed acyclic graph G, its vertices can be output in topologically sorted order using $O(m + n)$ time using $m + 3n + o(n + m)$ bits of space. The algorithm can also detect if G is not acyclic.*

Theorem 10 (♠). *Given a d-degenerate graph G, its vertices can be output in d-degenerate order using $m + 3n + o(m + n)$ bits and $O(m + n)$ time. The algorithm can also detect if the given graph is not d-degenerate.*

5   DFS and Its Applications Using $O(m + n)$ Bits

In this section, we prove the following.

Theorem 11 (♠). *A DFS traversal of a directed or undirected graph G can be performed in $O(m + n)$ time using $O(m + n)$ bits. Using this, given a connected undirected graph G, in $O(m + n)$ time and $O(m + n)$ bits of space we can determine whether·G is 2-vertex (and/or edge) connected. If G is 2-edge (or vertex) connected, in the same time and space we can compute ear (open) decomposition. If not, in the same amount of time and space, we can compute all the bridges and cut vertices of the graph. Also, within same time and space bound, we can output 2-vertex (and edge) connected components.*

Proof (sketch:). We use the unary degree sequence encoding of the input graph G to store the DFS tree, and use succinct rank/select structure to navigate G in depth first manner especially to backtrack. By rerunning DFS and some bookkeeping, we compute the chain decomposition of G, and obtain space efficient implementation of the algorithm of Schmidt [25] for the applications of DFS mentioned in the theorem.   □

6 DFS and Its Applications Using $O(n)$ Bits

Building on Elmasry et al. [15] who gave space efficient implementation of DFS taking $O(n)$ bits and $O(m \lg \lg n)$ time, we provide the following space efficient implementations of some of the classical applications of DFS.

Theorem 12 (♠). *Using $O(n)$ bits, we can*

- *perform a topological sort of the vertices of a directed acyclic graph G in the same $O(m \lg \lg n)$ time,*
- *determine the strongly connected components of a directed graph in $O(m \lg n \lg \lg n)$ time, and*
- *find a sparse ($O(n)$ edges) spanning biconnected subgraph of an undirected biconnected graph in $O(m \lg \lg n)$ time.*

7 Conclusions and Open Problems

We end with the following interesting open problems.

- Can we perform BFS using $n + o(n)$ bits and $O(m \lg^c n)$ time for some constant c?
- Can we test 2-vertex (and/or edge) connectivity using $O(n)$ bits?
- Brandes [9] obtained an *s-t-numbering* from the ear decompostion of a graph, in linear time using $O(m+n)$ words. Can we improve the space bound to $O(n)$ bits or even $O(m + n)$ bits?

Acknowledgement. The authors thank Saket Saurabh for suggesting the exploration of algorithms using $O(m+n)$ bits for DFS and Anish Mukherjee for helpful discussions.

References

1. Arora, S., Barak, B.: Computational Complexity - A Modern Approach. Cambridge University Press, Cambridge (2009)
2. Asano, T., et al.: Depth-first search using O(n) bits. In: Ahn, H.-K., Shin, C.-S. (eds.) ISAAC 2014. LNCS, vol. 8889, pp. 553–564. Springer, Heidelberg (2014)
3. Asano, T., Kirkpatrick, D., Nakagawa, K., Watanabe, O.: $\widetilde{O}(\sqrt{n})$-space and polynomial-time algorithm for planar directed graph reachability. In: Ésik, Z., Csuhaj-Varjú, E., Dietzfelbinger, M. (eds.) MFCS 2014, Part II. LNCS, vol. 8635, pp. 45–56. Springer, Heidelberg (2014)
4. Asano, T., Mulzer, W., Rote, G., Wang, Y.: Constant-work-space algorithms for geometric problems. JoCG **2**(1), 46–68 (2011)
5. Banerjee, N., Chakraborty, S., Raman, V., Roy, S., Saurabh, S.: Time-space trade-offs for dynamic programming algorithms in trees and bounded treewidth graphs. In: Xu, D., Du, D., Du, D. (eds.) COCOON 2015. LNCS, vol. 9198, pp. 349–360. Springer, Heidelberg (2015)
6. Barba, L., Korman, M., Langerman, S., Sadakane, K., Silveira, R.I.: Space-time trade-offs for stack-based algorithms. Algorithmica **72**(4), 1097–1129 (2015)

7. Barnes, G., Buss, J., Ruzzo, W., Schieber, B.: A sublinear space, polynomial time algorithm for directed s-t connectivity. SICOMP **27**(5), 1273–1282 (1998)
8. Batagelj, V., Zaversnik, M.: An $O(m)$ algorithm for cores decomposition of networks. CoRR cs.DS/0310049 (2003)
9. Brandes, U.: Eager st-ordering. In: Möhring, R.H., Raman, R. (eds.) ESA 2002. LNCS, vol. 2461, pp. 247–256. Springer, Heidelberg (2002)
10. Brodnik, A., Carlsson, S., Demaine, E.D., Munro, J.I., Sedgewick, R.D.: Resizable arrays in optimal time and space. In: Dehne, F., Gupta, A., Sack, J.-R., Tamassia, R. (eds.) WADS 1999. LNCS, vol. 1663, pp. 37–48. Springer, Heidelberg (1999)
11. Chakraborty, D., Pavan, A., Tewari, R., Vinodchandran, N.V., Yang, L.: New time-space upperbounds for directed reachability in high-genus and H-minor-free graphs. In: FSTTCS, pp. 585–595 (2014)
12. Clark, D.: Compact pat trees. Ph.D. thesis, University of Waterloo, Canada (1996)
13. Cormen, T.H., Leiserson, C.E., Rivest, R.L., Stein, C.: Introduction to Algorithms, 3rd edn. MIT Press, Cambridge (2009)
14. Darwish, O., Elmasry, A.: Optimal time-space tradeoff for the 2D convex-hull problem. In: Schulz, A.S., Wagner, D. (eds.) ESA 2014. LNCS, vol. 8737, pp. 284–295. Springer, Heidelberg (2014)
15. Elmasry, A., Hagerup, T., Kammer, F.: Space-efficient basic graph algorithms. In: 32nd STACS, pp. 288–301 (2015)
16. Eppstein, D., Loffler, M., Strash, D.: Listing all maximal cliques in large sparse real-world graphs. ACM J. Exp. Algorithmics **18**, 1–3 (2013)
17. Frederickson, G.N.: Upper bounds for time-space trade-offs in sorting and selection. J. Comput. Syst. Sci. **34**(1), 19–26 (1987)
18. Gupta, A., Hon, W.-K., Shah, R., Vitter, J.S.: A framework for dynamizing succinct data structures. In: Arge, L., Cachin, C., Jurdziński, T., Tarlecki, A. (eds.) ICALP 2007. LNCS, vol. 4596, pp. 521–532. Springer, Heidelberg (2007)
19. Hagerup, T., Kammer, F.: Succinct choice dictionaries. CoRR abs/1604.06058 (2016)
20. Hon, W.K., Sadakane, K., Sung, W.K.: Succinct data structures for searchable partial sums with optimal worst-case performance. Theor. Comput. Sci. **412**(39), 5176–5186 (2011)
21. Munro, J.I.: Tables. In: Chandru, V., Vinay, V. (eds.) FSTTCS 1996. LNCS, vol. 1180, pp. 37–42. Springer, Heidelberg (1996)
22. Munro, J.I., Paterson, M.: Selection and sorting with limited storage. Theor. Comput. Sci. **12**, 315–323 (1980)
23. Munro, J.I., Raman, V.: Selection from read-only memory and sorting with minimum data movement. Theor. Comput. Sci. **165**(2), 311–323 (1996)
24. Reingold, O.: Undirected connectivity in log-space. J. ACM **55**(4), 1–17 (2008)
25. Schmidt, J.M.: A simple test on 2-vertex- and 2-edge-connectivity. Inf. Process. Lett. **113**(7), 241–244 (2013)

Metric 1-Median Selection: Query Complexity vs. Approximation Ratio

Ching-Lueh Chang[(✉)]

Department of Computer Science and Engineering,
Yuan Ze University, Taoyuan, Taiwan
clchang@saturn.yzu.edu.tw

Abstract. Consider the problem of finding a point in a metric space $(\{1, 2, \ldots, n\}, d)$ with the minimum average distance to other points. We show that this problem has no deterministic $o(n^{1+1/(h-1)})$-query $(2h-\epsilon)$-approximation algorithms for any constants $h \in \mathbb{Z}^+ \setminus \{1\}$ and $\epsilon > 0$.

1 Introduction

The METRIC 1-MEDIAN problem asks for a point in an n-point metric space with the minimum average distance to other points. It has a Monte-Carlo $O(n/\epsilon^2)$-time $(1 + \epsilon)$-approximation algorithm for all $\epsilon > 0$ [8,9]. In $\mathbb{R}^D$, Kumar et al. [11] give a Monte-Carlo $O(2^{\text{poly}(1/\epsilon)}D)$-time $(1 + \epsilon)$-approximation algorithm for 1-median selection and another algorithm for k-median selection, where $D \geq 1$ and $\epsilon > 0$. Any Monte-Carlo $O(1)$-approximation algorithm for metric k-median selection takes $\Omega(nk)$ time [7,12]. Algorithms for k-median and k-means clustering abound [1,6,7,9,10,12].

Chang [4], Wu [13] and Chang [2] show that METRIC 1-MEDIAN has a deterministic nonadaptive $O(n^{1+1/h})$-time $(2h)$-approximation algorithm for all constants $h \in \mathbb{Z}^+ \setminus \{1\}$. Furthermore, Chang [5] shows the nonexistence of deterministic $o(n^2)$-query $(4-\Omega(1))$-approximation algorithms for METRIC 1-MEDIAN. This paper generalizes his result to show that METRIC 1-MEDIAN has no deterministic $o(n^{1+1/(h-1)})$-query $(2h-\epsilon)$-approximation algorithms for any constants $h \in \mathbb{Z}^+ \setminus \{1\}$ and $\epsilon > 0$.

As in the previous lower bounds for deterministic algorithms [3,5], we use an adversarial method. Our proof proceeds as follows:

(i) Design an adversary Adv for answering the distance queries of any deterministic algorithm A with query complexity $q(n) = o(n^{1+1/(h-1)})$.
(ii) Show that A's output, denoted z, has a large average distance to other points, according to Adv's answers to A.
(iii) Construct a distance function with respect to which a certain point $\hat{\alpha}$ has a small average distance to other points.

C.-L. Chang—Supported in part by the Ministry of Science and Technology of Taiwan under grant 103-2221-E-155-026-MY2.

T.N. Dinh and M.T. Thai (Eds.): COCOON 2016, LNCS 9797, pp. 131–142, 2016.
DOI: 10.1007/978-3-319-42634-1_11

(iv) Construct the final distance function $d(\cdot,\cdot)$ similar to and bounded by that in item (iii).
(v) Show that $d(\cdot,\cdot)$ is a metric.
(vi) Show the consistency of $d(\cdot,\cdot)$ with Adv's answers.
(vii) Compare $\hat{\alpha}$ in item (iii) with z in item (ii) to establish our lower bound on A's approximation ratio.

Chang [3] does item (ii) by answering the distance between two distinct points as 2 if they both involve in only a few queries and as 3 otherwise. For a stronger lower bound, he expands the range of answers from $\{2,3\}$ to $\{2,3,4\}$ [5]. In contrast, our high-level idea for item (ii) is to answer the queries according to the distances on a graph with small degrees only (note that every vertex in such a graph has a large average distance to other points).

All our constructions and analyses are built on two novel graph sequences, $\{H^{(i)}\}_{i=0}^{q(n)}$ and $\{G^{(i)}\}_{i=0}^{q(n)}$, in Sec. 3. But at a high level, we share the following paradigm for item (iii) with Chang [5]:

- Keep a small set S of points whose distances to other points are answered by Adv as large values during A's execution.
- Then set a point $\hat{\alpha} \in S$ involved in only a few queries to have a small average distance to other points.

Below is the rationale of this paradigm: If Adv answers the $\hat{\alpha}$-v distance as a small value for some point v with a large average distance to other points, then the average distance from $\hat{\alpha}$ to other points will have to be large by the triangle inequality, a bad news for item (iii). So we want Adv to answer as large values the distances from $\hat{\alpha}$ to other points during A's execution.

The exact constructions in items (iii)–(iv) combine $G^{(q(n))}$ and $H^{(q(n))}$ with the planting of small $\hat{\alpha}$-v distances for many points v in a rather technical way. Their careful design eases the remaining items.

Sects. 3.1, 3.2 and 3.3 correspond to items (ii), (iii) and (iv)–(vii), respectively. The full version of this paper is at http://arxiv.org/abs/1509.05662.

2   Definitions and Preliminaries

For $n \in \mathbb{N}$, $[n] \equiv \{1,2,\ldots,n\}$. An algorithm A is c-approximate for METRIC 1-MEDIAN if $A^d(1^n)$ outputs a c-approximate 1-median of $([n], d)$ for each finite metric space $([n], d)$, where $c \geq 1$. For simplicity, abbreviate $A^d(1^n)$ as A^d.

Fact 1 [2,4,13]. *For each constant $h \in \mathbb{Z}^+ \setminus \{1\}$, METRIC 1-MEDIAN has a deterministic nonadaptive $O(n^{1+1/h})$-time $(2h)$-approximation algorithm.*

For a predicate P, let $\chi[P] = 1$ if P is true and $\chi[P] = 0$ otherwise. A weighted undirected graph $G = (V, E, w)$ has a finite vertex set V, an edge set E and a weight function $w \colon E \to (0,\infty)$. When the domain of w is a superset of E, interpret (V, E, w) simply as $(V, E, w|_E)$, where $w|_E$ denotes the restriction of w on E. For all $S \subseteq V$, $N_G(S) \equiv \bigcup_{v \in S} N_G(v)$. For all $s, t \in V$, the shortest s-t distance in G, denoted $d_G(s,t)$, is the infimum of the weights (w.r.t. w) of all s-t paths in G. So $d_G(s,t) = \infty$ if there are no s-t paths.

3   Query Complexity vs Approximation Ratio

Throughout this section, (1) $n \in \mathbb{Z}^+$, (2) $\delta \in (0,1)$ and $h \in \mathbb{Z}^+ \setminus \{1\}$ are constants (i.e., they are independent of n), (3) A is a deterministic $o(n^{1+1/(h-1)})$-query algorithm for METRIC 1-MEDIAN, and (4) $S = [\lfloor \delta n \rfloor] \subseteq [n]$.

All pairs in $[n]^2$ are assumed to be unordered in this section. By padding queries, assume W.L.O.G. that A will have queried for the distances between its output and all other points when halting. Denote A's (worst-case) query complexity by

$$q(n) = o\left(n^{1+1/(h-1)}\right).$$

By padding queries, assume the number of queries of A to be exactly $q(n)$. W.L.O.G., forbid to make the same query twice or to query for the distance from a point to itself, where the queries for $d(x,y)$ and $d(y,x)$ are considered to be the same for $x, y \in [n]$. Furthermore, let n be sufficiently large to satisfy

$$\delta n^{1/(h-1)} > 3, \tag{1}$$

$$\frac{2q(n)}{|S| - 1} \leq \delta n^{1/(h-1)}. \tag{2}$$

By (2),

$$q(n) \leq \delta n^{1+1/(h-1)}. \tag{3}$$

Define two unweighted undirected graphs $G^{(0)}$ and $H^{(0)}$ by

$$E_G^{(0)} \equiv \{(u,v) \mid (u,v \in [n] \setminus S) \wedge (u \neq v)\}, \tag{4}$$

$$G^{(0)} \equiv \left([n], E_G^{(0)}\right), \tag{5}$$

$$E_H^{(0)} \equiv \emptyset, \tag{6}$$

$$H^{(0)} \equiv \left([n], E_H^{(0)}\right). \tag{7}$$

Algorithm Adv in Fig. 1 answers A's queries. In particular, for all $i \in [q(n)]$, the ith iteration of the loop of Adv answers the ith query of A, denoted $(a_i, b_i) \in [n]^2$. It constructs three unweighted undirected graphs, $G^{(i)} = ([n], E_G^{(i)})$, $H^{(i)} = ([n], E_H^{(i)})$ and $Q^{(i)}$. As $G^{(i-1)}$ is unweighted for all $i \in [q(n)]$, P_i in line 5 of Adv is an a_i-b_i path in $G^{(i-1)}$ with the minimum number of edges. By line 16 of Adv, the edges of $Q^{(i)}$ are precisely the first i queries of A.

An Intuitive Exposition. Line 17 of Adv in Fig. 1 answers the ith query of A according to $H^{(i)}$. So, to make the output of A have a large average distance to other points, we want $H^{(\cdot)}$ to have small degrees only. For this purpose, line 8 forms $G^{(i)}$ by removing the edges having a large-degree endpoint in $H^{(i)}$. Once an edge is absent in $G^{(i)}$, it cannot be inserted into $H^{(i+1)}$ by lines 5–6 of the next iteration, thus keeping the degrees small in $H^{(\cdot)}$. Lines 5–6 and 17 suggest that the answer to the ith query of A is just the length of P_i. So Adv should "remember"

```
1: Let E_G^{(0)}, G^{(0)}, E_H^{(0)} and H^{(0)} be as in (4)-(7);
2: for i = 1, 2, ..., q(n) do
3:     Receive the ith query of A, denoted (a_i, b_i);
4:     if d_{G^{(i-1)}}(a_i, b_i) ≤ h then
5:         Find a shortest a_i-b_i path P_i in G^{(i-1)};
6:         E_H^{(i)} ← E_H^{(i-1)} ∪ {e | e is an edge on P_i};
7:         H^{(i)} ← ([n], E_H^{(i)});
8:         E_G^{(i)} ← E_G^{(i-1)} \ {(u,v) ∈ E_G^{(i-1)} \ E_H^{(i)} | (deg_{H^{(i)}}(u) ≥ δn^{1/(h-1)} - 2) ∨
           (deg_{H^{(i)}}(v) ≥ δn^{1/(h-1)} - 2)};
9:         G^{(i)} ← ([n], E_G^{(i)});
10:    else
11:        E_H^{(i)} ← E_H^{(i-1)};
12:        H^{(i)} ← ([n], E_H^{(i)});
13:        E_G^{(i)} ← E_G^{(i-1)};
14:        G^{(i)} ← ([n], E_G^{(i)});
15:    end if
16:    Q^{(i)} ← ([n], {(a_j, b_j) | j ∈ [i]});
17:    Output min{d_{H^{(i)}}(a_i, b_i), h - (1/2) · χ[∃v ∈ {a_i, b_i}, (v ∈ S) ∧ (deg_{Q^{(i)}}(v) ≤
         δn^{1/(h-1)})]} as the answer to the ith query of A;
18: end for
```

Fig. 1. Algorithm Adv for answering A's queries

P_i *to be able to answer the future queries consistently with its answer to the ith query. This is why line 8 preserves all the edges in* $E_H^{(i)}$ *(including those of* P_i *by line 6) when forming* $G^{(i)}$.

Roughly, Adv *works as follows: Answer each query by the length of a shortest path. Mark the edges of that path by adding them to* $H^{(\cdot)}$. *Once a vertex is incident to too many marked edges, remove its incident unmarked edges to keep its degree small in* $H^{(\cdot)}$. *Preserve all marked edges for consistency among answers.*

Lemma 1.

$$E_H^{(0)} \subseteq E_H^{(1)} \subseteq \cdots \subseteq E_H^{(q(n))} \subseteq E_G^{(q(n))} \subseteq E_G^{(q(n)-1)} \subseteq \cdots \subseteq E_G^{(0)}.$$

Proof. By lines 6 and 11 of Adv in Fig. 1, $E_H^{(i-1)} \subseteq E_H^{(i)}$ for all $i \in [q(n)]$. By lines 8 and 13, $E_G^{(i)} \subseteq E_G^{(i-1)}$ for all $i \in [q(n)]$.

To show that $E_H^{(q(n))} \subseteq E_G^{(q(n))}$, we shall prove the stronger statement that $E_H^{(i)} \subseteq E_G^{(i)}$ for all $i \in \{0, 1, \ldots, q(n)\}$ by mathematical induction. By (6), $E_H^{(0)} \subseteq E_G^{(0)}$. Assume as the induction hypothesis that $E_H^{(i-1)} \subseteq E_G^{(i-1)}$. The following shows that $E_H^{(i)} \subseteq E_G^{(i-1)}$ by examining each $e \in E_H^{(i)}$:

Case 1: $e \in E_H^{(i-1)}$. By the induction hypothesis, $e \in E_G^{(i-1)}$.

Case 2: $e \notin E_H^{(i-1)}$. As $e \in E_H^{(i)} \setminus E_H^{(i-1)}$, lines 6 and 11 show that e is on P_i (and that the ith iteration of the loop of Adv runs line 6 rather than line 11). By line 5, each edge on P_i is in $E_G^{(i-1)}$. In particular, $e \in E_G^{(i-1)}$.

Having shown that $E_H^{(i)} \subseteq E_G^{(i-1)}$, lines 8 and 13 will both result in $E_H^{(i)} \subseteq E_G^{(i)}$, completing the induction step. $\qquad\square$

Lemma 2. *For all $i \in [q(n)]$ with $d_{G^{(i-1)}}(a_i, b_i) \leq h$,*

$$d_{H^{(i)}}(a_i, b_i) = d_{H^{(q(n))}}(a_i, b_i) = d_{G^{(q(n))}}(a_i, b_i) = d_{G^{(i-1)}}(a_i, b_i).$$

Proof. By line 4 of Adv, the ith iteration of the loop runs lines 5–9. Lines 5–7 put (the edges of) a shortest a_i-b_i path in $G^{(i-1)}$ into $H^{(i)}$; hence

$$d_{H^{(i)}}(a_i, b_i) \leq d_{G^{(i-1)}}(a_i, b_i).$$

This and Lemma 1 complete the proof. $\qquad\square$

3.1   The Average Distance from A's Output to Other Points

This subsection shows that the output of A^{Adv} has a large average distance to other points, according to the answers of Adv.

Lemma 3. *For all $i \in [q(n)]$ and $v \in [n]$, we have $deg_{H^{(i)}}(v) \leq deg_{H^{(i-1)}}(v) + 2$.*

Proof. If the ith iteration of the loop of Adv runs lines 11–14 but not 5–9, then $H^{(i)} = H^{(i-1)}$, proving the lemma. So assume otherwise. Being shortest, P_i in line 5 does not repeat vertices. Therefore, v is incident to at most two edges on P_i, which together with lines 6–7 complete the proof. $\qquad\square$

Lemma 4. *For all $v \in [n]$, $deg_{H^{(q(n))}}(v) < \delta n^{1/(h-1)}$.*

Proof. Assume otherwise. Then prove the existence of $i \in [q(n)]$ satisfying

$$\deg_{H^{(i-1)}}(v) < \delta n^{1/(h-1)} - 2, \tag{8}$$
$$\deg_{H^{(i)}}(v) \geq \delta n^{1/(h-1)} - 2. \tag{9}$$

As $H^{(i-1)} \neq H^{(i)}$ by (8)–(9), the ith iteration of the loop of Adv runs lines 5–9 but not 11–14. By (9) and line 8 of Adv,

$$\left\{ u \in [n] \mid (u,v) \in E_G^{(i)} \right\} = \left\{ u \in [n] \mid (u,v) \in E_G^{(i-1)} \setminus \left(E_G^{(i-1)} \setminus E_H^{(i)} \right) \right\}. \tag{10}$$

By Lemma 1, $E_H^{(i)} \subseteq E_G^{(i-1)}$. So by (10), $N_{G^{(i)}}(v) = N_{H^{(i)}}(v)$, implying $\deg_{G^{(i)}}(v) = \deg_{H^{(i)}}(v)$. By (8) and Lemma 3, $\deg_{H^{(i)}}(v) < \delta n^{1/(h-1)}$. Finally, $\deg_{H^{(q(n))}}(v) \leq \deg_{G^{(i)}}(v)$ by Lemma 1. Summarize the above. $\qquad\square$

Lemma 5. *For all $v \in [n]$,*

$$|\{u \in [n] \mid d_{H^{(q(n))}}(v, u) < h\}| \leq 2\delta^{h-1} n.$$

Proof. Use Lemma 4. $\qquad\square$

Denote the output of A^{Adv} by z. Furthermore,

$$I \equiv \{ j \in [q(n)] \mid z \in \{a_j, b_j\} \}. \tag{11}$$

The following lemma analyzes the sum of the distances, as answered by line 17 of Adv, from z to other points.

Lemma 6.

$$\sum_{i \in I} \min \left\{ d_{H^{(i)}}(a_i, b_i), h - \frac{1}{2} \cdot \chi \left[\exists v \in \{a_i, b_i\}, (v \in S) \wedge \left(deg_{Q^{(i)}}(v) \leq \delta n^{1/(h-1)} \right) \right] \right\}$$
$$\geq n \cdot \left(h - 2h\delta^{h-1} - o(1) - \delta \right).$$

Proof. By Lemma 1, $d_{H^{(i)}}(a_i, b_i) \geq d_{H^{(q(n))}}(a_i, b_i)$ for all $i \in [q(n)]$. Now use Lemma 5 to bound $\sum_{i \in I} d_{H^{(q(n))}}(a_i, b_i)$ from below. The rest is not hard.   □

3.2   Planting a Point with a Small Average Distance to Other Points

This subsection constructs a distance function with respect to which a certain point has an average distance of approximately $1/2$ to other points.

Lemma 7. $|E_H^{(q(n))}| \leq h \cdot q(n)$.

Proof. By lines 4–5 of Adv, P_i in line 5 has at most h edges. So by lines 6 and 11, $|E_H^{(i)}| \leq |E_H^{(i-1)}| + h$. Finally, there are $q(n)$ queries.   □

Lemma 8.

$$\left| \left\{ u \in [n] \mid deg_{H^{(q(n))}}(u) \geq \delta n^{1/(h-1)} - 2 \right\} \right| = \frac{h}{\delta} \cdot o(n).$$

Proof. By Lemma 7, the average degree in $H^{(q(n))}$ is at most $2h \cdot q(n)/n = h \cdot o(n^{1/(h-1)})$. Use the averaging argument.   □

By (1), $S \setminus \{z\} \neq \emptyset$. Define

$$\hat{\alpha} \equiv \underset{\alpha \in S \setminus \{z\}}{\operatorname{argmin}} \ deg_{Q^{(q(n))}}(\alpha), \tag{12}$$

breaking ties arbitrarily.

Lemma 9. *For all* $i \in [q(n)]$, $deg_{Q^{(i)}}(\hat{\alpha}) \leq \delta n^{1/(h-1)}$.

Proof. Use (12) and the averaging argument.   □

Inductively, let

$$V_0 \equiv \{\hat{\alpha}\}, \tag{13}$$
$$V_1 \equiv N_{Q^{(q(n))}}(\hat{\alpha}) \setminus V_0, \tag{14}$$
$$V_{j+1} \equiv N_{H^{(q(n))}}(V_j) \setminus \left(\bigcup_{i=0}^{j} V_i \right) \tag{15}$$

for all $j \in [h-2]$. Furthermore,

$$V_h \equiv [n] \setminus \left(\bigcup_{i=0}^{h-1} V_i \right). \tag{16}$$

The following lemma is not hard to see from (13)–(16).

Lemma 10. $(V_0, V_1, \ldots, V_h)$ is a partition of $[n]$, i.e., $\bigcup_{k=0}^{h} V_k = [n]$ and $V_i \cap V_j = \emptyset$ for all distinct $i, j \in \{0, 1, \ldots, h\}$.

An Intuitive Exposition. By Lemma 9, $|V_1|$ is small. Because we have seen that $H^{(q(n))}$ has small degrees only, $|V_j|$ grows slowly as j increases from 1 to $h-1$. Consequently, $\sum_{j=1}^{h-1} |V_j|$ should be small. In fact, $|V_h| \approx n$. So if we connect $\hat{\alpha}$ to each point in V_h by an edge of weight $1/2$, then $\hat{\alpha}$ will have an average distance of approximately $1/2$ to other points. Technicalities complicate the exact constructions, though.

Define

$$B \equiv \left\{ u \in [n] \mid \deg_{H^{(q(n))}}(u) \geq \delta n^{1/(h-1)} - 2 \right\}, \tag{17}$$

$$\mathcal{E} \equiv \left[E_G^{(q(n))} \setminus \left(\bigcup_{i,j \in \{0,1,\ldots,h\}, |i-j| \geq 2} V_i \times V_j \right) \right] \cup (\{\hat{\alpha}\} \times (V_h \setminus (B \cup S))). \tag{18}$$

By (12), $\hat{\alpha} \in S$, implying $\hat{\alpha} \notin V_h \setminus (B \cup S)$. So $\mathcal{E}$ does not contain a self-loop. For all distinct $u, v \in [n]$,

$$w(u,v) \equiv \begin{cases} 1/2, & \text{if one of } u \text{ and } v \text{ is } \hat{\alpha} \text{ and the other is in } V_h \setminus (B \cup S), \\ 1, & \text{otherwise.} \end{cases} \tag{19}$$

Furthermore, define

$$\mathcal{G} \equiv ([n], \mathcal{E}, w) \tag{20}$$

to be a weighted undirected graph.

Lemma 11.

$$\sum_{j=1}^{h-1} |V_j| \leq 2\delta^{h-1} n.$$

Proof. By Lemma 9 and (14), we have $|V_1| \leq \delta n^{1/(h-1)}$. By Lemma 4 and (15), $|V_{j+1}| \leq |V_j| \cdot \delta n^{1/(h-1)}$ for $j \in [h-2]$. Now bound $\sum_{j=1}^{h-1} |V_j|$ by a geometric series.   □

Lemma 12.

$$|V_h \setminus (B \cup S)| \geq n \left(1 - 2\delta^{h-1} - \frac{h}{\delta} \cdot o(1) - \delta \right).$$

Proof. By Lemma 8 and (17), $|B| = (h/\delta) \cdot o(n)$. Furthermore,

$$|V_h| \overset{\text{Lemmas 10–11}}{\geq} n - 2\delta^{h-1}n - |V_0| \overset{(13)}{=} n - 2\delta^{h-1}n - 1.$$

$\square$

The following lemma says that $\hat{\alpha}$ has an average distance of approximately $1/2$ to other points w.r.t. the distance function $\min\{d_{\mathcal{G}}(\cdot, \cdot), h\}$.

Lemma 13.

$$\sum_{v \in [n]} \min\{d_{\mathcal{G}}(\hat{\alpha}, v), h\} \leq n \cdot \left(\frac{1}{2} + 2h\delta^{h-1} + \frac{h^2}{\delta} \cdot o(1) + h\delta\right).$$

Proof. By (18)–(20), $d_{\mathcal{G}}(\hat{\alpha}, v) \leq 1/2$ for all $v \in V_h \setminus (B \cup S)$. This and Lemma 12 complete the proof (note that $\min\{d_{\mathcal{G}}(\cdot, \cdot), h\} \leq h$). $\square$

3.3   A Metric Consistent with Adv's answers

This subsection constructs a metric $d \colon [n]^2 \to [0, \infty)$ consistent with Adv's answers in line 17. So Lemma 6 will require z, which is the output of A^{Adv}, to have an average distance (w.r.t. d) of at least approximately h to other points. Although $d(\cdot, \cdot)$ will not be exactly $\min\{d_{\mathcal{G}}(\cdot, \cdot), h\}$, Lemma 13 will forbid $\sum_{v \in [n]} d(\hat{\alpha}, v)/n$ to exceed $1/2$ by too much. Comparing z with $\hat{\alpha}$ yields our lower bound.

An Intuitive Exposition. *Suppose that A ever queries for the $\hat{\alpha}$-v_1, v_1-v_2 and v_2-v_3 distances, where $v_1 \in [n]$, $v_2, v_3 \in V_h \setminus (B \cup S)$ and $v_2 \neq v_3$. By (14), $v_1 \in V_1$. All of Adv's answers are clearly based on $\{H^{(i)}\}_{i=0}^{q(n)}$ and $\{G^{(i)}\}_{i=0}^{q(n)}$, which, unlike $\mathcal{G}$, do not have edges of weight $1/2$. So for variants of $\min\{d_{\mathcal{G}}(\cdot, \cdot), h\}$ to be used as the final metric, we need to prevent the edges of $\mathcal{G}$ with weight $1/2$ from creating "shortcuts" that, together with the triangle inequality, violate Adv's answers.*

By (19), $\min\{d_{\mathcal{G}}(\hat{\alpha}, v_2), h\} = 1/2$. So for $\min\{d_{\mathcal{G}}(\cdot, \cdot), h\}$ to be consistent with Adv's answers, the $\hat{\alpha}$-v_1 and v_1-v_2 distances returned by Adv must differ by at most $1/2$ in absolute value by the triangle inequality. From the picking of $\hat{\alpha}$, it will be easy to show that Adv answers the $\hat{\alpha}$-v_1 distance with $h - 1/2$. Consequently, the v_1-v_2 distance returned by Adv should be at least $(h - 1/2) - 1/2 = h - 1$. As $v_1 \in V_1$ and $v_2 \in V_h$, we turn to prove that every point in V_1 has a distance of at least $h - 1$ to every point in V_h.

Again by (19), $\min\{d_{\mathcal{G}}(\hat{\alpha}, v_j), h\} = 1/2$ for all $j \in \{2, 3\}$. So for $\min\{d_{\mathcal{G}}(\cdot, \cdot), h\}$ to be consistent with Adv's answers, the v_2-v_3 distance returned by Adv must not exceed $1/2 + 1/2 = 1$ by the triangle inequality. For this purpose, we just need to prove the existence of a v_2-v_3 edge or, more generally, an edge between any two distinct points in $[n] \setminus (B \cup S)$.

The above descriptions are somewhat inaccurate. E.g., the final metric is not exactly $\min\{d_{\mathcal{G}}(\cdot, \cdot), h\}$.

Recall that $H^{(i)}$ and $G^{(i)}$ are unweighted for all $i \in \{0, 1, \ldots, q(n)\}$. They can be treated as having the weight function w while preserving $d_{H^{(i)}}(\cdot, \cdot)$ and $d_{G^{(i)}}(\cdot, \cdot)$, as shown by the lemma below.

Lemma 14. *For all* $i \in \{0, 1, \ldots, q(n)\}$, *each path* P *in* $H^{(i)}$ *or* $G^{(i)}$ *has exactly* $w(P)$ *edges.*

Proof. As $\hat{\alpha} \in S$ by (12), (19) implies $w(u, v) = 1$ for all distinct u, $v \in [n] \setminus S$. This and (4) imply that all edges in $E_G^{(0)}$ have weight 1 w.r.t. w. So by Lemma 1, the edges in $E_H^{(i)} \cup E_G^{(i)}$ have weight 1 w.r.t. w. $\qquad\square$

Lemma 15.

$$E_H^{(q(n))} \cap \left(\bigcup_{i,j \in \{0,1,\ldots,h\}, |i-j| \geq 2} V_i \times V_j \right) = \emptyset.$$

Proof. Use (15). $\qquad\square$

Lemma 16. $E_H^{(q(n))} \subseteq \mathcal{E}.$

Proof. By Lemma 15 and (18), $E_G^{(q(n))} \cap E_H^{(q(n))} \subseteq \mathcal{E}$. Now invoke Lemma 1. $\square$

Lemma 17. *Let* P *be a path in* $\mathcal{G}$ *that visits no edges in* $\{\hat{\alpha}\} \times (V_h \setminus (B \cup S))$. *If the first and the last vertices of* P *are in* V_h *and* V_1, *respectively, then* $w(P) \geq h - 1$.

Proof. Because P visits no edges in $\{\hat{\alpha}\} \times (V_h \setminus (B \cup S))$, no edges on P are in $V_i \times V_j$ for any i, $j \in \{0, 1, \ldots, h\}$ with $|i - j| \geq 2$ by (18) and (20). This forces P to visit at least one edge in $V_{i+1} \times V_i$ for each $i \in [h - 1]$. As $\hat{\alpha} \notin \bigcup_{i=1}^{h} V_i$ by (13)–(16), (19) gives $w(u, v) = 1$ for all $(u, v) \in \bigcup_{i=1}^{h-1} (V_{i+1} \times V_i)$. $\qquad\square$

Lemma 18. *Let* P *be a shortest* a_i-b_i *path in* $\mathcal{G}$, *where* $i \in [q(n)]$. *If* P *visits exactly one edge in* $\{\hat{\alpha}\} \times (V_h \setminus (B \cup S))$ *and* $\hat{\alpha} \in \{a_i, b_i\}$, *then* $w(P) \geq h - 1/2$.

Proof. Assume $\hat{\alpha} = a_i$ by symmetry. Decompose P into an edge $(\hat{\alpha}, v)$, where $v \in V_h \setminus (B \cup S)$, and a v-b_i path $\tilde{P}$ in $\mathcal{G}$ that visits no edges in $\{\hat{\alpha}\} \times (V_h \setminus (B \cup S))$. Clearly, $b_i \in V_1$. In summary, $\tilde{P}$ is a V_h-V_1 path in $\mathcal{G}$ visiting no edges in $\{\hat{\alpha}\} \times (V_h \setminus (B \cup S))$. By Lemma 17, $w(\tilde{P}) \geq h - 1$, implying $w(P) \geq h - 1/2$. $\square$

Lemma 19. *For all* $i \in [q(n)]$ *with* $\hat{\alpha} \in \{a_i, b_i\}$,

$$\chi \left[\exists v \in \{a_i, b_i\}, (v \in S) \wedge \left(deg_{Q^{(i)}}(v) \leq \delta n^{1/(h-1)} \right) \right] = 1.$$

Proof. By (12), $\hat{\alpha} \in S$. This and Lemma 9 complete the proof. $\qquad\square$

Lemma 20. *For all distinct* u, $v \in [n] \setminus (B \cup S)$, *we have* $(u, v) \in E_G^{(q(n))}$.

Proof. As u, $v \in [n] \setminus B$, (17) implies

$$deg_{H^{(i)}}(u) < \delta n^{1/(h-1)} - 2, \tag{21}$$
$$deg_{H^{(i)}}(v) < \delta n^{1/(h-1)} - 2 \tag{22}$$

when $i = q(n)$. So by Lemma 1, (21)–(22) hold for all $i \in [q(n)]$.

As $u, v \in [n] \setminus S$ and $u \neq v$, we have $(u, v) \in E_G^{(0)}$ by (4). By lines 8 and 13 of Adv,

$$E_G^{(i-1)} \setminus \left\{ (x, y) \in [n]^2 \mid \left(\deg_{H^{(i)}}(x) \geq \delta n^{1/(h-1)} - 2 \right) \vee \left(\deg_{H^{(i)}}(y) \geq \delta n^{1/(h-1)} - 2 \right) \right\} \subseteq E_G^{(i)} \quad (23)$$

for all $i \in [q(n)]$. By (21)–(23), $(u, v) \in E_G^{(i)}$ if $(u, v) \in E_G^{(i-1)}$, for all $i \in [q(n)]$. The proof is complete by mathematical induction on i. □

Lemma 21. *Let P be a shortest a_i-b_i path in $\mathcal{G}$, where $i \in [q(n)]$. If P visits exactly two edges in $\{\hat{\alpha}\} \times (V_h \setminus (B \cup S))$, then $G^{(q(n))}$ has an a_i-b_i path with exactly $w(P)$ edges.*

Proof. Clearly, the two edges of P in $\{\hat{\alpha}\} \times (V_h \setminus (B \cup S))$, denoted $(u, \hat{\alpha})$ and $(\hat{\alpha}, v)$, are consecutive on P. Replace the subpath $(u, \hat{\alpha}, v)$ of P by the edge (u, v) to yield an a_i-b_i path $\tilde{P}$. Except for the two edges of P in $\{\hat{\alpha}\} \times (V_h \setminus (B \cup S))$, all edges of P are in $E_G^{(q(n))}$ by (18). As $u, v \in V_h \setminus (B \cup S)$ and $u \neq v$, $(u, v) \in E_G^{(q(n))}$ by Lemma 20. In summary, all the edges of $\tilde{P}$ are in $E_G^{(q(n))}$. Consequently, $\tilde{P}$ is an a_i-b_i path in $G^{(q(n))} = ([n], E_G^{(q(n))})$. So we are left only to prove that $\tilde{P}$ has exactly $w(P)$ edges, which, by Lemma 14, is equivalent to proving $w(\tilde{P}) = w(P)$.

Note that $\hat{\alpha} \notin V_h \setminus (B \cup S)$ by (12). By the construction of $\tilde{P}$ and recalling that $u, v \in V_h \setminus (B \cup S)$ and $u \neq v$,

$$w\left(\tilde{P}\right) = w(P) - w(u, \hat{\alpha}) - w(\hat{\alpha}, v) + w(u, v) \overset{(19)}{=} w(P) - \frac{1}{2} - \frac{1}{2} + 1 = w(P).$$

□

Lemma 22. *Every simple path in $\mathcal{G}$ visiting exactly one edge in $\{\hat{\alpha}\} \times (V_h \setminus (B \cup S))$ either starts or ends at $\hat{\alpha}$.*

Proof. By (12), $\hat{\alpha} \in S$. So by (4) and Lemma 1, $\hat{\alpha}$ is incident to no edges in $E_G^{(q(n))}$. Consequently, the set of all edges of $\mathcal{G}$ incident to $\hat{\alpha}$ is $\{\hat{\alpha}\} \times (V_h \setminus (B \cup S))$ by (18). The lemma is now easy to see. □

Lemma 23. *For all $i \in [q(n)]$,*

$$\min \left\{ d_{H^{(i)}}(a_i, b_i), h - \frac{1}{2} \cdot \chi \left[\exists v \in \{a_i, b_i\}, (v \in S) \wedge \left(\deg_{Q^{(i)}}(v) \leq \delta n^{1/(h-1)} \right) \right] \right\}$$

$$\leq \min \left\{ d_{\mathcal{G}}(a_i, b_i), h - \frac{1}{2} \cdot \chi \left[\exists v \in \{a_i, b_i\}, (v \in S) \wedge \left(\deg_{Q^{(i)}}(v) \leq \delta n^{1/(h-1)} \right) \right] \right\}. \quad (24)$$

Proof. Clearly, we may assume $d_{\mathcal{G}}(a_i, b_i) < \infty$. Pick an a_i-b_i path P in $\mathcal{G}$ with

$$w(P) = d_{\mathcal{G}}(a_i, b_i). \quad (25)$$

We deal only with the hardest case that P visits exactly two edges in $\{\hat{\alpha}\} \times (V_h \setminus (B \cup S))$. In this case, Lemma 21 implies

$$d_{G^{(q(n))}}(a_i, b_i) \leq w(P). \quad (26)$$

If $d_{G^{(i-1)}}(a_i, b_i) \leq h$, then $d_{H^{(i)}}(a_i, b_i) = d_{G^{(q(n))}}(a_i, b_i)$ by Lemma 2, which together with (25)–(26) completes the proof. Otherwise, $d_{G^{(q(n))}}(a_i, b_i) > h$ by Lemma 1, which together with (25)–(26) shows $d_{\mathcal{G}}(a_i, b_i) > h$ and thus completes the proof.   □

Define $d \colon [n]^2 \to [0, \infty)$ by

$$d(a_i, b_i) = d(b_i, a_i)$$
$$\equiv \min\left\{ d_{\mathcal{G}}(a_i, b_i), h - \frac{1}{2} \cdot \chi\left[\exists v \in \{a_i, b_i\}, (v \in S) \wedge \left(\deg_{Q^{(i)}}(v) \leq \delta n^{1/(h-1)} \right) \right] \right\}, \quad (27)$$
$$d(u, v)$$
$$\equiv \min\{ d_{\mathcal{G}}(u, v), h \} \tag{28}$$

for all $i \in [q(n)]$ and $(u, v) \in [n]^2 \setminus \{(a_j, b_j) \mid j \in [q(n)]\}$.

Lemma 24. $([n], d)$ *is a metric space.*

Proof. We only prove the triangle inequality for $d(\cdot, \cdot)$. Clearly, $d_{\mathcal{G}}(\cdot, \cdot)$ is a metric and $d_{\mathcal{G}}(\cdot, \cdot) \notin (0, 1/2)$. By (27)–(28), $d(\cdot, \cdot)$ truncates $d_{\mathcal{G}}(\cdot, \cdot)$ to within either $h - 1/2$ or h, preserving the triangle inequality.   □

Lemma 25. *For all* $i \in [q(n)]$, $d_{H^{(i)}}(a_i, b_i) \geq d_{\mathcal{G}}(a_i, b_i)$.

Proof. Take a shortest a_i-b_i path P in the unweighted graph $H^{(i)} = ([n], E_H^{(i)})$. So by Lemma 14, $d_{H^{(i)}}(a_i, b_i) = w(P)$. By Lemma 1, P's edges are in $E_H^{(q(n))}$. So by Lemma 16, P is a path in $\mathcal{G} = ([n], \mathcal{E}, w)$, implying $d_{\mathcal{G}}(a_i, b_i) \leq w(P)$.   □

The following lemma says that line 17 of Adv answers queries consistently with $d(\cdot, \cdot)$.

Lemma 26. *For all* $i \in [q(n)]$,

$$\min\left\{ d_{H^{(i)}}(a_i, b_i), h - \frac{1}{2} \cdot \chi\left[\exists v \in \{a_i, b_i\}, (v \in S) \wedge \left(\deg_{Q^{(i)}}(v) \leq \delta n^{1/(h-1)} \right) \right] \right\} = d(a_i, b_i).$$

Proof. Use Lemmas 23 and 25 and (27).   □

Theorem 1. METRIC 1-MEDIAN *has no deterministic* $o(n^{1+1/(h-1)})$-*query* $(2h - \epsilon)$-*approximation algorithms for any constants* $h \in \mathbb{Z}^+ \setminus \{1\}$ *and* $\epsilon > 0$.

Proof. By Lemma 26 and line 17 of Adv, Adv answers A's queries consistently with $d(\cdot, \cdot)$. By Lemma 24, $([n], d)$ is a metric space. Now use Lemma 13 to bound $\sum_{v \in [n]} d(\hat{\alpha}, v)$ from the above. Then use Lemmas 6 and 26 to bound $\sum_{v \in [n]} d(z, v)$ from below. Finally, pick δ to be sufficiently small.   □

References

1. Arya, V., Garg, N., Khandekar, R., Meyerson, A., Munagala, K., Pandit, V.: Local search heuristics for k-median and facility location problems. SIAM J. Comput. **33**(3), 544–562 (2004)
2. Chang, C.-L.: A deterministic sublinear-time nonadaptive algorithm for metric 1-median selection. To appear in Theoretical Computer Science
3. Chang, C.-L.: Some results on approximate 1-median selection in metric spaces. Theor. Comput. Sci. **426**, 1–12 (2012)
4. Chang, C.-L.: Deterministic sublinear-time approximations for metric 1-median selection. Inf. Process. Lett. **113**(8), 288–292 (2013)
5. Chang, C.-L.: A lower bound for metric 1-median selection. Technical report. arXiv:1401.2195 (2014)
6. Chen, K.: On coresets for k-median and k-means clustering in metric and Euclidean spaces and their applications. SIAM J. Comput. **39**(3), 923–947 (2009)
7. Guha, S., Meyerson, A., Mishra, N., Motwani, R., O'Callaghan, L.: Clustering data streams: theory and practice. IEEE Trans. Knowl. Data Eng. **15**(3), 515–528 (2003)
8. Indyk, P.: Sublinear time algorithms for metric space problems. In: Proceedings of the 31st Annual ACM Symposium on Theory of Computing, pp. 428–434 (1999)
9. Indyk, P.: High-Dimensional Computational Geometry. Ph.D. thesis, Stanford University (2000)
10. Jaiswal, R., Kumar, A., Sen, S.: A simple D^2-sampling based PTAS for k-means and other clustering problems. In: Proceedings of the 18th Annual International Conference on Computing and Combinatorics, pp. 13–24 (2012)
11. Kumar, A., Sabharwal, Y., Sen, S.: Linear-time approximation schemes for clustering problems in any dimensions. J. ACM **57**(2), 5 (2010)
12. Mettu, R.R., Plaxton, C.G.: Optimal time bounds for approximate clustering. Mach. Learn. **56**(1–3), 35–60 (2004)
13. Wu, B.-Y.: On approximating metric 1-median in sublinear time. Inf. Process. Lett. **114**(4), 163–166 (2014)

Frequent-Itemset Mining
Using Locality-Sensitive Hashing

Debajyoti Bera[1]([✉]) and Rameshwar Pratap[2]

[1] Indraprastha Institute of Information Technology-Delhi (IIIT-D), New Delhi, India
dbera@iiitd.ac.in
[2] TCS Innovation Labs, New Delhi, India
rameshwar.pratap@gmail.com

Abstract. The Apriori algorithm is a classical algorithm for the frequent itemset mining problem. A significant bottleneck in Apriori is the number of I/O operation involved, and the number of candidates it generates. We investigate the role of LSH techniques to overcome these problems, without adding much computational overhead. We propose randomized variations of Apriori that are based on asymmetric LSH defined over Hamming distance and Jaccard similarity.

1 Introduction

Mining *frequent itemsets* in a transactions database appeared first in the context of analyzing supermarket transaction data for discovering association rules [1,2], however this problem has, since then, found applications in diverse domains like finding correlations [13], finding episodes [9], clustering [14]. Mathematically, each transaction can be regarded as a subset of the items ("itemset") those that present in the transaction. Given a database $\mathcal{D}$ of such transactions and a support threshold $\theta \in (0, 1)$, the primary objective of frequent itemset mining is to identify θ-frequent itemsets (denoted by FI, these are subsets of items that appear in at least θ-fraction of transactions).

Computing FI is a challenging problem of data mining. The question of deciding if there exists any FI with k items is known to be NP-complete [7] (by relating it to the existence of bi-cliques of size k in a given bipartite graph) but on a more practical note, simply checking support of any itemset requires reading the transaction database – something that is computationally expensive since they are usually of an extremely large size. The state-of-the-art approaches try to reduce the number of candidates, or not generate candidates at all. The best known approach in the former line of work is the celebrated Apriori algorithm [2].

Apriori is based on the *anti-monotonicity property of* partially-ordered sets which says that no superset of an infrequent itemset can be frequent. This algorithm works in a bottom-up fashion by generating itemsets of size l in level l, starting at the first level. After finding frequent itemsets at level l they are joined pairwise to generate $l + 1$-sized *candidate itemsets*; FI are identified among the candidates by computing their support explicitly from the data. The algorithm

© Springer International Publishing Switzerland 2016
T.N. Dinh and M.T. Thai (Eds.): COCOON 2016, LNCS 9797, pp. 143–155, 2016.
DOI: 10.1007/978-3-319-42634-1_12

terminates when no more candidates are generated. Broadly, there are two downsides to this simple but effective algorithm. The first one is that the algorithm has to compute support[1] of every itemset in the candidate, even the ones that are highly infrequent. Secondly, if an itemset is infrequent, but all its subsets are frequent, Apriori doesn't have any easy way of detecting this without reading every transaction of the candidates.

A natural place to look for fast algorithms over large data are randomized techniques; so we investigated if LSH could be of any help. An earlier work by Cohen *et al.* [5] was also motivated by the same idea but worked on a different problem (see Sect. 1.2). LSH is explained in Sect. 2, but roughly, it is a randomized hashing technique which allows efficient retrieval of approximately "similar" elements (here, itemsets).

1.1 Our Contribution

In this work, we propose LSH-Apriori – a basket of three explicit variations of Apriori that uses LSH for computing FI. LSH-Apriori handles both the above mentioned drawbacks of the Apriori algorithm. First, LSH-Apriori significantly cuts down on the number of infrequent candidates that are generated, and further due to its dimensionality reduction property saves on reading every transaction; secondly, LSH-Apriori could efficiently filter our those infrequent itemset without looking every candidate. The first two variations essentially reduce computing FI to the approximate nearest neighbor (cNN) problem for Hamming distance and Jaccard similarity. Both these approaches can drastically reduce the number of false candidates without much overhead, but has a non-zero probability of error in the sense that some frequent itemset could be missed by the algorithm. Then we present a third variation which also maps FI to elements in the Hamming space but avoids the problem of these false negatives incurring a little cost of time and space complexity. Our techniques are based on asymmetric LSH [12] and LSH with one-sided error [10] which are proposed very recently.

1.2 Related Work

There are a few hash based heuristic to compute FI which outperform the Apriori algorithm and PCY [11] is one of the most notable among them. PCY focuses on using hashing to efficiently utilize the main memory over each pass of the database. However, our objective and approach both are fundamentally different from that of PCY.

The work that comes closest to our work is by Cohen *et al.* [5]. They developed a family of algorithms for finding interesting associations in a transaction database, also using LSH techniques. However, they specifically wanted to avoid any kind of filtering of itemsets based on itemset support. On the other hand, our problem is the vanilla frequent itemset mining which requires filtering itemsets satisfying a given minimum support threshold.

---

[1] Note that computing support is an I/O intensive operation and involves reading every transaction.

1.3 Organization of the Paper

In Sect. 2, we introduce the relevant concepts and give an overview of the problem. In Sect. 3, we build up the concept of LSH-Apriori which is required to develop our algorithms. In Sect. 4, we present three specific variations of LSH-Apriori for computing FI. Algorithms of Subsects. 4.1 and 4.2 are based on Hamming LSH and Minhashing, respectively. In Subsect. 4.3, we present another approach based on CoveringLSH which overcomes the problem of producing false negatives. In Sect. 5, we summarize the whole discussion. Proofs are omitted due to space constraint and can be found in the full-version [3].

2 Background

The input to the classical frequent itemset mining problem is a database $\mathcal{D}$ of n transactions $\{T_1, \ldots, T_n\}$ over m items $\{i_1, \ldots, i_m\}$ and a support threshold $\theta \in (0,1)$. Each transaction, in turn, is a subset of those items. Support of itemset $I \subseteq \{i_1, \ldots, i_m\}$ is the number of transactions that contain I. The objective of the problem is to determine *every itemset* with support at least θn. We will often identify an itemset I with its transaction vector $\langle I[1], I[2], \ldots, I[n] \rangle$ where $I[j]$ is 1 if I is contained in T_j and 0 otherwise. An equivalent way to formulate the objective is to find itemsets with at least θn 1's in their transaction vectors. It will be useful to view $\mathcal{D}$ as a set of m transaction vectors, one for every item.

Notations			
$\mathcal{D}$	Database of transactions: $\{t_1, \ldots, t_n\}$	n	Number of transactions
$\mathcal{D}_l$	FI of level-l: $\{I_1, \ldots I_{m_l}\}$	θ	Support threshold, $\theta \in (0,1)$
α_l	Maximum support of any item in $\mathcal{D}_l$	m	Number of items
ε	Error tolerance in LSH, $\varepsilon \in (0,1)$	m_l	Number of FI of size l
δ	Probability of error in LSH, $\delta \in (0,1)$	$|v|$	Number of 1's in v

2.1 Locality Sensitive Hashing

We first briefly explain the concept of locality sensitive hashing (LSH).

Definition 1 (Locality Sensitive Hashing [8]). *Let S be a set of m vectors in $\mathbb{R}^n$, and U be the hashing universe. Then, a family $\mathcal{H}$ of functions from S to U is called as $(S_0, (1-\varepsilon)S_0, p_1, p_2)$-sensitive (with $\varepsilon \in (0,1]$ and $p_1 > p_2$) for the similarity measure $Sim(.,.)$ if for any $x, y \in S$:*

- *if $Sim(x,y) \geq S_0$, then $\Pr_{h \in \mathcal{H}}[h(x) = h(y)] \geq p_1$,*
- *if $Sim(x,y) \leq (1-\varepsilon)S_0$, then $\Pr_{h \in \mathcal{H}}[h(x) = h(y)] \leq p_2$.*

Not all similarity measures have a corresponding LSH. However, the following well-known result gives a sufficient condition for existence of LSH for any *Sim*.

Lemma 1. *If Φ is a strict monotonic function and a family of hash function $\mathcal{H}$ satisfies $\mathrm{Pr}_{h \in \mathcal{H}}[h(x) = h(y)) = \Phi(Sim(x,y)]$ for some $Sim : \mathbb{R}^n \times \mathbb{R}^n \to \{0,1\}$, then the conditions of Definition 1 are true for Sim for any $\varepsilon \in (0,1)$.*

The similarity measures that are of our interest are *Hamming* and *Jaccard* over binary vectors. Let $|x|$ denote the Hamming weight of a binary vector x. Then, for vectors x and y of length n, Hamming distance is defined as $\mathrm{Ham}(x,y) = |x \oplus y|$, where $x \oplus y$ denotes a vector that is element-wise Boolean XOR of x and y. Jaccard similarity is defined as $\langle x, y \rangle / |x \vee y|$, where $\langle x, y \rangle$ indicates inner product, and $x \vee y$ indicates element-wise Boolean OR of x and y. LSH for these similarity measures are simple and well-known [4,6,8]. We recall them below; here I is some subset of $\{1, \ldots, n\}$ (or, n-length transaction vector).

Definition 2 (Hash Function for Hamming Distance). *For any particular bit position i, we define the function $h_i(I) := I[i]$. We will use hash functions of the form $g_J(I) = \langle h_{j_1}(I), h_{j_2}(I), \ldots, h_{j_k}(I) \rangle$, where $J = \{j_1, \ldots, j_k\}$ is a subset of $\{1, \ldots, n\}$ and the hash values are binary vectors of length k.*

Definition 3 (Minwise Hash Function for Jaccard Similarity). *Let π be some permutations over $\{1, \ldots, n\}$. Treating I as a subset of indices, we will use hash functions of the form $h_\pi(I) = \arg\min_i \pi(i)$ for $i \in I$.*

The probabilities that two itemsets hash to the same value for these hash functions are related to their Hamming distance and Jaccard similarity, respectively.

2.2   Apriori Algorithm for Frequent Itemset Mining

As explained earlier, Apriori works level-wise and in its l-th level, it generates all θ-frequent itemsets with l-items each; for example, in the first level, the algorithm simply computes support of individual items and retains the ones with support

---

Input: Transaction database $\mathcal{D}$, support threshold θ;
Result: θ-frequent itemsets;
1 $l = 1$ /* level */;
2 $F = \big\{\{x\} \mid \{x\}$ is θ-frequent in $\mathcal{D}\big\}$ /* frequent itemsets in level-1 */ ;
3 Output F;
4 **while** F *is not empty* **do**
5 $\quad$ $l = l + 1$;
6 $\quad$ $C = \{I_a \cup I_b \mid I_a \in F,\ I_b \in F,\ I_a$ and I_b are compatible$\}$;
7 $\quad$ $F = \emptyset$;
8 $\quad$ **for** *itemset I in C* **do**
9 $\quad\quad$ Add I to F if support of I in $\mathcal{D}$ is at least θn /* reads database*/ ;
10 $\quad$ **end**
11 $\quad$ Output F;
12 **end**

Algorithm 1. Apriori algorithm for frequent itemset mining

at least θn. Apriori processes each level, say level-$(l+1)$, by joining all pairs of θ-frequent *compatible itemsets* generated in level-l, and further filtering out the ones which have support less than θn. Here, two candidate itemsets (of size l each) are said to be compatible if their union has size exactly $l+1$. A high-level pseudocode of Apriori is given in Algorithm 1. All our algorithms rely on a good implementation of set whose runtime cost is not included in our analysis.

3   LSH-Apriori

The focus of this paper is to reduce the computation of processing all pairs of itemsets at each level in line 6 (which includes computing support by going through $\mathcal{D}$). Suppose that level l outputs m_l frequent itemsets. We will treat the output of level l as a collection of m_l transaction vectors $\mathcal{D}_l = \{I_1, \ldots I_{m_l}\}$, each of length n and one for each frequent itemset of the l-th level. Our approach involves defining appropriate notions of similarity between itemsets (represented by vectors) in $\mathcal{D}_l$ similar to the approach followed by Cohen *et al.* [5]. Let I_i, I_j be two vectors each of length n. Then, we use $|I_i, I_j|$ to denote the number of bit positions where both the vectors have a 1.

Definition 4. *Given a parameter $0 < \varepsilon < 1$, we say that $\{I_i, I_j\}$ is θ-frequent (or similar) if $|I_i, I_j| \geq \theta n$ and $\{I_i, I_j\}$ is $(1-\varepsilon)\theta$-infrequent if $|I_i, I_j| < (1-\varepsilon)\theta n$. Furthermore, we say that I_j is similar to I_i if $\{I_i, I_j\}$ is θ-frequent.*

Let I_q be a frequent itemset at level $l-1$. Let $\text{FI}(I_q, \theta)$ be the set of itemsets I_a such that $\{I_q, I_a\}$ is θ-frequent at level l. Our main contributions are a few randomized algorithms for identifying itemsets in $\text{FI}(I_q, \theta)$ with high-probability.

Definition 5 ($\text{FI}(I_q, \theta, \varepsilon, \delta)$). *Given a θ-frequent itemset I_q of size $l-1$, tolerance $\varepsilon \in (0,1)$ and error probability δ, $\text{FI}(I_q, \theta, \varepsilon, \delta)$ is a set F' of itemsets of size l, such that with probability at least $1-\delta$, F' contains every I_a for which $\{I_q, I_a\}$ is θ-frequent.*

It is clear that $\text{FI}(I_q, \theta) \subseteq \text{FI}(I_q, \theta, \varepsilon, \delta)$ with high probability. This motivated us to propose LSH-Apriori, a randomized version of Apriori, that takes δ and ε as additional inputs and essentially replaces line **6** by LSH operations to combine every itemset I_q with only similar itemsets, unlike Apriori which combines all pairs of itemsets. This potentially creates a significantly smaller C without missing out too many frequent itemsets. The modifications to Apriori are presented in Algorithm 2 and the following lemma, immediate from Definition 5, establishes correctness of LSH-Apriori.

Input: $\mathcal{D}_l = \{I_1, \ldots, I_{m_l}\}$, θ, (Additional) error probability δ, tolerance ε;
6a (Pre-processing) Initialize hash tables and add all items $I_a \in \mathcal{D}_l$;
6b (Query) Compute $\text{FI}(I_q, \theta, \varepsilon, \delta)$ $\forall I_q \in \mathcal{D}_l$ by hashing I_q and checking collisions;
6c $C \leftarrow \{I_q \cup I_b \mid I_q \in \mathcal{D}_l, \ I_b \in \text{FI}(I_q, \theta, \varepsilon, \delta)\}$;

Algorithm 2. LSH-Apriori level $l+1$ (only modifications to Apriori line: 6)

Lemma 2. *Let I_q and I_a be two θ-frequent compatible itemsets of size $(l-1)$ such that the itemset $J = I_q \cup I_a$ is also θ-frequent. Then, with probability at least $1 - \delta$, $\mathrm{FI}(I_q, \theta, \varepsilon, \delta)$ contains I_a (hence C contains J).*

In the next section we describe three LSH-based randomized algorithms to compute $\mathrm{FI}(I_q, \theta, \varepsilon, \delta)$ for all θ-frequent itemset I_q from the earlier level. The input to these subroutines will be $\mathcal{D}_l$, the frequent itemsets from earlier level, and parameters $\theta, \varepsilon, \delta$. In the *pre-processing stage* at level l, the respective LSH is initialized and itemsets of $\mathcal{D}_l$ are hashed; we specifically record the itemsets hashing to every bucket. LSH guarantees (*w.h.p.*) that pairs of similar items hash into the same bucket, and those that are not hash into different buckets. In the *query stage* we find all the itemsets that any I_q ought to be combined with by looking in the bucket in which I_q hashed, and then combining the compatible ones among them with I_q to form C. Rest of the processing happens à la Apriori.

The internal LSH subroutines may output false-positives – itemsets that are not θ-frequent, but such itemsets are eventualy filtered out in line 9 of Algorithm 1. Therefore, the output of LSH-Apriori does not contain any false positives. However, some frequent itemsets may be missing from its output (false negatives) with some probability depending on the parameter δ as stated below in Theorem 3 (proof follows from the union bound).

Theorem 3 (Correctness). *LSH-Apriori does not output any itemset which is not θ-infrequent. If X is a θ-frequent itemset of size l, then the probability that LSH-Apriori does not output X is at most $\delta 2^l$.*

The tolerance parameter ε can be used to balance the overhead from using hashing in LSH-Apriori with respect to its savings because of reading fewer transactions. Most LSH, including those that we will be using, behave somewhat like dimensionality reduction. As a result, the hashing operations do not operate on all bits of the vectors. Furthermore, the pre-condition of similarity for joining ensure that (w.h.p.) most infrequent itemsets can be detected before verifying them from $\mathcal{D}$. To formalize this, consider any level l with m_l θ-frequent itemsets $\mathcal{D}_l$. We will compare the computation done by LSH-Apriori at level $l+1$ to what Apriori would have done at level $l+1$ given the same frequent itemsets $\mathcal{D}_l$. Let c_{l+1} denote the number of candidates Apriori would have generated and m_{l+1} the number of frequent itemsets at this level (LSH-Apriori may generate fewer).

Overhead: Let $\tau(LSH)$ be the time required for hashing an itemset for a particular LSH and let $\sigma(LSH)$ be the space needed for storing respective hash values. The extra overhead in terms of space will be simply $m_l \sigma(LSH)$ in level $l+1$. With respect to overhead in running time, LSH-Apriori requires hashing each of the m_l itemsets twice, during pre-processing and during querying. Thus total time overhead in this level is $\vartheta(LSH, l+1) = 2m_l \tau(LSH)$.

Savings: Consider the itemsets in $\mathcal{D}_l$ that are compatible with any $I_q \in \mathcal{D}_l$. Among them are those whose combination with I_q do not generate a θ-frequent itemset for level $l+1$; call them as *negative* itemsets and denote their number by $r(I_q)$. Apriori will have to read all n transactions of $\sum_{I_q} r(I_q)$ itemsets in order to reject them. Some of these negative itemsets will be added to FI by LSH-Apriori – we will call them *false positives* and denote their count by $FP(I_q)$; the rest those which correctly not added with I_q – lets call them as *true nega-tives* and denote their count by $TN(I_q)$. Clearly, $r(I_q) = TN(I_q) + FP(I_q)$ and $\sum_{I_q} r(I_q) = 2(c_{l+1} - m_{l+1})$. Suppose $\phi(LSH)$ denotes the number of transactions a particular LSH-Apriori reads for hashing any itemset; due to the dimensionality reduction property of LSH, $\phi(LSH)$ is always $o(n)$. Then, LSH-Apriori is able to reject all itemsets in TN by reading only ϕ transactions for each of them; thus for itemset I_q in level $l + 1$, a particular LSH-Apriori reads $(n - \phi(LSH)) \times TN(I_q)$ fewer transactions compared to a similar situation for Apriori. Therefore, total savings at level $l + 1$ is $\varsigma(LSH, l + 1) = (n - \phi(LSH)) \times \sum_{I_q} TN(I_q)$.

In Sect. 4, we discuss this in more detail along with the respective LSH-Apriori algorithms.

4 FI via LSH

Our similarity measure $|I_a, I_b|$ can also be seen as the inner product of the binary vectors I_a and I_b. However, it is not possible to get any LSH for such similarity measure because for example there can be three items I_a, I_b and I_c such that $|I_a, I_b| \geq |I_c, I_c|$ which implies that $\Pr(h(I_a) = h(I_b)) \geq \Pr(h(I_c) = h(I_c)) = 1$, which is not possible. Noting the exact same problem, Shrivastava *et al.* introduced the concept of *asymmetric LSH* [12] in the context of binary inner product similarity. The essential idea is to use two different hash functions (for pre-processing and for querying) and they specifically proposed extending MinHashing by padding input vectors before hashing. We use the same pair of padding functions proposed by them for n-length binary vectors in a level l: $P_{(n, \alpha_l)}$ for preprocessing and $Q_{(n, \alpha_l)}$ for querying are defined as follows.

- In $P(I)$ we append $(\alpha_l n - |I|)$ many 1's followed by $(\alpha_l n + |I|)$ many 0's.
- In $Q(I)$ we append $\alpha_l n$ many 0's, then $(\alpha_l n - |I|)$ many 1's, then $|I|$ 0's.

Here, $\alpha_l n$ (at LSH-Apriori level l) will denote the maximum number of ones in any itemset in $\mathcal{D}_l$. Therefore, we always have $(\alpha_l n - |I|) \geq 0$ in the padding functions. Furthermore, since the main loop of Apriori is not continued if no frequent itemset is generated at any level, $(\alpha_l - \theta) > 0$ is also ensured at any level that Apriori is executing.

We use the above padding functions to reduce our problem of finding similar itemsets to finding nearby vectors under Hamming distance (using Hamming-based LSH in Subsect. 4.1 and Covering LSH in Subsect. 4.3) and under Jaccard similarity (using MinHashing in Subsect. 4.2).

4.1 Hamming Based LSH

In the following lemma, we relate Hamming distance of two itemsets I_x and I_y with their $|I_x, I_y|$.

Lemma 4. *For two itemsets I_x and I_y, $\mathrm{Ham}(P(I_x), Q(I_y)) = 2(\alpha_l n - |I_x; I_y|)$.*

Therefore, it is possible to use an LSH for Hamming distance to find similar itemsets. We use this technique in the following algorithm to compute $\mathrm{FI}(I_q, \theta, \varepsilon, \delta)$ for all itemset I_q. The algorithm contains an optimization over the generic LSH-Apriori pseudocode (Algorithm 2). There is no need to separately execute lines:7–10 of Apriori; one can immediately set $F \leftarrow C$ since LSH-Apriori computes support before populating FI.

Input: $\mathcal{D}_l = \{I_1, \ldots, I_{m_l}\}$, query item I_q, threshold θ, tolerance ε, error δ.
Result: $\mathrm{FI}_q = \mathrm{FI}(I_q, \theta, \varepsilon, \delta)$ for every $I_q \in \mathcal{D}_l$.

6a Preprocessing step: Setup hash tables and add vectors in $\mathcal{D}_l$;

  i   Set $\rho = \frac{\alpha_l - \theta}{\alpha_l - (1-\varepsilon)\theta}$, $k = \log_{\left(\frac{1+2\alpha_l}{(1+2(1-\varepsilon)\theta)}\right)} m_l$ and $L = m_l^\rho \log\left(\frac{1}{\delta}\right)$;

  ii   Select functions $g_1, \ldots, g_L$ *u.a.r.*;

  iii For every $I_a \in \mathcal{D}_l$, pad I_a using $P()$ and then hash $P(I_a)$ into buckets $g_1(P(I_a)), \ldots, g_L(P(I_a))$;

6b Query step: For every $I_q \in \mathcal{D}_l$, we do the following ;

  i $S \leftarrow$ all I_q-compatible itemsets in all buckets $g_i(Q(I_q))$, for $i = 1 \ldots L$;

  ii   **for** $I_a \in S$ **do**
        If $|I_a, I_q| \geq \theta n$, then add I_a to FI_q /* reads database*/;
        (*) If no itemset similar to I_q found within $\frac{L}{\delta}$ tries, then break loop;
   end

Algorithm 3. LSH-Apriori (only lines 6a,6b) using Hamming LSH

Correctness of this algorithm is straightforward. Also, $\rho < 1$ and the space required and overhead of reading transactions is $\theta(kLm_l) = o(m_l^2)$. It can be further shown that $\mathbb{E}[FP(I_q)] \leq L$ for $I_q \in \mathcal{D}_l$ which can be used to prove that $\mathbb{E}[\varsigma] \geq (n - \phi)(2(c_{l+1} - m_{l+1}) - m_l L)$ where $\phi = O(kL) = \tilde{O}(m_l^\rho) = o(m_l)$. Details of these calculations including proof of the next lemma are omitted.

Lemma 5. *Algorithm 3 correctly outputs $\mathrm{FI}(I_q, \theta, \varepsilon, \delta)$ for all $I_q \in \mathcal{D}_l$. Additional space required is $o(m_l^2)$, which is also the total time overhead. The expected savings can be bounded by $\mathbb{E}[\varsigma(l+1)] \geq (n - o(m_l))((c_{l+1} - 2m_{l+1}) + (c_{l+1} - o(m_l^2)))$.*

Expected savings outweigh time overhead if $n \gg m_l$, $c_{l+1} = \theta(m_l^2)$ and $c_{l+1} > 2m_{l+1}$, i.e., in levels where the number of frequent itemsets generated are fewer compared to the number of transactions as well as to the number of candidates generated. The additional optimisation (*) essentially increases the savings when all $l + 1$-extensions of I_q are $(1 - \varepsilon)\theta$-infrequent — this behaviour will be predominant in the last few levels. It is easy to show that in this case, $FP(I_q) \leq \frac{L}{\delta}$ with probability at least $1 - \delta$; this in turn implies that $|S| \leq \frac{L}{\delta}$. So, if we did not find any similar I_a within first $\frac{L}{\delta}$ tries, then we can be sure, with reasonable probability, that there are no itemsets similar to I_q.

4.2 Min-Hashing Based LSH

Cohen et al. had given an LSH-based randomized algorithm for finding interesting itemsets without any requirement for high support [5]. We observed that their Minhashing-based technique [4] cannot be directly applied to the high-support version that we are interested in. The reason is roughly that Jaccard similarity and itemset similarity (w.r.t. θ-frequent itemsets) are not monotonic to each other. Therefore, we used padding to monotonically relate Jaccard similarity of two itemsets I_x and I_y with their $|I_x, I_y|$.

Lemma 6. *For two padded itemsets I_x and I_y, $\mathrm{JS}(P(I_x), Q(I_y)) = \frac{|I_x, I_y|}{2\alpha_l n - |I_x, I_y|}$.*

Once padded, we follow similar steps (as [5]) to create a *similarity preserving summary* $\hat{D}_l$ of D_l such that the Jaccard similarity for any column pair in D_l is approximately preserved in $\hat{D}_l$, and then explicitly compute $\mathrm{FI}(I_q, \theta, \varepsilon, \delta)$ from $\hat{D}_l$. $\hat{D}_l$ is created by using λ independent minwise hashing functions (see Definition 3). λ should be carefully chosen since a higher value increases the accuracy of estimation, but at the cost of *large summary vectors* in $\hat{D}_l$. Let us define $\hat{\mathrm{JS}}(I_i, I_j)$ as the fraction of rows in the summary matrix in which min-wise entries of columns I_i and I_j are identical. Then by Theorem 1 of Cohen et al. [5], we can get a bound on the number of required hash functions:

Theorem 7 (Theorem 1 of [5]). *Let $0 < \epsilon, \delta < 1$ and $\lambda \geq \frac{2}{\omega \epsilon^2} \log \frac{1}{\delta}$. Then for all pairs of columns I_i and I_j following are true with probability at least $1 - \delta$:*

- *If $\mathrm{JS}(I_i, I_j) \geq s* \geq \omega$, then $\hat{\mathrm{JS}}(I_i, I_j) \geq (1 - \epsilon)s*$,*
- *If $\mathrm{JS}(I_i, I_j) \leq \omega$, then $\hat{\mathrm{JS}}(I_i, I_j) \leq (1 + \epsilon)\omega$.*

Input: D_l, query item I_q, threshold θ, tolerance ε, error δ
Result: $\mathrm{FI}_q = \mathrm{FI}(I_q, \theta, \varepsilon, \delta)$ for every $I_q \in D_l$.

6a Preprocessing step: Prepare $\hat{D}_l$ via MinHashing;

i   Set $\omega = \frac{(1-\varepsilon)\theta}{2\alpha_l - (1-\varepsilon)\theta}$, $\epsilon = \frac{\alpha_l \varepsilon}{\alpha_l + (\alpha_l - \theta)(1-\varepsilon)}$ and $\lambda = \frac{2}{\omega \epsilon^2} \log \frac{1}{\delta}$;

ii   Choose λ many independent permutations (see Theorem 7);

iii   For every $I_a \in D_l$, pad I_a using $P()$ and then hash $P(I_a)$ using λ independent permutations;

6b Query step: For every $I_q \in D_l$, we do the following ;

i   Hash $Q(I_q)$ using λ independent permutations;

ii   **for** *compatible* $I_a \in D_l$ **do**

      If $\hat{\mathrm{JS}}(P(I_a), Q(I_q)) \geq \frac{(1-\epsilon)\theta}{2\alpha_l - \theta}$ for some I_a, then add I_a to FI_q;

    end

Algorithm 4. LSH-Apriori (only lines 6a,6b) using Minhash LSH (This algorithm can be easily boosted to $o(\lambda m_l)$ time by applying banding technique (see Section 4 of [5]) on the minhash table.)

Lemma 8. *Algorithm 4 correctly computes $\mathrm{FI}(I_q, \theta, \varepsilon, \delta)$ for all $I_q \in D_l$. Additional space required is $O(\lambda m_l)$, and the total time overhead is $O((n + \lambda)m_l)$. The expected savings is given by $\mathbb{E}[\varsigma(l + 1)] \geq 2(1 - \delta)(n - \lambda)(c_{l+1} - m_{l+1})$.*

The proof is omitted due to space constraints. Note that λ depends on α_l but is independent of n. This method should be applied only when $\lambda \ll n$. And in that case, for levels with number of candidates much larger than the number of frequent itemsets discovered (i.e., $c_{l+1} \gg \{m_l, m_{l+1}\}$), time overhead would not appear significant compared to expected savings.

4.3   Covering LSH

Due to their probabilistic nature, the LSH-algorithms presented earlier have the limitation of producing false positives and more importantly, false negatives. Since the latter cannot be detected unlike the former, these algorithms may miss some frequent itemsets (see Theorem 3). In fact, once we miss some FI at a particular level, then all the FI which are *"supersets"* of that FI (in the subsequent levels) will be missed. Here we present another algorithm for the same purpose which overcomes this drawback. The main tool is a recent algorithm due to Pagh [10] which returns approximate nearest neighbors in the Hamming space. It is an improvement over the seminal LSH algorithm by Indyk and Motwani [8], also for Hamming distance. Pagh's algorithm has a small overhead over the latter; to be precise, the query time bound of [10] differs by at most $\ln(4)$ in the exponent in comparison with the time bound of [8]. However, its big advantage is that it generates no false negatives. Therefore, this LSH-Apriori version also does not miss any frequent itemset.

The LSH by Pagh is with respect to Hamming distance, so we first reduce our FI problem into the Hamming space by using the same padding given in Lemma 4. Then we use this LSH in the same manner as in Subsect. 4.1. Pagh coined his hashing scheme as coveringLSH which broadly mean that given a threshold r and a tolerance $c > 1$, the hashing scheme guaranteed a collision for every pair of vectors that are within radius r. We will now briefly summarize coveringLSH for our requirement; refer to the paper [10] for full details.

Similar to HammingLSH, we use a family of Hamming projections as our hash functions: $\mathcal{H}_{\mathcal{A}} := \{x \mapsto x \wedge a \mid a \in \mathcal{A}\}$, where $\mathcal{A} \subseteq \{0,1\}^{(1+2\alpha_l)n}$. Now, given a query item I_q, the idea is to iterate through all hash functions $h \in \mathcal{H}_{\mathcal{A}}$, and check if there is a collision $h(P(I_x)) = h(Q(I_q))$ for $I_x \in \mathcal{D}_l$. We say that this scheme doesn't produce false negative for the threshold $2(\alpha_l - \theta)n$, if at least one collision happens when there is an $I_x \in \mathcal{D}_l$ when $\text{Ham}(P(I_x), Q(I_q)) \leq 2(\alpha_l - \theta)n$, and the scheme is efficient if the number of collision is not too many when $\text{Ham}(P(I_x), Q(I_q)) > 2(\alpha_l - (1-\varepsilon)\theta)n$ (proved in Theorems 3.1, 4.1 of [10]). To make sure that all pairs of vector within distance $2(\alpha_l - \theta)n$ collide for some h, we need to make sure that some h map their "mismatching" bit positions (between $P(I_x)$ and $Q(I_q)$) to 0. We describe construction of hash functions next.

n'	θ'	t	c	ϵ	ν
$(1+2\alpha_l)n$	$2(\alpha_l-\theta)n$	$\left\lceil \dfrac{\ln m_l}{2(\alpha_l-(1-\varepsilon)\theta)n} \right\rceil$	$\dfrac{\alpha_l-(1-\varepsilon)\theta}{\alpha_l-\theta}$	$\epsilon \in (0,1)$ s.t. $\dfrac{\ln m_l}{2(\alpha_l-(1-\varepsilon)\theta)n} + \epsilon \in \mathbb{N}$	$\dfrac{t+\epsilon}{ct}$

CoveringLSH: The parameters relevant to LSH-Apriori are given above. Notice that after padding, dimension of each item is n', threshold is θ' (i.e., min-support is θ'/n'), and tolerance is c. We start by choosing a random function $\varphi : \{1, \ldots, n'\} \to \{0,1\}^{t\theta'+1}$, which maps bit positions of the padded itemsets to bit vectors of length $t\theta' + 1$. We define a family of bit vectors $a(v) \in \{0,1\}^{n'}$, where $a(v)_i = \langle \varphi(i), v \rangle$, for $i \in \{1, \ldots, n'\}$, $v \in \{0,1\}^{t\theta'+1}$ and $\langle m(i), v \rangle$ denotes the inner product over $\mathbb{F}_2$. We define our hash function family $\mathcal{H}_\mathcal{A}$ using all such vectors $a(v)$ except $a(\mathbf{0})$: $\mathcal{A} = \left\{ a(v) | v \in \{0,1\}^{t\theta'+1}/\{\mathbf{0}\} \right\}$.

Pagh described how to construct $\mathcal{A}' \subseteq \mathcal{A}$ [10, Corollary 4.1] such that $\mathcal{H}_{\mathcal{A}'}$ has a very useful property of no false negatives and also ensuring very few false positives. We use $\mathcal{H}_{\mathcal{A}'}$ for hashing using the same manner of Hamming projections as used in Subsect. 4.1. Let ψ be the expected number of collisions between any itemset I_q and items in $\mathcal{D}_l$ that are $(1-\varepsilon)\theta$-infrequent with I_q. The following Theorem captures the essential property of coveringLSH that is relevant for LSH-Apriori, described in Algorithm 5. It also bounds the number of hash functions which controls the space and time overhead of LSH-Apriori. Proof of this theorem follows from Theorem 4.1 and Corollary 4.1 of [10].

Theorem 9. *For a randomly chosen φ, a hash family $\mathcal{H}_{A'}$ described above and distinct $I_x, I_q \in \{0,1\}^n$:*

- *If* $\mathrm{Ham}\big(P(I_x), Q(I_q)\big) \leq \theta'$, *then there exists* $h \in \mathcal{H}_{A'}$ *s.t.* $h(P(I_x)) = h(Q(I_q))$,
- *Expected number of false positives is bounded by* $\mathbb{E}[\psi] < 2^{\theta'\varepsilon+1} m_l^{\frac{1}{c}}$,
- $|\mathcal{H}_{A'}| < 2^{\theta'\varepsilon+1} m_l^{\frac{1}{c}}$.

Input: $\mathcal{D}_l$, query item I_q, threshold θ, tolerance ε, error δ.
Result: $\mathrm{FI}_q = \mathrm{FI}(I_q, \theta, \varepsilon, \delta)$ for every $I_q \in \mathcal{D}_l$.
6a Preprocessing step: Setup hash tables according to $\mathcal{H}_{A'}$ and add items;
  i   For every $I_a \in \mathcal{D}_l$, hash $P(I_a)$ using all $h \in \mathcal{H}_{A'}$;
6b Query step: For every $I_q \in \mathcal{D}_l$, we do the following ;
  i $S \leftarrow$ all itemsets that collide with $Q(I_q)$;
  ii   **for** $I_a \in S$ **do**
        If $|I_a, I_q| \geq \theta n$, then add I_a to FI_q /* reads database*/;
        (*) If no itemset similar to I_q found within $\frac{\psi}{\delta}$ tries, break loop;
      end

Algorithm 5. LSH-Apriori (only lines 6a,6b) using Covering LSH

Lemma 10. *Algorithm 5 outputs all θ-frequent itemsets and only θ-frequent itemsets. Additional space required is $O\big(m_l^{1+\nu}\big)$, which is also the total time overhead. The expected savings is given by $\mathbb{E}[\varsigma(l+1)] \geq 2\left(n - \frac{\log m_l}{c} - 1\right)\big((c_{l+1} - m_{l+1}) - m_l^{1+\nu}\big)$.*

The (*) line is an additional optimisation similar to what we did for HammingLSH Sect. 4.1; it efficiently recognizes those frequent itemsets of the earlier level none of whose extensions are frequent. The guarantee of not missing any valid itemset comes with a heavy price. Unlike the previous algorithms, the conditions under which expected savings beats overhead are quite stringent, namely, $c_{l+1} \in \{\omega(m_l^2), \omega(m_{l+1}^2)\}$, $\frac{2^n}{5} > m_l > 2^{n/2}$ and $\epsilon < 0.25$ (since $1 < c < 2$, these bounds ensure that $\nu < 1$ for later levels when $\alpha_l \approx \theta$).

5 Conclusion

In this work, we designed randomized algorithms using locality-sensitive hashing (LSH) techniques which efficiently outputs almost all the frequent itemsets with high probability at the cost of a little space which is required for creating hash tables. We showed that time overhead is usually small compared to the savings we get by using LSH.

Our work opens the possibilities for addressing a wide range of problems that employ on various versions of frequent itemset and sequential pattern mining problems, which potentially can efficiently be randomized using LSH techniques.

References

1. Agrawal, R., Imielinski, T., Swami, A.N.: Mining association rules between sets of items in large databases. In: Proceedings of the 1993 ACM SIGMOD International Conference on Management of Data, Washington, D.C., 26–28 May 1993, pp. 207–216 (1993)
2. Agrawal, R., Srikant, R.: Fast algorithms for mining association rules in large databases. In: Proceedings of 20th International Conference on Very Large Data Bases, 12–15 September 1994, Santiago de Chile, Chile, pp. 487–499 (1994)
3. Bera, D., Pratap, R.: Frequent-itemset mining using locality-sensitive hashing. CoRR, abs/1603.01682 (2016)
4. Broder, A.Z., Charikar, M., Frieze, A.M., Mitzenmacher, M.: Min-wise independent permutations. J. Comput. Syst. Sci. **60**(3), 630–659 (2000)
5. Cohen, E., Datar, M., Fujiwara, S., Gionis, A., Indyk, P., Motwani, R., Ullman, J.D., Yang, C.: Finding interesting associations without support pruning. IEEE Trans. Knowl. Data Eng. **13**(1), 64–78 (2001)
6. Gionis, A., Indyk, P., Motwani, R.: Similarity search in high dimensions via hashing. In: VLDB 1999, Proceedings of 25th International Conference on Very Large Data Bases, 7–10 September 1999, Edinburgh, Scotland, UK, pp. 518–529 (1999)
7. Gunopulos, D., Khardon, R., Mannila, H., Saluja, S., Toivonen, H., Sharma, R.S.: Discovering all most specific sentences. ACM Trans. Database Syst. **28**(2), 140–174 (2003)
8. Indyk, P., Motwani, R.: Approximate nearest neighbors: towards removing the curse of dimensionality. In: Proceedings of the Thirtieth Annual Symposium on the Theory of Computing, Dallas, Texas, USA, 23–26 May 1998, pp. 604–613 (1998)
9. Mannila, H., Toivonen, H., Verkamo, A.I.: Discovery of frequent episodes in event sequences. Data Min. Knowl. Discov. **1**(3), 259–289 (1997)
10. Pagh, R.: Locality-sensitive hashing without false negatives. In: Proceedings of the Twenty-Seventh Annual ACM-SIAM Symposium on Discrete Algorithms, SODA 2016, Arlington, VA, USA, 10–12 January 2016, pp. 1–9 (2016)

11. Park, J.S., Chen, M., Yu, P.S.: An effective hash based algorithm for mining association rules. In: Proceedings of the ACM SIGMOD International Conference on Management of Data, San Jose, California, 22–25 May, pp. 175–186 (1995)
12. Shrivastava, A., Li, P.: Asymmetric minwise hashing for indexing binary inner products and set containment. In: Proceedings of the 24th International Conference on World Wide Web, 2015, Florence, Italy, 18–22 May 2015, pp. 981–991 (2015)
13. Silverstein, C., Brin, S., Motwani, R.: Beyond market baskets: generalizing association rules to dependence rules. Data Min. Knowl. Discov. **2**(1), 39–68 (1998)
14. Wang, H., Wang, W., Yang, J., Yu, P.S.: Clustering by pattern similarity in large data sets. In: Proceedings of the 2002 ACM SIGMOD International Conference on Management of Data, Madison, Wisconsin, 3–6 June 2002, pp. 394–405 (2002)

Computational Complexity

On the Hardness of Switching to a Small Number of Edges

Vít Jelínek[1], Eva Jelínková[2(✉)], and Jan Kratochvíl[2]

[1] Computer Science Institute, Faculty of Mathematics and Physics,
Charles University, Malostranské nám. 25,
118 00 Praha, Czech Republic
jelinek@iuuk.mff.cuni.cz
[2] Department of Applied Mathematics, Faculty of Mathematics and Physics,
Charles University, Malostranské nám. 25, 118 00 Praha, Czech Republic
{eva,honza}@kam.mff.cuni.cz

Abstract. Seidel's switching is a graph operation which makes a given vertex adjacent to precisely those vertices to which it was non-adjacent before, while keeping the rest of the graph unchanged. Two graphs are called switching-equivalent if one can be made isomorphic to the other one by a sequence of switches.

Jelínková et al. [DMTCS 13, no. 2, 2011] presented a proof that it is NP-complete to decide if the input graph can be switched to contain at most a given number of edges. There turns out to be a flaw in their proof. We present a correct proof.

Furthermore, we prove that the problem remains NP-complete even when restricted to graphs whose density is bounded from above by an arbitrary fixed constant. This partially answers a question of Matoušek and Wagner [Discrete Comput. Geom. 52, no. 1, 2014].

Keywords: Seidel's switching · Computational complexity · Graph density · Switching-minimal graphs · NP-completeness

1 Introduction

Seidel's switching is a graph operation which makes a given vertex adjacent to precisely those vertices to which it was non-adjacent before, while keeping the rest of the graph unchanged. Two graphs are called switching-equivalent if one can be made isomorphic to the other one by a sequence of switches. The class of graphs that are pairwise switching-equivalent is called a switching class.

Hage in his PhD thesis [4, p. 115, Problem 8.5] posed the problem to characterize the graphs that have the maximum (or minimum) number of edges in their switching class. We call such graphs *switching-maximal* and *switching-minimal*, respectively.

V. Jelínek and J. Kratochvíl—Supported by CE-ITI project GACR P202/12/G061.
E. Jelínková—Supported by the grant SVV-2016-260332.

© Springer International Publishing Switzerland 2016
T.N. Dinh and M.T. Thai (Eds.): COCOON 2016, LNCS 9797, pp. 159–170, 2016.
DOI: 10.1007/978-3-319-42634-1_13

Some properties of switching-maximal graphs were studied by Kozerenko [7]. He proved that any graph with sufficiently large minimum degree is switching-maximal, and that the join of certain graphs is switching-maximal. Further, he gave a characterization of triangle-free switching-maximal graphs and of non-hamiltonian switching-maximal graphs.

It is easy to observe that a graph is switching-maximal if and only if its complement is switching-minimal. We call the problem to decide if a graph is switching-minimal SWITCH-MINIMAL.

Jelínková et al. [6] studied the more general problem SWITCH-FEW-EDGES – the problem of deciding if a graph can be switched to contain at most a certain number of edges. They presented a proof that the problem is NP-complete. Unfortunately, their proof is not correct.

In this paper, we provide a different proof of the NP-hardness of SWITCH-FEW-EDGES, based on a reduction from a restricted version of MAX-CUT. Furthermore, we strengthen this result by proving that for any $c > 0$, SWITCH-FEW-EDGES is NP-complete even if we require that the input graph has density at most c. We also prove that if the problem SWITCH-MINIMAL is co-NP-complete, then for any $c > 0$, the problem is co-NP-complete even on graphs with density at most c.

We thus partially answer a question of Matoušek and Wagner [10] posed in connection with properties of simplicial complexes – they asked if deciding switching-minimality was easy for graphs of bounded density. Our results also indicate that it might be unlikely to get an easy characterization of switching-minimal (or switching-maximal) graphs, which contributes to understanding Hage's question [4].

1.1 Formal Definitions and Previous Results

Let G be a graph. Then the *Seidel's switch* of a vertex subset $A \subseteq V(G)$ is denoted by $S(G, A)$ and is defined by

$$S(G, A) = (V(G), E(G) \triangle \{xy : x \in A,\ y \in V(G) \setminus A\}),$$

where $\triangle$ denotes the symmetric difference of the sets. $S(G, A)$ is also the graph obtained from G by consecutive switching of the vertices of A (in any order).

We say that two graphs G and H are *switching-equivalent* (denoted by $G \sim H$) if there is a set $A \subseteq V(G)$ such that $S(G, A)$ is isomorphic to H. The set $[G] = \{S(G, A) : A \subseteq V(G)\}$ is called the *switching class* of G.

We say that a graph G is $(\leq k)$-*switchable* if there is a set $A \subseteq V(G)$ such that $S(G, A)$ contains at most k edges. Analogously, a graph G is $(\geq k)$-*switchable* if there is a set $A \subseteq V(G)$ such that $S(G, A)$ contains at least k edges.

It is easy to observe that a graph G is $(\leq k)$-switchable if and only if the complement $\overline{G}$ is $(\geq (\binom{n}{2} - k))$-switchable. We may, therefore, focus on $(\leq k)$-switchability only.

We examine the following problems.

SMALL-CAPS SWITCH-FEW-EDGES
Input: A graph $G = (V, E)$, an integer k
Question: Is G ($\leq k$)-switchable?

SWITCH-MINIMAL
Input: A graph $G = (V, E)$
Question: Is G switching-minimal?

We say that a graph is *switching-reducible* if G is *not* switching-minimal, in other words, if there is a set $A \subseteq V(G)$ such that $S(G, A)$ contains fewer edges than G. For further convenience, we also define the problem SWITCH-REDUCIBLE.

SWITCH-REDUCIBLE
Input: A graph $G = (V, E)$
Question: Is G switching-reducible?

Let $G = (V, E)$ be a graph. We say that a partition V_1, V_2 of V is a *cut* of G. For a cut V_1, V_2, the set of edges that have exactly one end-vertex in V_1 is denoted by $\operatorname{cutset}(V_1)$, and the edges of $\operatorname{cutset}(V_1)$ are called *cut-edges*. We let $\delta(V_1)$ denote the size of $\operatorname{cutset}(V_1)$. When there is no danger of confusion, we also say that a single subset $V_1 \subseteq V$ is a cut (meaning the partition V_1, $V \setminus V_1$).

1.2 Easy Cases

In this subsection we present several results about easy special cases of the problems that we focus on. This complements our hardness results.

The following theorem was proved by Ehrenfeucht et al. [2] and also independently (in a slightly weaker form) by Kratochvíl [8].

Theorem 1. *Let $\mathcal{P}$ be a graph property that can be decided in time $\mathcal{O}(n^a)$ for an integer a. Let every graph with $\mathcal{P}$ contain a vertex of degree at most $d(n)$. Then the problem if an input graph is switching-equivalent to a graph with $\mathcal{P}$ can be decided in time $\mathcal{O}(n^{d(n)+1+\max(a,2)})$.*

The proof of Theorem 1 also gives an algorithm that works in the given time. Hence, it also provides an algorithm for SWITCH-FEW-EDGES: in a graph with at most k edges all vertex degrees are bounded by k. Hence, we can use $d(n) = k$ and $a = 2$ and get an $\mathcal{O}(n^{k+3})$-time algorithm. It was further proved by Jelínková et al. [6] that SWITCH-FEW-EDGES is fixed-parameter tractable; it has a kernel with $2k$ vertices, and there is an algorithm running in time $\mathcal{O}(2.148^k \cdot n + m)$, where m is the number of edges of the input graph. In Sect. 2, we provide a corrected NP-completeness proof.

The following proposition states a basic relation of switching-minimality and graph degrees.

Proposition 1 (Folklore). *Every switching-minimal graph $G = (V, E)$ on n vertices has maximum degree at most $\lfloor (n-1)/2 \rfloor$.*

Proof. Clearly, if G contains a vertex v of degree greater than $\lfloor (n-1)/2 \rfloor$, then $S(G, \{v\})$ has fewer edges than G, showing that G is not switching-minimal. $\square$

We remark that for a given graph G we can efficiently construct a switch whose maximum degree is at most $\lfloor (n-1)/2 \rfloor$; one by one, we switch vertices whose degree exceeds this bound (in this way, the number of edges is decreased in each step). However, the graph constructed by this procedure is not necessarily switching-minimal.

The next proposition is an equivalent formulation of Lemma 2.5 of Kozerenko [7], strengthening Proposition 1.

Proposition 2. *A graph G on n vertices is switching-minimal if and only if for every $A \subseteq V(G)$, we have $2\delta(A) \leq |A|(n - |A|)$.*

We derive the following consequence.

Proposition 3. *Let G be a graph with n vertices. If the maximum vertex degree in G is at most $\frac{n}{4}$, then G is switching-minimal.*

Proof. Let A be any subset of $V(G)$. We observe that $\delta(A) = \delta(V(G) \setminus A)$; hence we can assume without loss of generality that $|A| \leq n/2$, and thus $n - |A| \geq n/2$.

Further, as $\delta(A) \leq \sum_{v \in A} \deg(v)$, we have that $\delta(A) \leq |A| \frac{n}{4}$. Hence, $2\delta(A) \leq |A|(n - |A|)$, and the condition of Proposition 2 is fulfilled. $\square$

Proposition 3 implies that SWITCH-FEW-EDGES and SWITCH-MINIMAL are trivially solvable in polynomial time for graphs on n vertices with maximum degree at most $\frac{n}{4}$.

We note that in Proposition 3, the bound $\frac{n}{4}$ in general cannot be improved. To see this, consider an arbitrary bipartite k-regular graph G on n vertices, with partition classes X and $Y = V(G) \setminus X$, and with $k > \frac{n}{4}$. Since G can be switched to a $(\frac{n}{2} - k)$-regular bipartite graph $S(G, X)$, G is not switching-minimal.

2   NP-Completeness of SWITCH-FEW-EDGES

Jelínková et al. [6] presented a proof that the problem SWITCH-FEW-EDGES is NP-complete. Unfortunately, there is an error in their proof (see Remark 1 on page 8). We present another proof here. The core of the original proof is a reduction from the MAX-CUT problem. Our reduction works in a similar way. However, we need the following more special version of MAX-CUT (we prove the NP-completeness of LARGE-DEG-MAX-CUT in Sect. 3).

LARGE-DEG-MAX-CUT
Input: A graph G with $2n$ vertices such that the minimum vertex degree of G is $2n - 4$ and the complement of G does not contain triangles; an integer j
Question: Does there exist a cut V_1 of $V(G)$ with at least j cut-edges?

Proposition 4. *Let G be a graph with 2n vertices such that the minimum vertex degree of G is 2n − 4 and the complement of G does not contain triangles. In polynomial time, we can find a graph G′ such that $|V(G')| = 4|V(G)|$ and the following statements are equivalent for every integer j:*

(a) *There is a cut in G with at least j cut-edges,*
(b) *There exists a set $A \subseteq V(G')$ such that $S(G', A)$ contains at most $|E(G')| - 16j$ edges.*

Proof. We first describe the construction of the graph G'. For each vertex u of G we create a corresponding four-tuple $\{u', u'', u''', u''''\}$ of pairwise non-adjacent vertices in G'. An edge of G is then represented by a complete bipartite graph interconnecting the two four-tuples, and a non-edge in G is represented by 8 edges that form a cycle that alternates between the two four-tuples (see Fig. 1).

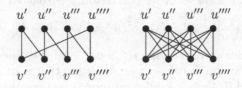

Fig. 1. The representation of non-edges and edges of G

A vertex four-tuple in G' corresponding to a vertex of G is called an *o-vertex*. A pair of o-vertices corresponding to an edge of G is called an *o-edge* and a pair of o-vertices corresponding to an non-edge of G is called an *o-non-edge*. Where there is no danger of confusion, we identify o-vertices with vertices of G, o-edges with edges of G and o-non-edges with non-edges of G.

We now prove that the statements (a) and (b) are equivalent. First assume that there is a cut V_1 of $V(G)$ with j' cut-edges. Let V_1' be the set of vertices u', u'', u''', u'''' for all $u \in V_1$. We prove that $S(G', V_1')$ contains at most $|E(G')| - 16j'$ edges.

We say that a non-edge *crosses the cut* V_1 if the non-edge has exactly one vertex in V_1. It is clear that G' contains 16 edges per every o-edge and 8 edges per every o-non-edge. In $S(G', V_1')$, every o-edge corresponding to an edge that is not a cut-edge is unchanged by the switch, because its end-o-vertices are either both contained in V_1' or both in $V(G) \setminus V_1'$; hence, the o-edge yields 16 edges. Similarly, every o-non-edge corresponding to a non-edge that does not cross the cut yields 8 edges.

Figure 2 illustrates the switches of o-non-edges and o-edges that have exactly one end-o-vertex in V_1. We can see that every o-non-edge corresponding to a non-edge that crosses the cut yields 8 edges in $S(G', V_1')$, and that every o-edge corresponding to a cut-edge yields 0 edges. Altogether, $S(G', V_1')$ has $|E(G')| - 16j'$ edges, which we wanted to prove.

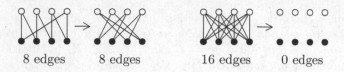

8 edges   8 edges   16 edges   0 edges

Fig. 2. Switches of an o-non-edge and of an o-edge

Now assume that there exists a set $A \subseteq V(G')$ such that $S(G', A)$ contains at most $|E(G')| - 16j$ edges. We want to find a cut in G with at least j cut-edges.

We say that an o-vertex u of G' is *broken in A* if A contains exactly one, two or three vertices out of u', u'', u''', u''''; otherwise, we say that u is *legal in A*. We say that an o-edge or o-non-edge $\{u, v\}$ is *broken in A* if at least one of the o-vertices u, v is broken. Otherwise, we say that $\{u, v\}$ is *legal in A*.

If all vertices of G are legal in A, we say that A is *legal*. Legality is a desired property, because for a legal set A we can define a subset V_A of $V(G)$ such that

$$V_A = \{u \in V(G) : \{u', u'', u''', u''''\} \subseteq A\}.$$

The set V_A then defines a cut in G. If a set is not legal, we proceed more carefully to get a cut from it. For any vertex subset A, we say that a set A' is a *legalization* of A if A' is legal and if A' and A differ only on o-vertices that are broken in A.

We want to show that for every illegal set A, there exists its legalization A' such that the number of edges in $S(G', A')$ is not much higher than in $S(G', A)$. To this end, we give the algorithm Legalize which for a set A finds such a legalization A'. During the run of the algorithm, we keep a set A''. In the beginning we set $A'' := A$ and in each step we change A'' so that more o-vertices are legal.

We define some notions needed in the algorithm. Let v be an o-vertex and consider the o-vertices that are adjacent to v (through an o-edge); we call them *o-neighbors* of v. The o-neighbors of v are four-tuples of vertices and some of those vertices are in A'', some of them are not. We define $\mathrm{dif}(v)$ as the number of such vertices that are in A'' minus the number of such vertices that are not in A''. (Note that $\mathrm{dif}(v)$ is always an even number, because the total number of vertices in o-neighbors is even. If all o-neighbors were legal, then $\mathrm{dif}(v)$ would be divisible by four.)

The algorithm is given in Fig. 3. As in the last step the algorithm legalizes all remaining broken o-vertices, it is clear that the set A'' output by the algorithm is a legalization of A. We prove that $|E(S(G', A''))| - |E(S(G', A))| \leq 7$.

We need to introduce more terminology. A pair of vertices of G' which belong to the same o-vertex is called a *v-pair*. A pair of vertices of G' which belong to different o-vertices that are adjacent (in G) is called an *e-pair*. A pair of vertices of G' which belong to different o-vertices that are non-adjacent (in G) is called an *n-pair*. It is easy to see that any edge of G' or $S(G', A'')$ is either a v-pair, an e-pair or an n-pair. We call such edges *v-edges*, *e-edges* and *n-edges*, respectively.

We say that a broken o-vertex v is *asymmetric* if it contains an odd number of vertices of A''; we say that a broken o-vertex is *symmetric* if it contains two vertices out of A''.

Algorithm Legalize(A)

Set $A'' := A$; do the following while any of the cases applies.

Case 1. There exists a broken o-vertex v such that $|\text{dif}(v)| \geq 4$. If $\text{dif}(v) \geq 4$, set
$A'' := A'' \setminus \{v', v'', v''', v''''\}$. Otherwise, set $A'' := A'' \cup \{v', v'', v''', v''''\}$.

Case 2. Case 1 does not apply and there exists an asymmetric broken vertex v such
that v contains exactly one vertex from A'' and $\text{dif}(v) = 2$. Set $A'' :=$
$A'' \setminus \{v', v'', v''', v''''\}$.

Case 3. Case 1 does not apply and there exists an asymmetric broken vertex v such
that v contains exactly three vertices from A'' and $\text{dif}(v) = -2$. Set $A'' :=$
$A'' \cup \{v', v'', v''', v''''\}$.

Case 4. None of Cases 1, 2, 3 applies and there exist two adjacent broken o-vertices
u and v. Set

$$A_1 := (A'' \cup \{u', u'', u''', u''''\}) \setminus \{v', v'', v''', v''''\},$$

$$A_2 := (A'' \cup \{v', v'', v''', v''''\}) \setminus \{u', u'', u''', u''''\}.$$

If $|E(S(G', A_1))| < |E(S(G', A_2))|$ then set $A'' := A_1$, otherwise $A'' := A_2$.

Case 5. None of the above cases applies. Then legalize the remaining broken o-
vertices arbitrarily (without changing the legal ones), output A'' and STOP.

Fig. 3. The algorithm legalize

To measure how the number of edges of $S(G', A'')$ changes during the run of
the algorithm, we define a variable $c(A'')$ which we call the *charge* of the graph
$S(G', A'')$. Before the first step we set $c(A'') := |E(S(G', A))|$. After a step of
the algorithm, we update $c(A'')$ in the following way.

- For every v-pair or e-pair that was an edge of $S(G', A'')$ before the step and
  is no longer an edge of $S(G', A'')$ after the step, we decrease $c(A'')$ by one.
- For every v-pair or e-pair that was not an edge of $S(G', A'')$ before the step
  and that has become an edge of $S(G', A'')$ after the step, we increase $c(A'')$
  by one.
- For every o-vertex v that was legalized in the step and is incident to an o-non-
  edge, we change $c(A'')$ in the following way:
  - If v was symmetric, we increase $c(A'')$ by 2.5 for every o-non-edge incident
    to v;
  - If v was asymmetric, we increase $c(A'')$ by 1.5 for every o-non-edge inci-
    dent to v.

To explain the last two points, we observe how the number of n-edges
increases after legalizing an o-vertex. By analyzing all cases of o-non-edges with
one or two broken end-o-vertices, we get that there are four cases where the
o-non-edges have less than 8 n-edges before legalization: either 6 or 4 n-edges. In
these cases, both end-o-vertices are broken. If there are only 4 n-edges, at least
one of the end-o-vertices is symmetric. After one end-o-vertex is legalized, the
number of n-edges increases by 2 or 4. When the second end-o-vertex is legalized,
the number of n-edges does not increase for this particular o-non-edge.

After both end-o-vertices are legalized, the charge has been changed in the following way: if both end-o-vertices were symmetric, we have increased the charge by 5. If one of them was symmetric and the other one was asymmetric, we have increased the charge by 4. Finally, if both were asymmetric, we have increased the charge by 3. In all these cases, the increase is an upper bound on the number of contributed n-edges.

Further, every v-edge or e-edge that has appeared or disappeared during the run of the algorithm is counted immediately after the corresponding step. Hence, we have proved the following claim.

Claim 1. *At the end of the algorithm we have that* $c(A'') \geq |E(S(G', A''))|$.

Claim 2. *After every step of the algorithm except for the last one, the charge* $c(A'')$-*is decreased. After the last step, the charge is increased by at most 7. Hence,* $c(A'') \leq |E(S(G', A))| + 7$.

To prove Claim 2, we count how the charge changes after each step. Due to space limitations, the proof of Claim 2 is omitted; to give an insight into the proof, we remark that thanks to the assumptions on the input graph G, each o-vertex is incident to at most three o-non-edges, which limits the charge increase – without the assumption, the charge could increase largely due to a single step of the algorithm. Further, thanks to the assumption that the complement of G does not contain triangles, the number of o-vertices legalized in Step 5 (and hence the charge increase) is bounded easily.

By Claims 1 and 2 we have that $|E(S(G', A''))| \leq |E(S(G', A))| + 7$, and hence A'' is the sought legalization of A.

We continue the proof of Proposition 4. We have already argued that a legal set A'' defines a subset $V_{A''}$ of $V(G)$, and hence a cut in G. Assume that cutset($V_{A''}$) has j' edges. From the proof of the first implication of Proposition 4 we know that the number of edges in $S(G', A'')$ can be expressed as $|E(G')| - 16j'$.

On the other hand, we have proved that the number of edges in $S(G', A'')$ is at most $|E(G')| - 16j + 7$. We get that $|E(G')| - 16j' \leq |E(G')| - 16j + 7$, and hence $j' \geq j - 7/16$. As both j and j' are integers, we have that $j' \geq j$. Hence, cutset($V_{A''}$) has at least j edges, and Proposition 4 is proved.   □

Remark 1. As we noted before, our proof is a corrected version of an erroneous proof presented in [6]. The argument of the original (attempted) proof was based on a similar construction as our proof of Proposition 4, except that each o-vertex was formed by two vertices rather than four, an o-edge was represented by a copy of $K_{2,2}$ rather than $K_{4,4}$, and an o-non-edge was represented by two disjoint edges. It was then incorrectly claimed in [6, Lemma 4.3] that for any graph G, there is a switching-minimal legal switch of the corresponding graph G'. The claim is false, as can be seen e.g. by taking the graph G consisting of two disjoint triangles, where G' has 42 edges, the optimal legal switch has 26 edges, but there are illegal switches with just 18 edges.

Theorem 2. SWITCH-FEW-EDGES *is NP-complete.*

Proof. Theorem 3 in the next section gives the NP-completeness of LARGE-DEG-MAX-CUT. Further, by Proposition 4, an instance (G, j) of LARGE-DEG-MAX-CUT can be transformed into an instance (G', j') of SWITCH-FEW-EDGES such that there is a cut in G with at least j cut-edges if and only if G' is $(\leq j')$-switchable. The transformation works in polynomial time.

Finally, it is clear that the problem SWITCH-FEW-EDGES is in NP.   □

3   The NP-Completeness of LARGE-DEG-MAX-CUT

Let G be a graph with $2n$ vertices. A *bisection of G* is a partition S_1, S_2 of $V(G)$ such that $|S_1| = |S_2| = n$ (hence, a bisection is a special case of a cut). The size of cutset(S_1) is called the *size* of the bisection S_1, S_2. A *minimum bisection of G* is a bisection of G with minimum size.

Garey et al. [3] proved that, given a graph G and an integer b, the problem to decide if G has a bisection of size at most b is NP-complete (by a reduction of MAX-CUT). Their formulation is slightly different from ours – two distinguished vertices must be each in one part of the partition, and the input graph does not have to be connected. However, their reduction from MAX-CUT (see [3, pp. 242–243]) produces only connected graphs as instances of the bisection problem, and it is immediate that the two distinguished vertices are not important in the proof. Hence, their proof gives also the NP-completeness of the following version of the problem.

CONNECTED-MIN-BISECTION
Input: A connected graph G with $2n$ vertices, an integer b
Question: Is there a bisection S_1, S_2 of $V(G)$ such that cutset(S_1) contains at most b edges?

From the NP-completeness of MIN-BISECTION, Bui et al. [1] proved the NP-completeness of MIN-BISECTION restricted to 3-regular graphs (as a part of a more general result, see [1, proof of Theorem 2]). We use their result to prove the NP-completeness of LARGE-DEG-MAX-CUT.

LARGE-DEG-MAX-CUT
Input: A graph G with $2n$ vertices such that the minimum vertex degree of G is $2n - 4$ and the complement of G is connected and does not contain triangles; an integer j
Question: Does there exist a cut V_1 of G with at least j cut-edges?

Lemma 1. *Let G be a connected 3-regular graph on $2n$ vertices. Let b be the size of the minimum bisection in G and let c be the size of the maximum cut in $\overline{G}$. Then $b = n^2 - c$.*

Due to space constraints, the proof of Lemma 1 is omitted.

Theorem 3. LARGE-DEG-MAX-CUT *is NP-complete.*

Proof. Let (G, b) be an instance of CONNECTED-MIN-BISECTION. We use the construction of Bui et al. [1, proof of Theorem 2]. Their first step is to construct from an instance (G, b) of MIN-BISECTION a 3-regular graph G^* such that G has a minimum bisection of size b if and only if G^* has a minimum bisection of size b. Further, it is immediate from their construction that G^* contains no triangles, and if G is connected, then G^* is connected as well. Moreover, G^* has an even number of vertices.

We see that $\overline{G^*}$ fulfills the conditions of an instance of LARGE-DEG-MAX-CUT. By Lemma 1 we know that G^* has a minimum bisection of size b if and only if $\overline{G^*}$ has a maximum cut of size $m^2 - b$.

Altogether, G has a minimum bisection of size b if and only if $\overline{G^*}$ has a maximum cut of size $m^2 - b$. Hence, $(\overline{G^*}, m^2 - b)$ is an equivalent instance of LARGE-DEG-MAX-CUT. To finish the proof that LARGE-DEG-MAX-CUT is NP-complete, we observe that LARGE-DEG-MAX-CUT is in NP.   □

4 Switching of Graphs with Bounded Density

The *density* of a graph G is defined as $D(G) = |E(G)| / \binom{|V(G)|}{2}$.

In connection with properties of simplicial complexes, Matoušek and Wagner [10] asked if deciding switching-minimality was easy for graphs of bounded density. We give a partial negative answer by proving that the problem SWITCH-FEW-EDGES stays NP-complete even for graphs of density bounded by an arbitrarily small constant. This is in contrast with Proposition 3, which shows that any graph G with maximum degree at most $|V(G)|/4$ is switching-minimal. The core of our argument is the following proposition.

Proposition 5. *Let G be a graph, let k be an integer, and let c be a fixed constant in $(0, 1)$. In polynomial time, we can find a graph G' and an integer k' such that*

1. *$D(G') \leq c$,*
2. *G' is $(\leq k')$-switchable if and only if G is $(\leq k)$-switchable,*
3. *G' is switching-minimal if and only if G is switching-minimal, and*
4. *$|V(G')| = O(|V(G)|)$.*

Proof. Let $n = |V(G)|$ and let $N = \max\left\{n, \left\lceil \frac{3n}{4c} \right\rceil\right\}$. We construct the graph G' in the following way. Let $V = V(G)$. Then $V(G') = V \cup Y \cup Z$, where Y is a set of N vertices and Z is a set of N more vertices, and $E(G') = \{\{v_1, v_2\} : v_1 \in Y, v_2 \in V\} \cup E(G)$.

We prove that G' fulfills the conditions of Proposition 5. It is easy to see that Condition 4 holds and that G' can be obtained in polynomial time. We prove that Conditions 2 and 3 hold, too.

Assume that G is switching-reducible, i.e., there exists a set $A \subseteq V$ such that $S(G, A)$ contains fewer edges than G. Let us count the number of edges in $S(G', A)$.

It is easy to see that if we switch a subset of V in G', the number of edges whose one endpoint is outside V is unchanged, and the number of edges with both endpoints outside V remains zero. We also observe that $S(G', A)[V]$ (the induced subgraph of $S(G', A)$ on the vertex subset V) is equal to $S(G, A)$. Hence, $S(G', A)$ has fewer edges than G', showing that G' is switching-reducible.

Moreover, if $S(G, A)$ has l edges for an integer l, then $S(G', A)$ has $l + nN$ edges. Thus, if G is $(\leq k)$-switchable, we have that G' is $(\leq k + nN)$-switchable.

Now assume that G' is switching-reducible, i.e., there exists a set $A \subseteq V(G')$ such that $S(G', A)$ has fewer edges than G'. If $A \subseteq V$, we have that $S(G, A)$ has fewer edges than G, and Condition 3 is satisfied. On the other hand, if $A \not\subseteq V$, we use the following claim.

Claim 3. *Let A be a subset of $V(G')$ and let $A' = A \cap V$. Then the number of edges in $S(G', A')$ is less than or equal to the number of edges in $S(G', A)$.*

Due to space constraints, the proof of Claim 3 is omitted. As a consequence of Claim 3, if G' is switching-reducible, then it can be reduced by switching a set $A' \subseteq V$. The same set A' then reduces G, and Condition 3 of the proposition holds. Analogically, if G' can be switched to contain L edges for an integer L, then G can be switched to contain $L - nN$ edges. Hence, we have proved Condition 2 with $k' = k + nN$.

It remains to check Condition 1. By definition, the density of G' is

$$D(G') = \frac{2|E(G')|}{(2N + n)(2N + n - 1)} \leq \frac{2\left(\binom{n}{2} + nN\right)}{(2N + n)(2N + n - 1)}$$
$$\leq \frac{n^2 + 2nN}{4N^2} \leq \frac{3nN}{4N^2} = \frac{3n}{4N} \leq c.$$

This completes the proof.   □

Proposition 5 allows us to state a stronger version of Theorem 2 for the special case of graphs with bounded density.

Theorem 4. *For every $c > 0$, the problem* SWITCH-FEW-EDGES *is NP-complete for graphs of density at most c.*

Proof. This follows from Theorem 2 and Proposition 5.   □

5   Concluding Remarks

5.1. We have not yet proved that the problem SWITCH-REDUCIBLE is NP-complete (and hence, SWITCH-MINIMAL is co-NP-complete). Note however, that if SWITCH-REDUCIBLE is NP-complete, then by Proposition 5 it remains NP-complete on graphs of bounded density.

5.2. Lindzey [9] noticed that it is possible to speed-up several graph algorithms using switching to a lower number of edges – he obtained up to super-polylogarithmic speed-ups of algorithms for diameter, transitive closure, bipartite maximum matching and general maximum matching. However, he focuses on switching digraphs (with a definition somewhat different to Seidel's switching in undirected graphs), where the situation is in sharp contrast with our results – a digraph with the minimum number of edges in its switching-class can be found in $O(n+m)$ time.

5.3. It has been observed (cf. e.g. [2]) that for a graph property $\mathcal{P}$, the complexity of deciding $\mathcal{P}$ is independent of the complexity of deciding if an input graph can be switched to a graph possessing the property $\mathcal{P}$. Switching to few edges thus adds another example of a polynomially decidable property (counting the edges is easy) whose switching version is hard. Previously known cases are the NP-hardness of deciding switching-equivalence to a regular graph [8] and deciding switching-equivalence to an H-free graph for certain specific graphs H [5].

5.4. Let $d > 0$ be a constant. What can we say about the complexity of SWITCH-REDUCIBLE and SWITCH-FEW-EDGES on graphs of maximum degree at most dn? If $d \leq \frac{1}{4}$, the two problems are trivial by Proposition 3. On the other hand, for $d \geq \frac{1}{2}$ the restriction on maximum degree becomes irrelevant, in view of Proposition 1. For any $d \in (\frac{1}{4}, \frac{1}{2})$, the complexity of the two problems on instances of maximum degree at most dn is open.

References

1. Bui, T.N., Chaudhuri, S., Leighton, F.T., Sipser, M.: Graph bisection algorithms with good average case behavior. Combinatorica **7**(2), 171–191 (1987)
2. Ehrenfeucht, A., Hage, J., Harju, T., Rozenberg, G.: Complexity issues in switching of graphs. In: Ehrig, H., Engels, G., Kreowski, H.-J., Rozenberg, G. (eds.) TAGT 1998. LNCS, vol. 1764, pp. 59–70. Springer, Heidelberg (2000)
3. Garey, M.R., Johnson, D.S., Stockmeyer, L.: Some simplified NP-complete graph problems. Theor. Comput. Sci. **1**(3), 237–267 (1976)
4. Hage, J.: Structural Aspects of Switching Classes. Ph.D. thesis, Leiden Institute of Advanced Computer Science (2001)
5. Jelínková, E., Kratochvíl, J.: On switching to H-free graphs. J. Graph Theor. **75**(4), 387–405 (2014)
6. Jelínková, E., Suchý, O., Hliněný, P., Kratochvíl, J.: Parameterized problems related to Seidel's switching. Discrete Math. Theor. Comput. Sci. **13**(2), 19–42 (2011)
7. Kozerenko, S.: On graphs with maximum size in their switching classes. Comment. Math. Univ. Carol. **56**(1), 51–61 (2015)
8. Kratochvíl, J.: Complexity of hypergraph coloring and Seidel's switching. In: Bodlaender, H.L. (ed.) WG 2003. LNCS, vol. 2880, pp. 297–308. Springer, Heidelberg (2003)
9. Lindzey, N.: Speeding up graph algorithms via switching classes. In: Kratochvíl, J., Miller, M., Froncek, D. (eds.) IWOCA 2014. LNCS, vol. 8986, pp. 238–249. Springer, Heidelberg (2015)
10. Matoušek, J., Wagner, U.: On Gromov's method of selecting heavily covered points. Discrete Comput. Geom. **52**(1), 1–33 (2014)

On Hard Instances of Non-Commutative Permanent

Christian Engels[1] and B.V. Raghavendra Rao[2(✉)]

[1] Tokyo Institute of Technology, Tokyo, Japan
engels@is.titech.ac.jp
[2] IIT Madras, Chennai, India
bvrr@cse.iitm.ac.in

Abstract. Recent developments on the complexity of the non-commutative determinant and permanent [Chien et al. STOC 2011, Bläser ICALP 2013, Gentry CCC 2014] have settled the complexity of non-commutative determinant with respect to the structure of the underlying algebra. Continuing the research further, we look to obtain more insights on hard instances of non-commutative permanent and determinant.

We show that any Algebraic Branching Program (ABP) computing the Cayley permanent of a collection of disjoint directed two-cycles with distinct variables as edge labels requires exponential size. For graphs where every connected component contains at most six vertices, we show that evaluating the Cayley permanent over any algebra containing 2×2 matrices is #P complete.

Further, we obtain efficient algorithms for computing the Cayley permanent/determinant on graphs with bounded component size, when vertices within each component are not far apart from each other in the Cayley ordering. This gives a tight upper and lower bound for size of ABPs computing the permanent of disjoint two-cycles. Finally, we exhibit more families of non-commutative polynomial evaluation problems that are complete for #P.

Our results demonstrate that apart from the structure of underlying algebras, relative ordering of the variables plays a crucial role in determining the complexity of non-commutative polynomials.

1 Introduction

Background. The study of algebraic complexity theory was initiated by Valiant in his seminal paper [18] where he showed that computing the permanent of an integer matrix is #P complete. Since then, separating the complexities of permanent and determinant has been the focal point of this research area which led to the development of several interesting results and techniques. (See [7,17] for good surveys on these topics.)

The underlying ring plays an important role in algebraic complexity theory. While the research focused mainly on the permanent vs determinant problem over fields and commutative rings there has also been an increasing amount of interest over non-commutative algebras. Nisan [16] was the first to consider the complexity of these two polynomials over non-commutative algebras. He showed

© Springer International Publishing Switzerland 2016
T.N. Dinh and M.T. Thai (Eds.): COCOON 2016, LNCS 9797, pp. 171–181, 2016.
DOI: 10.1007/978-3-319-42634-1_14

that any non-commutative arithmetic formula over the free $\mathbb{K}$ algebra computing the permanent or determinant of an $n \times n$ matrix requires size $2^{\Omega(n)}$ where $\mathbb{K}$ is any field. Later on, this was generalized to other classes of algebras in [8]. More recently, Limaye, Malod and Srinivasan [14] generalized Nisan's technique to prove lower bounds against more general classes of non-commutative circuits. Nisan's work left the problem of determining the arithmetic circuit complexity of non-commutative determinant as an open question.

In a significant breakthrough, Arvind and Srinivasan [3] showed that computing the Cayley determinant is #P hard over certain matrix algebras. Finally this question was settled by Bläser [6] who classified such algebras. Further, Gentry [13] simplified these reductions.

Motivation. Though the studies in [3,6] highlight the role of the underlying algebra in determining the complexity of the non-commutative determinant they do not shed much light on the combinatorial structure of non-commutative polynomials that are #P hard. One could ask: *Does the hardness stem from the underlying algebra or are there inherent properties of polynomials that make them #P hard in the non-commutative setting?* Our results in this paper indicate that relative ordering among the variable in the monomials constituting a polynomial f plays an important role in the hardness of certain non-commutative polynomials.

As a first step, we look for polynomials that are easier to compute than the determinant in the commutative setting and whose non-commutative versions are #P hard. Natural candidate polynomials are the elementary symmetric polynomials and special cases of determinant/permanent. One way to obtain special cases of determinant/permanent would be to restrict the structure of the underlying graph. For example, let G be a directed graph consisting of n cycles $(0,1), (2,3), \ldots,$ $(2n-2, 2n-1)$ of length two with self loops where each edge is labeled by a distinct variable. The permanent of G, $\mathrm{perm}(G)$, is given by $\prod_{i=0}^{n-1}(x_{2i,2i}x_{2i+1,2i+1} + x_{2i,2i+1}x_{2i+1,2i})$ where $x_{i,j}$ is the variable labeling of the edge (i,j). This is one of the easiest to compute but non trivial special case of permanent.

Our Results. We study the complexity of the Cayley permanent (C-perm) on special classes of graphs. The Cayley permanent (Cayley determinant) are given by $\sum_{\sigma \in S_n} x_{1,\sigma(1)} \cdots x_{n,\sigma(n)}$ ($\sum_{\sigma \in S_n} \mathrm{sgn}(\sigma) x_{1,\sigma(1)} \cdots x_{n,\sigma(n)}$) respectively. We exhibit a family of graphs G_n (consisting of a collection of disjoint two-cycles) for which any algebraic branching program (ABP) computing the C-perm must have size at least $2^{\Omega(n)}$ (Corollary 2). Further, we exhibit a parameter $\mathrm{cut}(G)$ (see Sect. 4 for the definition) for a collection G of disjoint two-cycles on n vertices such that any ABP computing C-perm(G) has size $2^{\Theta(\mathrm{cut}(G))}$ (Theorem 3). This makes the lower bound in Corollary 2 tight up to a constant factor in the exponent. It should be noted that our results also hold for the case of the Cayley determinant (C-det) on such graphs. We also observe that for graphs of component size greater or equal to six, evaluating C-perm is #P complete (Theorem 5).

On the positive side, for graphs where each strongly connected component has at most c vertices we obtain an ABP of size $n^{O(c)} c^{\mathrm{near}(G)}$ computing the

C-perm (Theorem 1) where $near(G)$ is a parameter (see Definition 1) depending on the labeling of vertices on the graph.

We demonstrate a non-commutative variant of the elementary symmetric polynomial that is #P hard over certain algebras (Theorem 7).[1] Finally, we show that computing C-perm on rank one matrices is #P hard.

Related Results. The study of commutative permanent on special classes of matrices was initiated by Barvinok [5] who gave a polynomial time algorithm for computing the permanent of rank one matrices over a field. More recently, Flarup et al. [11] showed that computing the permanent of bounded tree-width graphs can be done by polynomial size formulas. This was further extended by Flarup and Lyaudet [12] to other width measures on graphs. Datta et al. [9] showed that computing the permanent on planar graphs is as hard as the general case.

Comparison to Other Results. Results reported in [3,6,13] highlight the importance of the underlying algebra and characterizes algebras for which C-det is #P hard. In contrast, our results shed light on the role played by the order in which vertices are labeled in a graph. For example, the commutative permanent of disjoint two-cycles has a depth three formula given by $\prod_{i=0}^{n-1}(x_{2i,2i}x_{2i+1,2i+1} + x_{2i,2i+1}x_{2i+1,2i})$ whereas C-perm on almost all orderings of vertices requires exponential size ABPs.

All proofs that have been omitted due to space restrictions can be found in the full version of the paper [10].

2   Preliminaries

For definitions of complexity classes the reader is referred to any of the standard text books on Computational Complexity Theory, e.g., [1]. Let $\mathbb{K}$ be a field and $S = \mathbb{K}[x_1,\ldots,x_n]$ be the ring of polynomials over $\mathbb{K}$ in n variables. Let R denote a non-commutative ring with identity and associativity property. Unless otherwise stated, we assume that R is an algebra over $\mathbb{K}$ and contains the algebra of $n \times n$ matrices with entries from $\mathbb{K}$ as a subalgebra.

An *arithmetic circuit* is a directed acyclic graph where every vertex has an in-degree either zero or two. Vertices of zero in-degree are called *input* gates and are labeled by elements in $R \cup \{x_1 \ldots, x_n\}$. Vertices of in-degree two are called *internal* gates and have their labels from $\{\times, +\}$. An arithmetic circuit has at least one vertex of out degree zero called an *output* gate. We assume that an arithmetic circuit has exactly one output gate. A polynomial p_g in $R[x_1,\ldots,x_n]$ can be associated with every gate g of an arithmetic circuit defined in an inductive fashion. Input gates compute their label. Let g be an internal gate with left child f and right child h, then $p_g = p_f$ op p_h where op is the label of g. The polynomial computed by the circuit is the polynomial at one of the output gates and denoted by p_C. The size of an arithmetic circuit is the number of gates in it and is denoted by $size(C)$.

---

[1] One of the anonymous reviewers suggested that this result follows from a folklore fact. However since there is no explicit reference for this folklore fact, we have included the proof for completeness.

We restrict ourselves to circuits where coefficients of the polynomials computed at every gate can be represented in at most $\mathsf{poly}(\mathsf{size}(C))$ bits.

An *algebraic branching program* (ABP) is a directed acyclic graph with two special nodes s, t and edges labeled by variables or constants in R. The weight of a path is the product of the weights of its edges. The polynomial computed by an ABP P is the sum of the weights of all $s \rightsquigarrow t$ paths in P, and is denoted by p_P. We denote by the size of an ABP the number of vertices.

Over a non-commutative ring, there are many possibilities for defining the determinant/permanent of a matrix depending on the ordering of the variables (see for example [4]). We will use the well known definitions of the *Cayley determinant* and *Cayley permanent*. Let $X = (x_{i,j})_{1 \leq i,j \leq n}$ be an $n \times n$ matrix with distinct variables $x_{i,j}$. Then

$$\mathsf{C\text{-}det}(X) = \sum_{\sigma \in S_n} \mathrm{sgn}(\sigma) x_{1,\sigma(1)} \cdots x_{n,\sigma(n)}; \quad \text{and} \quad \mathsf{C\text{-}perm}(X) = \sum_{\sigma \in S_n} x_{1,\sigma(1)} \cdots x_{n,\sigma(n)}.$$

In the above, S_n denotes the set of all permutations on n symbols. Note that C-det and C-perm can also be seen as functions taking $n \times n$ matrices with entries from R as input. Given a weighted directed graph G on n vertices with weight $x_{i,j}$ for the edge $(i, j) \in E(G)$, the Cayley permanent of G denoted by $\mathsf{C\text{-}perm}(G)$ is the permanent of the weighted adjacency matrix of G. It is known that [7] $\mathsf{C\text{-}perm}(G)$ is the sum of the Cayley weights of all cycle covers of G.

The tensor product of two matrices $A, B \in \mathbb{K}^{n \times n}$ with entries $a_{i,j}, b_{i,j}$ is denoted by $A \otimes B$ and is given by

$$A \otimes B = \begin{pmatrix} a_{1,1}B & \cdots & a_{1,n}B \\ \vdots & \ddots & \vdots \\ a_{n,1}B & \cdots & a_{n,n}B \end{pmatrix}.$$

Let P be an ABP over disjoint sets of variables $X \cup Y$, with $|X| = n$ and $|Y| = m$. Let $p_P(X, Y)$ be the polynomial computed by P. P is said to be *read once certified* [15] in Y if there are numbers $0 = i_0 < i_1 < \cdots < i_m$ where i_m is at most the length of P and there is a permutation $\pi \in S_m$ such that between layers from i_j to i_{j+1} no variable other than $y_{\pi(j+1)}$ from the set Y appears as a label. We use the following result from [15]. The proof given in [15] works only in the commutative setting, see [10] for a proof of the non-commutative case.

Proposition 1 [15]. *Let P be an ABP on $X \cup Y$ read-once certified in Y. Then the polynomial $\sum_{(e_1, e_2, \ldots, e_m) \in \{0,1\}^m} p_P(X, e_1, \ldots, e_m)$ can be computed by an ABP of size $2\mathsf{size}(P)$).*

Let A be a non-deterministic s-space bounded algorithm that uses non-deterministic bits in a read-once fashion and outputs a monomial on each of the accepting paths. We assume that a non-commutative monomial is output as a string in a write-only tape and non-deterministic paths are represented by binary strings $e \in \{0,1\}^m$, $m \leq 2^{O(s)}$. The polynomial p_A computed by A is the sum of the monomial output on each of the accepting paths of A, i.e., $p(x_1, \ldots, x_n) = \sum_e A(x_1, \ldots, x_n, e)$, where the sum is taken over all accepting paths e of A, and $A(x_1, \ldots, x_n, e)$ denotes the monomial output along path represented by e.

Proposition 2 (Folklore). *Let $A(X)$ be an s-space bounded non-deterministic algorithm as above. There is a non-commutative ABP P of size $2^{O(s)}$ that computes the polynomial $p_A(X)$, the polynomial computed by $A(X)$.*

3   An Algorithm for Cayley Permanent

In this section, we give an algorithm for C-perm that is parameterized by the maximum difference between labelings of vertices in individual components.

In what follows, we identify the vertices of a graph with the set $[n]$. A directed graph G on n vertices is said to have *component size bounded by c* if every strongly connected component of G contains at most c vertices where $c > 0$. We assume that edges of G are labeled by distinct variables. Firstly, we define a parameter that measures the closeness of labelings in each component.

Definition 1. *Let G be a directed graph. The nearness parameter $\mathsf{near}(C)$ of a strongly connected component C of G is defined as $\mathsf{near}(C) = \max_{i,j \in C} |i - j|$. The nearness parameter of G is defined as $\mathsf{near}(G) = \max_C \mathsf{near}(C)$, where the maximum is taken over the set of all strongly connected components in G.*

Theorem 1. *Let G be a directed graph with component size bounded by c and edges labeled by distinct variables. Then there exists an ABP of size $n^{O(c)} c^{\mathsf{near}(G)}$ computing the Cayley permanent of the adjacency matrix of G.*

Proof. For an edge $(i, j) \in E(G)$, let $x_{i,j}$ denote the variable label on (i, j). Let A_G be the weighted adjacency matrix of G. Note that, the Cayley permanent of A_G equals the sum of weights of cycle covers in G where the weight of a cycle cover γ is the product of labels of edges in γ multiplied in the Cayley order.

We describe a non-deterministic small-space bounded procedure P that guesses a cycle cover γ in G and outputs the product of weights of γ with respect to the Cayley ordering as a string of variables. Additionally, we ensure that the algorithm P uses the non-deterministic bits in a read-once fashion, and by the closure property of ABP under read-once exponential sums (c.f. Proposition 1), we obtain the required ABP. Suppose $C_1, \ldots, C_r$ are the strongly connected components of G, sorted in the ascending order of the smallest vertex in each component. Then any cycle cover γ of G can be decomposed into cycle cover γ_i of the component C_i. The only difficulty in computing the weight of γ is the Cayley ordering of the variables. However, with a careful implementation, we show that this can be done in space $O(\log c \cdot \mathsf{near}(G) + \log n)$. We represent a cycle cover in G as a permutation γ where $\gamma(i)$ is the successor of vertex i in the cycle cover represented by γ. We begin with the description of the non-deterministic procedure P. Let T represent the set of vertices v in the partial cover that is being built by the procedure where the weight of the edge going out of v is not yet output, and pos the current position going from 1 to n. Let $\mathsf{Acc}(G)$ be the sum of the terms output by the following algorithm on all accepting paths.

1. Initialize pos := 1, $T := \emptyset$, $\gamma :=$ the cycle cover of the empty graph, $f = 1$.
2. For $1 \leq i \leq r$ repeat steps 3 & 4.
3. Non-deterministically guess a cycle cover γ' in C_i, and set $\gamma = \gamma \uplus \gamma'$, $T = T \cup V(C_i)$ where $V(C_i)$ is the set vertices in C_i.
4. While there is a vertex $k \in T$ with $k = $ pos do the following:
   Set $f = f \cdot x_{k,\gamma(k)}$; pos := pos + 1; and $T := T \setminus \{k\}$.
5. If pos $= n$, then output f and accept.

Claim. $\mathsf{Acc}(G) = \mathsf{C\text{-}perm}(G)$. Moreover, the algorithm P uses $O(\log c \cdot \mathsf{near}(G) + \log n)$ space, and is read-once on the non-deterministic bits.

Proof (of the Claim). Recall that a permutation $\gamma \in S_n$ is a cycle cover of G if and only if it can be decomposed into vertex disjoint cycle covers $\gamma_1, \ldots, \gamma_r$ of the strongly connected components $C_1, \ldots, C_r$ in G. Thus Step 3 enumerates all possible cycle covers in G. Also, the weights output at every accepting path are in the Cayley order.

We have $T = \{k \mid$ pos $< k$ and k occurs in the components already explored $\}$. Firstly, we argue that at any point in time in the algorithm, $|T| \leq \mathsf{near}(G) + c$. Suppose the algorithm has processed components up to C_i and is yet to process C_{i+1}. Let $\mu = \max_{v \in T} v$. Since the components are in ascending order with respect to the smallest vertex in them, the component C_j with $\mu \in C_j$ must have $\mathsf{near}(C_j) \geq \mu - $ pos. Thus $\mu - $ pos $\leq \mathsf{near}(G)$. Also, just before step 3 in any iteration, for any $v \in T$, we have pos $< v \leq \mu$ and hence $|T| \leq \mu - pos+c \leq \mathsf{near}(G) + c$.

Note that it is enough to store the labels of the vertices in T and the choice $\gamma(v)$ made during the non-deterministic guess for each $v \in T$ and hence $O(|T| \log n)$ additional bits of information needs to be stored. However, we will show that it is possible to implement the algorithm without explicitly remembering the vertices in T and using only $O(|T| \log c)$ additional bits in memory. Suppose that the vertices in T are ordered as they appear in $C_1, C_2, \ldots, C_r$ where vertices within a component are considered in the ascending order of their labels. Let B be a vector of length $\mathsf{near}(G) + c$ where each entry B_j is $\log c$ bits long which indicates the neighbour of the jth vertex in T. Now, we show how to implement step 4 in the procedure using B as a data structure for T. To check if there is a $k \in T$ with $k = $ pos, we can scan the components from $C_1, \ldots, C_i$ and check if the vertex assigned to pos occurs in one of the components. Remember that $\gamma(k)$ is the successor of k in the cycle cover γ. To obtain $\gamma(k)$ from B, we need to know the number j of vertices v that appear in components $C_1, \ldots, C_i$ such that $v \geq $ pos and that occur before k. Then $\gamma(k) = B_j$. Once B_j is used, we remove B_j from B and shift the array $B_{j+1}, \ldots B_{\mathsf{near}(G)+c}$ by one index towards the left. Further, we can implement step 3 by simply appending the information for $V(C_i)$ given by γ' to the right of the array B. We require at most $O(c \log n)$ bits of space guessing a cycle cover γ_i for component C_i which can be re-used after the non-deterministic guessing of γ_i is complete. Thus the overall space requirement of the algorithm is bounded by $O(\log c \cdot (\mathsf{near}(G) + c) + c \log n)$.

By Proposition 2, we get an ABP P computing a polynomial $p_G(X, Y)$ such that $\mathsf{C\text{-}perm}(G) = \sum_{e_1, \ldots, e_m \in \{0,1\}} p_G(X, e)$, $m = O(c \log n)$. Combining the above algorithm with the closure property of algebraic branching programs over read-once variables given by Proposition 1, we get a non-commutative arithmetic branching program computing $\mathsf{C\text{-}perm}(G)$. It can be seen that size of the resulting branching program is at most $2^{O((c\log c + \log c \cdot \mathsf{near}(G)) + c\log n)} = n^{O(c)} \cdot c^{\mathsf{near}(G)}$ for large enough n.

Corollary 1. *Let G be as in Theorem 1. There is an ABP of size $n^{O(c)}c^{\mathsf{near}(G)}$ computing the Cayley determinant of G.*

4 Unconditional Lower Bound

We now show that any branching program computing the non-commutative permanent of directed graphs with component size 2 must be of exponential size. This shows that the upper bound in Theorem 1 is tight up to a constant factor in the exponent, however, with a different but related parameter. All our lower bound results hold for free algebras over any field $\mathbb{K}$.

Our proof crucially depends on Nisan's [16] partial derivative technique. We begin with some notations following his proof. Let f be a non-commutative degree d polynomial in n variables. Let $B(f)$ denote the smallest size of a non-commutative ABP computing f. For $k \in \{0, \ldots, d\}$ let $M_k(f)$ be the matrix with rows indexed by all possible sequences containing k variables and columns indexed by all possible sequences containing $d - k$ variables (repetitions allowed). Hence the matrix has dimension $n^k \times n^{d-k}$. The entry of $M_k(f)$ at $(x_{i_1} \ldots x_{i_k}, x_{j_1} \ldots x_{j_{d-k}})$ is the coefficient of the monomial $x_{i_1} \cdots x_{i_k} \cdot x_{j_1} \cdots x_{j_{d-k}}$ in f. Nisan established the following result:

Theorem 2 [16]. *For any homogeneous polynomial f of degree d, we have $B(f) = \sum_{k=0}^{d} \mathsf{rank}(M_k(f))$.*

We prove lower bounds for the Cayley permanent of graphs with every strongly connected component of size exactly 2, i.e., each strongly connected component being a two-cycle with self loops on the vertices. Note that any collection of $n/2$ vertex disjoint two-cycles can be viewed as a permutation $\pi \in S_n$ consisting of disjoint transpositions and that π is an involution. Conversely, any involution π on n elements represents a graph G_π with connected component size 2.

For a permutation $\pi \in S_n$ let the *cut at i* denoted by $C_i(\pi)$ be the set of pairs $(j, \pi(j))$ that cross i, i.e., $C_i(\pi) = \{(j, \pi(j)) \mid i \in [j, \pi(j)] \cup [\pi(j), j]\}$. The *cut* parameter $\mathsf{cut}(\pi)$ of π is defined as $\mathsf{cut}(\pi) = \max_{1 \le k \le n} |C_k(\pi)|$. Let G be a collection of vertex disjoint 2-cycles denoted by $(a_1, b_1), \ldots, (a_{n/2}, b_{n/2})$ where n is even. The corresponding involution is $\pi_G = (a_1, b_1) \cdots (a_{n/2}, b_{n/2})$. By abusing the notation a bit, we let $\mathsf{cut}(G) = \mathsf{cut}(\pi_G)$. Without loss of generality, assume that $a_i < b_i$, and $a_1 < a_2 < \cdots < a_{n/2}$. Firstly, we note that $\mathsf{cut}(\pi)$ is bounded by $\mathsf{near}(G)$.

Lemma 1. *For any collection of disjoint 2-cycles G on n vertices, $\mathsf{cut}(\pi) \le \mathsf{near}(G)$ where π is the involution represented by G.*

Further, we note that the upper bound given in Theorem 1 holds true even if we consider $\mathsf{cut}(G)$ instead of $\mathsf{near}(G)$.

Lemma 2. *Let G be a collection of disjoint 2-cycles and self loops where every edge is labeled by a distinct variable or a constant from R. Then there is an ABP of size $2^{O(\mathsf{cut}(G))}n^2$ computing the Cayley permanent of G.*

Lemma 3. *Let G be a collection of ℓ disjoint two-cycles described by the involution π and self loops at every vertex with edge labeled by distinct variables. Then $M_\ell(\text{C-perm}(G))$ contains $I_2^{\otimes t}$ as a sub-matrix where $t = \max_k |C_k(\pi)|$, $A^{\otimes t}$ is the tensor product of A with itself t times and I_2 is the 2×2 identity matrix.*

Proof. Let $k \in [\ell]$, and $r = |C_k(\pi)| \leq \ell$. Let $C_k(\pi) = \{(a_{i_1}, b_{i_1}), \ldots, (a_{i_r}, b_{i_r})\}$ be such that $a_{i_j} \leq k \leq b_{i_j}$ for all j. Let G_k be the graph restricted to involutions in $C_k(\pi)$. By induction on m, we argue that $M_r(\text{C-perm}(G_k))$ contains $I_2^{\otimes r}$ as a sub-matrix. The lemma would then follow since $M_r(\text{C-perm}(G_k))$ is itself a sub-matrix of $M_\ell(\text{C-perm}(G))$.

We begin with $r = 1$ as the base case. Consider the transposition (a_{i_j}, b_{i_j}), with $a_{i_j} \leq k \leq b_{i_j}$. The corresponding two cycle has four edges. Let f_{i_j} be the Cayley permanent of this graph then $M_1(f_{i_j})$ has the 2×2 identity matrix as a sub-matrix. Let us dwell on this simple part. For ease of notation let the variables corresponding to the self loops be given by x_a, x_b for (a_{i_j}, a_{i_j}) and (b_{i_j}, b_{i_j}) respectively and the edge (a_{i_j}, b_{i_j}) by $x_{(a,b)}$ and the edge (b_{i_j}, a_{i_j}) by $x_{b,a}$. Now our matrix has monomials $x_a, x_{a,b}$ as rows and $x_b, x_{b,a}$ as columns. We can ignore the other orderings as these will always be zero. As the valid cycle covers are given by $x_a x_b$ and $x_{a,b} x_{b,a}$ the proof is clear.

For the induction step, suppose $r > 1$. Suppose $a_1 < a_2 < \cdots < a_r$. Let G'_k be the graph induced by $C_k(\pi) \setminus (a_1, b_1)$. Let $M' = M_{r-1}(\text{C-perm}(G'_k))$. The rows of M' are labeled by monomials consisting of variables with first index $\leq k$ and the columns of M' are labeled by monomials consisting only of variables with first index $> k$. Let $M = M_r(\text{C-perm}(G_k))$. M can be obtained from M' as follows: Make two copies of the row labels of M', the first one with monomials pre-multiplied by x_{a_1,a_1}, and the second pre-multiplied by x_{a_1,b_1}. Similarly, make two copies of the columns of M', the first by inserting x_{b_1,b_1} to the column labels of M' at appropriate position, and then inserting x_{b_1,a_1} similarly. Now, the matrix M can be viewed as two copies of M' that are placed along the diagonal. Thus $M = M' \otimes I_2$, combining this with Induction Hypothesis completes the proof.

Remark 1. *It should be noted that the ordering of the variables is crucial in the above argument. If $a_1, b_1 < k$ in the above, then $\mathsf{rank}(M) = \mathsf{rank}(M')$.*

Theorem 3. *Let G be a collection of disjoint two cycles described by the involution π and self loops at every vertex, with edges labeled by distinct variables. Then any non-commutative ABP computing the Cayley permanent on G has size at least $2^{\Omega(\mathsf{cut}(G))}$.*

Let $\pi = (a_1, b_1) \cdots (a_{n/2}, b_{n/2})$, $a_1 < a_2 < \cdots < a_{n/2}$ be an involution. Then G_π is the set of 2-cycles $(a_1, b_1), \ldots, (a_{n/2}, b_{n/2})$ and self loops at every vertex.

Corollary 2. *Let G be a collection of disjoint two cycles described by the involution π and self loops at every vertex, with edges labeled by distinct variables. Then $B(\mathsf{C\text{-}perm}(G)) \in 2^{\Theta(\mathsf{cut}(G))}$. Further, there exists a graph G with $\mathsf{cut}(G) = \Theta(n)$.*

Finally, we have,

Theorem 4. *For all but a $1/\sqrt{n}$ fraction of graphs G with connected component size 2, any ABP computing the $\mathsf{C\text{-}perm}$ on G requires size $2^{\Omega(n)}$.*

5   #P Completeness

In this section, we show multiple hardness results for simple polynomials over certain classes of non-commutative algebras. We give a #P completeness result for specific graphs of component size at most six. The completeness result is obtained by a careful analysis of the parameters and a small modification of the reduction from $\#SAT$ to non-commutative determinant given recently by Gentry [13].

Theorem 5. *Let R be a division algebra over a field $\mathbb{K}$ of characteristic zero containing the algebra of 2×2 matrices over $\mathbb{K}$. Computing the Cayley Permanent on graphs with component size 6 with edges labeled from R is #P complete.*

It is known that computing the commutative permanent of the weighted adjacency matrix of a planar graph is as hard as the general case [9]. We observe that the reduction in [9] extends to the non-commutative case.

Theorem 6. $\mathsf{C\text{-}perm} \leq_m^p \mathsf{planar\text{-}C\text{-}perm}$; *and* $\mathsf{C\text{-}det} \leq_m^p \mathsf{planar\text{-}C\text{-}det}$. *Moreover, the above reductions work over any non-commutative algebra.*

We demonstrate some more families of polynomials whose commutative variants are easy but certain non-commutative variants are as hard as the permanent polynomial. We begin with a non-commutative variant of the elementary symmetric polynomial. The elementary symmetric polynomial of degree d, $\mathsf{Sym}_{n,d}$ is given by $\mathsf{Sym}_{n,d}(x_1,\ldots,x_n) = \sum_{S \subseteq [n],\ |S|=d} \prod_{i \in S} x_i$. There are several non-commutative variants of the above polynomial. The first one is analogous to the Cayley permanent, i.e., $\mathsf{Cayley\text{-}Sym}_{n,d} = \sum_{S=\{i_1 < i_2 < \cdots < i_d\}} \prod_{j=1}^{d} x_{i_j}$. It is not hard to see that the above mentioned non-commutative version of $\mathsf{Cayley\text{-}Sym}_{n,d}$ can be computed by depth 3 non-commutative circuits for every value of $d \in [n]$. However, the above definition is not satisfactory, since it is not invariant under permutation of variables, which is the inherent property of elementary symmetric polynomials. We define a variant of non-commutative elementary symmetric polynomial which is invariant under the permutation of variables.

$$\mathsf{nc\text{-}Sym}_{n,d}(x_1,\ldots,x_n) \stackrel{\triangle}{=} \sum_{\{i_1,\ldots,i_d\} \subseteq [n]} \sum_{\sigma \in S_d} \prod_{j=1}^{d} x_{i_{\sigma(j)}}.$$

We show that with coefficients from the algebra of $n \times n$ matrices allowed, $\mathsf{nc\text{-}Sym}_{n,d}$ cannot be computed by polynomial size circuits unless $\mathsf{VP} = \mathsf{VNP}$. We need the following definition introduced in [2,3].

Definition 2. *The* Hadamard product *between two polynomials* $f = \sum_m \alpha_m m$ *and* $g = \sum_m \beta_m m$, *written as* $f \odot g$, *is defined as* $f \odot g = \sum_m \alpha_m \beta_m m$.

Theorem 7. *Over any* $\mathbb{K}$ *algebra* R *containing the* $n \times n$ *matrices as a subalgebra,* nc$-$Sym$_{n,n}$ *does not have polynomial size arithmetic circuits unless* perm$_n \in$ VP.

Proof. Suppose that nc$-$Sym$_{n,n}$ has a circuit C of size polynomial in n. We need to show that perm $\in$ VP. Let $X = (x_{i,j})_{1 \le i,j \le n}$ be matrix of variables, and $y_1, \ldots, y_n$ be distinct variables different from $x_{i,j}$. In the commutative setting, it was observed in [19] that perm(X) equals the coefficient of $y_1 \cdots y_n$ in the polynomial

$$P(X,Y) \stackrel{\triangle}{=} \prod_{i=1}^{n} \left(\sum_{j=1}^{n} x_{i,j} y_j \right) \tag{1}$$

over the polynomial ring $\mathbb{K}[x_{1,1}, \ldots, x_{n,n}]$. However, the same cannot be said in the case of non-commuting variables. If $x_{i,j} y_k = y_k x_{i,j}$ for $i, j, k \in [n]$, then in the non-commutative development of (1), the sum of coefficients of *all* permutations of the monomial $y_1 \cdots y_n$ equals perm(X) i.e., the commutative permanent. Hence the value perm(X) can be extracted using a Hadamard product with nc$-$Sym$_{n,n}(y_1, \ldots, y_n)$ and then substituting $y_1 = 1, \ldots, y_n = 1$. However, we cannot assume $x_{i,j} y_k = y_k x_{i,j}$, since the Hadamard product may not be computable under this assumption. Let $\ell = \sum_{i,j} x_{i,j}$. Now we argue that perm$(X) =$ (nc$-$Sym$_{n,n}(\ell y_1, \ldots, \ell y_n) \odot P)(y_1 = 1, \ldots, y_n = 1)$. Given a permutation $\sigma \in S_n$, there is a unique monomial $m_\sigma = x_{1,\sigma(1)} y_{\sigma(1)} \cdots x_{n,\sigma(n)} y_{\sigma(n)}$ in P containing the variables $y_{\sigma(1)}, \ldots, y_{\sigma(n)}$ in that order. Thus taking Hadamard product with P filters out all monomials but m_σ from the term $\prod_{i=1}^{n} \ell y_{\sigma(i)}$. The monomials where a y_j occurs more than once are eliminated by nc$-$Sym$_{n,n}(\ell y_1, \ldots, \ell y_n)$. Thus the only monomials that survive in the Hadamard product are of the form m_σ, $\sigma \in S_n$. Now substituting $y_i = 1$ for $i \in [n]$ we get perm$(X) =$ (nc$-$Sym$_{n,n}(\ell y_1, \ldots, \ell y_n) \odot P)(y_1 = 1, \ldots, y_n = 1)$.

Note that the polynomial $P(X,Y)$ can be computed by an ABP of size $O(n^2)$. Then, by [2,3], we obtain an arithmetic circuit D of size $O(n^2 \text{size}(C))$ that computes the polynomial nc$-$Sym$_{n,n} \odot P$. Substituting $y_1 = 1, \ldots, y_n = 1$ in D gives the required arithmetic circuit for perm(X).

Barvinok [5] showed that computing the permanent of an integer matrix of constant rank can be done in strong polynomial time. In a similar spirit, we explore the complexity of computing the Cayley permanent of bounded rank matrices with entries from $\mathbb{K} \cup \{x_1, \ldots, x_n\}$. We consider the following notion of rank for matrices with variable entries. Let $A \in (\mathbb{K} \cup \{x_1, \ldots, x_n\})^{n \times n}$. Then row-rank$(A) = \max_{a_1, \ldots, a_n \in \mathbb{K}} \text{rank}(A|_{x_1=a_1, \ldots, x_n=a_n})$. The column rank of A is defined analogously. As opposed to the case of the commutative permanent, for any algebra R containing the algebra of $n \times n$ matrices over $\mathbb{K}$, we have:

Corollary 3. C-perm *and* C-det *of rank one matrices with entries from* $\mathbb{K} \cup \{x_1, \ldots, x_n\}$ *over any* $\mathbb{K}$ *algebra does not have polynomial size arithmetic circuits unless* perm $\in$ VP.

Acknowledgements. The authors like to thank V. Arvind and Markus Bläser for helpful discussions and pointing out specific problems to work on. The authors also thank anonymous referees for their comments which helped in improving the presentation. This work was partially done while the first author was visiting IIT Madras sponsored by the Indo-Max-Planck Center for Computer Science.

References

1. Arora, S., Barak, B.: Computational Complexity: A Modern Approach. Cambridge University Press, Cambridge (2009)
2. Arvind, V., Joglekar, P.S., Srinivasan, S.: Arithmetic circuits and the hadamard product of polynomials. In: FSTTCS, pp. 25–36 (2009)
3. Arvind, V., Srinivasan, S.: On the hardness of the noncommutative determinant. In: STOC, pp. 677–686 (2010)
4. Aslaksen, H.: Quaternionic determinants. Math. Int. **18**(3), 57–65 (1996)
5. Barvinok, A.I.: Two algorithmic results for the traveling salesman problem. Math. Oper. Res. **21**(1), 65–84 (1996)
6. Bläser, M.: Noncommutativity makes determinants hard. In: Fomin, F.V., Freivalds, R., Kwiatkowska, M., Peleg, D. (eds.) ICALP 2013, Part I. LNCS, vol. 7965, pp. 172–183. Springer, Heidelberg (2013)
7. Bürgisser, P.: Completeness and Reduction in Algebraic Complexity Theory. Springer, Heidelberg (2000)
8. Chien, S., Sinclair, A.: Algebras with polynomial identities and computing the determinant. SIAM J. Comput. **37**(1), 252–266 (2007)
9. Datta, S., Kulkarni, R., Limaye, N., Mahajan, M.: Planarity, determinants, permanents, and (unique) matchings. ToCT **1**(3), 10 (2010)
10. Engels, C., Raghavendra Rao, B.V.: New Algorithms and Hard Instances for Non-Commutative Computation. ArXiv e-prints, September 2014
11. Flarup, U., Koiran, P., Lyaudet, L.: On the expressive power of planar perfect matching and permanents of bounded treewidth matrices. In: Tokuyama, T. (ed.) ISAAC 2007. LNCS, vol. 4835, pp. 124–136. Springer, Heidelberg (2007)
12. Flarup, U., Lyaudet, L.: On the expressive power of permanents and perfect matchings of matrices of bounded pathwidth/cliquewidth. ToCS **46**(4), 761–791 (2010)
13. Gentry, C.: Noncommutative determinant is hard: a simple proof using an extension of barrington's theorem. In: CCC, pp. 181–187, June 2014
14. Limaye, N., Malod, G., Srinivasan, S.: Lower bounds for non-commutative skew circuits. In: Electronic Colloquium on Computational Complexity (ECCC), vol. 22, p. 22 (2015)
15. Mahajan, M., Rao, B.V.R.: Small space analogues of valiant's classes and the limitations of skew formulas. Comput. Complex. **22**(1), 1–38 (2013)
16. Nisan, N.: Lower bounds for non-commutative computation (extended abstract). In: STOC, pp. 410–418 (1991)
17. Shpilka, A., Yehudayoff, A.: Arithmetic circuits: a survey of recent results and open questions. FTTS **5**(3–4), 207–388 (2010)
18. Valiant, L.G.: Completeness classes in algebra. In: STOC 1979, pp. 249–261 (1979)
19. von zur Gathen, J.: Feasible arithmetic computations: Valiant's hypothesis. J. Symb. Comput. **4**(2), 137–172 (1987)

The Effect of Range and Bandwidth on the Round Complexity in the Congested Clique Model

Florent Becker[1], Antonio Fernández Anta[2], Ivan Rapaport[3(✉)], and Eric Rémila[4]

[1] LIFO (EA 4022), Université d'Orléans, Orléans, France
[2] IMDEA Networks Institute, Madrid, Spain
[3] DIM-CMM (UMI 2807 CNRS), Universidad de Chile, Santiago, Chile
rapaport@dim.uchile.cl
[4] Univ. Lyon, UJM Saint-Etienne, Saint-Etienne, France

Abstract. The congested clique model is a message-passing model of distributed computation where k players communicate with each other over a complete network. Here we consider synchronous protocols in which communication happens in rounds (we allow them to be randomized with public coins). In the *unicast* communication mode, each player i has her own n-bit input x_i and may send $k-1$ different b-bit messages through each of her $k-1$ communication links in each round. On the other end is the *broadcast* communication mode, where each player can only broadcast a single message over all her links in each round. The goal of this paper is to complete our Brief Announcement at PODC 2015, where we initiated the study of the space that lies between the two extremes. For that purpose, we parametrize the congested clique model by two values: the *range* r, which is the maximum number of different messages a player is allowed to send in each round, and the *bandwidth* b, which is the maximum size of these messages. We show that the space between the unicast and broadcast congested clique models is very rich and interesting. For instance, we show that the round complexity of the pairwise set-disjointness function PWDISJ is completely sensitive to the range r. This translates into a $\Omega(k)$ gap between the unicast ($r = k-1$) and the broadcast ($r = 1$) modes. Moreover, provided that $r \geq 2$ and $rb/\log r = O(k)$, the round complexity of PWDISJ is $\Theta(n/k \log r)$. On the other hand, we also prove that the behavior of PWDISJ is exceptional: almost every boolean function f has maximal round complexity $\Theta(n/b)$. Finally, we prove that $\min\left(\left\lceil \frac{b'}{\lceil \log r \rceil} \right\rceil, \left\lceil \frac{r'}{r-1} \right\rceil \left\lceil \frac{b'}{b} \right\rceil\right)$ is an upper bound for the gap between the round complexities with parameters (b, r) and parameters (b', r') of any boolean function.

Supported in part by the ANR project QuasiCool (ANR-12-JS02-011-01), MINECO grant TEC2014- 55713-R, Regional Government of Madrid (CM) grant Cloud4BigData (S2013/ICE-2894, co-funded by FSE & FEDER), NSF of China grant 61520106005, EC H2020 grants ReCred and NOTRE, CONICYT via Basal in Applied Mathematics, Núcleo Milenio Información y Coordinación en Redes ICM/FIC RC130003, Fondecyt 1130061.

T.N. Dinh and M.T. Thai (Eds.): COCOON 2016, LNCS 9797, pp. 182–193, 2016.
DOI: 10.1007/978-3-319-42634-1_15

1   Introduction

In this paper we study a synchronous, message-passing model of distributed computation where the underlying communication network is a complete graph. Therefore, the only obstacle to perform any task is due to *congestion*. In fact, the main theoretical purpose of this model, known as *congested clique*, is to serve as a basic model for understanding the role played by congestion in distributed computation [14,15,21,25,27,28]. (Besides this, there are interesting connections between the congested clique model and popular systems such as MapReduce [20].)

The model is defined as follows. There are k players. Each player has her own n-bit input x_i and they all collaborate in order to compute a joint boolean function $f(x_1, \ldots, x_k)$. They communicate with each other in synchronous rounds. More precisely, each of the k players may send up to $k - 1$ different b-bit messages through each of her $k - 1$ communication links. A protocol that computes f stops when every player knows the output. We use the number of rounds as the goodness metric to be minimized. The absolute minimum of this parameter is what we call *round complexity*. In this paper all protocols are allowed to be randomized with public coins. More precisely, the k players have access to a common infinite string of independent random bits. Protocols may return the wrong answer with probability at most ϵ, for some fixed, small $\epsilon > 0$.

Most work on this (unicast) congested clique model considers the joint input as a graph G by giving to each player i the boolean vector $x_i \in \{0,1\}^n$, which is the indicator function of her neighborhood in G. Note that in this case $n = k$ and, therefore, the total number of bits exchanged in each round is bn^2. Unfortunately, due to the huge number of bits transmitted globally per round (even for $b = 1$), no lower bound is known for this model. Drucker et al. gave in [15] an explanation for this difficulty. They proved that in this model it is possible to simulate powerful classes of bounded-depth circuits (and therefore lower bounds in the congested clique would yield lower bounds in circuit complexity). The intrinsic power of the (synchronous) congested clique model has allowed some authors [10,14,19,21] to provide extremely fast protocols for some natural graph problems (assuming always that $b = \log n$, following the spirit of the $\mathcal{CONGEST}$ model [29]).

In the broadcast version of the congested clique model, each player can only broadcast a single b-bit message over all her links in each round [15]. This setting is equivalent to the multi-party, number-in-hand computation model, where communication takes place in a shared blackboard [1,2,5–7,15]. In fact, writing a message $\mathcal{M}$ on the blackboard is equivalent to broadcasting $\mathcal{M}$. In this setting, the number of transmitted bits per round decreases from bn^2 to bn. Therefore, obtaining lower bounds using communication complexity reductions becomes possible. For instance, detecting deterministically a triangle in the input graph G requires $\Omega(n/(e^{\mathcal{O}(\sqrt{\log n})}b))$ rounds [15]. On the other hand, fast protocols are also known in the broadcast congested clique model [1,2,18,23].

There is a particular boolean function that we are going to use throughout this paper. This function, that we call *pairwise set-disjointness*, is defined below.

Definition 1. *Let $k = 2k'$. Let $x = (x_1 \ldots x_k) \in (\{0,1\}^n)^k$. Each x_i is the indicator vector of a subset $X_i \subseteq \{1, \ldots, n\}$. Function pairwise set-disjointness* PWDISJ *is defined by:* PWDISJ$(x) = 1$ *if* $\forall 1 \leq i \leq k', X_i \cap X_{i+k'} = \emptyset$; *and* PWDISJ$(x) = 0$ *otherwise.*

Our goal is to complete the work of [4], where we initiated the study of the round complexity of boolean functions according to two parameters of the model:

- The *range r*: the maximum number of different messages a player can send over her links in one round.
- The *bandwidth b*: the maximum size, in bits, of each of these messages.

By analogy with the notation introduced in [15], we denote this model by CLIQUE-RCAST$_{r \times b}$. Note that the two extreme cases $r = 1$ and $r = k - 1$, which correspond to the broadcast and the unicast communication modes, are the cases already considered in the literature. More precisely,

$$\text{CLIQUE-RCAST}_{(k-1) \times b} = \text{CLIQUE-UCAST}_b,$$
$$\text{CLIQUE-RCAST}_{1 \times b} = \text{CLIQUE-BCAST}_b.$$

Note also that, if the available bandwidth b is too small, then having a big range r becomes useless, since the number of possible different messages with a bandwidth b is 2^b. More precisely,

$$\forall r \geq 2^b, \text{CLIQUE-RCAST}_{r \times b} = \text{CLIQUE-RCAST}_{2^b \times b} = \text{CLIQUE-UCAST}_b.$$

Thus, in the sequel, we will assume that $r \leq 2^b$. We denote by ROUND$_{r \times b}(f)$ the round complexity of function f. That is, ROUND$_{r \times b}(f)$ denotes the minimal number of rounds needed by any k-player protocol in CLIQUE-RCAST$_{r \times b}$ for computing f. We also denote,

$$\mathbf{U}\text{ROUND}_b(f) = \text{ROUND}_{(k-1) \times b}(f),$$
$$\mathbf{B}\text{ROUND}_b(f) = \text{ROUND}_{1 \times b}(f).$$

A protocol in CLIQUE-RCAST$_{r \times b}$ is said to be a *broadcasting protocol* if it consists of every player broadcasting its complete input. Obviously, for any function f, there exists a broadcasting protocol which computes f, and we get the universal bound ROUND$_{r \times b}(f) \leq \mathbf{B}\text{ROUND}_b(f) \leq \lceil n/b \rceil$. In order to understand the role played by the range r and the bandwidth b in the round complexity of the congested clique model we define the following ratio.

$$\Gamma_{r' \times b'}^{r \times b}(f) = \frac{\text{ROUND}_{r \times b}(f)}{\text{ROUND}_{r' \times b'}(f)}.$$

The values above obviously depend on k, n and ϵ. But we omit them in order to avoid heavy notation. Finally, by taking the uniform probability over $\{0,1\}^{\{0,1\}^{kn}}$, we also consider what happens with random boolean functions. For instance, we compute probabilities such us $\Pr\{\Gamma_{r' \times b'}^{r \times b}(f) = \alpha\}$, for fixed α.

1.1 Our Results

In Sect. 2 we compare the broadcast model and the unicast model. For that purpose we consider the pairwise set-disjointness function PWDISJ. We prove that $\mathbf{U}\text{ROUND}_b(\text{PWDISJ}) = \mathcal{O}(n/kb)$ while $\mathbf{B}\text{ROUND}_b(\text{PWDISJ}) = \Omega(n/b)$. In other words, $\Gamma^{1\times b}_{(k-1)\times b}(\text{PWDISJ}) = \Omega(k)$. This gives a large gap between the unicast and broadcast congested clique models, that grows at least linearly with k.

In Sect. 3 we prove that the round complexity of PWDISJ is completely sensitive to the range r even in the intermediate values between unicast and broadcast. More precisely, we prove that for k sufficiently large and for $r \geq 2$ such that $rb/\log r = \mathcal{O}(k)$ the following holds: $\text{ROUND}_{r\times b}(\text{PWDISJ}) = \Theta(n/k\log r)$. Then, we give some interpretations to this result. In particular, we conclude that $\Gamma^{r\times \log k}_{r'\times \log k}(\text{PWDISJ}) = \Theta(\log r'/\log r)$ for every $r' \geq r \geq 2$. Note that the logarithmic bandwidth is the most studied case in the congested clique model, and this result yields a hierarchy of models of different computational power according to the range r for this case.

In Sect. 4 we prove that almost every boolean function f satisfies that $\mathbf{U}\text{ROUND}_b(f) = \mathbf{B}\text{ROUND}_b(f) = \lceil n/b \rceil$, provided that k is sufficiently large and that $0 \leq \epsilon \leq 0.2$. In other words, $\Gamma^{1\times b}_{(k-1)\times b}(f) = 1$ for almost every f. This means that the gap we found in Sect. 2 for function PWDISJ is exceptional and that the power given by having $r > 1$ is almost always useless. Nevertheless, as pointed out by Drucker et al. [15], finding for $k = n$ an explicit boolean function f with the behavior $\mathbf{U}\text{ROUND}_b(f) = \omega(1)$ is (equivalent to solving) a long-standing open problem in circuit complexity theory.

The goal of Sect. 5 is to compare models with different combinations of range and bandwidth for arbitrary boolean functions f. For doing this we analyze the ratio $\Gamma^{r\times b}_{r'\times b'}(f)$. We make the following observation: for almost every function f we have $\Gamma^{r\times b}_{r'\times b'}(f) = \Theta(b'/b)$. Moreover, if $r \geq r'$ or $r = 2^b$ then $\Gamma^{r\times b}_{r'\times b'}(f) \leq \lceil b'/b \rceil$ for every boolean function f. The general upper bound we obtain is the following $\Gamma^{r\times b}_{r'\times b'}(f) \leq \min\left(\left\lceil \frac{b'}{\lceil \log r \rceil} \right\rceil, \left\lceil \frac{r'}{r-1} \right\rceil \left\lceil \frac{b'}{b} \right\rceil\right)$, for $r \geq 2$.

1.2 Related Work: The Asynchronous Case

The congested clique model with bandwidth $b = 1$ –that is, the multiplayer, number-in-hand, message passing model– was introduced by Dolev and Feder [13]. The main difference with our setting is that the original model was *asynchronous*. Hence, protocols, instead of being designed to minimize the number of rounds, were designed to minimize the number of exchanged bits. The first communication complexity lower bounds were obtained by Ďuriš and Rolim [16].

Recently, new techniques and new results have been developed, and tight bounds for the communication complexity of different functions have been obtained. In [30] the authors introduced the symmetrization technique and were able to prove tight $\Omega(nk)$ lower bounds for several direct-sum-like functions such as coordinate-wise AND or coordinate-wise OR. These lower bounds also apply in the blackboard communication mode, where players write messages on a blackboard, visible to everybody. (Note that, in the asynchronous setting, the communication complexity in the blackboard mode gives stronger lower bounds than the communication complexity in the message-passing, point-to-point mode.) This symmetrization technique has been used and developed by other authors as well [26, 31].

It is important to point out that there exists a strict separation between the blackboard communication mode and the message-passing communication mode. For instance, the communication complexity for computing the multiparty set-disjointness function is $\Theta(n \log k + k)$ in the blackboard communication mode [9] and it is $\Theta(nk)$ in the message-passing communication mode [8]. These results on set-disjointness were obtained by using information complexity, a theory introduced in [11]. Information complexity turned out to be an extremely useful theory for proving communication complexity lower bounds [3, 12, 17].

2   A Gap in the Round Complexity of Broadcast Versus Unicast

The first question we would like to answer is the following: How much do we gain if, instead of broadcasting, we have the possibility of sending at least two different messages in each round? This seems to be a simple question. But it is a fundamental one if we want to understand the role played by the range in the congested clique model. For answering this we use the pairwise set-disjointness function PWDISJ defined in Sect. 1.

Theorem 1. $\mathbf{U}\text{ROUND}_b(\text{PWDISJ}) = \mathcal{O}(n/kb)$.

Proof. We prove that $\mathbf{U}\text{ROUND}_b(\text{PWDISJ}) \leq \left\lceil \frac{\lceil n/k \rceil}{b} \right\rceil + 1$. The protocol is as follows. Let $T = \left\lceil \frac{\lceil n/k \rceil}{b} \right\rceil$. For every $1 \leq t \leq T$, let

$$w_{i,j,t} = (x_i)_{(j-1)\lceil n/k \rceil + (t-1)b+1}, \dots, (x_i)_{(j-1)\lceil n/k \rceil + tb}.$$

Round $1 \leq t \leq T$. Each player i sends to each player j (including itself) the b bits of $w_{i,j,t}$.

Round $T + 1$. Each player j broadcasts 1 if at all rounds t, all its incoming messages from player $1 \leq i \leq k'$ were disjoints with all its incoming messages from player $i + k'$.

Clearly, after T rounds, player j receives $(x_i)_{(j-1)\lceil n/k \rceil+1}, \dots, (x_i)_{j\lceil n/k \rceil}$ from every i. Hence, PWDISJ$(x) = 0$ if and only if a 0 is broadcasted by some player in the last round. Therefore, every player will know the answer after the last round. $\square$

Theorem 2. $\mathbf{B}\text{ROUND}_b(\text{PWDISJ}) = \Omega(n/b)$.

Proof. It is well-known that, in the two party case $k = 2$, the round complexity of set-disjointness with error probability ϵ is $\Omega(n/b)$ [22]. If $k > 2$ we get the same bound for (PWDISJ by considering the instance where $x_1 = x \in \{0,1\}^n$ is given to player 1, $x_{1+k'} = y \in \{0,1\}^n$ is given to player 2, and the empty set ϕ, represented by $(0,\ldots,0)^T$, is given to all the other $k-2$ players. $\square$

Corollary 1. *Let* $k = n$. *Then,* $\mathbf{U}\text{ROUND}_1(\text{PWDISJ}) = 2$ *and* $\mathbf{B}\text{ROUND}_b(\text{PWDISJ}) = \Omega(n/b)$.

Corollary 2. $\Gamma_{(k-1)\times b}^{1\times b}(\text{PWDISJ}) = \Omega(k)$.

3   A Hierarchy of Models According to the Range

In previous section we proved that the broadcast ($r = 1$) and the unicast ($r = k - 1$) models are fundamentally different in their power to solve one particular problem. These two models are the two ends of the spectrum of values of the range r. In this section we prove that the sensitivity to the range is more general. In particular, we show that the round complexity of PWDISJ is completely sensitive to the range.

Lemma 1. $\text{ROUND}_{r\times b}(\text{PWDISJ}) = \Omega\left(\frac{n}{\min(kb, rb+\lceil\log r\rceil k)}\right)$.

Proof. We use a reduction from the two-party communication problem $\text{DISJ}_{k'n}$, where instances are pairs (x,y) of boolean vectors, each of length $k'n$. The communication complexity (bits to be exchanged) of $\text{DISJ}_{k'n}$ is $\Theta(k'n)$ [22]. We transform an instance of $\text{DISJ}_{k'n}$ into an instance of PWDISJ in the direct way. From (x,y) we define the input $(x_1,\ldots,x_k)$ of function PWDISJ as follows: $x = x_1\cdots x_{k'}$ and $y = x_{k'+1}\cdots x_k$. Obviously, $\text{DISJ}_{k'n}(x,y) = 1 \iff \text{PWDISJ}(x_1,\ldots,x_k) = 1$.

Let us consider any protocol P that solves PWDISJ in T_P rounds. If we group players 1 to k' into a global player A and players $k'+1$ to k into a global player B, protocol P would yield a protocol for solving $\text{DISJ}_{k'n}$. So the question is the following: How many bits are exchanged between A and B? Let us derive an upper bound for this.

Consider a player i in A. Player i sends one message of length b to each player in B, thus he sends $k'b$ bits. However, since $r \le 2^b$, the messages sent by player i to players in B can be compressed as follows. Since player i can send up to r different messages, one can consider that she sends to each player $j \in B$ a message numbered from the set $\{0,1,\ldots,r-1\}$ that identifies the message $m(i,j)$ sent to player j. These numbers, of $\lceil\log r\rceil$ bits each, can be used to obtain the actual message from a table that contains the r messages, of b bits each, sent by i. Hence, the total number of bits sent by i to B is upper bounded by the length of the k' numbers, $\lceil\log r\rceil k'$ bits, and the size of the message table, br bits; a total of $rb + \lceil\log r\rceil k'$ bits.

Let us define $\beta = \min(bk', rb + \lceil\log r\rceil k')$. In each round, the number of bits exchanged between A and B is upper bounded by $k\beta$. Therefore, considering that the communication complexity of $\text{DISJ}_{k'n}$ is $\Theta(k'n)$, it follows that $T_P k\beta = \Omega(k'n)$. Therefore, $\text{ROUND}_{r\times b}(\text{PWDISJ}) = \Omega(\frac{n}{2\beta})$, as claimed. $\square$

Lemma 2. $\text{ROUND}_{r \times b}(\text{PWDISJ}) \leq \left\lceil \frac{n}{k \lfloor \log r \rfloor} \right\rceil + 1$.

Proof. Consider the same protocol used in the proof of Theorem 1 but with messages of $\lfloor \log r \rfloor \leq b$ bits. $\qquad \square$

Putting these together, we get the following theorem.

Theorem 3. *For k sufficiently large and for $r \geq 2$ such that $\frac{rb}{\log r} = O(k)$,*

$$\text{ROUND}_{r \times b}(\text{PWDISJ}) = \Theta\left(\frac{n}{k \log r}\right).$$

Proof. The upper bound follows from the previous lemma. For the lower bound, it follows from Lemma 1 that

$$\text{ROUND}_{r \times b}(\text{PWDISJ}) \geq \frac{n}{k \min(b, \lceil \log r \rceil (1 + \frac{2rb}{\lceil \log r \rceil k}))} \geq \frac{n}{k \lceil \log r \rceil (1 + \frac{2rb}{\lceil \log r \rceil k})},$$

where the last inequality follows from $\lceil \log r \rceil \leq b$ and $\frac{2rb}{\lceil \log r \rceil k} > 0$. Since $\frac{rb}{\log r} = O(k)$, we deduce that, for k sufficiently large, there is a constant $\Delta > 0$ such that $\frac{2rb}{\lceil \log r \rceil k} \leq \Delta$, and hence

$$\text{ROUND}_{r \times b}(\text{PWDISJ}) \geq \frac{n}{k \lceil \log r \rceil (1 + \Delta)} = \Omega\left(\frac{n}{k \lceil \log r \rceil}\right).$$

$\qquad \square$

The natural way to interpret Theorem 3 is to parametrize everything by k. Following the spirit of the $\mathcal{CONGEST}$ model [29], we are going to restrict both the bandwidth and the range by taking $b = \log k$ and varying r from 2 to $k - 1$. Observe that, when $b = \log k$ and $r \leq k - 1$, it always holds that $\frac{rb}{\log r} = O(k)$. Hence, the next corollaries are direct consequences of Theorem 3.

Corollary 3. *For every n and every constant integer $c \geq 2$, we have*

$$\text{ROUND}_{\log k \times \log k}(\text{PWDISJ}) = \Theta\left(\frac{n}{k \log \log k}\right) \quad \text{and} \quad \text{ROUND}_{c \times \log k}(\text{PWDISJ}) = \Theta\left(\frac{n}{k}\right)$$

In other words, $\Gamma_{\log k \times \log k}^{c \times \log k}(\text{PWDISJ}) = \Theta(\log \log k)$.

In general, we can state the following corollary.

Corollary 4. *For every n and every $r' \geq r \geq 2$, we have*

$$\text{ROUND}_{r' \times \log k}(\text{PWDISJ}) = \Theta\left(\frac{n}{k \log r'}\right) \quad \text{and} \quad \text{ROUND}_{r \times \log k}(\text{PWDISJ}) = \Theta\left(\frac{n}{k \log r}\right)$$

In other words, $\Gamma_{r' \times \log k}^{r \times \log k}(\text{PWDISJ}) = \Theta\left(\frac{\log r'}{\log r}\right)$.

4   Most Functions Have Maximal Round Complexity

From the results presented in the previous sections one may be tempted to conclude that, in general, increasing the range r increases the power of the protocols. In particular, one may conclude that the unicast congested clique model has much more power than the broadcast congested clique model (even if in the first we restrict the bandwidth to $b = 1$ while in the latter we allow it to be $b = o(n)$). We show here that this fact, which holds for function PWDISJ, holds for very few other functions. More precisely, we are going to prove that for almost every boolean function f, the broadcasting protocol is optimal. We start by considering deterministic decision protocols that compute functions f correctly (i.e., they make no mistake). (Some proofs are omitted.)

Lemma 3. *The number of T-round deterministic decision protocols in the unicast congested clique model* CLIQUE-UCAST$_b$ *is at most* $2^{N(T)}$, *where*

$$N(T) = 2^{T(k-1)b+n}(1 + \frac{(k+1)(k-1)b}{2^{(k-1)b}}).$$

Now, we still consider deterministic protocols, but now we allow them to make mistakes. We say that a deterministic protocol P computes f with error $\epsilon \geq 0$ if it outputs $f(x)$ for at least $(1 - \epsilon)2^{nk}$ of the inputs x of f.

Lemma 4. *Let P be a deterministic decision protocol and let $P(x)$ denote the output of P with input $x \in \{0,1\}^{nk}$. Let $M_\epsilon(P)$ be the number of functions f which are computed by P with an error $\epsilon > 0$. We have,*

$$M_\epsilon(P) \leq \left(\frac{2e}{\epsilon}\right)^{\epsilon 2^{nk}} = 2^{\log(\frac{2e}{\epsilon})\epsilon 2^{nk}}.$$

We show now that a deterministic protocol P that computes a function f chosen uniformly at random with error ϵ requires the maximal number of rounds $\lceil n/b \rceil$ with high probability. Let us extend our notation, so that $\text{UROUND}_b^\epsilon(f)$ is the round complexity of function f when protocols are deterministic and error ϵ is allowed.

Theorem 4. *For k sufficiently large and for every n, and $\epsilon > 0$ such that $1 - \log(\frac{2e}{\epsilon})\epsilon > 0$, we have*

$$\Pr\{\text{UROUND}_b^\epsilon(f) = \lceil n/b \rceil\} \geq 1 - 2^{-2^{kn}(\frac{1-\log(\frac{2e}{\epsilon})\epsilon}{2})}.$$

For $\epsilon = 0$ (i.e. the case without error), we have

$$\Pr\{\text{UROUND}_b^0(f) = \lceil n/b \rceil\} \geq 1 - 2^{-2^{kn}\,0.5}.$$

Proof. Since there are $2^{2^{kn}}$ different functions $f : \{0,1\}^{kn} \to \{0,1\}$, we have

$$\Pr\{\text{UROUND}_b^\epsilon(f) \leq T\} \leq \frac{2^{N(T)}\max_P M_\epsilon(P)}{2^{2^{kn}}}.$$

From Lemmas 3 and 4, for $\epsilon > 0$, we have

$$\Pr\{\mathbf{Uround}_b^\epsilon(f) \le T\} \le 2^{2^{T(k-1)b+n}(1+\frac{(k+1)(k-1)b}{2^{(k-1)b}})} 2^{\log(\frac{2\epsilon}{\epsilon})\epsilon 2^{nk}} 2^{-2^{kn}}$$

$$\le 2^{-2^{kn}(1-\log(\frac{2\epsilon}{\epsilon})\epsilon - 2^{T(k-1)b+n-kn}(1+\frac{(k+1)(k-1)b}{2^{(k-1)b}}))}.$$

For k sufficiently large, the quantity $1 + \frac{(k+1)(k-1)b}{2^{(k-1)b}}$ can be upper bounded (by 2 for example). Now let us take $T = \lceil n/b \rceil - 1$. Then, we have $Tb - n \le -b$ and, thus

$$2^{T(k-1)b+n-kn} = 2^{(k-1)(Tb-n)} \le 2^{-b(k-1)} \le 2^{-k}$$

Thus, for k sufficiently large, the term, $2^{T(k-1)b+n-kn}(1 + \frac{(k+1)(k-1)b}{2^{(k-1)b}})$ can be upper bounded by any positive value, in particular by $\frac{1-\log(\frac{2\epsilon}{\epsilon})\epsilon}{2}$. Thus, we get that

$$\Pr\{\mathbf{Uround}_b^\epsilon(f) \le \lceil n/b \rceil - 1\} \le 2^{-2^{kn}(\frac{1-\log(\frac{2\epsilon}{\epsilon})\epsilon}{2})}.$$

which is the result since, for any f, one trivially has $\mathbf{Uround}_b^\epsilon(f) \le \mathbf{Uround}_b^0(f) \le \lceil n/b \rceil$.

For $\epsilon = 0$ we proceed on the same way, after noticing that $\max_P M_0(P) = 1$. □

Theorem 5. *For k sufficiently large, for every n, and $0 \le \epsilon \le 0.2$, there exists a positive constant $c(\epsilon) > 0$ such that $\Pr\{\mathbf{Uround}_b^\epsilon(f) = \lceil n/b \rceil\} \ge 1 - 2^{-2^{kn}c(\epsilon)}$.*

Recall that $\mathbf{Uround}_b(f)$ is the round complexity of computing function f with randomized protocols, which may use public coins, with success probability $1 - \epsilon$. From the previous results we can prove that most functions have round complexity $\lceil n/b \rceil$.

Corollary 5. *For k sufficiently large and for every n, and $0 \le \epsilon \le 0.2$, there exists a positive constant $c(\epsilon) > 0$ such that,*

$$\Pr\{\mathbf{Uround}_b(f) = \lceil n/b \rceil\} \ge 1 - 2^{-2^{kn}c(\epsilon)}.$$

Proof. The result follows from Theorem 5, and Theorem 3.20 at [24] using the uniform distribution as the distribution μ of the inputs. □

The following bound is obvious for any function f.

$$\Gamma_{(k-1)\times b}^{1\times b}(f) = \frac{\mathbf{Bround}_b(f)}{\mathbf{Uround}_b(f)} \ge \min_f \Gamma_{(k-1)\times b}^{1\times b}(f) \ge 1.$$

Next corollary, which is a direct consequence of Corollary 5, says that previous inequality is in fact an equality for almost every boolean function.

Corollary 6. *For k sufficiently large and for every n, and $0 \le \epsilon \le 0.2$, there exists a positive constant $c(\epsilon) > 0$ such that,*

$$\Pr\{\Gamma_{(k-1)\times b}^{1\times b}(f) = 1\} \ge 1 - 2^{-2^{kn}c(\epsilon)}.$$

5   Comparing Models with Different Combinations of Range and Bandwidth for Arbitrary Boolean Functions

In this section we explore the relative round complexities of different modes of the congested clique model with various combinations of range and bandwidth $\Gamma_{r' \times b'}^{r \times b}(f)$ for arbitrary boolean functions f. The first result shows that for *most* boolean functions f, $\Gamma_{r' \times b'}^{r \times b}(f) = \Theta(b'/b)$.

Theorem 6. *For k sufficiently large and for every n, there is a positive constant $c(\epsilon) > 0$ such that*

$$\Pr\left\{\Gamma_{r' \times b'}^{r \times b}(f) = \lceil n/b \rceil / \lceil n/b' \rceil\right\} \geq 1 - 2^{-2^{kn} c(\epsilon) + 1}.$$

Proof. From Corollary 5, a function f *simultaneously* satisfies $\mathrm{ROUND}_{r \times b}(f) = \lceil n/b \rceil$ and $\mathrm{ROUND}_{r' \times b'}(f) = \lceil n/b' \rceil$ with probability at least $1 - 2^{-2^{kn} c(\epsilon) + 1}$. $\square$

Now, we show that in fact the typical case shown in the previous theorem is not far from the worst case, studied in the following sequence of results.

Theorem 7. *Let r be such that $r \geq r'$ or $r = 2^b$. Then, for every function f, $\Gamma_{r' \times b'}^{r \times b}(f) \leq \lceil b'/b \rceil$.*

Proof. Let P' be a T-round protocol in $\mathrm{CLIQUE\text{-}RCAST}_{r' \times b'}$. From P' we construct the protocol P in $\mathrm{CLIQUE\text{-}RCAST}_{r \times b}$ as follows. Consider the message $m_t(i, j)$ sent by player i to player j in round t of P'. For each $1 \leq \ell \leq \lceil b'/b \rceil$, let $\mathrm{block}_t^\ell(i, j)$ be the ℓth block of length b of $m_t(i, j)$. The last block is padded with 0s. For each ℓ and i, we have: $|\{\mathrm{block}_t^\ell(i, j), 1 \leq j \leq k \in N\}| \leq \min\{r', 2^b\} \leq r$.

Then, during round number $(t - 1) \lceil b'/b \rceil + \ell$ of P, player i sends to player j the b bits of $\mathrm{block}_t^i(u, v)$. The inequalities above ensure that P is a well-defined protocol in $\mathrm{CLIQUE\text{-}RCAST}_{r \times b}$. Since P knows the bandwidth b' it can discard the padding bits. The total number of rounds executed by P is $T \lceil b'/b \rceil$. $\square$

Theorem 8. *Let $b \leq b' \leq n$, and k sufficiently large. Then, there exists a function f such that: $\Gamma_{r' \times b'}^{r \times b}(f) = \lceil b'/b \rceil$.*

Proof. Let $b' = n$. In this case, every function $f : (\{0,1\}^n)^k \to \{0,1\}$ can be solved in one round in the model $\mathrm{CLIQUE\text{-}RCAST}_{r' \times b'}$. On the other hand, from Corollary 5, almost every function $f : (\{0,1\}^n)^k \to \{0,1\}$ satisfies $\mathrm{ROUND}_{r \times b}(f) = \lceil n/b \rceil = \lceil b'/b \rceil$. When $n > b'$, let us define $n' = b'$. From Corollary 5, almost every function $f' : (\{0,1\}^{n'})^k \to \{0,1\}$ satisfies $\mathrm{ROUND}_{r \times b}(f') = \left\lceil \frac{n'}{b} \right\rceil$. Let us take one such function f', and define a new function $f : (\{0,1\}^n)^k \to \{0,1\}$ as follows: $f(x_1, x_2 .... x_k) = f'(y_1, y_2, ..., y_k)$, where each y_i is the vector formed with the n' first bits of x_i. Hence, $\mathrm{ROUND}_{r \times b}(f) = \mathrm{ROUND}_{r \times b}(f') = \lceil n'/b \rceil = \lceil b'/b \rceil$ while $\mathrm{ROUND}_{r' \times b'}(f) = 1$. $\square$

Remark 1. When $b|b'$ is a multiple of b and $b'|n$, we have $\lceil n/b \rceil / \lceil n/b' \rceil = (n/b)/(n/b') = b'/b = \lceil b'/b \rceil$. When $n = b'$, we also have $\lceil n/b \rceil / \lceil n/b' \rceil = \lceil b'/b \rceil$. Thus, in the previous cases, for $r \geq r'$ of $r = 2^b$, the maximal value $\lceil b'/b \rceil$ for the value of $\Gamma_{r' \times b'}^{r \times b}(f)$ is reached with high probability. On the other hand, in some cases, there exists a small but intriguing gap between the maximal value $\lceil b'/b \rceil$ and the value $\lceil n/b \rceil / \lceil n/b' \rceil$ reached with high probability. For example, take $b = 2$, $b' = 3$. For $n = 4$, we have $\lceil b'/b \rceil = 2$ and $\lceil n/b \rceil / \lceil n/b' \rceil = 1$.

Note that Theorem 7 holds when $r' \leq r$ or $r = 2^b$. Without this hypothesis we only get the following weaker, general bound.

Theorem 9. *Let $r \geq 2$ and $r' \geq 1$. Then, for every function f,*

$$\Gamma_{r' \times b'}^{r \times b}(f) \leq \min \left(\left\lceil \frac{b'}{\lfloor \log r \rfloor} \right\rceil, \left\lceil \frac{r'}{r-1} \right\rceil \left\lceil \frac{b'}{b} \right\rceil \right).$$

Observe that the two values of the minimum are complementary, since none implies the other.

References

1. Ahn, K.J., Guha, S., McGregor, A.: Analyzing graph structure via linear measurements. In: Proceedings of SODA 2012, pp. 459–467 (2012)
2. Ahn, K.J., Guha, S., McGregor, A.: Graph sketches: sparsification, spanners, and subgraphs. In: Proceedings of PODS 2012, pp. 5–14 (2012)
3. Bar-Yossef, Z., Jayram, T.S., Kumar, R., Sivakumar, D.: An information statistics approach to data stream and communication complexity. In: Proceedings of FOCS 2002, pp. 209–218 (2002)
4. Becker, F., Fernández Anta, A., Rapaport, I., Rémila, E.: Brief announcement: a hierarchy of congested clique models, from broadcast to unicast. In: Proceedings of PODC 2015, pp. 167–169 (2015)
5. Becker, F., Kosowski, A., Nisse, N., Rapaport, I., Suchan, K.: Allowing each node to communicate only once in a distributed system: shared whiteboard models. In: Proceedings of SPAA 2012, pp. 11–17 (2012)
6. Becker, F., Matamala, M., Nisse, N., Rapaport, I., Suchan, K., Todinca, I.: Adding a referee to an interconnection network: what can (not) be computed in one round. In: Proceedings of IPDPS 2011, pp. 508–514 (2011)
7. Becker, F., Montealegre, P., Rapaport, I., Todinca, I.: The simultaneous number-in-hand communication model for networks: private coins, public coins and determinism. In: Halldórsson, M.M. (ed.) SIROCCO 2014. LNCS, vol. 8576, pp. 83–95. Springer, Heidelberg (2014)
8. Braverman, M., Ellen, F., Oshman, R., Pitassi, T., Vaikuntanathan, V.: A tight bound for set disjointness in the message-passing model. In: Proceedings of FOCS 2013, pp. 668–677 (2013)
9. Braverman, M., Oshman, R.: On information complexity in the broadcast model. In: Proceedings of PODC 2015, pp. 355–364 (2015)
10. Censor-Hillel, K., Kaski, P., Korhonen, J.H., Lenzen, C., Paz, A., Suomela, J.: Algebraic methods in the congested clique. In: Proceedings of PODC 2015, pp. 143–152

11. Chakrabart, A., Shi, Y., Wirth, A., Yao, A.: Informational complexity and the direct sum problem for simultaneous message complexity. In: Proceedings of FOCS 2001, pp. 270–278. IEEE (2001)
12. Chattopadhyay, A., Mukhopadhyay, S.: Tribes is hard in the message passing model. In: Proceedings of STACS 2009, pp. 224–237 (2015)
13. Dolev, D., Feder, T.: Multiparty communication complexity. In: Proceedings of FOCS 1989, pp. 428–433 (1989)
14. Dolev, D., Lenzen, C., Peled, S.: "Tri, Tri Again": finding triangles and small subgraphs in a distributed setting. In: Aguilera, M.K. (ed.) DISC 2012. LNCS, vol. 7611, pp. 195–209. Springer, Heidelberg (2012)
15. Drucker, A., Kuhn, F., Oshman, R.: On the power of the congested clique model. In: Proceedings of PODC 2014, pp. 367–376 (2014)
16. Duriš, P., Rolim, J.D.: Lower bounds on the multiparty communication complexity. J. Comput. Syst. Sci. **56**(1), 90–95 (1998)
17. Gronemeier, A.: Asymptotically optimal lower bounds on the NIH-multi-party information complexity of the AND-function and disjointness. In: Proceedings of STACS 2009, pp. 505–516 (2009)
18. Guha, S., McGregor, A., Tench, D.: Vertex and hyperedge connectivity in dynamic graph streams. In: Proceedings of PODS 2015, pp. 241–247 (2015)
19. Hegeman, J.W., Pandurangan, G., Pemmaraju, S.V., Sardeshmukh, V.B., Scquizzato, M.: Toward optimal bounds in the congested clique: graph connectivity and MST. In: Proceedings of PODC 2015, pp. 91–100 (2015)
20. Hegeman, J.W., Pemmaraju, S.V.: Lessons from the congested clique applied to MapReduce. In: Halldórsson, M.M. (ed.) SIROCCO 2014. LNCS, vol. 8576, pp. 149–164. Springer, Heidelberg (2014)
21. Hegeman, J.W., Pemmaraju, S.V., Sardeshmukh, V.B.: Near-constant-time distributed algorithms on a congested clique. In: Kuhn, F. (ed.) DISC 2014. LNCS, vol. 8784, pp. 514–530. Springer, Heidelberg (2014)
22. Kalyanasundaram, B., Schintger, G.: The probabilistic communication complexity of set intersection. SIAM J. Discrete Math. **5**(4), 545–557 (1992)
23. Kari, J., Matamala, M., Rapaport, I., Salo, V.: Solving the induced subgraph problem in the randomized multiparty simultaneous messages model. In: Scheideler, C. (ed.) SIROCCO 2015. LNCS, vol. 9439, pp. 370–384. Springer, Heidelberg (2015)
24. Kushilevitz, E., Nisan, N.: Communication Complexity. Cambridge University Press, Cambridge (2006)
25. Lenzen, C.: Optimal deterministic routing and sorting on the congested clique. In: Proceedings of PODC 2013, pp. 42–50 (2013)
26. Li, Y., Sun, X., Wang, C., Woodruff, D.P.: On the communication complexity of linear algebraic problems in the message passing model. In: Kuhn, F. (ed.) DISC 2014. LNCS, vol. 8784, pp. 499–513. Springer, Heidelberg (2014)
27. Lotker, Z., Pavlov, E.: MST construction in $O(\log \log n)$ communication rounds. In: Proceedings of SPAA 2003, pp. 94–100 (2003)
28. Patt-Shamir, B., Teplitsky, M.: The round complexity of distributed sorting. In: Proceedings of PODC 2011, pp. 249–256 (2011)
29. Peleg, D.: Distributed Computing: A Locality-Sensitive Approach. Society for Industrial and Applied Mathematics, Philadelphia (2000)
30. Phillips, J.M., Verbin, E., Zhang, Q.: Lower bounds for number-in-hand multiparty communication complexity, made easy. In: Proceedings of SODA 2012, pp. 486–501
31. Woodruff, D.P., Zhang, Q.: An optimal lower bound for distinct elements in the message passing model. In: Proceedings of SODA 2014, pp. 718–733 (2014)

Minimum Cost Homomorphisms
with Constrained Costs

Pavol Hell and Mayssam Mohammadi Nevisi[✉]

School of Computing Science, Simon Fraser University, Burnaby, Canada
{pavol,mayssamm}@sfu.ca

Abstract. The minimum cost homomorphism problem is a natural optimization problem for homomorphisms to a fixed graph H. Given an input graph G, with a cost associated with mapping any vertex of G to any vertex of H, one seeks to minimize the sum of costs of the assignments over all homomorphisms of G to H. The complexity of this problem is well understood, as a function of the target graph H. For bipartite graphs H, the problem is polynomial time solvable if H is a proper interval bigraph, and is NP-complete otherwise. In many applications, the costs may be assumed to be the same for all vertices of the input graph. We study the complexity of this restricted version of the minimum cost homomorphism problem. Of course, the polynomial cases are still polynomial under this restriction. We expect the same will be true for the NP-complete cases, i.e., that the complexity classification will remain the same under the restriction. We verify this for the class of trees. For general graphs H, we prove a partial result: the problem is polynomial if H is a proper interval bigraph and is NP-complete when H is not chordal bipartite.

Keywords: Homomorphisms · NP-completeness · Dichotomy

1 Introduction

Suppose G and H are graphs (without loops or multiple edges). A *homomorphism* $f : G \to H$ is a mapping $V(G) \to V(H)$ such that $f(u)f(v) \in E(H)$ whenever $uv \in E(G)$. For a fixed graph H, a number of computational problems have been considered. In the *homomorphism problem*, one asks whether or not an input graph G admits a homomorphism to H. It is known that this problem is polynomial time solvable if H is bipartite, and is NP-complete otherwise [1]. In the *list homomorphism problem*, the input graph G is equipped with lists (sets) $L(x) \subseteq V(H)$, for all $x \in V(G)$, and one asks whether or not there exists a homomorphism $f : G \to H$ with $f(x) \in L(x)$ for all $x \in V(G)$. This problem is known to be polynomial time solvable if H is an interval bigraph,

---

P. Hell and M.M. Nevisi—Both authors were supported by the NSERC Discovery Grant of the second author, who was additionally supported by the grant ERCCZ LL 1201. Also, part of this work was done while the second author was visiting the Simons Institute for the Theory of Computing.

© Springer International Publishing Switzerland 2016
T.N. Dinh and M.T. Thai (Eds.): COCOON 2016, LNCS 9797, pp. 194–206, 2016.
DOI: 10.1007/978-3-319-42634-1_16

and is NP-complete otherwise [2]. (An *interval bigraph* is a bipartite graph H with parts X and Y such that there exist intervals $I_x, x \in X$, and $J_y, y \in Y$, for which $xy \in E(H)$ if and only if $I_x \cap J_y \neq \emptyset$.) In this paper we address the *minimum cost homomorphism problem*, in which the input graph is equipped with a cost function $c : V(G) \times V(H) \to \mathbb{N}$ and one tries to minimize the total cost $\sum_{u \in V(G)} c(u, f(u))$. Minimum cost homomorphism problems were introduced in [3]. They were motivated by an application in repair and maintenance scheduling; however, the problem arises in numerous other contexts, e.g. in the minimum colour sum problem and the optimum cost chromatic partition problem [4,5]. To state it as a decision problem, the input includes an integer k, and one asks whether or not there exists a homomorphism of total cost at most k. This problem is known to be polynomial time solvable if H is a proper interval bigraph, and is NP-complete otherwise [6]. (An interval bigraph is a *proper interval bigraph* if the above two families of intervals $I_x, x \in X$, and $J_y, y \in Y$ can be chosen to be inclusion-free, i.e., no I_x properly contains another $I_{x'}$ and similarly for the J_y's.)

These results are *dichotomies* in the sense that for each H the problem is polynomial time solvable or NP-complete. They have subsequently been studied in more general contexts, for graphs with possible loops, for digraphs, and for general relational structures (in the context of constraint satisfaction problems). In particular, there is a dichotomy for the homomorphism problem for graphs with possible loops [1], but dichotomy is only conjectured for digraphs (and more general structures) [7,8]. A dichotomy for list homomorphism problems for graphs with possible loops was established in [2,9], then a general dichotomy was proved for all relational systems in [10]. (A more structural dichotomy classification for digraphs was given in [12].) For minimum cost homomorphism problems, a dichotomy for graphs with possible loops is given in [6]. A structural dichotomy classification for digraphs was conjectured in [3], and proved in [11] (cf. [13,14]). Then a general dichotomy for all relational systems was proved in [15]. Even more general dichotomy results are known, for so-called finite valued constraint satisfaction problems [16].

It is easy to see that minimum cost homomorphism problems generalize list homomorphism problems, which in turn generalize homomorphism problems. Minimum cost homomorphism problems also generalize two graph optimization problems, the minimum colour sum problem, and the optimum cost chromatic partition problem [4,5]. In the former, the cost function has only two values, 0 and 1 (and $k = 0$). In the latter, the cost function is assumed to be constant across $V(G)$, i.e., $c(x, u) = c(u)$ for all $x \in V(G)$. This restriction, that costs only depend on vertices of H, appears quite natural even for the general minimum cost homomorphism problems, and appears not have been studied. In this paper we take the first steps in investigating its complexity.

Let H be a fixed graph. The *minimum constrained cost homomorphism problem* for H has as input a graph G, together with a cost function $c : V(H) \to \mathbb{N}$, and an integer k, and asks whether there is a homomorphism $f : G \to H$ of total cost $cost(f) = \sum_{u \in V(G)} c(f(u)) \leq k$.

It appears that the added constraint on the cost function may leave the dichotomy classification from [6] unchanged; in fact, we can show it does not change it for trees H (and in some additional cases, cf. Lemma 8 below).

Theorem 1. *Let H be a fixed tree. Then the minimum constrained cost homomorphism problem to H is polynomial time solvable if H is a proper interval bigraph, and is NP-complete otherwise.*

We believe the same may be true for general graphs H. We have obtained the following partial classification.

Theorem 2. *Let H be a fixed graph. Then the minimum constrained cost homomorphism problem to H is polynomial time solvable if H is a proper interval bigraph, and is NP-complete if H is not a chordal bipartite graph.*

Of course, the first statement of the theorem follows from [6]. Only the second claim, the NP-completeness, needs to be proved. A bipartite graph H is *chordal bipartite* if it does not contain an induced cycle of length greater than four. Both chordal bipartite graphs and proper interval bigraphs can be recognized in polynomial time [17,18]. Proper interval bigraphs are a subclass of chordal bipartite graphs, and Lemma 8 below gives a forbidden subgraph characterization of proper interval bigraphs within the class of chordal bipartite graphs.

Our NP-completeness reductions in the proofs of Theorems 1 and 2 use a shorthand, where vertices v of the input graph G have *weights* $w(v)$. Adding polynomially bounded vertex weights does not affect the time complexity of our problems. Let G, H be graphs, and, for every $v \in V(G)$ and every $i \in V(H)$, let $c_i(v)$ denote the cost of mapping v to i. Let $w : V(G) \to \mathbb{N}$ be a weight function. The weighted cost of a homomorphism $f : G \to H$ is $cost(f) = \sum_{v \in V(G)} w(v).c_{f(v)}(v)$. In the *weighted minimum cost homomorphism problem* for a fixed graph H, the input is a graph G, together with cost functions $c_i : V(G) \to \mathbb{N}$ (for all $i \in V(H)$), vertex weights $w : V(G) \to \mathbb{N}$, and an integer k; and the question is if there is a homomorphism of G to H of weighted cost at most k.

The variant with constrained costs is defined similarly: the *weighted minimum constrained cost homomorphism problem* for H has as input a graph G, cost function $c : V(H) \to \mathbb{N}$, vertex weights $w : V(G) \to \mathbb{N}$, and an integer k, and it asks if there is a homomorphism $f : G \to H$ with cost $\sum_{v \in V(G)} w(v).c(f(v)) \leq k$.

Clearly, when w is a polynomial function, the weighted minimum cost homomorphism problem and the minimum cost homomorphism problem are polynomially equivalent. It turns out that this is also the case for the problems with constrained costs.

Theorem 3. *Let H be a fixed graph. The minimum constrained cost homomorphism problem to H and the weighted minimum constrained cost homomorphism problem to H with polynomial weights are polynomially equivalent.*

2   Chordal Bipartite Graphs

In this section, we investigate the minimum constrained cost homomorphism problem for graphs H with induced even cycles of length at least six. First we treat the case of hexagon, then we handle longer cycles.

Lemma 4. *Let H be a graph which contains hexagon as an induced subgraph. Then, the weighted minimum constrained cost homomorphism problem to H is NP-complete.*

For a fixed graph H, the *pre-colouring extension problem* to H takes as input a graph G in which some vertices v have been pre-assigned to images $f(v) \in V(H)$ (we say v is *pre-coloured* by $f(v)$), and asks whether or not there exists a homomorphism $f : G \to H$ that extends this pre-assignment. This can be viewed a special case of the list homomorphism problem to H (all lists are either singletons or the entire set $V(H)$), and has been studied under the name of One-Or-All list homomorphism problem, denoted OAL-HOM(H) [2]. Here we adopt the abbreviation OAL-HOM(H) for the pre-colouring extension problem.

The problem OAL-HOM(H) was first studied in [2,19].

Lemma 5 [2]. *Let C be a cycle of length $2k$ with $k \geq 3$. Then the pre-colouring extension problem to C is NP-complete.*

We can now present the proof of Lemma 4.

Proof. The membership in NP is clear. Let $C = 1, 2, \cdots, 6$ denote the hexagon and $h_1 h_2 \cdots h_6$ be an induced subgraph of H which is isomorphic to C. We reduce from the pre-colouring extension homomorphism problem to C.

Let (G, L) be an instance of OAL-HOM(C), i.e., G is a bipartite graph with $n \geq 2$ vertices and $m \geq 1$ edges, and some vertices v of G have been pre-assigned to $f(v) \in V(C)$. We construct an instance (G', c, w, T) of the weighted minimum constrained cost homomorphism problem to H as follows. The graph G' is a bipartite graph obtained from a copy of G, by adding, for every vertex $v \in V(G)$ pre-coloured k, a gadget that is the cartesian product of v and the hexagon, using six new vertices $(v, 1), (v, 2), \cdots, (v, 6)$, and six new edges $(v, 1)(v, 2), (v, 2)(v, 3), \cdots, (v, 6)(v, 1)$. We also connect v to exactly two neighbours of (v, k) in its corresponding gadget. A vertex v and its corresponding gadget is illustrated in Fig. 1.

We define the vertex weight function w as follows.

- for every vertex v in the copy of G, let $w(v) = 1$
- for every pre-coloured vertex $v \in V(G)$:
    - $w((v, 1)) = w((v, 4)) = 5 \times 36n^3 + 1$,
    - $w((v, 2)) = w((v, 5)) = 1$,
    - $w((v, 3)) = 36n^2$,
    - $w((v, 6)) = 6n$.

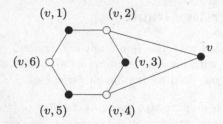

Fig. 1. A gadget in G' for a vertex $v \in V(G)$ pre-coloured by 3

We define the homomorphism cost function c as follows.

- $c(h_1) = c(h_4) = 0$,
- $c(h_2) = c(h_5) = 36n^2$,
- $c(h_3) = 1$,
- $c(h_6) = 6n$,
- $c(h_i) = 5 \times 36n^3 + 1$ for all other vertices $h_i \in V(H)$.

Finally, we set $T = 5 \times 36n^3 = 180n^3$.

We now claim that there is an extension of the pre-colouring f to a homomorphism of G to C if and only if there is a homomorphism of G' to H with weighted cost at most T.

First, assume that the pre-colouring can be extended to a homomorphism $f : G \to C$. We define a homomorphism $g : G' \to H$ as follows.

- $g(u) = h_i$ iff $f(u) = i$ for every vertex $u \in V(G)$ and every $1 \le i \le 6$,
- $g((u, i)) = h_i$ for every vertex $u \in V(G)$ pre-coloured k and every $1 \le i \le 6$.

Claim. The function g is a homomorphism of G' to H. Moreover, it only maps vertices of G' to the copy of C in G, i.e., g only uses vertices $h_1, h_2, \cdots, h_6$.

To prove the above claim, we distinguish three types of edges in G'.

1. Edges uv corresponding to the edges in G ($u, v \in V(G)$): These are clearly mapped to edges in H by g as $g(u) = f(u)$ for all vertices $u \in V(G)$ and f is a homomorphism of G to C.
2. Edges $(u, i)(u, i + 1)$ that connect two vertices of the gadgets: These edges map to the corresponding edge $h_i h_{i+1}$ by definition of g (indices modulo 6).
3. Edges that connect a vertex $u \in V(G)$ to two vertices in its corresponding gadget: Notice that there is a gadget for u in G' only when u is pre-coloured i. So, we have $f(u) = i$. This further implies that $g(u) = h_i$. Also, notice that $g((u, i - 1)) = h_{i-1}$ and $g((u, i + 1)) = h_{i+1}$ by the definition of g (again, all indices modulo 6). Hence, edges $u(u, i - 1)$ and $u(u, i + 1)$ also map to edges $h_{i-1}h_i$ and $h_i h_{i+1}$, respectively.

This completes the proof of the above Claim. We now show that the cost of g is at most $T = 180n^3$.

- For every vertex $u \in V(G)$, $w(u) = 1$ and $c(g(u)) \leq 36n^2$. Also, there are exactly n such vertices in G'. This contributes at most $36n^3$ to the cost of the homomorphism.
- For every pre-coloured vertex $u \in V(G)$, its corresponding gadget contributes exactly $4 \times 36n^2$:
  - vertices $(u, 1)$ and $(u, 4)$ do not contribute, as $c(h_1) = c(h_4) = 0$,
  - vertices $(u, 2)$ and $(u, 5)$ each contribute $36n^2$,
  - vertices $(u, 3)$ and $(u, 6)$ each contributes $36n^2 = 6n \times 6n = 36n^2 \times 1$.

There are at most n gadgets in G' (one for every vertex $u \in V(G)$), and so, the total contribution of all vertices of the gadgets is at most $4 \times 36n^3$. Therefore, the cost of g is at most $5 \times 36n^3 = 180n^3 = T$.

Conversely, let g be a homomorphism of G' to H which costs at most T. We prove that there is a homomorphism $f : G \to C$ extending the pre-colouring. First, we show that g has the following two properties.

- It only maps vertices of G' to the vertices of the hexagon $h_1, h_2, \cdots, h_6$,
- all gadgets are mapped identically to the hexagon in H, that is, for all pre-coloured vertices $u \in V(G)$ and for every $1 \leq i \leq 6$, $g((u, i)) = h_i$.

The first property holds because $c(a) > T$ for every vertex $a \in V(H)$ other than the vertices of the hexagon (and the fact that, by definition, all vertex weights are positive integers). In fact, we must have $w(u) \times c(g(u)) \leq T$, or equivalently, $c(g(u)) < \frac{(T+1)}{w(u)}$, for every vertex $u \in V(G')$. This restricts possible images of vertices with large vertex weights. Consider vertices in the gadget of a vertex $u \in V(G')$. For instance, every $(u, 4)$ must map to either h_1 or h_4. Similarly, none of the $(u, 3)$ vertices can map to any vertex other than h_1, h_3, or h_4. Given that $(u, 3)$ and $(u, 4)$ are adjacent in G', their images must also be adjacent in H. This enforces $f((u, 3)) = h_3$ and $f((u, 4)) = h_4$ (for every u that has a gadget in G'). Similar to $(u, 4)$, g must also map every $(u, 1)$ to either h_1 or h_4, but $g((u, 1)) = h_4$ is not feasible as it does not leave any options for the image of $(u, 2)$. Hence, $g((u, 1)) = h_1$. This further implies that $g((u, 6)) = h_6$ (as it is adjacent to $(u, 1)$), and finally, $g((u, 2)) = h_2$ and $g((u, 5)) = h_5$.

It is now easy to verify that for every vertex $u \in V(G)$ pre-coloured j, we always have $g(u) = h_j$. This is because u is adjacent to $(u, j - 1)$ and $(u, j + 1)$ in G' and the only vertex in H that is adjacent to the $g((u, j - 1)) = h_{j-1}$ and $g((u, j + 1)) = h_{j+1}$ and the cost of mapping to it is less than or equal to T is h_j. This completes the proof as we can define a homomorphism $f : G \to C$ extending the pre-colouring by setting $f(v) = i \iff g(v) = h_i$. ∎

A shorthand of the construction used in the above proof is shown in Fig. 2. We now extend Lemma 4 to larger even cycles.

Lemma 6. *Let H be a bipartite graph which contains a cycle of length at least eight as an induced subgraph. Then the weighted minimum constrained cost homomorphism problem to H is NP-complete.*

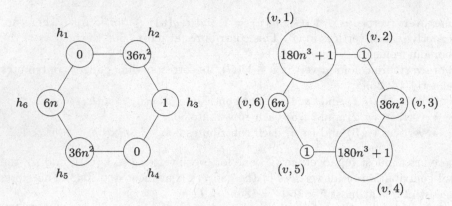

Fig. 2. A hexagon in H together with associated homomorphism costs (left), and a gadget in G' together with vertex weights (right).

Proof Sketch. The proof is similar to the proof of Lemma 4. We only discuss the reduction here. Let $C = 1, 2, \cdots, 2k$ be an even cycle, and $h_1 h_2 \cdots h_{2k}$ be an induced subgraph of H which is isomorphic to C ($k \geq 4$). Again, we reduce from OAL-HOM(C). We take an instance of the OAL-HOM(C), i.e., a graph G with $n \geq 2$ vertices and $m \geq 1$ edges, with some vertices of G pre-coloured by vertices of C. We construct a corresponding instance (G', c, w, T) of the weighted minimum constrained cost homomorphism problem to H.

The graph G' is constructed exactly as before: we start with a copy of G and for every vertex v pre-coloured by t, we add the cartesian product of v and C using $2k$ new vertices and $2k$ new edges. Finally, make v adjacent to two vertices in its corresponding gadget, $(v, t-1)$ and $(v, t+1)$ (all indices modulo $2k$).

We define the vertex weight function w as follows.

- for every vertex v in the copy of G, let $w(v) = 1$
- for every pre-coloured vertex $v \in V(G)$:
  - $w((v, 1)) = w((v, 4)) = 50kn^2$,
  - $w((v, 2)) = w((v, 3)) = w((v, 5)) = 1$,
  - $w((v, i)) = 9n$ for all $6 \leq i \leq 2k$

We define the homomorphism cost function c as follows.

- $c(h_1) = c(h_4) = 0$,
- $c(h_2) = c(h_3) = c(h_5) = 8kn$,
- $c(h_i) = 1$ for all $6 \leq i \leq 2k$,
- $c(h_i) = 50kn^2$ otherwise.

Finally, we set $T = 50kn^2 - 1$. As in the proof of Lemma 4, we argue that there is a homomorphism of G to C extending the pre-colouring if and only if there is a homomorphism of G' to H with cost at most T.

This completes the proof of Theorem 2, as chordal bipartite graphs have no induced cycles of length greater than four.

We note that Theorem 2 gives only a partial dichotomy for the minimum constrained cost homomorphism problem, as there is a gap between the class of chordal bipartite graphs and the class of proper interval bigraphs. Specifically, the following result clarifies the gap.

Lemma 7 [20]. *A chordal bipartite graph H is a proper interval bigraph if and only if it does not contain a bipartite claw, a bipartite net, or a bipartite tent (Fig. 3).*

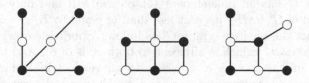

Fig. 3. The bipartite claw, net and tent

3   The Dichotomy for Trees

In this section, we prove an extension of Theorem 2 to graphs H that contain a bipartite claw. As in the case of large cycles, we focus on the weighted version of the problem and show that it is NP-complete when the target graph H contains a bipartite claw. As a corollary we will obtain our dichotomy classification for trees, Theorem 1.

Lemma 8. *Let H be a fixed graph containing the bipartite claw as an induced subgraph. Then the weighted minimum constrained cost homomorphism problem to H is NP-complete.*

It is well known that the problem of finding a maximum independent set in a graph is NP-complete. Alekseev and Lozin citelozin proved that the problem is still NP-complete even when the input is restricted to be a 3-partite graph, cf. Gutin et al. [6].

Theorem 9 [6,21]. *The problem of finding a maximum independent set in a 3-partite graph G, even given the three partite sets, in NP-complete.*

The main idea of the proof of Lemma 8 is similar to the proofs of Lemmas 4 and 6. We show that finding an independent set of size at least k in an arbitrary 3-partite graph G is equivalent to finding a homomorphism of cost at most k' in an auxiliary graph G' together with constrained costs c and vertex weights w. To construct G', we start by adding a fixed number of placeholder vertices; vertices that, with the appropriate weights and costs, always map to the same specific vertices of the target graph H in any homomorphism of G' to H of minimum cost. We then use these placeholder vertices in our construction to ensure that the vertices corresponding to each part of the input graph G are only mapped to certain vertices of H.

Proof. The membership in NP is clear. To show that the problem is NP-hard, we reduce from the problem of finding a maximum independent set in a 3-partite graph, stated in Theorem 9. Let G be a 3-partite graph in which we seek an independent set of size k, with parts V_1, V_2, and V_3, and denote by and n and m the number of vertices and edges in G, respectively. We assume that G is non-empty. Without loss of generality, we can assume that $|V_1| \geq 1$. We construct an instance $(G', c, w, T_{G,k})$ of the weighted minimum cost graph homomorphism and show that G has an independent set of size k if and only if there is a homomorphism of G' to H with cost less than or equal to $T_{G,k}$.

We construct the bipartite graph G' as follows. Subdivide every edge e in G using a new vertex d_e (which is adjacent to both ends of e). Add three vertices b_1, b_2 and b_3 and make each b_i adjacent to all vertices in V_i for $i = 1, 2, 3$. Finally, add three more vertices c_0, c_1 and c_2. Make c_0 adjacent to b_1, b_2 and b_3, c_1 adjacent to b_1 and c_2 adjacent to b_2. A 3-partite graph G together with its corresponding G' is depicted in Fig. 4. For future reference, we denote the set $\{b_1, b_2, b_3, c_0, c_1, c_2\}$ by V_4.

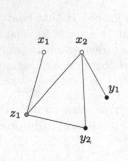 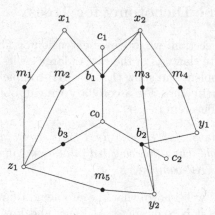

Fig. 4. A 3-partite graph G with parts $V_1 = \{x_1, x_2\}$, $V_2 = \{y_1, y_2\}$, $V_3 = \{z_1\}$ (left) and its corresponding bipartite graph G' (right)

Let $H' = (X, Y)$ be an induced subgraph of H which is isomorphic to a bipartite claw with parts $X = \{v_0, v_1, v_2, v_3\}$ and $Y = \{u_1, u_2, u_3\}$, and edge set

$$E' = \{u_1v_1, u_2v_2, u_3v_3, u_1v_0, u_2v_0, u_3v_0\}.$$

Define the homomorphism cost function c as follows (see Fig. 5).

- $c(v_0) = 4$
- $c(v_1) = c(u_1) = 1$
- $c(u_2) = c(v_3) = 3$
- $c(v_2) = c(u_3) = 0$
- $c(u) = 160n(m + n)$ for every other vertex $u \notin X \cup Y$

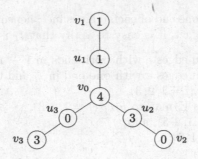

Fig. 5. A bipartite claw, with homomorphism costs

Define the vertex weights of G' as follows.

- $w(b_1) = w(c_1) = 50n(m+n)$
- $w(b_3) = w(c_2) = 160n(m+n)$
- $w(b_2) = w(c_0) = 1$
- $w(u) = 4(m+n)$ for every vertex $u \in V_1$
- $w(u) = 3(m+n)$ for every vertex $u \in V_2$
- $w(u) = 12(m+n)$ for every vertex $u \in V_3$

Finally, let $T_{G,k}$ be the sum of the following values.

- $T_{G,k}^1 = 16(m+n)|V_1|$,
- $T_{G,k}^2 = 12(m+n)|V_2|$,
- $T_{G,k}^3 = 48(m+n)|V_3|$,
- $T_{G,k}^4 = 2 \times 50n(m+n) + 4 + 3$,
- $T_{G,k}^e = 3m$, and,
- $T_{G,k}^I = -12(m+n)k$.

Equivalently, $T_{G,k} = 100n(m+n) + 7 + 3m + (4|V_1| + 36|V_3|)(m+n) + 12(m+n)(n-k)$, We prove that G has an independent set of size k if and only if there is a homomorphism of G' to H of cost less than or equal to $T_{G,k}$.

First, assume that I is an independent set of size k in G with parts $I_1 \subset V_1$, $I_2 \subset V_2$, and $I_3 \subset V_3$. Let k_i denote $|I_i|$ $(i = 1, 2, 3)$. Define the homomorphism f_I as follows.

- $f_I(u) = v_i$ for all vertices $u \in I_i$ $(i = 1, 2, 3)$,
- $f_I(u) = v_0$ for all vertices $u \in V(G) - I$,
- $f_I(d_e) = u_j$ for every edge e with one end in I_j $(j = 1, 2, 3)$,
- $f_I(d_e) = u_3$ for every edge e with both ends in $V - I$,
- $f_I(b_j) = u_j$ for $j = 1, 2, 3$, and finally,
- $f_I(c_k) = v_k$ for $k = 0, 1, 2$.

Notice that at most one end of each edge is in I, hence, the above assignment is indeed a function. In fact, it is easy to verify that f_I is a homomorphism.

- edges subdivided from edges e with both ends in $V - I$ map to $v_0 u_3$,
- edges subdivided from edges e with one end in I_i and the other end in $V - I$ map to $u_i v_i$ and $u_i v_0$ $(i = 1, 2, 3)$,
- edges connecting b_i to V_i map to $u_i v_i$ $(i = 1, 2, 3)$,
- $c_0 b_i$ map to $v_0 u_i$ $(i = 1, 2, 3)$, and,
- $b_i c_i$ map to $v_i u_i$ $(i = 1, 2)$.

We now compute the cost of f_I and show that it does not exceed $T_{G,k}$.

- The vertices in V_1 contribute exactly $(|V_1| - k_1) \times 16(m + n) + k_1 \times 4(m + n)$, or, $T^1_{G,k} - 12k_1(m + n)$,
- the vertices in V_2 contribute exactly $(|V_2| - k_2) \times 12(m + n) + k_1 \times 0$, or, $T^2_{G,k} - 12k_2(m + n)$,
- the vertices in V_3 contribute exactly $(|V_3| - k_3) \times 48(m + n) + k_3 \times 36(m + n)$, or, $T^3_{G,k} - 12k_3(m + n)$,
- the vertices in V_4 contribute a total of $100n(m + n) + 7 = T^4_{G,k}$ (see Table 1),
- the vertices d_e contribute at most $3m = T^e_{G,k}$.

Notice that $k = k_1 + k_2 + k_3$, hence, the cost of f_I is at most $T_{G,k}$.

Table 1. Contribution of vertices in V_4 to the cost of homomorphism f_I

Vertex v	$w(v)$	$f_I(v)$	$c(f_I(v))$	Contributed cost of v
b_1	$50n(m + n)$	u_1	1	$50n(m + n)$
b_2	1	u_2	3	3
b_3	$160n(m + n)$	u_3	0	0
c_0	1	v_0	4	4
c_1	$50n(m + n)$	v_1	1	$50n(m + n)$
c_2	$160n(m + n)$	v_2	0	0

Conversely, assume that f is a homomorphism of G' to H which costs less than or equal to $T_{G,k}$. Note that $T_{G,k} < 150n(m + n)$. This prevents any vertex v to map to a vertex a when $c(v, a) \times w(v) \geq T_{G,k}$. In particular, b_1 and c_1 can only map to vertices a with $c(a) < 3$, i.e., v_1, u_1, v_2, u_3. But b_1 and c_1 are adjacent and the only edge in H among these four vertices is $u_1 v_1$. Similarly, b_3 and c_2 can only map to u_3 or v_2. Observe that $f(b_3) = v_2$ is not feasible, as it implies $f(c_0) = u_2$ and hence $f(b_1) \in \{v_0, v_2\}$. Thus, we have $f(b_3) = u_3$, $f(b_1) = u_1$, $f(c_1) = v_1$, $f(c_0) = v_0$, $f(c_2) = v_2$, and finally $f(b_2) = u_2$.

This restricts possible images of vertices in V. Specifically, all vertices in V_1 are adjacent to b_1, thus, f can only map them to v_1 or v_0, the neighbourhood of $u_1 = f(b_1)$. Similarly, each vertex in V_2 will only map to v_2 or v_0, and each vertex in V_3 will only map to v_3 or v_0.

Let I denote the set of vertices of G that f maps to v_1, v_2 or v_3. Notice that I is an independent set in G. This is because any two adjacent vertices in G are of distance two in G' but the shortest path between v_1 and v_2, or between v_2 and v_3, or between v_3 and v_1 in H' has length 4.

We complete the proof by showing that $|I| \geq k$. Let $|I| = k'$ and assume for a contradiction that $k' < k$. Let f_I denote the homomorphism of G' to H constructed from I as described in the first part of the proof with $cost(f_I) \leq T_{G,k'}$. Observe that f and f_I are identical for every vertex $v \in V_i$ ($i = 1, 2, 3, 4$). Hence, $|cost(f) - cost(f_I)| \leq 3m$. This implies that $cost(f_I) \leq cost(f) + 3m$. Also, note that $cost(f_I) \geq T_{G,k'} - 3m$, hence, we have $T_{G,k'} - 3m \leq T_{G,k} + 3m$, or equivalently, $T_{G,k'} - T_{G,k} \leq 6m$. But this is a contradiction because:

$$T_{G,k'} - T_{G,k} = T_{G,k'}^I - T_{G,k}^I = 12(m+n)(k-k') \geq 12(m+n).$$

∎

We can now apply Theorem 3 and derive the same conclusion for the problem without vertex weights.

Theorem 10. *Let H be a fixed graph containing the bipartite claw as an induced subgraph. Then the minimum constrained cost homomorphism problem to H is NP-complete.*

Note that Lemma 7 implies that for trees, a chordal bipartite H is a proper interval bigraph if and only if it does not contain an induced bipartite claw. Thus we obtain Theorem 1 as a corollary.

4   Conclusion

We left open the complexity of the minimum constrained cost graph homomorphism problems in general. In particular, it remains to check whether the problem is NP-complete also for graphs H that contain a bipartite net or a bipartite tent.

References

1. Hell, P., Nešetřil, J.: On the complexity of H-coloring. J. Comb. Theory Ser. B **48**, 92–110 (1990)
2. Feder, T., Hell, P., Huang, J.: List homomorphisms and circular arc graphs. Combinatorica **19**, 487–505 (1999)
3. Gutin, G., Rafiey, A., Yeo, A., Tso, M.: Level of repair analysis and minimum cost homomorphisms of graphs. Discrete Appl. Math. **154**, 881–889 (2006)
4. Bar-Noy, A., Kortsarz, G.: Minimum color sum of bipartite graphs. J. Algorithms **28**, 339–365 (1998)
5. Supowit, K.: Finding a maximum planar subset of a set of nets in a channel. IEEE Trans. Comput.-Aided Des. **6**, 93–94 (1987)
6. Gutin, G., Hell, P., Rafiey, A., Yeo, A.: A dichotomy for minimum cost homomorphisms. Eur. J. Comb. **29**, 900–911 (2008)

7. Feder, T., Vardi, M.Y.: The computational structure of monotone monadic SNP and constraint satisfaction: a study through datalog and group theory. SIAM J. Comp. **28**, 57–104 (1998)
8. Bulatov, A., Jeavons, P., Krokhin, A.: Classifying the complexity of constraints using finite algebras. SIAM J. Comput. **34**(3), 720–742 (2005)
9. Feder, T., Hell, P., Huang, J.: Bi-arc graphs and the complexity of list homomorphisms. J. Graph Theory **42**, 61–80 (2003)
10. Bulatov, A.: Complexity of conservative constraint satisfaction problems. ACM Trans. Comput. Logic **12**, 24:1–24:66 (2011)
11. Hell, P., Rafiey, A.: The dichotomy of minimum cost homomorphism problems for digraphs. SIAM J. Discrete Math. **26**(4), 1597–1608 (2012)
12. Hell, P., Rafiey, A.: The dichotomy of list homomorphisms for digraphs. In: Proceedings of the Symposium on Discrete Algorithms, SODA 2011, pp. 1703–1713 (2011)
13. Hell, P., Rafiey, A.: Duality for min-max orderings and dichotomy for min cost homomorphisms. arXiv preprint arXiv:0907.3016 (2009)
14. Hell, P., Rafiey, A.: Minimum cost homomorphism problems to smooth and balanced digraphs. Manuscript (2007)
15. Takhanov, R.: A dichotomy theorem for the general minimum cost homomorphism problem. In: 27th International Symposium on Theoretical Aspects of Computer Science, vol. 5, pp. 657–668 (2010)
16. Kolmogorov, V., Živný, S.: The complexity of conservative valued CSPs. J. ACM **60**, 10:1–10:38 (2013)
17. Müller, H.: Recognizing interval digraphs and interval bigraphs in polynomial. Discrete Appl. Math. **78**, 189–205 (1997)
18. Spinrad, J., Brandstädt, A., Stewart, L.: Bipartite permutation graphs. Discrete Appl. Math. **18**, 279–292 (1987)
19. Feder, T., Hell, P.: List homomorphism to reflexive graphs. J. Comb. Theory B **72**, 236–250 (1998)
20. Hell, P., Huang, J.: Interval bigraphs and circular arc graphs. J. Graph Theory **46**, 313–327 (2004)
21. Alekseev, V.E., Lozin, V.V.: Independent sets of maximum weight in (p, q)-colorable graphs. Discrete Math. **265**, 351–356 (2003)

Approximation Algorithms

An Improved Constant-Factor Approximation Algorithm for Planar Visibility Counting Problem

Sharareh Alipour[1]([✉]), Mohammad Ghodsi[1,2], and Amir Jafari[1]

[1] Sharif University of Technology, Tehran, Iran
Sharareh.alipour@gmail.com
[2] Institute for Research in Fundamental Sciences (IPM), Tehran, Iran

Abstract. Given a set S of n disjoint line segments in $\mathbb{R}^2$, the visibility counting problem (VCP) is to preprocess S such that the number of segments in S visible from any query point p can be computed quickly. This problem can trivially be solved in logarithmic query time using $O(n^4)$ preprocessing time and space. Gudmundsson and Morin proposed a 2-approximation algorithm for this problem with a trade-off between the space and the query time. They answer any query in $O_\epsilon(n^{1-\alpha})$ with $O_\epsilon(n^{2+2\alpha})$ of preprocessing time and space, where α is a constant $0 \leq \alpha \leq 1, \epsilon > 0$ is another constant that can be made arbitrarily small, and $O_\epsilon(f(n)) = O(f(n)n^\epsilon)$.

In this paper, we propose a randomized approximation algorithm for VCP with a tradeoff between the space and the query time. We will show that for an arbitrary constants $0 \leq \beta \leq \frac{2}{3}$ and $0 < \delta < 1$, the expected preprocessing time, the expected space, and the query time of our algorithm are $O(n^{4-3\beta} \log n)$, $O(n^{4-3\beta})$, and $O(\frac{1}{\delta^3} n^\beta \log n)$, respectively. The algorithm computes the number of visible segments from p, or m_p, exactly if $m_p \leq \frac{1}{\delta^3} n^\beta \log n$. Otherwise, it computes a $(1+\delta)$-approximation m_p' with the probability of at least $1 - \frac{1}{\log n}$, where $m_p \leq m_p' \leq (1+\delta)m_p$.

Keywords: Computational geometry · Visibility · Randomized algorithm · Approximation algorithm · Graph theory

1 Introduction

Problem Statement: Let $S = \{s_1, s_2, \ldots, s_n\}$ be a set of n disjoint closed line segments in the plane contained in a bounding box, $\mathbb{B}$. Two points p and q in the bounding box are visible to each other with respect to S, if the open line segment $\overline{pq}$ does not intersect any segments of S. A segment $s_i \in S$ is also said to be visible from a point p, if there exists a point $q \in s_i$ such that q is visible from p. *The visibility counting problem (VCP) is to find m_p, the number of segments of S visible from a query point p. We know that the visibility polygon of a given point $p \in \mathbb{B}$ is defined as $VP_S(p) = \{q \in \mathbb{B} : p$ and q are visible$\}$, and the visibility polygon of a given segment s_i is defined as $VP_S(s_i) = \bigcup_{q \in s_i} VP_S(q)$.*

© Springer International Publishing Switzerland 2016
T.N. Dinh and M.T. Thai (Eds.): COCOON 2016, LNCS 9797, pp. 209–221, 2016.
DOI: 10.1007/978-3-319-42634-1_17

Consider the $2n$ end-points of the segments of S as vertices of a geometric graph. Add a straight-line-edge between each pair of visible vertices. The result is *the visibility graph of S* or $VG(S)$. We can extend each edge of $VG(S)$ in both directions to the points that the edge hits some segments in S or the bounding box. This creates at most two new vertices and two new edges. Adding all these vertices and edges to $VG(S)$ results in a new geometric graph called *the extended visibility graph of S* or $EVG(S)$. $EVG(S)$ reflects all the visibility information from which the visibility polygon of any segment $s_i \in S$ can be computed [9].

Related Work: $VP_S(p)$ can be computed in $O(n \log n)$ time using $O(n)$ space [3,13]. Vegter proposed an output sensitive algorithm that reports $VP_S(p)$ in $O(|VP_S(p)| \log(\frac{n}{|VP_S(p)|}))$ time, by preprocessing the segments in $O(m \log n)$ time using $O(m)$ space, where $m = O(n^2)$ is the number of edges of $VG(S)$ and $|VP_S(p)|$ is the number of vertices of $VP_S(p)$ [14].

$EVG(S)$ can be used to solve VCP. $EVG(S)$ can optimally be computed in $O(n \log n + m)$ time [7]. If a vertex is assigned to any intersection point of the edges of $EVG(S)$, we have a planar graph, which is called the planar arrangement of the edges of $EVG(S)$. All points in any face of this arrangement have the same number of visible segments and this number can be computed for each face in the preprocessing step [9]. Since there are $O(n^4)$ faces in the planar arrangement of $EVG(S)$, a point location structure of size $O(n^4)$ can answer each query in $O(\log n)$ time. But, $O(n^4)$ preprocessing time and space is high. Also, for any query point p, by computing $VP_S(p)$, m_p can be computed in $O(n \log n)$ with no preprocessing. This has led to several results with a tradeoff between the preprocessing cost and the query time [2,4,8,12,15].

There are two approximation algorithms for VCP by Fischer *et al.* [5,6]. One of these algorithms uses a data structure of size $O((m/r)^2)$ to build a (r/m)-cutting for $EVG(S)$ by which the queries are answered in $O(\log n)$ time with an absolute error of r compared to the exact answer ($1 \leq r \leq n$). The second algorithm uses the random sampling method to build a data structure of size $O((m^2 \log^{O(1)} n)/l)$ to answer any query in $O(l \log^{O(1)} n)$ time, where $1 \leq l \leq n$. In the latter method, the answer of VCP is approximated up to an absolute value of δn for any constant $\delta > 0$ (δ affects the constant factor of both data structure size and the query time).

In [13], Suri and O'Rourke represent the visibility polygon of a segment by a union of set of triangles. Gudmundsson and Morin [9] improved the covering scheme of [13]. Their method builds a data structure of size $O_\epsilon(m^{1+\alpha}) = O_\epsilon(n^{2(1+\alpha)})$ in $O_\epsilon(m^{1+\alpha}) = O_\epsilon(n^{2(1+\alpha)})$ preprocessing time, from which each query is answered in $O_\epsilon(m^{(1-\alpha)/2}) = O_\epsilon(n^{1-\alpha})$ time, where $0 < \alpha \leq 1$. This algorithm returns m_p' such that $m_p \leq m_p' \leq 2m_p$. The same result can be achieved from [1,11]. In [1], it is proven that the number of visible end-points of the segments in S, denoted by ve_p, is a 2-approximation of m_p, that is $m_p \leq ve_p \leq 2m_p$.

Our Results: In this paper, we present a randomized $(1 + \delta)$-approximation algorithm, where $0 < \delta \leq 1$. The expected preprocessing time and space of our algorithm are $O(m^{2-3\beta/2} \log m)$ and $O(m^{2-3\beta/2})$ respectively, and our query time is $O(\frac{1}{\delta^3} m^{\beta/2} \log m)$, where $0 \le \beta \le \frac{2}{3}$ is chosen arbitrarily in the pre-processing time. In this algorithm, a graph $G(p)$ is associated to each query point p; the construction of $G(p)$ is explained in Sect. 2. It will be shown that $G(p)$ has a planar embedding and this formula holds: $m_p = n - F(G(p)) + 1$ or $n - F(G(p)) + 2$, where $F(G(p))$ is the number of faces of $G(p)$. Using Euler's formula for planar graphs, we will show that if p is inside a bounded face of $G(p)$, then $m_p = ve_p - C(G(p)) + 1$, otherwise $m_p = ve_p - C(G(p))$, where $C(G(p))$ is the number of connected components of $G(p)$. In Sects. 3 and 4, we will present algorithms to approximate ve_p and $C(G(p))$. This leads to an overall approximation for m_p. Some detail of our algorithm is as follows: First, we try to calculate $VP_S(p)$ by running the algorithm presented in [14] for $\frac{1}{\delta^3} m^{\beta/2} \log m$ steps. If this algorithm terminates, the exact value of m_p is calculated, which is obviously less than $\frac{1}{\delta^3} m^{\beta/2} \log m$. Otherwise, our algorithm instead returns m_p', such that $m_p \le m_p' \le (1+\delta) m_p$ with the probability of at least $1 - \frac{1}{\log n}$. Table 1 compares the performance of our algorithm with the best known result for this problem. Note that if we choose a constant number $0 < \delta < 1$, then our query time is better than [9], however our algorithm returns a $(1 + \delta)$-approximation of the answer with a high probability.

Table 1. Comparison of our method and the best known result for VCP. Note that β $(0 \le \beta \le \frac{2}{3})$ is chosen in the preprocessing time and $1 + \delta$ $(0 < \delta \le 1)$ is the approximation factor of the algorithm which affects the query time and $O_\epsilon(f(n)) = O(f(n)n^\epsilon)$, where ϵ is a constant number that can be arbitrary small.

Reference	Preprocessing time	Space	Query	Approx-Factor
[9]	$O_\epsilon(m^{2-3\beta/2})$	$O_\epsilon(m^{2-3\beta/2})$	$O_\epsilon(m^{3\beta/4})$	2
Our result	$O(m^{2-3\beta/2} \log m)$	$O(m^{2-3\beta/2})$	$O(\frac{1}{\delta^3} m^{\beta/2} \log m)$	$1 + \delta$

2 Definitions and the Main Theorem

For each point $a' \in s_i$, let $\overrightarrow{pa'}$ be the ray emanating from the query point p toward a' and let $a = pr(a')$ be the first intersection point of $\overrightarrow{pa'}$ and a segment in S or the bounding box right after touching a'. We say that $a = pr(a')$ is covered by a' or the projection of a' is a. Also, suppose that $\overline{x'y'}$ is a subsegment of s_i and $\overline{xy}$ is a subsegment of s_j, such that $pr(x') = x$ and $pr(y') = y$ and for any point $z' \in \overline{x'y'}$, $pr(z') \in \overline{xy}$, then we say that $\overline{xy}$ is covered by $\overline{x'y'}$.

For each query point p, we construct a graph denoted by $G(p)$ as follows: a vertex v_i is associated to each segment $s_i \in S$, and an edge (v_i, v_j) is put if s_j covers one end-point of s_i (or vice-versa; that is, if s_i covers one end-point of s_j). Obviously, there are two edges between v_i and v_j, if s_j (or s_i) covers both end-points of s_i (or s_j). As an example, refer to Fig. 1(a) and (d). Note that the bounding box is not considered here.

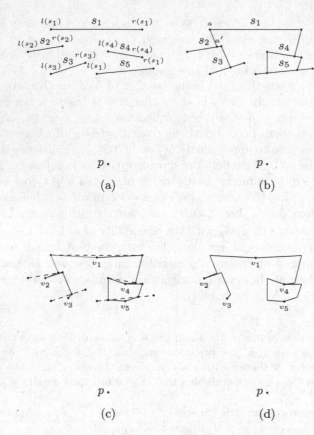

Fig. 1. The steps to draw a planar embedding of $G(p)$. (a) The segments are $s_1, \ldots, s_5$ with their left and right end-points and a given query point is p. (b) For each end-point $a \in s_i$ not visible to p, if $a' \in s_j$ such that $pr(a') = a$, we draw $\overline{aa'}$. (c) Put a vertex v_i for each segment s_i in a distance sufficiently close to the middle of s_i. For each a and a' (described in (b)), connect a to v_i and a' to v_j. This creates an edge between v_i and v_j shown in red (d) Remove the segments and the remaining is the planar embedding of $G(p)$. Note that the final embedding has 5 vertices and 5 edges and each edge is drown as 3 consequence straight lines. (Color figure online)

For any segment $s \in S$, let $l(s)$ and $r(s)$ be the first and second end-points of s, respectively swept by a ray around p in clockwise order (Fig. 1(a)).

Lemma 1. *$G(p)$ has a planar embedding.*

Proof. Here is the construction. For each end-point $a \in s_i$ not visible from p, let $a' \in s_j$ such that $pr(a') = a$. Draw the straight-line $\overline{aa'}$. Doing this, we have a collection of non-intersecting straight-lines. For each s_i, we put a vertex v_i located very close to the mid-point of s_i. Also, for each segment $\overline{aa'}$, we connect a to v_i and a' to v_j. This creates an edge consisting of three consecutive straight-lines $\overline{v_i a}$, $\overline{aa'}$, and $\overline{a'v_j}$ that connects v_i to v_j. Obviously, none of these edges intersect. Finally, all the original segments are removed. The remaining is the vertices and edges of a planar embedding of $G(p)$ (See Fig. 1).

From now on, we use $G(p)$ as the planar embedding of the graph $G(p)$. As we know the Euler's formula for any non-connected planar graph G with multiple edges is: $V(G) - E(G) + F(G) = 1 + C(G)$, where $E(G), V(G), F(G)$, and $C(G)$ are the number of edges, vertices, faces, and connected components of G, respectively. We have the following theorem to calculate m_p, using $G(p)$.

Theorem 1. *The number of segments not visible from p is equal to $F(G(p)) - 2$ if p is inside a bounded face of $G(p)$, or is equal to $F(G(p)) - 1$, otherwise.*

Proof. We construct a bijection ϕ between the segments not visible from p to the faces of $G(p)$ except the unbounded face and the face that contains p. This will complete the proof of our theorem.

Suppose that s_i is a segment not visible from p. Then, we can partition s_i into k subsegments, $\overline{q_0 q_1}, \overline{q_1 q_2}, \ldots, \overline{q_{k-1} q_k}$ such that $q_0 = l(s_i)$, $q_k = r(s_i)$, and for each $\overline{q_i q_{i+1}}$, there is a subsegment $\overline{q_i' q_{i+1}'} \in s_j$ that covers $\overline{q_i q_{i+1}}$. Let $s_1', s_2', \ldots, s_k'$ be the set of segments such that $\overline{xy} \in s_{i+1}'$ covers $\overline{q_i q_{i+1}}$ (note that some segments may appear more than once in the above sequence) (Fig. 2). We claim that the vertices $v_i, v_1', v_2', \ldots, v_k'$ form a bounded face of $G(p)$ that does not contain p. In ϕ, we associate this face to s_i. Since v_1' is the vertex associated to the first segment that covers $\overline{q_0 q_1}$, s_1' will cover $l(s_i)$ and hence v_i is adjacent to v_1'. Similarly, since s_k' covers $r(s_i)$, hence v_i is adjacent to v_k'. The next subsegment that covers a subsegment of s_i comes from s_2'. This means that $r(s_1')$ is covered by s_2' or $l(s_2')$ is covered by s_1'. This implies that v_1' is adjacent to v_2'. Similarly, we can show that v_i' is adjacent to v_{i+1}' for all $1 \le i < k$. To complete the construction, we show that the closed path formed

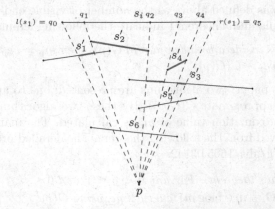

Fig. 2. s_i is not visible from p. It can be partitioned into 5 subsegments $\overline{q_0 q_1}, \overline{q_1 q_2}, \overline{q_2 q_3}, \overline{q_3 q_4}$, and $\overline{q_4 q_5}$, each is covered respectively by subsegment of s_1', s_2', s_3', s_4', and s_3' shown above.

by $v_i \to v'_1 \to v'_2, \ldots \to v'_k \to \dot{v}_i$ is a bounded face not containing p. Consider a ray around p in clockwise order. The area that this ray touches under s_i and above $s'_1, \ldots, s'_k$ is bounded by $v_i, v'_1, v'_2, \ldots, v'_k$. So, p is not inside this region.

Now, we show that our map ϕ is one-to-one and onto. If $\phi(s_i) = \phi(s_j)$, then according to the construction of ϕ, a subsegment of s_i covers a subsegment of s_j and a subsegment of s_j covers a subsegment of s_i. This is a contradiction since these segments do not intersect. To prove the onto-ness, we need to show for any bounded face f that does not contain p, there is a vertex v_i corresponding to a segment s_i that is not visible to p such that $\phi(s_i) = f$.

To find s_i, we use the sweeping ray around p. Since f is assumed to be bounded and not containing p, the face f is between two rays from p; one from the left and the other from the right. If we start sweeping from left to right, there is a segment corresponding to the vertices of f whose end-point is the first to be covered by the other segments corresponding to the vertices of f. We claim that s_i is the desired segment i.e. s_i is not visible to p and $\phi(s_i) = f$. For example in Fig. 2, the closed path $v_i \to v'_1 \to v'_2 \to v'_3 \to v'_4, \to v'_3, \to v_i$ forms a face and s_i is the first segment among $\{s_i, s'_1, s'_2, s'_3, s'_4\}$ such that $l(s_i)$ is covered by one of the segments in $\{s_i, s'_1, s'_2, s'_3, s'_4\}$.

Obviously, $l(s_i)$ is not visible from p. v'_1 is adjacent to v_i which means that a subsegment of s'_1 covers a subsegment of s_i. Since v'_1 and v'_2 are adjacent, this means that a subsegment of s'_2 consecutively covers the next subsegment of s_i right after s'_1. Continuing this procedure, we conclude that a subsegment of each s'_i covers some subsegment of s_i continuously right after s'_{i-1}. v'_k and v_i are also adjacent, so $r(s_i)$ is not visible from p. We conclude that subsegments of $s'_1, s'_2 \ldots, s_k$ completely cover s_i and hence s_i is not visible from p.

So, if p is in the unbounded face of $G(p)$, the number of segments which are not visible from p is $F(G(p)) - 1$, otherwise it is $F(G(p)) - 2$.

The Euler's formula is used to compute $F(G(p))$. Obviously, $V(G(p))$ is n. For each end-point not visible from p, an edge is added to $G(p)$; therefore, $E(G(p))$ is $2n - ve_p$ (ve_p was defined above as the number of visible end-points from p). The Euler's formula and Theorem 1 indicate the following lemma.

Lemma 2. *If p is inside a bounded face of $G(p)$, then $m_p = ve_p - C(G(p)) + 1$, otherwise, $m_p = ve_p - C(G(p))$.*

In the rest of this paper, two algorithms are presented; one to approximate ve_p and the other to approximate $C(G(p))$. By these two algorithms and applying Lemma 2, an approximation value of m_p is calculated. The main result of this paper is thus derived from the following theorem. The detailed proof is presented in http://arxiv.org/abs/1605.03542.

Theorem 2 *(Main theorem). For any $0 < \delta \le 1$ and $0 \le \beta \le \frac{2}{3}$, VCP can be approximated in $O(\frac{1}{\delta^3} m^{\beta/2} \log m)$ query time using $O(m^{2-3\beta/2} \log m)$ expected preprocessing time and $O(m^{2-3\beta/2})$ expected space. This algorithm returns a value m'_p such that with the probability at least $1 - \frac{1}{\log m}$, $m_p \le m'_p \le (1+\delta)m_p$ when $m_p \ge \frac{1}{\delta^3} m^{\beta/2} \log m$ and returns the exact value when $m_p < \frac{1}{\delta^3} m^{\beta/2} \log m$.*

3   An Approximation Algorithm to Compute ve_p

In this section, we present an algorithm to approximate ve_p, the number of visible end-points. In the preprocessing phase, we build the data structure of the algorithm presented in [14] which calculates $VP_S(p)$ in $O(|VP_S(p)|\log(n/|VP_S(p)|))$ time, where $|VP_S(p)|$ is the number of vertices of $VP_S(p)$. In [14], the algorithm for computing $VP_S(p)$, consists of a rotational sweep of a line around p. During the sweep, the subsegments visible from p along the sweep-line are collected. In the preprocessing phase, we choose a fixed parameter β, where $0 \le \beta \le \frac{2}{3}$. In the query time we also choose a fixed parameter $0 < \delta \le 1$ which is the value of approximation factor of the algorithm.

We use the algorithm presented in [14] to find the visible end-points, but for any query point, we stop the algorithm if more than $\frac{2}{\delta^3} m^{\beta/2} \log m$ of the visible end-points are found.

If the sweep line completely sweeps around p before counting $\frac{1}{\delta^3} m^{\beta/2} \log m$ of the visible end-points, then we have completely computed $VP_S(p)$ and we have $|VP_S(p)| \le \frac{2}{\delta^3} m^{\beta/2} \log m$. In this case, the number of visible segments can be calculated exactly in $O(\frac{1}{\delta^3} m^{\beta/2} \log m)$ time. Otherwise, $ve_p > \frac{2}{\delta^3} m^{\beta/2} \log m$ and the answer is calculated in the next step of algorithm, that we now explain.

The visibility polygon of an end-point a is a star shaped polygon consisting of $m_a = O(n)$ non-overlapping triangles [3,13], which are called *the visibility triangles of a* denoted by $VT_S(a)$. Notice that m_a is the number of edges of $EVG(S)$ incident to a. The query point p is visible to an end-point a, if and only if it lies inside one of the visibility triangles of a. Let VT_S be the set of visibility triangles of all the end-points of the segments in S. Then, the number of visible end-points from p is the number of triangles in VT_S containing p. We can construct VT_S in $O(m \log m) = O(n^2 \log n)$ time using $EVG(S)$ and $|VT_S| = O(m) = O(n^2)$ [9].

We can preprocess a given set of triangles using the following lemma to count the number of triangles containing any query point.

Lemma 3. *Let Δ be a set of n triangles. There exists a data structure of size $O(n^2)$, such that in the preprocessing time of $O(n^2 \log n)$, the number of triangles containing a query point p can be calculated in $O(\log n)$ time.*

Proof. Consider the planar arrangement of the edges of the triangles in Δ as a planar graph. Let f be a face of this graph. Then, for any pair of points p and q in f, the number of triangles containing p and q are equal. Therefore, we can compute these numbers for each face in a preprocessing phase and then, for any query point locate the face containing that point. There are $O(n^2)$ faces in the planar arrangement of Δ, so a point location structure of size $O(n^2)$ can answer each query in $O(\log n)$ time [10]. Note that the number of triangles containing a query point differs in 1 for any pair of adjacent faces.

3.1   The Algorithm

Here, we present an algorithm to approximate ve_p. We use this algorithm when $m_p > \frac{1}{\delta^3} m^{\beta/2} \log m$. In the preprocessing phase we take a random subset $RVT_1 \subset VT_S$ such that each member of VT_S is chosen with the probability of $\frac{1}{m^\beta}$.

Lemma 4. $E(|RVT_1|) = O(m^{1-\beta})$.

Proof. Let $VT_S = \{\Delta_1, \Delta_2, \dots, \Delta_{m'}\}$, where $m' = O(m) = O(n^2)$ and $X_i = 1$ if $\Delta_i \in RTV_1$, and $X_i = 0$ otherwise. We have,

$$E(|RVT_1|) = E(\sum\nolimits_{i=1}^{m'} X_i) = \sum\nolimits_{i=1}^{m'} E(X_i) = \sum\nolimits_{i=1}^{m'} \frac{1}{m^\beta} = \frac{m'}{m^\beta} = O(m^{1-\beta}).$$

Suppose that in the preprocessing time, we choose $m^{\beta/2}$ independent random subsets $RVT_1, \dots, RVT_{m^{\beta/2}}$ of VT_S. By Lemma 3, for any query point p, the number of triangles of each RVT_i containing p denoted by $(ve_p)_i$, is calculated in $O(\log m)$ time by $O(m^{2-2\beta} \log m)$ expected preprocessing time and $O(m^{2-2\beta})$ expected space. Then, $ve'_p = m^\beta \frac{\sum_{i=1}^{m^{\beta/2}} (ve_p)_i}{m^{\beta/2}}$ is returned as the approximation value of ve_p.

3.2   Analysis of Approximation Factor

Lemma 5. Let $X_i = m^\beta (ve_p)_i$, we have $E(X_i) = ve_p$.

Proof. Suppose that $VT(p) = \{\Delta'_1, \Delta'_2, \dots, \Delta'_{ve_p}\} \subset VT_S$ be the set of all triangles containing p. Let $Y_j = 1$ if $\Delta'_j \in RVT_i$, and $Y_j = 0$ otherwise. So, $(ve_p)_i = \sum_{j=1}^{ve_p} Y_j$ and $E((ve_p)_i) = E(\sum_{j=1}^{ve_p} Y_j) = \frac{ve_p}{m^\beta}$. $E(X_i) = E(m^\beta (ve_p)_i) = m^\beta E((ve_p)_i) = m^\beta \frac{ve_p}{m^\beta} = ve_p$.

In addition, we can conclude the following lemma:

Lemma 6. $E(\frac{\sum_{i=1}^{m^{\beta/2}} X_i}{m^{\beta/2}}) = ve_p$.

So, $X_1, X_2, \dots, X_{m^{\beta/2}}$ are random variables with $E(X_i) = ve_p$. According to Chebyshev's Lemma the following lemma holds

Lemma 7 *(Chebyshev's Lemma). Given $X_1, X_2, \dots, X_n$ sequence of i.i.d.'s random variables with finite expected value $E(X_1) = E(X_2) = \dots = \mu$, we have,*

$$P((|\frac{X_1 + \dots + X_n}{n} - \mu|) > \varepsilon_1) \le \frac{Var(X)}{n\varepsilon_1^2}.$$

Lemma 8. *With a probability at least $1 - \frac{1}{\log m}$ we have, $(1-\delta)ve_p \le ve'_p \le (1+\delta)ve_p$.*

Proof. Using Lemma 7, we choose $\varepsilon_1 = \delta ve_p$. Here, δ indicates the approximation factor of the algorithm. Obviously, $Var(X_i) = m^{2\beta}(ve_p)(1 - \frac{1}{m^\beta})\frac{1}{m^\beta}$. So,

$$\mathbb{P} = P(|ve'_p - ve_p| > \delta ve_p) \leq \frac{m^\beta ve_p}{m^{\beta/2}\delta^2(ve_p)^2}.$$

We know that $ve_p \geq \frac{1}{\delta^2}m^{\beta/2}\log m$, so $\mathbb{P} = P(|ve'_p - ve_p| > \delta ve_p) \leq \frac{1}{\log m}$. With the probability of at least $1 - \mathbb{P}$, we have, $(1 - \delta)ve_p \leq ve'_p \leq (1 + \delta)ve_p$. Also, for a large m, we have $\mathbb{P} \sim 0$.

3.3   Analysis of Time and Space Complexity

In the first step of the query time, we run the algorithm of [14]. The preprocessing time and space for constructing the data structure of [14] are $O(m \log m)$ and $O(m)$, respectively, which computes $VP_S(p)$ in $O(|VP_S(p)| \log(n/|VP_S(p)|))$ time. As we run this algorithm for at most $\frac{1}{\delta^3}m^{\beta/2}\log m$ steps, the query time of the first step is $O(\frac{1}{\delta^3}m^{\beta/2}\log m)$.

According to Lemma 4, $E(|RVT_i|) = O(m^{1-\beta})$. Using Lemma 3, the expected preprocessing time and space for each RVT_i are $O(m^{2-2\beta}\log m)$ and $O(m^{2-2\beta})$ respectively, such that in $O(\log m)$ we can calculate $(ve_p)_i$. So, the expected preprocessing time and space are $m^{\beta/2}O(m^{2-2\beta}\log m) = O(m^{2-\frac{3}{2}\beta}\log m)$ and $m^{\beta/2}O(m^{2-2\beta}) = O(m^{2-\frac{3}{2}\beta})$ respectively. In the second step, for each RVT_i the value of $(ve_p)_i$ is calculated in $O(\log m)$. Therefore, the query time is $O(\frac{1}{\delta^3}m^{\beta/2}\log m) + O(m^{\beta/2}\log m)$. So, we have the following lemma.

Lemma 9. *There exists an algorithm that for any query point p, approximates ve_p in $O(\frac{1}{\delta^3}m^{\beta/2}\log m)$ query time using $O(m^{2-3\beta/2}\log m)$ expected preprocessing time and $O(m^{2-3\beta/2})$ expected space $(0 \leq \beta \leq \frac{2}{3})$. This algorithm returns the exact value of ve_p when $ve_p < \frac{1}{\delta^2}m^{\beta/2}\log m$. Otherwise, a value of ve'_p is returned such that with the probability of at least $1 - \frac{1}{\log m}$, we have $(1 - \delta)ve_p \leq ve'_p \leq (1 + \delta)ve_p$.*

4   An Approximation Algorithm for Computing the Number of Components of $G(p)$

Now, we explain an algorithm to compute the number of connected components of $G(p)$, each is simply called a component of $G(p)$. Let c be a component such that p is not inside any of its faces. Without loss of generality we can assume that p lies below c. Obviously, there exist rays emanating from p that do not intersect any segments corresponding to the vertices of c. We start sweeping one of these rays in a clockwise direction. Let $l(c)$ (left end-point of c) be the first end-point of a segment of c and $r(c)$ (right end-point of c) be the last end-point of a segment of c that are crossed by this ray (Fig. 3). This way every component

c has $l(c)$ and $r(c)$ except the component containing p. Also, note that $r(c)$ and $l(c)$ do not depend on the choice of the starting ray. As said, the bounding box is not a part of $G(p)$, but $G(p)$ is contained in the bounding box.

Lemma 10. *For each component c, except the one containing p, the projections of $l(c)$ and $r(c)$ both belong either to the same segment or the bounding box.*

Proof. Assume that $pr(l(c))$ belongs to a segment $s \in S$. Since $l(s)$ is on the left of $l(c)$, s can not be among the segments of c. We claim that $r(s)$ is on the right of $r(c)$. Obviously, if this claim is true then, if $pr(r(c)) \in s'$, then $l(s')$ is on the left of $l(c)$. Clearly, if $s \neq s'$, then these two should intersect, which is impossible. Also, this implies that if $pr(l(c))$ is on the bounding box, then $pr(r(c))$ should to be on the bounding box as well. The claim is proven by contradiction. Assume that $r(s)$ is on the left of $r(c)$. Since, $r(s)$ is not visible from p, then there should exist a segment s' that covers $r(s)$. Since, s is not in c and s' is connected to s, s' can not be in c, so $l(s')$ is to the right of $l(c)$ and hence is not visible. Therefore, there should exist a different segment s'' that covers $l(c)$ and with the same argument s'' can not be in c and $l(s'')$ should be covered by another segment. This process can not be continued indefinitely since the number of segments is finite and therefore we will reach a contradiction.

Let s'_1, s'_2, s'_3, and s'_4 be the segments of the bounding box. According to Lemma 10, we can associate a pair of adjacent visible subsegments or a connected visible part of the bounding box for each component of $G(p)$. For example, in Fig. 3, s_1 has two visible subsegments which are associated to the component composed of s_3 and s_4. If we can count the number of visible subsegments of each segment and the number of visible parts of the bounding box, then we can compute the exact value of $C(G(p))$. Because each pair of consecutive visible

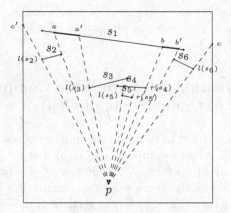

Fig. 3. $\overline{aa'}$ and $\overline{bb'}$ are the visible subsegments of s_1. The bounding box has one visible part from c to c'. $G(p)$ has three components; $\{s_1, s_2, s_6\}, \{s_3, s_4\}$, and $\{s_5\}$. $l(s_2), l(s_3)$, and $l(s_5)$ are the left end-points of these components, respectively. $r(s_6), r(s_4)$, and $r(s_5)$ are the right end-points of these components, respectively.

subsegments of a segment and each visible part of the bounding box are associated to a component. Let c' be the number of visible parts of the bounding box. If $c' > 0$, then p is in the unbounded face. So, if each segment s_i has c_i visible subsegments, then $C(G(p)) = c' + \sum_{i=1}^{n} \max\{(c_i - 1), 0\}$. For example in Fig. 3, $c_1 = 2$, $c_2 = 1$, $c_3 = 1$, $c_4 = 2$, $c_5 = 1$ and $c_6 = 1$, also $c' = 1$. This implied that $C(G(p)) = 3$. If $c' = 0$, then p is in a bounded face and this face is contained in a component with no left and right end-point, so in this case $C(G(p)) = 1 + \sum_{i=1}^{n} \max\{(c_i - 1), 0\}$.

In the following we propose an algorithm to approximate the number of visible subsegments of each segment $s_i \in S \cup \{s'_1, s'_2, s'_3, s'_4\}$.

4.1 Algorithm

According to [9], it is possible to cover the visibility region of each segment $s_i \in S \cup \{s'_1, s'_2, s'_3, s'_4\}$ with $O(m_{s_i})$ triangles denoted by $VT(s_i)$. Here, $|VT(s_i)| = O(m_{s_i})$, where m_{s_i} is the number of edges of $EVG(S)$ incident on s_i. Note that the visibility triangles of s_i may overlap. If we consider the visibility triangles of all segments, then there is a set $VT_S = \{\Delta_1, \Delta_2, \dots\}$ of $|VT_S| = O(m)$ triangles. We say Δ_i is related to s_j if and only if $\Delta_i \in VT(s_j)$. For a given query point p, m''_p, the number of triangles in VT_S containing p, is between m_p and $2m_p$. So, m''_p gives a 2-approximation factor solution for VCP [9]. Since the visibility triangles of each segment may overlap, some of the segments are counted repeatedly. In [9], it is shown that each segment s_i is counted c_i times, where c_i is the number of visible subsegments of s_i. In other words, there are c_i triangles related to s_i in VT_S which contain p.

A similar approach can be used to approximate $C(G(p))$. A random subset $RVT_1 \subset VT_S$ is chosen such that each member of VT_S is chosen with probability $\frac{1}{m^\beta}$. For a given query point p, let $c'_{i,1} \geq 1$ be the number of triangles related to s_i in RVT_1 containing p. We report $C_1 = \sum_{i=1}^{n}(m^\beta c'_{i,1} - 1)$ as the approximated value of $C(G(p))$ received by RVT_1. We choose $m^{\beta/2}$ random subsets $RVT_1, \dots, RVT_{m^{\beta/2}}$ of VT_S. Let p be the given query point, for each RVT_j, $C_j = \sum_{i=1}^{n}(m^\beta c'_{i,j} - 1)$ is calculated. At last, $C'_p = \frac{\sum_{j=1}^{m^{\beta/2}} C_j}{m^{\beta/2}}$ is reported as the approximation value of $C(G(p))$.

4.2 Analysis of Approximation Factor

We show that with the probability at least $\frac{1}{\log m}$, if $C(G(p)) > \frac{1}{\delta^2} m^{\beta/2} \log m$, then C'_p is a $(1 + \delta)$-approximation of $C(G(p))$.

Lemma 11. $E(C_j) = C(G(p))$.

Proof. $E(C_j) = E(\sum_{i=1}^{n} m^\beta c'_{i,j} - 1) = \sum_{i=1}^{n} E(m^\beta c'_{i,j} - 1) = \sum_{i=1}^{n} c_i - 1 = C(G(p))$.

Using Lemma 7, we have, $\mathbb{P} = P(|\frac{C_1+\cdots+C_{m^{\beta/2}}}{m^{\beta/2}} - C(G(p))| > \delta C(G(p))) \leq \frac{Var(C_i)}{m^{\beta/2}\delta^2 C(G(p))^2}$. $Var(C_i) = m^{2\beta}C(G(p))(\frac{1}{m^\beta})(1-\frac{1}{m^\beta})$. Since we have, $C(G(p)) > \frac{1}{\delta^2}m^{\beta/2}\log m$, then $\mathbb{P} = P(|\frac{C_1+\cdots+C_{m^{\beta/2}}}{m^{\beta/2}} - C(G(p))| > \delta C(G(p))) \leq \frac{1}{\log m}$.

So, with the probability at least $1-\mathbb{P}$, $(1-\delta)C(G(p)) \leq C'_p \leq (1+\delta)C(G(p))$. And for a large m, we have, $\mathbb{P} \sim 0$.

4.3   Analysis of Time and Space Complexity

By Lemma 3, for each RVT_i, a data structure of expected preprocessing time and size of $O(m^{2-2\beta}\log m)$ and $O(m^{2-2\beta})$ is needed. RVT_i returns C_i in $O(\log m)$ for each query point p. So, the expected space for all $m^{\beta/2}$ data structures is $O(m^{2-2\beta+\beta/2}\log m)$ and the query time for calculating C'_p is $O(m^{\beta/2}\log m)$.

Lemma 12. *There exists an algorithm that approximates $C(G(p))$ in $O(\frac{1}{\delta^2}m^{\beta/2}\log m)$ query time by using $O(m^{2-3\beta/2})$ expected preprocessing time and $O(m^{2-3\beta/2})$ expected space $(0 \leq \beta \leq \frac{2}{3})$. For each query p, this algorithm returns a value C'_p such that with probability at least $1 - \frac{1}{\log m}$, $(1-\delta)C(G(p)) \leq C'_p \leq (1+\delta)C(G(p))$ when $C(G(p)) > \frac{1}{\delta^2}m^{\beta/2}\log m$.*

References

1. Alipour, S., Zarei, A.: Visibility testing and counting. In: Atallah, M., Li, X.-Y., Zhu, B. (eds.) FAW-AAIM 2011. LNCS, vol. 6681, pp. 343–351. Springer, Heidelberg (2011)
2. Aronov, B., Guibas, L.J., Teichmann, M., Zhang, L.: Visibility queries and maintenance in simple polygons. Discret. Comput. Geom. **27**, 461–483 (2002)
3. Asano, T.: An efficient algorithm for finding the visibility polygon for a polygonal region with holes. IEICE Trans. **68**(9), 557–589 (1985)
4. Bose, P., Lubiw, A., Munro, J.I.: Efficient visibility queries in simple polygons. Comput. Geom. Theory Appl. **23**(7), 313–335 (2002)
5. Fischer, M., Hilbig, M., Jahn, C., Meyer auf der Heide F., Ziegler M.: Planar visibility counting. CoRR, abs/0810.0052 (2008)
6. Fischer, M., Hilbig, M., Jahn, C., Meyer auf der Heide F., Ziegler M.: Planar visibility counting. In: Proceedings of the 25th European Workshop on Computational Geometry (EuroCG 2009), pp. 203–206 (2009)
7. Ghosh, S.K., Mount, D.: An output sensitive algorithm for computing visibility graphs. SIAM J. Comput. **20**, 888–910 (1991)
8. Ghosh, S.K.: Visibility Algorithms in the Plane. Cambridge University Press, Cambridge (2007)
9. Gudmundsson, J., Morin, P.: Planar visibility: testing and counting. In: Annual Symposium on Computational Geometry, pp. 77–86 (2010)
10. Kirkpatrick, D.: Optimal search in planar subdivisions. SIAM J. Comput. **12**(1), 28–35 (1983)
11. Nouri, M., Ghodsi, M.: Space/query-time tradeoff for computing the visibility polygon. Comput. Geom. **46**(3), 371–381 (2013)
12. Pocchiola, M., Vegter, G.: The visibility complex. Int. J. Comput. Geom. Appl. **6**(3), 279–308 (1996)

13. Suri, S., O'Rourke, J.: Worst-case optimal algorithms for constructing visibility polygons with holes. In: Proceedings of the Second Annual Symposium on Computational Geometry (SCG 86), pp. 14–23 (1986)
14. Vegter, G.: The visibility diagram: a data structure for visibility problems and motion planning. In: Gilbert, J.R., Karlsson, R. (eds.) SWAT 90. LNCS, vol. 447, pp. 97–110. Springer, Heidelberg (1990)
15. Zarei, A., Ghodsi, M.: Efficient computation of query point visibility in polygons with holes. In: Proceedings of the 21st Annual ACM Symposium on Computational Geometry (SCG 2005) (2005)

Approximation Algorithms for the Star k-Hub Center Problem in Metric Graphs

Li-Hsuan Chen[1], Dun-Wei Cheng[2], Sun-Yuan Hsieh[2], Ling-Ju Hung[2(✉)], Chia-Wei Lee[2], and Bang Ye Wu[1]

[1] Department of Computer Science and Information Engineering,
National Chung Cheng University, Chiayi 62102, Taiwan
{clh100p,bangye}@cs.ccu.edu.tw
[2] Department of Computer Science and Information Engineering,
National Cheng Kung University, Tainan 701, Taiwan
dunwei.ncku@gmail.com, hsiehsy@mail.ncku.edu.tw,
hunglc@cs.ccu.edu.tw, cwlee@csie.ncku.edu.tw

Abstract. Given a metric graph $G = (V, E, w)$ and a center $c \in V$, and an integer k, the STAR k-HUB CENTER PROBLEM is to find a depth-2 spanning tree T of G rooted by c such that c has exactly k children and the diameter of T is minimized. Those children of c in T are called hubs. The STAR k-HUB CENTER PROBLEM is NP-hard. (Liang, *Operations Research Letters*, 2013) proved that for any $\epsilon > 0$, it is NP-hard to approximate the STAR k-HUB CENTER PROBLEM to within a ratio $1.25 - \epsilon$. In the same paper, a 3.5-approximation algorithm was given for the STAR k-HUB CENTER PROBLEM. In this paper, we show that for any $\epsilon > 0$, to approximate the STAR k-HUB CENTER PROBLEM to a ratio $1.5 - \epsilon$ is NP-hard. Moreover, we give 2-approximation and $\frac{5}{3}$-approximation algorithms for the same problem.

1 Introduction

Hub location problems have been well studied in the literatures since they have various applications in transportation and telecommunication systems (see the two survey papers [1,3]). Suppose that we have a set of demand nodes that want to communicate with each other through some hubs in a network. If a demand node can be served by several hubs, then this kind of hub location problem is called *multi-allocation*. A hub location problem is called *single allocation* if each demand node can be served by exactly one hub. The goal of classical hub location problems is to minimized the total cost of routing in the network. In this

This research is partially supported by the Ministry of Science and Technology of Taiwan under grants MOST 103–2218–E–006–019–MY3, MOST 103–2221–E–006–135–MY3, MOST 103–2221–E–006–134–MY2.

Ling-Ju Hung is supported by the Ministry of Science and Technology of Taiwan under grant MOST 104–2811–E–006–056.

Chia-Wei Lee is supported by the Ministry of Science and Technology of Taiwan under grant MOST 104-2811-E-006-037.

© Springer International Publishing Switzerland 2016
T.N. Dinh and M.T. Thai (Eds.): COCOON 2016, LNCS 9797, pp. 222–234, 2016.
DOI: 10.1007/978-3-319-42634-1_18

paper, we consider the STAR k-HUB CENTER PROBLEM introduced by Yaman and Elloumi [11]. It is classified as single allocation. Unlike those classical hub location problems, STAR k-HUB CENTER PROBLEM is used to design a two level telecommunications network with the optimization criterion that the poorest service quality is minimized. Suppose that we have a set of demand nodes located in a metric space, each of them would like to communicate with all the others in a two-level tree structure network. In the two-level network, there is a given central hub c and we want to pick k nodes among the set of demand nodes as hubs and to connect them with the central hub c. Then each of the remaining demand nodes is connected to exactly one of the k chosen hubs such that the longest path in the tree structure network is minimized.

Let u, v be two vertices in a tree T. Use $d_T(u, v)$ to denote the distance between u, v in T. Define $D(T) = \max_{u,v \in T} d_T(u, v)$ called the diameter of T. For a vertex v in a tree T, we use $N_T(v)$ to denote the set of vertices adjacent to v in T. In this paper, we consider a graph $G = (V, E, w)$ with a distance function $w(\cdot, \cdot)$ being a metric on V such that $w(v, v) = 0$, $w(u, v) = w(v, u)$, and $w(u, v) + w(v, r) \geq w(u, r)$ for all $u, v, r \in V$. We give the formal definition of the STAR k-HUB CENTER PROBLEM as follows.

STAR k-HUB CENTER PROBLEM (SkHCP)
Input: A metric graph $G = (V, E, w)$, a center vertex $c \in V$, and a positive integer k.
Output: A depth-2 spanning tree T^* rooted by c called the central hub such that c has exactly k children (called hubs) and the diameter of T^*, $D(T^*)$, is minimized.

Yaman and Elloumi [11] showed the NP-hardness of STAR k-HUB CENTER PROBLEM and proposed two integer programming formulations of the same problem. Liang [8] showed that the STAR k-HUB CENTER PROBLEM does not admit a $(1.25 - \epsilon)$-approximation algorithm for any $\epsilon > 0$ unless $P = NP$ and gave a 3.5-approximation algorithm.

A similar problem of the STAR k-HUB CENTER PROBLEM called the SINGLE ALLOCATION p-HUB CENTER PROBLEM was introduced in [2, 10] and further studied in [6, 7, 9]. The difference between the two problems is that the SINGLE ALLOCATION p-HUB CENTER PROBLEM assumes that hubs are fully interconnected. Thus, for SINGLE ALLOCATION p-HUB CENTER PROBLEM, the communication between hubs is not necessary to go through a specified central hub c.

In this paper, we answer the open problem proposed by Liang [8] that for the STAR k-HUB CENTER PROBLEM whether we can bridge the gap between the lower bound $(1.25 - \epsilon)$ and the upper bound 3.5 of the approximation ratio. We show that for any $\epsilon > 0$, to approximate the STAR k-HUB CENTER PROBLEM to a ratio $(1.5 - \epsilon)$ is NP-hard. Moreover, we give a 2-approximation algorithm running in time $O(n)$ and a $\frac{5}{3}$-approximation algorithms for the same problem running in time $O(kn^4)$.

2   Inapproximability

In this section, we show that for any $\epsilon > 0$, if there exists a $(1.5 - \epsilon)$-approximation algorithm for STAR k-HUB CENTER PROBLEM running in polynomial time, then SET COVER can be approximated to within a ratio 3 in polynomial time.

SET COVER
Input: A universe $\mathcal{U}$ of elements, $|\mathcal{U}| = n$ and a collection $\mathcal{S}$ of subsets of $\mathcal{U}$.
Output: $\mathcal{S}' \subseteq \mathcal{S}$ of minimum cardinality such that $\bigcup_{s_i \in \mathcal{S}'} s_i = \mathcal{U}$.

The SET COVER problem is a well-known NP-hard problem. Dinur and Steurer [5] showed that for any $\epsilon > 0$, to approximate SET COVER to within a factor $(1 - \epsilon) \ln n$ is NP-hard.

Lemma 1. *For any $\epsilon > 0$, if* STAR k-HUB CENTER PROBLEM *can be approximated to a ratio $(1.5 - \epsilon)$ in polynomial time, then* SET COVER *admits a 3-approximation algorithm running in polynomial time.*

Proof. Let $(\mathcal{U}, \mathcal{S})$ be an input instance of SET COVER. We construct a metric graph $G = (V_1 \cup V_2 \cup \mathcal{S}_1 \cup \mathcal{S}_2 \cup \{c, x_1, x_2, y\}, E, w)$ of the STAR k-HUB CENTER PROBLEM according to $(\mathcal{U}, \mathcal{S})$ where c is the specified center. Let $V_1 = \mathcal{U}$ and $V_2 = \mathcal{U}$. For each set $s_i \in \mathcal{S}$ create a vertex in $\mathcal{S}_1$ and a vertex in $\mathcal{S}_2$. In the following, we define the cost of edges in G.

- $w(c, v) = 2$ if $v \in V_1 \cup V_2 \cup \{y\}$ and $w(c, z) = 1$ if $z \in \mathcal{S}_1 \cup \mathcal{S}_2 \cup \{x_1, x_2\}$.
- For $v_1 \in V_1$,
  - $w(v_1, v_1') = 2$ if $v_1' \in V_1$;
  - $w(v_1, v_2') = 4$ if $v_2' \in V_2$;
  - $w(v_1, q) = 1$ if v_1 is an element of $q \in \mathcal{S}$ where $q \in \mathcal{S}_1$ represents the set $q \in \mathcal{S}$; otherwise $w(v_1, q) = 2$;
  - $w(v_1, q') = 3$ if $q' \in \mathcal{S}_2$;
  - $w(v_1, x_1) = 2$; $w(v_1, x_2) = 3$; $w(v_1, y) = 4$.
- For $v_2 \in V_2$,
  - $w(v_2, v_2') = 2$ if $v_2' \in V_2$;
  - $w(v_2, q) = 3$ if $q \in \mathcal{S}_1$;
  - $w(v_2, q') = 1$ if v_2 is an element of $q' \in \mathcal{S}$ where $q' \in \mathcal{S}_2$ represents the set $q' \in \mathcal{S}$; otherwise $w(v_2, q') = 2$;
  - $w(v_2, x_1) = 3$; $w(v_2, x_2) = 2$; $w(v_2, y) = 4$.
- For $p \in \mathcal{S}_1$,
  - $w(p, q) = 2$ if $q \in \mathcal{S}_1$;
  - $w(p, q') = 2$ if $q' \in \mathcal{S}_2$;
  - $w(p, x_1) = 1$; $w(p, x_2) = 2$; $w(p, y) = 3$.
- For $p' \in \mathcal{S}_2$,
  - $w(p', q') = 2$ if $q' \in \mathcal{S}_2$;
  - $w(p', x_1) = 2$; $w(p', x_2) = 1$; $w(p', y) = 3$.
- $w(x_1, x_2) = 2$ and $w(x_1, y) = 3$.
- $w(x_2, y) = 3$.

Table 1. The cost of edges in G where $v_1, v_1' \in V_1$, $v_2, v_2' \in V_2$, $p, q \in \mathcal{S}_1$, $p', q' \in \mathcal{S}_2$, and $w(v_1, q)$, $w(v_2, q')$, $w(v_1', p)$, $w(v_2', p')$ are either 1 or 2.

$w(u,v)$	c	v_1'	v_2'	q	q'	x_1	x_2	y
c	0	2	2	1	1	1	1	2
v_1	2	2	4	$w(v_1,q)$	3	2	3	4
v_2	2	4	2	3	$w(v_2,q')$	3	2	4
p	1	$w(v_1',p)$	3	2	2	1	2	3
p'	1	3	$w(v_2',p')$	2	2	2	1	3
x_1	1	2	3	1	2	0	2	3
x_2	1	3	2	2	1	2	0	3
y	2	4	4	3	3	3	3	0

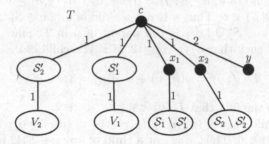

Fig. 1. A solution of Star $(2t+3)$-Hub Center Problem in G

In Table 1, we list the edge cost of all edges (u,v) in G specifically. It is not hard to see that any three vertices u, v, r in G satisfy $w(u,v) + w(v,r) \geq w(u,r)$. Thus G is a metric graph.

Let T^* be an optimal solution of Star k-Hub Center Problem in G. Suppose that $\mathcal{S}' \subseteq \mathcal{S}$ is an optimal solution of Set Cover, $|\mathcal{S}'| = t$ and $k = 2t + 3$. W.l.o.g., assume that $2t < |\mathcal{U}| < |\mathcal{S}|$. We now construct a solution T (see Fig. 1) of Star $(2t+3)$-Hub Center Problem in G. Let $\mathcal{S}_1' = \mathcal{S}_2' = \mathcal{S}'$. Let $N_T(c) = \mathcal{S}_1' \cup \mathcal{S}_2' \cup \{x_1, x_2, y\}$. Notice that $\mathcal{S}'$ is an optimal solution of Set Cover. For $v_1 \in V_1$, there exists $s \in \mathcal{S}_1'$ such that v_1 is an element in the set $s \in \mathcal{S}'$. Let v_1 be a child of s. For $v_2 \in V_2$, there exists $s' \in \mathcal{S}_2'$ such that v_2 is an element in the set $s' \in \mathcal{S}'$. Let v_2 be a child of s'. Let all vertices in $\mathcal{S}_1 \setminus \mathcal{S}_1'$ be children of x_1 in T. Let all vertices in $\mathcal{S}_2 \setminus \mathcal{S}_2'$ be children of x_2 in T. We see that $D(T) = 4$ and $D(T^*) \leq 4$.

Now we show that $D(T^*) > 3$. Suppose that $D(T^*) = 3$. We see that y must be a child of c; otherwise $d_{T^*}(y,c) \geq 3$ and $D(T^*) > 3$. If y has a child v, then $d_{T^*}(v,c) > 3$. Thus y has no children. Since $2t < |\mathcal{U}|$, there exists $v \in V_1 \cup V_2$ that is not a child of c. We see that $d_{T^*}(v,y) > 3$, a contradiction to the assumption that $D(T^*) = 3$. This shows that $D(T^*) > 3$. Hence $D(T^*) = 4$.

Suppose that there exists an approximation algorithm that finds a solution T of STAR $(2t+3)$-HUB CENTER PROBLEM in G and $D(T) < 6$. Let $N_T(c) = V_1' \cup V_2' \cup S_1' \cup S_2' \cup X$ where $V_1' \subset V_1$, $V_2' \subset V_2$, $S_1' \subset S_1$, $S_2' \subset S_2$, and $X \subseteq \{x_1, x_2, y\}$.

CLAIM 1. y *must be a child of* c *in* T.

PROOF OF CLAIM. Suppose that y is not a child of c. If y is a child of $v \in V_1' \cup V_2'$, $d_T(y, c) = 6$, a contradiction. If y is a child of $v \in S_1' \cup S_2' \cup X$, then $d_T(y, c) = 4$. Suppose that $v \in S_1' \cup \{x_1\}$ is the parent of y in T. Since $2t < |\mathcal{U}|$, there exists $v' \in V_2 \setminus V_2'$ such that $d_T(v', c) \geq 2$. If v' is a child of v,

$$d_T(v', y) = w(v', v) + w(v, y) = 3 + 3 = 6,$$

a contradiction to the fact that $D(T) < 6$.

If v' is not a child of v, $d_T(v', y) = d_T(v', c) + d_T(y, c) \geq 6$, a contradiction to the fact that $D(T) < 6$. Thus y is not a child of any $v \in S_1' \cup \{x_1\}$.

Suppose that $v \in S_2' \cup \{x_2\}$ is the parent of y in T. Since $2t < |\mathcal{U}|$, there exists $v' \in V_1 \setminus V_1'$ such that $d_T(v', c) \geq 2$. If v' is a child of v,

$$d_T(v', y) = w(v', v) + w(v, y) = 3 + 3 = 6,$$

a contradiction to the fact that $D(T) < 6$. If v' is not a child of v, $d_T(v', y) = d_T(v', c) + d_T(y, c) \geq 6$, a contradiction to the fact that $D(T) < 6$. Thus, y is not a child of any $v \in S_2' \cup \{x_2\}$. This shows that y is a child of c in T. ∎

CLAIM 2. y *has no children in* T.

PROOF OF CLAIM. If y has a child v, then $d_T(c, v) \geq 5$. For $u \in V_1' \cup V_2' \cup S_1' \cup S_2'$, $d_T(u, v) = d_T(u, c) + d_T(v, c) \geq 6$, a contradiction. Thus y has no children. ∎

According to Claims 1 and 2, in T, y is a child of c and y has no children. Since $D(T) < 6$, $d_T(y, c) = w(y, c) = 2$, and y has no children in T, for $v \in V_1 \cup V_2 \cup S_1 \cup S_2 \cup \{x_1, x_2\}$, $d_T(v, c) < 4$.

CLAIM 3. *If for all* $v \in V_1 \setminus V_1'$, $d_T(v, c) = 2$, *then* $S_1' \cup S_1''$ *is a set cover of* $\mathcal{U}$, $|S_1' \cup S_1''| \leq 2t + 2$ *where* $S_1'' \subset S$ *satisfying that for each* $u \in V_1'$ *there is exactly one set in* S_1'' *containing* u.

PROOF OF CLAIM. Since for all $v \in V_1 \setminus V_1'$, $d_T(v, c) = 2$, the element v must be a child of $s \in S_1$ satisfying that $v \in s$. We see that S_1' is a set cover of $V_1 \setminus V_1'$. For each $u \in V_1'$, we pick exactly one set in S that contains u, call the collection of sets S_1''. It is easy to see that $|S_1''| = |V_1'|$ and S_1'' is a set cover of V_1'. Thus, $S_1' \cup S_1''$ is a set cover of $V_1 = \mathcal{U}$ satisfying $|S_1' \cup S_1''| \leq 2t + 2$ that can be found in polynomial time. ∎

CLAIM 4. *If there exists a solution* T *of the* STAR $(2t+3)$-HUB CENTER PROBLEM *in* G *such that* $D(T) < 6$, *then* $S_2' \cup S_2''$ *is a set cover of* $\mathcal{U}$, $|S_2' \cup S_2''| \leq 2t + 2$ *where* $S_2'' \subset S$ *satisfying that for each* $u \in V_2'$ *there is exactly one set in* S_2'' *containing* u.

PROOF OF CLAIM. Suppose that the condition of Claim 3 is not true, there exists $v \in V_1 \setminus V_1'$ at distance $d_T(v,c) = 3$ from c in T. If the parent of v is in $V_1' \cup V_2' \cup S_2' \cup \{x_2\}$, we see that $d_T(v,c) > 3$, a contradiction to the fact that $d_T(v,c) = 3$. Thus v must be a child of $u \in S_1' \cup \{x_1\}$ and $d_T(v,c) = w(v,u) + w(u,c) = 3$ where $w(v,u) = 2$ and $w(u,c) = 1$.

For each $v' \in V_2 \setminus V_2'$, if v' is a child of $u' \in V_1' \cup V_2' \cup S_1' \cup \{x_1\}$, then $d_T(v',y) \geq 6$, a contradiction to the fact $D(T) < 6$. Thus $u' \in S_2 \cup \{x_2\}$. If $u' = x_2$, we see that $d_T(v',v) = d_T(v',c) + d_T(v,c) = 3 + 3 = 6$, a contradiction to the fact that $D(T) < 6$. If $u' \in S_2$ that $d_T(v',u') = w(v',u') = 2$, then $d_T(v',v) = d_T(v',c) + d_T(v,c) = 3 + 3 = 6$, a contradiction to the fact that $D(T) < 6$. This implies that for $v' \in V_2 \setminus V_2'$, $d_T(v',c) = 2$. Since for all $v' \in V_2 \setminus V_2'$, $d_T(v',c) = 2$, the element v' must be a child of $s \in S_2$ satisfying that $v' \in s$. We see that S_2' is a set cover of $V_2 \setminus V_2'$. For each $z \in V_2'$, we pick exactly one set in S that contains z, call the collection of sets S_2''. It is easy to see that $|S_2''| = |V_2'|$ and S_2'' is a set cover of V_2'. Thus, $S_2' \cup S_2''$ is a set cover of $V_2 = \mathcal{U}$ satisfying $|S_2' \cup S_2''| \leq 2t + 2$ that can be found in polynomial time. ∎

By Claims 3 and 4, if $D(T) < 6$, then SET COVER has a 3-approximation algorithm running in polynomial time. Notice that $D(T^*) = 4$. Thus, for any $\epsilon > 0$, if there exists an approximation algorithm that finds a $(1.5 - \epsilon)$ approximate solution of STAR $(2t + 3)$-HUB CENTER PROBLEM in G in polynomial time, then SET COVER has a 3-approximation algorithm running in polynomial time. □

Theorem 1. *For any $\epsilon > 0$, to approximate* STAR k-HUB CENTER PROBLEM *to a ratio $(1.5 - \epsilon)$ is NP-hard.*

Proof. By Lemma 1, if STAR k-HUB CENTER PROBLEM can be approximated to a ratio $(1.5 - \epsilon)$ in polynomial time, then there exists a 3-approximate solution of SET COVER that can be found in polynomial time. This contradicts to that for any $\epsilon > 0$ to approximate SET COVER to within factor $(1 - \epsilon) \ln n$ is NP-hard [5]. Thus, for any $\epsilon > 0$, to approximate STAR k-HUB CENTER PROBLEM to within a factor $(1.5 - \epsilon)$ is NP-hard. This completes the proof. □

3   New Approximation Algorithms

Let $\mathcal{T}$ be the collection depth-2 trees rooted by c satisfying that c has exact k children. We see that $T^* = \arg\min_{T \in \mathcal{T}} \{D(T)\}$ is an optimal tree of the STAR k-HUB CENTER PROBLEM. For $v \in T^*$, let $f^*(v)$ denote the parent of v in T^*. We use $C^* = N_{T^*}(c)$ to denote the set of children of c in T^* and $N_{T^*}[c] = C^* \cup \{c\}$. We call C^* the set of vertices in the first layer of T^* and $V \setminus N_{T^*}[c]$ the set of vertices in the second layer of T^*.

3.1   A 2-Approximation Algorithm

We give a 2-approximation algorithm for the STAR k-HUB CENTER PROBLEM.

Algorithm BasicAPX$_{SkHCP}$

Step 1: Pick k vertices $\{v_1, v_2, \ldots, v_k\}$ closest to c. Let $N_T(c) = \{v_1, v_2, \ldots, v_k\}$,
   w.l.o.g., assume that $w(v_1, c) \leq w(v_2, c) \leq \cdots \leq w(v_k, c)$.
Step 2: Connect to all vertices in $V \setminus \{c, v_1, v_2, \ldots, v_k\}$ to v_1 in T.
Step 3: Return the depth-2 tree T.

Theorem 2. *There is a 2-approximation algorithm for the* STAR k-HUB CEN-
TER PROBLEM *running in time* $O(n)$ *where n is the number of vertices in the
input graph.*

Proof. Note that picking the kth vertex closest to c can be done by a linear
time selection algorithm [4]. It is not hard to see that in time $O(n)$ Algorithm BasicAPX$_{SkHCP}$ returns a depth-2 spanning tree T rooted by c satisfying
that c has exact k children. Let T^* be an optimal tree of the STAR k-HUB CEN-
TER PROBLEM. Now we show that the approximation ratio is 2 by showing that
$D(T) \leq 2 \cdot D(T^*)$.

Since $w(v_1, c) \leq w(v_2, c) \leq \cdots \leq w(v_k, c)$, for $u \in N_T(c)$, $w(u, c) \leq w(v_k, c)$.
Since $D(T^*) \geq w(v_1, c) + w(v_k, c)$, it is easy to see that for $u \in N_T(c)$,

$$d_T(u, v_1) = w(u, c) + w(c, v_1) \leq D(T^*).$$

For any vertex $v \in V \setminus \{c, v_1, v_2, \ldots, v_k\}$,

$$d_T(v, v_1) = w(v, v_1) \leq D(T^*).$$

For any u, v in $V \setminus \{c\}$, we have the following three cases.

- Both u, v are adjacent to v_1. Then

$$d_T(u, v) = d_T(u, v_1) + d_T(v_1, v) \leq 2 \cdot D(T^*).$$

- The vertex u is adjacent to v_1 and v is adjacent to c. Then

$$d_T(u, v) = d_T(u, v_1) + d_T(v, v_1) \leq 2 \cdot D(T^*).$$

- Both u, v are adjacent to c. Then

$$d_T(u, v) = d_T(u, c) + d_T(v, c) \leq 2 \cdot D(T^*).$$

Since for any $u, v \in T \setminus \{c\}$, $d_T(u, v) \leq 2 \cdot D(T^*)$, we see that $D(T) \leq 2 \cdot D(T^*)$.
This completes the proof. □

3.2 A $\frac{5}{3}$-Approximation Algorithm

In this section, we give a 5/3-approximation algorithm for the STAR k-HUB
CENTER PROBLEM.

Let T^* be an optimal tree. Let $x = \arg\max_{v \in V \setminus N_{T^*}[c]} d_{T^*}(v, c)$ be a farthest vertex from c in the second layer of T^* and $m_1 = f^*(x)$. We use
$\ell = \max_{v \in V \setminus N_{T^*}[c]} \{w(v, f^*(v))\}$ to denote the cost of a longest edge with one
end vertex in the second layer of T^* and the other end vertex in $C^* = N_{T^*}(c)$.

Algorithm APX_{SkHCP}

Step 1: Run Algorithm APX1.
Step 2: Run Algorithm APX2.
Step 3: Return the best solution found by Algorithms APX1 and APX2.

Algorithm APX1

Let $U := V \setminus \{c\}$. For each $v \in U$, let $m_1 = v$ and for $p, q \in U$, let $\ell = w(p, q)$, do the following steps to find a depth-2 spanning tree T of G rooted by c. Let M be the set of children of c in T, initialize $M = \emptyset$. Keep a tree T found by the following steps having the minimum diameter.

Step 1: Add edge (m_1, c) in the tree T, let $M := M \cup \{m_1\}$, and let $U := U \setminus \{m_1\}$.
Step 2: For each $v \in U$, if $v \in U$ and $w(m_1, v) \leq \ell$, we add edges (v, m_1) in T and
      let $U := U \setminus \{v\}$.
Step 3: While $i = |M| + 1 \leq k$ and $U \neq \emptyset$,
    – choose $v \in U$, let $m_i = v$, add edge (m_i, c) in T, let $U := U \setminus \{v\}$,
      and let $M := M \cup \{m_i\}$;
    – for $u \in U$, if $w(u, m_i) \leq 2\ell$, then add edge (u, m_i) in T and
      $U := U \setminus \{u\}$.
Step 4: If $|M| < k$ and $U = \emptyset$, we change the shape of T by selecting $k - |M|$
      vertices closest to c from the second layer to be the children of c, call the
      new tree T'; otherwise let $T' := T$.

Algorithm APX2

Let $U = V \setminus \{c\}$. For $y \in U$ and for $z \in U \setminus \{y\}$, do the following steps to find a depth-2 spanning tree T'' of G rooted by c.

Step 1: Let $\ell = w(y, z)$ and let y be the child of c in T^*.
Step 2: Pick $(k - 1)$ vertices $\{v_1, v_2, \ldots, v_{k-1}\}$ closest to c from $U \setminus \{y, z\}$.
      Let $N_{T''}(c) = \{y, v_1, v_2, \ldots, v_{k-1}\}$, w.l.o.g., assume that $w(v_1, c) \leq$
      $w(v_2, c) \leq \cdots \leq w(v_{k-1}, c)$.
Step 3: Let all vertices in $U \setminus \{v_1, v_2, \ldots, v_{k-1}, y\}$ be the children of y.

Lemma 2. *Algorithm APX1 returns a $(1 + 4\delta)$-approximation solution of* STAR k-HUB CENTER PROBLEM *in time* $O(kn^4)$ *where* $\delta = \frac{\ell}{D(T^*)}$ *and* n *is the number of vertices in* G.

Proof. Suppose that T^* is an optimal tree. Let $C^* = N_{T^*}(c) = \{s_1, s_2, \ldots, s_k\}$ be the set of children of c in T^*. Let $S_1, S_2, \ldots, S_k$ be components of $T^* \setminus \{c\}$. Note that each component S_i is a star in $T^* \setminus \{c\}$ with the center s_i. Let $x = \arg\max_{v \in V \setminus N_{T^*}[c]} d_{T^*}(v, c)$ be a farthest vertex from c in the second layer of T^* and $m_1 = f^*(x)$. Let $\ell = \max_{v \in V \setminus N_{T^*}[c]} \{w(v, f^*(v))\}$ be the cost of a longest edge with one end vertex in the second layer of T^* and the other end vertex in C^*. Suppose that the algorithm guesses the correct m_1 and ℓ. We may assume that $m_1 \in S_1$. W.l.o.g., we assume that $s_1 = m_1$. Since the algorithm adds edges (v, m_1) if $w(v, m_1) \leq \ell$, we see that $S_1 \subset N_T[m_1]$. Notice that for each vertex

$v \in S_j, j = 1, \ldots, k$, we see that $w(v, f^*(v)) \leq \ell$ by the definition of ℓ. For $u, v \in S_j \setminus \{s_j\}, j = 1, \ldots, k, f^*(u) = f^*(v) = s_j$ and $w(u, v) \leq w(u, s_j) + w(v, s_j) \leq 2\ell$. Thus, for each $S_j, j > 1$, if there is a vertex $x \in S_j$ specified as $m_i \in M, i > 1$, then all the other vertices in S_j are children of one of $m_1, m_2, \ldots, m_i$ in T. Moreover, we see that for each $m_i, 1 < i \leq |M|$, there exists $S_j, 1 < j \leq k$, such that $m_i \in S_j$ and $S_j \cap M = \{m_i\}$. If there exists $S_j, 1 < j \leq k, S_j \cap M = \emptyset$, then all vertices of S_j are children of one vertex of M in T and $|M| < k$.

It is easy to see that if $|M| = k$, then T is a depth-2 spanning tree of G satisfying that c has exactly k children. Suppose that $|M| < k$ and we change the shape of T by taking $k - |M|$ vertices closest to c from the leaves to be new children of c, call the new tree T' and let M' be the set of new children of c. Notice that in T', vertices in M' have no children. We see that T' is a depth-2 spanning tree of G satisfying that c has exactly k children.

CLAIM 1. *For $u, v \in M$, $d_{T'}(u, v) \leq D(T^*)$.*

PROOF OF CLAIM. Since u, v are in different components of $T^* \setminus \{c\}$, we have

$$d_{T'}(u, v) = w(u, c) + w(v, c) \leq d_{T^*}(u, c) + d_{T^*}(v, c) \leq D(T^*).$$

∎

CLAIM 2. *For $u \in M$ and $v \in M'$, $d_{T'}(u, v) \leq D(T^*) + \ell$.*

PROOF OF CLAIM. For $v \in M'$, let $f(v)$ denote the parent of v in T. Notice that for $v \in M'$, $f(v) \in M$ and $w(v, f(v)) \leq 2\ell$. Suppose that v is a child of c in T^*. If $f^*(u) \neq v$, we have

$$d_{T'}(u, v) = w(u, c) + w(v, c) \leq w(u, f^*(u)) + w(f^*(u), c) + w(v, c) \leq D(T^*).$$

If $f^*(u) = v$, then

$$\begin{aligned} d_{T'}(u, v) &= w(u, c) + w(v, c) \leq w(u, v) + w(v, c) + w(v, c) \\ &\leq \ell + w(m_1, c) + w(v, c) \leq \ell + d_{T^*}(m_1, v) \\ &\leq D(T^*) + \ell. \end{aligned}$$

Suppose that v is not a child of c in T^*. Since vertices in M' are selected from the vertices in the second layer of T that are closest to c, there exists v' in the second layer of T' that is a child of c in T^* satisfying that $w(v', c) \geq w(v, c)$. Thus,

$$\begin{aligned} d_{T'}(u, v) &= w(u, c) + w(v, c) \leq w(u, c) + w(v', c) \\ &= w(u, f^*(u)) + w(f^*(u), c) + w(v', c) \\ &\leq \ell + w(m_1, c) + w(v', c) = \ell + d_{T^*}(m_1, v') \\ &\leq D(T^*) + \ell. \end{aligned}$$

∎

CLAIM 3. *For $u, v \in M'$, $d_{T'}(u, v) \leq D(T^*) + 3\ell$.*

PROOF OF CLAIM. Since $u \in M'$, in T the parent of u, called $f(u)$, must be in M. We see that $w(u, f(u)) \leq 2\ell$. By Claim 2, $d_{T'}(f(u), v) \leq D(T^*) + \ell$. Hence

$$\begin{aligned} d_{T'}(u, v) &= w(u, c) + w(v, c) \\ &\leq w(u, f(u)) + w(f(u), c) + w(v, c) \\ &\leq 2\ell + d_{T'}(f(u), v) \\ &\leq 3\ell + D(T^*). \end{aligned}$$

Thus, for $u, v \in M'$, $d_{T'}(u, v) \leq D(T^*) + 3\ell$.   ∎

CLAIM 4. *For $u, v \in V \setminus (M \cup M' \cup \{c\})$, $d_{T'}(u, v) \leq D(T^*) + 4\ell$.*

PROOF OF CLAIM. Note that for vertices in $V \setminus (M \cup M' \cup \{c\})$, their parents in T' are the same as their parents in T. Let $f(u)$ and $f(v)$ be parents of u and v in both T and T', respectively. Since $u, v \in V \setminus (M \cup M' \cup \{c\})$, we see that $f(u), f(v) \in M$, $w(u, f(u)) \leq 2\ell$, and $w(v, f(v)) \leq 2\ell$.

Suppose that $f(u) \neq f(v)$. By Claim 1, $d_{T'}(f(u), f(v)) \leq D(T^*)$.

$$\begin{aligned} d_{T'}(u, v) &= w(u, f(u)) + w(f(u), c) + w(v, f(v)) + w(f(v), c) \\ &\leq 4\ell + d_{T'}(f(u), f(v)) \\ &\leq 4\ell + D(T^*). \end{aligned}$$

Suppose that $f(u) = f(v)$. It is easy to see that $d_{T'}(u, v) \leq 4\ell$. This shows that for $u, v \in V \setminus (M \cup M' \cup \{c\})$, $d_{T'}(u, v) \leq D(T^*) + 4\ell$.   ∎

CLAIM 5. *For $u \in V \setminus (M \cup M' \cup \{c\})$ and $v \in M'$, $d_{T'}(u, v) \leq D(T^*) + 3\ell$.*

PROOF OF CLAIM. Let $f'(u)$ be the parent of u in T'. We see that $f(u) = f'(u)$ and $f(u) \in M$ where $f(u)$ is the parent of u in T. Since $f(u) \in M$ and $v \in M'$, by Claim 2, $d_{T'}(f(u), v) \leq D(T^*) + \ell$. Thus,

$$\begin{aligned} d_{T'}(u, v) &= w(u, f(u)) + w(f(u), c) + w(v, c) \\ &\leq 2\ell + d_{T'}(f(u), v) \leq 2\ell + \ell + D(T^*) \\ &= 3\ell + D(T^*). \end{aligned}$$

This completes the proof of the claim.   ∎

Thus, by Claims 1–5, $D(T') \leq D(T^*) + 4\ell$. We obtain that

$$\frac{D(T')}{D(T^*)} \leq \frac{D(T^*) + 4\ell}{D(T^*)} = (1 + 4\delta)$$

where $\delta = \frac{\ell}{D(T^*)}$.

The algorithm guesses m_1 and ℓ, there are $O(n)$ possibilities of m_1 and $O(n^2)$ possibilities of ℓ. In Algorithm APX1, there are $O(n^3)$ depth-2 spanning trees constructed. It is not hard to see that it takes $O(kn)$ time to construct a tree T'. The running time of Algorithm APX1 is $O(kn^4)$. This completes the proof.   □

Lemma 3. *Algorithm APX2 either returns an optimal solution or a $(2 - 2\delta)$-approximation solution of* STAR k-HUB CENTER PROBLEM *in time $O(kn^2)$ where $\delta = \frac{\ell}{D(T^*)} < 1/2$ and n is the number of vertices in G.*

Proof. Suppose that T^* is an optimal tree. Let (y, z) be the longest edge in the second layer of T^* such that $f^*(y) = c$ and $w(y, z) = \ell$. For $u \in V$, we use $f''(u)$ to denote the parent of u in T''.

For $v \in V \setminus \{z\}$, we show that $d_{T''}(v, y) \leq D(T^*) - \ell$.

There are two cases.

Case 1: If $f''(v) = y$, we see that

$$d_{T''}(v, y) = w(v, y) \leq d_{T^*}(v, y) = d_{T^*}(v, z) - \ell \leq D(T^*) - \ell.$$

Case 2: Suppose that $f''(v) = c$. We have two subcases.

Case 2.1: If $f^*(v) = c$, we see that

$$d_{T''}(v, y) = w(v, c) + w(c, y) = d_{T^*}(v, y)$$
$$= d_{T^*}(v, z) - \ell \leq D(T^*) - \ell.$$

Case 2.2: If $f^*(v) \neq c$, then there exists $v' \in N_{T''}(y) \setminus \{c, y, z\}$ such that $f^*(v') = c$ and $w(v', c) \geq w(v, c)$. We see that

$$d_{T''}(v, y) = w(v, c) + w(c, y) \leq w(v', c) + w(c, y)$$
$$= d_{T^*}(v', y) = d_{T^*}(v', z) - \ell$$
$$\leq D(T^*) - \ell.$$

Now we show that for $v \in V \setminus \{z\}$, $d_{T''}(z, v) \leq D(T^*)$.

$$d_{T''}(z, v) \leq d_{T''}(z, y) + d_{T''}(y, v) \leq \ell + D(T^*) - \ell = D(T^*).$$

For $u, v \in V \setminus \{z\}$, we see that

$$d_{T''}(u, v) \leq d_{T''}(u, y) + d_{T''}(v, y) \leq 2D(T^*) - 2\ell.$$

Note that $D(T'') \geq D(T^*)$. If $D(T^*) \leq 2\ell$, then

$$D(T'') \leq 2D(T^*) - 2\ell \leq 2D(T^*) - D(T^*) = D(T^*).$$

Thus, if $D(T^*) \leq 2\ell$, Algorithm APX2 returns an optimal solution.

Suppose that $D(T^*) > 2\ell$. We see that

$$\frac{D(T'')}{D(T^*)} \leq \frac{2D(T^*) - 2\ell}{D(T^*)} = 2 - 2\delta$$

where $\delta = \frac{\ell}{D(T^*)} < 1/2$.

Algorithm APX2 guesses y and z, there are $O(n)$ possibilities of y and $O(n)$ possibilities of z. In Algorithm APX2, there are $O(n^2)$ depth-2 spanning trees constructed. Suppose that in advance the algorithm takes $O(kn)$ time to compute $u_1, u_2, \ldots, u_{k+1} \in V \setminus \{c\}$ that are closest to c such that $w(u_1, c) \leq w(u_2, c) \leq \cdots \leq w(u_k, c) \leq w(u_{k+1}, c)$. In Algorithm APX2, it takes $O(k)$ time to pick $v_1, v_2, \ldots, v_{k-1} \in V \setminus \{c, y, z\}$ that are closest to c to be the children of c. The running time of Algorithm APX2 is $O(kn^2 + kn) = O(kn^2)$. This completes the proof. $\qquad\square$

By Lemmas 2 and 3, we have the following theorem.

Theorem 3. *There is a $\frac{5}{3}$-approximation algorithm for the* STAR k-HUB CEN- *TER* PROBLEM *running in time* $O(kn^4)$ *where n is the number of vertices in the input graph.*

Proof. By Lemma 2, Algorithm APX1 runs in time $O(kn^4)$ and finds a $(1 + 4\delta)$-approximate solution where $\delta = \frac{\ell}{D(T^*)}$. By Lemma 3, Algorithm APX2 runs in time $O(kn^2)$ and either finds an optimal solution or a $(2 - 2\delta)$-approximation solution, $\delta < 1/2$. In Step 3 of Algorithm APX$_{SkHCP}$, it takes $O(1)$ time to return the best solution found by Algorithm APX1 and Algorithm APX2. We see that the worst approximate ratio happens when $1 + 4\delta = 2 - 2\delta$ and $\delta = \frac{1}{6}$. This shows that the approximation ratio is $\frac{D(T)}{D(T^*)} \leq \frac{5}{3}$ and the running time of Algorithm APX$_{SkHCP}$ is $O(kn^4)$. $\qquad\square$

4 Concluding Remarks

In this paper, we reduce the gap between the upper and lower bounds of approximability for the STAR k-HUB CENTER PROBLEM. For the future work, it is interesting to see whether there exists an α-approximation algorithm and $\alpha < 5/3$ or to prove that for any $\epsilon > 0$, it is NP-hard to approximate the STAR k-HUB CENTER PROBLEM to a ratio $\frac{5}{3} - \epsilon$.

References

1. Alumur, S.A., Kara, B.Y.: Network hub location problems: the state of the art. Netw. Hub Location Probl.: State Art **190**, 1–21 (2008)
2. Campbell, J.F.: Integer programming formulations of discrete hub location problems. Eur. J. Oper. Res. **72**, 387–405 (1994)
3. Campbell, J.F., Ernst, A.T.: Hub location problems. In: Drezner, Z., Hamacher, H.W. (eds.) Facility Location: Applications and Theory, pp. 373–407. Springer, Berlin (2002)
4. Cormen, T.H., Leiserson, C.E., Rivest, R.L., Stein, C.: Introduction to Algorithms. The MIT Press, Cambridge (2009)
5. Dinur, I., Steurer, D.: Analytical approach to parallel repetition. In: Proceedings of STOC 2014, pp. 624–633 (2014)
6. Ernst, A.T., Hamacher, H., Jiang, H., Krishnamoorthy, M., Woeginger, G.: Uncapacitated single and multiple allocation p-hub center problem. Comput. Oper. Res. **36**, 2230–2241 (2009)

7. Kara, B.Y., Tansel, B.Ç.: On the single-assignment p-hub center problem. Eur. J. Oper. Res. **125**, 648–655 (2000)
8. Liang, H.: The hardness and approximation of the star p-hub center problem. Oper. Res. Lett. **41**, 138–141 (2013)
9. Meyer, T., Ernst, A., Krishnamoorthy, M.: A 2-phase algorithm for solving the single allocation p-hub center problem. Comput. Oper. Res. **36**, 3143–3151 (2009)
10. O'Kelly, M.E., Miller, H.J.: Solution strategies for the single facility minimax hub location problem. Pap. Reg. Sci. **70**, 376–380 (1991)
11. Yaman, H., Elloumi, S.: Star p-hub center problem and star p-hub median problem with bounded path length. Comput. Oper. Res, **39**, 2725–2732 (2012)

Balls and Funnels: Energy Efficient Group-to-Group Anycasts

Jennifer Iglesias[1], Rajmohan Rajaraman[2], R. Ravi[1], and Ravi Sundaram[2($\boxtimes$)]

[1] Carnegie Mellon University, Pittsburgh, PA, USA
{jiglesia,ravi}@andrew.cmu.edu
[2] Northeastern University, Boston, MA, USA
{rraj,koods}@ccs.neu.edu

Abstract. We introduce group-to-group anycast (g2g-anycast), a network design problem of substantial practical importance and considerable generality. Given a collection of groups and requirements for directed connectivity from source groups to destination groups, the solution network must contain, for each requirement, an omni-directional down-link broadcast, centered at any node of the source group, called the ball; the ball must contain some node from the destination group in the requirement and all such destination nodes in the ball must aggregate into a tree directed towards the source, called the funnel-tree. The solution network is a collection of balls along with the funnel-trees they contain. g2g-anycast models DBS (Digital Broadcast Satellite), Cable TV systems and drone swarms. It generalizes several well known network design problems including minimum energy unicast, multicast, broadcast, Steiner-tree, Steiner-forest and Group-Steiner tree. Our main achievement is an $O(\log^4 n)$ approximation, counterbalanced by an $\log^{(2-\epsilon)} n$ hardness of approximation, for general weights. Given the applicability to wireless communication, we present a scalable and easily implemented $O(\log n)$ approximation algorithm, Cover-and-Grow for fixed-dimensional Euclidean space with path-loss exponent at least 2.

Keywords: Network design · Wireless · Approximation

1 Introduction

1.1 Motivation

Consider a DBS (Digital Broadcast Satellite) system such as Dish or DIRECTV in the USA (see Fig. 1). The down-link is an omni-directional broadcast from constellations of satellites to groups of apartments or neighborhoods serviced by one or more dish installations. The up-link is sometimes a wired network but in remote areas it is usually structured as a tree consisting of point-to-point wireless links directed towards the network provider's head-end (root). The high availability requirement of such services are typically satisfied by having multiple head-ends and anycasting to them. The same architecture is found in CATV

© Springer International Publishing Switzerland 2016
T.N. Dinh and M.T. Thai (Eds.): COCOON 2016, LNCS 9797, pp. 235–246, 2016.
DOI: 10.1007/978-3-319-42634-1_19

(originally Community Antenna TV), or cable TV distribution systems as well as sensor networks where an omni-directional broadcast from a beacon is used to activate and control the sensors; the sensors then funnel their information back using relays. Moreover, this architecture is also beginning to emerge in drone networks, for broadcasting the Internet, by companies such as Google [10] and Facebook's Connectivity Labs [8]. The Internet is to be broadcast from drones flying fixed patterns in the sky to a collection of homes on the ground. The Internet up-link from the homes is then aggregated using wireless links organized as a tree to be sent back to the drones. Anycasting is an integral part of high-availability services such as Content Delivery Networks (CDNs) where reliable connectivity is achieved by reaching some node in the group. What is the common architecture underlying all these applications and what is the constraining resource that is driving their form?

The various distribution systems can be abstractly seen to consist of a down-link *ball* and an up-link *funnel-tree* (see Fig. 1). The ball is an omni-directional broadcast from the publisher or content-producer to a large collection of subscribers or content-consumers. At the same time, the consumers have information that they need to dynamically send back to the publisher in order to convey their preferences and requirements. The funnel-tree achieves this up-link efficiently in terms of both time and energy. Aggregation of information and use of relays uses less energy as compared to omni-directional broadcasts by each node back to the publisher and also avoids

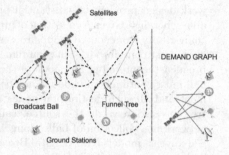

Fig. 1. Pictogram of Digital Broadcast Satellite System with 2 satellite groups and 4 ground station groups on left with associated demand graph on the right. The broadcast balls are denoted by dotted black lines, and the funnel trees by solid yellow lines. (Color figure online)

the scheduling needed to avoid interference. In this work, we focus primarily on total energy consumption. The application scenarios mentioned in the opening paragraph are all energy sensitive. Sensor networks [11] and drone fleets [12] are particularly vulnerable to energy depletion. For the purpose of energy conservation, generally each wireless node can dynamically adjust its transmitting power based on the distance of the receiving nodes and background noise. In the most common power-attenuation model [14], the signal power falls as $\frac{1}{r^\kappa}$ where r is the distance from the transmitter to the receiver and κ is the path-loss exponent - a constant between 2 and 4 dependent on the wireless environment. A key implication of non-linear power attenuation is that relaying through an intermediate node can sometimes be more energy efficient than transmitting directly - a counter-intuitive violation of the triangle inequality - e.g., in a triangle ABC with obtuse angle ABC, where $d_{AB}^2 + d_{BC}^2 < d_{AC}^2$.

1.2   Problem Formulation and Terminology

In this paper, we consider a general formulation that encompasses a wide variety of scenarios: given a collection of groups (of nodes) along with a directed demand graph over these groups the goal is to design a collection of balls and associated funnel-trees of lowest cost so that every demand requirement is met - meaning that if there is an arc from a source group to a destination group then the solution must have a ball centered at a node of the source group that includes a funnel-tree containing a node of the destination group.

Formally, we define the *group-to-group anycast* problem, or *g2g-anycast*, as follows: as input we are given n nodes along with a collection of source groups $S_1, S_2, \ldots, S_p$ and a collection of destination groups $T_1, T_2, \ldots, T_q$ which are subsets of these nodes; a demand graph on these groups consisting of directed arcs from source groups S_i to destination groups T_j. A nonnegative cost c_{uv} is specified between every pair of nodes; when a node u incurs a cost C in doing an omni-directional broadcast it reaches all nodes v such that $c_{uv} \leq C$. A metric d_{uv} is also specified between every pair of nodes and when a node u connects to node v in the funnel-tree using a point-to-point link it incurs a cost d_{uv}. A solution consists of a broadcast ball around every source node s (we give a radius which the source can broadcast to), and a funnel tree rooted at s. A demand S_i, T_j is satisfied if there is a broadcast ball from some $s \in S_i$ which contains some $t \in T_j$ and the funnel tree of s also includes t. The cost of the solution is the sum of the ball-radii around the source nodes (under the broadcast costs c) and the sum of the costs of the funnel trees (under the funnel metric d) that connect all terminal-nodes used to cover the demands to the source nodes within whose balls they lie. We do not allow funnel trees to share edges (even if they are going to the same source group), and will pay for each copy of an edge used.

- First, the bipartite demand graph is no less general than an arbitrary demand graph since a given group can be both a source group and destination group.
- Second, since funnel trees sharing the same edge pay separately, solutions to the problem decompose across the sources and it is sufficient to solve the case where we have exactly one source group $S = \{s_1, s_2, \ldots, s_k\}$ and destination groups $T_1, T_2, \ldots, T_q$ (i.e. the demand graph is a star consisting of all arcs $(S, T_j), 1 \leq j \leq q$). This observation also enables parallelized implementations.
- Lastly, there is no loss of generality in assuming a metric d_{uv} for funnel-tree costs; even if the costs were arbitrary their metric completion is sufficient for determining the optimal funnel-tree.

We refer collectively to the (ball) costs c_{uv} and (funnel-tree) metric distances d_{uv} as *weights*. In this paper we consider two cases - one, the general case where the weights can be arbitrary and two, the special case where the nodes are embedded in a Euclidean space and all weights are induced from the embedding.

1.3   Our Contributions

Our main results on the minimum energy g2g-anycast problem are as follows (Fig. 2):

	g2g, any metric	g2s, any metric	g2g, ℓ_2^2 norm
Upper	$O(\log^4 n)$	$2\ln n$	$O(\log n)$
Lower	$\Omega(\log^{2-\epsilon} n)$	$\Omega(\log n)$	$(1-o(1))\ln n$

Fig. 2. A summary of upper and lower bounds achieved in the different problems. The lower bound holds for every fixed $\epsilon > 0$

1. We present a polynomial-time $O(\log^4 n)$ approximation algorithm for the g2g-anycast problem on n nodes with general weights. We complement this with an $\Omega(\log^{2-\epsilon} n)$ hardness of approximation, for any $\epsilon > 0$ (Sect. 2).
2. One scenario with practical application is where every destination group is a singleton set while source groups continue to have more than one node; we refer to this special case of g2g-anycast as *g2s anycast*. We present a *tight* logarithmic approximation result for g2s-anycast (Sect. 3).
3. For the realistic scenario where the nodes are embedded in a 2-D Euclidean plane with path-loss exponent $\kappa \geq 2$, we design an efficient $O(\log n)$-approximation algorithm Cover-and-Grow, and also establish a matching logarithmic hardness of approximation result (Sect. 4).
4. Lastly, we compare Cover-and-Grow with 4 alternative heuristics on random 2-D Euclidean instances; we discover that Cover-and-Grow does well in a wide variety of practical situations in terms of both running time and quality, besides possessing provable guarantees. This makes Cover-and-Grow a go-to solution for designing near-optimal data dissemination networks in the wireless infrastructure space (Sect. 5).

1.4   Related Work

A variety of power attenuation models for wireless networks have been studied in the literature [14]. Though admittedly coarse, the model based on the path loss exponent (varying from 2, in free space to 4, in lossy environments) is the standard way of characterizing attenuation [13]. The problems of energy efficient multicast and broadcast in this model have been extensively studied [9,16–18]. Two points worth mentioning in this context are: one, we consider the funnel-tree as consisting of point-to-point directional transmissions rather than an omni-directional broadcast since the nonlinear cost of energy makes it more economical to relay through an intermediate node, and two, we consider only energy spent in transmission but not in reception.

Network design problems are notoriously NP-hard. Over time sophisticated approximation techniques have been developed, ranging from linear programming and randomized rounding to metric embeddings [19]. The g2g-anycast problem with general weights is a substantial generalization including problems such as minimum spanning trees, multicast trees, broadcast trees, Steiner trees and Steiner forests. Even the set cover problem can be seen as a special case where the destination groups are singletons. The g2g-anycast also generalizes the much harder group Steiner tree problem [4,5].

2 Approximating g2g-anycast

In this section, we present an $O(\log^4 n)$-approximation for the g2g-anycast problem with general weights by a reduction to the generalized set-connectivity problem. We then give a reduction from the group Steiner tree problem that demonstrates that there is no polynomial-time $\log^{2-\epsilon} n$-approximation algorithm for g2g-anycast unless $P = NP$.

2.1 Approximation Algorithm for g2g-anycast with General Weights

The generalized set-connectivity problem [2] takes as input an edge-weighted undirected graph $G = (V, E)$, and collection of demands $\{(S_1, T_1), \ldots, (S_k, T_k)\}$, each pair are disjoint vertex sets. The goal is to find a minimum-weight subgraph that contains a path from any node in S_i to any node in T_i for every $i \in \{1, \ldots, k\}$. Without loss of generality, the edge weights can be assumed to form a metric. Chekuri et al. [2] present an $O(\log^2 n \log^2 k)$-approximation for this problem using minimum density junction trees.

We show a reduction from the g2g-anycast problem with general weights to the generalized set-connectivity problem. Recall that without loss of generality, we may assume that in the g2g problem, we are given a single source group S, a collection of destination groups $T_1, \ldots, T_q$, nonnegative (broadcast) costs c_{uv}, and (funnel-tree) metric costs d_{uv}.

2.2 The Reduction

The main idea of the reduction is to overload the broadcast cost of the ball radius around each node in the source group S into a larger single metric in which we use the generalized set-connectivity algorithm. In particular, for every source node $s_i \in S$, we sort the nodes in $T_1 \cup \ldots \cup T_q$ in increasing order of broadcast cost from s_i to get the sorted order, say $t_1^i, \ldots, t_r^i$ where t_j^i is at distance c_{ij} from s_i, and we have $c_{i1} \le c_{i2} \ldots \le c_{ir}$, where $|T_1 \cup$

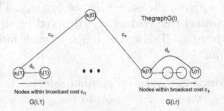

Fig. 3. A connected component $G(i)$ in the reduction of the g2g-anycast problem with general weights to the generalized set connectivity problem.

$\ldots \cup T_q| = r$. We now build r different graphs $G(i, 1), \ldots, G(i, r)$ where $G(i, j)$ is a copy of the metric completion of G under the funnel tree costs d induced on the node set $\{s_i, t_1^i, \ldots, t_j^i\}$, with the copies denoted as $\{s_i(j), t_1^i(j), \ldots, t_j^i(j)\}$. (Note that the terminal node t_a^i appears in copies a through r.) Finally, we take the r copies of the node s_i denoted $s_i(1), s_i(2), \ldots, s_i(r)$ and connect them to a new node $s_i(0)$ where the cost of the edge from $s_i(j)$ to $s_i(0)$ is c_{ij}. Thus these r different copies $G(i, 1), \ldots, G(i, r)$ all connected to the new node $s_i(0)$

together form one connected component $G(i)$. We now repeat this process for every source node s_i for $i \in \{1, \ldots, k\}$ to get k different graphs $G(1), \ldots, G(k)$ (Fig. 3).

We are now ready to define the generalized set connectivity demands. We define a new super source set $SS = \{s_1(0), s_2(0), \ldots, s_k(0)\}$. For each of the destination groups T_x, we define the terminal set TT_x to be the union of the copies of all corresponding terminal nodes in any of the copies $G(i)$. More precisely $TT_x = \{\cup_i t_a^i(j) | a \leq j \leq r, t_a^i \in T_x\}$. The final demand pairs for the set connectivity problem are $\{(SS, TT_1), \ldots, (SS, TT_q)\}$.

Lemma 1. *Given an optimal solution to the g2g-anycast problem, there is a solution to the resulting set connectivity problem described above of the same cost.*

Proof. Suppose the solution of the g2g problem involved picking broadcast ball radii $c_1, \ldots, c_k$ from source nodes $s_1, \ldots, s_k$ respectively. We also have funnel trees $H_1, \ldots, H_k$ that connect terminals $T(H_1), \ldots, T(H_k)$ to $s_1, \ldots, s_k$ respectively. Note that all terminals in $T(H_x)$ are within the thresholds that receive the broadcast from s_x, i.e. for every such terminal $t \in H_x$, the broadcast cost of the edge between s_x and t is at most the radius threshold c_x at which s_x is broadcasting.

Consider the tree H_x with terminals $T(H_x)$ connected to the root s_x, so that c_x is the largest weight of any of the edges from s_x to any terminal in $T(H_x)$. (If all of them were even closer, we can reduce the broadcast cost c_x of broadcasting from s_x and reduce the cost of the g2g solution.) Let the terminal in the funnel tree with this broadcast cost be $t(x)$ and in the sorted order of weights from s_x let the rank of $t(x)$ be p. We now consider the graph copy $G(x, p)$ and take a copy of the funnel tree H_x in this copy. To this we add an edge from the root $s_x(p)$ to the node $s_x(0)$ of cost c_{xp}. The total cost of this tree thus contains the funnel tree cost of H_x (denoted by $d(H_x)$) as well as the broadcast cost of c_{xp} from s_x. Taking the union of such funnel trees over all the copies gives the lemma.

Lemma 2. *Given an optimal solution to the set connectivity problem described above, there is a solution to the g2g-anycast problem from which it was derived of the same total weight.*

Proof. In the other direction, consider each copy $G(x)$ in turn and consider the set of edges in the tree containing the source node $s_x(0)$ in the solution to the generalized set-connectivity instance. First notice that it contains at most one of the edges to a copy $s_x(q)$ for some q. Indeed if we have edges to two different copies $s_x(p)$ and $s_x(q)$ from $s_x(0)$ for $p < q$, then since $G(x, p) \subset G(x, q)$, we can consider the tree edges in $G(x, p)$ and buy them in $G(x, q)$ where they also occur to cover the same set of terminals at smaller cost. In this way, we can save the broadcast cost of the copy of the edge from $s_x(0)$ to $s_x(p)$ contradicting the optimality of the solution. Now that we have only one of the edges, say to $s_x(q)$ from $s_x(0)$, we can consider all the edges of the tree in the copy $G(x, q)$ and include these edges in a funnel tree H'_x. The distance of the edge from $s_x(0)$ to

$s_x(q)$ pays for the broadcasting cost from s_x in the original instance and the cost of the rest of the tree is the same as the funnel tree cost of H'_q (Note that our observation above implies that edges in the metric completion in the tree can be converted to paths in the graph and hence connect all the nodes in the tree).

Since every terminal superset TT_j is connected to some source node of SS, all the demands of the g2g problem must be satisfied in the collection of funnel trees H'_x constructed in this way giving a solution to the g2g problem of the same cost.

The above two lemmas with the result of [2] gives us the following result.

Theorem 1. *The general weights version of the g2g-anycast problem with k destination groups admits a polynomial-time approximation algorithm with performance ratio $O(\log^2(k) \log^2 n)$ in an n-node graph.*

2.3   Hardness of Approximating g2g-anycast

We observe that the g2g-anycast problem with general weights can capture the group Steiner tree problem which is known to be $\log^{2-\epsilon} n$-hard to approximate unless NP is contained in quasi-polynomial time [6].

In the group Steiner tree problem, we are given an undirected graph with metric edge costs, a root s and a set of subsets of nodes called groups, say $T_1, \ldots, T_g$, and the goal is to find a minimum cost tree that connects the root with at least one node from each group. We can easily define this as a g2g-anycast problem with a singleton source group $S = \{s\}$ with the single root node. The terminal sets for the g2g-anycast problem are the groups $T_1, \ldots, T_g$, with the demand graph $(S, T_1), \ldots, (S, T_g)$. We can set the broadcast costs of any node in the graph from s to be zero; we use the given metric costs in the group Steiner problem as the funnel tree costs to capture the cost of the group Steiner tree. Any solution to the resulting g2g-anycast problem is a single tree connecting s to at least one node in each of the groups as required and its total weight is just its funnel tree cost that reflects precisely the cost of this feasible group Steiner tree solution. The hardness follows from this approximation-preserving reduction.

3   Approximating g2s-anycast

In this section, we consider g2s-anycast, a special case of the g2g-anycast, in which each destination group is a singleton set (i.e., has exactly one terminal). Let S denote the source-set and $t_1, \ldots, t_q$ denote the terminals.

The desired solution is a collection of broadcast balls and funnel trees T_v, each rooted at a source node v, so that for every demand (S, t_j), there exists at least one node v in S such that $t_j \in T_v$.

We now present a $\Theta(\log n)$-approximation algorithm for g2s-anycast problem. Our algorithm iteratively computes an approximation to a minimum density assignment, which assigns a subset of as yet unassigned terminals to a source node, and then combines these assignments to form the final solution.

Minimum Density Assignment. We seek a source s and a tree T_s rooted at s that connects s to a subset of terminals, such that the ratio $(c(T_s) + d(T_s))/|T_s|$ is minimized among all choices of s and T_s (here $c(T_s)$ denotes the minimum broadcast cost for s to reach the terminals in T_s, while $d(T_s)$ denotes the funnel-tree cost, i.e. the sum of the metric distances d_{uv} over all edges $uv \in T_s$). We present a constant-approximation to the problem, using a constant factor approximation algorithm for the rooted k-MST problem, which is defined as follows: given a graph G with weights on edges and a root node, determine a tree of minimum weight that spans at least k vertices. The best known approximation factor for the k-MST problem [15] is 2 [3]. We now present our algorithm for minimum density assignment.

- For each source $s \in S$, integer $k \in [1, n]$, and integer r drawn from the set $\{c_{st_j} | 1 \leq j \leq q\}$:
  - Let G' denote the graph with vertex set $\{s\} \cup \{t_j | c_{st_j} \leq r\}$, and edge weights given by d.
  - Compute a 2-approximation $T'(s, r, k)$ to the k-MST problem over the graph G' with s being the root.
- Among all trees computed in the above iterations, return a tree that minimizes $\min_{s,r,k}(d(T'(s, r, k)) + r)/k$.

Lemma 3. *The above algorithm is a polynomial-time 2-approximation algorithm for the minimum density assignment problem.*

Proof. We first show that the algorithm is polynomial time. The number of different choices for the source equals the size of the source set, the number of choices for k is n, and the number of different values for r is the number of different broadcast costs, which is at most n. Thus the number of iterations in the for loop is at most n^3. Consider an optimal solution T to the minimum density assignment problem, rooted at source s. It is a valid solution to the k-MST problem in the iteration given by $s, r = c(T), k = |T|$. For this particular iteration, the tree $T'(s, r, k)$ satisfies $(d(T'(s, r, k) + r)/k \leq (2d(T) + r)/k \leq 2 \cdot (d(T) + r)/k$. Since our algorithm returns the tree that has the best density, we have a 2-approximation for the minimum density assignment.

Approximation Algorithm for g2s-Anycast. Our algorithm is a greedy iterative algorithm, in which we repeatedly compute an approximation to the minimum density assignment problem, and return an appropriate union of all of the trees computed.

- For each source s, set T_s to $\{s\}$.
- While all terminals are not assigned:
  - Compute a 2-approximation T to the minimum density assignment problem using any source s and the unassigned terminals.
  - If T is rooted at source s, then set T_s to be the minimum spanning tree of the union of the trees T and T_s.
- Return the collection $\{T_s\}$.

Theorem 2. *The greedy algorithm yields an approximation algorithm with performance ratio $2 \ln n$ to the g2s-anycast problem.*

Proof. Let OPT denote the cost of the optimal solution to the problem. Any solution is composed of at most m trees, one for each of the sources, with each singleton group being included as a node in one of these trees. Let T_s^* denote the tree rooted at source s in an optimal solution.

Consider any iteration i of our algorithm. Let n_i denote the number of unassigned terminals at the start of the iteration i. By an averaging argument, we know there exists a source s such that

$$\frac{d(T_s^*) + c(T_s^*)}{|T_s^*|} \leq \frac{OPT}{n_i},$$

By Lemma 3, it follows that in the ith iteration of the greedy algorithm, if T_i is the tree computed in the step, then

$$\frac{d(T_i) + c(T_i)}{|T_i|} \leq \frac{2 \cdot OPT}{n_i},$$

Adding over all steps, we obtain that the total cost is

$$\sum_i (d(T_i) + c(T_i)) \leq 2 \cdot OPT \cdot \sum_i \frac{|T_i|}{n_i} \leq 2 \cdot OPT \cdot H_n \leq 2OPT \ln n.$$

Hardness of Approximation. We complement the positive result with a matching inapproximability result which shows that the above problem is as hard as set cover.

Theorem 3. *Unless $NP = P$ there is no polynomial-time $\alpha \ln n$ approximation to the g2s-anycast problem, for a suitable constant $\alpha > 0$.*

We defer the proof of this theorem to Appendix A of the full version [7].

4   Euclidean g2g-anycast

In this section, we present a $\Theta(\log n)$-approximation for the more realistic version of the g2g-anycast problem in the 2-D Euclidean plane. We achieve our results by a reduction to an appropriately defined set cover problem.

In detail, all the points in both the source group S and destination groups $T_1, \ldots, T_q$ lie in the 2-D Euclidean plane. The cost of an edge (u, v) is the Euclidean distance between u and v raised to the path loss exponent κ. For the rest of this section, we assume that $\kappa = 2$. (The corresponding results for $\kappa > 2$ follow with very simple modifications.) First we show that even this special case of the g2g-anycast problem does not permit an approximation algorithm with ratio $(1 - \epsilon) \ln n$ on an instance with n nodes unless NP is in quasi-polynomial time. Next, we present *Cover-and-Grow*, an $O(\log n)$-approximation algorithm that applies a greedy heuristic to an appropriately defined instance of the set covering problem.

Hardness of 2-D g2g-anycast. Again we can prove a hardness via a reduction from set cover.

Theorem 4. *The 2-D Euclidean version of the g2g problem on n nodes does not permit a polynomial-time $(1 - o(1)) \ln n$ approximation algorithm unless $NP = P$.*

The proof of this can be found in Appendix B of the full version [7].

4.1 Cover-and-Grow

We now describe a matching $O(\log n)$-approximation for the problem. For this we first need the following property of minimum spanning trees of points in the 2-D Euclidean plane within a unit square, when the costs of any edge in the tree are the squared Euclidean distances between the edge's endpoints.

Theorem 5 [1]. *The weight of a minimum spanning tree of a finite number of points in the 2-D Euclidean plane within a unit square, where the weight of any edge is the square of the Euclidean distance between its endpoints, is at most 3.42.*

We can apply this theorem to bound the cost of the funnel trees within any demand ball in the solution within a factor of at most 3.42 of the cost of the ball. Indeed, by scaling the diameter of the demand ball to correspond to unit distance, the above theorem shows that for any finite set of terminal nodes (i.e. nodes in the destination group) within the ball, a funnel tree which is an MST that connects these terminal nodes to the center of the ball has total cost at most 3.42. The cost of the demand ball is the square of the Euclidean distance of the ball radius which, in the scaled version, has cost $(\frac{1}{2})^2 = \frac{1}{4}$. This shows that the funnel tree has cost at most 13.68 times the cost of the funnel ball. This motivates an algorithm that uses balls of varying radii around each source node as a "set" that has cost equal to the square of the ball radius (the ball cost) and covers all the terminal nodes within this ball (which can be connected in a funnel tree of cost at most 13.68 times that of the demand ball).

Algorithm Cover-and-Grow

1. Initialize the solution to be empty.
2. While there is still an unsatisfied demand edge
   - For every source node s_i, for every possible radius at which there is a terminal node belonging to some destination group T for which the demand (S, T) is yet unsatisfied, compute the ratio of the square of the Euclidean radius of the ball to the number of *as yet unsatisfied* destination groups whose terminal nodes lie in the ball.
   - Pick the source node and ball radius whose ratio is minimum among all the available balls, and add it to the solution (both the demand ball around this node and a funnel tree from one node of each destination group whose demand is unsatisfied at this point). Update the set of unsatisfied demands accordingly.

Theorem 6. *Algorithm Cover-and-grow runs in polynomial time and gives an $O(\log n)$-approximate solution for the 2-D g2g-anycast problem in an n-node graph.*

Proof. We will use a reduction from the given 2-D g2g-anycast problem to an appropriate set cover problem as described in the algorithm. The elements of the set cover problem are the terminal sets T_j such that the demand graph has the edge (S, T_j). For every source node $s_i \in S$, and for every possible radius r at which there is a terminal node belonging to some destination group T for which there is a demand (S, T), we consider a set $X(s_i, r)$ that contains all the destination groups T_j such that some node of T_j lies within this ball. The cost of this set is r^2.

First, we argue that an optimal solution for the 2-D g2g-anycast problem of cost C^* gives a solution of cost at most C^* to this set cover problem. Next, we show how any feasible solution to the set cover problem of cost C gives a feasible solution to the 2-D g2g-anycast problem of cost at most $14.68C$. These two observations give us the result since the algorithm we describe is the standard greedy approximation algorithm for set cover.

To see the first observation, given an optimal solution for the 2-D g2g-anycast problem of cost C^*, we pick the sets corresponding to the demand balls in the solution for the set cover problem. Since these demand balls are a feasible solution to the anycast problem, they together contain at least one terminal from each of the destination groups T_j for which there is a demand edge (S, T_j). These balls form a solution to the set cover problem and the demand ball costs of the anycast solution alone pay for the corresponding costs of the set cover problem. Hence this feasible set cover solution has cost at most C^*.

For the other direction, given any feasible solution to the set cover problem of cost C, note that this pays for the demand balls around the source nodes in this set cover solution. Now we can use the implication in the paragraph following Theorem 5 to construct a funnel tree for each of these demand balls that connects all the terminals within these balls to the source node at the center of the ball with cost at most 13.68 times the cost of the demand ball around the source node. Summing over all such balls in the solution gives the result.

5 Empirical Results

We conducted simulations comparing Cover-and-Grow with four different natural heuristics for points embedded in a unit square in the 2-D Euclidean plane. These simulations allow us gain perspective on the real-world utility of Cover-and-Grow vis a vis alternatives that do not possess provable guarantees but yet have the potential to be practical. The specifics of the simulation and the details of the results are discussed in Appendix C of the full version [7]. Cover-and-Grow performs comparably to the heuristics in performance; and the runtime of Cover-and-Grow was better than the heuristics except for the T-centric approach.

References

1. Aichholzer, O., Allen, S., Aloupis, G., Barba, L., Bose, P., de Varufel, J.L., Iacono, J., Langerman, S., Souvaine, D., Taslakian, P., Yagnatinsky, M.: Sum of squared edges for MST of a point set in a unit square. In: Japanese Conference on Discrete and Computational Geometry (JCDCG) (2013)
2. Chekuri, C., Even, G., Gupta, A., Segev, D.: Set connectivity problems in undirected graphs and the directed Steiner network problem. ACM Trans. Algorithms **7**(2), 18:1–18:17 (2011)
3. Garg, N.: Saving an epsilon: a 2-approximation for the k-MST problem in graphs. In: ACM Theory of Computing, pp. 396–402 (2005)
4. Garg, N., Konjevod, G., Ravi, R.: A polylogarithmic approximation algorithm for the group Steiner tree problem. J. Algorithms **37**(1), 66–84 (2000)
5. Halperin, E., Kortsarz, G., Krauthgamer, R., Srinivasan, A., Wang, N.: Integrality ratio for group Steiner trees and directed Steiner trees. SIAM J. Comput. **36**(5), 1494–1511 (2007)
6. Halperin, E., Krauthgamer, R.: Polylogarithmic inapproximability. In: ACM Theory of Computing, pp. 585–594 (2003)
7. Iglesias, J., Rajaraman, R., Ravi, R., Sundaram, R.: Balls and funnels: energy efficient group-to-group anycasts. In: Dinh, T.N., Thai, M.T. (eds.) COCOON 2016. LNCS, vol. 558, pp. 235–246. Springer, Heidelberg (2016). CoRR http://arxiv.org/abs/1605.07196
8. Lapowsky, I.: Facebook lays out its roadmap for creating internet-connected drones. Wired (2014). http://www.wired.com/2014/09/facebook-drones-2/
9. Li, D., Liu, Q., Hu, X., Jia, X.: Energy efficient multicast routing in ad hoc wireless networks. Comput. Commun. **30**(18), 3746–3756 (2007)
10. McNeal, G.: Google wants Internet broadcasting drones, plans to run tests in New Mexico. Forbes (2014). http://www.forbes.com/sites/gregorymcneal/2014/09/19/google-wants-internet-broadcasting-drones-plans-to-run-tests-in-new-mexico/
11. Milyeykovski, V., Segal, M., Shpungin, H.: Location, location, location: using central nodes for efficient data collection in WSNs. In: WiOpt, pp. 333–340, May 2013
12. Olsson, P.M.: Positioning algorithms for surveillance using unmanned aerial vehicles. Licentiate thesis, Linköpings universitet (2011)
13. Path Loss Wikipedia. http://en.wikipedia.org/wiki/Path_loss
14. Rappaport, T.: Wireless Communications: Principles and Practice, 2nd edn. Prentice Hall PTR, Upper Saddle River (2001)
15. Ravi, R., Sundaram, R., Marathe, M.V., Rosenkrantz, D.J., Ravi, S.S.: Spanning trees short or small. In: ACM-SIAM Discrete Algorithms, SODA 1994, pp. 546–555. SIAM (1994)
16. Wan, P.J., Călinescu, G., Li, X.Y., Frieder, O.: Minimum-energy broadcasting in static ad hoc wireless networks. Wirel. Netw. **8**(6), 607–617 (2002)
17. Wieselthier, J.E., Nguyen, G.D., Ephremides, A.: On the construction of energy-efficient broadcast and multicast trees in wireless networks. In: INFOCOM, pp. 585–594 (2000)
18. Wieselthier, J.E., Nguyen, G.D., Ephremides, A.: Algorithms for energy-efficient multicasting in static ad hoc wireless networks. MONET **6**(3), 251–263 (2001)
19. Williamson, D.P., Shmoys, D.B.: The Design of Approximation Algorithms, 1st edn. Cambridge University Press, New York (2011)

Assigning Proximity Facilities for Gatherings

Shin-ichi Nakano[✉]

Gunma University, Kiryu 376-8515, Japan
nakano@cs.gunma-u.ac.jp

Abstract. In this paper we study a recently proposed variant of the problem the r-gathering problem. An r-*gathering* of customers C to facilities F is an assignment A of C to *open facilities* $F' \subset F$ such that r or more customers are assigned to each open facility. (Each open facility needs enough number of customers.) Then the cost of an r-gathering is $\max\{\max_{i \in C}\{co(i, A(i))\}, \max_{j \in F'}\{op(j)\}\}$, and the r-*gathering problem* finds an r-gathering having the minimum cost.

Assume that F is a set of locations for emergency shelters, $op(f)$ is the time needed to prepare a shelter $f \in F$, and $co(c, f)$ is the time needed for a person $c \in C$ to reach assigned shelter $A(c) \in F$. Then an r-gathering corresponds to an evacuation plan such that each opened shelter serves r or more people, and the r-gathering problem finds an evacuttion plan minimizing the evacuation time span.

However in a solution above some person may be assigned to a farther open shelter although it has some closer open shelter. It may be difficult for the person to accept such an assignment for an emergency situation. Therefore Armon considered the problem with one more additional constraint, that is, each customer should be assigned to a closest open facility, and gave a 9-approximation algorithm for the problem.

In this paper we give a simple 3-approximation algorithm for the problem.

1 Introduction

The facility location problem and many of its variants are studied [4,5].

In the basic facility location problem we are given (1) a set C of customers, (2) a set F of facilities, (3) an opening cost $op(f)$ for each $f \in F$, and (4) a connecting cost $co(c, f)$ for each pair of $c \in C$ and $f \in F$, then we open a subset $F' \subset F$ of facilities and find an assignment A from C to F' so that a designated cost is minimized. A typical *max* version of the cost of an assignment A is $\max\{\max_{i \in C}\{co(i, A(i))\}, \max_{j \in F'}\{op(j)\}\}$. We assume that co satisfies the triangle inequality.

In this paper we study a recently proposed variant of the problem, called the r-gathering problem [2].

An r-*gathering* of customers C to facilities F is an assignment A of C to *open facilities* $F' \subset F$ such that r or more customers are assigned to each open facility. (Each open facility needs enough number of customers.) We assume $|C| \geq r$ holds. Then max version of the cost of an r-gathering is

© Springer International Publishing Switzerland 2016
T.N. Dinh and M.T. Thai (Eds.): COCOON 2016, LNCS 9797, pp. 247–253, 2016.
DOI: 10.1007/978-3-319-42634-1_20

$\max\{\max_{i \in C}\{co(i, A(i))\}, \max_{j \in F'}\{op(j)\}\}$. Then the min-max version of the *r-gathering problem* finds an *r*-gathering having the minimum cost. (For the min-sum version see the brief survey in [2].)

Assume that F is a set of locations for emergency shelters, $op(f)$ is the time needed to prepare a shelter $f \in F$, and $co(c, f)$ is the time needed for a person $c \in C$ to reach assigned shelter $A(c) \in F$. Then an *r*-gathering corresponds to an evacuation plan such that each opened shelter serves r or more people, and the *r*-gathering problem finds an evacuttion plan minimizing the evacuation time span.

Armon [2] gave a simple 3-approximation algorithm for the problem and proved that with assumption $P \neq NP$ the problem cannot be approximated within a factor of less than 3 for any $r \geq 3$.

However in a solution above some person may be assigned to a farther open shelter although it has some closer open shelter. It may be difficult for the person to accept such an assignment for an emergency situation. Therefore Armon [2] also considered the problem with one more additional constraint, that is, each customer should be assigned to a closest open facility, and gave a 9-approximation algorithm for the problem. We call the problem *the proximity r-gathering problem*.

In this paper we give a simple 3-approximation algorithm for the proximity *r*-gathering problem.

The remainder of this paper is organized as follows. Section 2 contains our main algorithm for the proximity *r*-gathering problem. Section 3 considers a case with outliers. Section 4 gives a slightly improved algorithm for the original *r*-gathering problem. (This part will be appear in [1], but in Japanese.) Section 5 contains a conclusion and an open problem.

A preliminary version of the paper is presented at (unrefereed) meeting [6].

2 Algorithm

We need some preparation.

A lower bound $lb(i, j)$ of the cost assigning $i \in C$ to $j \in F$ in any *r*-gathering is derived as follows. Let $N(j)$ be the set of r customers having up to *r*-th smallest connection costs to facility $j \in F$. If $i \in N(j)$ then define $lb(i, j) = max\{op(j), co(k, j)\}$, where k is the customer having the *r*-th smallest connection cost to j. Otherwise $lb(i, j) = max\{op(j), co(i, j)\}$. Then a lower bound $lb(i)$ of the cost for $i \in C$ in any *r*-gathering is derived as $lb(i) = min_{j \in F}\{lb(i, j)\}$. Since we need to assign $i \in C$ to some facility, $lb(i)$ is also a lower bound for the cost of the solution of the proximity *r*-gathering problem. Let $bestf(i)$ for $i \in C$ be a facility $j \in F$ attaining cost $lb(i)$. Let $mates(i)$ for $i \in C$ be $N(bestf(i))$ if $i \in N(bestf(i))$, and $N(bestf(i)) \cup \{i\} - \{k\}$ otherwise. Thus if we assign $mates(i)$ to $bestf(i) \in F$ then the cost of the part is at least $lb(i)$.

We regard $co(f, f') = min_{i \in C}\{co(i, f) + co(i, f')\}$ for $f, f' \in F$. We define by opt the cost of the solution, that is $min_A max\{max_{i \in C}\{co(i, A(i))\}, max_{j \in F'}\{op(j)\}\}$, where $F' \subset F$ is the set of opened facilities. Clearly $opt \geq lb(i)$ holds for any $i \in C$.

Algorithm 1. Best-or-factor3

for all $i \in C$ **do**
  Compute $lb(i)$, $bestf(i)$ and $mates(i)$
end for
Sort C in the non-increasing order of $lb(i)$
for all $i \in C$ in the non-increasing order of $lb(i)$ **do**
  if $bestf(i)$ is not assigned to yet, and none of $mates(i)$ has been assigned yet **then**
    Open $bestf(i)$
    for all $k \in mates(i)$ **do**
      Assign k to $bestf(i)$ /* Best-Assignment */
    end for
    for all f such that $co(f, bestf(i)) \le 2 \cdot lb(i)$ **do**
      Shut down f
    end for
  end if
end for
for all unassigned $k \in C$ **do**
  Assign k to a closest open facility   /* Factor3-Assignment */
end for

Now we give our algorithm to solve the proximity r-gathering problem.

Clearly Algorithm **Best-or-factor3** finds an r-gathering. (Since whenever we newly open a facility we always assign r customers.)

The algorithm is similar to algorithm **Best-or-rest** in [2] for the original r-gathering problem, except (1) our algorithm has the "shut down f "operation, and (2) sorts C in the non-increasing order, while **Best-or-rest** [2] sorts C in the non-decreasing order. Actually we can modify algorithm **Best-or-rest** for the original r-gathering problem so that it does not need the sort. We show this in the later section.

We have the following lemma.

Lemma 1. *Algorithm* **Best-or-factor3** *finds an r-gathering such that each customer is assigned to a closest open facility.*

Proof. Assume otherwise for a contradiction. Since Factor3-Assignment never open any facility and always assign a customer to a closest open facility, we only consider for Best-Assignment. Then some $i' \in mates(i)$ assigned to facility $bestf(i)$ has a closer open facility, say $bestf(k)$ for some $k \in C$. We have two cases based on the opening order of $bestf(i)$ and $bestf(k)$.

If $bestf(k)$ opens earlier than $bestf(i)$ then $lb(k) \ge lb(i)$ holds, then

$$
\begin{aligned}
co(bestf(i), bestf(k)) &\le co(i', bestf(i)) + co(i', bestf(k)) \\
&< co(i', bestf(i)) + co(i', bestf(i)) \\
&\le 2 \cdot lb(i) \\
&\le 2 \cdot lb(k)
\end{aligned}
$$

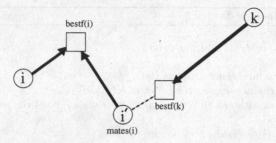

Fig. 1. Illustration for the proof of Lemma 1.

which contradicts to the fact that after we open facility $bestf(k)$ we shut down every surrounding facility with connection cost at most $2 \cdot lb(k)$. We need the sort for the last inequality (Fig. 1).

Otherwise $bestf(i)$ opens earlier than $bestf(k)$ and $lb(i) \geq lb(k)$ holds, so

$$co(bestf(k), bestf(i)) \leq co(i', bestf(k)) + co(i', bestf(i))$$
$$\leq co(i', bestf(i)) + co(i', bestf(i))$$
$$\leq 2 \cdot lb(i)$$

which contradicts to the fact that after we open facility $bestf(i)$ we shut down every surrounding facility within connection cost at most $2 \cdot lb(i)$.   □

We have the following two theorems.

Theorem 1. *The cost of an r-gathering found by Algorithm* **Best-or-factor3** *is at most* $3 \cdot opt$.

Proof. Consider the cost for each assignment of $i \in C$. For Best-Assignment the cost is $lb(i) \leq opt$. So we need to consider only for $Factor3$-$Assignment$.

Each $i \in C$ assigned in $Factor3$-$Assignment$ was not assigned to $bestf(i)$ in Best-Assignment but later assigned to its closest already opened facility. So we consider only for connection costs.

Assume we assign $i \in C$ in $Factor3$-$Assignment$. We show that i always has an open facility with the connection cost at most $3\,opt$. We have two cases based on the reason why i was not assigned in $Best$-$Assignment$.

Case 1(a): Some $i' \in mate(i)$ is already assigned to some $bestf(k)$ since $i' \in mates(k)$ also holds. See Fig. 2(a).

The connection cost $co(i, bestf(k))$ is at most

$$co(i, bsetf(i)) + co(i', bestf(i)) + co(i', bestf(k)) \leq lb(i) + lb(i) + lb(k)$$
$$\leq 3\,opt$$

Thus $i \in C$ has an open facility with a connection cost at most $3\,opt$.

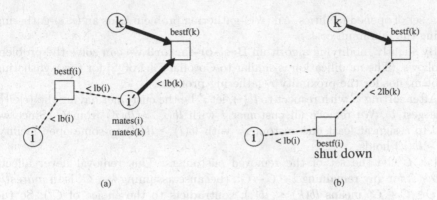

Fig. 2. Illustration for the proof of Theorem 1

Case 1(b): $bestf(i)$ is already shut down just after some $bestf(k)$ is opened. See Fig. 2(b).

The connection cost $co(i, bestf(k))$ is at most

$$co(i, bsetf(i)) + co(bestf(i), bestf(k)) \leq lb(i) + 2 \cdot lb(k)$$
$$\leq 3\, opt$$

Thus $i \in C$ has an open facility with the connection cost at most $3\, opt$.   □

Theorem 2. *Algorithm* **Best-or-factor3** *runs in time* $O(r|C| + |C||F|^2 + |C|\log|C|)$.

Proof. For each $j \in F$ by using a linear time selection algorithm [3] [p. 220] find the r-th closest customer to j, then choosing closer customers we can compute the set of up to $(r-1)$-th closest customers to j in $O(|C|)$ time. Thus we need $O(|C||F|)$ time in total to compute such a customer and a set of customers for all $j \in F$.

Then we can compute $lb(i), bestf(i), mates(i)$ for all $i \in C$ in $O(|C||F|)$ time. We also compute $co(f, f')$ for every $f, f' \in F$ in $O(|C||F|^2)$ time.

We need $O(|C|\log|C|)$ time for the sort.

Then **Best-Assignment** part runs in $O(r|C| + |C||F|^2)$ time, and **Factor3-Assignment** part runs in $O(|C||F|)$ time.

Thus in total the algorithm runs in $O(r|C| + |C||F|^2 + |C|\log|C|)$ time.   □

3 Outlier

An (r, ϵ)-gathering of C to F is an r-gathering of $C - C'$ to F, where C' is any subset of C with size at most $\epsilon|C|$. Intuitively we can ignore at most $\epsilon|C|$ (outlier) customers for the assignment. The cost of an (r, ϵ)-gathering A is defined naturally, that is $\max\{\max_{i \in C-C'}\{co(i, A(i))\}, \max_{j \in F'}\{op(j)\}\}$, where $F' \subset F$

is the set of opened facilities. An (r, ϵ)-gathering problem finds an (r, ϵ)-gathering having the minimum cost.

By slightly modifying algorithm **Best-or-factor3** we can solve the problem as follows. (The modification is similar to Corollary 3.4 of [2] for the r-gathering problem, not for the proximity r-gathering problem.)

After sorting C with respect to $lb(i)$, let i' be the customer having the $\lceil \epsilon |C| \rceil$-th largest lb. We remove all customer i with $lb(i) > lb(i')$ from C. Since we need to assign at least one customer i with $lb(i) \geq lb(i')$ to some open facility, $opt \geq lb(i')$ holds.

Let C' be the set of the removed customers. This removal never affects $mates(i)$ for any remaining $i \in C - C'$, (because assuming $k \in C'$ is in $mates(i)$ for $i \in C - C'$ means $lb(k) \leq lb(i)$, contradicts to the choice of C'). So the removal also never affects $lb(i)$ and $bestf(i)$ for any remaining $i \in C - C'$.

Thus for the remaining customers algorithm **Best-or-factor3** computes an r-gathering with cost at most $3lb(i') \leq 3opt$. Now we have the following theorem.

Theorem 3. *One can find an (r, ϵ)-gathering with cost at most $3 \cdot opt$ in $O(r|C| + |C||F|^2 + |C| \log |C|)$ time.*

4 r-Gathering Without Sort

The following algorithm **Best-or-rest** is a 3-approximate algorithm for the original r-gathering problem which is basically derived from [2] by just removing the sort of C.

Algorithm 2. Best-or-rest

---
for all $i \in C$ do
  Compute $lb(i)$, $bestf(i)$ and $mates(i)$
end for
for all $i \in C$ do
  if $bestf(i)$ is not assigned to yet, and all $mates(i)$ are not assigned yet **then**
    Open $bestf(i)$
    for all $k \in mates(i)$ do
      Assign k to $bestf(i)$ /* Best-Assignment */
    end for
  end if
end for
for all unassigned $k \in C$ do
  Assign k to a closest open facility   /* Rest-Assignment */
end for

We have the following theorems.

Theorem 4. *The cost of an r-gathering found by Algorithm **Best-or-rest** is at most $3 \cdot opt$.*

Proof. The proof is just a subset of the proof of Theorem 1.

Consider the cost for each assignment of $i \in C$. For *Best-Assignment* the cost is $lb(i) \leq opt$. So we need to consider only for *Rest-Assignment*.

Each $i \in C$ assigned in *Rest-Assignment* was not assigned to $bestf(i)$ but later assigned to its closest already opened facility. So we consider only for connection costs.

Assume we assign $i \in C$ in *Rest-Assignment*. The reason why i was not assigned in *Best-Assignment* is some $i' \in mate(i)$ is already assigned to some $bestf(k)$ since $i' \in mates(k)$ also holds.

The connection cost $co(i, bestf(k))$ is at most

$$co(i, bsetf(i)) + co(i', bestf(i)) + co(i', bestf(k)) \leq lb(i) + lb(i) + lb(k)$$
$$\leq 3\, opt$$

Thus $i \in C$ has an open facility with a connection cost at most $3\, opt$.   □

We can prove the running time of the algorithm is $O(|C||F| + r|C|)$, by a similar way to the proof of Theorem 2. While in [2] the running time was $O(|C||F| + r|C| + |C| \log |C|)$ since it needs a sort of $|C|$.

5   Conclusion

In this paper we provided a simple approximation algorithm to solve the proximity r-gathering problem. The approximation ratio is 3, which improve the former result [2] of 9.

The algorithm can solve a slightly more general problem in which each $f \in F$ has a distinct minimum number r_f of customers needed to open. The algorithm also runs in $O(r|C| + |C||F|^2 + |C| \log |C|)$ time. We assume $r > r_f$ holds for all $f \in F$.

Can we design an approximation algorithm for the min-sum version of the proximity r-gathering problem?

References

1. Akagi, T., Arai, R., Nakano, S.: Faster min-max r-gatherings. IEICE Trans. Fundam., vol. E99-A, (2016, Accepted). (in Japanese)
2. Armon, A.: On min-max r-gatherings. Theor. Comput. Sci. **412**, 573–582 (2011)
3. Cormen, T.H., Leiserson, C.E., Rivest, R.L., Stein, C.: Introduction to Algorithms, 3rd edn. MIT Press, Cambridge (2009)
4. Drezner, Z.: Facility Location: A Survey of Applications and Methods. Springer, New York (1995)
5. Drezner, Z., Hamacher, H.W.: Facility Location: Applications and Theory. Springer, Heidelberg (2004)
6. Nakano, S.: Assigning proximity facilities for gatherings, IPSJ SIG Technical report, 2015-AL-151-5 (2015)

Cryptography

Combiners for Chosen-Ciphertext Security

Cong Zhang[1], David Cash[1], Xiuhua Wang[2], Xiaoqi Yu[3],
and Sherman S.M. Chow[2([⊠])]

[1] Department of Computer Science, Rutgers University, New Brunswick, NJ, USA
[2] Department of Information Engineering, The Chinese University of Hong Kong,
Sha Tin, N.T., Hong Kong
sherman@ie.cuhk.edu.hk
[3] Department of Computer Science, The University of Hong Kong,
Pok Fu Lam, Hong Kong

Abstract. Security against adaptive chosen-ciphertext attack (CCA) is
a *de facto* standard for encryption. While we know how to construct
CCA-secure encryption, there could be pragmatic issues such as black-
box design, software mis-implementation, and lack of security-oriented
code review which may put the security in doubt. On the other hand,
for double-layer encryption in which the two decryption keys are held
by different parties, we expect the scheme remains secure even when
one of them is compromised or became an adversary. It is thus desirable
to combine two encryption schemes, where we cannot be assured that
which one is really CCA-secure, to a new scheme that is CCA-secure. In
this paper we propose new solutions to this problem for symmetric-key
encryption and public-key encryption. One of our result can be seen as
a new application of the detectable CCA notion recently proposed by
Hohenberger *et al.* (Eurocrypt 2012).

Keywords: Encryption · Chosen-ciphertext security · Robust combiners

1 Introduction

Secure systems are usually complex and involve multiple components. If a component turns out to be problematic, the whole system may become totally insecure. For security-critical applications, a prudent practice is to have a robust
design, such that the system remains secure even if a component is insecure. Of
course, if one could identify which component is insecure, the designer can simply
replace it with a secure one. Yet, it is notoriously difficult to ensure that a system
component is secure in general. One example is that a component primitive is
implemented as a black-box which the combiner cannot assert its security. On
the other hand, even if the source code (of the software) or the circuit footprint
(of the hardware) were available, asserting its security depends on the rigor and
quality of the corresponding security-oriented review.

Sherman S.M. Chow is supported in part by the Early Career Award and the grants
(CUHK 439713 & 14201914) of the Research Grants Council, Hong Kong.

© Springer International Publishing Switzerland 2016
T.N. Dinh and M.T. Thai (Eds.): COCOON 2016, LNCS 9797, pp. 257–268, 2016.
DOI: 10.1007/978-3-319-42634-1_21

In this paper, we look into a basic cryptographic tool which is encryption. We consider both public-key encryption (PKE) and symmetric-key encryption (SKE). The work of Herzberg [7] motivated the need of robust design of cryptosystem, which features a combination of multiple instantiation of the same primitives (*e.g.*, one may consider ElGamal encryption and RSA encryption as examples), such that as long as one of them ensures a certain level of security, the same security guarantee is preserved by the combined design, without knowing beforehand which one is that. *Robust combiner* achieving this property can ensure security even if there is doubt in the security of the component primitives [6]. It is also termed as tolerant cryptographic schemes [7] or cryptanalysis-tolerant schemes [5] in the literature.

Herzberg [7] proposed combiners that are secure against chosen-plaintext attack (CPA) or chosen-ciphertext attack. However, it is hard to achieve security against *adaptive* chosen-ciphertext attack (CCA in this paper[1]) if one of the component schemes turns out to be malleable. In the CCA attack, the adversary can query to a decryption oracle even after the adversary has obtained the challenge ciphertext, and the only disallowed query is the challenge ciphertext itself. Hence, if a part of the ciphertext is malleable, an adversary can simply maul it and obtain the plaintext from the decryption oracle. Dodis and Katz [5] proposed a cryptanalysis-tolerant CCA-secure encryption scheme, which remains secure when only an unknown one of the component schemes is CCA-secure.

Another usage of such a combiner is to achieve security for cryptosystems in which the decryption requires two private keys held by different parties. Security remains preserved when one of the parties is compromised by the adversary. An application is to support revocation via a security-mediator, a party whom needs to help the non-revoked users in every decryption request. Immediate revocation can be achieved once it is instructed to stop entertaining any further (partial) decryption request of the revoked user. For example, Chow *et al.* [2] proposed a CCA-secure security-mediated certificateless encryption scheme, combining an identity-based encryption with a public-key encryption generically. Without a combiner, a specific ad-hoc construction is probably needed [3].

Our Results. In this paper, we give two other cryptanalysis-tolerant CCA-secure encryption schemes, one for PKE and one for SKE. Our PKE combiner matches well with the notion of detectable chosen-ciphertext attack (DCCA) proposed by Hohenberger *et al.* [8] recently. Intuitively, DCCA is a weaker version of CCA, where "dangerous" ciphertexts are not allowed to be queried to the decryption oracle. Here, whether a ciphertext is dangerous can be checked by a polynomial-time function. Our combiner aims to achieve indistinguishability against DCCA attack, by detecting whether a query is originated from the challenge ciphertext of a component scheme. If so, such decryption query is disallowed. This gives a conceptually simple combiner with an elementary security proof. Furthermore, it illustrates yet another application of this DCCA notion.[2]

---

[1] We remark that it is called CCA2 in the literature when the adaptiveness matters.

[2] While the original paper has discussed the application of DCCA in ruling out some known implementation bug of a "sloppy" encryption scheme [8], our combiner does not assume the bug from the component scheme can be easily detected.

Yet, our combiner is downgrading the security of the component scheme since one of them is CCA-secure, but the resulting scheme is only DCCA-secure. For getting CCA-security, we resort back to the result of Hohenberger *et al.* [8]. Their work showed that we can construct a CCA-secure encryption scheme by a nested encryption approach, taking a DCCA-secure scheme, a CPA-secure scheme, and a 1-bounded CCA-secure scheme [4]. A q-bounded CCA-secure encryption system is secure against q chosen ciphertext queries, which can be constructed via a CPA-secure encryption primitive [4].

We then propose another combiner to directly obtain an SKE scheme with CCA security, by taking two SKE schemes in which only one of them is CCA-secure. This is different from our combiner for PKE. Note that an SKE scheme with security against chosen-plaintext attack and integrity of the ciphertext implies that this scheme is also CCA-secure [1]. For this combiner, our strategy is to work on these two properties instead, by taking two component schemes where an unknown one of them possesses of both properties.

Finally, we review in appendix the nested encryption technique of Hohenberger *et al.* [8] for obtaining CCA security.

2 Preliminaries

2.1 CCA Security for PKE

Definition 1 (Public-Key Encryption). *A public-key encryption scheme $\mathcal{PKE}$ consists of the following three probabilistic polynomial-time (PPT) algorithms* (KeyGen, Enc, Dec).

- $(EK, DK) \leftarrow$ KeyGen(1^λ): *the algorithm outputs a pair of keys consisting of the public encryption key EK and the private decryption key DK, according to the input security parameter 1^λ.*
- $C \leftarrow$ Enc(EK, m): *the algorithm takes a public key EK and a plaintext m as inputs, and outputs a ciphertext C.*
- $m \leftarrow$ Dec(DK, C): *the algorithm uses the private key DK to decrypt a ciphertext C to recover the plaintext m, or to output $\perp$ denoting C is invalid.*

When the context is clear, we may put the input key as a subscript instead, or simply omit it.

We recall the definition of CCA security. Consider the following experiment $Exp_{\mathcal{A},\mathcal{PKE}}^{\mathsf{cca}}(1^\lambda)$ for $\mathcal{PKE}$:

- *Setup*: The challenger $\mathcal{C}$ takes a security parameter 1^λ and runs KeyGen to output keys (EK, DK). It gives $\mathcal{A}$ EK, and keeps DK to itself.
- *Query Phase 1*: $\mathcal{A}$ is given full access to the decryption oracle Dec$(DK, \cdot)$. When the adversary $\mathcal{A}$ decides to terminate the query phase, it outputs a pair of messages m_0, m_1 of the same length.
- *Challenge*: The challenger $\mathcal{C}$ randomly picks a bit $b \leftarrow \{0, 1\}$, computes $C^* \leftarrow$ Enc(EK, m_b) and sends C^* to $\mathcal{A}$.

- *Query Phase 2*: $\mathcal{A}$ continues to have access to $\mathsf{Dec}(DK, \cdot)$, but is not allowed to request for a decryption of C^*. Finally $\mathcal{A}$ outputs a bit b'.
- *Output*: The output of the experiment is defined to be 1 if $b' = b$, otherwise 0.

A PKE scheme $\mathcal{PKE} = (\mathsf{KeyGen}, \mathsf{Enc}, \mathsf{Dec})$ is CCA secure if for all PPT adversaries $\mathcal{A}$, there exists a negligible function $\mathsf{negl}()$ such that:

$$\Pr[Exp^{\mathsf{cca}}_{\mathcal{A},\mathcal{PKE}}(1^\lambda) = 1] \leq \frac{1}{2} + \mathsf{negl}(\lambda).$$

2.2   Detectable Chosen Ciphertext Security

Detectable chosen ciphertext attack (DCCA) is an attack mode against PKE introduced by Hohenberger *et al.* [8], which is weaker than the standard CCA notion. Considering a DCCA-secure PKE (or detectable encryption) suggests a new way to build CCA-secure encryption scheme. Their results show that one can construct a CCA-secure PKE scheme by applying nested encryption techniques on three primitives that are DCCA-secure, 1-bounded CCA-secure, and CPA-secure respectively.

A detectable encryption scheme is defined by $\Pi = (\mathsf{KeyGen}, \mathsf{Enc}, \mathsf{Dec}, F)$, where KeyGen, Enc, and Dec behave as those in traditional encryption schemes, but with an additional efficient boolean function $F()$ available, which is designed to detect "dangerous" ciphertext. Specifically, $F()$ will be applied before any decryption query in Phase 2 of the original CCA game. When the queried ciphertext C "is related to" the challenge ciphertext C^*, meaning that adversary can infer "useful" information about C^* from the decryption query of C, $F()$ will return 1 and the query is rejected; else the decryption result of C will be returned to the adversary. Definition 2 formally describes the syntax of a detectable encryption scheme.

Definition 2 (Detectable Encryption). *A detectable encryption scheme consists of the following PPT algorithms* $(\mathsf{KeyGen}, \mathsf{Enc}, \mathsf{Dec}, F)$.

- $\mathsf{KeyGen}, \mathsf{Enc}, \mathsf{Dec}$ *are defined as those in a regular PKE scheme.*
- $\{0, 1\} \leftarrow F(EK, C, C^*)$*: The detecting function* F *takes as inputs a public key* EK *and two ciphertexts* C *and* C^**, and outputs 1 if* C *and* C^* *has some relations, else outputs 0.*

The definition of $F()$ above is at its full generality. We may omit the input of EK from $F()$ when the function $F()$ does not need it.

Correctness is defined as in a regular encryption scheme. A DCCA-secure scheme must satisfy unpredictability for F and indistinguishability under DCCA.

Unpredictability of the Detecting Function [8]. Intuitively, it is hard for the adversary to find a useful ciphertext C, given the detectable function $F()$ and a public key EK. This is formally defined via the game $Exp^{\mathsf{unp}}_{\mathcal{A},\Pi}(1^\lambda)$ for a detectable scheme $\Pi = (\mathsf{KeyGen}, \mathsf{Enc}, \mathsf{Dec}, F)$ played by an adversary $\mathcal{A}$.

- *Setup*: The challenger $\mathcal{C}$ takes a security parameter 1^λ and runs KeyGen to output keys (EK, DK). It gives EK to $\mathcal{A}$, and keeps DK to itself.
- *Query*: $\mathcal{A}$ can fully access the decryption oracle $\mathsf{Dec}(DK, \cdot)$. When $\mathcal{A}$ concludes the query phase, it outputs a message m and a ciphertext C.
- *Challenge*: The challenger $\mathcal{C}$ outputs a ciphertext $C^* \leftarrow \mathsf{Enc}(EK, m)$.
- *Output*: The experiment outputs $F(EK, C, C^*)$.

A detectable encryption scheme Π is said to have unpredictability for F if, for any PPT adversary $\mathcal{A}$, we have $\Pr[Exp_{\mathcal{A},\Pi}^{\mathsf{unp}} = 1] \leq \mathsf{negl}(\lambda)$.

One can formulate a stronger version of the above game, in which the adversary is given the decryption key instead of the oracle [8]. This implies the basic version of undetectability since the adversary can simulate the decryption oracle when given DK.

Indistinguishability under DCCA [8]. Now we formalize the confidentiality guarantee according to the following experiment $Exp_{\mathcal{A},\Pi}^{\mathsf{dcca}}(1^\lambda)$:

- *Setup*: The challenger $\mathcal{C}$ takes a security parameter 1^λ and runs KeyGen to output keys (EK, DK). It gives $\mathcal{A}$ EK, and keeps DK to itself.
- *Query Phase 1*: $\mathcal{A}$ is given full access to the decryption oracle $\mathsf{Dec}(DK, \cdot)$. When the adversary $\mathcal{A}$ decides that the query phase ends, it outputs messages m_0, m_1 of the same length.
- *Challenge*: The challenger $\mathcal{C}$ randomly picks a bit $b \leftarrow \{0, 1\}$, computes $C^* \leftarrow \mathsf{Enc}(EK, m_b)$ and sends C^* to $\mathcal{A}$.
- *Query Phase 2*: $\mathcal{A}$ continues to have access to $\mathsf{Dec}(DK, \cdot)$, but is not allowed to issue a decryption query such that $F(EK, C, C^*) = 1$.
- *Output*: $\mathcal{A}$ wins the game and the experiment outputs 1 if and only if $b' = b$.

A detectable encryption scheme Π is said to have indistinguishability under DCCA, if we have $\Pr[Exp_{\mathcal{A},\Pi}^{\mathsf{dcca}} = 1] \leq \frac{1}{2} + \mathsf{negl}(\lambda)$ for any PPT adversary $\mathcal{A}$.

2.3 Authenticated (Symmetric-Key) Encryption

Definition 3 (Symmetric-Key Encryption). *A symmetric-key encryption scheme $\mathcal{SKE}$ consists of the following three probabilistic polynomial-time (PPT) algorithms* (KeyGen, Enc, Dec).

- $SK \leftarrow \mathsf{KeyGen}(1^\lambda)$: *the algorithm outputs a secret key SK according to the input security parameter 1^λ.*
- $C \leftarrow \mathsf{Enc}(SK, m)$: *the algorithm takes a secret key SK and a plaintext m as inputs, and outputs a ciphertext C.*
- $m \leftarrow \mathsf{Dec}(SK, C)$: *the algorithm decrypts a ciphertext C to the corresponding plaintext m, or outputs $\perp$, by using the secret key SK.*

Confidentiality. We recall the definition of CPA security and CCA security. Consider the following experiment, $Exp^{atk}_{\mathcal{A},\mathcal{SKE}}(1^\lambda)$ for $\mathcal{SKE}$:

- *Setup*: The challenger $\mathcal{C}$ runs KeyGen(1^λ) and obtains the secret key SK.
- *Query Phase 1*: $\mathcal{A}$ is given full access to the encryption oracle Enc($SK,\cdot$) for atk = cpa, and an additional decryption oracle Dec($SK,\cdot$) for atk = cca. When the adversary $\mathcal{A}$ decides to terminate the query phase, it outputs a pair of messages m_0, m_1 of the same length.
- *Challenge*: The challenger $\mathcal{C}$ randomly picks a bit $b \leftarrow \{0,1\}$, computes $C^* \leftarrow$ Enc(SK, m_b) and sends C^* to $\mathcal{A}$.
- *Query Phase 2*: $\mathcal{A}$ continues to have access to Enc($SK,\cdot$) for atk = cpa. For atk = cca, the adversary also has access to the decryption oracle, but is not allowed to request for a decryption of C^*. Finally $\mathcal{A}$ outputs a bit b'.
- *Output*: The output of the experiment is defined to be 1 if $b' = b$, otherwise 0.

An SKE scheme $\mathcal{SKE}$ = (KeyGen, Enc, Dec) is CPA/CCA-secure if for all PPT adversaries $\mathcal{A}$, there exists a negligible function negl() such that:

$$\Pr[Exp^{atk}_{\mathcal{A},\mathcal{SKE}}(1^\lambda) = 1] \leq \frac{1}{2} + \mathsf{negl}(\lambda)$$

for atk = cpa or atk = cca respectively.

Integrity. *Integrity of ciphertexts* (INT) or *integrity of plaintexts* (INT-PTXT) is formally defined via the following experiment, $Exp^{atk}_{\mathcal{A},\mathcal{SKE}}(1^\lambda)$, played by an adversary $\mathcal{A}$ for an SKE scheme $\mathcal{SKE}$, where atk = int or atk = int-ptxt respectively.

- *Setup*: The challenger $\mathcal{C}$ runs KeyGen(1^λ) and obtains the secret key SK.
- *Query*: $\mathcal{A}$ is given full access to the encryption oracle Enc($SK,\cdot$) and the decryption oracle Dec($SK,\cdot$).
- *Challenge*: When the adversary $\mathcal{A}$ decides to terminate the query phase, it outputs a forgery C^*.
- *Output*: The challenger $\mathcal{C}$ decrypts C^* to obtain M^*.
  - For atk = int, if $M^* \neq \perp$ and C^* has never appeared as a response by the challenger to any encryption oracle query of $\mathcal{A}$, $\mathcal{A}$ is considered to have won the game, and the experiment outputs 1; otherwise, outputs 0.
  - For atk = int-ptxt, if $M^* \neq \perp$ and M^* has never appeared in any encryption oracle query, $\mathcal{A}$ is considered to have won the game, and the experiment outputs 1; otherwise, outputs 0.

An SKE scheme $\mathcal{SKE}$ = (KeyGen, Enc, Dec) is INT-secure/INT-PTXT-secure if for all PPT adversaries $\mathcal{A}$, there exists a negligible function negl() such that:

$$\Pr[Exp^{atk}_{\mathcal{A},\mathcal{SKE}}(1^\lambda) = 1] \leq \mathsf{negl}(\lambda)$$

for atk = int or atk = int-ptxt respectively.

An INT-secure scheme is also INT-PTXT-secure [1].

3   Combiner from CCA Security to DCCA Security

Given two public-key encryption schemes, $\mathcal{PKE}_1$ and $\mathcal{PKE}_2$ where $\mathcal{PKE}_1 = (\mathsf{KeyGen}_1, \mathsf{Enc}_1, \mathsf{Dec}_1)$, $\mathcal{PKE}_2 = (\mathsf{KeyGen}_2, \mathsf{Enc}_2, \mathsf{Dec}_2)$ such that only one of them is CCA-secure, we can build a detectable public-key encryption scheme $\mathcal{PKE} = (\mathsf{KeyGen}, \mathsf{Enc}, \mathsf{Dec}, F)$ that achieves DCCA security.

- $(EK, DK) \leftarrow \mathsf{KeyGen}(1^\lambda)$: $\mathsf{KeyGen}(1^\lambda)$ executes the key generation algorithm of $\mathcal{PKE}_1$ and $\mathcal{PKE}_2$: $(EK_1, DK_1) \leftarrow \mathcal{PKE}_1.\mathsf{KeyGen}_1(1^\lambda)$, $(EK_2, DK_2) \leftarrow \mathcal{PKE}_2.\mathsf{KeyGen}_2(1^\lambda)$; and outputs $EK = (EK_1, EK_2)$ and $DK = (DK_1, DK_2)$.
- $C \leftarrow \mathsf{Enc}(m)$: this algorithm chooses a random string r which is as long as the message m, and sets $C = (C_1, C_2) = (\mathcal{PKE}_1.\mathsf{Enc}_1(r), \mathcal{PKE}_2.\mathsf{Enc}_2(r \oplus m))$.
- $m \leftarrow \mathsf{Dec}(DK, C)$: this algorithm returns $m = \mathsf{Dec}_{DK_1}(C_1) \oplus \mathsf{Dec}_{DK_2}(C_2)$.
- $\{0, 1\} \leftarrow F(EK, C, C^*)$: Let $C^* = (C_1^*, C_2^*)$ be the challenge ciphertext. Similarly, parse C as (C_1, C_2). We define the detecting function $F(EK, C, C^*)$ to output 1 if and only if:

$$C_1 = C_1^* \quad \text{or} \quad C_2 = C_2^*;$$

otherwise, *i.e.*, $C_1 \neq C_1^*$ and $C_2 \neq C_2^*$, it outputs 0.

In the following, we will show that this construction achieves DCCA security.

Lemma 4. *The detecting function F satisfies unpredictability.*

Proof. Since both $\mathcal{PKE}_1, \mathcal{PKE}_2$ are probabilistic schemes, without receiving the challenge ciphertext and with no decryption key, no adversary can output a ciphertext C such that $F(EK, C, C^*) = 1$ with non-negligible probability. Thus the unpredictability of the detecting function F for the combiner scheme is satisfied. Next we prove indistinguishability of encryptions, which will then complete the proof of DCCA security. $\square$

Lemma 5. *If $\mathcal{PKE}_1$ is CCA-secure, then $\mathcal{PKE}$ is DCCA-secure.*

Proof. Since $\mathcal{PKE}$ satisfies unpredictability of the detecting function F, it suffices to show that $\mathcal{PKE}$ is indistinguishable. If there is an adversary $\mathcal{A}$ which can break the indistinguishability experiment of $\mathcal{PKE}$ with non-negligible probability ϵ, then we can construct a simulator $\mathcal{B}$ to break the CCA experiment of $\mathcal{PKE}_1$ with probability ϵ.

Given EK_1 of $\mathcal{PKE}_1$, $\mathcal{B}$ calls $\mathsf{KeyGen}_2()$ of $\mathcal{PKE}_2$, and sends $EK = (EK_1, EK_2)$ to $\mathcal{A}$. $\mathcal{B}$ can simulate the decryption oracle in Phase 1, by using that of $\mathcal{PKE}_1$ and the private decryption key DK_2.

During the challenge phase, $\mathcal{A}$ submits m_0, m_1 (of the same length) to $\mathcal{B}$. $\mathcal{B}$ chooses a randomness r_0, calculates $r_1 = r_0 \oplus m_0 \oplus m_1$, and sends r_0, r_1 to the challenger $\mathcal{C}$. Then the challenger $\mathcal{C}$ chooses $b \xleftarrow{\$} \{0, 1\}$ and sends $\mathsf{Enc}_1(r_b)$ to $\mathcal{B}$. Receiving $\mathsf{Enc}_1(r_b)$, $\mathcal{B}$ computes $\mathsf{Enc}_2(r_0 \oplus m_0)$ and sends the challenge ciphertext $C^* = (C_1^*, C_2^*)$, where $C_1^* = \mathsf{Enc}_1(r_b)$, $C_2^* = \mathsf{Enc}_2(r_0 \oplus m_0)$ to $\mathcal{A}$.

If $b = 0$, this challenge ciphertext is correctly distributed for m_0. If $b = 1$, we view r_b as $r_1 = r_0 \oplus m_0 \oplus m_1$, then r_1 is randomly distributed, and $r_0 \oplus m_0 = r_1 \oplus m_1$, so the challenge ciphertext is also correctly distributed for m_1.

In Phase 2, the adversary is only allowed to submit C to the decryption oracle if $F(EK, C, C^*) = 0$. Parsing C as (C_1, C_2) and C^* as (C_1^*, C_2^*). Under this condition, we have $C_1 \neq C_1^*$. $\mathcal{B}$ can submit C_1 to the decryption oracle of its own challenger. Thus, $\mathcal{B}$ can also properly simulate the decryption oracle for $\mathcal{A}$ in Phase 2.

Finally, $\mathcal{A}$ outputs its guess b'. $\mathcal{B}$ simply forwards b' as its own guess. When $\mathcal{A}$ guesses it is $b' = 0$, it means $C_1^* = \mathsf{Enc}_1(r_0)$. When $\mathcal{A}$ guesses it is $b' = 1$, it means $C_1^* = \mathsf{Enc}_1(r_1)$. The messages just match with those chosen by $\mathcal{B}$ in its own game. Hence, the probability of $\mathcal{B}$ to win its game is also ϵ. $\qquad\square$

Since our encryption scheme does not look symmetric regarding the choice of $\mathcal{PKE}_1, \mathcal{PKE}_2$ (versus the ordering of $\mathcal{PKE}_2$ then $\mathcal{PKE}_1$), we will now consider the case that $\mathcal{PKE}_2$ is CCA-secure.

Lemma 6. *If $\mathcal{PKE}_2$ is CCA-secure, then $\mathcal{PKE}$ is DCCA-secure.*

Proof. In this case, the challenger $\mathcal{C}$ is for $\mathcal{PKE}_2$. Our goal is to construct a simulator $\mathcal{B}$ to win the CCA experiment of $\mathcal{PKE}_2$ with the help of adversary $\mathcal{A}$.

Given EK_2 of $\mathcal{PKE}_2$, $\mathcal{B}$ calls $\mathsf{KeyGen}_1()$ of $\mathcal{PKE}_1$, and sends $EK = (EK_1, EK_2)$ to $\mathcal{A}$. Simulation of the decryption oracle in Phase 1 is similar to the corresponding treatment in the previous proof.

In the challenge phase, $\mathcal{A}$ submits m_0, m_1 (of the same length) to $\mathcal{B}$. $\mathcal{B}$ choose a randomness r, calculates $r_0 = r \oplus m_0$, $r_1 = r \oplus m_1$, and sends r_0, r_1 to the challenger $\mathcal{C}$. Then the challenger $\mathcal{C}$ chooses $b \xleftarrow{\$} \{0, 1\}$ and sends $\mathsf{Enc}_2(r_b)$ to $\mathcal{B}$. Receiving $\mathsf{Enc}_2(r_b)$, $\mathcal{B}$ computes $\mathsf{Enc}_1(r)$ and sends the challenge ciphertext $C^* = (C_1^*, C_2^*)$, where $C_1^* = \mathsf{Enc}_1(r)$, $C_2^* = \mathsf{Enc}_2(r_b)$ to the adversary $\mathcal{A}$. The challenge ciphertext is correctly distributed since $r_b = r \oplus m_b$ for $b \in \{0, 1\}$.

In Phase 2, the adversary is only allowed to query the decryption for C where $F(EK, C, C^*) = 0$. Parsing C as (C_1, C_2) and C^* as (C_1^*, C_2^*), we thus have $C_2 \neq C_2^*$. $\mathcal{B}$ can then submit C_2 to the decryption oracle in its own game.

Finally, $\mathcal{B}$ outputs what $\mathcal{A}$ outputs. Similar to our analysis of the distribution of the challenge ciphertext, the guess of $\mathcal{A}$ just matches with the guess of $\mathcal{B}$. Therefore $\mathcal{B}$ can win with probability ϵ. $\qquad\square$

Combining Lemmas 4, 5 and 6, we can see that $\mathcal{PKE}$ is DCCA-secure if one of $\mathcal{PKE}_1, \mathcal{PKE}_2$ is CCA-secure. We can then construct a CCA-secure scheme by using the nested encryption technique [8], which we review in Appendix.

4   Combiner for Secret Key Encryption

Similar to our analysis above, we can get a combiner for SKE if the nested encryption scheme of Hohenberger *et al.* [8] also applied on SKE. However, apart from the two component schemes, it also requires a 1-bounded CCA-secure scheme and another CPA-secure scheme. This appears to be complicated for an SKE scheme. We thus give a direct CCA-combiner for SKE below.

One of the advantages that SKE possesses over PKE is that the encryption needs the secret key, so that an adversary might not easily obtain by itself a ciphertext which decrypts to a valid message. This property is called integrity (INT). An SKE which is both CPA-secure and INT-secure is called authenticated encryption, and is CCA-secure [1]. Below we aim to propose a combiner for authenticated encryption, which achieves both CPA and INT security.

We require the ciphertext produced by $\mathsf{Enc}(SK, m; r)$ of the $\mathcal{SKE}$ schemes, is in the form of $C = (r, \mathsf{E}_{SK}(m, r))$ where r is the randomness used in the implicitly-defined deterministic key-ed function $\mathsf{E}_{SK}(m, r)$. We also assume that they are perfectly correct, $i.e.$, $\Pr[\mathsf{Dec}(SK, \mathsf{Enc}(SK, m)) = m] = 1$, $\forall SK \leftarrow \mathsf{KeyGen}(1^\lambda)$.

Given two encryption schemes $\mathcal{SKE}_1, \mathcal{SKE}_2$, one of which is CPA+INT-secure, and both have the stated ciphertext form and perfect correctness, our combiner for CPA+INT-secure SKE $\mathcal{SKE} = (\mathsf{KeyGen}, \mathsf{Enc}, \mathsf{Dec})$ is as follows.

- $\mathsf{KeyGen}(1^\lambda)$: takes the security parameter 1^λ, sets the $SK = (SK_1, SK_2)$, where SK_i is the secret key generated by $\mathcal{SKE}_i.\mathsf{KeyGen}_i(1^\lambda)$ for $i \in \{1, 2\}$.
- $\mathsf{Enc}(SK, m)$: chooses randomness R_1, R_2 such that $R_1 \oplus R_2 = m$ and sets

$$C_1 = (r_1, \mathsf{E}_1(R_1, r_1))$$
$$C_2 = (r_2, \mathsf{E}_2(R_2, r_2))$$
$$C_3 = (r_3, \mathsf{E}_1(C_1\|C_2\|r_4, r_3))$$
$$C_4 = (r_4, \mathsf{E}_2(C_2\|C_1\|r_3, r_4))$$

where r_1, r_2, r_3, r_4 are the randomness used in the corresponding encryption algorithm. Outputs $C = (C_1, C_2, C_3, C_4)$.
- $\mathsf{Dec}(SK, C)$: firstly parses C into (C_1, C_2, C_3, C_4), checks if both of $C_3 = (r_3, \mathsf{E}_1(C_1\|C_2\|r_4, r_3))$ and $C_4 = (r_4, \mathsf{E}_2(C_2\|C_1\|r_3, r_4))$ hold. If no, outputs $\bot$. Otherwise, gets $R_1' = \mathsf{Dec}(SK_1, C_1)$ and $R_2' = \mathsf{Dec}(SK_2, C_2)$. If none of them is $\bot$, returns $R_1' \oplus R_2'$.

The two lemmas below assert the security of this combiner scheme $\mathcal{SKE}$.

Lemma 7. $\mathcal{SKE}$ is CPA-secure.

Proof. Since $\mathcal{SKE}$ is symmetric for $\mathcal{SKE}_1$ and $\mathcal{SKE}_2$, without loss of the generality, we suppose $\mathcal{SKE}_1$ is CPA+INT-secure and $\mathcal{SKE}_2$ is an arbitrary encryption scheme.

If $\mathcal{SKE}$ is not CPA-secure, we can construct a simulator $\mathcal{B}$ to win the CPA game of $\mathcal{SKE}_1$. Given the encryption oracle of $\mathcal{SKE}_1$, $\mathcal{B}$ generates the parameters of $\mathcal{SKE}_2$ and runs $\mathsf{Enc}_2()$ by itself. In Phase 1, when $\mathcal{A}$ issues an encryption query for message m, $\mathcal{B}$ randomly picks R_i calls the $\mathsf{Enc}_1()$ oracle to get $C_1 = (r_1, \mathsf{E}_1(m \oplus R_i, r_1))$, then $\mathcal{B}$ randomly chooses r_2 and r_4, computes $C_2 = (r_2, \mathsf{E}_2(R_i, r_2))$ and sends $C_1\|C_2\|r_4$ to the encryption oracle. Receiving

$C_3 = (r_3, \mathsf{E}_1(C_1 \| C_2 \| r_4, r_3))$, the simulator $\mathcal{B}$ also calculates C_4 with r_3 and r_4. Note that the messages encrypted in C_1 and C_2 follows the distribution of the real scheme, and C_3 and C_4 simply follow the construction as in the real scheme.

In the challenge phase, the adversary $\mathcal{A}$ submits m_0, m_1 to $\mathcal{B}$, and $\mathcal{B}$ randomly picks R and sends $m'_0 = m_0 \oplus R$ and $m'_1 = m_1 \oplus R$ to the challenger $\mathcal{C}$. Then the challenger $\mathcal{C}$ flips a coin $b \in \{0,1\}$, encrypts m'_b to obtain $C'_b = (r_1, \mathsf{E}_1(m'_b, r_1))$, and sends it to $\mathcal{B}$. $\mathcal{B}$ then randomly picks r_2, r_4, and computes $C^*_2 = (r_2, \mathsf{E}_2(R, r_2))$ and submits $C'_b \| C^*_2 \| r_4$ to the encryption oracle. After receiving $C^*_3 = (r_3, \mathsf{E}_1(C'_b \| C^*_2 \| r_4))$, the simulator calculates $C^*_4 = (r_4, \mathsf{E}_2(C'_b \| C^*_2 \| r_3))$ and sends $(C'_b, C^*_2, C^*_3, C^*_4)$ as the challenge ciphertext for $\mathcal{A}$. It is a well-distributed challenge ciphertext, following the same analysis as that for the simulation of encryption oracle by $\mathcal{B}$.

In Phase 2, the simulator acts exactly as in Phase 1, and outputs what the adversary $\mathcal{A}$ outputs. We can see that the guess of $\mathcal{A}$ is correct if and only that of $\mathcal{B}$ is correct. Thus if $\mathcal{A}$ can guess correctly with non-negligible advantage over $\frac{1}{2}$, so does $\mathcal{B}$ in breaking the CPA security of $\mathcal{SKE}_1$. $\qquad\square$

Lemma 8. *$\mathcal{SKE}$ is INT-secure.*

Proof. Without loss of the generality, we suppose that $\mathcal{SKE}_1$ is CPA+INT-secure. If $\mathcal{SKE}$ is not INT-secure, we construct a simulator $\mathcal{B}$ that breaks the INT-security of $\mathcal{SKE}_1$.

$\mathcal{B}$ is given the encryption oracle $\mathsf{Enc}_1()$ of $\mathcal{SKE}_1$, and runs $\mathsf{Enc}_2()$ normally by itself. In Phase 1, when the adversary $\mathcal{A}$ queries for an encryption of message m, $\mathcal{B}$ randomly picks an R and calls the $\mathsf{Enc}_1()$ oracle to gets $C'_1 = (r'_1, \mathsf{E}_1(R, r'_1))$, then $\mathcal{B}$ randomly picks r'_2 and r'_4, computes $C'_2 = (r'_2, \mathsf{E}_2(M \oplus R, r'_2))$ and sends $C'_1 \| C'_2 \| r'_4$ to $\mathsf{Enc}_1()$ oracle. Receiving $C'_3 = (r'_3, \mathsf{E}_1(C'_1 \| C'_2 \| r'_4, r'_3))$, the simulator $\mathcal{B}$ also calculates C'_4 with r'_3 and r'_4. Finally $\mathcal{B}$ returns $C' = (C'_1, C'_2, C'_3, C'_4)$ as the response. Note that C'_1 and C'_3 obtained by $\mathcal{B}$ from its own encryption oracle $\mathsf{Enc}_1()$ are always directly forwarded to $\mathcal{A}$ as is.

In the challenge phase, the adversary $\mathcal{A}$ returns the forgery C^*. $\mathcal{B}$ parses $C^* = (C^*_1, C^*_2, C^*_3, C^*_4)$. If C^*_1 or C^*_3 has not been forwarded by $\mathcal{B}$ to $\mathcal{A}$ before, $\mathcal{B}$ returns it to break the INT security of $\mathcal{SKE}_1$.

Consider to the contrary that both C^*_1 and C^*_3 directly came from the response of the encryption oracle due to some queries of $\mathcal{B}$. Note that for any valid forgery, $C^*_3 = (r^*_3, \mathsf{E}_1(C^*_1 \| C^*_2 \| r^*_4, r^*_3))$, which is ensured by the validity checking of the decryption algorithm. From the perfect correctness of $\mathcal{SKE}_1$, every valid ciphertext can only be decrypted to a single message. Since C^*_3 was created from $\mathsf{Enc}_1()$, the corresponding query supplied to $\mathsf{Enc}_1()$, and hence the decryption result of C^*_3, must be $C^*_1 \| C^*_2 \| r^*_4$, Also note that every $\mathsf{Enc}_1()$ oracle query $\mathcal{B}$ has ever made is for returning a ciphertext $C' = (C'_1, C'_2, C'_3, C'_4)$ to $\mathcal{A}$. When C^*_3 was returned by $\mathsf{Enc}_1()$, it must be triggered by an encryption oracle query by $\mathcal{A}$ which leads to the creation of $C' = (C^*_1, C^*_2, C^*_3, C'_4)$, where $C'_4 = (r^*_4, \mathsf{E}_2(C^*_2 \| C^*_1 \| r^*_3, r^*_4)) = C^*_4$. So the forgery $C^* = (C^*_1, C^*_2, C^*_3, C^*_4)$ is exactly the same as C' given by $\mathcal{B}$ to $\mathcal{A}$ before, violating the rule of the game played by $\mathcal{A}$. Contradiction occurs and this concludes the proof. $\qquad\square$

5   Conclusion

We show two provably-secure robust combiners for ensuring security against chosen-ciphertext attack (CCA). Our robust combiner for public-key encryption (PKE) is inspired by the detectable chose-ciphertext attack (DCCA) notion proposed by Hohenberger *et al.* [8]. Instead of directly obtaining a combiner to get a CCA-secure PKE from two possibly CCA-secure PKE schemes, our goal is to devise a combiner for DCCA security, given that only one of the schemes is CCA-secure. A CCA-secure scheme can thus be obtained following the nested encryption approach proposed by Hohenberger *et al.* [8]. For our robust combiner for symmetric-key encryption (SKE), instead of directly working on CCA-security, we work on the CPA-security and the ciphertext integrity, in which a combination of both implies CCA-security of an SKE scheme [1].

It is our future work to build a more efficient SKE combiner.

A   CCA Security from DCCA Security

CCA-secure PKE can be obtained by combining DCCA PKE, 1-bounded CCA PKE, and CPA PKE [8]. We remark that the same technique also works in identity-based encryption (IBE), attribute-based encryption (ABE), and threshold PKE/IBE. The same holds true for our combiner in Sect. 3.

We use Π_{DCCA}, Π_{CPA}, and Π_{qb} to denote the encryption primitives which are DCCA-secure, CPA-secure, and q-bounded-CCA-secure (where $q = 1$) respectively. For a probabilistic algorithm $\mathsf{Enc}(\cdot)$, we can transform it to a deterministic one $\mathsf{Enc}(\cdot; r)$ where r is a well-distributed random value.

We describe the CCA-secure encryption scheme in the context of IBE. It can easily degenerated to SKE/PKE, or extended into threshold PKE/IBE or ABE.

A.1   Syntax of IBE

In IBE, any user can request for a secret key SK_{ID} related to her identity ID from a trusted private key generator. The secret key SK_{ID} can decrypt the ciphertext encrypted for ID correctly. An IBE scheme is defined as follows.

- $(MPK, MSK) \leftarrow \mathsf{Setup}(1^\lambda)$: This algorithm takes as the security parameter 1^λ and returns a master public key MPK and a master secret key MSK. MPK is omitted from the input of the rest of the algorithms.
- $SK_{ID} \leftarrow \mathsf{Extract}(MSK, ID)$: This algorithm takes as inputs the master security key MSK and an user identity ID, and it returns a user secret key SK_{ID}.
- $C \leftarrow \mathsf{Enc}(ID, m)$: This algorithm takes as inputs a user identity ID, and a message m, it then returns a ciphertext C encrypting m for ID.
- $m \leftarrow \mathsf{Dec}(ID, SK_{ID}, C)$: It takes as inputs a secret key SK_{ID} corresponding to the identity ID, and a ciphertext C. It returns m or an invalid symbol $\bot$.

A.2 CCA-Secure Construction

- $(MPK, MSK) \leftarrow \mathsf{Setup}(1^\lambda)$: Run all the underlying IBE setup algorithms: $\mathsf{Setup}_{\mathrm{DCCA}}(1^\lambda)$ to get $(MPK_{\mathrm{DCCA}}, MSK_{\mathrm{DCCA}})$, then $\mathsf{Setup}_{\mathrm{CPA}}(1^\lambda)$ to obtain $(MPK_{\mathrm{CPA}}, MSK_{\mathrm{CPA}})$ and $\mathsf{Setup}_{\mathrm{qb}}(1^\lambda)$ to get $(MPK_{\mathrm{qb}}, SK_{\mathrm{qb}})$. Keep $MSK = (MSK_{\mathrm{DCCA}}, MSK_{\mathrm{CPA}}, MSK_{\mathrm{qb}})$ in secret and output the master public key as $MPK = (MPK_{\mathrm{DCCA}}, MPK_{\mathrm{CPA}}, MPK_{\mathrm{qb}})$.
- $SK_{ID} \leftarrow \mathsf{Extract}(ID)$: Run $\mathsf{Extract}_{\mathrm{DCCA}}(ID)$ to obtain $SK_{\mathrm{DCCA}.ID}$, then $\mathsf{Extract}_{\mathrm{CPA}}(ID)$ to obtain $SK_{\mathrm{CPA}.ID}$, and $\mathsf{Extract}_{\mathrm{qb}}(ID)$ to obtain $SK_{\mathrm{qb}.ID}$. Finally, output $SK_{ID} = (SK_{\mathrm{DCCA}.ID}, SK_{\mathrm{CPA}.ID}, SK_{\mathrm{qb}.ID})$.
- $C \leftarrow \mathsf{Enc}(ID, m)$: First pick three random values $r_{\mathrm{DCCA}}, r_{\mathrm{CPA}}, r_{\mathrm{qb}} \in \{0,1\}^\lambda$, encrypt two of them with the message m in C_{DCCA} using r_{DCCA} as the encryption randomness, i.e., $\mathsf{Enc}_{\mathrm{DCCA}}(ID, (r_{\mathrm{CPA}}||r_{\mathrm{qb}}||m); r_{\mathrm{DCCA}})$; then compute two more encryption of it via $C_{\mathrm{qb}} = \mathsf{Enc}_{\mathrm{qb}}(ID, C_{\mathrm{DCCA}}; r_{\mathrm{qb}})$ and $C_{\mathrm{CPA}} = \mathsf{Enc}_{\mathrm{CPA}}(ID, C_{\mathrm{DCCA}}; r_{\mathrm{CPA}})$. Finally, we set $C = (C_{\mathrm{CPA}}, C_{\mathrm{qb}})$.
- $m \leftarrow \mathsf{Dec}(ID, SK_{ID}, C)$: Parse C into $(C_{\mathrm{CPA}}, C_{\mathrm{qb}})$. Decrypt the second ciphertext $\mathsf{Dec}_{\mathrm{qb}}(ID, SK_{ID}, C_{\mathrm{qb}})$ to obtain C_{DCCA}. Then decrypt it to obtain $(r_{\mathrm{CPA}}||r_{\mathrm{qb}}||m)$. Check that both $C_{\mathrm{qb}} = \mathsf{Enc}_{\mathrm{qb}}(ID, C_{\mathrm{DCCA}}; r_{\mathrm{qb}})$ and $C_{\mathrm{CPA}} = \mathsf{Enc}_{\mathrm{CPA}}(ID, C_{\mathrm{DCCA}}; r_{\mathrm{CPA}})$ holds. If so, output m; otherwise output $\bot$.

References

1. Bellare, M., Namprempre, C.: Authenticated encryption: relations among notions and analysis of the generic composition paradigm. In: Okamoto, T. (ed.) ASIACRYPT 2000. LNCS, vol. 1976, pp. 531–545. Springer, Heidelberg (2000). doi:10.1007/3-540-44448-3_41
2. Chow, S.S.M., Boyd, C., Nieto, J.M.G.: Security-mediated certificateless cryptography. In: Public Key Cryptography (PKC), pp. 508–524 (2006). http://dx.doi.org/10.1007/11745853_33
3. Chow, S.S.M., Roth, V., Rieffel, E.G.: General certificateless encryption and timed-release encryption. In: Ostrovsky, R., De Prisco, R., Visconti, I. (eds.) SCN 2008. LNCS, vol. 5229, pp. 126–143. Springer, Heidelberg (2008). doi:10.1007/978-3-540-85855-3_9
4. Cramer, R., Hanaoka, G., Hofheinz, D., Imai, H., Kiltz, E., Pass, R., Shelat, A., Vaikuntanathan, V.: Bounded CCA2-secure encryption. In: Kurosawa, K. (ed.) ASIACRYPT 2007. LNCS, vol. 4833, pp. 502–518. Springer, Heidelberg (2007). doi:10.1007/978-3-540-76900-2_31
5. Dodis, Y., Katz, J.: Chosen-ciphertext security of multiple encryption. In: Kilian, J. (ed.) TCC 2005. LNCS, vol. 3378, pp. 188–209. Springer, Heidelberg (2005). doi:10.1007/978-3-540-30576-7_11
6. Harnik, D., Kilian, J., Naor, M., Reingold, O., Rosen, A.: On robust combiners for oblivious transfer and other primitives. In: Cramer, R. (ed.) EUROCRYPT 2005. LNCS, vol. 3494, pp. 96–113. Springer, Heidelberg (2005). doi:10.1007/11426639_6
7. Herzberg, A.: Folklore, practice and theory of robust combiners. J. Comput. Secur. 17(2), 159–189 (2009). doi:10.3233/JCS-2009-0336
8. Hohenberger, S., Lewko, A., Waters, B.: Detecting dangerous queries: a new approach for chosen ciphertext security. In: Pointcheval, D., Johansson, T. (eds.) EUROCRYPT 2012. LNCS, vol. 7237, pp. 663–681. Springer, Heidelberg (2012). doi:10.1007/978-3-642-29011-4_39

Homomorphic Evaluation of Lattice-Based Symmetric Encryption Schemes

Pierre-Alain Fouque[1,3](✉), Benjamin Hadjibeyli[2], and Paul Kirchner[3]

[1] Institut Universitaire de France, Université de Rennes 1, Rennes, France
[2] École normale supérieure de Lyon, Lyon, France
Benjamin.Hadjibeyli@ens-lyon.fr
[3] École normale supérieure, Paris, France
{Pierre-Alain.Fouque,Paul.Kirchner}@ens.fr

Abstract. Optimizing performance of Fully Homomorphic Encryption (FHE) is nowadays an active trend of research in cryptography. One way of improvement is to use a hybrid construction with a classical symmetric encryption scheme to transfer encrypted data to the Cloud. This allows to reduce the bandwidth since the expansion factor of symmetric schemes (the ratio between the ciphertext and the plaintext length) is close to one, whereas for FHE schemes it is in the order of 1,000 to 1,000,000. However, such a construction requires the decryption circuit of the symmetric scheme to be easy to evaluate homomorphically. Several works have studied the cost of homomorphically evaluating classical block ciphers, and some recent works have suggested new homomorphic oriented constructions of block ciphers or stream ciphers. Since the multiplication gate of FHE schemes usually squares the noise of the ciphertext, we cannot afford too many multiplication stages in the decryption circuit. Consequently, FHE-friendly symmetric encryption schemes have a decryption circuit with small multiplication depth.

We aim at minimizing the cost of the homomorphic evaluation of the decryption of symmetric encryption schemes. To do so, we focus on schemes based on learning problems: Learning With Errors (LWE), Learning Parity with Noise (LPN) and Learning With Rounding (LWR). We show that they have lower multiplicative depth than usual block ciphers, and hence allow more FHE operations before a heavy bootstrapping becomes necessary. Moreover, some of them come with a security proof. Finally, we implement our schemes in HElib. Experimental evidence shows that they achieve lower amortized and total running time than previous performance from the literature: our schemes are from 10 to 10,000 more efficient for the time per bit and the total running time is also reduced by a factor between 20 to 10,000. Of independent interest, the security of our LWR-based scheme is related to LWE and we provide an efficient security proof that allows to take smaller parameters.

1 Introduction

Fully Homomorphic Encryption (FHE) is nowadays one of the most active trend of research in cryptography. In a nutshell, a FHE scheme is an encryption scheme

© Springer International Publishing Switzerland 2016
T.N. Dinh and M.T. Thai (Eds.): COCOON 2016, LNCS 9797, pp. 269–280, 2016.
DOI: 10.1007/978-3-319-42634-1_22

that allows evaluation of arbitrarily complex programs on encrypted data. This idea has been introduced by Rivest et al. [26] in 1978, while the first plausible construction has been given by Gentry [16] in 2009. Since, numerous papers have focused on improving the efficiency of the constructions. Even if there still remains works before FHE becomes practical, it arouses more and more interest and the scope of application goes from genomics to finance [24].

One way of improvement has been introduced in [24]. It focuses on minimizing the communication complexity of the scheme. The idea is to use a "hybrid" encryption scheme: some parts of the scheme are replaced by a symmetric encryption scheme. Instead of encrypting the data under the FHE scheme, the client will only encrypt its symmetric key under the FHE scheme, and encrypt its data under the symmetric scheme. The cloud will then homomorphically evaluate the decryption of the symmetric scheme on the symmetrically encrypted data and the homomorphically encrypted symmetric key, to get a ciphertext corresponding to a homomorphic encryption of the data. Clearly, such a construction has low communication complexity, since the only online data transfer is made under the symmetric scheme. However, the cloud might pay a huge cost at the homomorphic evaluation of the symmetric decryption. Thus, one can look for the most "FHE-friendly" symmetric encryption scheme to use in the hybrid.

Being "FHE-friendly" consists in optimizing several criteria. First, as the application we gave suggests, we want a scheme with a small expansion factor, so that the communication complexity stays low. Then, other criteria depend on the FHE construction we are building upon. All current FHE schemes are based on variants of Gentry's initial idea: ciphertext consists of encryption of data with noise, and homomorphic operations increase this noise. When the upper bound of noise is reached, one has to "bootstrap", to reduce the noise to its initial level. Typically, functions are represented as arithmetic circuits and multiplications have a far higher cost than additions in terms of noise. Thus, we will want to minimize the multiplicative depth of the decryption circuit of our symmetric scheme. In addition, we will also take into account the total running time of our homomorphic evaluation step. This metric highly depends on the chosen FHE scheme, but multiplications often happen to be the main bottleneck again.

Our Contributions. In this paper, we focus on symmetric schemes having shallow decryption circuits. We build secure schemes with constant or small decryption circuit, namely with small multiplication depth. Contrary to the direction followed by many recent work, that tweak block ciphers or stream ciphers [3,9], our approach is related to provable security. Indeed, we notice that one can construct lattice-based schemes with very small decryption circuit and then, we evaluate the performances of our schemes using HElib to compare them with other symmetric ciphers. Finally, we try to use HElib features (full packing and parallelization) in order to achieve better performances. We describe two kinds of ciphers: the first family has its security related to the difficulty of solving the LPN problem in specific instances, while the second family has a *security proof* based on the LWE problem. The first construction is similar to "symmetric cryptography" since we do not have a clean security proof and consequently, we provide a more thorough security analysis. However, the security seems to be easier to understand than ad-hoc constructions usually used in symmetric cryptography, since the security problem on which the scheme is based can be formally stated. We present a very efficient construction specifically tailored to this problem to secure our construction from Arora-Ge type of attack on LPN. The performance of the schemes from this family can be 10 times more efficient than the most efficient previous cipher. For the second family, we have a rigorous security proof related to LWE, while the scheme is based on LWR. The performance of the second family can be very efficient, about 10,000 times faster, but the caveat is that the decrypted plaintext contains random bits in the least significant bits if we do not compute homomorphically the truncation using the costly ExtractDigits function. Therefore, if we want to remove the erroneous bits, the performances become equivalent to previous ciphers, while being more efficient than AES. In some cases, we can compute with such noise.

We notice that contrary to what is claimed in many works [24], it is not necessary to re-encrypt the symmetrically-encrypted ciphertext using the FHE scheme when the server receives the data. We show that the evaluation of the homomorphic decryption procedure gives ciphertexts encrypted with FHE. This improves the performance of the scheme, since we homomorphically evaluate the function that maps the key K to the $\mathsf{Dec}(K, c)$, given the ciphertext c and some multiplications in Dec will be simplified once c is known.

Then, we describe our efficient FHE-friendly symmetric schemes based on lattices, and more precisely on learning problems. Our results show that we can get circuits with very small multiplication depth for the decryption algorithms of these schemes. In addition, their security relies on hard problems or on hard instances of lattice problems in the worst cases, as opposed to usual block ciphers.

We present a scheme whose security is based on the Learning Parity With Noise problem (LPN) introduced in [18]. We have to specify an error correcting code (ECC) for this scheme so that the decryption circuit is small. We choose to use a repetition code in order to simplify the decoding and reduce its circuit in term of multiplications. More complex ECC exist with constant decoding such as [19] but they are only interesting from an asymptotic point of view. However, prohibiting decryption failures makes the scheme vulnerable to the Arora-Ge [6] attack and to avoid its most efficient variant [2] using Gröbner basis algorithms, we use a very efficient transformation, similar to random locally function [5], which increases the algebraic degree of the polynomials system. We provide a detailed analysis of this attack. The function we propose is also very similar to [1] and we can show that our construction achieves better influence parameters, but it has higher complexity class since we need a logarithmic depth circuit.

Then, we introduce another scheme whose security is based on the Learning With Rounding problem (LWR) and a very similar version whose security relies directly on the Learning With Errors (LWE) problem. In order to encrypt many bits using small parameters, we provide a direct proof from LWE to the security of the scheme. We do not rely on any reduction from LWE to LWR since the first reduction given by Banerjee et al. [7] requires exponential parameters and the one by Alwen et al. [4] requires parameter linear in the number of samples. Here, our reduction is only logarithmic in the number of samples. Furthermore, we extend both schemes to their ring versions. In this case, we optimized the number of multiplications using a FFT algorithm to compute the polynomial multiplications. Finally, we extend them to their module versions, which generalizes standard and ring versions.

Along with a theoretical analysis, we give a homomorphic evaluation in HElib to make practical comparisons. While the homomorphic evaluation of AES went down to 11 ms per bit [17] and LowMC, a block cipher designed to be FHE-friendly (and whose security has recently been analyzed [11,12]), went down to 3 ms per bit [3], which was the best so far, we go under a millisecond per bit (with the module version of our LPN scheme). In some scenario, our performance for the scheme based on LWR are drastically better if we allow FHE-encrypted plaintexts to contain noise in the least significant bits. Moreover, our schemes are a lot more flexible, in the sense that they need smaller FHE parameters, and while these performance were amortized over a computation taking several minutes, the evaluation of our schemes takes only from a second to a minute.

Related Work. Many papers have presented homomorphic evaluations of block ciphers. It has started in [17], where AES has been chosen as a benchmark for measuring the performance of HElib. Then, performance has been improved in [23]. AES has then been used as benchmark for comparing FHE schemes in [10,13]. Similarly, Simon has been used to compare FHE schemes [21]. Recently, the problem has been taken the other way round, with works trying to find the most FHE-friendly block cipher. First, a lightweight block cipher like Prince has been suggested and evaluated [14]. Then, a new block cipher, LowMC, has been designed specifically for this kind of application [3], as well as for multiparty computations. Finally, using stream ciphers has also been proposed [9].

Organization of the Paper. In Sect. 2 we recall definitions about symmetric encryption and Lattice problems in Cryptography. In Sect. 3, we explain how we use homomorphic operation more efficiently. Then we introduce in Sect. 4 our symmetric schemes based on learning problems: LPN, LWR and LWE. The security and performance analysis of the schemes are proved in the final version.

2   Preliminaries

Symmetric Encryption. We will say that a function of k (from positive integers to positive real numbers) is negligible if it approaches zero faster than any inverse polynomial, and noticeable if it is larger than some inverse polynomial (for infinitely many values of k).

Definition 1. *A symmetric encryption scheme is a tuple* (Gen, Enc, Dec) *of Probabilistic Polynomial-time (PPT) algorithms as follows:*

- Gen(1^λ): *given a security parameter λ, output a secret key k;*
- Enc(k, m): *given a key k and a message m, output a ciphertext c;*
- Dec(k, c): *given a key k and a ciphertext c, output a message m';*

and which satisfies the correctness property: if $k := \mathsf{Gen}(1^\lambda)$, then for all messages m, $\Pr[\mathsf{Dec}(k, \mathsf{Enc}(k, m)) \neq m]$ is negligible (in λ).

For the sake of clarity, we will often write the key as a subscript and the scheme name as a superscript of our algorithms, like in Enc_k^S. Semantic security is implied by the following property, which will be satisfied by our schemes.

Definition 2. *A symmetric encryption scheme S has pseudo-random ciphertexts (ciphertexts indistinguishable from random) if no PPT $\mathcal{A}$ can distinguish between ciphertexts from the scheme and the uniform distribution, i.e. for all PPT $\mathcal{A}$ and for all messages m, it holds that $\big| \Pr[\mathcal{A}(\mathsf{Enc}_k^S(m), 1^\lambda) = 1] - \Pr[\mathcal{A}(r, 1^\lambda) = 1] \big|$ is negligible (in λ), where r is drawn randomly over the ciphertext space.*

Learning Problems. Given a finite set S and a probability distribution D on S, $s \leftarrow D$ denotes the drawing of an element of S according to D and $s \leftarrow S$ the random drawing of an element of S endowed with uniform probability.

Learning with Errors. The Gaussian distribution with standard deviation σ is defined on $\mathbb{R}$ by the density function $\frac{1}{\sqrt{2\pi}\sigma} \exp(-\frac{1}{2}(\frac{x}{\sigma})^2)$. The Learning With Errors problem (LWE) has been introduced in [25]. For $s \in \mathbb{Z}_q^k$, the LWE distribution $D_{s,\chi}^{LWE}$ is defined over $\mathbb{Z}_q^k \times \mathbb{Z}_q$ and consists in samples $(a, \langle a, s \rangle + e)$ where $a \leftarrow \mathbb{Z}_q^k$ and $e \leftarrow \chi$ for some distribution χ over $\mathbb{Z}_q$. Typically, χ is taken to be some integral Gaussian distribution when assuming that LWE is hard. As in most works [25], we will consider here rounded Gaussian distributions: it basically consists in sampling a Gaussian distribution, reducing the result modulo 1, multiplying it by q and rounding it to the nearest integer. LWE consists, for s chosen according to some distribution over $\mathbb{Z}_q^k$ (typically, the uniform distribution), in distinguishing between any desired number of samples from $D_{s,\chi}^{LWE}$ and the same number of samples drawn from the uniform distribution over $\mathbb{Z}_q^k \times \mathbb{Z}_q$. For rounded Gaussian distributions, LWE is usually considered to be hard when the standard deviation σ verifies $\sigma > \sqrt{k}$ [25].

LWE can be extended into a ring version RLWE [22]. Let $R = \mathbb{Z}[X]/(P(X))$ for a monic irreducible polynomial P of degree k, and let $R_q = R/qR$. Generally, P is chosen to be some power-of-two cyclotomic polynomial, which are of the form $X^{2^z} + 1$. For an element $s \in R_q$, we define the RLWE distribution $D_{s,\chi}^{RLWE}$ over $R_q \times R_q$ by samples $(a, a.s + e)$ where $a \leftarrow R_q$ and $e \leftarrow \chi^k$ where χ^k consists in k independent samples from χ and e is interpreted as an element of R_q. The Ring-LWE problem (RLWE) consists, for s drawn according to some distribution over R_q, in distinguishing $D_{s,\chi}^{RLWE}$ from the uniform distribution over $R_q \times R_q$.

We will also use the Module-LWE problem (MLWE). It has been introduced in [8] under the name of GLWE, for General LWE. However, we will call it MLWE as in [20], because it indeed corresponds to introducing a module structure over LWE. For an element $s \in R_q^k$, where the underlying ring polynomial has degree d, we define the MLWE distribution $D_{s,\chi}^{MLWE}$ over $R_q^k \times R_q$ by samples $(a, \langle a.s \rangle + e)$ where $a \leftarrow R_q^k$ and $e \leftarrow \chi^d$ is interpreted as an element of R_q. MLWE generalizes LWE and RLWE: LWE is when $d = 1$ and RLWE when $k = 1$.

By a standard hybrid argument, LWE can be extended to several secrets. It can be shown that the problem which consists in distinguishing samples $(a, \langle a, s_1 \rangle + e_1, \ldots, \langle a, s_n \rangle + e_n) \in \mathbb{Z}_q^k \times \mathbb{Z}_q^n$ from the uniform distribution over $\mathbb{Z}_q^k \times \mathbb{Z}_q^n$, where each $s_j \in \mathbb{Z}_q^k$ is chosen independently for any $n = poly(k)$, is at least as hard as LWE for a single secret s. An analogous statement can be shown for RLWE and MLWE. Finally, the LWE [25], RLWE [22] and MLWE [20] hardness assumptions have been reduced to standard lattice assumptions. The security of MLWE seems to be intermediate between that of LWE based on hardness results in arbitrary lattices and the security of RLWE in ideal lattices.

Learning Parity with Noise (LPN). We denote by $\mathcal{B}_\eta$ the Bernoulli distribution of parameter $\eta \in [0, 1]$, i.e. a bit $b \leftarrow \mathcal{B}_\eta$ is chosen such that $\Pr[b = 1] = \eta$ and $\Pr[b = 0] = 1 - \eta$. The LPN problem consists in LWE for $q = 2$. The distribution χ chosen over $\mathbb{Z}_2$ corresponds to a Bernoulli distribution. We extend LPN to RLPN and MPLN. The only difference is that the underlying polynomial will not be cyclotomic anymore, but some irreducible polynomial modulo 2. Similarly, these problems are also extended to a polynomial number of secrets. The main difference between LWE and LPN is that the security of LPN remains heuristic because no reduction has been made so far to lattice problems.

Learning with Rounding (LWR). The LWR problem has been introduced in [7] as a derandomization of LWE. The idea is to replace the addition of a random noise by a rounding function. Let k be the security parameter and moduli $q \geq p \geq 2$ be integers. We define the function $\lfloor . \rceil_p : \mathbb{Z}_q \rightarrow \mathbb{Z}_p$ by $\lfloor x \rceil_p = \lfloor (p/q).\bar{x} \rceil$, where $\bar{x}$ is an integer congruent to $x \bmod q$. We extend $\lfloor . \rceil_p$ component-wise to vectors and matrices over $\mathbb{Z}_q$. Let R denote the cyclotomic polynomial ring $R = \mathbb{Z}[z]/(z^k + 1)$ for k a power of two. For any modulus q, we define the quotient ring $R_q = R/qR$ and extend $\lfloor . \rceil_p$ coefficient-wise to it. Note that we can use any common rounding method, like the floor or ceiling functions. In our implementations, we use the floor, because it is equivalent to dropping the least-significant digits in base 2 when q and p are both powers of 2.

For a vector $s \in \mathbb{Z}_q^k$, the LWR distribution D_s^{LWR} is defined over $\mathbb{Z}_q^k \times \mathbb{Z}_p$ by elements $(a, \lfloor \langle a, s \rangle \rceil_p)$ with $a \leftarrow \mathbb{Z}_q^k$. For a vector $s \in R_q$, the ring-LWR (RLWR) distribution D_s^{RLWR} is defined over $R_q \times R_p$ by elements $(a, \lfloor a.s \rceil_p)$ with $a \leftarrow R_q$. And for a vector $s \in R_q^k$, the module-LWR (MLWR) distribution D_s^{MLWR} is defined over $R_q^k \times R_p$ by elements $(a, \lfloor \langle a.s \rangle \rceil_p)$ with $a \leftarrow R_q^k$. For a given distribution D over $s \in \mathbb{Z}_q^k$, LWR consists in distinguishing between any desired number of independent samples from D_s^{LWR} and the same number of samples drawn uniformly and independently from $\mathbb{Z}_q^k \times \mathbb{Z}_p$. RLWR and MLWR are defined analogously. All these problems can be extended to several secrets, as stated for LWE. The LWR has been reduced to LWE when q/p is exponential in k [7], and when q/p is poly(k) and linear in the number of samples by [4].

3   Fully-Homomorphic Encryption (FHE)

While classical encryption preserves the privacy of information, homomorphic encryption aims also at making some computation on the encrypted data.

FHE Definitions. Formally, we have a message space M with a set of functions f we would like to compute on messages, and we want an algorithm which efficiently computes functions f' on the ciphertext space C such that $\mathsf{Dec}(f'(\{c_i\}_i)) = f(\{\mathsf{Dec}(c_i)\}_i)$. Thus, we want the decryption function to be a homomorphism from C to M for these functions f. This notion, originally called a privacy homomorphism, was introduced in [26]. Here is a formal definition of a homomorphic scheme, sometimes referred to as "somewhat homomorphism".

Definition 3. *Let $\mathcal{F}$ be a set of functions. A $\mathcal{F}$-homomorphic encryption (HE) scheme is a tuple of PPT algorithms* $(\mathsf{Gen}, \mathsf{Encrypt}, \mathsf{Decrypt}, \mathsf{Eval})$ *as follows:*

- $\mathsf{Gen}(1^\lambda)$: *given a security parameter λ, output a public key pk, a secret key sk and an evaluation key ek;*
- $\mathsf{Enc}(pk, m)$: *given the public key pk and a message m, output a ciphertext c;*
- $\mathsf{Dec}(sk, c)$: *given the secret key sk and a ciphertext c, output a message m';*
- $\mathsf{Eval}(ek, f, \Psi = (c_1, \ldots, c_l))$: *given the evaluation key, a function f and a tuple Ψ of l ciphertexts, where l is the arity of f, output a ciphertext c';*

satisfying the correctness property: for all functions $f \in \mathcal{F}$ and messages $\{m_i\}_{i \leq l}$, where l is the arity of f, if $(pk, sk, ek) := \mathsf{Gen}(1^\lambda)$ and $c_i := \mathsf{Enc}(pk, m_i)$ for all i, then $\Pr[\mathsf{Dec}(sk, \mathsf{Eval}(ek, f, (c_1, \ldots, c_l))) \neq f(m_1, \ldots, m_l)]$ is negligible (in λ).

Homomorphic Evaluation of Symmetric Encryption Schemes. We now give a more precise description of the scenario where a symmetric encryption scheme is used to improve FHE performance, as described in [24], and on which we will to rely to analyse the performance of our schemes.

Optimizing Communication with the Cloud. Consider the setting where a client uploads its data encrypted under a FHE scheme on a cloud service and wants the cloud to compute on this data and return encrypted outputs. Typically, FHE schemes come with an expansion factor of the order of 1,000 to 1,000,000. To mitigate this problem, the client will send its data encrypted under some semantically secure symmetric encryption scheme (which, by itself, is not homomorphic at all) along with the homomorphic encryption of its symmetric key. Then, the steps of symmetric decryption can all be carried out on homomorphically encrypted entries. Thus, the cloud can obtain the data encrypted under the FHE scheme by homomorphically evaluating the decryption circuit.

Here is a formal description of the protocol. Let $H = (\mathsf{Gen}^H, \mathsf{Enc}^H, \mathsf{Dec}^H, \mathsf{Eval}^H)$ be a FHE scheme and let $S = (\mathsf{Gen}^S, \mathsf{Enc}^S, \mathsf{Dec}^S)$ be a symmetric encryption scheme. Let λ be the security parameter and m be the data the client wants to send to the cloud. Let $(pk, sk, ek) := \mathsf{Gen}^H(1^\lambda)$ and $k := \mathsf{Gen}^S(1^\lambda)$.

- The client sends messages $c_1 := \mathsf{Enc}^H_{pk}(k)$ and $c_2 := \mathsf{Enc}^S_k(m)$ to the cloud.
- Given a couple of ciphertexts (c_1, c_2) received from the client, the cloud computes (either at the reception or just before further computing) $x = \mathsf{Enc}^H_{pk}(c_2)$ and then $c = \mathsf{Eval}_{ek}(\mathsf{Dec}^S, c_1, x)$: this is why we need an efficient homomorphic evaluation of the decryption circuit of our symmetric scheme.

– Now, the cloud possesses a FHE-encrypted ciphertext c, which means that $\mathsf{Dec}_{sk}^{H}(c) = m$. Furthermore, it can now homomorphically evaluate any function f: for all f, $\mathsf{Dec}_{sk}^{H}(\mathsf{Eval}_{ek}^{H}(f, c)) = f(c)$.

Indeed, if the evaluation algorithm allows constant arguments, i.e. arguments which are not homomorphically encrypted, this scenario can be optimized further, simply by noticing that c_2 does not have to be homomorphically encrypted. Thus, when receiving c_1 and c_2, the cloud will just directly compute $c = \mathsf{Eval}_{ek}(\mathsf{Dec}^{S}, c_2, c_1)$. It still has to homomorphically evaluate the decryption circuit of the symmetric scheme, but it saves a homomorphic encryption, and operations with constants might be faster. This can also be seen as evaluating the function $K \mapsto Dec^{S}(K, m)$, which depends on m.

All the symmetric encryption schemes we will use are based on lattices, and, more precisely, on learning problems. Some of them will rely on the LPN problem, while the others will rely on the LWR or on the LWE problem. Our initial goal is to construct efficient FHE-friendly encryption schemes. Symmetric encryptions are used in FHE scenario in order to transfer the cloud. Here, we first describe a much more efficient scenario for symmetric and homomorphic encryption than the classical scenario described in [24].

4   FHE-Friendly Symmetric Encryption Based on Learning

An Encryption Scheme Based on MLPN. Our first encryption scheme is a generalization of the scheme introduced in [18], under the name of LPN − C by Gilbert *et al.* A $[n, m, d]$ linear binary (error-correcting) code C is a linear subspace of $\mathbb{F}_2^n$ with dimension m such that d is the minimum ℓ^1 distance between two elements of the code. We associate it with an encoding function $E : \mathbb{F}_2^m \rightarrow C$ and a decoding function $D : C \rightarrow \mathbb{F}_2^m$. LPN − C is a symmetric encryption scheme whose security can be reduced to the hardness of LPN. Let E and D be respectively the encoding and decoding functions of a $[n, m, d]$ linear binary code.

Here, we describe the more general version of our scheme. Similarly, we can define LPN − C and RLPN − C based on LPN and RLPN problems.

Definition 4 (MLPN − C). *Let d, k and n be polynomials in λ. Let consider $\mathbb{F}_{2^d}$ a finite field defined by an irreducible polynomial P of degree d. The symmetric encryption scheme MLPN − C is defined as follows: $\mathsf{Gen}(1^\lambda)$: output $S \leftarrow \mathbb{F}_{2^d}^{k \times n}$; $\mathsf{Enc}_S(x)$: output $(a, E(x) \oplus a.S \oplus e)$, where $a \leftarrow \mathbb{F}_{2^d}^k$ and $e \leftarrow \mathcal{B}_\eta^{d \times n}$ is interpreted as an element of $\mathbb{F}_{2^d}^n$; and $\mathsf{Dec}_S(a, y)$: output $D(y \oplus a.S)$.*

For a message of m bits, this scheme produces a ciphertext of $n + k$ bits. Indeed, the expansion factor can tend to the one of the linear code we are using, which is n/m, since n is any polynomial in k. Furthermore, one can consider that a does not have to be sent, and can be replaced, for example, by the seed used in order to generate it.

We choose the 3-repetition code to have a small multiplication depth circuit of degree 2. We define the encoding scheme for $a \in \mathbb{F}_{2^d}$ as $(a^{2^{d-1}}, a^{2^{d-1}}, a^{2^{d-1}})$. In order to decode a code word $(a, b, c) \in (\mathbb{F}_{2^d})^3$, we compute $ab + bc + ac$. (The normal encoding with would be (a, a, a) and the decoding $(ab + ac + bc)^{2^{d-1}}$, but we prefer to incorporate the power 2^{d-1} in the encoding in order to make the homomorphic part more efficient.)

Proposition 1 [18]. LPN − C *(resp.* RLPN − C, MLPN − C*) is semantically secure as soon as the corresponding LPN (resp. RLPN, MLPN) problem is hard.*

As it stands, if we do not bound the number of errors sent along with a message, this scheme will produce decryption failures. They will happen when the Hamming weight of the noise vector e is greater than the correction capacity of the error-correcting code. We study the probability of decryption failures and we can choose the noise parameter η so that this probability is very low. However, in this case, more efficient attacks than BKW algorithm $O(2^{k/\log(k/-\log(1-2\eta))})$ can be used to recover the secret in time $O(k^3/(1-\eta)^k)$. To thwart attacks, we will increase their complexity using *delinearization steps* described later.

An important point is the choice of the error-correcting code used in the scheme. In our context, we would like a code with shallow decoding circuit, and indeed, codes with shallow decoding circuits are quite rare. For example, linear codes, which have really simple encoding circuits, have complicated decoding circuits. We would like to use a 3-repetition code, which has decoding depth 1. We will keep using such a code in practice as it leads to very efficient homomorphic performance, but the particular structure given to the noise requires a careful analysis of the security, that we will do in the following section. Consequently, in order to also thwart this attack, the delinearization steps can be useful.

Delinearization Steps. In order to counter the Arora-Ge attack, we choose to add some noise on our values after computing the scalar product. In practice, the ciphertext we send consists of $(a, E(x) \oplus F(a.S) \oplus e)$ where F is some function involving enough layers of multiplication so that the Arora-Ge attack does not work. Of course, this step increases the parameters we have to choose for homomorphically evaluating our scheme, however, a few steps (3 in order to have a sufficient security parameter) are needed in order to prevent the attacks. We admit that such techniques are far away from provable security and come from symmetric cryptography since F is a kind of cheap non-linear operation. However, contrary to the symmetric setting, here the adversary cannot control the inputs to this function and many well-known chosen plaintext attacks are then prohibited and only known plaintext attacks need to be studied. It is easy to see how to adapt the decryption process. The function F we choose works as follows: on a vector V, it consecutively applies several transformations T_i, for $i \leq d$, such that $[T_i(V)]_j = V_j + V_{x_{ij}} * V_{y_{ij}}$, where the set of indices x_{ij} and y_{ij} is chosen so that monomials do not cancel. The degree of F in the inputs is 2^d. We estimated the number of applications of such transformations needed in order to counteract the most efficient variant of the Arora-Ge attack and for $n = 512$, three steps seem reasonable.

Even though, our scheme has a security proof, the parameters we choose do not allow us to use the reduction. Indeed, we pick either a structured noise (and Arora-Ge algorithms must be taken into account) or a very small noise to reduce the decryption failure. Therefore the delinearization steps increase the complexity of these attacks and a thorough security analysis is needed. These steps are similar to local random functions [1,5] and we can show similar security.

An Encryption Scheme Based on LWR. We present a symmetric encryption scheme whose security can be reduced to the LWE problem. We describe the more general MLWR − SYM and we can similarly define LWR − SYM and RLWR − SYM.

Definition 5 (MLWR − SYM). *Let d, k and n be polynomials in λ. Let consider R, with underlying polynomial P of degree d. The symmetric encryption scheme MLWR − SYM is defined as follows: $\mathsf{Gen}(1^\lambda)$: output $S \leftarrow R_q^k$; $\mathsf{Enc}_S(x)$: output $(a, x + \lfloor a.S \rceil_p)$, where $a \leftarrow R_q^k$ and $\mathsf{Dec}_S(a, y)$: output $(y - \lfloor a.S \rceil_p)$.*

For a message of size n over $\mathbb{Z}_p$, this scheme produces a ciphertext consisting of a random vector of length k over $\mathbb{Z}_q$ and a vector of length n of $\mathbb{Z}_p$. Thus, the expansion factor is $1 + \frac{\log q}{\log p}\frac{k}{n}$. Now, for the same reasons as for LPN − C, this expansion factor can basically be considered as 1. The decryption circuit has depth one plus the depth of the rounding function. When using the floor function and if q and p are power of two, then the rounding consists in dropping the least significant bits of the result. The most efficient FHE-friendly encryption scheme works in only adding the plaintext on the $\log p$ most significant bits of $\lfloor a.S \rceil_p$ to avoid the costly ExtractDigits homomorphic function and the returned plaintext contains noise.

Proposition 2. LWR − SYM *(resp. RLWR − SYM, MLWR − SYM) is semantically secure as soon as the corresponding LWR (resp. RLWR, MLWR) problem is hard.*

An Encryption Scheme Based on LWE. We can adapt LWR − SYM so that its security proof relies directly on the LWE assumption. This new scheme will basically be the same as the previous one, except that the vector a will be chosen according to some biased distribution D_S. The distribution D_S we will use is defined on $\mathbb{Z}_q^n$ and depends on some matrix $S \in \mathbb{Z}_q^{k \times n}$ and a distribution χ.

We will quickly present it in the case where $k = 1$. It verifies the property that $\Pr[D_s = a]$ is proportional to $\Pr[||\lfloor \frac{p}{q}.(\langle a, s \rangle + e)\rceil] - \frac{p}{q}.(\langle a, s \rangle + e)| < \frac{1}{4}]$, where $e \leftarrow \chi$ and $\overline{\langle a, s \rangle + e}$ means that $\langle a, s \rangle + e$ is interpreted as an element of $\mathbb{Z}$. This basically means that we want the value $(p/q).(\overline{\langle a, s \rangle + e})$ to be close to its rounding for our samples a. One can efficiently sample according to this distribution D_s: sample a uniformly, and output it if and only if, when sampling e according to χ, the value $(p/q).(\overline{\langle a, s \rangle + e})$ is at distance less than $1/4$ from its rounding. Since the distribution of $\langle a, s \rangle + e$ is indistinguishable from uniform, the probability that a vector a gets rejected is (around) $1/2$. To extend this distribution to a matrix S, we will take a distance of $1/2 - 1/4n$ instead of $1/4$.

Definition 6 (LWE − SYM). *Let k and n be polynomials in λ. The symmetric encryption scheme LWE − SYM is defined as follows: $\mathsf{Gen}(1^\lambda)$: output $S \leftarrow \mathbb{Z}_q^{k \times n}$; $\mathsf{Enc}_S(x)$: output $(a, x + \lfloor a.S \rceil_p)$, where $a \leftarrow D_S$ and $\mathsf{Dec}_S(a, y)$: output $y - \lfloor a.S \rceil_p$.*

Our scheme relying on RLWR and MLWR can also be adapted to schemes called RLWE − SYM and MLWE − SYM in a similar way, that we do not explicit here. We now show that the security of LWE − SYM (resp. RLWE − SYM, MLWE − SYM) directly reduces to the LWE (resp. RLWE, MLWE) hardness assumption. Our reduction is better than previous ones in the case of one secret. We introduce the problem LWR_D (resp. $RLWR_D$, $MLWR_D$) as the same problem as LWR (resp. RLWR, MLWR) except that a is drawn according to the distribution D. To choose secure parameters for LWR, we picked $k = 128$ and $p < \sqrt{q}$ according to [15].

Proposition 3. LWE − SYM *(resp.* RLWE − SYM, MLWE − SYM*) is semantically secure as soon as the corresponding LWE (resp. RLWE, MLWE) problem is hard.*

Since σ is usually chosen to be at least $\sqrt{k}$ in LWE, our modulus-to-error ratio q/p verifies $q/p > O(n\sqrt{k}\log m)$, which is an improvement compared to previous reductions which depend on m rather than $\log m$.

The schemes LWR − SYM and LWE − SYM are similar, except that LWE − SYM involves some checking when generating vectors a. Thus, LWE − SYM has exactly the same efficiency as LWR − SYM for the homomorphic part. The only difference of performance lies in the symmetric encryption, because the generation of the vectors a is a constant factor longer. Thus, we only present the implementation of LWR − SYM, because we are only interested in the homomorphic part.

References

1. Akavia, A., Bogdanov, A., Guo, S., Kamath, A., Rosen, A.: Candidate weak pseudorandom functions in AC0 ∘MOD₂. In: Innovations in Theoretical Computer Science, ITCS 2014, Princeton, NJ, USA, 12–14 January 2014, pp. 251–260 (2014)
2. Albrecht, M.R., Cid, C., Faugère, J., Fitzpatrick, R., Perret, L.: Algebraic algorithms for LWE problems. In: IACR Cryptology ePrint Archive 2014, p. 1018 (2014)
3. Albrecht, M.R., Rechberger, C., Schneider, T., Tiessen, T., Zohner, M.: Ciphers for MPC and FHE. In: Oswald, E., Fischlin, M. (eds.) EUROCRYPT 2015. LNCS, vol. 9056, pp. 430–454. Springer, Heidelberg (2015)
4. Alwen, J., Krenn, S., Pietrzak, K., Wichs, D.: Learning with rounding, revisited - new reduction, properties and applications. In: Canetti, R., Garay, J.A. (eds.) CRYPTO 2013, Part I. LNCS, vol. 8042, pp. 57–74. Springer, Heidelberg (2013)
5. Applebaum, B.: Cryptographic hardness of random local functions - survey. In: Electronic Colloquium on Computational Complexity (ECCC), vol. 22, p. 27 (2015)
6. Arora, S., Ge, R.: New algorithms for learning in presence of errors. In: Aceto, L., Henzinger, M., Sgall, J. (eds.) ICALP 2011, Part I. LNCS, vol. 6755, pp. 403–415. Springer, Heidelberg (2011)
7. Banerjee, A., Peikert, C., Rosen, A.: Pseudorandom functions and lattices. In: Pointcheval, D., Johansson, T. (eds.) EUROCRYPT 2012. LNCS, vol. 7237, pp. 719–737. Springer, Heidelberg (2012)
8. Brakerski, Z., Vaikuntanathan, V.: Efficient fully homomorphic encryption from (standard) LWE. In: Ostrovsky, R. (ed.) 52nd FOCS, pp. 97–106. IEEE Computer Society Press, October 2011
9. Canteaut, A., Carpov, S., Fontaine, C., Lepoint, T., Naya-Plasencia, M., Paillier, P., Sirdey, R.: How to compress homomorphic ciphertexts. In: IACR Cryptology ePrint Archive 2015, p. 113 (2015)

10. Cheon, J.H., Coron, J.-S., Kim, J., Lee, M.S., Lepoint, T., Tibouchi, M., Yun, A.: Batch fully homomorphic encryption over the integers. In: Johansson, T., Nguyen, P.Q. (eds.) EUROCRYPT 2013. LNCS, vol. 7881, pp. 315–335. Springer, Heidelberg (2013)

11. Dinur, I., Liu, Y., Meier, W., Wang, Q.: Optimized interpolation attacks on LowMC. Cryptology ePrint Archive, Report 2015/418 (2015). http://eprint.iacr.org/

12. Dobraunig, C., Eichlseder, M., Mendel, F.: Higher-order cryptanalysis of LowMC. Cryptology ePrint Archive, Report 2015/407 (2015). http://eprint.iacr.org/

13. Doroz, Y., Hu, Y., Sunar, B.: Homomorphic AES evaluation using NTRU. Cryptology ePrint Archive, Report 2014/039 (2014). http://eprint.iacr.org/2014/039

14. Doröz, Y., Shahverdi, A., Eisenbarth, T., Sunar, B.: Toward practical homomorphic evaluation of block ciphers using prince. In: Böhme, R., Brenner, M., Moore, T., Smith, M. (eds.) FC 2014 Workshops. LNCS, vol. 8438, pp. 208–220. Springer, Heidelberg (2014)

15. Duc, A., Tramèr, F., Vaudenay, S.: Better algorithms for LWE and LWR. In: Oswald, E., Fischlin, M. (eds.) EUROCRYPT 2015. LNCS, vol. 9056, pp. 173–202. Springer, Heidelberg (2015)

16. Gentry, C.: Fully homomorphic encryption using ideal lattices. In: Mitzenmacher, M. (ed.) 41st ACM STOC, pp. 169–178. ACM Press, May/June 2009

17. Gentry, C., Halevi, S., Smart, N.P.: Homomorphic evaluationof the AES circuit. In: Safavi-Naini, R., Canetti, R. (eds.) CRYPTO 2012. LNCS, vol. 7417, pp. 850–867. Springer, Heidelberg (2012)

18. Gilbert, H., Robshaw, M., Seurin, Y.: How to encrypt with the LPN problem. In: Aceto, L., Damgård, I., Goldberg, L.A., Halldórsson, M.M., Ingólfsdóttir, A., Walukiewicz, I. (eds.) ICALP 2008, Part II. LNCS, vol. 5126, pp. 679–690. Springer, Heidelberg (2008)

19. Goldwasser, S., Gutfreund, D., Healy, A., Kaufman, T., Rothblum, G.N.: Verifying and decoding in constant depth. In: Johnson, D.S., Feige, U. (eds.) 39th ACM STOC, pp. 440–449. ACM Press, June 2007

20. Langlois, A., Stehlé, D.: Worst-case to average-case reductions for module lattices. Des. Codes Cryptogr. **75**(3), 565–599 (2015)

21. Lepoint, T., Naehrig, M.: A comparison of the homomorphic encryption schemes FV and YASHE. In: Pointcheval, D., Vergnaud, D. (eds.) AFRICACRYPT 2014. LNCS, vol. 8469, pp. 318–335. Springer, Heidelberg (2014)

22. Lyubashevsky, V., Peikert, C., Regev, O.: A toolkit for ring-LWE cryptography. In: Johansson, T., Nguyen, P.Q. (eds.) EUROCRYPT 2013. LNCS, vol. 7881, pp. 35–54. Springer, Heidelberg (2013)

23. Mella, S., Susella, R.: On the homomorphic computation of symmetric cryptographic primitives. In: Stam, M. (ed.) IMACC 2013. LNCS, vol. 8308, pp. 28–44. Springer, Heidelberg (2013)

24. Naehrig, M., Lauter, K., Vaikuntanathan, V.: Can homomorphic encryption be practical? In: Proceedings of the 3rd ACM Workshop on Cloud Computing Security Workshop, CCSW 2011, pp. 113–124. ACM, New York (2011)

25. Regev, O.: On lattices, learning with errors, random linear codes, and cryptography. In: Gabow, H.N., Fagin, R. (eds.) 37th ACM STOC, pp. 84–93. ACM Press (2005)

26. Rivest, R.L., Adleman, L., Dertouzos, M.L.: On data banks and privacy homomorphisms. Found. Secure Comput. **4**, 169–179 (1978). Academia Press

Four-Round Zero-Knowledge Arguments of Knowledge with Strict Polynomial-Time Simulation from Differing-Input Obfuscation for Circuits

Ning Ding[1,2]($\boxtimes$), Yanli Ren[3], and Dawu Gu[1]

[1] Department of Computer Science and Engineering,
Shanghai Jiao Tong University, Shanghai, China
{dingning,dwgu}@sjtu.edu.cn
[2] State Key Laboratory of Cryptology, Beijing, China
[3] School of Communication and Information Engineering,
Shanghai University, Shanghai, China
renyanli@shu.edu.cn

Abstract. In this paper we present a 4-round zero-knowledge argument of knowledge for **NP** with strict-polynomial-time simulation and expected polynomial-time extraction based on differing-input obfuscation for some circuit samplers and other reasonable assumptions.

1 Introduction

Zero-knowledge (ZK) proof and argument systems, introduced in [6,15], are a fundamental notion in cryptography. Since the introduction, there are many works constructing ZK protocols that satisfy various properties such as constant rounds, proof of knowledge and strict/expected polynomial-time simulation and extraction etc. As for constant-round constructions, Goldreich and Kahan [14] presented a 5-round ZK proof, and Lindell [18] presented a 5-round ZK proof of knowledge and Feige and Shamir [11] gave a 4-round ZK argument of knowledge (ZKAOK). The simulators (resp. extractor if there is) of these protocols use verifier's code (resp. prover's code) in a black-box way and run in expected polynomial-time. Barak [2] presented a constant-round public-coin non-black-box ZK argument with strict polynomial-time simulation, which admits a 6-round implementation shown in [19].

Recently, cryptography community has witnessed a new breakthrough of program obfuscation and its applications. Some works applied differing-input obfuscation [1] to get new results on the exact round complexity of ZK. Informally, a differing-input obfuscator $di\mathcal{O}$ for a circuit/machine sampler is one such that for each pair of circuits/machines and an auxiliary input output by the sampler, if it is hard to find an input on which the two circuits/machines do not agree even given the auxiliary input, their obfuscated programs output by $di\mathcal{O}$ are indistinguishable for any adversary having this auxiliary input.

© Springer International Publishing Switzerland 2016
T.N. Dinh and M.T. Thai (Eds.): COCOON 2016, LNCS 9797, pp. 281–292, 2016.
DOI: 10.1007/978-3-319-42634-1_23

Pandey *et al.* [20] presented a 4-round (concurrent) ZK argument with strict polynomial-time simulation from $di\mathcal{O}$ for machines. Ding [8] presented a 4-round ZKAOK with strict polynomial-time simulation and extraction from this kind of $di\mathcal{O}$. We note that the usage of $di\mathcal{O}$ for machines in [8,20] cannot be replaced by $di\mathcal{O}$ for circuits because the programs in [8,20] that should be obfuscated have unbounded running-time and cannot be implemented by fixed polynomial-size circuits. Since $di\mathcal{O}$ for machines is stronger than $di\mathcal{O}$ for circuits, Ding [9] considered to reduce the exact round complexity of ZKAOK from $di\mathcal{O}$ for circuits and accordingly put forward a 6-round ZKAOK with strict polynomial-time simulation and extraction based on this kind of $di\mathcal{O}$.

Thus for ZKAOK with strict polynomial-time simulation, the protocol achieving best round complexity without using $di\mathcal{O}$ so far is the 6-round one in [2]. While allowed to use $di\mathcal{O}$ for machines, the best round complexity is achieved by the 4-round one in [8]. So a natural question arises: can we reduce the rounds for such ZKAOK based on the weaker assumption of $di\mathcal{O}$ for circuits? This paper will focus on this question.

1.1 Our Results

We present a 4-round ZKAOK with strict polynomial-time simulation and expected polynomial-time exaction from $di\mathcal{O}$ for circuits and other reasonable assumptions. Garg *et al.* [13] showed that the existence of general $di\mathcal{O}$ conflicts with some special obfuscation and thus one of them cannot exist. In their construction, the auxiliary input sampled by the sampler is contrived, which plays a key role in the proof. However, the auxiliary inputs randomly generated by the samplers in this paper are quite natural, e.g. public coins, perfectly-hiding commitments and a transcript of Blum's proof for Hamilton cycles in [5]. As suggested in [16] that the notion of $di\mathcal{O}$ is plausible for samplers which auxiliary-input outputs are public coins, we further take $di\mathcal{O}$ in [1] as a candidate obfuscator for the samplers in this paper. Our result is as follows.

Theorem 1 *(Informal). Assuming the existence of $di\mathcal{O}$ for some circuit samplers and other reasonable assumptions, there exists a 4-round ZKAOK for* **NP** *with strict polynomial-time simulation and expected polynomial-time extraction.*

Our Techniques. Recall that the protocol in [9] is constructed in two steps. First it presents a 8-round statistically ZKAOK with strict-polynomial-time simulation and extraction and then modifies the protocol by additionally at Step 6 letting the prover obfuscate its original next-message function of Step 8 with $di\mathcal{O}$ for circuits and send the obfuscated program as well as the original message of Step 6 to the verifier. Thus the number of the total rounds is reduced to 6. We note that the key condition for applying $di\mathcal{O}$ is that the simulator's strategy of Step 8 can be implemented by a fixed polynomial-size circuit so that the round-compressing method can be performed.

Our idea for further reducing rounds is to apply this method in some earlier step e.g. Step 4. That is, we would like to let the prover obfuscate its original next-message function of both Steps 6 and 8 at Step 4 and send the obfuscated program as well as the original message of Step 4 to the verifier, and thus reduce the number of the total rounds to 4. Note that this idea requires the prover's/simulator's strategy of the last two prover-steps can be implemented in fixed polynomial-size.

Unfortunately, for the protocol [9], the simulator's strategy of Steps 6, 8 cannot be implemented by a fixed polynomial-size circuit, so the round-compressing method cannot be applied. This essentially attributes to the usage of a 4-round universal argument in the protocol. In simulation, the simulator needs to make use of verifier's code and its running-time of the universal argument is a polynomial in verifier's running-time which can be an arbitrarily polynomial. Moreover, it can only finish its computation of the universal argument at Step 6, which implies that the round-compressing method cannot be applied at Step 4.

We modify the execution of the universal argument by letting the verifier send all its messages of two steps to the prover in one step: the first message is sent in plaintext while the second message is sent in encryption using fully homomorphic encryption. Thus the prover/simulator can generate the first prover's message in plaintext and second prover's message in encryption in one step too. It can be seen that this modification makes the simulator finish its computation of the universal argument earlier than before.

Concretely, our protocol basically consists of two parts. The first part uses 4 rounds, which is a compressed version of Barak's non-black-box ZK in [2], where the verifier sends its messages of the universal argument in the above manner. Thus the simulator, having verifier's code, can generate all its messages of the universal argument in the corresponding manner at Step 4.

The second part is used to ensure the soundness, or even stronger, realize extraction. This part also consists of 4 rounds. Since the verifier sends its two messages of the universal argument in one step, a cheating prover may violate the soundness of the universal argument. Thus we let the verifier re-send the second message of the universal argument in encryption followed by a proof that the second messages in the two encryptions sent by it are indeed same (or it knows a witness of a witness-hiding protocol), and then let the prover reply with the answer in encryption too. In extraction we let the extractor send a different second message in encryption, also followed by a valid proof where the extractor proves that it knows the witness of the witness-hiding protocol. When getting two answers, the extractor can extract either a witness for the public input or some piece of the witness for the statement that the universal argument is proving (which can be further used to recover the whole witness). Since the latter cannot happen, what is exacted is the witness for the public input.

Then since the simulator's strategy of the second part can be implemented by a fixed polynomial-size circuit, we apply the round-compressing method at Step 4 and thus reduce the number of the total rounds to 4.

Organizations. For lack of space we omit the preliminaries but will point out the literature for them. In Sect. 2 we present our uncompressed protocol, which is the main body of this paper. In Sect. 3 we apply the round-compressing method to reduce the round number.

2   The Uncompressed Protocol

In this section we present the uncompressed protocol which admits the key property that its simulator's strategy of Steps 6 and 8 can be implemented in fixed polynomial-size. In Sect. 2.1 we present some building blocks and overview of the protocol. In Sect. 2.2 we show it in detail.

2.1   Overview of the Protocol

The protocol still employs those primitives used in [9] as follows.

HCom: a 2-round trapdoor perfectly-hiding commitment scheme, of which the binding property holds against $n^{O(\log \log n)}$-size algorithms, referred to [12]. HCom satisfies that when given the coins in generating a commitment, the committed message can be retrieved from the commitment. For simplicity let (msg, HCom) denote the two messages of the scheme HCom and Trapdoor denote the trapdoor.

Com: a non-interactive perfectly-binding commitment scheme [4].

$\mathcal{H}_n$: a collision-resistent hash function family and each $h \in \mathcal{H}_n$ maps arbitrarily polynomially long strings to n-bit strings. The collision resistance of $\mathcal{H}_n$ holds against $n^{O(\log \log n)}$-size algorithms.

LS: the Lapidot-Shamir 3-round public-coin WI argument of knowledge [17], which enjoys a key property that the first two messages are independent of the witness and the public input. When instantiated with HCom, it is perfectly WI. If ignoring the first message msg of HCom, LS uses 3 rounds and let $(\mathsf{LS}_1, \mathsf{LS}_2, \mathsf{LS}_3)$ denote the 3 messages.

LS′, LS″: two independent running instances of LS for proving different statements, which are instantiated with Com. Let $(\mathsf{LS}_1', \mathsf{LS}_2', \mathsf{LS}_3')$ denote the 3 messages of LS′, $(\mathsf{LS}_1'', \mathsf{LS}_2'', \mathsf{LS}_3'')$ denote the 3 messages of LS″.

UA: the 4-round public-coin universal argument of knowledge in [3] constructed from $\mathcal{H}_n$. Let $(\mathsf{UA}_1, \mathsf{UA}_2, \mathsf{UA}_3, \mathsf{UA}_4)$ denote the 4 messages of UA.

FHE $=$ (KeyGen, Enc, Dec, Evaluate): a fully homomorphic encryption scheme satisfying that any two encryptions output by Enc and Evaluate corresponding to a same decryption are statistically close. For instance, the scheme in [7] can be adapted to satisfy the requirements (e.g. choose x_0 in the public key as an exact multiple of p and allow reducing modulo x_0 during Evaluate and re-randomize the output of Evaluate in a similar way in Enc when the parameters are appropriately chosen in the somewhat homomorphic scheme and assume the weak circular security for the bootstrappable scheme). We employ any FHE satisfying these requirements in this paper.

Let L be in **NP** and (x, w) is an instance-witness pair of L. Our protocol for L basically runs as follows.

1. P and V interact of Barak's preamble, in which the verifier first samples a random hash function h and then the prover responds with a commitment Z using HCom and lastly the verifier sends a random r.

Public input: x (statement to be proved is "$x \in L$");
Prover's auxiliary input: w, (a witness for $x \in L$).

---

1. $V \rightarrow P$: Send $h \in \mathcal{H}_n$, msg.
2. $P \rightarrow V$: Send $f(r_1), f(r_2), \alpha, Z \leftarrow \mathsf{HCom}(h(0^n))$.
3. $V \rightarrow P$: Send $\beta, r \in \{0,1\}^n$, $\mathsf{UA}_1, pk, X_{\mathsf{UA}_3} \leftarrow \mathsf{Enc}(pk, \mathsf{UA}_3), C_{\mathsf{LS}_2} \leftarrow \mathsf{Com}(\mathsf{LS}_2)$, $\mathsf{LS}_1', \mathsf{LS}_1''$.
4. $P \rightarrow V$: Send $\gamma, C_{\mathsf{UA}_2} \leftarrow \mathsf{HCom}(0^{|\mathsf{UA}_2|}), \mathsf{LS}_1, \mathsf{LS}_2', \mathsf{LS}_2''$.
5. $V \rightarrow P$: Send $X_{\mathsf{UA}_3}' \leftarrow \mathsf{Enc}(pk, \mathsf{UA}_3), \mathsf{LS}_3'$.
6. $P \rightarrow V$: Send $X_{\mathsf{HCom}(0^{|\mathsf{UA}_4|})}$.
7. $V \rightarrow P$: Send $\mathsf{UA}_3, C_{\mathsf{UA}_4} \leftarrow \mathsf{Dec}(sk, X_{\mathsf{HCom}(0^{|\mathsf{UA}_4|})}), \mathsf{LS}_2, \mathsf{LS}_3''$.
8. $P \rightarrow V$: Send LS_3.

Protocol 1. The zero-knowledge argument of knowledge $(P; V)$ for L.

2. P sends $f(r_1), f(r_2)$ to V for uniformly random r_1, r_2 where f is any one-way function and then proves to V the knowledge of one pre-image of $f(r_1), f(r_2)$ via Blum's proof. Note that this proof is witness indistinguishable and also witness hiding in this scenario.
3. V samples (sk, pk) of FHE and sends UA_1 and the encryption of UA_3 to P (which replies with a commitment).
4. V re-computes a fresh encryption of UA_3 and sends it to P. Then V proves to P in LS' that either it knows a pre-image of $f(r_1)$, $f(r_2)$ or the two encryptions correspond to the same UA_3.
5. P sends an encryption of some commitment to V, which replies with UA_3 and the decryption to P's encryption that is the commitment. Then V proves to P in LS'' that either it knows a pre-image of $f(r_1)$, $f(r_2)$ or UA_3 is the message encrypted in the two encryptions and the decryption is correct.
6. P proves to V in LS using w as witness that either $x \in L$ or if letting $\mathsf{UA}_2, \mathsf{UA}_4$ denote the messages committed in the two commitments, $\mathsf{UA}_1, \mathsf{UA}_2, \mathsf{UA}_3, \mathsf{UA}_4$ are a valid transcript of UA for proving that there is a program Π satisfying $h(\Pi)$ is the message committed in Z and Π can output r.

2.2   Actual Construction

The actual construction of the protocol follows the above overview but with intensive parallel implementations of different phases, shown in Protocol 1. We present the detailed specification as follows.

1. $V \rightarrow P$: Sample $h \in \mathcal{H}_n$ and $(msg, \mathsf{Trapdoor})$ of HCom where Trapdoor is the trapdoor of HCom corresponding to msg. Send h, msg to P.
2. $P \rightarrow V$: Sample $r_1, r_2 \in \{0,1\}^n$ and compute $f(r_1), f(r_2)$ and α. Compute $Z \leftarrow \mathsf{HCom}(h(0^n))$. Send $f(r_1), f(r_2), \alpha$ and Z to V.

- (α, β, γ) is a transcript of Blum's proof for that the prover knows a pre-image of $f(r_1), f(r_2)$. So P can compute α using either r_1 or r_2. The remainder messages β, γ will be generated later.

3. $V \rightarrow P$: Sample $\beta, r \in \{0,1\}^n, \mathsf{UA}_1, \mathsf{UA}_3, \mathsf{LS}_2$. Sample $(sk, pk) \leftarrow \mathsf{KeyGen}(n)$ and compute $X_{\mathsf{UA}_3} \leftarrow \mathsf{Enc}(pk, \mathsf{UA}_3)$. Compute $C_{\mathsf{LS}_2} \leftarrow \mathsf{Com}(\mathsf{LS}_2)$ and $\mathsf{LS}_1', \mathsf{LS}_1''$. Send $\beta, r, \mathsf{UA}_1, pk, X_{\mathsf{UA}_3}, C_{\mathsf{LS}_2}, \mathsf{LS}_1', \mathsf{LS}_1''$ to P.
   - The UA system is to prove that there are a program Π of size less than $n^{\log\log n}$ and some coins such that $h(\Pi)$ and the coins are an opening of Z and Π on input the message of Step 2 outputs r in $n^{\log\log n}$ steps.
   - The LS, LS' and LS'' systems will be described below.

4. $P \rightarrow V$: Compute γ and then erase all information used in generating α, γ. Compute $C_{\mathsf{UA}_2} \leftarrow \mathsf{HCom}(0^{|\mathsf{UA}_2|})$ and LS_1. Sample $\mathsf{LS}_2', \mathsf{LS}_2''$. Send them to P.
   - The LS system is to prove that either there is w for $x \in L$ or there are openings of $C_{\mathsf{UA}_2}, C_{\mathsf{UA}_4}$, in which let $\mathsf{UA}_2, \mathsf{UA}_4$ denote the committed messages and then $(\mathsf{UA}_1, \mathsf{UA}_2, \mathsf{UA}_3, \mathsf{UA}_4)$ is a valid transcript of UA. Here P uses witness w for $x \in L$ in LS.

5. $V \rightarrow P$: Compute $X'_{\mathsf{UA}_3} \leftarrow \mathsf{Enc}(pk, \mathsf{UA}_3)$ independently. Compute LS_3'. Send them to P.
   - The LS' system is to prove the knowledge of a pre-image of $f(r_1), f(r_2)$ or that there are $\mathsf{Trapdoor}, \mathsf{UA}_3, sk$ satisfying $\mathsf{Trapdoor}$ is the trapdoor corresponding to msg and UA_3 is the message encrypted in $X_{\mathsf{UA}_3}, X'_{\mathsf{UA}_3}$ using sk, and (sk, pk) is generated by KeyGen with some coins. Here V uses the witness for the second statement of LS' (similarly for LS'').

6. $P \rightarrow V$: Compute $X_{\mathsf{HCom}(0^{|\mathsf{UA}_4|})} \leftarrow \mathsf{Enc}(pk, \mathsf{HCom}(0^{|\mathsf{UA}_4|}))$ and send it to V.

7. $V \rightarrow P$: Compute $C_{\mathsf{UA}_4} \leftarrow \mathsf{Dec}(sk, X_{\mathsf{HCom}(0^{|\mathsf{UA}_4|})})$ (where sk is the secret key to pk). Compute LS_3''. Send $\mathsf{UA}_3, C_{\mathsf{UA}_4}, \mathsf{LS}_2, \mathsf{LS}_3''$ to P.
   - The LS'' system is to prove the knowledge of a pre-image of $f(r_1), f(r_2)$ or that UA_3 is the message encrypted in X_{UA_3} and X'_{UA_3}, and C_{UA_4} is generated as specified and LS_2 is the message committed in C_{LS_2}.

8. $P \rightarrow V$: Compute LS_3 using witness w for $x \in L$. Send LS_3 to V. Finally V accepts x iff $(\mathsf{LS}_1, \mathsf{LS}_2, \mathsf{LS}_3)$ is convincing.

Claim 1. *Protocol 1 is an interactive argument for L.*

Proof. The completeness is obvious and the soundness follows from Claim 3. $\square$

Claim 2. *Assuming the existence of $\mathsf{HCom}, \mathsf{FHE}$, Protocol 1 is statistical zero-knowledge.*

Proof (Sketch). We construct a simulator S for any polynomial-size verifier V^* and $x \in L$. $S(x, V^*)$ works as follows.

1. Emulate V^* to send out h, msg. Then generate $f(r_1), f(r_2), \alpha$ honestly and compute $Z \leftarrow \mathsf{HCom}(h(\Pi))$ where Π denotes the remainder strategy of V^*. Send $f(r_1), f(r_2), \alpha, Z$ to V^*.

2. Emulate V^* to send out $\beta, r, \mathsf{UA}_1, pk, X_{\mathsf{UA}_3}, C_{\mathsf{LS}_2}, \mathsf{LS}'_1, \mathsf{LS}''_1$. Compute γ using r_1 or r_2 as witness and then erase all information used in generating α, γ. Compute UA_2 with witness Π (and the coins) and further $C_{\mathsf{UA}_2} \leftarrow \mathsf{HCom}(\mathsf{UA}_2)$. Generate LS_1 and sample $\mathsf{LS}'_2, \mathsf{LS}''_2$. Send them to V^*.

Let Circuit denote the circuit that having some coins hardwired, on input UA_3 computes UA_4 (adopting the prover strategy of UA) and $\mathsf{HCom}(\mathsf{UA}_4)$ using the hardwired coins. Generate $X_{\mathsf{HCom}(\mathsf{UA}_4)} \leftarrow \mathsf{Evaluate}(pk, \mathsf{Circuit}, X_{\mathsf{UA}_3})$ and keep it. (S will not postpone the computation of $X_{\mathsf{HCom}(\mathsf{UA}_4)}$ to Step 6 and thus its running-time in Steps 6 and 8 is fixed polynomial.)

3. Emulate V^* to send out $X'_{\mathsf{UA}_3}, \mathsf{LS}'_3$. If $(\mathsf{LS}'_1, \mathsf{LS}'_2, \mathsf{LS}'_3)$ is convincing, send $X_{\mathsf{HCom}(\mathsf{UA}_4)}$ to V. Otherwise, abort the simulation.

4. Emulate V^* to send out $\mathsf{UA}_3, C_{\mathsf{UA}_4}$ (that is supposed to be $\mathsf{HCom}(\mathsf{UA}_4)$), $\mathsf{LS}_2, \mathsf{LS}''_3$. If $(\mathsf{LS}''_1, \mathsf{LS}''_2, \mathsf{LS}''_3)$ is convincing, retrieve UA_4 from C_{UA_4} using the knowledge of the coins previously hardwired in Circuit. Then compute LS_3 using as witness $\mathsf{UA}_2, \mathsf{UA}_4$ and send it to V^*. Otherwise, abort the simulation.

Since Π is a witness for the public input of UA, S can use it to finish the interaction and run in polynomial-time. The statistical ZK property follows from the perfectly-hiding property of HCom and the perfectly WI property of LS, and the statistical indistinguishability of two encryptions output by Enc and Evaluate of FHE corresponding to a same decryption. It is noticeable that at Step 6 the prover's message is $X_{\mathsf{HCom}(0^{|\mathsf{UA}_4|})}$ which decryption is $\mathsf{HCom}(0^{|\mathsf{UA}_4|})$, while S's message is $X_{\mathsf{HCom}(\mathsf{UA}_4)}$ which decryption is $\mathsf{HCom}(\mathsf{UA}_4)$. Due to the perfectly-hiding property of HCom, the decryptions of the prover's and simulator's messages are of same distribution. For each same decryption, $X_{\mathsf{HCom}(0^{|\mathsf{UA}_4|})}$ (output by Enc) and $X_{\mathsf{HCom}(\mathsf{UA}_4)}$ (output by Evaluate) are statistically close, due to the property of FHE. So are they for independently generated $\mathsf{HCom}(0^{|\mathsf{UA}_4|})$ and $\mathsf{HCom}(\mathsf{UA}_4)$.

At Step 5 V^* may send a fake X'_{UA} that is not an encryption of UA_3 but it can send a valid LS''_3 at Step 7. If this happens, S's message of Step 8 differs from P's. In the following we show this occurs with negligible probability. Thus S's messages and P's are statistically close. (Actually, the output of S is identically distributed to a real view of V^* if ignoring the messages of Steps 6 and 8.)

Suppose, on the contrary, V^* can send a valid LS''_3 with noticeable probability for an invalid X'_{UA_3}. Then by running the extractor of LS'' we can generate a witness for the public input of LS'' that must be a pre-image of $f(r_1)$ or $f(r_2)$. This is impossible because V^* cannot get this pre-image due to the witness hiding property of (α, β, γ). Thus the statistical zero-knowledge property holds. $\square$

Claim 3. *Assuming the existence of $\mathcal{H}_n, \mathsf{FHE}, \mathsf{Com}, \mathsf{HCom}$, Protocol 1 is an argument of knowledge.*

Proof (Sketch). We show there is an extractor E such that if P' is a polynomial-size prover that can convince V of $x \in L$ with noticeable probability ϵ, $E(P', x)$ outputs a witness for x with probability $\epsilon - \mathsf{neg}(n)$. E works as follows.

1. Sample h, msg. Send them to P'. Then emulate P' to send out f_1, f_2 (supposed to be $f(r_1), f(r_2)$), α, Z.

2. Sample $\beta, r, \mathsf{UA}_1, \mathsf{UA}_3, \mathsf{LS}_2$ honestly and $(sk, pk) \leftarrow \mathsf{KeyGen}(n)$. Compute $X_{\mathsf{UA}_3} \leftarrow \mathsf{Enc}(pk, \mathsf{UA}_3)$, $C_{\mathsf{LS}_2} \leftarrow \mathsf{Com}(\mathsf{LS}_2)$, $\mathsf{LS}_1', \mathsf{LS}_1''$. Send them to P' and emulate it to send out $\gamma, C_{\mathsf{UA}_2}, \mathsf{LS}_1, \mathsf{LS}_2', \mathsf{LS}_2''$.

   If (α, β, γ) is not convincing, abort the running. Otherwise, run the extractor of the (α, β, γ) system in expected polynomial-time to extract a pre-image of f_1 or f_2, denoted r^*.

3. Randomly choose UA_3' and compute $X_{\mathsf{UA}_3}' \leftarrow \mathsf{Enc}(pk, \mathsf{UA}_3')$. Compute LS_3' using r^* as witness. Send them to P and emulate it to output an encryption, denoted X, that is supposed to be $X_{\mathsf{HCom}(0^{|\mathsf{UA}_4|})}$.

4. Compute $C_{\mathsf{UA}_4} \leftarrow \mathsf{Dec}(sk, X)$. Sample a fresh LS_2 and compute LS_3'' using r^* as witness. Send $\mathsf{UA}_3', C_{\mathsf{UA}_4}, \mathsf{LS}_2, \mathsf{LS}_3''$ to P' and emulate it to output LS_3.

5. If $(\mathsf{LS}_1, \mathsf{LS}_2, \mathsf{LS}_3)$ is not convincing, abort the running. Otherwise, run the exactor of LS in expected polynomial-time to gain a witness and output it.

Since P' can convince V with probability ϵ, we have that P''s messages interacting with E is convincing with probability $\epsilon - \mathsf{neg}(n)$. In fact, V uses the witnesses for the second statements of $\mathsf{LS}', \mathsf{LS}''$ respectively, while E uses r^*, a pre-image of f_1, f_2, as witness in the two systems. Due to the IND-CPA security of FHE and the WI property of $\mathsf{LS}', \mathsf{LS}''$, P' cannot tell the difference with noticeable probability.

So what we need to show is that this witness is for $x \in L$ with probability $\epsilon - \mathsf{neg}(n)$. Suppose this is not the case. Then the witness is for the second statement of LS. Thus it contains $\mathsf{UA}_2, \mathsf{UA}_4$ satisfying $\mathsf{UA}_1, \mathsf{UA}_2, \mathsf{UA}_3', \mathsf{UA}_4$ is a valid transcript of UA. By adopting the exaction strategy of UA shown in [2], we can generate a program Π, a witness used in UA, in $n^{O(\log \log n)}$-time. Then with a similar argument in [2], we can run the extraction process twice from Step 3 in which the two r are different and obtain two programs, denoted Π_1, Π_2, with noticeable probability. Since the two r are different, $\Pi_1 \neq \Pi_2$, which either breaks the collision-resistance of h or breaks the binding property of HCom. This is impossible. So what is extracted is indeed a witness for $x \in L$. $\square$

So combining the three claims, we have proved the following proposition.

Proposition 1. *Assuming all the underlying primitives, Protocol 1 is a 8-round statistically ZKAOK for* **NP** *with strict-polynomial-time simulation. Further, its simulator can simulate the messages of Steps 6, 8 in fixed polynomial-time.*

3   Compressing the Last Five Rounds with Obfuscation

In this section we adopt the round-compressing method to reduce rounds. In Sect. 3.1 we present the analysis of the obfuscation of the prover's/simualtor's strategy of last two steps. In Sect. 3.2 we apply the round-compressing method to reduce the round number of Protocol 1 to 4.

3.1 Obfuscating P/S's Next-Message Functions of Last Two Prover Steps

We adopt the notations and route in [9]. Let $P_{x,w,u}$ denote the honest prover $P(x,w)$ with randomness u hardwired and $S_{x,V^*,v}$ denote $S(x,V^*)$ with randomness v hardwired. For each u used by the prover, parse $u = (u_1, u_2)$, where u_1 is used for computing the messages of Steps 2, 4 and u_2 is used for Step 6. Note that S uses v in Steps 2, 4. Then we divide a view of V^* to two parts $\mathsf{view}_1 \circ \mathsf{view}_2$, where view_1 denotes V^*'s view up to Step 4 and view_2 denotes its view of Steps 6, 8.

Due to Claim 2 and its proof, for each fixed view_1 P and S have the equal-probability to generate it. Let P_{view_1, u_2} be $P_{x,w,u}$'s next-message function of last two prover steps and Q_{view_1} be $S_{x,V^*,v}$'s next-message function of last two prover steps. That is, P_{view_1, u_2} on input the message of Step 5 outputs the message of Step 6 and on input the messages of Steps 5 and 7 first generates the message of Step 6 and then outputs the message of Step 8 (in the consecutive execution). Similarly for Q_{view_1}.

Since P has a witness w for $x \in L$, P_{view_1, u_2} can be computed by a fixed polynomial-size circuit. As for Q_{view_1}, recall that S computes UA_2, C_{UA_2} and $X_{\mathsf{HCom}(\mathsf{UA}_4)}$ at Step 4, and keeps them for the remainder interaction. So S's strategy of Steps 6 and 8 can also be computed by a fixed polynomial-size circuit. So this means that P_{view_1, u_2} and Q_{view_1} can be implemented in same polynomial-size. In the following we first present a sampler $\mathsf{Sampler}$ that can output P_{view_1, u_2} and Q_{view_1} as well as view_1 in which v relies on u such that it is hard to find a differing-input for P_{view_1, u_2} and Q_{view_1} even having view_1.

Algorithm 1. The circuit sampler $\mathsf{Sampler}$ which has (V^*, x, w) hardwired.
Input: the system parameter n.
Output: $(P_{\mathsf{view}_1, u_2}, Q_{\mathsf{view}_1}, \mathsf{view}_1)$.

– Sample u and invoke an interaction between $P_{x,w,u}$ and V^* to generate the joint view. Let $\mathsf{view}_1, \mathsf{view}_2, P_{\mathsf{view}_1, u_2}$ be defined as above.

– Run the extractor of LS' to extract a witness for the public input of LS' in expected polynomial-time. Due to the witness hiding property of (α, β, γ), the witness cannot be a pre-image of $f(r_1), f(r_2)$. So it contains $\mathsf{Trapdoor}, sk$.

– Since view_1 consists of the prover's messages of Steps 2 and 4, $\mathsf{Sampler}$ first adopts S's strategy to generate $h(\Pi), \mathsf{UA}_2$, and then computes some coins such that these coins and $h(\Pi), \mathsf{UA}_2$ are also openings of Z, C_{UA_2} using the knowledge of $\mathsf{Trapdoor}$. (Thus $\mathsf{Sampler}$ can find coins corresponding to u satisfying that S also outputs the same view_1.)

– Knowing the coins used in generating $\mathsf{HCom}(0^{|\mathsf{UA}_4|})$, $\mathsf{Sampler}$ finds UA_4 and some coins satisfying they are also an opening of $\mathsf{HCom}(0^{|\mathsf{UA}_4|})$ in the following way: decrypt $X_{\mathsf{HCom}(0^{|\mathsf{UA}_4|})}$ with sk to get $\mathsf{HCom}(0^{|\mathsf{UA}_4|})$; compute UA_4 corresponding to UA_3; finally generate the coins using the knowledge of $\mathsf{Trapdoor}$ such that these coins and UA_4 are an opening of $\mathsf{HCom}(0^{|\mathsf{UA}_4|})$.

(Since $\mathsf{HCom}(0^{|\mathsf{UA_4}|}) = \mathsf{HCom}(\mathsf{UA_4})$ which is the decryption of $X_{\mathsf{HCom}(0^{|\mathsf{UA_4}|})}$ in view_2, due to the property of FHE in Sect. 2.1, there exist some coins such that in S's running, Evaluate when using them as internal coins also outputs $X_{\mathsf{HCom}(\mathsf{UA_4})} = X_{\mathsf{HCom}(0^{|\mathsf{UA_4}|})}$. But Simpler does not need to compute these coins explicitly.) Compute the right messages and coins corresponding to the new witness $\mathsf{UA_2}, \mathsf{UA_4}$ of the LS system such that these coins and the messages result in the same transcript of LS in view_2.

- Let Q_{view_1} denote the circuit that has view_1, $X_{\mathsf{HCom}(0^{|\mathsf{UA_4}|})}$ (that is $X_{\mathsf{HCom}(\mathsf{UA_4})}$ when the implicit right coins are used by Evaluate) as well as other necessary messages needed for computing S's strategy of last two steps hardwired. When receiving V^*'s valid message of Step 5, it directly sends $X_{\mathsf{HCom}(\mathsf{UA_4})}$ to V^* and when receiving V^*'s valid message of Step 7, it runs with S's strategy of Step 8. (Thus both P_{view_1, u_2} and Q_{view_1} output the same view_2 except for negligible probability.)

- Output $(P_{\mathsf{view}_1, u_2}, Q_{\mathsf{view}_1}, \mathsf{view}_1)$.

Assume $di\mathcal{O}$ [1] works for Sampler with any (V^*, x, w). Then we have the following claim. (Claim 4 and Proposition 2 are literally almost same as those in [9]. But due to the difference of the protocols and the sampler strategies, their meanings and proofs are different.)

Claim 4. *Let* $(P_{\mathsf{view}_1, u_2}, Q_{\mathsf{view}_1}, \mathsf{view}_1) \leftarrow \mathsf{Sampler}(1^n)$, $\mathcal{P}_{\mathsf{view}_1, u_2} \leftarrow di\mathcal{O}$ $(P_{\mathsf{view}_1, u_2})$, $\mathcal{Q}_{\mathsf{view}_1} \leftarrow di\mathcal{O}(Q_{\mathsf{view}_1})$. *Then* $\mathcal{P}_{\mathsf{view}_1, u_2}$ *and* $\mathcal{Q}_{\mathsf{view}_1}$ *are indistinguishable for any polynomial-size distinguisher even having* view_1.

Thus based on Claim 4, with a similar argument in [9], we have the following proposition that the obfuscation of the two functions are still indistinguishable even when P and S are independently executed.

Proposition 2. *For any* $V^*, x \in L$, *any polynomial-size distinguisher* D, *independently random* u, v *(resulting in independent* view_1), $|\Pr[D(\mathsf{view}_1, \mathcal{P}_{\mathsf{view}_1, u_2}) = 1] - \Pr[D(\mathsf{view}_1, \mathcal{Q}_{\mathsf{view}_1}) = 1]| = neg(n)$, *where the probabilities are taken over all values of* u, v *and* $di\mathcal{O}$'s *independent coins in generating* $\mathcal{P}_{\mathsf{view}_1, u_2}$ *and* $\mathcal{Q}_{\mathsf{view}_1}$.

3.2 Achieving the Final Protocol

We compress the rounds of Protocol 1 by at Step 4 letting P additionally compute $\mathcal{P}_{\mathsf{view}_1, u_2}$ for random u_2 and send this obfuscation as well as the original message to V, which then adopts the remainder honest verifier's strategy of Protocol 1 to interact with $\mathcal{P}_{\mathsf{view}_1, u_2}$ locally. The final protocol is shown in Protocol 2. Employing the argument in [9], we restate the main theorem as follows.

Theorem 2. *Assuming the existence of* $\mathsf{HCom}, di\mathcal{O}, \mathcal{H}_n, \mathsf{FHE}, \mathsf{Com}$, *Protocol 2 is a 4-round ZKAOK for* **NP** *with strict polynomial-time simulation and expected polynomial-time exaction.*

Public input: x (statement to be proved is "$x \in L$");
Prover's auxiliary input: w, (a witness for $x \in L$).

---

1. $V \to P$: Send $h \in \mathcal{H}_n, msg$.
2. $P \to V$: Send $f(r_1), f(r_2), \alpha, Z \leftarrow \mathsf{HCom}(h(0^n))$.
3. $V \to P$: Send $\beta, r \in \{0,1\}^n$, UA_1, pk, $X_{\mathsf{UA}_3} \leftarrow \mathsf{Enc}(pk, \mathsf{UA}_3)$, $C_{\mathsf{LS}_2} \leftarrow \mathsf{Com}(\mathsf{LS}_2)$, $\mathsf{LS}_1', \mathsf{LS}_1''$.
4. $P \to V$: Send $\gamma, C_{\mathsf{UA}_2} \leftarrow \mathsf{HCom}(0^{|\mathsf{UA}_2|})$, $\mathsf{LS}_1, \mathsf{LS}_2', \mathsf{LS}_2''$, $\mathcal{P}_{\mathsf{view}_1, u_2}$.

Protocol 2. The final 4-round protocol for L.

Acknowledgments. We are grateful to the reviewers of COCOON 2016 for their useful comments. This work is supported by the National Natural Science Foundation of China (Grant No. 61572309) and Major State Basic Research Development Program (973 Plan) of China (Grant No. 2013CB338004) and Research Fund of Ministry of Education of China and China Mobile (Grant No. MCM20150301).

References

1. Ananth, P., Boneh, D., Garg, S., Sahai, A., Zhandry, M.: Differing-inputs obfuscation and applications. In: IACR Cryptology ePrint Archive 2013, p. 689 (2013)
2. Barak, B.: How to go beyond the black-box simulation barrier. In: FOCS, pp. 106–115 (2001)
3. Barak, B., Goldreich, O.: Universal arguments and their applications. In: IEEE Conference on Computational Complexity, pp. 194–203 (2002)
4. Blum, M.: Coin flipping by telephone. In: Gersho, A. (ed.) CRYPTO, pp. 11–15, U. C. Santa Barbara, Dept. of Elec. and Computer Eng., ECE Report No. 82-04 (1981)
5. Blum, M.: How to prove a theorem so no one else can claim it. In: Proceedings of the International Congress of Mathematicians, pp. 1444–1451 (1987)
6. Brassard, G., Chaum, D., Crépeau, C.: Minimum disclosure proofs of knowledge. J. Comput. Syst. Sci. **37**(2), 156–189 (1988)
7. van Dijk, M., Gentry, C., Halevi, S., Vaikuntanathan, V.: Fully homomorphic encryption over the integers. In: Gilbert, H. (ed.) EUROCRYPT 2010. LNCS, vol. 6110, pp. 24–43. Springer, Heidelberg (2010). http://dx.doi.org/10.1007/978-3-642-13190-5
8. Ding, N.: Obfuscation-based non-black-box extraction and constant-round zero-knowledge arguments of knowledge. In: Chow, S.S.M., Camenisch, J., Hui, L.C.K., Yiu, S.M. (eds.) ISC 2014. LNCS, vol. 8783, pp. 120–139. Springer, Heidelberg (2014). http://dx.doi.org/10.1007/978-3-319-13257-0_8
9. Ding, N.: On zero-knowledge with strict polynomial-time simulation and extraction from differing-input obfuscation for circuits. In: Lehmann, A., Wolf, S. (eds.) Information Theoretic Security. LNCS, vol. 9063, pp. 51–68. Springer, Heidelberg (2015). http://dx.doi.org/10.1007/978-3-319-17470-9_4

10. Dodis, Y., Nielsen, J.B. (eds.): TCC 2015. LNCS, vol. 9015. Springer, Heidelberg (2015). http://dx.doi.org/10.1007/978-3-662-46497-7

11. Feige, U., Shamir, A.: Witness indistinguishable and witness hiding protocols. In: STOC, pp. 416–426. ACM (1990)

12. Fischlin, M.: Trapdoor commitment schemes and their applications. Ph.D. thesis, Fachbereich Mathematik Johann Wolfgang Goethe-Universit at Frankfurt am Main (2001)

13. Garg, S., Gentry, C., Halevi, S., Wichs, D.: On the implausibility of differing-inputs obfuscation and extractable witness encryption with auxiliary input. In: Garay, J.A., Gennaro, R. (eds.) CRYPTO 2014, Part I. LNCS, vol. 8616, pp. 518–535. Springer, Heidelberg (2014). http://dx.doi.org/10.1007/978-3-662-44371-2_29

14. Goldreich, O., Kahan, A.: How to construct constant-round zero-knowledge proof systems for NP. J. Cryptol. 9(3), 167–190 (1996)

15. Goldwasser, S., Micali, S., Rackoff, C.: The knowledge complexity of interactive proof systems. SIAM J. Comput. 18(1), 186–208 (1989)

16. Ishai, Y., Pandey, O., Sahai, A.: Public-coin differing-inputs obfuscation and its applications. In: Dodis and Nielsen [10], pp. 668–697. http://dx.doi.org/10.1007/978-3-662-46497-7_26

17. Lapidot, D., Shamir, A.: Publicly verifiable non-interactive zero-knowledge proofs. In: Menezes, A., Vanstone, S.A. (eds.) CRYPTO 1990. LNCS, vol. 537, pp. 353–365. Springer, Heidelberg (1991)

18. Lindell, Y.: A note on constant-round zero-knowledge proofs of knowledge. J. Cryptol. 26(4), 638–654 (2013)

19. Ostrovsky, R., Visconti, I.: Simultaneous resettability from collision resistance. In: Electronic Colloquium on Computational Complexity (ECCC), vol. 19, p. 164 (2012). http://dblp.uni-trier.de/db/journals/eccc/eccc19.html#OstrovskyV12

20. Pandey, O., Prabhakaran, M., Sahai, A.: Obfuscation-based non-black-box simulation and four message concurrent zero knowledge for NP. In: Dodis and Nielsen [10], pp. 638–667. http://dx.doi.org/10.1007/978-3-662-46497-7_25

Inferring Sequences Produced by a Linear Congruential Generator on Elliptic Curves Using Coppersmith's Methods

Thierry Mefenza[1,2]([✉])

[1] ENS, CNRS, INRIA, PSL, Paris, France
[2] Department of Mathematics, University of Yaounde 1, Yaoundé, Cameroon
thierrymefenza@yahoo.fr

Abstract. We analyze the security of the Elliptic Curve Linear Congruential Generator (EC-LCG). We show that this generator is insecure if sufficiently many bits are output at each iteration. In 2007, Gutierrez and Ibeas showed that this generator is insecure given a certain amount of most significant bits of some consecutive values of the sequence. Using the Coppersmith's methods, we are able to improve their security bounds.

Keywords: Elliptic Curve Linear Congruential Generator · Lattice reduction · Coppersmith's methods · Elliptic curves

1 Introduction

In cryptography, a pseudo-random number generator is a deterministic algorithm which takes as input a short random seed and outputs a long sequence which is indistinguishable in polynomial time from a truly random sequence. Pseudo-random numbers have found a number of applications in the literature. For instance they are useful in cryptography for key generation, encryption and signature. In 1994, Hallgren [Hal94] proposed a pseudo-random number generator based on a subgroup of points of an elliptic curve defined over a prime finite field. This generator is known as the Linear Congruential Generator on Elliptic Curves (EC-LCG). Let E be an elliptic curve defined over a prime finite field $\mathbb{F}_p$, that is a rational curve given by the following Weierstrass equation

$$E : y^2 = x^3 + ax + b$$

for some $a, b \in \mathbb{F}_p$ with $4a^3 + 27b^2 \neq 0$. It is well known that the set $E(\mathbb{F}_p)$ of $\mathbb{F}_p$-rational points (including the special point O at infinity) forms an Abelian group with an appropriate composition rule (denoted $\oplus$) where O is the neutral element. For a given point $G \in E(\mathbb{F}_p)$, the EC-LCG is a sequence U_n of points defined by the relation:

$$U_n = U_{n-1} \oplus G = nG \oplus U_0, \quad n \in \mathbb{N}$$

© Springer International Publishing Switzerland 2016
T.N. Dinh and M.T. Thai (Eds.): COCOON 2016, LNCS 9797, pp. 293–304, 2016.
DOI: 10.1007/978-3-319-42634-1_24

where $U_0 \in E(\mathbb{F}_p)$ is the initial value or seed. We refer to G as the *composer* of the generator. The EC-LCG provides a very attractive alternative to linear and non-linear congruential generators and it has been extensively studied in the literature [Shp05, HS02, GL01, GBS00, MS02, BD02]. In cryptography, we want to use the output of the generator as a stream cipher. One can notice that if two consecutive values U_n, U_{n+1} of the generator are revealed, it is easy to find U_0 and G. So, we output only the most significant bits of each coordinate of U_n, $n \in \mathbb{N}$ in the hope that this makes the resulting output sequence difficult to predict. In this paper, we show that the EC-LCG is insecure if sufficiently many bits are output at each stage. Therefore a secure use of this generator requires to output fewer bits at each iteration and the efficiency of the schemes is thus degraded. Our attacks used the well-known Coppersmith's methods for finding small roots on polynomial equations. These methods have been introduced in 1996 by Coppersmith for polynomial of one or two variables [Cop96a, Cop96b] and have been generalized to many variables. These methods have been used to infer many pseudorandom generators and to cryptanalyze many schemes in cryptography (see [BCTV16, BVZ12] and the references therein). In this paper we used such techniques to improve the previous bounds known on the security of the EC-LCG in the literature. Our improvements are theoretical since in practice, the performance of Coppersmith's method in our case is bad because of large dimension of the lattice.

Prior Work. In the cryptography setting, the initial value U_0 and the constants G, a and b may be kept secret. Gutierrez and Ibeas [GI07] consider two cases: the case where the *composer* G is known and a, b are kept secret and the case where the *composer* G is unknown and a, b are kept secret. In the first case, they showed that the EC-LCG is insecure if a proportion of at most 1/6 of the least significant bits of two consecutive values of the sequence is hidden. When the *composer* is unknown, they showed heuristically that the EC-LCG is insecure if a proportion of at most 1/46 of the least significant bits of three consecutive values of the sequence is hidden. Their result is based on a lattice basis reduction attack, using a certain linearization technique. In some sense, their technique can be seen as a special case of the problem of finding small solutions of multivariate polynomial congruences. The Coppersmith's methods also tackle the problem of finding small solutions of multivariate polynomial congruences. Gutierrez and Ibeas due to the special structure of the polynomials involved claimed that "the Coppersmith's methods does not seem to provide any advantages", and that "It may be very hard to give any precise rigorous or even convincing heuristic analysis of this approach". Our purpose in this paper is to tackle this issue.

Our Contributions. We infer the EC-LCG sequence using Coppersmith's method for calculating the small roots of multivariate polynomials modulo an integer. The method for multivariate polynomials is heuristic since it is not proven and may fail (but in practice it works most of the time). At the end of the Coppersmith's methods we use the methods from [BCTV16] to analyze the success condition. In the case where the *composer* is known, we showed that the EC-LCG is insecure if a proportion of at most 1/5 of the least significant bits of two consecutive values U_0 and U_1 of the sequence is hidden. This improves the previous bound $1/6$ of Gutierrez and Ibeas. We further improve this result by considering several consecutive values of the sequence. We showed that the EC-LCG is insecure if a proportion of at most $3/11$ of the least significant bits of these values is hidden. In the case where the *composer* is unknown, we showed that the EC-LCG is insecure if a proportion of at most $1/24$ of the least significant bits of two consecutive values U_0 and U_1 of the sequence is hidden. This improves the previous bound $1/46$ of Gutierrez and Ibeas. We further improve this result by considering sufficiently many consecutive values of the sequence. We showed that the EC-LCG is insecure if a proportion of at most $1/8$ of the least significant bits of these values is hidden.

The table below gives a comparison between our results and those of Gutierrez and Ibeas. It gives the bound of the proportion of least significant bits hidden from each consecutive values necessary to break the EC-LCG in (heuristic) polynomial time. The basic proportion corresponds to the case where the adversary knows bits coming from the minimum number of intermediate values leading to a feasible attack; while the asymptotic proportion corresponds to the case when the bits known by the adversary knows bits coming from arbitrary number of values.

	Basic proportion		Asymptotic proportion	
	Prior result	Our result	Prior result	Our result
Known *composer*	1/6	1/5	None	3/11
Unknown *composer*	1/46	1/24	None	1/8

2   Preliminaries

For some $\Delta > 0$, we say that $W = (x_W, y_W) \in \mathbb{F}_p^2$ is a Δ-approximation to $U = (x_U, y_U) \in \mathbb{F}_p^2$ if there exists integers e, f satisfying:

$$|e|, |f| \leqslant \Delta, \quad x_W + e = x_U, \quad y_W + e = y_U.$$

Throughout the paper, $\Delta < p^\delta$, with $0 < \delta < 1$, corresponds to the situation where a proportion of at most δ of the least significant bits of the output sequence remain hidden.

2.1   The Group Law on Elliptic Curves

In this subsection, we recall the group law $\oplus$ on elliptic curves defined by the Weierstrass equation (for more details on elliptic curves, see [BSS99,Was08]), since our pseudorandom generator is defined recursively by adding a fixed composer G to the previous value. Let $E/\mathbb{F}_p : y^2 = x^3 + ax + b$ be an elliptic curve over $\mathbb{F}_p$. For two points $P = (x_P, y_P)$ and $Q = (x_Q, y_Q)$, with $P, Q \neq O$ the addition law $\oplus$ is defined as follows:

$$P \oplus Q = R = (x_R, y_R),$$

– If $x_P \neq x_Q$, then

$$x_R = m^2 - x_P - x_Q, \quad y_R = m(x_P - x_R) - y_P, \quad \text{where, } m = \frac{y_Q - y_P}{x_Q - x_P} \quad (1)$$

– If $x_P = x_Q$ but ($y_P \neq y_Q$ or $y_P = y_Q = 0$), then $R = O$
– If $P = Q$ and $y_P \neq 0$, then

$$x_R = m^2 - 2x_P, \quad y_R = m(x_P - x_R) - y_P, \quad \text{where, } m = \frac{3x_Q^2 + a}{2y_P}$$

2.2 Coppersmith's Methods

In this section, we give a short description of Coppersmith's method for solving a multivariate modular polynomial system of equations modulo an integer N. We refer the reader to [JM06] for details and proofs.

Problem Definition. Let $f_1(y_1, \ldots, y_n), \ldots, f_s(y_1, \ldots, y_n)$ be irreducible multivariate polynomials defined over $\mathbb{Z}$, having a root $(x_1, \ldots, x_n)$ modulo a known integer N, namely $f_i(x_1, \ldots, x_n) \equiv 0 \mod N$. We want this root to be *small* in the sense that each of its components is bounded by a known value X_i.

Polynomials Collection. In a first step, one generates a collection $\mathfrak{P}$ of polynomials $\{\tilde{f}_1, \ldots, \tilde{f}_r\}$ linearly independent having $(x_1, \ldots, x_n)$ as a root modulo powers of N. Usually, multiples and powers of products of f_i, $i \in \{1, \ldots, s\}$ are chosen , namely $\tilde{f}_\ell = y_1^{\alpha_{1,\ell}} \cdots y_n^{\alpha_{n,\ell}} f_1^{k_{1,\ell}} \cdots f_s^{k_{s,\ell}}$ for some integers $\alpha_{1,\ell}, \ldots, \alpha_{n,\ell}$, $k_{1,\ell}, \ldots, k_{s,\ell}$ for $\ell \in \{1, \ldots, r\}$. Such polynomials satisfy the relation $\tilde{f}_\ell(x_1, \ldots, x_n) \equiv 0 \mod N^{\sum_{i=1}^{s} k_{i,\ell}}$, i.e., there exists an integer c_i such that $\tilde{f}_l(x_1, \ldots, x_n) = c_i N^{k_\ell}$, $k_\ell = \sum_{j=1}^{s} k_{j,\ell}$.

Monomials. We denote $\mathfrak{M}$ the set of monomials appearing in collection of polynomials $\mathfrak{P}$. Then each polynomial $\tilde{f}_i$ can be expressed as a vector with respect to a chosen order on $\mathfrak{M}$. We construct a matrix $\mathcal{M}$ and we define $\mathcal{L}$ the lattice generated by its rows. From that point, one computes an LLL-reduction on the lattice $\mathcal{L}$ and computes the Gram-Schmidt's orthogonalized basis of the LLL output basis. Extracting the coefficients appearing in the obtained vectors, one can construct polynomials defined over $\mathbb{Z}$ such that $\{p_1(x_1, \ldots, x_n) = 0, \ldots, p_n(x_1, \ldots, x_n) = 0\}$. Under the (heuristic) assumption that all created polynomials define an algebraic variety of dimension 0, the previous system can be solved (e.g., using elimination techniques such as Groebner basis) and the desired root recovered in polynomial time.

The conditions on the bounds X_i that make this method work are given by the following (simplified) inequation (see [JM06] for details):

$$\prod_{y_1^{k_1} \cdots y_n^{k_n} \in \mathfrak{M}} X_1^{k_1} \cdots X_n^{k_n} < N^{\sum_{\ell=1}^{r} \sum_{i=1}^{s} k_{i,\ell}}. \quad (2)$$

For such techniques, the most complicated part is the choice of the collection of polynomials, what could be a really intricate task when working with multiple polynomials.

2.3   Analytic Combinatorics

In the following, we recall the analytic combinatorics methods [FS09] to count the exponents of the bounds $X_1, \ldots, X_n$ and of the modulo N on the monomials and polynomials appearing in the inequality (2) in Coppersmith's methods. Those methods can be used to compute the cardinalities of the sets $\mathfrak{P}$ and $\mathfrak{M}$. We used the same notations as in [BCTV16] and for more details of the methods the reader is referred to that paper. We see $\mathfrak{P}$ (respectively $\mathfrak{M}$) as a combinatorial class with size function $S(\tilde{f}_\ell) = \deg(\tilde{f}_\ell)$ (respectively $S(y_\mathbf{k}) = \deg(y_\mathbf{k})$, where $y_\mathbf{k} \in \mathfrak{M}$). We recall that a combinatorial class is a finite or countable set on which a size function is defined, satisfying the following conditions: (i) the size of an element is a non-negative integer and (ii) the number of elements of any given size is finite. We define another function χ, called a *parameter* function, such that $\chi(\tilde{f}_\ell) = k_\ell$ (respectively $\chi(y_\mathbf{k}) = k_i$, where k_i is the degree of the variable y_i in $y_\mathbf{k}$). This allows us to compute for some non negative integer t, ψ (respectively α_i) as:

$$\psi = \chi_{<t}(\mathfrak{P}) = \sum_{a \in \mathfrak{P}: S(a)<t} \chi(a) \quad \alpha_i = \chi_{<t}(\mathfrak{M}) = \sum_{a \in \mathfrak{P}: S(a)<t} \chi(a).$$

To do so we should be able to compute given a combinatorial class $\mathfrak{A}$ ($\mathfrak{A} = \mathfrak{P}$ or $\mathfrak{A} = \mathfrak{M}$) with size function S and the parameter function χ,

$$\chi_{\leqslant p}(\mathfrak{A}) = \sum_{a \in \mathfrak{A}: S(a) \leqslant p} \chi(a).$$

We proceed as follows:

1. We give another description of $\mathfrak{A}$ with respect to S and χ. This description associates to the combinatorial class an ordinary generating function (OGF) $F(z, u)$ (using Table 1, see [BCTV16] for details). When the class contains elements of different sizes (such as variables of degree 1 and polynomials of degree e), the variables in the OGF are represented by the atomic element $\mathcal{Z}$ and the polynomials by the element $\mathcal{Z}^e$, in order to take into account the degree of these polynomials. Then we "mark" the element useful for the parameter, with a new variable u. At this level we only know how to compute $\sum_{a \in \mathfrak{A}: S(a)=p} \chi(a)$. An easier way to compute $\chi_{\leqslant p}(\mathfrak{A})$ is to force all elements a of size less than or equal to p to be of size exactly p by adding enough times a *dummy* element y_0 such that $\chi(y_0) = 0$. In our context of polynomials, the aim of the dummy variable y_0 is to homogenize the polynomial.

2. We have:

$$\chi_{\leqslant}(\mathfrak{A})(z) = \sum_{p=0}^{+\infty} \chi_{\leqslant p}(\mathfrak{A}) z^p = \frac{\partial F(z, u)}{\partial u}\bigg|_{u=1},$$

Table 1. Combinatorics constructions and their OGF

	Construction	OGF
Atomic class	$\mathcal{Z}$	$Z(z) = z$
Neutral class	ε	$E(z) = 1$
Disjoint union	$\mathcal{A} = \mathcal{B} + \mathcal{C}$ (when $\mathcal{B} \cap \mathcal{C} = \emptyset$)	$A(z) = B(z) + C(z)$
Complement	$\mathcal{A} = \mathcal{B} \setminus \mathcal{C}$ (when $\mathcal{C} \subseteq \mathcal{B}$)	$A(z) = B(z) - C(z)$
Cartesian product	$\mathcal{A} = \mathcal{B} \times \mathcal{C}$	$A(z) = B(z) \cdot C(z)$
Cartesian exponentiation	$\mathcal{A} = \mathcal{B}^k = \mathcal{B} \times \cdots \times \mathcal{B}$	$A(z) = B(z)^k$
Sequence	$\mathcal{A} = \mathrm{SEQ}(\mathcal{B}) = \varepsilon + \mathcal{B} + \mathcal{B}^2 + \ldots$	$A(z) = \frac{1}{1-B(z)}$

3. Since Coppersmith's method is usually used in an asymptotic way, singularity analysis enables us to find the asymptotic value of the coefficients in an simple way by using the following theorem (see [FS09], page 392):

Theorem 1 (Transfer Theorem). *Let* $\mathfrak{A}$ *be a combinatorial class with an ordinary generating function* F *regular enough such that there exists a value* c *verifying*

$$F(z) = \sum_{n=0}^{+\infty} F_n z^n \underset{z \to 1}{\sim} \frac{c}{(1-z)^{\alpha}}$$

for a non-negative integer α. *The asymptotic value of the coefficient* F_n *is*

$$F_n \underset{n \to \infty}{\sim} (cn^{\alpha-1})/(\alpha-1)!.$$

3   Predicting EC-LCG Sequences for Known *Composer*

In the cryptographic setting, the initial value $U_0 = (x_0, y_0)$ and the constants G, a and b are supposed to be the secret key. In the following, we infer the EC-LCG sequence in the case where the *composer* G is known and the curve parameters are kept secret. We show that the generator is insecure if at least a proportion of $4/5$ of the most significant bits of two consecutive values U_0 and U_1 of the sequence is output.

Theorem 2 *(two consecutive outputs).* *Given* Δ-*approximations* W_0, W_1 *to two consecutive affine value* U_0, U_1 *produced by the EC-LCG, and given the value of the composer* $G = (x_G, y_G)$. *Under the heuristic assumption that all created polynomials we get by applying Coppersmiths method with the polynomial set* $\mathfrak{P}$ *below define an algebraic variety of dimension* 0, *one can recover the seed* U_0 *in heuristic polynomial time in* $\log p$ *as soon as* $\Delta < p^{\delta}$, *with* $\delta < 1/5$.

Proof We suppose without loss of generality that $U_0 \notin \{-G, G\}$. Then, clearing denominators in (1), we can translate

$$U_1 = U_0 \oplus G$$

into the following identities in the field $\mathbb{F}_p$:

$$L_1 = L_1(x_0, y_0, x_1) = 0 \bmod p, \quad L_2 = L_2(x_0, y_0, x_1, y_1) = 0 \bmod p$$

where $U_0 = (x_0, y_0)$, $U_1 = (x_1, y_1)$ and

$$L_1 = x_G^3 + x_1 x_G^2 - x_0 x_G^2 - 2x_1 x_G x_0 - x_G x_0^2 + x_0^3 + 2y_G y_0 + x_1 x_0^2 - y_G^2 - y_0^2,$$

$$L_2 = y_1 x_G - y_1 x_0 - y_G x_0 + y_G x_1 - y_0 x_1 + y_0 x_G.$$

We denote $W_0 = (\alpha_0, \beta_0)$ and $W_1 = (\alpha_1, \beta_1)$. Then using the equalities $x_j = \alpha_j + e_j$ and $y_j = \beta_j + f_j$, for $j \in \{0, 1\}$, where $|e_j|, |f_j| < \Delta$ leads to the following polynomial system:

$$\begin{cases} f(e_0, e_1, f_0) = 0 \bmod p \\ g(e_0, e_1, f_0, f_1) = 0 \bmod p \ . \end{cases}$$

where $f(z_1, z_2, z_3) = A_1 z_1 + A_2 z_2 + A_3 z_3 + A_4 z_1^2 + A_5 z_1 z_2 + z_1^3 + z_1^2 z_2 - z_3^3 + A_6$ and $g(z_1, z_2, z_3, z_4) = B_1 z_1 + B_2 z_2 + B_3 z_3 + B_4 z_4 + z_1 z_4 + z_2 z_3 + B_5$ are polynomials whose coefficients A_i's and B_i's are functions of x_G, and the approximations values $\alpha_0, \alpha_1, \beta_0, \beta_1$. If we set $u_1 = z_1^3 + z_1^2 z_2 - z_3^3$ and $v_1 = z_1 z_4 + z_2 z_3$, then the polynomial f becomes $f_1(z_1, z_2, z_3, u_1) = A_1 z_1 + A_2 z_2 + A_3 z_3 + A_4 z_1^2 + A_5 z_1 z_2 + u_1 + A_6$ and g becomes $g_1(z_1, z_2, z_3, z_4, v_1) = B_1 z_1 + B_2 z_2 + B_3 z_3 + B_4 z_4 + v_1 + B_5$.

Description of the Attack. The adversary is therefore looking for the small solutions of the following modular multivariate polynomial system:

$$\begin{cases} f_1(z_1, z_2, z_3, u_1) = 0 \bmod p \\ g_1(z_1, z_2, z_3, z_4, v_1) = 0 \bmod p \ . \end{cases}$$

With $|z_j| < \Delta$, $|u_1| < X = \Delta^3$ and $|v_1| < Y = \Delta^2$. The attack consists in applying Coppersmith's methods for multivariate polynomials. From now, we use the following collection of polynomials (parameterized by some integer $t \in \mathbb{N}$):

$$\mathfrak{P} = \left\{ z_1^{j_1} \ldots z_4^{j_4} f_1^{i_1} g_1^{i_2} \bmod p^{i_1 + i_2} : i_1 + i_2 > 0 \text{ and } j_1 + \cdots + j_4 + 2i_1 + i_2 < 2t \right\}$$

The list of monomials appearing within this collection can be described as:

$$\mathfrak{M} = \left\{ z_1^{i_1} z_2^{i_2} z_3^{i_3} z_4^{i_4} u_1^{i_5} v_1^{i_6} \bmod \Delta^{i_1 + i_2 + i_3 + i_4} X^{i_5} Y^{i_6} : i_1 + \cdots + i_4 + 2i_5 + i_6 < 2t \right\}.$$

If we use for instance the lexicography order on monomials, (with $z_1 < z_2 < z_3 < z_4 < u_1 < v_1$) on the set of monomials, then the leading monomial (denoted LM) of f_1 is $LM(f_1) = u_1$ and $LM(g_1) = v_1$. Then the polynomials in $\mathfrak{P}$ are linearly independent since we have prohibited the multiplication by u_1 and v_1.

Bounds for the Polynomials Modulo p. We consider the set $\mathfrak{P}$ as a combinatorial class, with the size function $S(z_1^{j_1} \ldots z_4^{j_4} f_1^{i_1} g_1^{i_2}) = j_1 + \cdots + j_4 + 2i_1 + i_2$

and the parameter function $\chi(z_1^{j_1} \ldots z_4^{j_4} f_1^{i_1} g_1^{i_2}) = i_1 + i_2$. The degree of each variable z_i, u_1, v_1 is 1, whereas the degree of f_1 is 2 and the degree of g_1 is 1. For the sake of simplicity, we can consider $0 \leqslant i_1 + i_2$, since the parameter function equals 0 for elements $z_1^{j_1} \ldots z_4^{j_4} f_1^{i_1} g_1^{i_2}$ with $i_1 + i_2 = 0$.

We can described $\mathfrak{P}$ as: $\prod_{i=1}^{4} \mathrm{SEQ}(Z) \times \mathrm{SEQ}(uZ^2) \times \mathrm{SEQ}(uZ) \times \mathrm{SEQ}(Z)$, where the last term is for the *dummy* value z_0.

This leads to the generating function:

$$F(z,u) = \left(\frac{1}{1-z}\right)^5 \times \frac{1}{1-uz^2} \times \frac{1}{1-uz}.$$

As $z \to 1$, $1 - z^n \sim n(1-z)$ leads to:

$$\left.\frac{\partial F}{\partial u}(u,z)\right|_{u=1} \underset{z \to 1}{\sim} \frac{3(1-z)}{4(1-z)^9} \sim \frac{3}{4(1-z)^8},$$

since $2t \sim 2t - 1$, this leads to: $\chi_{<2t}(\mathfrak{P}) \sim \frac{3}{4} \times \frac{(2t)^7}{7!}$

Bounds for the Monomials Modulo Δ. We consider the set $\mathfrak{M}$ as a combinatorial class, with the size function $S(z_1^{i_1} \ldots z_4^{i_4} u^{i_5} v^{i_6}) = i_1 + \cdots + i_4 + 2i_5 + i_6$ and the parameter function $\chi(z_1^{i_1} \ldots z_4^{i_4} u^{i_5} v^{i_6}) = i_1 + \cdots + i_4$. As z_1, z_2, z_3, z_4, u, v "count for" $1, 1, 1, 1, 2$ and 1 respectively in the condition of the set, we can described $\mathfrak{M}$ as: $\mathrm{SEQ}(Z^2) \times \mathrm{SEQ}(Z) \times \prod_{i=1}^{4} \mathrm{SEQ}(uZ) \times \mathrm{SEQ}(Z)$, where the last term is for the *dummy* value z_0.

Which leads to the generating function: $F(z,u) = \frac{1}{(1-z^2)(1-z)^2} \times \left(\frac{1}{1-uz}\right)^4$. As previously, we obtain $\chi_{<2t,\Delta}(\mathfrak{M}) \sim \frac{2(2t)^7}{7!}$.

Bounds for the Monomials Modulo X (Respectively Modulo Y). We consider the set $\mathfrak{M}$ as a combinatorial class, with the size function $S(z_1^{i_1} \ldots z_4^{i_4} u^{i_5} v^{i_6}) = i_1 + \cdots + i_4 + 2i_5 + i_6$ and the parameter function $\chi(z_1^{i_1} \ldots z_4^{i_4} u^{i_5} v^{i_6}) = i_5$ (respectively $\chi(z_1^{i_1} \ldots z_4^{i_4} u^{i_5} v^{i_6}) = i_6$). As z_1, z_2, z_3, z_4, u, v "count for" $1, 1, 1, 1, 2$ and 1 respectively in the condition of the set, we can described $\mathfrak{M}$ as: $\prod_{i=1}^{5} \mathrm{SEQ}(Z) \times \mathrm{SEQ}(uZ^2) \times \mathrm{SEQ}(Z)$ (respectively $\prod_{i=1}^{4} \mathrm{SEQ}(Z) \times \mathrm{SEQ}(Z^2) \times \mathrm{SEQ}(uZ) \times \mathrm{SEQ}(Z)$) where the last one is for the *dummy* value z_0.

Which leads to the generating function: $F(z,u) = \frac{1}{(1-z)^6} \times \frac{1}{1-uz^2}$ (respectively $F(z,u) = \frac{1}{(1-z)^5(1-z^2)} \times \frac{1}{1-uz}$). This leads to: $\chi_{<2t,X}(\mathfrak{M}) \sim \frac{(2t)^7}{4 \times 7!}$ (respectively $\chi_{<2t,Y}(\mathfrak{M}) \sim \frac{(2t)^7}{2 \times 7!}$).

Condition. We denote by $\nu_1 = \chi_{<2t,\Delta}(\mathfrak{M})$, $\nu_2 = \chi_{<2t,X}(\mathfrak{M})$, $\nu_3 = \chi_{<2t,Y}(\mathfrak{M})$ and $\varepsilon = \chi_{<2t}(\mathfrak{P})$. The inequality (2) is $p^{\varepsilon} > \Delta^{\nu_1} X^{\nu_2} Y^{\nu_3}$, i.e. $\Delta < p^{\frac{\varepsilon}{\nu_1 + 3\nu_2 + 2\nu_3}}$, where:

$$\frac{\varepsilon}{\nu_1 + 3\nu_2 + 2\nu_3} \sim \frac{\chi_{<2t}(\mathfrak{P})}{\chi_{<2t,\Delta}(\mathfrak{M}) + 3\chi_{<2t,X}(\mathfrak{M}) + 2\chi_{<2t,Y}(\mathfrak{M})} \sim \frac{1}{5},$$

this leads to the claimed bound: $\Delta < p^{\frac{1}{5}}$. $\square$

This bound improves the known bound $\Delta < p^{1/6}$. Next we further improve the previous bound and we show that the generator is insecure if at least a proportion of 8/11 of the most significant bits of an arbitrary large number of consecutive values U_i of the sequence is output.

Theorem 3 (more consecutive outputs). *Given Δ-approximations W_0, $W_1,\ldots,W_n$ (for some integer $n > 1$) to $n + 1$ consecutive affine values U_0, $U_1,\ldots,U_n$ produced by the EC-LCG, and given the value of the composer $G = (x_G, y_G)$. Under the heuristic assumption that all created polynomials we get by applying Coppersmiths method with the polynomial set $\mathfrak{P}$ below define an algebraic variety of dimension 0, one can recover the seed U_0 in polynomial time in $\log p$ as soon as $\Delta < p^\delta$, with $\delta < \frac{3n}{11n+4}$.*

Proof (Sketch) We can generalize the previous proof by considering n couples of consecutive values (U_i, U_{i+1}), $i \in \{0,\ldots,n-1\}$ and the same variable change to get n couple of polynomials f_{i+1}, g_{i+1} of the same shape as f_1 and g_1. We then apply the method to the following collection of polynomials:

$$\mathfrak{P} = \left\{ \begin{array}{l} z_0^{j_0} \cdots z_{2n+1}^{j_{2n+1}} f_1^{i_1} \cdots f_n^{i_n} g_1^{l_1} \cdots g_n^{l_n} \bmod p^{i_1+l_1\cdots+i_n+l_n} \\ \text{s.t. } i_1 + l_1 + \cdots + i_n + l_n > 0 \\ \text{and } j_0 + \cdots + j_{2n+1} + 2(i_1 + \cdots + i_n) + l_1 + \cdots + l_n < 2t \end{array} \right\},$$

and the following set of monomials:

$$\mathfrak{M} = \left\{ \begin{array}{l} z_0^{j_0} \cdots z_{2n+1}^{j_{2n+1}} u_1^{i_1} v_1^{l_1} \cdots u_n^{i_n} v_n^{l_n} \bmod \Delta^{j_0+\cdots+j_{2n+1}} X^{i_0+\cdots+i_n} Y^{l_0+\cdots+l_n} \\ \text{s.t. } j_0 + \cdots + j_{2n+1} + 2(i_1 + \cdots + i_n) + l_1 + \cdots + l_n < 2t \end{array} \right\},$$

to get the result (see the full version of the paper for the complete proof). $\square$

4   Predicting EC-LCG Sequences for Unknown Composer

In this section, we infer the EC-LCG sequence in the case where the *composer G* is unknown and the curve parameters are kept secret. In the following, We show that the generator is insecure if at least a proportion of 23/24 of the most significant bits of three consecutive values U_0 and U_1 and U_2 of the sequence is output.

Theorem 4 (three consecutive outputs). *Given Δ-approximations W_0, W_1, W_2 to three consecutive affine values U_0, U_1, U_2 produced by the EC-LCG. Under the heuristic assumption that all created polynomials we get by applying Coppersmiths method with the polynomial set $\mathfrak{P}$ below define an algebraic variety of dimension 0, one can recover the seed U_0 and the composer G in polynomial time in $\log p$ as soon as $\Delta < p^\delta$ with $\delta < 1/24$.*

Proof We set $U_0 = (x_0, y_0)$, $U_1 = (x_1, y_1)$, $U_2 = (x_2, y_2)$, $W_0 = (\alpha_0, \beta_0)$, $W_1 = (\alpha_1, \beta_1)$ and $W_2 = (\alpha_2, \beta_2)$. We then have the equalities:

$$x_i = \alpha_i + e_i, \; y_j = \beta_j + f_j, \quad \text{where} \quad |e_i|, |f_i| < \Delta, \, i \in \{0,1,2\}. \tag{3}$$

We also have:

$$\begin{cases} y_0^2 = x_0^3 + ax_0 + b \\ y_1^2 = x_1^3 + ax_1 + b \\ y_2^2 = x_2^3 + ax_2 + b \ . \end{cases}$$

Eliminating the curve parameters a, b and assuming without loss of generality that $U_2 \neq \pm U_1$ (that is, $x_2 \neq x_1$), we obtain the following equation:

$$y_2^2(x_0 - x_1) + x_2^3(x_1 - x_0) + x_0^3(x_2 - x_1) + y_0^3(x_1 - x_2) + x_1^3(x_0 - x_2) + y_1^2(x_2 - x_0) = 0$$

Using the equalities (3), leads to the equation:

$$f(e_0, e_1, e_2, f_0, f_1, f_2) = 0 \bmod p$$

where f is a polynomial of degree 4 whose coefficients are functions of $\alpha_0, \alpha_1, \alpha_2$, β_0, β_2, and β_2.

Description of the Attack. The adversary is therefore looking for the solutions smaller than Δ of the following modular multivariate polynomial equation:

$$f(z_1, \ldots, z_6) = 0 \bmod p$$

The attack consists in applying Coppersmith's methods as in the former subsection. If we consider monomials with respect to lexicographic order, then the leading monomial of f is $z_1^3 z_2$. From now on, we use the following collection of polynomials:

$$\mathfrak{P} = \{ \tilde{f}_{j_1, \ldots, j_6, i} = z_1^{j_1} \ldots z_6^{j_6} f^i \bmod p^i : i > 0 \quad \text{and} \quad j_1 + \cdots + j_6 + 4i < 4t$$
$$\text{and} \ (0 \leqslant j_1 < 3 \vee j_2 = 0) \},$$

One can check that the polynomials $\tilde{f}_{j_1, \ldots, j_6, i}$ are linearly independent since $LM(f) \neq z_1^{j_1} \ldots z_6^{j_6}$ for each $\tilde{f}_{j_1, \ldots, j_6, i}$ from $\mathfrak{P}$. The list of monomials appearing within this collection can be described as:

$$\mathfrak{M} = \left\{ z_1^{j_1} \ldots z_6^{j_6} \bmod \Delta^{j_1 + \cdots + j_6} : j_1 + \cdots + j_6 < 4t \right\}.$$

Bounds for the Polynomials Modulo p. We consider the set $\mathfrak{P}$ as a combinatorial class, with the size function $S(\tilde{f}_{j_1, \ldots, j_6, i}) = j_1 + \cdots + j_6 + 4i$ and the parameter function $\chi(\tilde{f}_{j_1, \ldots, j_6, i}) = i$. Since the degree of each variable z_i is 1 and the degree of f is 4, we can described $\mathfrak{P}$ as:

$$\prod_{i=1}^{4} \text{SEQ}(Z) \times \text{SEQ}(uZ^4) \times \left(\underbrace{(\varepsilon + Z + Z^2)}_{z_1} \underbrace{(\varepsilon + Z\text{SEQ}(Z))}_{z_2} + \underbrace{Z^3 \text{SEQ}(Z)}_{z_1} \right) \times \text{SEQ}(Z),$$

where the last term is for the *dummy* value z_0. This leads to the generating function:

$$F(z, u) = \left(\frac{1}{1-z} \right)^5 \times \frac{1}{1 - uz^4} \times \left((1 + z + z^2)(1 + z/(1-z)) + \frac{z^3}{1-z} \right).$$

This leads to: $\chi_{<4t}(\mathfrak{P}) \sim \frac{1}{4} \times \frac{(4t)^7}{7!}$

Bounds for the Monomials Modulo Δ. We consider the set $\mathfrak{M}$ as a combinatorial class, with the size function $S(z_1^{j_1} \ldots z_6^{j_6}) = j_1 + \cdots + j_6$ and the parameter function $\chi(z_1^{j_1} \ldots z_6^{j_6}) = j_1 + \cdots + j_6$. Since the degree of each z_i is 1, we can then described $\mathfrak{M}$ as: $\prod_{i=1}^{6} \mathrm{SEQ}(uZ) \times \mathrm{SEQ}(Z)$, where the last term is for the *dummy* value z_0. Which leads to the generating function: $F(z, u) = \left(\frac{1}{1-uz}\right)^6 \times \frac{1}{1-z}$. We then obtain: $\chi_{<4t}(\mathfrak{M}) \sim \frac{6(3t)^7}{7!}$

Condition. If we denote by $\nu = \chi_{<4t}(\mathfrak{P})$, and $\varepsilon = \chi_{<4t}(\mathfrak{M})$, the inequality (2) is $p^\nu > \Delta^\varepsilon$, i.e. $\Delta < p^{\frac{\nu}{\varepsilon}}$, where: $\frac{\nu}{\varepsilon} \sim \frac{\chi_{<4t}(\mathfrak{P})}{\chi_{<4t}(\mathfrak{M})} \sim \frac{1}{24}$, this leads to the claimed bound: $\Delta < p^{\frac{1}{24}}$. $\qquad \square$

This bound improves the known bound $\Delta < p^{1/46}$. Next, we further improve the previous bound and we show that the generator is insecure if at least a proportion of 7/8 of the most significant bits of an arbitrary large number of consecutive values U_i of the sequence is output.

Theorem 5 (more consecutive outputs). *Given Δ-approximations W_0, $W_1, \ldots, W_{n+1}$ (for some integer $n > 1$) to $n + 2$ consecutive affine values U_0, $U_1, \ldots, U_{n+1}$ produced by the EC-LCG. Under the heuristic assumption that all created polynomials we get by applying Coppersmiths method with the polynomial set $\mathfrak{P}$ below define an algebraic variety of dimension 0, one can recover the seed U_0 and the composer G in polynomial time in $\log p$ as soon as $\Delta < p^\delta$ with $\delta < n/4(2n + 4)$.*

Proof See the full version of the paper. $\qquad \square$

5   Conclusion

We analyzed the security of the Elliptic Curve Linear Congruential Generator (EC-LCG). In the case where the *composer* is known, we showed that this generator is insecure if at least a proportion of 8/11 of the most significant bits of an arbitrary large number of consecutive values U_i of the sequence is output. We also consider the cryptographic setting where the *composer* is unknown and we showed that this generator is insecure if at least a proportion of 7/8 of the most significant bits of an arbitrary large number of consecutive values U_i of the sequence is output. Our results are theoretical since in practice, the performance of Coppersmith's method in our attacks is bad because of large dimension of the constructed lattice but they are good evidences of the weaknesses of this generator. This generator should then be used with great care.

Acknowledgments. The author was supported in part by the French ANR JCJC ROMAnTIC project (ANR-12-JS02-0004) and by the Simons foundation Pole PRMAIS. I would like to thank anonymous referees for their helpful comments.

References

[BCTV16] Benhamouda, F., Chevalier, C., Thillard, A., Vergnaud, D.: Easing Coppersmith methods using analytic combinatorics: applications to public-key cryptography with weak pseudorandomness. In: Cheng, C.-M., Chung, K.-M., Persiano, G., Yang, B.-Y. (eds.) PKC 2016. LNCS, vol. 9615, pp. 36–66. Springer, Heidelberg (2016). doi:10.1007/978-3-662-49387-8_3

[BD02] Beelen, P., Doumen, J.: Pseudorandom sequences from elliptic curves. In: Mullen, G.L., Stichtenoth, H., Tapia-Recillas, H. (eds.) Finite Fields with Applications to Coding Theory, Cryptography and Related Areas, pp. 37–52. Springer, Berlin (2002)

[BSS99] Blake, I.F., Seroussi, G., Smart, N.P.: Elliptic Curves in Cryptography. Cambridge University Press, Cambridge (1999)

[BVZ12] Bauer, A., Vergnaud, D., Zapalowicz, J.-C.: Inferring sequences produced by nonlinear pseudorandom number generators using Coppersmith's methods. In: Fischlin, M., Buchmann, J., Manulis, M. (eds.) PKC 2012. LNCS, vol. 7293, pp. 609–626. Springer, Heidelberg (2012)

[Cop96a] Coppersmith, D.: Finding a small root of a univariate modular equation. In: Maurer, U.M. (ed.) EUROCRYPT 1996. LNCS, vol. 1070, pp. 155–165. Springer, Heidelberg (1996)

[Cop96b] Coppersmith, D.: Finding a small root of a bivariate integer equation; factoring with high bits known. In: Maurer, U.M. (ed.) EUROCRYPT 1996. LNCS, vol. 1070, pp. 178–189. Springer, Heidelberg (1996)

[FS09] Flajolet, P., Sedgewick, R.: Analytic Combinatorics. Cambridge University Press, Cambridge (2009)

[GBS00] Gong, G., Berson, T.A., Stinson, D.R.: Elliptic curve pseudorandom sequence generators. In: Heys, H.M., Adams, C.M. (eds.) SAC 1999. LNCS, vol. 1758, pp. 34–49. Springer, Heidelberg (2000)

[GI07] Gutierrez, J., Ibeas, A.: Inferring sequences produced by a linear congruential generator on elliptic curves missing high-order bits. Des. Code Crypt. **45**, 199–212 (2007)

[GL01] Gong, G., Lam, C.C.Y.: Linear recursive sequences over elliptic curves. In: Helleseth, T., Kumar, P.V., Yang, K. (eds.) Proceedings of the International Conference on Sequences and Their Applications, Bergen, pp. 182–196. Springer, London (2001)

[Hal94] Hallgren, S.: Linear congruential generators over elliptic curves. Preprint CS-94-143, Dept. of Comp. Sci. (1994)

[HS02] Hess, F., Shparlinski, I.E.: On the linear complexity and multidimensional distribution of congruential generators over elliptic curves. Des. Code Crypt. **35**, 111–117 (2005)

[JM06] Jochemsz, E., May, A.: A strategy for finding roots of multivariate polynomials with new applications in attacking RSA variants. In: Lai, X., Chen, K. (eds.) ASIACRYPT 2006. LNCS, vol. 4284, pp. 267–282. Springer, Heidelberg (2006)

[MS02] Mahassni, E., Shparlinski, I.E.: On the uniformity of distribution of congruential generators over elliptic curves. In: Helleseth, T., Kumar, P.V., Yang, K. (eds.) Proceedings of International Conference on Sequences and Their Applications, Bergen, pp. 257–264. Springer, London (2001, 2002)

[Shp05] Shparlinski, I.E.: Pseudorandom points on elliptic curves over finite fields (2005). Preprint

[Was08] Washington, L.C.: Elliptic Curves Number Theory and Cryptography, 2nd edn. Chapman and Hall/CRC, Boca Raton (2008)

Network and Algorithms

The Routing of Complex Contagion in Kleinberg's Small-World Networks

Wei Chen[1]([⊠]), Qiang Li[2]([⊠]), Xiaoming Sun[2]([⊠]), and Jialin Zhang[2]([⊠])

[1] Microsoft Research, Beijing, China
[2] Institute of Computing Technology, Chinese Academy of Sciences, Beijing, China
liqiang01@ict.ac.cn

Abstract. In Kleinberg's small-world network model, strong ties are modeled as deterministic edges in the underlying base grid and weak ties are modeled as random edges connecting remote nodes. The probability of connecting a node u with node v through a weak tie is proportional to $1/|uv|^{\alpha}$, where $|uv|$ is the grid distance between u and v and $\alpha \geq 0$ is the parameter of the model. Complex contagion refers to the propagation mechanism in a network where each node is activated only after $k \geq 2$ neighbors of the node are activated.

In this paper, we propose the concept of routing of complex contagion (or *complex routing*), where at each time step we can select one eligible node (nodes already having two active neighbors) to activate, with the goal of activating the pre-selected target node in the end. We consider decentralized routing scheme where only the links connected to already activated nodes are known to the selection strategy. We study the routing time of complex contagion and compare the result with simple routing and complex diffusion (the diffusion of complex contagion, where all eligible nodes are activated immediately in the same step with the goal of activating all nodes in the end).

We show that for decentralized complex routing, the routing time is lower bounded by a polynomial in n (the number of nodes in the network) for all range of α both in expectation and with high probability (in particular, $\Omega(n^{\frac{1}{\alpha+2}})$ for $\alpha \leq 2$ and $\Omega(n^{\frac{\alpha}{2(\alpha+2)}})$ for $\alpha > 2$ in expectation). Our results indicate that complex routing is exponentially harder than both simple routing and complex diffusion at the sweetspot of $\alpha = 2$.

Keywords: Computational social science · Complex contagion · Diffusion · Decentralized routing · Small-world networks · Social networks

1 Introduction

Social networks are known to be the medium for spreading disease, information, ideas, innovations, and other types of behaviors. Social scientists have been

The work is partially supported by National Natural Science Foundation of China (61222202, 61433014, 61502449) and the China National Program for support of Top-notch Young Professionals.

T.N. Dinh and M.T. Thai (Eds.): COCOON 2016, LNCS 9797, pp. 307–318, 2016.
DOI: 10.1007/978-3-319-42634-1_25

studying social networks and diffusions in the networks for decades, and many of the research results are inspirational to researches in the intersection of social science, economics, and computation on modeling social networks and diffusions in them.

In the seminal work [15,17], Granovetter classified relationships in a social network as strong ties and weak ties. Strong ties represent close relationships, such as family members and close friends, while weak ties represent acquaintance relationship that people casually maintain. The surprising result in this study is that people often obtain important job referrals leading to their current jobs through weak ties instead of strong ties, which leads to the popular term *the strength of weak ties*. His research demonstrated the importance of weak ties in information diffusion in social networks. Another famous experiment related to information diffusion is Milgram's small-world experiment [22], in which Milgram asked subjects to forward a letter to their friends in order for the letter to reach a person not known to the initiator of the letter. The result showed that on average it takes only six hops to connect two people in U.S. unknown to each other, hence the famous term of *six-degree of separation*.

The above studies motivated the modeling of small-world networks [20,26]. Watts and Strogatz modeled the small-world network as a ring where nodes close to one another in ring distance are connected representing strong ties, and some strong ties are rewired to connect to other random nodes on the ring, which represent weak ties [26]. They also proposed short diameter (the distance between any pair of nodes is small) and high clustering coefficient (the probability that two friends of a node are also friends of each other) as two characteristics of small-world networks. Kleinberg [20] improved the model of Watts and Strogatz by building a small-world network on top of a base grid, where grid edges representing strong ties, and each node u initiating a weak tie connecting to another node v with probability proportional to $1/|uv|^{\alpha}$, where $|uv|$ is the grid distance between u and v and α is the small-world parameter. Kleinberg showed that when α equals the dimension of the grid, the decentralized greedy routing, where in each routing step the current node routes the message to its neighbor with grid distance closest to the target node, achieves efficient routing performance [20]. This efficient decentralized routing behavior qualitatively matches the result of Milgram's small-world experiment. Kleinberg further showed that when α is not equal to the grid dimension, no decentralized routing scheme could be efficient, and in particular, the small-world model of Newman and Watts [24] corresponds to the one-dimensional Kleinberg's model with $\alpha = 0$. Kleinberg's small-world network model is the one we use in this paper.

In another work [16], Granovetter proposed the threshold model to characterize diffusions of rumors, innovations, or riot behaviors. An individual in a social network is activated by a certain behavior only when the number of her neighbors already adopting the behavior exceeds a threshold. This threshold model motivated the linear threshold, fixed threshold, and general threshold models proposed by Kempe et al. [18], and is directly related to the model of complex contagion we use in this paper.

More recently, Centola and Macy [5] classified the threshold model into simple contagion and complex contagion. Simple contagion refers to diffusion models with threshold being one on every node, which means that a node can be activated as long as there is one active neighbor. Simple contagion corresponds to diffusions of virus or simple information, where one can get activated by simply receiving the virus and information. Complex contagion, on the other hand, refers to diffusion models with threshold at least two, meaning that a node can be activated only after multiple of its neighbors are activated. Complex contagion corresponds to diffusions requiring complex decision process by individuals, such as adopting a costly new product, adopting a disruptive innovation, etc., where people usually need multiple independent sources of confirmation about the utility of the new product or new innovation before taking the action. The important point Centola and Macy argued is that, while weak ties are effective in transmitting information quickly across a long range in a network, they may not be as effective in complex contagion. This is because for complex contagions to spread quickly in a network, it requires weak ties forming not only long bridges connecting different regions of the network but also *wide* bridges in the sense that many weak ties can work together to bring the contagion from one region of the network to another region of the network.

Motivated by the above work, Ghasemiesfeh et al. provided the first analytical study of complex contagion in small-world networks [14]. They studied the diffusion of k-complex contagion (or *k-complex diffusion*), where all nodes have threshold k and all nodes with at least k active neighbors are activated right away. They showed that the *diffusion time*, which is the time for the diffusion to activate all nodes in a network starting from k initial seed nodes connected with strong ties, is polylogarithmic to the size of the network when $\alpha = 2$. Ebrahimi et al. [11] further generalized the results and proved that the diffusion time for k-complex diffusion has polylogarithmic upper bound when $\alpha \in (2, \frac{2(k^2+k+1)}{k+1})$ in Kleinberg's grid model. They also show that in Kleinberg's model with α outside this range, the diffusion time is lower bounded by a polynomial in n.

In this paper, we go beyond the diffusion of complex contagion (or *complex diffusion*), to study a new propagation phenomenon closer to decentralized routing in [20], which we call the *routing of complex contagion* (or *complex routing*). In complex routing, we model weak ties as directed edges as in [20], and study the time for two seed nodes connected by a strong tie to activate a target node t farthest on the grid (we call it the *routing time*). At each step only one new node can be activated, and the decision of which node to activate is decentralized which means it is only based on the current activated nodes and their outgoing weak tie neighbors as well as the underlying grid, same as decentralized routing in [20]. Such decentralized routing behavior corresponds to real-world phenomenon where a group of people want to influence a target person by influencing intermediaries between the source group and the target person, and influencing these intermediaries requires effort and thus has to be carried out one at a time. Active friending [27] is an application similar to the above scenario recently proposed in the context of online social networks such as Facebook for increasing the chance of a target user accepting the friending request from the source.

1.1 Our Results

In this paper, we show that, unlike simple routing or complex diffusion, in complex routing problem for any $k \geq 2$, for the entire range of α, the routing time is polynomial in n both in expectation and with high probability for any decentralized routing algorithm. Compared with simple routing or complex diffusion, the results at the sweetspot of $\alpha = 2$ are the most interesting: simple routing has routing time $O(\log^2 n)$ in expectation [20] and complex diffusion has an upper bound of $O(\log^{k+1.5} n)$ in expected diffusion time [14], while complex routing has a lower bound of $\Omega(n^{\frac{1}{4}})$ in expected routing time, for any $k \geq 2$. This exponentially wide gap indicates intrinsic difference between complex routing and simple routing or complex diffusion. We further show that if we allow activating m nodes in one step, the routing time is lower bounded by $\Omega(n^{\frac{1}{4}}/m)$, which means that to get a polylogarithmic upper bound on the routing time m has to be $\Omega(n^{\frac{1}{4}}/\log^c n)$ for some constant c.

Our main contribution is that we propose the study of complex routing, and prove that the routing time has polynomial lower bound in the entire range of α for complex routing. Our results indicate that complex routing is much harder than complex diffusion and the routing time of complex contagion differs exponentially compared to simple contagion at sweetspot.

1.2 Additional Related Work

Social and information networks and network diffusions have been extensively studied, and a comprehensive coverage has been provided by recent textbooks such as [10, 25]. In this section, we provide most related work in addition to the ones already discussed in the introduction.

Since the proposal of the small-world network models by [20, 26], many extensions and variants have been studied. For example, Kleinberg proposed a small-world model based on tree structure [21], Fraigniaud and Giakkoupis extended the model to allow power-law degree distribution [12] or arbitrary base graph structure [13].

In terms of network diffusion, a line of research initiated in [18, 19] studied the maximization problem of finding a set of small seeds to maximize the influence spread, usually under a stochastic diffusion model. For Chen et al. [8] provided efficient influence maximization algorithms for large-scale networks, while Chen [7] proved that minimizing the size of the seed set for a given coverage in the fixed threshold model is hard to approximate to any polylogarithmic factor.

Threshold behavior is also studied in bootstrap percolation [1], where all nodes have the same threshold and initial seeds are randomly selected. Bootstrap percolation focuses on the study of the critical fraction f of the seed nodes selected so that the entire network is infected in the end. The network structures investigated for bootstrap percolation include grid [6], trees [3], random regular graphs [2], complex networks [4] etc.

The rest of the paper is organized as follows. Section 2 provides the technical model and problem definitions. Section 3 presents the results and analyses on complex routing. We conclude the paper in Sect. 4.

2 Model and Problem Definitions

We now provide the precise definitions of the network model, the propagation model, and the problems we are studying in this paper.

2.1 Kleinberg's Small-World Networks

The Kleinberg's small-world network model defines a random graph based on a set V of n nodes organized in a $\sqrt{n} \times \sqrt{n}$ two-dimensional grid [20]. For convenience, we connect the top boundary nodes of the grid with the corresponding bottom boundary nodes, and connect the left boundary nodes with the corresponding right boundary nodes, creating a two-dimensional torus, in which the positions of all nodes are symmetric. For nodes u and v on the torus, the Manhattan distance $|uv|$ between them is the shortest distance from u to v (or v to u) using grid edges.

There are two types of edges in this random graph: *strong ties* and *weak ties*. Strong ties refer to the undirected edges between any pair of nodes with Manhattan distance no more than p, where $p \geq 1$ is a universal constant. Weak ties refer to random edges connecting any node u with other possibly remote nodes v in the grid. Each node u has q weak tie connections created independently from one another, and the i-th weak tie initiated by u has endpoint v with probability proportional to $1/|uv|^\alpha$, where $\alpha \geq 0$ is a parameter of the model. In order to get the probability distribution of weak ties, we multiply $1/|uv|^\alpha$ by the normalizing factor $\mathcal{Z} = 1/\sum_{v \in V} |uv|^{-\alpha}$ (on a torus, this value is the same for any $u \in V$). For a node u in the network, u's *grid-neighbors* are nodes linked with u through strong ties while *weak-neighbors* are nodes linked with u through weak ties.

The original network model by Kleinberg [20] considers the weak tie from u to v as a directed edge, and we call it the *directed Kleinberg's small-world network model*, while some work including [14] considers the weak ties as undirected edges. Define random graph $G(n, k, \alpha)$ as directed Kleinberg's small-world network with n nodes and parameter α and $p = q = k$. We only consider directed network models in this paper.

2.2 Routing of Complex Contagion

We model the propagation of information, disease, or innovations in a network as a *contagion*. Each node in a network has three possible states — *inactive*, *exposed*, *infected* (or *activated*), and a node can transformed from the inactive state to the exposed state and then to the infected state, but not in the reverse direction.

A contagion proceeds in discrete time steps $0, 1, 2, \ldots$. At time $t \geq 1$, a node becomes exposed if at time $t-1$ at least k of its neighbors (or in-neighbors in the case of directed networks) are infected. An exposed node may become infected immediately or at a later step, which will be specified later. A *simple contagion* refers to the contagion with $k = 1$, that is, one infected neighbor is enough to expose (and potentially infect) the node, while a *complex contagion* refers to the case of $k \geq 2$, that is, at least two infected neighbors are needed to infect a new node. We refer the complex contagion with $k \geq 2$ as k-complex contagion.

We study a different propagation phenomenon closer to the decentralized routing behavior studied in [20] originally for the small-world network model, which we call *routing of complex contagion*, or simply *complex routing*.

To study k-complex routing, at time 0, we set k consecutive nodes on the grid in one dimension as infected initially, which we refer as *seed nodes*. For convenience, we also set $p = k$. When $p = k$, the k-complex routing is guaranteed to infect all nodes eventually through strong ties only. In complex routing, we have a target node t besides the set of k initial seed nodes.

The task is to infect or activate node t as fast as possible. We can only select one exposed node to activate at each time step. Moreover, when selecting the node to activate at time i, one only knows the out-neighbors of already activated nodes since decentralized routing is applied. This corresponds to the situation where a group of people try to influence a target by gradually growing their allies in the social network towards the target, and they only know the friends of their allies and try to recruit one of them into the allies at the next time step. Note that when $k = 1$, k-complex routing is essentially the decentralized simple routing studied in [20].

To study how fast the routing could be successful, we define the *routing time* as the number of time steps needed to activate the farthest target node t from the seed node in terms of the Manhattan distance.

3   Results on Complex Routing

When studying complex routing, we use the directed Kleinberg's small-world network model, same as the model originally proposed by Kleinberg in [20] for decentralized routing. As described in the model, we consider decentralized routing in which a node can only send activation to its out-neighbors. Hence only when a node is pointed to by edges from k different activated nodes it becomes exposed. For the strong tie, we still treat them as undirected or bi-directional. In each time step, we only have the knowledge of the current activated nodes and the out-neighbors of the current activated nodes. This allows us to apply the Principle of Deferred Decisions [23] in the same way as applied in [20], which means that the weak ties of a node u are defined and known only when u is activated. Initial seeds set is a set of k consecutive nodes, so the k-complex routing will eventually activate target t when we set $p = k$ in Kleinberg's small-world network model.

We consider a 2-complex routing task from a pair of grid neighbor nodes $S_0 = \{s_0^1, s_0^2\}$ to a destination t where s_0^1, s_0^2 have Manhattan distance of 1 on the grid. In this paper, we discuss the routing with initial grid distance of $|s_0^1 t| = \Theta(\sqrt{n})$. The strategy of activating nodes from exposed nodes set is not restricted. A special scheme is choosing the node with smallest Manhattan distance to t in each time step, which is the greedy algorithm. But our result holds for any decentralized node selection schemes, even randomized ones. The following theorem provides the lower bound result on the routing time.

Theorem 1. *For any decentralized routing schemes (even randomized ones), the routing time of 2-complex routing in $G(n, 2, \alpha)$ has the following lower bounds based on the parameter α, for any small $\varepsilon > 0$:*

1. *For $\alpha \in [0, 2)$, the routing time is $\Omega(n^{\frac{1-\varepsilon}{\alpha+2}})$ with probability at least $1 - O(n^{-\varepsilon})$ and the expected routing time is $\Omega(n^{\frac{1}{\alpha+2}})$.*
2. *For $\alpha = 2$, the routing time is $\Omega(n^{\frac{1}{4}})$ with probability at least $1 - O(\frac{1}{\log n})$ and the expected routing time is $\Omega(n^{\frac{1}{4}})$.*
3. *For $\alpha \in (2, +\infty)$, the routing time is $\Omega(n^{\frac{\alpha-2\varepsilon}{2(\alpha+2)}})$ with probability at least $1 - O(n^{-\varepsilon})$ and the expected routing time is $\Omega(n^{\frac{\alpha}{2(\alpha+2)}})$.*

First we give some necessary definitions. For a set of nodes S, define $\mathcal{E}(S)$ to be the set of exposed nodes for the current activated set S, namely $\mathcal{E}(S) = \{x \notin S \mid x \text{ has at least two in-neighbors in set } S\}$. In a routing protocol, let S_i be the set of the current activated nodes in time i. In time step i, we can choose at most *one* node $u \in \mathcal{E}(S_{i-1})$, and activate u (which means we add u to S_{i-1} in time i and obtain S_i). From the definition of $\mathcal{E}(S)$ we know that complex routing proceeds following the direction of edges in directed Kleinberg's small-world model.

3.1   Proof of Deterministic Scheme

We consider deterministic decentralized routing schemes first. Due to the page limit, the proofs of lower bounds for randomized schemes are omitted, and they are included in our full version [9]. First we discuss routing time for $\alpha = 2$.

Suppose $S_0, S_1, \cdots, S_\ell$ is the sequence of activated sets of nodes in routing where S_i is the set of current activated nodes in time step i. The initial seeds are $\{s_0^1, s_0^2\}$ so $S_0 = \{s_0^1, s_0^2\}$, $S_i = \{s_0^1, s_0^2, s_1, \cdots, s_i\}$ and in time $i \geq 1$ we add a new node s_i selected from $\mathcal{E}(S_{i-1})$, particularly $s_l = t$. Let $d_i = d(S_i \cup \mathcal{E}(S_i), t)$, where $d(S, u)$ is the minimum Manhattan distance between node $v \in S$ and u. It is easy to observe that d_i is a non-increasing sequence and $d_{\ell-1} = d_\ell = 0$. For convenience, we write s_0^1 as S_{-1} and define that $d_{-1} = |s_0^1 t| = \sqrt{n}$. We then prove that when the parameter $\alpha = 2, \Pr(\forall\ 0 \leq i < cn^{\frac{1}{4}}, d_{i-1} - d_i \leq n^{\frac{1}{4}})$ is high enough, where $c < 1$ is a positive constant we will set later. Define event $\chi = \{\forall 0 \leq i < cn^{\frac{1}{4}}, d_{i-1} - d_i \leq n^{\frac{1}{4}}\}$. Event χ means from time step 0 to $cn^{1/4} - 1$, the Manhattan distance between the current activated set and target t decrease at most $n^{\frac{1}{4}}$ in each time step.

Lemma 1. *For decentralized 2-complex routing in directed Kleinberg's small-world network $G(n, 2, \alpha)$ with $\alpha = 2$, given the initial seeds $\{s_0^1, s_0^2\}$ and farthest target t with $|s_0^1 t| = \Theta(\sqrt{n})$, then for some suitable constant $c \in (0, 1)$,*

$$\Pr(\forall\, 0 \le i < cn^{\frac{1}{4}}, d_{i-1} - d_i \le n^{\frac{1}{4}}) \ge 1 - O(\frac{1}{\log n}).$$

Proof. Let $u_i = \arg\min_x \{d(x, t) | x \in S_i \cup \mathcal{E}(S_i)\}$, so u_i is the node that is closest to node t and can be activated by set S_i or belong to S_i. Since u_{i-1} is the node that with the shortest Manhattan distance to t among $\mathcal{E}(S_{i-1}) \cup S_{i-1}$ and $s_i \in \mathcal{E}(S_{i-1})$, $|s_i t| \ge |u_{i-1} t| = d_{i-1}$. Thus if $d_{i-1} - d_i > 0$, we know that s_i is not the node closest to t among $S_i \cup \mathcal{E}(S_i)$ since $|s_i t| \ge d_{i-1}$. Besides, we can also get that $u_i \in \mathcal{E}(S_i) \setminus \mathcal{E}(S_{i-1})$ and s_i activate u_i together with another node in S_{i-1}. Combining with the definition that $|u_i t| = d_i$, we know $|s_i u_i| \ge |s_i t| - |u_i t| = d_{i-1} - d_i$. Hence we have the following conclusions:

If $d_{i-1} - d_i > n^{1/4}$ for $i \ge 1$, then we can conclude (1) $|s_i u_i| > n^{\frac{1}{4}}$; (2) u_i is one of the out-neighbors of s_i, more specifically, s_i initiates a weak tie to u_i; (3) u_i is exposed exactly in time step i, so there is exactly one weak tie from some node in S_{i-1} to u_i. For $i = 0$, because $d(S_0, t) = d_{-1}$, so the gap between d_{-1} and d_0 is caused by $u_0 \in \mathcal{E}(S_0)$. The conclusions still hold.

We define the set of nodes that are the endpoints of the weak ties initiated by S_{i-1} as X_i. X_i is indeed the set of weak-neighbors in directed Kleinberg's small-world network. Apparently $u_i \in X_i$ according to assertion (3) above. If $d_{i-1} - d_i > n^{\frac{1}{4}}$ happens, u_i can be reached by s_i with a weak tie of distance at least $n^{\frac{1}{4}}$. Define $u \to v$ as node u initiates a weak tie with endpoint v. By union bound, we have:

$$
\begin{aligned}
&1 - \Pr(\forall\, 0 \le i < cn^{\frac{1}{4}}, d_{i-1} - d_i \le n^{\frac{1}{4}}) \\
&= \Pr(\exists\, 0 \le i < cn^{\frac{1}{4}}, d_{i-1} - d_i > n^{\frac{1}{4}}) \\
&\le \textstyle\sum_{i=0}^{cn^{\frac{1}{4}}-1} \Pr(d_{i-1} - d_i > n^{\frac{1}{4}}) \\
&\le \textstyle\sum_{i=0}^{cn^{\frac{1}{4}}-1} \Pr(s_i \to u_i, |s_i u_i| > n^{\frac{1}{4}}, u_i \in X_i) \\
&\le \textstyle\sum_{i=0}^{cn^{\frac{1}{4}}-1} \Pr(\exists\, x \in X_i, s_i \to x, |s_i x| > n^{\frac{1}{4}}).
\end{aligned}
\tag{1}
$$

Since there is $i + 1$ nodes in the set S_{i-1}, S_{i-1} initiate $q(i + 1)$ weak ties, which means that $|X_i| \le q(i + 1)$. Denote $\mathcal{H}_i \subseteq 2^V$ to be the set of all sets of nodes with size no more than $q(i+1)$. Then we fix the randomness of X_i and s_i:

$$
\begin{aligned}
&\Pr(\exists\, x \in X_i, s_i \to x, |s_i x| > n^{\frac{1}{4}}) \\
&\le \textstyle\sum_{C \in \mathcal{H}_i} \sum_{v \in V} \Pr\left((X_i = C) \wedge (s_i = v) \wedge (\exists\, x \in C, v \to x, |vx| > n^{\frac{1}{4}})\right) \\
&= \textstyle\sum_{C \in \mathcal{H}_i} \sum_{v \in V} \Pr\left((X_i = C) \wedge (s_i = v)\right) \Pr(\exists\, x \in C, v \to x, |vx| > n^{\frac{1}{4}}) \\
&\le \textstyle\sum_{C \in \mathcal{H}_i} \sum_{v \in V} \Pr\left((X_i = C) \wedge (s_i = v)\right) \sum_{x \in C} \Pr(v \to x, |vx| > n^{\frac{1}{4}}) \\
&\le \textstyle\sum_{C \in \mathcal{H}_i} \sum_{v \in V} \Pr\left((X_i = C) \wedge (s_i = v)\right) \cdot |C| \cdot 2\mathcal{Z}\frac{1}{n^{2 \cdot 1/4}} \\
&\le q(i+1) \cdot 2\frac{\mathcal{Z}}{n^{1/2}} \textstyle\sum_{C \in \mathcal{H}_i} \sum_{v \in V} \Pr\left((X_i = C) \wedge (s_i = v)\right) = 2q(i+1) \cdot \frac{\mathcal{Z}}{n^{1/2}}.
\end{aligned}
$$

By the property of decentralized routing, event $\{(X_i = C) \wedge (s_i = v)\}$ only depends on the random set S_{i-1} and the outgoing weak ties from S_{i-1}, and v

is not in S_{i-1}, while event $\{\exists x \in C, v \to x, |vx| > n^{\frac{1}{4}}\}$ only depends on the outgoing weak ties of the fixed node v. Thus event $\{(X_i = C) \wedge (s_i = v)\}$ is independent of event $\{\exists x \in C, v \to x, |vx| > n^{\frac{1}{4}}\}$. This gives us the first "=" in the equation. For a node x, if $|vx| \le n^{\frac{1}{4}}$, then $\Pr(v \to x, |vx| > n^{\frac{1}{4}}) = 0$; otherwise, $\Pr(v \to x, |vx| > n^{\frac{1}{4}}) \le 2p(v, x) \le 2\mathcal{Z}\frac{1}{n^{2 \cdot 1/4}}$. Hence we have the third "$\le$". Substitute it into Inequality (1):

$$1 - \Pr(\forall\, 0 \le i < cn^{\frac{1}{4}}, d_{i-1} - d_i \le n^{\frac{1}{4}})$$
$$\le \sum_{i=0}^{cn^{\frac{1}{4}}-1} \Pr(\exists\, x \in X_i, s_i \to x, |s_i x| > n^{\frac{1}{4}})$$
$$\le \sum_{i=0}^{cn^{\frac{1}{4}}-1} q(i+1) \cdot 2\frac{\mathcal{Z}}{n^{1/2}} \le cn^{\frac{1}{4}} \cdot qcn^{\frac{1}{4}} \cdot \Theta(\frac{1}{\log n})\frac{2}{n^{1/2}} = O(\frac{1}{\log n}).$$

Due to the above lemma, it is easy to see that the routing time is at least $cn^{\frac{1}{4}}$ with high probability for $\alpha = 2$.

Proof (of Theorem 1 (deterministic routing scheme)). Lemma 1 says, for $\alpha = 2$, in the first $cn^{\frac{1}{4}}$ steps, the grid distance between the current activated set and target t decreases at most $n^{\frac{1}{4}}$ in each step. Thus, for the first $cn^{\frac{1}{4}}$ steps, target t does not belong to the activated set and the routing procedure will continue. Hence with probability of $1 - O(\frac{1}{\log n})$, to activate the target t in $G(n, 2, \alpha)$ with $\alpha = 2$, decentralized 2-complex routing needs at least $cn^{\frac{1}{4}}$ time steps. The expected routing time is $cn^{\frac{1}{4}} \cdot (1 - O(\frac{1}{\log n})) = \Omega(n^{\frac{1}{4}})$.

When $\alpha \in [0, 2)$, like the proof in Lemma 1, we can prove for small $\varepsilon > 0$,

$$1 - \Pr(\forall\, 0 \le i < cn^{\frac{1-\varepsilon}{\alpha+2}}, d_{i-1} - d_i \le 2n^{\frac{\alpha+2\varepsilon}{2(\alpha+2)}})$$
$$\le \sum_{i=0}^{cn^{\frac{1-\varepsilon}{\alpha+2}}-1} \Pr(\exists\, x \in X_i, s_i \to x, |s_i x| > 2n^{\frac{\alpha+2\varepsilon}{2(\alpha+2)}})$$
$$\le \sum_{i=0}^{cn^{\frac{1-\varepsilon}{\alpha+2}}-1} 2(i+1) \cdot \mathcal{Z} \cdot 2(2n^{\frac{\alpha+2\varepsilon}{2(\alpha+2)}})^{-\alpha}$$
$$\le cn^{\frac{1-\varepsilon}{\alpha+2}} \cdot 2cn^{\frac{1-\varepsilon}{\alpha+2}} \cdot \Theta(\frac{1}{n^{1-\alpha/2}}) \cdot O(n^{-\frac{\alpha(\alpha+2\varepsilon)}{2(\alpha+2)}}) = O(n^{-\varepsilon}).$$

So the routing time is $\Omega(n^{\frac{1-\varepsilon}{\alpha+2}})$ with probability at least $1 - O(n^{-\varepsilon})$. By setting $\varepsilon = 0$ and adjusting the parameter c, the expected routing time can be obtained.

When $\alpha > 2$, we can prove that for small $\varepsilon > 0$,

$$\Pr(\forall\, 0 \le i < cn^{\frac{\alpha-2\varepsilon}{2(\alpha+2)}}, d_{i-1} - d_i \le n^{\frac{1+\varepsilon}{\alpha+2}}) \ge 1 - O(n^{-\varepsilon})$$

like the proof above. Hence with probability at least $1 - O(n^{-\varepsilon})$, we need $cn^{\frac{\alpha-2\varepsilon}{2(\alpha+2)}}$ time steps to find the target. Similarly we can get the bound for the expectation.

3.2 Discussion and Extension

We describe complex routing as the task of activating a node as fast as possible. Here we consider the task of activating a target node t that is n^γ grid distance away from the seeds. Similar hardness results can be inferred if we determine the step size and step number of routing cautiously. Here we just sate the theorem and do not provide the redundant proof.

Theorem 2. *For any decentralized routing schemes (even randomized ones), the routing time for 2-complex routing to activate a node with Manhattan distance $n^\gamma (0 < \gamma \le \frac{1}{2})$ away in $G(n, 2, \alpha)$ has the following lower bounds based on the parameter α, for any small $\varepsilon > 0$:*

1. *For $\alpha \in [0, 2)$, the routing time is $\Omega(n^{\frac{2\gamma(1-\varepsilon)}{\alpha+2}})$ with probability at least $1 - O(n^{-\varepsilon'})$ where $\varepsilon' = \max\{2\gamma\varepsilon, (1 - 2\gamma)(1 - \alpha/2)\}$ and the expected routing time is $\Omega(n^{\frac{2\gamma}{\alpha+2}})$.*
2. *For $\alpha = 2$, the routing time is $\Omega(n^{\frac{\gamma}{2}})$ with probability at least $1 - O(\frac{1}{\log n})$ and the expected routing time is $\Omega(n^{\frac{\gamma}{2}})$.*
3. *For $\alpha \in (2, +\infty)$, the routing time is $\Omega(n^{\frac{\gamma(\alpha-2\varepsilon)}{\alpha+2}})$ with probability at least $1 - O(n^{-\varepsilon})$ and the expected routing time is $\Omega(n^{\frac{\gamma\alpha}{\alpha+2}})$.*

We can obtain the same lower bound of routing time for k-complex routing. To ensure the success of complex routing, let $p = q = k$ for the Kleinberg's small-world network model and the size of seed nodes is k. The result is the same with 2-complex routing so we omit it.

Next, we extend our results to complex routing where at most m nodes can be activated in each time step. When $m = 1$, the result is what we covered in Theorem 1 for complex routing. When we do not restrict m, complex routing becomes complex diffusion. Thus a general m allows us to connect complex routing with diffusion, and see how large m is needed to bring down the polynomial lower bound in complex routing. From the theorem we know that we would not get polylogarithmic routing time for complex routing in $G(n, 2, 2)$ where m nodes can be activated in each step, unless $m = n^{\frac{1}{4}}/\log^{O(1)} n$.

Theorem 3. *In decentralized routing, for k-complex routing in $G(n, 2, \alpha)$, if at most m nodes can be activated in each time step, routing time has the following lower bounds based on the parameter α, for any small $\varepsilon > 0$:*

1. *For $\alpha \in [0, 2)$, the routing time is $\Omega(n^{\frac{1-\varepsilon}{\alpha+2}}/m)$ with probability at least $1 - O(n^{-\varepsilon})$ and the expected routing time is $\Omega(n^{\frac{1}{\alpha+2}}/m)$.*
2. *For $\alpha = 2$, the routing time is $\Omega(n^{\frac{1}{4}}/m)$ with probability at least $1 - O(\frac{1}{\log n})$ and the expected routing time is $\Omega(n^{\frac{1}{4}}/m)$.*
3. *For $\alpha \in (2, +\infty)$, the routing time is $\Omega(n^{\frac{\alpha-2\varepsilon}{2(\alpha+2)}}/m)$ with probability at least $1 - O(n^{-\varepsilon})$ and the expected routing time is $\Omega(n^{\frac{\alpha}{2(\alpha+2)}}/m)$.*

Proof. Assuming that S is the set of current activated nodes. In next time step, we can activate m nodes with the knowledge of the out-neighbors of S. But consider the original complex routing, we just activate one node in each step and we have m time steps to activate nodes. After each small step, we have the knowledge of the newly added node. Hence the method of activating m nodes with m time steps is more effective than infecting m nodes in just one time step. Therefore if we need T time steps to find the target with original complex-routing, the routing time with activating m nodes in each time step is at least $\frac{T}{m}$. The expected routing time is $\frac{T}{m} \cdot (1 - O(\frac{1}{\log n}))$. Then the theorem follows.

4   Conclusion

In this paper, we study the routing of complex contagion in Kleinberg's small-world networks. We show that for complex routing the routing time is lower bounded by a polynomial in the number of nodes in the network for the entire range of α, which is qualitatively different from the polylogarithmic upper bound in both complex diffusion and simple routing for $\alpha = 2$. Our results indicate that complex routing is much harder than both complex diffusion and simple routing at the sweetspot.

There are a number of future directions of this work. One may look into complex routing for undirected small-world networks or other variants of the small-world models. The qualitative difference between complex diffusion and complex routing for the case of $\alpha = 2$ may worth further investigation. For example, one may study if there is similar difference for a larger class of graphs, and under what network condition complex routing permits polylogarithmic solutions.

References

1. Adler, J.: Bootstrap percolation. Phys. A: Stat. Mech. Appl. **171**(3), 453–470 (1991)
2. Balogh, J., Pittel, B.G.: Bootstrap percolation on the random regular graph. Random Struct. Algorithms **30**(1–2), 257–286 (2007)
3. Balogh, J., Peres, Y., Pete, G.: Bootstrap percolation on infinite trees and non-amenable groups. Comb. Probab. Comput. **15**(5), 715–730 (2006)
4. Baxter, G.J., Dorogovtšev, S.N., Goltsev, A.V., Mendes, J.F.F.: Bootstrap percolation on complex networks. Phys. Rev. E **82**(1), 011103 (2010)
5. Centola, D., Macy, M.: Complex contagions and the weakness of long ties. Am. J. Sociol. **113**(3), 702–734 (2007)
6. Chalupa, J., Leath, P.L., Reich, G.R.: Bootstrap percolation on a Bethe lattice. J. Phys. C: Solid State Phys. **12**(1), L31 (1979)
7. Chen, N.: On the approximability of influence in social networks. In: Proceedings of the Nineteenth Annual ACM-SIAM Symposium on Discrete Algorithms, pp. 1029–1037 (2008)
8. Chen, W., Wang, Y., Yang, S.: Efficient influence maximization in social networks. In: Proceedings of the 15th ACM SIGKDD, pp. 199–208. ACM (2009)
9. Chen, W., Li, Q., Sun, X., Zhang, J.: The routing ofcomplex contagion in Kleinberg's small-world networks (2015). arXiv preprint: arXiv:1503.00448
10. Easley, D., Kleinberg, J.: Networks, Crowds, and Markets: Reasoning About a Highly Connected World, vol. 8. Cambridge University Press, Cambridge (2010)
11. Ebrahimi, R., Gao, J., Ghasemiesfeh, G., Schoenebeck, G.: Complex contagions in Kleinberg's small world model. In: Proceedings of the 2015 Conference on Innovations in Theoretical Computer Science, pp. 63–72. ACM (2015)
12. Fraigniaud, P., Giakkoupis, G.: The effect of power-law degrees on the navigability of small worlds: [extended abstract]. In: Proceedings of the 28th Annual ACM Symposium on Principles of Distributed Computing, pp. 240–249 (2009)
13. Fraigniaud, P., Giakkoupis, G.: On the searchability of small-world networks with arbitrary underlying structure. In: Proceedings of the 42nd ACM Symposium on Theory of Computing, pp. 389–398 (2010)

14. Ghasemiesfeh, G., Ebrahimi, R., Gao, J.: Complex contagion and the weakness of long ties in social networks: revisited. In: ACM Conference on Electronic Commerce, pp. 507–524 (2013)
15. Granovetter, M.: Getting a Job: A Study of Contacts and Careers. Harvard University Press, Cambridge (1974)
16. Granovetter, M.: Threshold models of collective behavior. Am. J. Sociol. **83**, 1420–1443 (1978)
17. Granovetter, M.S.: The strength of weak ties. Am. J. Sociol. **78**, 1360–1380 (1973)
18. Kempe, D., Kleinberg, J., Tardos, É.: Maximizing the spread of influence through a social network. In: Proceedings of the Ninth ACM SIGKDD, pp. 137–146. ACM (2003)
19. Kempe, D., Kleinberg, J.M., Tardos, É.: Influential nodes in a diffusion model for social networks. In: Caires, L., Italiano, G.F., Monteiro, L., Palamidessi, C., Yung, M. (eds.) ICALP 2005. LNCS, vol. 3580, pp. 1127–1138. Springer, Heidelberg (2005)
20. Kleinberg, J.: The small-world phenomenon: an algorithm perspective. In: Proceedings of the Thirty-Second Annual ACM Symposium on Theory of Computing, pp. 163–170. ACM (2000)
21. Kleinberg, J.M.: Small-world phenomena and the dynamics of information. In: NIPS, pp. 431–438 (2001)
22. Milgram, S.: The small world problem. Psychol. Today **2**(1), 60–67 (1967)
23. Motwani, R., Raghavan, P.: Randomized Algorithms. Cambridge University Press, Cambridge (1995)
24. Newman, M.E., Watts, D.J.: Scaling and percolation in the small-world network model. Phys. Rev. E **60**(6), 7332 (1999)
25. Newman, M.E.J.: Networks: An Introduction. Oxford University Press, Oxford (2010)
26. Watts, D.J., Strogatz, S.H.: Collective dynamics of small-world networks. Nature **393**(6684), 440–442 (1998)
27. Yang, D.-N., Hung, H.-J., Lee, W.-C., Chen, W.: Maximizing acceptance probability for active friending in online social networks. In: Proceedings of the 15th ACM SIGKDD International Conference on Knowledge Discovery and Data Mining, pp. 713–721 (2013)

The Maximum Disjoint Routing Problem

Farhad Shahmohammadi[1,2], Amir Sharif-Zadeh[1],
and Hamid Zarrabi-Zadeh[1(✉)]

[1] Department of Computer Engineering, Sharif University of Technology,
14588 89694 Tehran, Iran
{fshahmohammadi,asharifzadeh}@ce.sharif.edu, zarrabi@sharif.edu
[2] Department of Computer Science, University of California,
Los Angeles, CA 90095, USA

Abstract. Motivated by the bus escape routing problem in printed circuit boards, we revisit the following problem: given a set of n axis-parallel rectangles inside a rectangular region $\mathcal{R}$, find the maximum number of rectangles that can be extended toward the boundary of $\mathcal{R}$, without overlapping each other. We provide an efficient algorithm for solving this problem in $O(n^2 \log^3 n \log \log n)$ time, improving over the current best $O(n^3)$-time algorithm available for the problem.

1 Introduction

In the *maximum disjoint routing* problem, we are given a set of n axis-parallel rectangles inside a rectangular region $\mathcal{R}$, and the goal is to find a maximum number of rectangles that can be extended to the boundary of $\mathcal{R}$, without overlapping any other rectangle, whether it is extended or not. An instance of the problem is illustrated in Fig. 1.

The maximum disjoint routing problem is motivated by the *escape routing problem* in printed circuit boards (PCBs). The objective in the escape routing problem is to route the nets from their pins to the boundary of the enclosing component. There is a vast amount of work on this problem. In particular, the problem of routing a maximum number of nets to the boundary of component using disjoint paths on a grid has been solved efficiently using network flow algorithms [3,4]. Other flow-based solutions to PCB routing can be found in [5,6,14].

Most solutions available for PCB routing including the flow-based ones are net-centric, in the sense that they route the nets individually, without considering a top-level bus structure. However, recent work on escape routing has been shifted to the bus-level, where nets are grouped into buses, and the nets from each bus is required to be routed together [8–11,13]. In this model, the routing of a bus is obtained by projecting the bounding box of the bus onto one of the four sides of the bounding component. If we require to route a maximum number of buses in a single layer without any conflict, the problem becomes equivalent to the maximum disjoint routing problem, as defined above.

The first polynomial-time algorithm for the maximum disjoint routing problem was given by Kong *et al.* [8]. They presented an exact algorithm that solves

© Springer International Publishing Switzerland 2016
T.N. Dinh and M.T. Thai (Eds.): COCOON 2016, LNCS 9797, pp. 319–329, 2016.
DOI: 10.1007/978-3-319-42634-1_26

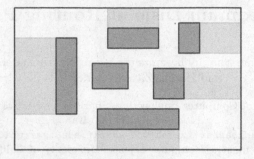

Fig. 1. An instance of the maximum disjoint routing problem. Input rectangles are shown in dark, and extended rectangles are shown in grey.

the problem in $O(n^6)$ time. Their algorithm indeed solves a more general problem of finding a maximum disjoint subset of boundary rectangles, where each boundary rectangle is attached to one of the four sides of the bounding box $\mathcal{R}$. Assadi *et al.* [2] improved this running time by providing an $O(n^4)$-time algorithm for the maximum disjoint routing problem. Ahmadinejad and Zarrabi-Zadeh [1] presented an $O(n^4)$-time algorithm that solves the more general problem of finding the maximum disjoint subset of boundary rectangles. Very recently, Keil *et al.* [7] improved this running time to $O(n^3)$ by presenting an algorithm that solves the maximum independent set problem on outerstring graphs.

Our Contribution. In this paper, we revisit the maximum disjoint routing problem, and present a new algorithm that solves the problem in $O(n^2 \operatorname{polylog}(n))$ time. This improves over the current best $O(n^3)$-time algorithms available for the problem [7]. The main ingredient of our improved result is an efficient solution for a special case of the problem in which each rectangle is a single point. We use a dynamic programming approach equipped with a geometric data structure to solve the point version of the problem efficiently. Our solution involves transforming the points into four dimensions, and using a geometric range searching structure to quickly query and compute subproblems. We then show how our solution for the point version can be extended to the general rectangle case, within the same time bound.

2   Preliminaries

Let S be a set of n axis-parallel rectangles located inside an axis-parallel rectangular region $\mathcal{R}$ in the plane. For each rectangle $r \in S$ and each direction $d \in \{up, down, left, right\}$, we denote by $\delta(r, d)$ the rectangle obtained by extending the rectangle r in direction d toward the boundary of $\mathcal{R}$. We call direction d a *free direction* for rectangle r, if $\delta(r, d)$ does not collide with the initial position of any other rectangle. By checking each pair of rectangles, we can find the free directions for all rectangles in $O(n^2)$ time.

For a rectangle r, we denote by left(r) and right(r) the x-coordinate of the left and the right side of r, respectively. Similarly, we denote by top(r) and bottom(r) the y-coordinate of the top and bottom side of r, respectively. Let $N = 2n + 2$. We define $V = \{v_1, \dots, v_N\}$ to be the set of all vertical lines obtained from extending the left and right sides of the rectangles in S, as well as the vertical sides of $\mathcal{R}$, sorted from left to right. Similarly, we define $H = \{h_1, \dots, h_N\}$ to be the set of all horizontal lines obtained from extending the top and bottom sides of the rectangles in S, as well as the horizontal sides of $\mathcal{R}$, sorted from top to bottom.

3   Subproblems

In order to solve the maximum disjoint routing problem, we first define three subproblems, and show how they can be solved efficiently. The three subproblems are the followings:

- OneWay$_d(i, j, k)$: where $1 \leqslant i < j \leqslant N$, $1 \leqslant k \leqslant N$, and d is one of the four possible directions. If $d \in \{up, down\}$, then OneWay$_d(i, j, k)$ is equal to the maximum number of rectangles lying completely in the area bounded by v_i, v_j, and h_k, that can be routed disjointly toward direction d. If $d \in \{left, right\}$, then we want to solve the same problem for the rectangles lying in the region bounded by h_i, h_j, and v_k.
- Parallel$_k(i, j)$: where $1 \leqslant i < j \leqslant N$, and $k \in \{horz, vert\}$. If $k = vert$, then the objective is to find the maximum number of rectangles lying completely in the area between v_i and v_j that can be routed disjointly toward up and down. If $k = horz$, then we want to solve the same problems for the rectangles between h_i and h_j that can be routed toward right and left.
- Corner$_k(i, j)$: where $1 \leqslant i, j \leqslant N$, and k is one of the four corners of $\mathcal{R}$. If k is the top-left corner, then the goal is to find the maximum number of rectangles lying completely in the top-left corner of $\mathcal{R}$ bounded by v_i and h_j, that can be routed toward up and left. The subproblem is defined analogously for the other three corners.

Whenever the subscripts d and k are clear from the context, we simply drop them in our notations.

Lemma 1. *Each instance of Parallel can be computed in $O(1)$ time, after $O(n^2)$ preprocessing time.*

Proof. Consider a vertical instance of Parallel (i.e., with $k = vert$). Suppose that direction $d \in \{up, down\}$ is free for a rectangle r. Then, for any other rectangle $t \neq r$, $\delta(r, d)$ will not collide with any of t, $\delta(t, up)$, and $\delta(t, down)$. Hence, the answer to Parallel(i, j) is equal to the number of rectangles between v_i and v_j, for which at least one of the directions in $\{up, down\}$ is free. Therefore, each instance of Parallel(i, j) can be solved in $O(n)$ time, leading to $O(n^3)$ overall time. To reduce the total processing time, we precompute and store the values of Parallel$(1, i)$, for all $1 \leqslant i \leqslant N$ in $O(n^2)$ time (note that v_1 represents the left side of $\mathcal{R}$). Now, the answer to Parallel(i, j) can be computed as Parallel$(1, j) -$ Parallel$(1, i - 1)$ in constant time. $\qquad\square$

Lemma 2. *Each instance of OneWay can be computed in $O(1)$ time, after $O(n^2)$ preprocessing time.*

Proof. Consider an instance of OneWay, toward the *up* direction. Similar to the previous lemma, the answer to OneWay(i, j, k) is equal to the number of rectangles in the specified region for which up is a free direction. Similar to Lemma 1, we first initialize the values of OneWay$(1, j, k)$ in $O(n^2)$ time. In order to perform the initialization, we obtain a list L containing all rectangles in S sorted by the y-coordinates of their bottom side in a decreasing order in $O(n \log n)$ time. For each $1 \leqslant j \leqslant N$, we then do the following: Let Q_j be a queue that contains all the rectangles lying between v_1 and v_j, ordered by the decreasing order of their bottom side. Each Q_j can be obtained from L in $O(n)$ time. We now loop through all values in H from top to bottom, and for each $h_k \in H$, we pop all the rectangles in front of the queue which lie completely between v_1, v_j and h_k. The answer to OneWay$(1, j, k)$ is equal to OneWay$(1, j, k-1)$ plus the number of popped rectangles for which the direction up is free. Therefore, we can solve all instances with $i = 1$ in $O(n^2)$ time. The answer to OneWay(i, j, k) can be computed from OneWay$(1, j, k) -$ OneWay$(1, i - 1, k)$ in constant time. $\qquad\square$

Lemma 3. *After $O(n^2)$ preprocessing time, each instance of Corner can be computed in $O(1)$ time.*

Proof. We reduce this subproblem to an instance of the maximum disjoint boundary rectangles (MDBR) problem [1], for which an $O(n^2)$-time solution is available. Assume, without loss of generality, that our instance is a top-left corner. We reduce it to an instance of MDBR as follows. For each rectangle $r \in S$, we replace r by at most two rectangles $\delta(r, d)$, for each direction $d \in \{up, left\}$ which is free for r. It is easy to verify that the maximum number of disjoint rectangles in this new instance is exactly equal to Corner(i, j). It is shown in [1] that after $O(n^2)$ preprocessing time, we can find the maximum number of disjoint rectangles bounded by v_i and h_j, for each pair (i, j) in $O(1)$ time. Therefore, the lemma follows. $\qquad\square$

4 Main Problem

Using Lemmas 1, 2 and 3, we can answer each instance of the OneWay, Parallel, and Corner subproblems in constant time, after $O(n^2)$ preprocessing time. In this section, we show how to solve the main problem efficiently using these subproblems. To this end, we first consider a special case of the problem in which each rectangle in S is a single point. We then generalize our algorithm for the point version to the normal rectangle case.

4.1   The Point Version

Here, we assume that S is a set of n points inside $\mathcal{R}$. Consider an optimal solution to the problem. Let r be the leftmost point in the optimal solution which is routed to right, and ℓ be the rightmost point which is routed to left. Similarly, we denote by t and b the bottom-most point routed upward and the topmost point routed downward, respectively. We distinguish between the following two cases:

- CASE 1: either ℓ is to the left of r (i.e., $\ell_x < r_x$), or b is below t (i.e., $b_y < t_y$)
- CASE 2: $r_x \leqslant \ell_x$ and $t_y \leqslant b_y$

Lemma 4. *In the first case, the optimal solution can be found in $O(n^2)$ time.*

Proof. Assume, w.l.o.g., that $\ell_x < r_x$. (The other case, $b_y < t_y$, can be handled similarly.) We divide $\mathcal{R}$ into five independent regions, as shown in Fig. 2. For each region, the directions to which the points in that region can be routed are shown by arrows.

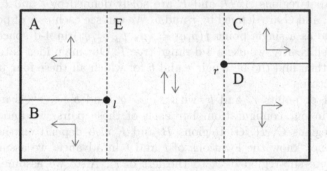

Fig. 2. The first case

The maximum number of points that can be routed in regions A, B, C, and D can be obtained from the Corner subproblems. For region E, the answer can be obtained from the Parallel subproblems. Therefore, the solution for this case can be obtained in $O(1)$ time. Since ℓ and r are not known in advance, we check all possible pairs (ℓ, r) in $O(n^2)$ time, and return the best solution.   □

Lemma 5. *If the second case holds, the optimal solution can be computed in $O(n^2 \log^3 n \log \log n)$ time.*

Proof. In this case, the four points ℓ, r, t, and b form a "wheel structure": ℓ is to the right of r, and t is below b. We assume, w.l.o.g., that t is to the left of b. The rectangle $\mathcal{R}$ is partitioned by this wheel structure into nine regions, as shown in Fig. 3. By the definition of the points forming the wheel structure, the central region is empty. In the other eight regions, labelled by A to H, the points can be routed toward the directions shown by arrows.

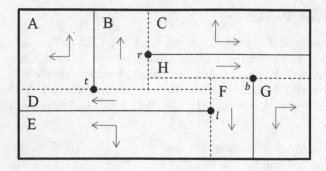

Fig. 3. The second case

Using subproblems OneWay and Corner, we can find answers to each of these regions in $O(1)$ time. Therefore, the optimal solution in this case can be simply obtained by checking all possible $O(n^4)$ configurations defined by ℓ, r, t, and b. In the following, we employ new ideas to perform this step more efficiently.

Observe that regions A, D, and E are solely defined by t and ℓ. Similarly, regions C, H, and G are defined by r and b. We can see each pair of points p and q in the plane as a single point $f(p,q) = (p_x, p_y, q_x, q_y)$ in 4-d space. To solve the problem efficiently, we use a 4-d range tree T. Our main idea is to fix points ℓ and t, and then find the best pair r and b for which all these four points form a wheel structure.

Each pair of points r' and b' with $r'_x \leqslant b'_x$ and $b'_y \leqslant r'_y$ is a candidate for r and b in our configuration. For each of these pairs, we know the best answer for regions C, H, G. Regions B and F also depend on these points. Since we do not know the locations of t and ℓ in advance, we assume that B is the complete sub-region of $\mathcal{R}$ to the left of r'. Also, we assume that F is the complete sub-region to the left side and below b'. After selecting ℓ' and t', some of the points currently in B and F will be out of these regions. We will fix this problem later. For now, we add to T the point $f(r', b')$ with value $\text{Corner}(C) + \text{Corner}(G) + \text{OneWay}(H) + \text{OneWay}(B) + \text{OneWay}(F)$. There are $O(n^2)$ such points. We use an augmented dynamic range tree for T that employs a dynamic version of fractional cascading to support range queries and point insertions/deletions in $O(\log^3 n \log \log n)$ time [12]. The tree itself can be built in $O(n^2 \log^3 n \log \log n)$ time.

Now we try to fix ℓ' and t' and use T to find the best pair (r', b') to form a wheel structure with a maximum possible answer. Each pair of points (t', ℓ') with $t'_x \leqslant \ell'_x$ and $\ell'_y \leqslant t'_y$ is a candidate for (t, ℓ). All the pairs of points (r', b') satisfying the following two conditions are candidates for (r, b) with respect to t' and ℓ':

- (r', b') forms a wheel structure with (t', ℓ'), i.e., $r'_x \leqslant \ell'_x$ and $t'_y \leqslant b'_y$.
- (r', b') is a valid candidate for r and b, i.e., $r'_x \leqslant b'_x$ and $b'_y \leqslant r'_y$.

These conditions together define a 4-d subspace in T. We can use a query on T to find the maximum value in this subspace in $O(\log^3 n \log \log n)$ time. The only problem is that for regions B and F, we are counting some points which are not part of those regions. These points have the following properties:

(i) The points below or on the left side of t' which are counted in region B.
(ii) The points on the left side of ℓ' which are counted in region F.

In order to get rid of these points, we use the following approach. First we loop through all the points as t'. Upon fixing t', we loop through all the points below or to the left of t'. None of these points must be considered in region B, regardless of where r and b are. For each such point, say p, first we check if direction up is free for it. If this direction is not free, then point p has no impact on the value of $OneWay(B)$. Hence, we do not need to take any action. If the direction up is free, then we must remove the impact of p in region B for all pairs of points r' and b' that counted p in B. It means that p must be to the left of r'. All these pairs form a 4-d subspace in T. Therefore, we can call a query on T to subtract one from the value of all pairs in this subspace. This action can be done in $O(n \log^3 n \log \log n)$ time for each point t'.

Now we only have to deal with the points in case (ii). After selecting t', all points below and to the right of t' are candidates for ℓ'. Since ℓ' must be on the right side of t', no point to the left of t' must be considered in region F, regardless of where r' and b' are. We can remove the impact of these points from region F exactly like what we did for the points in region B. Now, we sort all the candidates for ℓ' from left to right, and loop through them. We also set a vertical sweep line on t' and advance it to the left toward ℓ' when we change ℓ'. Whenever our sweep line hits a point like p, p is to the left of ℓ', and hence, it must not be counted in region F. We can remove its impact on F like what we did in the previous cases. Since our sweep line hits each point at most once, we perform at most one query on T for each point, and hence, the overall time is $O(n \log^3 n \log \log n)$. Therefore, after fixing points t' and ℓ', we can use our range tree T to find the points r' and b', yielding the best possible answer for regions B, C, D, F and G. The best answers for regions A, H and E only depend on points t' and ℓ'. Thus, we can find the best answer for all the eight regions using our subproblems and a query on T. After we found the best answer for a particular point t', we need the reset our range tree to its initial condition, so that we can use the same method for the next candidate t'. In order to do this, we can save all the -1 queries that we performed, and call $+1$ queries on the same regions, to return T to its initial state. As a result, we can check all the cases and return the best solution in $O(n^2 \log^3 n \log \log n)$ time. $\square$

The following is a corollary of Lemmas 4 and 5.

Theorem 1. *The point version of the maximum disjoint routing problem can be solved in $O(n^2 \log^3 n \log \log n)$ time.*

4.2   The Rectangle Version

Here, we consider the general rectangles version of the problem, and show how the algorithm described in the previous section for the point version can be extended to the rectangle case. Let S be a set of n rectangles inside $\mathcal{R}$. Consider an optimal solution to the problem. Let r be the rectangle with the leftmost left side routed to the right in the optimal solution, ℓ be the rectangle with the rightmost right side routed to the left, b be the rectangle with the topmost top side routed downward, and t be the rectangle with the bottom-most bottom side routed upward. Again, we consider the problem in two cases:

- CASE 1: either left(r) > right(ℓ), or bottom(t) > top(b).
- CASE 2: left(r) ⩽ right(ℓ) and bottom(t) ⩽ top(b).

Lemma 6. *In the first case, the optimal solution can be found in $O(n^2)$ time.*

Proof. Assume, w.l.o.g., that left(r) > right(ℓ). Similar to the point version, $\mathcal{R}$ is partitioned by r and ℓ into five regions, as shown in Fig. 4. The only difference here is that there might be some rectangles that reside in more than one region (i.e., do not completely reside in any region), and hence, they do not contribute to the solutions obtained for the subproblems. We call such rectangles the *shared* rectangles.

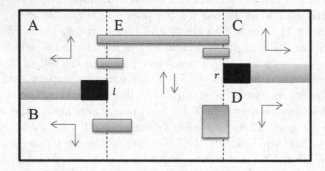

Fig. 4. The first case in the rectangle version.

For each shared rectangle, the restrictions for the intersecting regions also apply to the shared rectangle. It is easy to verify that each shared rectangle in the first case has at most one possible routing direction. We call this direction the *forced* direction for the rectangle. We claim that if the forced direction for a shared rectangle s is free, then in the optimal solution, s must be routed toward that forced direction.

To prove the claim, assume w.l.o.g. that s resides between regions A and E, and that the *up* direction is free for s. No rectangle in A can be routed right, and no rectangle in the regions to the right of A can be routed left. Moreover, no shared rectangle between A and E can be routed either left or right, because of the way ℓ and r are chosen. Therefore, there is no horizontally-routed rectangle that collides with $\delta(s, up)$. Since direction up is free for s, there is no rectangle above s, and hence, extending s to the up direction imposes no new restrictions on the other rectangles. Therefore, in the optimal solution, s must be routed upward, otherwise there would be a solution with a larger number of routed rectangles, contradicting the optimality of the solution. This completes the proof of the claim.

For each rectangle s, we can check if there exists a shared rectangle with a free direction in $\{up, down\}$, if s is selected as either r or ℓ. Note that for each rectangle and each direction, there is at most one shared rectangles which is free in that direction. Therefore, we can preprocess each rectangle and store free shared rectangles for that rectangle in $O(n)$ time. This preprocessing step takes $O(n^2)$ overall time. After that, for each pair of rectangles as r and ℓ, we can find the best solution, using the preprocessed subproblems and considering the shared rectangles, in $O(1)$ time.   □

Lemma 7. *In the second case, computing the optimal solution can be done in* $O(n^2 \log^3 n \log \log n)$ *time.*

Proof. The shared rectangles that arise in the second case are shown in Fig. 5. Each of these shared rectangles has only one forced direction, and hence, they can be treated in a same way as in the first case.

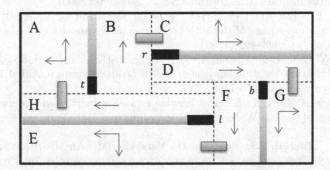

Fig. 5. The second case in the rectangle version

Note that any rectangle that is shared between regions other than those showed in Fig. 5 can not be routed toward any direction, and hence, they can be simply omitted. Therefore, like the first case and the point version, we can find the best solution using preprocessed subproblems and considering the shared rectangles.   □

The main result of this section, which is a corollary of Lemmas 6 and 7, is summarized as follows.

Theorem 2. *The maximum disjoint routing for a set of n rectangles can be computed in* $O(n^2 \log^3 n \log \log n)$ *time.*

5   Conclusions

In this paper, we presented an $O(n^2 \operatorname{polylog}(n))$-time algorithm for the maximum disjoint routing problem, improving over the current best $O(n^3)$-time algorithm available for the problem. Our algorithm simply generalizes to the weighted case, where each rectangle is assigned a weight, and the goal is to route a set of rectangles with maximum total weight. The polylog factor in the runtime of our algorithm is due to the cost of 4-d range queries. Using a more efficient data structure for answering range queries, one would be able to shave some of the log factors from the runtime. The decision version of the problem which asks whether "all" rectangles can be routed disjointly is interesting on its own, and remains open for further investigation.

References

1. Ahmadinejad, A., Zarrabi-Zadeh, H.: The maximum disjoint set of boundary rectangles. In: Proceedings of the 26th Canadian Conference on Computational Geometry, pp. 302–307 (2014)
2. Assadi, S., Emamjomeh-Zadeh, E., Yazdanbod, S., Zarrabi-Zadeh, H.: On the rectangle escape problem. In: Proceedings of the 25th Canadian Conference on Computational Geometry, pp. 235–240 (2013)
3. Chan, W.-T., Chin, F.Y.: Efficient algorithms for finding the maximum number of disjoint paths in grids. J. Algorithms **34**(2), 337–369 (2000)
4. Chan, W.-T., Chin, F.Y., Ting, H.-F.: A faster algorithm for finding disjoint paths in grids. In: Proceedings of the 10th International Symposium on Algorithms and Computation, pp. 393–402 (1999)
5. Fang, J.-W., Lin, I.-J., Chang, Y.-W., Wang, J.-H.: A network-flow-based RDL routing algorithm for flip-chip design. IEEE Trans. Comput. Aided Des. Integr. Circuits Syst. **26**(8), 1417–1429 (2007)
6. Hershberger, J., Suri, S.: Efficient breakout routing in printed circuit boards. In: Proceedings of the 7th Workshop on Algorithms and Data Structures, pp. 462–471 (1997)
7. Keil, J.M., Mitchell, J.S., Pradhan, D., Vatshelle, M.: An algorithm for the maximum weight independent set problem on outerstring graphs. In: Proceedings of the 27th Canadian Conference on Computational Geometry, pp. 2–7 (2015)
8. Kong, H., Ma, Q., Yan, T., Wong, M.D.F.: An optimal algorithm for finding disjoint rectangles and its application to PCB routing. In: Proceedings of the 47th ACM/EDAC/IEEE Design Automation Conference, pp. 212–217 (2010)
9. Kong, H., Yan, T., Wong, M.D.F., Ozdal, M.M.: Optimal bus sequencing for escape routing in dense PCBs. In: Proceedings of the 2007 IEEE/ACM International Conference on Computer-Aided Design, pp. 390–395 (2007)
10. Ma, Q., Wong, M.D.F.: NP-completeness and an approximation algorithm for rectangle escape problem with application to PCB routing. IEEE Trans. Comput. Aided Des. Integr. Circuits Syst. **31**(9), 1356–1365 (2012)
11. Ma, Q., Young, E., Wong, M.D.F.: An optimal algorithm for layer assignment of bus escape routing on PCBs. In: Proceedings of the 48th ACM/EDAC/IEEE Design Automation Conference, pp. 176–181 (2011)
12. Mehlhorn, K., Näher, S.: Dynamic fractional cascading. Algorithmica **5**(1–4), 215–241 (1990)

13. Wu, P.-C., Ma, Q., Wong, M.D.: An ILP-based automatic bus planner for dense PCBs. In: Proceedings of the 18th Asia South Pacific Design Automation Conference, pp. 181–186 (2013)
14. Yan, T., Wong, M.D.: A correct network flow model for escape routing. In: Proceedings of the 46th ACM/EDAC/IEEE Design Automation Conference, pp. 332–335 (2009)

Balanced Allocation on Graphs: A Random Walk Approach

Ali Pourmiri[✉]

School of Computer Science,
Institute for Research in Fundamental Sciences (IPM), Tehran, Iran
pourmiri@ipm.ir

Abstract. The standard balls-into-bins model is a process which randomly allocates m balls into n bins where each ball picks d bins independently and uniformly at random and the ball is then allocated in a least loaded bin in the set of d choices. When $m = n$ and $d = 1$, it is well known that at the end of process the maximum number of balls at any bin, the *maximum load*, is $(1 + o(1))\frac{\log n}{\log \log n}$ with high probability (With high probability refers to an event that holds with probability $1 - 1/n^c$, where c is a constant. For simplicity, we sometimes abbreviate it as whp). Azar et al. [3] showed that for the d-choice process, $d \geqslant 2$, provided ties are broken randomly, the maximum load is $\frac{\log \log n}{\log d} + \mathcal{O}(1)$.

In this paper we propose algorithms for allocating n sequential balls into n bins that are interconnected as a d-regular n-vertex graph G, where $d \geqslant 3$ can be any integer. Let l be a given positive integer. In each round t, $1 \leqslant t \leqslant n$, ball t picks a node of G uniformly at random and performs a non-backtracking random walk of length l from the chosen node. Then it allocates itself on one of the visited nodes with minimum load (ties are broken uniformly at random). Suppose that G has a sufficiently large girth and $d = \omega(\log n)$. Then we establish an upper bound for the maximum number of balls at any bin after allocating n balls by the algorithm, called *maximum load*, in terms of l with high probability. We also show that the upper bound is at most an $\mathcal{O}(\log \log n)$ factor above the lower bound that is proved for the algorithm. In particular, we show that if we set $l = \lfloor (\log n)^{\frac{1+\epsilon}{2}} \rfloor$, for every constant $\epsilon \in (0, 1)$, and G has girth at least $\omega(l)$, then the maximum load attained by the algorithm is bounded by $\mathcal{O}(1/\epsilon)$ with high probability. Finally, we slightly modify the algorithm to have similar results for balanced allocation on d-regular graph with $d \in [3, \mathcal{O}(\log n)]$ and sufficiently large girth.

1 Introduction

The standard balls-into-bins model is a process which randomly allocates m balls into n bins where each ball picks d bins independently and uniformly at random and the ball is then allocated in a least loaded bin in the set of d choices. When $m = n$ and $d = 1$, it is well known that at the end of process the maximum number of balls at any bin, the *maximum load*, is $(1 + o(1))\frac{\log n}{\log \log n}$ with high probability. Azar et al. [3] showed that for the d-choice process, $d \geqslant 2$,

© Springer International Publishing Switzerland 2016
T.N. Dinh and M.T. Thai (Eds.): COCOON 2016, LNCS 9797, pp. 330–341, 2016.
DOI: 10.1007/978-3-319-42634-1_27

provided ties are broken randomly, the maximum load is $\frac{\log\log n}{\log d} + \mathcal{O}(1)$. For a complete survey on the standard balls-into-bins process we refer the reader to [11]. Many subsequent works consider the settings where the choice of bins are not necessarily independent and uniform. For instance, Vöcking [13] proposed an algorithm called *always-go-left* that uses exponentially smaller number of choices and achieve a maximum load of $\frac{\log\log n}{d\phi_d} + \mathcal{O}(1)$ whp, where $1 \leqslant \phi_d \leqslant 2$ is an specified constant. In this algorithm, the bins are partitioned into d groups of size n/d and each ball picks one random bin from each group. The ball is then allocated in a least loaded bin among the chosen bins and ties are broken asymmetrically. In many applications selecting any random set of choices is costly. For example, in peer-to-peer or cloud-based systems balls (jobs, items,...) and bins (servers, processors,...) are randomly placed in a metric space (e.g., $\mathbb{R}^2$) and the balls have to be allocated on bins that are close to them as it minimizes the access latencies. With regard to such applications, Byer et al. [6] studied a model, where n bins (servers) are uniformly at random placed on a geometric space. Then each ball in turn picks d locations in the space and allocates itself on a nearest neighboring bin with minimum load among other d bins. In this scenario, the probability that a location close to a server is chosen depends on the distribution of other servers in the space and hence there is no a uniform distribution over the potential choices. Here, the authors showed the maximum load is $\frac{\log\log n}{\log d} + \mathcal{O}(1)$ whp. Later on, Kenthapadi and Panigrahy [10] proposed a model in which bins are interconnected as a Δ-regular graph and each ball picks a random edge of the graph. It is then placed at one of its endpoints with smaller load. This allocation algorithm results in a maximum load of $\log\log n + \mathcal{O}\left(\frac{\log n}{\log(\Delta/\log^4 n)}\right) + \mathcal{O}(1)$. Peres et al. [12] also considered a similar model where number of balls m can be much larger than n (i.e., $m \gg n$) and the graph is not necessarily regular. Then, they established upper bound $\mathcal{O}(\log n/\sigma)$ for the gap between the maximum and the minimum loaded bin after allocating m balls, where σ is the edge expansion of the graph. Following the study of balls-into-bins with correlated choices, Godfrey [9] generalized the model introduced by Kenthapadi and Panigrahy such that each ball picks an random edge of a hypergraph that has $\Omega(\log n)$ bins and satisfies some mild conditions. Then he showed that the maximum load is a constant whp. Recently, Bogdan et al. [5] studied a model where each ball picks a random node and performs a local search from the node to find a node with local minimum load, where it is finally placed on. They showed that when the graph is a constant degree expander, the local search guarantees a maximum load of $\Theta(\log\log n)$ whp.

Our Results. In this paper, we study balls-into-bins models, where each ball chooses a set of related bins. We propose allocation algorithms for allocating n sequential balls into n bins that are organized as a d-regular n-vertex graph G. Let l be a given positive integer. A non-backtracking random walk (NBRW) W of length l started from a node is a random walk in l steps so that in each step the walker picks a neighbor uniformly at random and moves to that neighbor with an additional property that the walker never traverses an edge twice in a row. Further information about NBRWs can be found in [1,2]. Our allocation algorithm, denoted by $\mathcal{A}(G, l)$, is based on a random sampling of bins from the neighborhood of a given node in G by a NBRW from the node. The algorithm proceeds as follows: In each round t, $1 \leqslant t \leqslant n$, ball t picks a node of G uniformly at random and performs a NBRW $W = (u_0, u_1 \ldots, u_l)$, called l-walk. After that the ball allocates itself on one of the visited nodes with minimum load and ties are broken randomly. Our result concerns bounding the maximum load attained by $\mathcal{A}(G, l)$, denoted by m^*, in terms of l. Note that if the balls are allowed to take NBRWs of length $l = \Omega(\log n)$ on a graph with girth at least l, then the visited nodes by each ball generates a random hyperedge of size $l+1$. Then applying the Godfrey's result [9] implies a constant maximum load whp. So, for the rest of the paper we focus on NBRWs of sub-logarithmic length (i.e., $l = o(\log_d n)$). We also assume that $l = \omega(1)$ and G is a d-regular n-vertex graph with girth at least $\omega(l \log \log n)$ and $d = \omega(\log n)$. However, when $l = \lfloor (\log n)^{\frac{1+\epsilon}{2}} \rfloor$, for any constant $\epsilon \in (0, 1)$, G with girth at least $\omega(l)$ suffices as well. It is worth mentioning that there exist several explicit families of n-vertex d-regular graph with arbitrary degree $d \geqslant 3$ and girth $\Omega(\log_d n)$ (see e.g. [8]).

In order to present the upper bound, we consider two cases:

I. If $l \geqslant 4\gamma_G$, where $\gamma_G = \sqrt{\log_d n}$, then we show that whp,

$$m^* = \mathcal{O}\left(\frac{\log \log n}{\log(l/\gamma_G)} \right).$$

Thus, for a given G satisfying the girth condition, if we set $l = \lfloor (\log_d n)^{\frac{1+\epsilon}{2}} \rfloor$, for any constant $\epsilon \in (0, 1)$, then we have $l/\gamma_G \geqslant (\log n)^{\epsilon/2}$ and by applying the above upper bound we have $m^* = \mathcal{O}(1/\epsilon)$ whp.

II. If $\omega(1) \leqslant l \leqslant 4 \cdot \gamma_G$, then we show that whp,

$$m^* = \mathcal{O}\left(\frac{\log_d n \cdot \log \log n}{l^2} \right).$$

In addition to the upper bound, we prove that whp,

$$m^* = \Omega\left(\frac{\log_d n}{l^2} \right)$$

If G is a d-regular graph with $d \in [3, \mathcal{O}(\log n)]$, then we slightly modify allocation algorithm $\mathcal{A}(G, l)$ and show the similar results for m^* in l. The algorithm $\mathcal{A}'(G, l)$ for sparse graphs proceeds as follows: Let us first define parameter

$$r_G = \lceil 2 \cdot \log_{d-1} \log n \rceil.$$

For each ball t, the ball takes a NBRW of size $l \cdot r_G$, say $(u_0, u_1, \cdots, u_{lr_G})$, and then a subset of visited nodes, $\{u_{j \cdot r_G} \mid 0 \leqslant j \leqslant l\}$, called $potential\ choices$, is selected and finally the ball is allocated on a least-loaded node of potential choices (ties are broken randomly). Provided G has sufficiently large girth, we show the similar upper and lower bounds as the allocation algorithm $\mathcal{A}(G, l)$ on d-regular graphs with $d = \omega(\log n)$.

Comparison with Related Works. The setting of our work is closely related to [5]. In that paper in each step a ball picks a node of a graph uniformly at random and performs a local search to find a node with local minimum load and finally allocates itself on it. They showed that with high probability the local search on expander graphs obtains a maximum load of $\Theta(\log \log n)$. In comparison to the mentioned result, our new protocol achieves a further reduction in the maximum load, while still allocating a ball close to its origin. Our result suggests a trade-off between allocation time and maximum load. In fact we show a constant upper bound for sufficient long walks (i.e., $l = (\log n)^{\frac{1+\epsilon}{2}}$, for any constant $\epsilon \in (0, 1)$). Our work can also be related to the one by Kenthapadi and Panigrahy where each ball picks a random edge from a $n^{\Omega(1/\log \log n)}$-regular graph and places itself on one of the endpoints of the edge with smaller load. This model results into a maximum load of $\Theta(\log \log n)$. Godfrey [9] considered balanced allocation on hypergraphs where balls choose a random edge e of a hypergraph satisfying some conditions, that is, first the size s s of each edge is $\Omega(\log n)$ and $\mathbf{Pr}\,[u \in e] = \Theta(\frac{s}{n})$ for any bin u. The latter one is called *balanced condition*. Berenbrink et al. [4] simplified Godfrey's proof and slightly weakened the balanced condition but since both analysis apply a Chernoff bound, it seems unlikely that one can extend the analysis for hyperedges of size $o(\log n)$. Our model can also be viewed as a balanced allocation on hypergraphs, because every l-walk is a random hyperedge of size $l + 1$ that also satisfies the balanced condition. By setting the right parameter for $l = o(\log n)$, we show that the algorithm achieves a constant maximum load with sub-logarithmic number of choices. In a different context, Alon and Lubetzky [2] showed that if a particle starts a NBRW of length n on n-vertex regualr expander graph with high-girth then the number of visits to nodes has a Poisson distribution. In particular they showed that the maximum visit to a node is at most $(1 + o(1)) \cdot \frac{\log n}{\log \log n}$. Our result can be also seen as an application of the mathematical concept of NBRWs to task allocation in distributed networks.

Techniques. To derive a lower bound for the maximum load we first show that whp there is a path of length l which is traversed by at least $\Omega(\log_d n/l)$ balls. Also, each path contains $l + 1$ choices and hence, by pigeonhole principle there is a node with load at least $\Omega(\log_d n/l^2)$, which is a lower bound for m^*. We establish the upper bound based on *witness graph* techniques. In our model, the potential choices for each ball are highly correlated, so the technique for building the witness graph is somewhat different from the one for standard balls-into-bins. Here we propose a new approach for constructing the witness graph. We also show a key property of the algorithm, called (α, n_1)-*uniformity*, that is useful for our proof technique. We say an allocation algorithm is (α, n_1)-uniform if the probability that, for every $1 \leqslant t \leqslant n_1$, ball t is placed on an arbitrary node is bounded by α/n, where $n_1 = \Theta(n)$ and $\alpha = \mathcal{O}(1)$. Using this property we conclude that for a given set of nodes of size $\Omega(\log n)$, after allocating n_1 balls, the average load of nodes in the set is some constant whp. Using witness graph method we show that if there is a node with load larger than some threshold then there is a collection of nodes of size $\Omega(\log n)$ where each of them has load larger than some specified constant. Putting these together implies that after allocating n_1 balls the maximum load, say m_1^*, is bounded as required whp. To derive an upper bound for the maximum load after allocation n balls, we divide the allocation process into n/n_1 phases and show that the maximum load at the end of each phase increases by at most m_1^* and hence $m^* \leqslant (n/n_1)m_1^*$ whp.

Discussion and Open Problems. In this paper, we proposed a balls-into-bins model, where each ball picks a set of nodes that are visited by a NBRW of length l and place itself on a visited node with minimum load. One may ask whether it is possible to replace a NBRW of length l by several parallel random walks of shorter length (started from the same node) and get the similar results?

In our result we constantly use the assumption that the graph locally looks like a d-ary tree. It is also known that cycles in random regular graph are restively far from each other (e.g., see [7]), so we believe that our approach can be extended for balanced allocation on random regular graphs.

Many works in this area (see e.g. [5,10]) assumed that the underlying networks is regular, it would be interesting to investigate random walk-based algorithms for irregular graphs.

Outline. In Sect. 2, we present notations and some preliminary results that are required for the analysis of the algorithm. In Sect. 3 we show how to construct a witness graph and then in Sect. 4 by applying the results we the upper bound for the maximum load.

2   Notations, Definitions and Preliminaries

In this section we provide notations, definitions and some preliminary results. A *non-backtracking random walk* (NBRW) W of length l started from a node is a simple random walk in l steps so that in each step the walker picks a neighbor uniformly at random and moves to that neighbor with an additional property that the walker never traverses an edge twice in a row. Throughout this paper we assume that $l \in [\omega(1), o(\log_d n)]$ is a given parameter and G is a d-regular graph with girth $10 \cdot l \cdot \log\log n$. Note that we will see that the condition on the girth can be relaxed to $\omega(l)$, for any l higher than $(\log_d n)^{\frac{1+\epsilon}{2}}$, where $\epsilon \in (0,1)$ is a constant.

It is easy to see that the visited nodes by a non-backtracking walk of length l on G induces a path of length l, which is called an l-*walk*. For simplicity, we use W to denote both the l-walk and the set of visited nodes by the l-walk. Also, we define $f(W)$ to be the number of balls in a least-loaded node of W. The *height* of a ball allocated on a node is the number balls that are placed on the node before the ball.

For every two nodes $u, v \in V(G)$, let $d(u,v)$ denote the length of shortest path between u and v in G. Since G has girth at least $\omega(l)$, every path of length at most l is specified by its endpoints, say u and v. So we denote the path by interval $[u,v]$. Also $V(H)$ denotes the vertex set of H. Note that due to the lack of space, proofs of the lemmas are omitted.

Definition 1 (Interference Graph). *For every given pair (G, l), the interference graph $\mathcal{I}(G, l)$ is defined as follows: The vertex set of $\mathcal{I}(G, l)$ is the set of all l-walks in G and two vertices W and W' of $\mathcal{I}(G, l)$ are connected if and only if $W \cap W' \neq \emptyset$. Note that if pair (G, l) is clear from the context, then the interference graph is denoted by $\mathcal{I}$.*

Now, let us interpret allocation process $\mathcal{A}(G, l)$ as follows:

For every ball $1 \leqslant t \leqslant n$, the algorithm picks a vertex of $\mathcal{I}(G, l)$, say W_t, uniformly at random and then allocates ball t on a least-loaded node of W_t (ties are broken randomly). Let $1 \leqslant n_1 \leqslant n$ be a given integer and assume that $\mathcal{A}(G, l)$ has allocated balls until the n_1-th ball. We then define $\mathcal{H}_{n_1}(G, l)$ to be the induced subgraph of $\mathcal{I}(G, l)$ by $\{W_t : 1 \leqslant t \leqslant n_1\} \subset V(\mathcal{I})$.

Definition 2. *Let λ and μ be given positive integers. We say rooted tree $T \subset \mathcal{I}(G, l)$ is a (λ, μ)-tree if T satisfies:*

(1) $|V(T)| = \lambda$,
(2) $|\cup_{W \in V(T)} W| \geqslant \mu$.

Note that the latter condition is well-defined because every vertex of T is an $(l+1)$-element subset of $V(G)$. A (λ, μ)-tree T is called c-loaded, if T is contained in $\mathcal{H}_{n_1}(G, l)$, for some $1 \leqslant n_1 \leqslant n$, and every node in $\cup_{W \in V(T)} W$ has load at least c.

2.1 Appearance Probability of a c-Loaded (λ, μ)-Tree

In this subsection we formally define the notion of (α, n_1)-uniformity for allocation algorithms, and then present our key lemma concerning the uniformity of $\mathcal{A}(G, l)$. By using this lemma we establish an upper bound for the probability that a c-loaded (λ, μ)-tree contained in $\mathcal{H}_{n_1}$ exists.

Definition 3. *Suppose that $\mathcal{B}$ be an algorithm that allocates n sequential balls into n bins. Then we say $\mathcal{B}$ is (α, n_1)-uniform if, for every $1 \leqslant t \leqslant n_1$ and every bin u, after allocating t balls we have that*

$$\mathbf{Pr}\left[\text{ball } t + 1 \text{ is allocated on } u\right] \leqslant \frac{\alpha}{n},$$

where α is some constant.

Lemma 2.1 (Key Lemma). *$\mathcal{A}(G, l)$ is an (α, n_1)-uniform allocation algorithm, where $n_1 = \lfloor n/(6e\alpha) \rfloor$.*

In the next lemma, we derive an upper bound for the appearance probability a c-loaded (λ, μ)-tree, whose proof is inspired by [10, Lemma 2.1].

Lemma 2.2. *Let λ, μ and c be positive integers. Then the probability that there exists a c-loaded (λ, μ)-tree contained in $\mathcal{H}_{n_1}(G, l)$ is at most*

$$n \cdot \exp(4\lambda \log(l + 1) - c\mu).$$

3   Witness Graph

In this section, we show that if there is a node whose load is larger than a threshold, then we can construct a c-loaded (λ, μ)-tree contained in $\mathcal{H}_{n_1}(G, l)$. Our construction is based on an iterative application of a 2-step procedure, called Partition-Branch. Before we explain the construction, we draw the reader's attention to the following remark:

Remark. The intersection (union) of two arbitrary graphs is a graph whose vertex set and edge set are the intersection (union) of the vertex and edge sets of those graphs. Let $\cap_g$ and $\cup_g$ denote the graphical intersection and union. Note that we use $\cap$ ($\cup$) to denote the set intersection (union) operation. Moreover, since G has girth $\omega(l)$, the graphical intersection of every two l-walks in G is either empty or a path (of length $\leqslant l$). Recall that W denotes both an l-walk and the set of nodes in the l-walk.

Partition-Branch. Let $k \geqslant 1$ and $\rho \geqslant 1$ be given integers and W be an l-walk with $f(W) \geqslant \rho + 1$. The Partition-Branch procedure on W with parameters ρ and k, denoted by $PB(\rho, k)$, proceeds as follows:

Partition: It partitions W into k edge-disjoint subpaths:

$$\mathcal{P}_k(W) = \{[u_i, u_{i+1}] \subset W, 0 \leqslant i \leqslant k-1\},$$

where $d(u_i, u_{i+1}) \in \{\lfloor l/k \rfloor, \lceil l/k \rceil\}$.

Branch: For a given $P_i = [u_i, u_{i+1}] \in \mathcal{P}_k(W)$, it finds (if exists) another l-walk W_{P_i} intersecting P_i that satisfies the following conditions:

(C1) $\emptyset \neq W_{P_i} \cap W \subseteq P_i \setminus \{u_i, u_{i+1}\}$.

(C2) $f(W_{P_i}) \geqslant f(W) - \rho$.

We say procedure $PB(\rho, k)$ on a given l-walk W is *valid*, if for every $P \in \mathcal{P}_k(W)$, W_P exists. We usually refer to W as the father of W_P. For a graphical view of the Partition-Branch procedure see Fig. 1.

Definition 4 (Event $\mathcal{N}_\delta$). *For any given $1 \leqslant \delta \leqslant l$, we say that event $\mathcal{N}_\delta$ holds, if after allocating at most n balls by $\mathcal{A}(G, l)$, every path of length δ is contained in less than $6 \log_{d-1} n/\delta$ l-walks that are randomly chosen by $\mathcal{A}(G, l)$.*

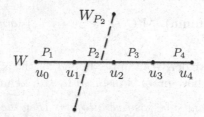

Fig. 1. The Partition step on W for $k = 4$ and the Branch step for P_2 that gives W_{P_2}, shown by dashed line.

For the sake of construction, let us define a set of parameters, depending on d, n, and l, which are used throughout the paper

$$k := \max\{4, \lfloor l/\sqrt{\log_d n}\rfloor\},$$
$$\delta := \lfloor \lfloor l/k \rfloor /4 \rfloor,$$
$$\rho := \lceil 6\log_d n/\delta^2 \rceil.$$

Lemma 3.1. *Suppose that event $\mathcal{N}_\delta$ holds and W be an l-walk with*

$$f(W) \geqslant \rho + 1.$$

Then the procedure $PB(\rho, k)$ on W is valid.

3.1 Construction of Witness Graph

In this subsection, we show how to construct a c-loaded (λ, μ)-tree contained in $\mathcal{H}_{n_1}$. Let $U_{n_1, l, h}$ denote the event that after allocating at most $n_1 \leqslant n$ balls by $\mathcal{A}(G, l)$ there is a node with load at least $h\rho + c + 1$, where $c = \mathcal{O}(1)$ and $h = \mathcal{O}(\log \log n)$ are positive integers that will be fixed later. Suppose that event $U_{n_1, l, h}$ conditioning on $\mathcal{N}_\delta$ happens. Then there is an l-walk R, called root, that corresponds to the ball at height $h\rho + c$ and has $f(R) \geqslant h\rho + c$. Applying Lemma 3.1 shows that $PB(\rho, k)$ on R is valid. So, let us define

$$\mathcal{L}_1 := \{W_P, P \in \mathcal{P}_k(R)\},$$

which is called the first level and R is the father of all l-walks in $\mathcal{L}_1$. (C2) in the Partition-Branch procedure ensures that for every $W \in \mathcal{L}_1$,

$$f(W) \geqslant (h-1)\rho + c.$$

Once we have the first level we recursively build the i-th level from the $(i-1)$-th level, for every $2 \leqslant i \leqslant h$. We know that each W except R is created by the Branch step on its father. Let us fix $W \in \mathcal{L}_{i-1}$ and its father W'. We then apply the Partition step on W and get $\mathcal{P}_k(W)$. We say $P \in \mathcal{P}_k(W)$ is a *free* subpath if it does not share any node with W'. By (C1), we have that $\emptyset \neq W \cap W' = [u, v] \subset P'$, for some $P' \in \mathcal{P}_k(W')$ and hence $d(u, v) \leqslant \lceil l/k \rceil$. So, $[u, v]$ shares node(s) with at most 2 subpaths in $\mathcal{P}_k(W)$ and thus $\mathcal{P}_k(W)$ contains at least $k - 2$ free subpaths. Let $\mathcal{P}_k^0(W) \subset \mathcal{P}_k(W)$ denote an arbitrary set of free subpaths of size $k - 2$. By (C2) and the recursive construction, we have that $f(W) \geqslant (h - i + 1)\rho + c$, for each $W \in \mathcal{L}_{i-1}$. Therefore, by Lemma 3.1, $PB(\rho, k)$ on W is valid. Now we define the i-th level as follows,

$$\mathcal{L}_i = \bigcup_{W \in \mathcal{L}_{i-1}} \{W_P, P \in \mathcal{P}_k^0(W)\}.$$

For a graphical view see Fig. 2. The following lemma guarantees that our construction gives a c-loaded (λ, μ)-tree in $\mathcal{H}_{n_1}$ with desired parameters.

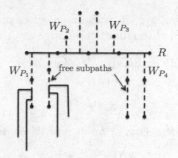

Fig. 2. The first level $\mathcal{L}_1 = \{W_{P_1}, W_{P_2}, W_{P_3}, W_{P_4}\}$ and the Branch step for free subpaths of $\mathcal{P}_k(W_{P_1})$.

Lemma 3.2. *Suppose that G has girth at least $10hl$ and $U_{n_1,l,h}$ conditioning on $\mathcal{N}_\delta$ happens. Then there exists a c-loaded (λ, μ)-tree $T \subset \mathcal{H}_{n_1}$, where $\lambda = 1 + k \sum_{j=0}^{h-1}(k-2)^j$ and $\mu = (l+1) \cdot k(k-2)^{h-1}$.*

4   Balanced Allocation on Dense Graphs

In this section we show the upper bound for the maximum load attained by $\mathcal{A}(G, l)$ for d-regular graph with $d = \omega(\log n)$. Let us recall the set of parameters for given G and l as follows,

$$k := \max\{4, \lfloor l/\sqrt{\log_d n}\rfloor\},$$
$$\delta := \lfloor \lfloor l/k\rfloor/4\rfloor,$$
$$\rho := \lceil 8 \log_d n/\delta^2\rceil,$$

and $U_{n_1,l,h}$ is the event that at the end of round n_1, there is a nodes with load at least $h\rho + c + 1$, where c is a constant and

$$h := \left\lceil \frac{\log\log n}{\log(k-2)} \right\rceil.$$

Note that when $l = (\log n)^{\frac{1+\epsilon}{2}}$ with constant $\epsilon \in (0, 1)$, then

$$k = \lfloor l/\sqrt{\log_d n}\rfloor \geqslant l/\sqrt{\log_3 n} \geqslant (\log n)^{\epsilon/3}.$$

Thus, $h = \left\lceil \frac{\log\log n}{\log(k-2)} \right\rceil$ is a constant. Therefore, in order to apply Lemma 3.2 for this case, it is sufficient that G has girth at least $10hl$ or $\omega(l)$. Also we have the following useful lemma.

Lemma 4.1. *With probability $1 - o(1/n)$, $\mathcal{N}_\delta$ holds.*

Theorem 4.2. *Suppose that G is a d-regular graph with girth at least $10hl$ and $d = \omega(\log n)$. Then, with high probability the maximum load attained by $\mathcal{A}(G, l)$, denoted by m^*, is bounded from above as follows:*

I. If $\omega(1) \leqslant l \leqslant 4\gamma_G$, where $\gamma_G = \sqrt{\log_d n}$. Then we have

$$m^* = \mathcal{O}\left(\frac{\log_d n \cdot \log\log n}{l^2}\right).$$

II. If $l \geqslant 4\gamma_G$, then we have

$$m^* = \mathcal{O}\left(\frac{\log\log n}{\log(l/\gamma_G)}\right).$$

Note that when $l = \Theta(\gamma_G)$, we get the maximum load $\mathcal{O}(\log\log n)$.

Proof. By Lemma 2.1 we have that $\mathcal{A}(G, l)$ is an (α, n_1)-uniform, where $n_1 = \lfloor n/(6e\alpha) \rfloor$. Let us divide the allocation process into s phases, where s is the smallest integer satisfying $sn_1 \geqslant n$. We now focus on the maximum load attained by $\mathcal{A}$ after allocating n_1 balls in the first phase, which is denoted by m_1^*. Let us assume that $U_{n_1,l,h}$ happens. Now, in order to apply Lemma 3.2, we only need that G has girth at least $10hl$. By Lemma 3.2, if $U_{n_1,l,h}$ conditioning on $\mathcal{N}_\delta$ happens, then there is a c-loaded (λ, μ)-tree T contained in $\mathcal{H}_{n_1}$, where $\lambda = 1 + k\sum_{j=0}^{h-1}(k-2)^j$ and $\mu \geqslant (l+1) \cdot k(k-2)^{h-1}$. Thus, we get

$$\mathbf{Pr}\left[U_{n_1,l,h} \mid \mathcal{N}_\delta\right] \mathbf{Pr}\left[\mathcal{N}_\delta\right] \leqslant \mathbf{Pr}\left[T \text{ exists} \mid \mathcal{N}_\delta\right] \mathbf{Pr}\left[\mathcal{N}_\delta\right]$$
$$= \mathbf{Pr}\left[T \text{ exists and} \mathcal{N}_\delta\right]$$
$$\leqslant \mathbf{Pr}\left[T \text{ exists}\right].$$

Therefore using the law of total probability and the above inequality we have

$$\mathbf{Pr}\left[U_{n_1,l,h}\right] = \mathbf{Pr}\left[U_{n_1,l,h} \mid \mathcal{N}_\delta\right] \mathbf{Pr}\left[\mathcal{N}_\delta\right] + \mathbf{Pr}\left[U_{n_1,l,h} \mid \neg\mathcal{N}_\delta\right] \mathbf{Pr}\left[\neg\mathcal{N}_\delta\right]$$
$$\leqslant \mathbf{Pr}\left[T \text{ exists}\right] + \mathbf{Pr}\left[\neg\mathcal{N}_\delta\right]$$
$$= \mathbf{Pr}\left[T \text{ exists}\right] + o(1/n). \tag{1}$$

where the last inequality follows from $\mathbf{Pr}\left[\neg\mathcal{N}_\delta\right] = o(1/n)$ by Lemma 4.1. By definition of h, we get

$$\lambda \leqslant 1 + k(1 + (k-2)^h) \leqslant 2k\log n$$

and

$$\mu = (l+1)k(k-2)^{h-1} \geqslant (l+1)(k-2)^h \geqslant (l+1)\log n.$$

It only remains to bound $\mathbf{Pr}\left[T \text{ exists}\right]$. By applying Lemma 2.2 and substituting μ and λ, we conclude that

$$\mathbf{Pr}\left[T \text{ exists}\right] \leqslant n\exp(4\lambda\log(l+1)) - c\mu \leqslant n\exp\{-z\log n\},$$

where $z = c(l+1) - 8k\log(l+1)$. Depending on k we consider two cases: First, $k = 4$. Then it is easy to see there exists a constant c such that $z \geqslant 2$. Second, $k = \lfloor l/\gamma_G \rfloor$. We know that $l < \log_d n$, so we have $l \leqslant \gamma_G^2$ and hence,

$$z \geqslant cl - 8l\log l/\gamma_G \geqslant l(c - 16\log \gamma_G/\gamma_G) = l(c - o(1)).$$

This yields that for some integer $c > 0$, $z = l(c - o(1)) > 2$ and hence in both cases we get $\mathbf{Pr}\,[T \text{ exists}] = o(1/n)$. Now, by Inequality (1) we infer that $m_1^* \leqslant h\rho + c + 1$ with probability $1 - o(1/n)$. In what follows we show the sub-additivity of the algorithm and concludes that in the second phase the maximum load increases by at most m_1^* whp. Assume that we have a copy of G, say G', whose nodes have load exactly m_1^*. Let us consider the allocation process of a pair of balls $(n_1 + t, t)$, for every $0 \leqslant t \leqslant n_1$, by $\mathcal{A}(G, l)$ and $\mathcal{A}(G', l)$. Let $X_u^{n_1+t}$ and Y_u^t, $t \geqslant 0$ denote the load of $u \in V(G) = V(G')$ after allocating balls $n_1 + t$ and t by $\mathcal{A}(G, l)$ and $\mathcal{A}(G', l)$, respectively. Now we show that for every integer $0 \leqslant t \leqslant n_1$ and $u \in V(G)$ we have that

$$X_u^{n_1+t} \leqslant Y_u^t. \tag{2}$$

When $t = 0$, clearly the inequality holds because $Y_u^0 = m_1^*$. We couple the both allocation processes $\mathcal{A}(G, l)$ and $\mathcal{A}(G', l)$ for a given pair of balls (n_1+t, t), $t \geqslant 0$, as follows. For every $1 \leqslant t \leqslant n_1$, the coupled process first picks a one-to-one labeling function $\sigma_t : V(G) \to \{1, 2, \dots, n\}$ uniformly at random. (Note that σ_t is also defined for G' as $V(G) = V(G')$.) Then it applies $\mathcal{A}(G, l)$ and selects l-walks W_{n_1+t} and its copy, say W_t', in G'. After that, balls $n_1 + t$ and t are allocated on least loaded nodes of W_{n_1+t} and W_t', respectively, and ties are broken in favor of nodes with minimum label. It is easily checked that the defined process is a coupling. Let us assume that Inequality (2) holds for every $t_0 \leqslant t$, then we show it for $t + 1$. Let $v \in W_{n_1+t+1}$ and $v' \in W_{t+1}'$ denote the nodes that are the destinations of pair $(n_1 + t + 1, t + 1)$. Now we consider two cases:

1. $X_v^{n_1+t} < Y_v^t$. Then allocating ball $n_1 + t + 1$ on v implies that

$$X_v^{n_1+t} + 1 = X_v^{n_1+t+1} \leqslant Y_v^t \leqslant Y_v^{t+1}.$$

So, Inequality (2) holds for $t + 1$ and every $u \in V(G)$.

2. $X_v^{n_1+t} = Y_v^t$. Since $W_{n_1+t+1} = W_{t+1}'$, $v \in W_{t+1}'$ and $v' \in W_{n_1+t+1}$. Also we know that v and v' are nodes with minimum load contained in W_{n+t+1} and W_{t+1}, So we have,

$$X_v^{n_1+t} \leqslant X_{v'}^{n_1+t} \leqslant Y_{v'}^t \leqslant Y_v^t.$$

Since $Y_v^t = X_v^{n_1+t}$, we have

$$Y_{v'}^t = Y_v^t = X_v^{n_1+t}.$$

If $v \neq v'$ and $\sigma_{t+1}(v') < \sigma_{t+1}(v)$, then it contradicts the fact that ball n_1+t+1 is allocated on v. Similarly, if $\sigma_{t+1}(v') > \sigma_{t+1}(v)$, it contradicts that ball t is allocated on v'. So, we have $v = v'$ and

$$X_v^{n_1+t} + 1 = X_v^{n+t+1} = Y_v^t + 1 = Y_v^{t+1}.$$

So in both cases, Inequality (2) holds for every $t \geqslant 0$. If we set $t = n_1$, then the maximum load attained by $A(G', l)$ is at most $2m_1^*$ whp. Therefore, by Inequality (2), $2m_1^*$ is an upper bound for the maximum load attained by $\mathcal{A}(G, l)$ in the second phase as well. Similarly, we apply the union bound and conclude that after allocating the balls in s phases, the maximum load m^* is at most sm_1^* with probability $1 - o(s/n) = 1 - o(1/n)$. $\square$

Acknowledgment. The author wants to thank Thomas Sauerwald for introducing the problem and several helpful discussions.

References

1. Alon, N., Benjamini, I., Lubetzky, E., Sodin, S.: Non-backtracking random walks mix faster. Commun. Contemp. Math. **9**, 585–603 (2007)
2. Alon, N., Lubetzky, E.: Poisson approximation for non-backtracking random walks. Israel J. Math. **174**(1), 227–252 (2009)
3. Azar, Y., Broder, A.Z., Karlin, A.R., Upfal, E.: Balanced allocations. SIAM J. Comput. **29**(1), 180–200 (1999)
4. Berenbrink, P., Brinkmann, A., Friedetzky, T., Nagel, L.: Balls into bins with related random choices. J. Parallel Distrib. Comput. **72**(2), 246–253 (2012)
5. Bogdan, P., Sauerwald, T., Stauffer, A., Sun, H.: Balls into bins via local search. In: Proceedings of the 24th Symposium Discrete Algorithms (SODA), pp. 16–34 (2013)
6. Byers, J.W., Considine, J., Mitzenmacher, M.: Geometric generalizations of the power of two choices. In: Proceedings of the 16th Symposium Parallelism in Algorithms and Architectures (SPAA), pp. 54–63 (2004)
7. Cooper, C., Frieze, A.M., Radzik, T.: Multiple random walks in random regular graphs. SIAM J. Discrete Math. **23**(4), 1738–1761 (2009)
8. Dahan, X.: Regular graphs of large girth and arbitrary degree. Combinatorica **34**(4), 407–426 (2014)
9. Godfrey, B.: Balls, bins with structure: balanced allocations on hypergraphs. In: Proceedings of the 19th Symposium Discrete Algorithms (SODA), pp. 511–517 (2008)
10. Kenthapadi, K., Panigrahy, R.: Balanced allocation on graphs. In: Proceedings of the 17th Symposium Discrete Algorithms (SODA), pp. 434–443 (2006)
11. Mitzenmacher, M., Richa, A.W., Sitaraman, R.: The power of two random choices: a survey of technique and results. Handb. Randomized Comput. **1**, 255–312 (2001)
12. Peres, Y., Talwar, K., Wieder, U.: Graphical balanced allocations and the $(1 + \beta)$-choice process. Random Struct. Algorithms (2014). doi:10.1002/rsa.20558
13. Vöcking, B.: How asymmetry helps load balancing. J. ACM **50**(4), 568–589 (2003)

Graph Theory and Algorithms

On the Power of Simple Reductions for the Maximum Independent Set Problem

Darren Strash[⊠]

Institute of Theoretical Informatics, Karlsruhe Institute of Technology,
Karlsruhe, Germany
strash@kit.edu

Abstract. Reductions—rules that reduce input size while maintaining the ability to compute an optimal solution—are critical for developing efficient maximum independent set algorithms in both theory and practice. While several simple reductions have previously been shown to make small domain-specific instances tractable in practice, it was only recently shown that advanced reductions (in a measure-and-conquer approach) can be used to solve real-world networks on millions of vertices [Akiba and Iwata, TCS 2016]. In this paper we compare these state-of-the-art reductions against a small suite of simple reductions, and come to two conclusions: just two simple reductions—vertex folding and isolated vertex removal—are sufficient for many real-world instances, and further, the power of the advanced rules comes largely from their initial application (i.e., kernelization), and not their repeated application during branch-and-bound. As a part of our comparison, we give the first experimental evaluation of a reduction based on maximum critical independent sets, and show it is highly effective in practice for medium-sized networks.

Keywords: Maximum independent set · Minimum vertex cover · Kernelization · Reductions · Exact algorithms

1 Introduction

Given a graph $G = (V, E)$, the maximum independent set problem asks us to compute a maximum cardinality set of vertices $I \subseteq V$ such that no vertices in I are adjacent to one another. Such a set is called a *maximum independent set* (MIS). The maximum independent set problem has applications in classification theory, information retrieval, computer vision [13], computer graphics [29], map labeling [17,32] and routing in road networks [20], to name a few. However, the maximum independent set problem is NP hard [16], and therefore, the currently-best-known algorithms take exponential time.

1.1 Previous Work

Most previous work has focused on the *maximum clique* problem and the *minimum vertex cover* problem, which are complementary to ours. That is, the maximum clique in the complement graph $\bar{G}$ is a maximum independent set in G, and if C is a minimum vertex cover in G, then $V \setminus C$ is a maximum independent set.

© Springer International Publishing Switzerland 2016
T.N. Dinh and M.T. Thai (Eds.): COCOON 2016, LNCS 9797, pp. 345–356, 2016.
DOI: 10.1007/978-3-319-42634-1_28

For computing a maximum clique, there are many branch-and-bound algorithms that are efficient in practice [26,27,31]. These algorithms achieve fast running times by prescribing the order to select vertices during search and by implementing fast-but-effective pruning techniques, such as those based on approximate graph coloring [31] or MaxSAT [24]. Among the fastest of these algorithms is the MCS algorithm by Tomita et al. [31], which is competitive in dense graphs even against algorithms that use bit parallelism [26]. Further priming these algorithms with a large initial solution obtained using local search [4] can be surprisingly effective at speeding up search [6].

Several techniques based on kernelization [1,15] have been very promising in solving both the maximum independent set and minimum vertex cover problems. In particular, Butenko et al. [10] showed that isolated vertex reductions disconnect medium-sized graphs derived from error-correcting codes into small connected components that can be solved optimally. Butenko and Trukhanov [11] introduced a reduction based on critical independent sets, finding exact maximum independent sets in graphs with up to 18,000 vertices generated with the Sanchis graph generator [28]. Though these works apply reduction techniques as a preprocessing step, further works apply reductions as a natural step of the algorithm. In the area of exact algorithms, it has long been clear that applying reductions in a measure-and-conquer approach can improve the theoretical running time of vertex cover and independent set algorithms [9,14]. However, few experiments have been conducted on the real-world efficacy of these techniques.

Recently, Akiba and Iwata [3] showed that applying advanced reductions with sophisticated branching rules in a measure-and-conquer approach is highly effective in practice. They show that an exact minimum vertex cover, and therefore an exact maximum independent set, can be found in many large complex networks with up to 3.2 million vertices in much less than a second. Further, on nearly all of their inputs, the state-of-the-art branch-and-bound algorithm MCS [31] fails to finish within 24 h. Thus, their method is orders of magnitude faster on these real-world graphs.

1.2   Our Results

While the results of Akiba and Iwata [3] are impressive, it is not clear how much their advanced techniques actually improve *search* compared to existing techniques. A majority of the graphs they tested have kernel size zero, and therefore no branching is required. We show that just 2 simple reduction rules—isolated vertex removal and vertex folding—are sufficient to make many of their test instances tractable with standard branch-and-bound solvers. We further provide the first comparison with another class of reductions that are effective on real-world complex networks: the critical independent set reduction of Butenko and Trukhanov [11] and the variant due to Larson [22], which computes a maximum critical independent set.

2 Preliminaries

We work with an undirected graph $G = (V, E)$ where V is a set of n vertices and $E \subset \{\{u, v\} \mid u, v \in V\}$ is a set of m edges. The open neighborhood of a vertex v, denoted $N(v)$, is the set of all vertices w such that $(v, w) \in E$. We further denote the closed neighborhood by $N[v] = N(v) \cup \{v\}$. We similarly define the open and closed neighborhoods of a set of vertices U to be $N(U) = \bigcup_{u \in U} N(u)$ and $N[U] = N(U) \cup U$, respectively. Lastly, for vertices $S \subseteq V$, the induced subgraph $G[S] \subseteq G$ is the graph on the vertices in S with edges in E between vertices in S.

2.1 Reduction Rules

There are several well-known reduction rules that can be applied to graphs for the minimum vertex cover problem (and hence the maximum independent set problem) to reduce the input size to its irreducible equivalent, the *kernel* [1]. Each reduction allows us to choose vertices that are in some MIS by following simple rules. If an MIS is found in the kernel, then undoing the reductions gives an MIS in the original graph. Reduction rules are typically applied as a preprocessing step. The hope is that the kernel is *small enough* to be solved by existing solvers in feasible time. If the kernel is empty, then a maximum independent set is found by simply undoing the reductions. We now briefly describe three classes of reduction rules that we consider here.

Simple Reductions. We first describe two simple reductions: *isolated vertex removal* and *vertex folding*.

An isolated vertex, also called a *simplicial vertex*, is a vertex v whose neighborhood forms a clique. That is, there is a clique C such that $V(C) \cap N(v) = N(v)$. Since v has no neighbors outside of the clique, it must be in *some* maximum independent set. Therefore, we can add v to the maximum independent set we are computing, and remove v and C from the graph. Isolated vertex removal was shown by Butenko et al. [10] to be highly effective in finding exact maximum independent sets on graphs derived from error-correcting codes [10]. This reduction is typically restricted to vertices of degree zero, one, and two in the literature. However, we consider vertices of any degree.

Vertex folding was first introduced by Chen et al. [12] to reduce the theoretical running time of exact branch-and-bound algorithms for the maximum independent set problem. This reduction is applied whenever there is a vertex v with degree 2 and non-adjacent neighbors u and w. Either v or both u and w are in some MIS. Therefore, we can contract u, v, and w to a single vertex v' and add the appropriate vertices to the MIS after finding an MIS in the kernel.

Critical Independent Set Reductions. One further reduction method shown to be effective in practice for sparse graphs is the critical independent set reduction by Butenko and Trukhanov [11]. A *critical set* is a set $U \subseteq V$ that

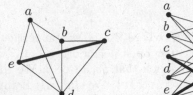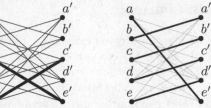

Fig. 1. A graph G (left) and its bi-double graph $B(G)$ (middle), illustrating that edges of G become two edges in $B(G)$. Right: a maximum matching (in this instance, a perfect matching) in $B(G)$.

maximizes $|U| - |N(U)|$ and $I_c = U \setminus N(U)$ is called a *critical independent set*. Butenko and Trukhanov show that every critical independent set is contained in some maximum independent set, and show that one can be found in polynomial time. Their algorithm works by repeatedly computing some critical independent set I_c and removing $N[I_c]$ from the graph, stopping when I_c is empty.

A critical set can be found by first computing the *bi-double graph* $B(G)$, then computing a maximum independent set in $B(G)$ [2,22,34]. $B(G)$ is a bipartite graph with vertices $V \cup V'$ where V' is a copy of V, and contains edge $(u, v') \subseteq V \times V'$ if and only if $(u, v) \in E$. Since $B(G)$ is bipartite, the maximum independent set in $B(G)$ can be solved by computing a maximum bipartite matching in polynomial time [18].

Butenko and Trukhanov [11] use the standard augmenting path algorithm to compute a maximum independent set in $B(G)$, and hence find a critical set in G, in $O(nm)$ time. One drawback of their approach is that the quality of the reduction depends on the maximum independent set found in the bi-double graph. As noted by Larson [22], if there is a perfect matching in $B(G)$ (such as in Fig. 1, right), then G has an empty critical empty set. However, in the experiments by Butenko and Trukhanov [11] these worst cases were not observed.

To prevent the worst-case, Larson [22] gave the first algorithm to find a maximum critical independent set, which accumulates vertices that are in *some* critical independent set and excludes their neighbors. He further gave a simple method to test if a vertex v is in a critical independent set: vertex v is in a critical independent set if and only if $\alpha(B(G)) = \alpha(B(G) - \{v, v'\} - N(\{v, v'\})) + 2$, where $\alpha(\cdot)$ is the *independence number*—the size of a maximum independent set. A naive approach would compute a new maximum matching from scratch to compute the independence number of each such bi-double graph, taking $O(n^2 m)$ time total (or $O(n^{3/2} m)$ time with the Hopcroft–Karp algorithm [18]). However, we can save the matching between executions to ensure only few augmenting paths are computed for each subsequent matching, giving $O(m^2)$ running time, which is better when $m = o(n^{3/2})$.

Advanced Reduction Rules. We list the advanced reduction rules from Akiba and Iwata [3]. Refer to Akiba and Iwata [3] for a more thorough discussion, including implementation details.

Firstly, they use vertex folding and degree-1 isolated vertex removal (also called *pendant* vertex removal), as described previously. They further test a full suite of other reductions from the literature, which we now briefly describe.

Linear Programming: A well-known [25] linear programming relaxation for the MIS problem with a half-integral solution (i.e., using only values 0, 1/2, and 1) can be solved using bipartite matching: maximize $\sum_{v \in V} x_v$ such that $\forall (u, v) \in E$, $x_u + x_v \leq 1$ and $\forall v \in V$, $x_v \geq 0$. Vertices with value 1 must be in some MIS and can thus be removed from G along with their neighbors. Akiba and Iwata [3] compute a solution whose half-integral part is minimal [19].

Unconfined [33]: Though there are several definitions of *unconfined* vertex in the literature, we use the simple one from Akiba and Iwata [3]. A vertex v is *unconfined* when determined by the following simple algorithm. First, initialize $S = \{v\}$. Then find a $u \in N(S)$ such that $|N(u) \cap S| = 1$ and $|N(u) \setminus N[S]|$ is minimized. If there is no such vertex, then v is confined. If $N(u) \setminus N[S] = \emptyset$, then v is unconfined. If $N(u) \setminus N[S]$ is a single vertex w, then add w to S and repeat the algorithm. Otherwise, v is confined. Unconfined vertices can be removed from the graph, since there always exists an MIS I with no unconfined vertices.

Twin [33]: Let u and v be vertices of degree 3 with $N(u) = N(v)$. If $G[N(u)]$ has edges, then add u and v to I and remove u, v, $N(u)$, $N(v)$ from G. Otherwise, some vertices in $N(u)$ may belong to some MIS I. We still remove u, v, $N(u)$ and $N(v)$ from G, and add a new gadget vertex w to G with edges to u's two-neighborhood (vertices at a distance 2 from u). If w is in the computed MIS, then none of u's two-neighbors are I, and therefore $N(u) \subseteq I$. Otherwise, if w is not in the computed MIS, then some of u's two-neighbors are in I, and therefore u and v are added to I.

Alternative: Two sets of vertices A and B are set to be *alternatives* if $|A| = |B| \geq 1$ and there exists an MIS I such that $I \cap (A \cup B)$ is either A or B. Then we remove A and B and $C = N(A) \cap N(B)$ from G and add edges from each $a \in N(A) \setminus C$ to each $b \in N(B) \setminus C$. Then we add either A or B to I, depending on which neighborhood has vertices in I. Two structures are detected as alternatives. First, if $N(v) \setminus \{u\}$ induces a complete graph, then $\{u\}$ and $\{v\}$ are alternatives (a *funnel*). Next, if there is a chordless 4-cycle $a_1 b_1 a_2 b_2$ where each vertex has at least degree 3. Then sets $A = \{a_1, a_2\}$ and $B = \{b_1, b_2\}$ are alternatives when $|N(A) \setminus B| \leq 2$, $|N(A) \setminus B| \leq 2$, and $N(A) \cap N(B) = \emptyset$.

Packing [3]: Given a non-empty set of vertices S, we may specify a *packing constraint* $\sum_{v \in S} x_v \leq k$, where x_v is 0 when v is in some MIS I and 1 otherwise. Whenever a vertex v is excluded from I (i.e., in the unconfined reduction), we remove x_v from the packing constraint and decrease the upper bound of the constraint by one. Initially, packing constraints are created whenever a vertex v is excluded from or included in the MIS. The simplest case for the packing reduction is when k is zero: all vertices must be in I to satisfy the constraint. Thus, if there is no edge in $G[S]$, S may be added to I, and S and $N(S)$ are removed from G. Other cases are much more complex. Whenever packing reductions are applied, existing packing constraints are updated and new ones are added.

3   Experimental Results

We first investigate the size of kernels computed by all kernelization techniques. We test four techniques: (1) using only isolated vertex removal and vertex folding (Simple), (2) using the critical independent set reduction rule due to Butenko and Trukhanov [11] (Critical), (3) the version of (2) by Larson [22] that always computes a maximum critical independent set (MaxCritical), and (4) the reductions tested by Akiba and Iwata [3] (Advanced). Note that we use the standard augmenting paths algorithm for computing a maximum bipartite matching (and not the Hopcroft–Karp algorithm [18]) to be consistent with the original experiments by Butenko and Trukhanov [11].

Next, we investigate the time to compute an exact solution on large instances. We test two algorithms: the full branch-and-reduce algorithm due to Akiba and Iwata [3] (B&R), and Simple kernelization followed by MCS, a state-of-the-art clique solver due to Tomita et al. [31] (Simple+MCS). We use our own implementation of MCS[1], since the code for the original implementation is not available and because we modify the MCS algorithm to solve the maximum independent set problem. We choose MCS because it is one of the leading solvers in practice, even competing with the bit-board implementations of San Segundo et al. [26,27].

Instances. We run our algorithms on synthetically-generated graphs, as well as a large corpus of real-world sparse data sets. For synthetic cases, we use graphs generated with the Sanchis graph generator [28]. For medium-sized real-world graphs, we consider small Erdős co-authorship networks from the Pajek data set [5] and biological networks from the Biological General Repository for Interaction Datasets v3.3.112 (BioGRID) [30]. We further consider large complex networks (including co-authorship networks, road networks, social networks, peer-to-peer networks, and Web crawl graphs) from the Koblenz Network Collection (KONECT) [21], the Stanford Large Network Dataset Repository (SNAP) [23], and the Laboratory for Web Algorithmics (LAW) [7,8].

3.1   Experimental Setup

All of our experiments were exclusively run on a machine with Ubuntu 14.04.3 and Linux kernel version 3.13.0-77. The machine has four Octa-Core Intel Xeon E5-4640 processors running at 2.4 GHz, 512 GB local memory, 420 MB L3-Cache, and 48256 KB L2-Cache. For Advanced reductions as well as B&R, we compiled and ran the original Java implementation of Akiba and Iwata[2] [3] with Java 8 update 60. We implemented all other algorithms[3] in C++11, and compiled them with gcc version 4.8.4 with optimization flag -O2. Each algorithm was run for one hour. All running times listed in our tables are in seconds, and we mark a data set with '-' when an algorithm does not finish within the time limit. We indicate the best solution, and the time to achieve it, by marking the value **bold**.

---

[1] https://github.com/darrenstrash/open-mcs.

[2] https://github.com/wata-orz/vertex_cover.

[3] https://github.com/darrenstrash/kernel-mis.

Table 1. We give the kernel size k and running time t for each reduction technique on synthetically-generated Sanchis data sets. We also list the data used to generate the graphs: the number of vertices n, number of edges m, and independence number $\alpha(G)$.

	Graph		Critical		MaxCritical		Advanced		Simple	
n	m	$\alpha(G)$	k	t	k	t	k	t	k	t
1 000	186 723	505	0	**0.06**	0	0.73	0	0.16	1 000	0.00
1 000	181 256	524	0	0.16	0	0.86	0	**0.15**	1 000	0.00
2 000	711 955	1 067	0	0.74	0	5.33	0	**0.37**	2 000	0.01
2 000	686 341	1 103	0	1.08	0	5.57	0	**0.34**	2 000	0.01
3 000	536 831	1 535	2 930	0.26	0	1.54	0	**0.02**	0	0.14
3 000	513 773	1 563	2 874	0.24	0	1.59	0	**0.02**	0	0.13
4 000	929 429	2 069	3 862	0.49	0	2.83	0	**0.03**	0	0.31
4 000	805 011	2 309	0	3.63	0	29.17	0	**0.47**	4 000	0.01
5 000	1 258 433	2 717	4 566	0.70	0	4.85	0	**0.03**	0	0.50
5 000	517 013	3 132	0	3.37	0	25.42	0	0.27	0	**0.18**
6 000	1 731 295	3 302	5 396	1.02	0	7.06	0	**0.04**	0	0.78
6 000	1 507 280	3 412	5 176	0.94	0	7.18	0	**0.04**	0	0.65
7 000	588 713	4 493	5 014	1.06	0	10.91	0	**0.03**	0	0.23
8 000	3 099 179	4 394	7 212	1.96	0	12.41	0	**0.06**	0	1.93
8 000	428 619	5 249	0	5.58	0	48.21	0	0.17	0	**0.29**
9 000	4 040 615	4 927	0	26.23	0	239.21	0	0.74	9 000	**0.04**
9 000	451 349	5 899	6 202	1.45	0	18.20	0	**0.02**	0	0.20
10 000	4 794 713	5 507	8 986	3.02	0	19.86	0	**0.07**	0	3.78
10 000	3 775 385	5 811	0	37.28	0	274.31	0	1.27	9 994	**0.06**
11 000	6 344 649	5 901	10 198	4.35	0	23.30	0	**0.09**	0	5.62
11 000	2 479 688	6 862	0	33.91	0	223.67	0	1.44	0	**1.10**
12 000	5 378 750	6 973	8 552	16.46	0	78.69	0	**0.07**	0	4.55
12 000	4 827 152	7 098	194	63.80	0	368.03	0	**0.07**	0	5.35
13 000	5 638 263	7 698	0	75.59	0	510.73	0	6.56	0	**4.50**
13 000	1 319 528	8 474	0	24.54	0	185.99	0	0.40	0	**1.15**
14 000	10 723 774	7 417	4	78.11	0	819.25	0	25.24	13 880	**0.67**
14 000	3 250 904	8 844	10 312	4.53	0	47.93	0	**0.05**	0	2.45
15 000	6 799 463	8 993	12 014	5.82	0	50.81	0	**0.10**	0	6.58
15 000	4 207 335	9 413	0	80.07	0	526.90	0	2.98	0	**3.37**
16 000	4 807 361	10 042	0	96.05	0	627.97	0	3.76	0	**3.44**
16 000	14 309 249	8 401	570	101.08	0	1 108.03	0	34.14	0	**29.35**
17 000	803 659	11 239	11 522	5.17	0	64.71	0	**0.03**	0	0.63
17 000	10 662 300	9 898	14 202	7.66	0	60.82	0	**0.14**	0	12.88
18 000	5 064 751	11 412	256	124.12	0	683.34	0	**0.08**	0	6.40
18 000	1 970 506	11 782	32	53.56	0	372.49	0	**0.05**	0	1.21

3.2 Kernel Sizes

First, we compare the kernel sizes computed by each reduction technique. We first run all algorithms on synthetically generated graphs, on which Critical was previously shown to be effective [11]. We generate instances with a known clique number using the Sanchis graph generator[4] [28], and then take the complement. Like Butenko and Trukhanov [11], we choose the clique number (and thus, the independence number in the complement graph) to be at least $n/2$.

As can be seen in Table 1, Critical succeeds in reducing the kernel to empty in many cases. During testing, we noticed that Critical did not enter a second iteration on most graphs. That is, in general, either the first critical independent set matched the size of a maximum independent set or the remaining graph had an empty critical set. It is unclear what causes this behavior, but we conjecture

---

[4] ftp://dimacs.rutgers.edu/pub/challenge/.

Table 2. We give the kernel size k and running time t for each reduction technique on Erdős and BioGRID graphs. We further give the number of vertices n and edges m for each graph.

Graph Name	n	m	Critical k	t	MaxCritical k	t	Advanced k	t	Simple k	t
Erdős Graphs										
erdos971	472	1 314	350	0.01	124	0.06	**0**	**0.00**	0	0.00
erdos972	5 488	8 972	46	0.20	0	13.83	**0**	**0.01**	0	0.00
erdos981	485	1 381	373	0.01	205	0.08	**0**	**0.00**	0	0.00
erdos982	5 822	9 505	44	0.22	0	15.65	**0**	**0.01**	0	0.00
erdos991	492	1 417	398	0.01	218	0.07	**0**	**0.01**	0	0.00
erdos992	6 100	9 939	42	0.22	0	17.11	**0**	**0.01**	0	0.00
BioGRID Graphs										
Arabidopsis-thaliana	7 225	17 223	1 534	1.16	188	19.18	**0**	**0.02**	31	0.00
Bos-taurus	389	357	109	0.01	3	0.05	**0**	**0.00**	0	0.00
Caenorhabditis-elegans	3 974	7 918	758	0.30	18	5.38	**0**	**0.01**	0	0.00
Candida-albicans-SC5314	379	371	66	0.00	14	0.05	**0**	**0.00**	0	0.00
Danio-rerio	238	249	71	0.00	11	0.02	**0**	**0.00**	0	0.00
Drosophila-melanogaster	8 229	39 086	3 479	2.78	973	28.01	**0**	**0.02**	30	0.01
Escherichia-coli	139	122	14	0.00	0	0.01	**0**	**0.00**	0	0.00
Gallus-gallus	336	343	81	0.00	3	0.04	**0**	**0.00**	0	0.00
Hepatitus-C-Virus	113	111	2	0.00	0	0.01	**0**	**0.00**	0	0.00
Homo-sapiens	19 592	169 285	5 675	18.70	1 629	210.89	**0**	**0.05**	150	0.03
Human-Herpesvirus-1	140	140	12	0.00	0	0.01	**0**	**0.00**	0	0.00
Human-Herpesvirus-4	219	217	2	0.00	0	0.02	**0**	**0.00**	0	0.00
Human-Herpesvirus-8	137	138	**0**	**0.00**	0	0.01	**0**	**0.00**	0	0.00
Human-HIV-1	1 030	1 186	**0**	**0.00**	0	0.36	**0**	**0.00**	0	0.00
Mus-musculus	8 567	19 265	1 377	1.39	51	27.56	**0**	**0.01**	0	0.00
Oryctolagus-cuniculus	183	168	28	0.00	0	0.02	**0**	**0.00**	0	0.00
Plasmodium-falciparum-3D7	1 224	2 443	336	0.04	0	0.40	**0**	**0.00**	0	0.00
Rattus-norvegicus	3 066	4 139	533	0.14	15	3.04	**0**	**0.01**	6	0.00
Saccharomyces-cerevisiae	6 660	228 752	5 732	2.48	5 180	212.87	**4 086**	**0.96**	4 575	0.05
Schizosaccharomyces-pombe	4 143	57 049	840	1.41	194	11.72	**0**	**0.01**	0	0.02
Xenopus-laevis	473	520	160	0.01	16	0.06	**0**	**0.00**	7	0.00

it could be due to how we compute the maximum matching: we use depth-first search in the bi-double graph. It is unclear which search strategy Butenko and Trukhanov [11] use in their experiments. MaxCritical always computes an empty kernel on these instances; however, it is significantly slower than Critical. This is because Critical computes only 2 maximum matchings on typical instances, while MaxCritical computes many more.

We now turn our attention to Simple and Advanced. Advanced is the clear winner on the Sanchis graphs. It always computes an empty kernel, and does so quickly. However, Simple also computes empty kernels on 28 of the instances. Even though Simple is only faster than Advanced on four instances, Simple still computes exact solutions on these instances within a few seconds. Therefore, the Advanced reductions are not required to make these instances tractable.

We further tested all algorithms on medium-sized real-world graphs. We ran all four reduction algorithms on Erdős collaboration graphs from the Pajek data set and on biological graphs from the BioGRID data set (we only show results on those graphs with 100 or more vertices). As seen in Table 2, MaxCritical still gives consistently smaller kernels than Critical, but unlike the Sanchis graphs, not all kernels are empty. However, the Simple reductions give consistently small kernels on these real-world instances, computing an empty kernel on all but 4 instances, and doing so as fast as Advanced. The size of three of these non-empty kernels is well within the range of feasibility of existing MIS solvers. Neither Simple nor Advanced can solve the `Saccharomyces-cerevisiae` data set exactly, and the Simple kernel is within 12 % of the Advanced kernel size.

3.3   Exact Solutions on Large-Scale Complex Networks

We now focus on computing an exact MIS for the larger instances considered by Akiba and Iwata [3]. We test the branch-and-reduce algorithm by Akiba and Iwata (B&R) that uses Advanced reductions with branching rules during recursion. We also run Simple to kernelize the graph, and then run MCS on the remaining connected components (Simple+MCS). Results are presented in Table 3. Since Critical and MaxCritical are slow on large instances and less effective on medium-sized real-world instances (see Table 2), we exclude them from these experiments.

Similar to the original experiments of Akiba and Iwata [3], B&R computes an exact MIS on 42 of these instances. However, surprisingly, Simple+MCS also computes exact solutions for 33 of these instances. In the remaining nine instances where B&R computes a solution but Simple+MCS does not, we see that the size k_{max} of the maximum connected component in the kernel is significantly smaller for B&R. For six of these instances, k_{max} is less than 600, which is within the range of traditional solvers. Therefore, we conclude that the speed of B&R is primarily due to the initial kernelization on these instances. However, the remaining three instances—`web-BerkStan`, `web-NotreDame` and `libimseti`—have kernels that are too large for traditional solvers. Therefore, these instances benefit the most from the branch-and-reduce paradigm.

4   Conclusion and Future Work

Although efficient in practice, the techniques used by Akiba and Iwata [3] are not necessary for computing a maximum independent set exactly in many large complex networks. Our results further suggest that the initial kernelization is far more effective than the techniques used in branch-and-bound. Further, while the critical independent set reduction due to Butenko and Trukhanov [11] and the variant due to Larson [22] compute small kernels in practice, they are too slow to compete with other reductions on real-world sparse graphs.

This leaves several open questions that are interesting for future research. In particular, we would like to understand the structure that causes branch-and-reduce techniques to be fast on some graphs, but slow on other (similar) instances. Is it possible to speed up branch-and-reduce algorithms by applying only simple kernelization techniques, and reserving advanced techniques for "difficult" portions of the graph? As we've seen, advanced rules are not always

Table 3. We give the size k_{max} of largest connected component in the kernel from each reduction technique and the running time t of each algorithm to compute an exact maximum independent set. We further give the number of vertices n and edges m for each graph.

Name	Graph n	m	B&R k_{max}	t	Simple+MCS k_{max}	t
LAW Graphs						
cnr-2000	325 557	2 738 969	2 404	-	17 626	-
dblp-2010	326 186	807 700	0	0.38	0	0.14
dblp-2011	986 324	3 353 618	0	1.03	6	0.62
eu-2005	862 664	16 138 468	51 864	-	313 797	-
hollywood-2009	1 139 905	56 375 711	0	22.01	9	21.38
hollywood-2011	2 180 759	114 492 816	0	47.50	17	44.66
in-2004	1 382 908	13 591 473	281	4.39	11 615	-
indochina-2004	7 414 866	150 984 819	8 246	-	509 355	-
uk-2002	18 520 486	261 787 258	9 408	-	2 043 389	-
SNAP Graphs						
as-Skitter	1 696 415	11 095 298	597	2 111.02	21 174	-
ca-AstroPh	18 772	198 050	0	0.07	0	0.02
ca-CondMat	23 133	93 439	0	0.04	0	0.01
ca-GrQc	5 242	14 484	0	0.04	0	0.00
ca-HepPh	12 008	118 489	0	0.05	7	0.01
ca-HepTh	9 877	25 973	0	0.05	0	0.01
email-Enron	36 692	183 831	0	0.11	6	0.03
email-EuAll	265 214	364 481	0	0.09	0	0.09
p2p-Gnutella04	10 876	39 994	0	0.01	7	0.01
p2p-Gnutella05	8 846	31 839	0	0.01	0	0.01
p2p-Gnutella06	8 717	31 525	0	0.01	0	0.01
p2p-Gnutella08	6 301	20 777	0	0.01	0	0.01
p2p-Gnutella09	8 114	26 013	0	0.02	0	0.01
p2p-Gnutella24	26 518	65 369	0	0.02	0	0.02
p2p-Gnutella25	22 687	54 705	0	0.02	0	0.01
p2p-Gnutella30	36 682	88 328	0	0.02	0	0.02
p2p-Gnutella31	62 586	147 892	0	0.05	0	0.03
roadNet-CA	1 965 206	2 766 607	10 807	-	89 667	-
roadNet-PA	1 088 092	1 541 898	5 834	-	35 780	-
roadNet-TX	1 379 917	1 921 660	4 102	-	49 143	-
soc-Epinions1	75 879	405 740	0	0.07	7	0.06
soc-LiveJournal1	4 847 571	42 851 237	295	8.09	28 037	-
soc-pokec	1 632 803	22 301 964	651 503	-	748 755	-
soc-Slashdot0811	77 360	469 180	0	0.07	8	0.13
soc-Slashdot0902	82 168	504 230	0	0.11	15	0.15
web-BerkStan	685 230	6 649 470	1 478	143.43	62 741	-
web-Google	875 713	4 322 051	70	1.23	770	1.51
web-NotreDame	325 729	1 090 108	3 548	12.27	3 578	-
web-Stanford	281 903	1 992 636	2 619	-	10 715	-
wiki-Talk	2 394 385	4 659 565	0	0.44	0	2.32
wiki-Vote	7 115	100 762	0	0.01	0	0.02
KONECT Graphs						
flickr-growth	2 302 925	22 838 276	9	1.60	139	31.59
flickr-links	1 715 255	15 555 041	9	1.10	68	17.04
libimseti	220 970	17 233 144	49 399	1 371.18	141 008	-
orkut-links	3 072 441	117 185 083	2 545 612	-	2 701 058	-
petster-carnivore	623 766	15 695 166	0	2.50	117	4.77
petster-cat	149 700	5 448 197	66	2.83	68 152	-
petster-dog	426 820	8 543 549	231	4.59	139 270	-
youtube-links	1 138 499	2 990 443	0	0.43	19	3.12
youtube-u-growth	3 223 643	9 376 594	0	1.51	33	17.59
baidu-internallink	2 141 300	17 014 946	10	1.02	71	36.99
baidu-relatedpages	415 641	2 374 044	492	1.59	11 458	-
hudong-internallink	1 984 484	14 428 382	79	1.89	1 546	45.92

necessary. Finally, much time is devoted to computing reductions in branch-and-reduce algorithms, perhaps more advanced (but slower) pruning techniques are now viable for these algorithms.

References

1. Abu-Khzam, N.F., Fellows, R.M., Langston, A.M., Suters, H.W.: Crown structures for vertex cover kernelization. Theor. Comput. Syst. **41**(3), 411–430 (2007)
2. Ageev, A.A.: On finding critical independent and vertex sets. SIAM J. Discrete Math. **7**(2), 293–295 (1994)
3. Akiba, T., Iwata, Y.: Branch-and-reduce exponential, FPT algorithms in practice: a case study of vertex cover. Theor. Comput. Sci. **609**(Part 1), 211–225 (2016)
4. Andrade, D.V., Resende, M.G., Werneck, R.F.: Fast local search for the maximum independent set problem. J. Heuristics **18**(4), 525–547 (2012)
5. Batagelj, V., Mrvar, A.: Pajek datasets (2006). http://vlado.fmf.uni-lj.si/pub/networks/data/
6. Batsyn, M., Goldengorin, B., Maslov, E., Pardalos, P.: Improvements to MCS algorithm for the maximum clique problem. J. Comb. Optim. **27**(2), 397–416 (2014)
7. Boldi, P., Rosa, M., Santini, M., Vigna, S.: Layered label propagation: a multiresolution coordinate-free ordering for compressing social networks. In: Srinivasan, S., Ramamritham, K., Kumar, A., Ravindra, M.P., Bertino, E., Kumar, R. (eds.) Proceedings of 20th International Conference on World Wide Web (WWW 2011), pp. 587–596. ACM Press (2011)
8. Boldi, P., Vigna, S.: The WebGraph framework I: compression techniques. In: Proceedings of 13th International Conference on World Wide Web (WWW 2004), pp. 595–601, Manhattan, USA, 2004. ACM Press
9. Bourgeois, N., Escoffier, B., Paschos, V.T., van Rooij, J.M.: Fast algorithms for max independent set. Algorithmica **62**(1–2), 382–415 (2012)
10. Butenko, S., Pardalos, P., Sergienko, I., Shylo, V., Stetsyuk, P.: Estimating the size of correcting codes using extremal graph problems. In: Pearce, C., Hunt, E. (eds.) Optimization. Springer Optimization and Its Applications, vol. 32, pp. 227–243. Springer, Heidelberg (2009)
11. Butenko, S., Trukhanov, S.: Using critical sets to solve the maximum independent set problem. Oper. Res. Lett. **35**(4), 519–524 (2007)
12. Chen, J., Kanj, I.A., Jia, W.: Vertex cover: further observations and further improvements. J. Algorithms **41**(2), 280–301 (2001)
13. Feo, T.A., Resende, M.G.C., Smith, S.H.: A greedy randomized adaptive search procedure for maximum independent set. Oper. Res. **42**(5), 860–878 (1994)
14. Fomin, F., Kratsch, D.: Exact Exponential Algorithms. Springer, Heidelberg (2010)
15. Gajarský, J., Hliněný, P., Obdržálek, J., Ordyniak, S., Reidl, F., Rossmanith, P., Sánchez Villaamil, F., Sikdar, S.: Kernelization using structural parameters on sparse graph classes. In: Bodlaender, H.L., Italiano, G.F. (eds.) ESA 2013. LNCS, vol. 8125, pp. 529–540. Springer, Heidelberg (2013)
16. Garey, M., Johnson, D.: Computers and Intractibility: A Guide to the Theory of NP-Completeness. W. H. Freeman, San Francisco (1979)
17. Gemsa, A., Nöllenburg, M., Rutter, I.: Evaluation of labeling strategies for rotating maps. In: Gudmundsson, J., Katajainen, J. (eds.) SEA 2014. LNCS, vol. 8504, pp. 235–246. Springer, Heidelberg (2014)

18. Hopcroft, J.E., Karp, R.M.: An $n^{5/2}$ algorithm for maximum matchings in bipartite graphs. SIAM J. Comput. **2**(4), 225–231 (1973)
19. Iwata, Y., Oka, K., Yoshida, Y.: Linear-time FPT algorithms via network flow. In: Proceedings of 25th ACM-SIAM Symposium on Discrete Algorithms, SODA 2014, pp. 1749–1761. SIAM (2014)
20. Kieritz, T., Luxen, D., Sanders, P., Vetter, C.: Distributed time-dependent contraction hierarchies. In: Festa, P. (ed.) SEA 2010. LNCS, vol. 6049, pp. 83–93. Springer, Heidelberg (2010)
21. Kunegis, J.: KONECT : the Koblenz network collection. In: Proceedings of 22nd International Conference on World Wide Web (WWW 2013), WWW 2013 Companion, pp. 1343–1350, New York, NY, USA, 2013. ACM
22. Larson, C.: A note on critical independence reductions. In: Bulletin of the Institute of Combinatorics and its Applications, vol. 51, pp. 34–46 (2007)
23. Leskovec, J., Krevl, A.: SNAP Datasets: Stanford large network dataset collection, June 2014. http://snap.stanford.edu/data
24. Li, C.-M., Fang, Z., Xu, K.: Combining MaxSAT reasoning and incremental upper bound for the maximum clique problem. In: Proceedings of IEEE 25th International Conference on Tools with Artificial Intelligence (ICTAI 2013), pp. 939–946, November 2013
25. Nemhauser, G., Trotter, J.: L.E. vertex packings: structural properties and algorithms. Math. Program. **8**(1), 232–248 (1975)
26. San Segundo, P., Matia, F., Rodriguez-Losada, D., Hernando, M.: An improved bit parallel exact maximum clique algorithm. Optim. Lett. **7**(3), 467–479 (2013)
27. San Segundo, P., Rodrguez-Losada, D., Jimnez, A.: An exact bit-parallel algorithm for the maximum clique problem. Comput. Oper. Res. **38**(2), 571–581 (2011)
28. Sanchis, L.A., Jagota, A.: Some experimental and theoretical results on test case generators for the maximum clique problem. INFORMS J. Comput. **8**(2), 87–102 (1996)
29. Sander, P.V., Nehab, D., Chlamtac, E., Hoppe, H.: Efficient traversal of mesh edges using adjacency primitives. ACM Trans. Graph. **27**(5), 144:1–144:9 (2008)
30. Stark, C., Breitkreutz, B., Reguly, T., Boucher, L., Breitkreutz, A., Tyers, M.: Biogrid: a general repository for interaction datasets. Nucleic Acids Res. **34**, D535–D539 (2006)
31. Tomita, E., Sutani, Y., Higashi, T., Takahashi, S., Wakatsuki, M.: A simple and faster branch-and-bound algorithm for finding a maximum clique. In: Rahman, M.S., Fujita, S. (eds.) WALCOM 2010. LNCS, vol. 5942, pp. 191–203. Springer, Heidelberg (2010)
32. Verweij, B., Aardal, K.: An optimisation algorithm for maximum independent set with applications in map labelling. In: Nešetřil, J. (ed.) ESA 1999. LNCS, vol. 1643, pp. 426–437. Springer, Heidelberg (1999)
33. Xiao, M., Nagamochi, H.: Confining sets and avoiding bottleneck cases: a simple maximum independent set algorithm in degree-3 graphs. Theor. Comput. Sci. **469**, 92–104 (2013)
34. Zhang, C.-Q.: Finding critical independent sets and critical vertex subsets are polynomial problems. SIAM J. Discrete Math. **3**(3), 431–438 (1990)

Deterministic Algorithms
for Unique Sink Orientations of Grids

Luis Barba[1,2], Malte Milatz[3], Jerri Nummenpalo[3], and Antonis Thomas[3(✉)]

[1] Carleton University, Ottawa, Canada
[2] Université Libre de Bruxelles, Brussels, Belgium
lbarbafl@ulb.ac.be
[3] Department of Computer Science, ETH Zürich, Zürich, Switzerland
{mmilatz,njerri,athomas}@inf.ethz.ch

Abstract. We study Unique Sink Orientations (USOs) of grids: Cartesian products of *two* complete graphs on n vertices, where the edges are oriented in such a way that each subgrid has a unique sink. We consider two different oracle models, the *edge query* and the *vertex query* model. An edge query provides the orientation of the queried edge, whereas a vertex query provides the orientation of all edges incident to the queried vertex. We are interested in bounding the number of queries to the oracle needed by an algorithm to find the sink. In the randomized setting, the best known algorithms find the sink using either $\Theta(n)$ edge queries, or $O(\log^2 n)$ vertex queries, in expectation. We prove that $O(n^{\log_4 7})$ edge queries and $O(n \log n)$ vertex queries suffice to find the sink in the deterministic setting. A deterministic lower bound for both models is $\Omega(n)$. Grid USOs are instances of LP-type problems and violator spaces for which derandomizations of known algorithms remain elusive.

Keywords: Unique sink orientation · LP-type problem · Violator spaces

1 Introduction

An (m, n)-*grid*, or simply a *grid*, is the Cartesian product $K_m \times K_n$ of two complete graphs with m and n vertices, respectively. An induced subgraph of a grid is called a *subgrid* if it is itself a grid. We identify the vertex set of an (m, n)-grid with the Cartesian product $[m] \times [n]$ where $[m] = \{1, \ldots, m\}$. Two vertices v, w in a grid are adjacent if and only if they differ in exactly one of their two coordinates. We call the edge vw a *horizontal edge* if v and w differ in the first coordinate, and we call it a *vertical edge* if they differ in the second coordinate. The *rows* and *columns* of the grid are defined accordingly; see Fig. 1.

A vertex in an oriented graph is called a *sink* if all its incident edges are incoming. An orientation of a grid is a *Unique Sink Orientation*, or *USO* for short, if all its non-empty subgrids have a unique sink. We call a grid with a unique sink orientation a grid USO. The SINK *problem* asks to find the sink of a grid USO by performing oracle queries which return the orientation of either

© Springer International Publishing Switzerland 2016
T.N. Dinh and M.T. Thai (Eds.): COCOON 2016, LNCS 9797, pp. 357–369, 2016.
DOI: 10.1007/978-3-319-42634-1_29

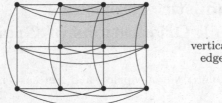 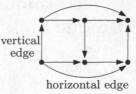

Fig. 1. Left: example of a $(4,3)$-grid (that means 4 columns and 3 rows) where a $(3,2)$-subgrid is shaded. Right: a unique sink orientation of a $(3,2)$-grid. This is the double-twist USO which we refer to in the text.

a single edge (*edge query*) or of all edges incident to a specified vertex (*vertex query*). We denote by SINK(m,n) the SINK problem on an (m,n)-grid USO. The goal of this paper is to provide bounds on the minimum number of queries needed to solve SINK(m,n) *deterministically*.

Previous Work. Unique sink orientations of grids, as presented in this paper, are a graph-theoretic model for a specific class of linear programs [6]. Namely, assume that the feasible region of a given linear program is a simple polytope of dimension d that has exactly $d+2$ facets. Every such $(d,d+2)$-*polytope* is a product of two simplices and hence its graph G is isomorphic to a grid [6]. The problem of minimizing a linear function on this polytope induces a grid USO and its unique sink corresponds to the optimum.

It is known that the above linear programming problem is equivalent to the *One line and n points* problem in the plane [8,16]: Given n points in general position in the plane and one vertical line ℓ, find the segment connecting a pair of points that has the lowest intersection with the given line. Gärtner et al. [8] proved that the grid USO arising from an instance of the one line and n points problem is *Holt-Klee*. The Holt-Klee property states that there are as many vertex-disjoint directed paths between source and sink as there are neighbors of the source (equivalently, the sink) in every subgrid [11]. Felsner et al. [6] have proved that, for grid USO, this is equivalent to a specific $(2,3)$-grid, the "double twist" (depicted in Fig. 1 right), not appearing as a subgraph of the grid USO. However, not every Holt-Klee grid USO comes from an instance of the one line and n points problem [6]. Analyzing *Random Edge* simplex algorithm for Holt-Klee grid USOs yields an algorithm to solve SINK that queries $\Theta(\log n \cdot \log m)$ vertices in expectation [6,8,15]. For general grid USOs, Gärtner et al. [9] exhibit a randomized algorithm that solves SINK(m,n) using $O(\log n \cdot \log m)$ vertex queries in expectation. The best lower bound (randomized) for this problem is $\Omega(\log n + \log m)$ (obtained from finding the sink of a single row or column). In the edge query model, they show that SINK(m,n) can be solved using $\Theta(m+n)$ edge queries in expectation. This yields an exponential gap between number of queries required in the two models. However, in the deterministic setting, it is still unclear whether such a gap exists.

Consider the *matrix USO optimization* problem: Find the global minimum of an injective real function defined on a two dimensional matrix USO, where a matrix is USO if each 2×2 sub-matrix has a unique local minimum. The matrix is given by an oracle that reveals the value of the queried entry. Since this matrix yields a grid USO when edges between entries in the same row or column are oriented from the larger to the smaller, the randomized algorithm from Gärtner et al. [9] solves the matrix USO optimization problem on a $m \times n$ matrix by querying $\Theta(m+n)$ entries in expectation. However, no deterministic counterpart to this algorithm is known. Similar optimization problems on matrices have been studied with stronger conditions and have several applications in geometric problems [1,5,7,13].

Connection to LP-type Problems. Gärtner et al. [9] hinted that SINK can be naturally written as an *LP-type problem*. An LP-type problem, originally defined by Sharir and Welzl [14], is a pair (S, w) where S is a finite set called the *constraints* and $w : 2^S \to \mathbb{R} \cup \{-\infty\}$ is a function subject to certain conditions: (1) Monotonicity: for every two sets $A \subseteq B \subseteq S$, we have $w(A) \leq w(B)$ and (2) Locality: for every two sets $A \subseteq B \subseteq S$ and every constraint $h \in S$, if $-\infty < w(A) = w(B) < w(B \cup \{h\})$ then $w(A) < w(A \cup \{h\})$. A set $B \subseteq S$ is called a *basis* if every proper subset of B has a smaller value for w than B itself. The *combinatorial dimension* of an LP-type problem is the maximum cardinality of a basis (for more information on LP-type problems refer to [12]). Solving an LP-type problem, means finding a basis with the same value under w as the whole set S.

To write $\text{SINK}(m, n)$ as an LP-type problem, we let the set of constraints S be the set of rows and columns and thus $|S| = m + n$. As any grid USO is acyclic (see Sect. 2), there is a topological ordering of its vertices with the sink having the highest rank in the ordering. Every subset of the rows and columns defines a subgrid H and we define $w(H)$ to be the rank of its sink in the topological ordering or $-\infty$ if the subgrid is empty. This definition satisfies both monotonicity and locality, yielding an LP-type problem. Moreover, a basis is simply a vertex defined by one row and one column. Solving $\text{SINK}(m, n)$ then corresponds to finding a basis of (S, w) with the highest value.

To solve an LP-type problem, there are linear-time (in $|S|$) randomized algorithms [12] which require access to certain primitive operations. For grid USOs these operations correspond directly to edge queries; hence, by the above construction we can solve $\text{SINK}(m, n)$ with the randomized algorithms of [12] using a linear number of edge queries. While derandomizations of algorithms for LP-type problems exist with the same performance [2,4], they make an extra assumption on the problem, which does not hold for grid USOs, namely the existence of a *subsystem oracle* (see Computational assumption 2 in [4]). Other problems where these assumptions do not hold have been studied [3], and it is an open question whether or not they can be solved deterministically with a linear number of primitive operations.

As a generalization of LP-type problems, Gärtner et al. [10] introduced *violator spaces*. Intuitively, instead of assigning a value to each set $A \subseteq S$ of constraints, we map A to a set of its *violators*, i.e., the set of constraints whose addition to A would change its basis. The goal is to find the basis with the same set of violators as S which can be done in linear expected number of primitive operations [10]. When modeling SINK as violator spaces, we assign to each vertex the subset of constraints to which it has an outgoing edge. Therefore, the vertex query model corresponds to an oracle that reveals the set of violators of a vertex.

Our Results. While randomized algorithms to solve SINK have been previously studied [6,8,9,15], to the best of our knowledge no non-trivial deterministic algorithm was known prior to this work. This is mainly due to the fact that the aforementioned subsystem oracle does not exist. In light of this fact, we aim to solve the problem using a different set of techniques. We present two deterministic algorithms to solve $\text{SINK}(m, n)$ using $O((m+n)\log(m+n))$ vertex queries and $O((m+n)^{\log_4 7})$ edge queries, respectively. Additionally, if the Holt-Klee property holds, we exhibit a deterministic algorithm to solve $\text{SINK}(m, n)$ using $O(m + n)$ vertex queries. Both query models exhibit a lower bound of $\Omega(m + n)$.

Outline. In Sect. 2 we state basic facts for grid USOs and define a way to partition the grid into subgrids. In Sect. 3 we focus on the vertex query model and describe our algorithms in this setting. Finally, Sect. 4 addresses the edge query model.

2   Grid USO Properties

Recall that an induced subgraph of a grid is called a *subgrid* if it is itself a grid. The subgrids of an (m, n)-grid are exactly those induced subgraphs whose vertex set is a Cartesian product $I \times J$, for some $I \subseteq [m]$ and $J \subseteq [n]$. We call the corresponding subgrid an $I \times J$-grid. If the original grid is oriented, the subgrids inherit this orientation. Our algorithms rely on two basic properties of grid USOs. The first is acyclicity, originally proved in [9].

Lemma 1 [9]. *Every (m, n)-grid USO is acyclic.*

The second ingredient is a partitioning strategy which we describe subsequently.

Induced USOs. For our algorithms we want to partition the grid into subgrids. Such a partition into subgrids induces itself a grid in a natural way, which inherits a unique sink orientation from the original grid USO; see Fig. 2 for an illustration. Formally, let G be an (m, n)-grid USO, and let $\mathcal{A} = \{A_1, \ldots, A_k\}$ and $\mathcal{B} = \{B_1 \ldots, B_l\}$ be partitions of $[m]$ and $[n]$, respectively. Let H be a (k, l)-grid whose vertex (i, j) we identify with the $A_i \times B_j$-subgrid of G for every $1 \leq i \leq k$

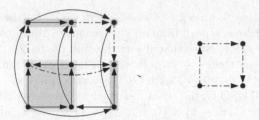

Fig. 2. Left: a $(3,3)$-grid USO whose coordinates have been partitioned into 2 parts of cardinalities two and one. Right: the induced $(2,2)$-grid USO; The edges contributing to the induced orientation are also shown in the original grid.

and for every $1 \leq j \leq l$. We define an orientation on H so that the edge between two adjacent vertices x and y of H is oriented towards y if the sink of x has at least one outgoing edge to a vertex of y in the original grid G. This orientation is well defined: If the sink of y also had an edge towards some vertex of x, there would be a cycle in G. Furthermore, there is always at least one such outgoing edge from one of the sinks as otherwise an appropriately chosen subgrid of G would have two sinks. We say that this orientation is *induced* on H by G and H is called the $\mathcal{A}$-$\mathcal{B}$-*partition grid*. The following lemma shows that the orientation induced on H by G is also a unique sink orientation. A similar result was proved by Gärtner et al. [9] in higher dimensional grid USOs.

Lemma 2. *Let G be an (m, n)-grid USO, and let $\mathcal{A}$ and $\mathcal{B}$ be partitions of $[m]$ and $[n]$, respectively. The orientation of the $\mathcal{A}$-$\mathcal{B}$-partition grid H induced by G is a unique sink orientation. Moreover, the sink of H is the subgrid of G that contains the sink of G.*

3   The Vertex Query Model

Recall that in the vertex query model a grid USO is given by a *vertex oracle* that reveals the orientations of the edges adjacent to the queried vertex. An adversary argument can be used to show that any deterministic algorithm needs $m + n - 1$ vertex queries in the worst case to solve $\text{SINK}(m, n)$. Our main result is Theorem 1 which describes a deterministic algorithm that needs $O((n + m) \log(n + m))$ vertex queries.

The Sink-Finding Algorithm. Before explaining the algorithm we introduce some notation. We define a reflexive partial order on the vertices of a grid USO G as follows. For any two vertices $v, w \in G$ we say that $w \succeq v$ if there is a directed path from w to v within the grid G (or if $w = v$). In words, we say that w is *larger* than v (or v is *smaller* than w). By Lemma 1, grid USOs are acyclic and, thus, this partial order is well defined. Moreover, the sink of the grid USO is the unique minimal vertex with respect to this ordering.

For a set of vertices V in a grid USO G let $join(V)$ be the set of vertices of the grid to which there is a path from all vertices in V. More formally $join(V) = \{w \in G \mid v \succeq w \; \forall v \in V\}$. Note that if s is the sink of the 2×2-grid containing two vertices v and w, then $s \in join(\{v, w\})$. Therefore, given a set of vertices V, we can compute $w \in join(V)$ with $O(|V|)$ vertex queries.

In an (m, n)-grid USO G, the *refined in-degree* of a vertex $v \in G$ is an ordered pair $[a_v, b_v] \in \{0, 1, \ldots, m-1\} \times \{0, 1, \ldots, n-1\}$ where a_v and b_v specify the number of incoming horizontal and vertical edges of v, respectively. We say that a vertex v with refined in-degree $[a_v, b_v]$ is a (k, t)-*vertex* if $a_v \geq k - 1$ while $b_v \geq t - 1$.

Consider the following algorithmic approach to solve $\textsc{Sink}(m, n)$. Assume that we are able to find in $O(m + n)$ time an $(\frac{m}{2}, \frac{n}{2})$-vertex v. We could then partition the whole grid into 4 $(\frac{m}{2}, \frac{n}{2})$-subgrids so that v is the sink of one of them. The next step would be to find the sink of the subgrid antipodal to the subgrid containing v in the $(2, 2)$-grid induced by this partition. By Lemma 2 we could thereafter discard one subgrid out of consideration and possibly still have to find the sink of one more subgrid. If one finds the sinks of the subgrids recursively, one would get the recursion:

$$T(m, n) = 2T\left(\frac{m}{2}, \frac{n}{2}\right) + O(m + n),$$

where $T(m, n)$ is the number of vertex queries needed by this algorithm to solve $\textsc{Sink}(m, n)$ which results in $T(m, n) = O((m + n)\log(m + n))$.

The main part of the algorithm description is to obtain an $(\frac{m}{2}, \frac{n}{2})$-vertex in linear time. For ease of presentation, we prove Lemmas 3 and 4 for the square (n, n)-grid. From those, we obtain Corollary 1, which shows how to compute an $(\frac{m}{8}, \frac{n}{8})$-vertex using a linear number of vertex queries. Using it as a black box, we manage to get an $(\frac{m}{2}, \frac{n}{2})$-vertex with only linearly many more queries in Lemma 5 and Corollary 2.

Lemma 3. *Let $D = \{v_1, \ldots, v_n\}$ be the set of diagonal vertices of an (n, n)-grid USO so that $v_i = (i, i)$. After sorting the vertices with respect to their indegree the j-th vertex has at least $j - 1$ incoming edges.*

Proof. Firstly, we evaluate all the vertices in D. We rename the coordinates such that for all $1 \leq i < j \leq n$ either $v_i \succeq v_j$ or v_i and v_j are incomparable. This defines a linear extension of the partial order $\succeq$ and we claim that v_j has at least $j - 1$ incoming edges for every $j = 1, \ldots, n$. To see this fix some v_j and consider the $(2, 2)$-subgrid $H_{i,j}$ containing both v_i and v_j for some $1 \leq i < j$. Because $H_{i,j}$ is a USO and since $v_i \succeq v_j$ (or v_i and v_j are incomparable), we know that $H_{i,j}$ contains no path from v_j to v_i. Therefore, v_j cannot be the source of $H_{i,j}$ which implies that v_j has at least one incoming edge in $H_{i,j}$. Since the edges in $H_{i,j}$ are disjoint from those of $H_{i',j}$ for $1 \leq i < i' < j$, we know that v_j has an incoming edge for each $1 \leq i < j$, i.e., v_j has at least $j - 1$ incoming edges. $\square$

Lemma 4. *Given an (n, n)-grid USO we can find an $(\frac{n}{4}, \frac{n}{4})$-vertex using $O(n)$ vertex queries.*

Proof. The algorithm we describe queries a linear number of vertices of the grid before finding a vertex with the required refined in-degree. After explaining the algorithm, we prove its correctness.

In the first phase, query the diagonal vertices $D = \{v_1, \ldots, v_n\}$ where $v_i = (i, i)$. If one of the vertices in D has the required refined in-degree, we are done. If not, sort and rename the vertices of D such that for $i < j$ vertex v_i has at most as many incoming edges as v_j (recall that in our model we only count the number of queries to the oracle).

In the second phase, let $V = \{v_{\lceil \frac{n}{2} \rceil}, \ldots, v_n\} \subseteq D$. Label the vertices in V as either *horizontal* or *vertical*, depending on whether the majority of incoming edges for the corresponding vertex are horizontal or vertical, respectively. Assume without loss of generality that there are more vertical vertices in V; otherwise change the role of the coordinates. Let $V' \subseteq V$ be the set of all vertical vertices in V and notice that $|V'| \geq |V|/2$. Then, find some $v \in join(V')$ by using at most $O(|V'|)$ additional queries and note that $v' \succeq v$ for every $v' \in V'$ (Fig. 3).

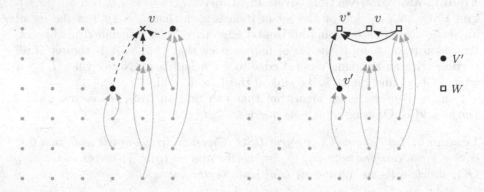

Fig. 3. An illustration for Lemma 4. The three black circles on the diagonal form V'. The dashed edges denote that there is a path from each vertex in V' to v. On the right hand side the vertices belonging to W are marked with a square.

Let I' be the set of indices containing the first coordinate of each vertex in V'. Assume that $v = (x_v, y_v)$ and let $W = I' \times \{y_v\}$ be a subset of vertices in the same row as v with $|W| = |V'| \geq |V|/2$. To conclude, the algorithm queries each vertex in W. It is clear from the description above that the number of vertices queried so far is $O(n)$. We claim that the sink of W will have the required refined in-degree. Because of the renaming, after the first phase and due to Lemma 3 we know that v_i has at least $i - 1$ incoming edges. Therefore, by the definition of V, we know that each vertex of $V' \subseteq V$ has at least $\lceil \frac{n}{2} \rceil - 1$ incoming edges. Moreover, each vertex in V' has at least $(\lceil \frac{n}{2} \rceil - 1)/2$ incoming vertical edges and $|V'| \geq \frac{n - \lceil \frac{n}{2} \rceil + 1}{2} \geq \frac{n}{4}$.

Let v^* be the sink of W obtained after querying each vertex in this set (including v). Let $[a, b]$ be the refined in-degree of v^*. We claim that $a, b \geq \frac{n}{4} - 1$.

Indeed, because v^* is smaller than each other vertex in W, we know that

$$a \geq |W| - 1 = |V'| - 1 = \frac{n - \lceil \frac{n}{2} \rceil + 1}{2} - 1 \geq \frac{n}{4} - 1.$$

Furthermore, there is a vertex $v' \in V'$ that is in the same column as v^*. Due to the *join* operation there is a path from v' to v and further to v^*. If $v' = v^*$ then the we know already that v^* has at least $\frac{n}{4} - 1$ incoming edges vertically. Otherwise, the edge between v' and v^* is oriented towards v^* and v^* has more incoming vertical edges than v', because of acyclicity. Therefore we establish that $b \geq \frac{\lceil \frac{n}{2} \rceil - 1}{2} + 1 \geq \frac{n}{4} \geq \frac{n}{4} - 1$ which shows that v^* is an $(\frac{n}{4}, \frac{n}{4})$-vertex. $\square$

The following corollary extends the previous lemma to non-square grids.

Corollary 1. *Given an (m, n)-grid USO we can find an $(\frac{m}{8}, \frac{n}{8})$-vertex using $O(m + n)$ vertex queries.*

Finding an $(\frac{m}{2}, \frac{n}{2})$-vertex. Given an (a, b)-vertex $v = (x_v, y_v)$ in an (m, n)-grid USO, let $I_v \subseteq [m]$ be the set of indices such that $I_v \times \{y_v\}$ is the set of all vertices with an outgoing horizontal edge to v, with v included in this set. Analogously, $J_v \subseteq [n]$ is the set of indices such that $\{x_v\} \times J_v$ is the set of all vertices with an outgoing vertical edge to v (including v). Notice that $|I_v| \geq a$ while $|J_v| \geq b$ and that v is the sink of the $I_v \times J_v$-grid.

An (α, β)-*oracle* is an algorithm that can find an $(\alpha m, \beta n)$-vertex v on a (m, n)-grid USO, using $O(m + n)$ queries.

Lemma 5. *Let G be an (m, n)-grid USO. Given an (α, β)-oracle such that $0 < \alpha, \beta < 1$, we can find both an $(\frac{m}{2}, \beta n)$-vertex and an $(\alpha m, \frac{n}{2})$-vertex in G using $O(1)$ oracle calls and $O(m + n)$ additional vertex queries.*

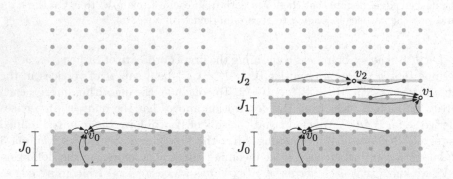

Fig. 4. An illustration of the first part of the proof of Lemma 5. On the left we depict the situation after the first call and on the right after the last call of the (α, β)-oracle.

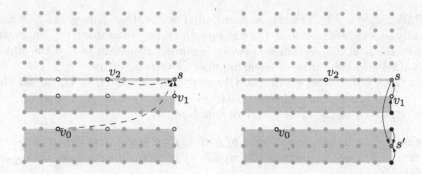

Fig. 5. The second part of the algorithm in Lemma 5. On the left we depict how to find s and on the right we depict how to find s'. The latter is the output of the algorithm.

Proof. We show how to find an $(\alpha m, \frac{n}{2})$-vertex. The procedure to find an $(\frac{m}{2}, \beta n)$-vertex is analogous. Let $B_0 = [n]$ and let G_0 be the $[m] \times B_0$-grid, i.e., $G = G_0$. Using an oracle call in G_i (initially $i = 0$), find an $(\alpha n, \beta n)$-vertex v_i in G_i. Recall that v_i is the sink of the $I_{v_i} \times J_{v_i}$-grid. Notice that $|I_{v_i}| \geq \alpha m$ while $|J_{v_i}| \geq \beta |B_i|$. Let $B_{i+1} = B_i \setminus J_{v_i}$ and let G_{i+1} be the $[m] \times B_{i+1}$-grid. Repeat this procedure with G_{i+1} as long as $|B_i| \geq \frac{n}{2}$; see Fig. 4.

Since $0 < \beta < 1$, after $k = O(1)$ iterations, the above procedure stops. Because the process stopped, we know that $|B_{k+1}| = |B_k \setminus J_{v_k}| < \frac{n}{2}$. Since $B_{k+1} = B_0 \setminus \cup_{i=0}^k J_{v_i}$, we know that $| \cup_{i=0}^k J_{v_i}| > \frac{n}{2}$.

Compute $s \in join(\{v_0, \ldots, v_k\})$ and note that s is smaller than v_i for each $1 \leq i \leq k$. As $k = O(1)$, s can be computed using $O(1)$ vertex queries. Assume that $s = (x_s, y_s)$ and let s' be the sink of the $\{x_s\} \times \cup_{i=0}^k J_{v_i}$-subgrid. Let $0 \leq h \leq k$ be an integer such that s' belongs to the $[m] \times J_{v_h}$-subgrid; see Fig. 5.

Let $[a, b]$ be the refined in-degree of s'. Recall that s is smaller than v_h, hence s' is also smaller than v_h. Because v_h is the sink of the $I_{v_h} \times J_{v_h}$-grid, s' is smaller than each vertex in the $I_{v_h} \times J_{v_h}$-grid. Therefore, at least $|I_{v_h}|$ vertices have outgoing edges to s' in the row containing s', i.e., $a \geq |I_{v_h}| \geq \alpha m$. Moreover, since s' is the sink of the $\{x_s\} \times \cup_{i=0}^k J_{v_i}$-subgrid, $b \geq | \cup_{i=0}^k J_{v_i}| > \frac{n}{2}$. Consequently, s' is an $(\alpha m, \frac{n}{2})$-vertex. □

Combining Lemma 5 with Corollary 1, we immediately get the following.

Corollary 2. *Let G be an (m, n)-grid USO. We can compute an $(\frac{m}{2}, \frac{n}{2})$-vertex using $O(m + n)$ vertex queries.*

Our main result can now be derived from Corollary 2.

Theorem 1. *The sink of an (m, n)-grid USO can be found after querying $O((m + n) \log(m + n))$ vertices.*

Holt-Klee Grid USOs in the Vertex Query Model. A grid USO is called *Holt-Klee* if it does not contain a specific forbidden subgraph called the "double twist" (depicted in Fig. 1 right). As mentioned before, Holt-Klee grid USOs arise naturally as a generalization of the one line and n points problem [8]. Improving on Theorem 1, in this section we provide a simple algorithm to find the sink of a Holt-Klee grid USO using a linear number of vertex queries.

Let G be a Holt-Klee grid USO. Given a $(2,2)$-subgrid H of G, we say that H is a *bow* if it contains a directed path (v_1, v_2, v_3, v_4) of length three. We refer to the edge v_2v_3 as the *middle edge* of this bow.

Lemma 6. *Let H be a bow subgrid of a Holt-Klee grid USO G. If g is the row or column of G which contains the middle edge of H, then g does not contain the sink of G.*

Proof. Let (v_1, v_2, v_3, v_4) be the directed path in the bow H and assume that the edge v_2v_3 is horizontal. Let w be any vertex in the row containing v_2v_3. Assume for a contradiction that w is the sink of G. Since w has all its edges incoming, the orientation of all the edges of the $(3,2)$-subgrid containing H and w is forced as shown in Fig. 6. The orientation of edge v_4v_5 is forced; otherwise the subgrid induced by the vertices v_3, v_4, v_5 and w would have two sinks. Similarly, the orientation of edge v_1v_5 is forced by acyclicity. This yields a double twist as a subgraph of G which is a contradiction. Thus, no vertex in the row containing v_2v_3 is the sink of G. □

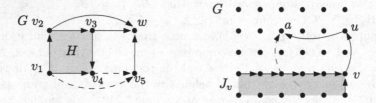

Fig. 6. Left: it is impossible to have the global sink in the same row of the middle edge of a bow. Right: illustration of case (3) in the algorithm for Theorem 2 which results in eliminating the column containing u and v.

Theorem 2. *The sink of a Holt-Klee (m,n)-grid USO G can be found with $O(m + n)$ vertex queries.*

Proof. Let r be an arbitrary row of G and query all of its vertices. With this information we can retrieve the total order of the vertices in this row. Let v be the sink of this row. Recall that v is the sink of the $[m] \times J_v$-grid. Therefore, if v is not the sink of G, then this sink lies in the $[m] \times ([n] \setminus J_v)$-grid.

Let u be an arbitrary vertex in the same column as v such that $v \succeq u$. After querying u, we have three cases: either (1) u is the sink of G, or (2) u has no horizontal outgoing edges, in which case the sink of G cannot lie in the row containing u, or (3) there is at least one outgoing edge from u to some vertex a. In this case, since u is the sink of its row, the $(2,2)$-subgrid containing u, v and a is necessarily a bow and hence, Lemma 6 implies that the sink of G cannot lie in the column containing u; see Fig. 6. Thus, regardless of the case, we can discard either a row or a column of G and update u or v accordingly to continue with this process. Consequently, after querying all vertices in the initial row r, with every additional query we can discard either a row or a column. That is, after $O(n+m)$ vertex queries the algorithm finds the sink of G.   □

4   The Edge Query Model

In the previous sections we have discussed the vertex query model. In the *edge query model* considered here, the algorithm accesses the grid by means of an *edge oracle* which can tell us the orientation of any given edge. One can show by an adversary argument that $\Omega(m+n)$ queries are necessary in the worst case (this lower bounds holds also for vertex queries). A trivial bound is $O(mn)$: find the sink of each row individually; then for each of these local sinks, query all incident edges to decide whether it is the global sink. The following theorem improves on this bound.

Theorem 3. *There exists a deterministic algorithm which finds the sink of an (m, n)-grid USO using $O((m+n)^{\log_4(7)})$ edge queries, where $\log_4(7) \approx 1.404$.*

To proof of Theorem 3 uses the following result for the *vertex* query model.

Lemma 7. *There exists a deterministic algorithm which finds the sink of a $(4, 4)$-grid USO using at most 7 vertex queries.*

To prove Lemma 7, we use a computer-based search to eliminate symmetrical situations and reduce the problem to three individually handled cases. We omit it from this paper as it does not offer any insights to the problem. Note, however, that the upper bound of 7 queries is exactly tight to our $n + m - 1$ lower bound.

Proof (of Theorem 3). Without loss of generality, we assume that m is a power of 4, and $m = n$. Otherwise, we can satisfy this property by adding at most linearly many rows and columns while maintaining the position of the sink.

Given an (m, m)-grid G, let $\mathcal{A}$ be a partition of the index set $[m]$ such that $|\mathcal{A}| = 4$ and $|A| = \frac{m}{4}$ for each $A \in \mathcal{A}$. Consider the induced unique sink orientation on the $\mathcal{A}$-$\mathcal{A}$-partition grid H.

Let v be a vertex of H. Since v corresponds to an $(\frac{m}{4}, \frac{m}{4})$-subgrid of G, we can find the sink s_v of this subgrid using a recursive call that takes $T(m/4)$ edge queries. Afterwards we can find out the orientations of the six edges adjacent to the vertex v in H by querying the $2m - 2$ edges adjacent to s_v in G. Thus we have implemented a vertex query in H using at most $T(m/4) + 2m - 2$ edge queries in G. Using Lemma 7, we find and query the sink s_H of H after
$$T(m) \leq 7 \cdot \Big(T(m/4) + 2m - 2\Big) \text{ edge queries in } G \text{ and hence } T(m) \in O(m^{\log_4(7)}).$$
In particular, querying s_H involves finding the sink of the subgrid s_H, which by Lemma 2 coincides with the sink of G.   □

5   Conclusions

In this paper, we have discussed the existence of deterministic algorithms to solve $\text{SINK}(m, n)$. We show that $O((m + n)\log(m + n))$ vertex queries and $O((m + n)^{1.404})$ edge queries suffice to solve $\text{SINK}(m, n)$. In the case of Holt-Klee grids, a linear number of vertex queries suffices. The obvious open problem is to close the gap between upper and lower bound in both vertex and edge query models. Following our approach, an equivalent statement to Lemma 4 in the edge query model would yield an algorithm using $O((m + n)\log(m + n))$ edge queries.

Acknowledgments. This work started at the 13th Gremo's Workshop on Open Problems (GWOP), Feldis (GR), Switzerland. We thank the organizers. In addition, we thank Bernd Gärtner and Emo Welzl for useful discussions.

References

1. Aggarwal, A., Klawe, M.M., Moran, S., Shor, P., Wilber, R.: Geometric applications of a matrix-searching algorithm. Algorithmica **2**(1–4), 195–208 (1987)
2. Chan, T.: Improved deterministic algorithms for linear programming in low dimensions. In: Proceedings of the SODA 2016 (2016, to appear)
3. Chan, T.M.: Deterministic algorithms for 2-d convex programming and 3-d online linear programming. J. Algorithms **27**(1), 147–166 (1998)
4. Chazelle, B., Matoušek, J.: On linear-time deterministic algorithms for optimization problems in fixed dimension. J. Algorithms **21**(3), 579–597 (1996)
5. Demaine, E.D., Langerman, S.: Optimizing a 2D function satisfying unimodality properties. In: Brodal, G.S., Leonardi, S. (eds.) ESA 2005. LNCS, vol. 3669, pp. 887–898. Springer, Heidelberg (2005)
6. Felsner, S., Gärtner, B., Tschirschnitz, F.: Grid orientations, (d, d + 2)-polytopes, and arrangements of pseudolines. Discrete Comput. Geom. **34**(3), 411–437 (2005)
7. Galil, Z., Park, K.: Dynamic programming with convexity, concavity and sparsity. Theoret. Comput. Sci. **92**(1), 49–76 (1992)
8. Gärtner, B., Tschirschnitz, F., Welzl, E., Solymosi, J., Valtr, P.: One line and n points. Random Struct. Algorithm **23**(4), 453–471 (2003)
9. Gärtner, B., Morris Jr., W.D., Rüst, L.: Unique sink orientations of grids. Algorithmica **51**(2), 200–235 (2008)
10. Gärtner, B., Matoušek, J., Rüst, L., Skovron, P.: Violator spaces: structure and algorithms. Discrete Appl. Math. **156**(11), 2124–2141 (2008)
11. Holt, F., Klee, V.: A proof of the strict monotone 4-step conjecture. Adv. Discrete Comput. Geom. Contemp. Math. **223**, 201–216 (1999)
12. Matoušek, J., Sharir, M., Welzl, E.: A subexponential bound for linear programming. Algorithmica **16**(4/5), 498–516 (1996)
13. Mityagin, A.: On the complexity of finding a local maximum of functions on discrete planar subsets. In: Alt, H., Habib, M. (eds.) STACS 2003. LNCS, vol. 2607, pp. 203–211. Springer, Heidelberg (2003)
14. Sharir, M., Welzl, E.: A combinatorial bound for linear programming and related problems. In: Finkel, A., Jantzen, M. (eds.) STACS 1992. LNCS, vol. 577, pp. 569–579. Springer, Heidelberg (1992)

15. Tschirschnitz, F.: LP-related properties of polytopes with few facets. Ph.D. thesis, ETH Zürich (2003)
16. Welzl, E.: Entering and leaving j-facets. Discrete Comput. Geom. **25**(3), 351–364 (2001)

From Graph Orientation
to the Unweighted Maximum Cut

Walid Ben-Ameur, Antoine Glorieux$^{(\boxtimes)}$, and José Neto

Samovar UMR 5157, Télécom SudParis, CNRS, Universit Paris-Saclay,
9 Rue Charles Fourier, 91011 Evry Cedex, France
{walid.benameur,antoine.glorieux,jose.neto}@telecom-sudparis.eu

Abstract. In this paper, starting from graph orientation problems, we introduce some new mixed integer linear programming formulations for the unweighted maximum cut problem. Then a new semidefinite relaxation is proposed and shown to be tighter than the Goemans and Williamson's semidefinite relaxation. Preliminary computational results are also reported.

1 Introduction

Let $G = (V, E)$ be a simple undirected graph, with node set V, edge set E, and let $(w_e)_{e \in E}$ denote nonnegative edge-weights. Given a node subset $S \subseteq V$, the *cut* defined by S, denoted by $\delta(S)$, is the subset of edges in E having exactly one endpoint in S, i.e. $\delta(S) = \{ij \in E \colon |S \cap \{i,j\}| = 1\}$. The *weight of the cut* defined by S, denoted by $w(\delta(S))$ is the sum of the weights of the edges belonging to the cut, i.e., $w(\delta(S)) = \sum_{e \in \delta(S)} w_e$. The maximum cut problem consists in finding a cut of maximum weight, denoted by $w^\star$, in the graph G: $\max\{w(\delta(S)) \colon S \subset V\}$. The cardinality of V will be denoted by n.

The maximum cut problem is a fundamental combinatorial optimization problem that emerges in several scientific disciplines: VLSI design [4], sparse matrix computation [2], parallel programming [9], statistical physics [4], quadratic programming [17], etc. A less known application is given by frequency assignment in networks where the weight of each edge represents the interference level between two nodes. Assuming that only two frequencies (resources) are available, assigning a frequency to each node such that the whole interference between nodes using the same frequency is minimized is a maximum cut problem.

The maximum cut problem is known to be NP-hard [20] in general, and not approximable within a ratio $\frac{16}{17} + \epsilon$ for any $\epsilon > 0$ unless $\mathsf{P} = \mathsf{NP}$ [18]. However, the problem may be polynomial for some instances. We know, for example, that the problem becomes easy when the underlying graph is weakly bipartite and the weights are nonnegative [16]. Other polynomial cases are reviewed, e.g., in [5].

One line of research to solve this problem has consisted in the development of (meta) heuristics, see, e.g. [6,12,27]. Another important line of research relies on linear programming formulations of the problem. This has namely led to deep

© Springer International Publishing Switzerland 2016
T.N. Dinh and M.T. Thai (Eds.): COCOON 2016, LNCS 9797, pp. 370–384, 2016.
DOI: 10.1007/978-3-319-42634-1_30

investigations on the polyhedral structure of the cut polytope: the convex hull of the incidence vectors of all the cuts of the graph, which has then been extensively used, e.g., in Branch and Cut algorithms [3,5,10]. More recently, essentially since the mid-1990's and the breakthrough paper by Goemans and Williamson [15], there has been a growing interest in semidefinite programming based algorithms. Goemans and Williamson's work presents a 0.87856-approximation algorithm for the maximum cut problem when the edge weights are nonnegative. Their method relies on the following semidefinite relaxation of the problem for a complete graph

$$
(\text{SDP0}) \begin{cases} \max \frac{1}{2} \sum_{i=1}^n \sum_{j=i+1}^n w_{ij}(1 - y_{ij}) \\ \text{s.t.} \\ y_{ii} = 1, \ \forall i \in [\![1,n]\!], \\ Y \succeq 0, \\ Y \in \mathbb{S}^n, \end{cases}
$$

where Y represents the matrix with entry y_{ij} in the ith row and jth column, $Y \succeq 0$ is the constraint that the matrix Y is positive semidefinite, and, for any given integer n, $\mathbb{S}^n$ denotes the set of symmetric matrices with order n.

To improve the quality of the bound $Z^\star_{\text{SDP0}}$ given by the formulation (SDP0) [15], different approaches have been proposed in the literature: namely by making use of polyhedral knowledge on the cut polytope and adding linear inequalities [13,19], or by means of lift-and-project techniques [1,22]. Another way to improve the upper bound given by the semidefinite relaxation is described in [7,8] where some spectral techniques are used leading to polynomial-time algorithms for some low rank weight matrices. This semidefinite approach of the problem also led to efficient solvers such as BiqMac [25] and BiqCrunch [21].

The reader can find in [5,10] and the references therein further results about the maximum cut problem including applications, polynomial cases, approximation algorithms, relationships with other combinatorial problems, polyhedral studies etc.

The paper is organized as follows. We introduce some new mixed integer linear programming formulations for the maximum cut problem based on graph orientations (Sect. 2). A semidefinite programming relaxation is then proposed. We show that the bound provided by this new SDP relaxation is stronger than the bound given by the relaxation (SDP0) introduced by Goemans and Williamson (Sect. 3). We also prove that the new bound is tight for complete graphs. We then introduce further Mixed Integer Programming formulations (Sect. 4). Several numerical experiments have been conducted showing the relevance of the SDP formulation and the performances of the new Mixed Integer Programming formulations (Sect. 5). A conclusion follows.

2   Mixed Integer Linear Programming Formulations

In this section, we gradually introduce our new formulation for the maximum cardinality cut problem, i.e. the maximum cut problem for the case when all the edge-weights are equal to 1: $w_e = 1, \forall e \in E$. For our purposes, we shall think of the original graph $G = (V, E)$ as directed (consider any arbitrary orientation) and let $B \in \{-1, 0, 1\}^{|V| \times |E|}$ stand for its incidence matrix, i.e., the column corresponding to the arc uv (or, equivalently, to the edge uv directed from node u to node v), has only nonzero entries in the rows corresponding to the nodes u and v: $B_{u,uv} = 1$ and $B_{v,uv} = -1$, respectively.

For clarity, we start introducing an auxiliary formulation. It involves two types of variables. The first type of variables $x \in \{-1, 1\}^{|E|}$ describes an orientation of the graph G and may be interpreted as follows. For each edge $ij \in E$ which is originally directed from node i to node j: if $x_{ij} = 1$ then ij is directed from i to j (i.e., the orientation is the same as the original one) and is directed from j to i otherwise (i.e., the edge is "reversed" with respect to the original orientation). The other variables are binary and denoted y_k^v, with $v \in V$ and $k \in [\![-d_v, d_v]\!]$, where d_v denotes the degree of the node v in G. They have the following interpretation: $y_k^v = 1$ if and only if $B_v x = k$, where B_v denotes the row of B corresponding to vertex v, so that the following equation trivially holds

$$\sum_{k=-d_v}^{d_v} k y_k^v = B_v x, \forall v \in V. \tag{1}$$

Notice that $B_v x = d_v^{(x,+)} - d_v^{(x,-)}$, where $d_v^{(x,+)}$ (resp. $d_v^{(x,-)}$) denotes the outdegree (resp. indegree) of the node v w.r.t. the orientation described by x.

Also, given the interpretation for the variables y, among those of the form y_k^v, for some fixed node $v \in V$, exactly one of them has value 1. Thus, the following contraints are satisfied

$$\sum_{k=-d_v}^{d_v} y_k^v = 1, \ \forall v \in V. \tag{2}$$

Then we can show the maximum cardinality cut problem may be formulated as the mixed-integer program

$$(\text{MIP1}) \begin{cases} Z_{\text{MIP1}}^\star = \frac{1}{2} \max \sum_{v \in V} \sum_{k=-d_v}^{d_v} |k| y_k^v \\ \text{s.t. } (1), (2), \\ x \in [-1; 1]^{|E|}, \\ y_k^v \in \{0, 1\}, \ \forall v \in V, \ \forall k \in [\![-d_v, d_v]\!]. \end{cases}$$

Proposition 1. *The optimal objective value of* (MIP1) *equals the maximum cardinality of a cut in the graph* G: $Z_{\text{MIP1}}^\star = w^\star$.

Proof. First, note that it is equivalent to take $x \in [-1; 1]^{|E|}$ or $x \in \{-1; 1\}^{|E|}$ since B is totally unimodular, the integrity of y implies the integrality of x. Observe that introducing sign variables $z \in \{-1, 1\}^{|V|}$ with the following interpretation: z_v is the sign of the only k for which $y_k^v \neq 0$, (MIP1) is equivalent to the following

$$\begin{cases} \max \frac{1}{2} \sum_{v \in V} z_v \sum_{k=-d_v}^{d_v} k y_k^v \\ \text{s.t. } (1), (2), \\ y_k^v \in \{0, 1\}, \ \forall v \in V, \ \forall k \in [\![-d_v, d_v]\!], \\ z \in \{-1, 1\}^{|V|}, \ x \in [-1; 1]^{|E|}. \end{cases} \Leftrightarrow \begin{cases} \max \frac{1}{2} \sum_{v \in V} z_v B_v x = \frac{1}{2} z^t B x \\ \text{s.t.} \\ z \in \{-1, 1\}^{|V|}, \ x \in \{-1, 0, 1\}^{|E|}. \end{cases}$$

$$\Leftrightarrow \begin{cases} \max \frac{1}{2}\sum_{uv\in E} x_{uv}(z_v - z_u) = \frac{1}{2}x^t B^t z \\ \text{s.t.} \\ z \in \{-1,1\}^{|V|}, \ x \in \{-1,1\}^{|E|}. \end{cases} \Leftrightarrow \begin{cases} \max \frac{1}{2}\sum_{uv\in E} |z_v - z_u| \\ \text{s.t.} \\ z \in \{-1,1\}^{|V|}. \end{cases}$$

Noting that, for any $z \in \{-1,1\}^{|V|}$, the quantity $\sum_{uv\in E} |z_v - z_u|$ equals twice the number of edges in $\delta(S)$, with $S = \{v \in V : z_v = 1\}$, the proposition follows. $\quad\square$

Remark 1. Given a cut $\delta(S)$ with maximum cardinality, we can associate to it the following feasible solution of (MIP1): the vector x corresponds to an orientation of all edges in the cut from S to $V \setminus S$, all other edges are not oriented ($x_{uv} = 0, \forall uv \in E \setminus \delta(S)$), and $y_k^v = 1$ if and only if k is equal to the outdegree minus the indegree w.r.t. the orientation given by x (ignoring edges which are not oriented). Then, for any edge $uv \in E$ which has the original orientation from u to v (i.e. the one given by the matrix B), the following equation holds: $x_{uv} = \sum_{k=1}^{d_u} y_k^u - \sum_{k=1}^{d_v} y_k^v$. From the latter we deduce (developing the expression $B_v x$):

$$\sum_{k=-d_v}^{d_v} k y_k^v = B_v x = \sum_{uv\in E}\left(\sum_{k=1}^{d_v} y_k^v - \sum_{k=1}^{d_u} y_k^u\right). \tag{3}$$

Observe also that each vertex v is incident to at least $\lceil\frac{d_v}{2}\rceil$ edges in the maximum cut. This implies that the variables y_k^v with $k \in \{1-\lceil\frac{d_v}{2}\rceil, \ldots, \lceil\frac{d_v}{2}\rceil - 1\}$ may be removed from formulation (MIP1), while Proposition 1 remains valid.

It follows that, in place of (MIP1), we may consider a formulation involving variables of the form y_k^v only: replace Eqs. (1) by (3). For each vertex v, we only consider y_k^v variables for $k \in [\![-d_v, -\lceil\frac{d_v}{2}\rceil]\!] \cup [\![\lceil\frac{d_v}{2}\rceil, d_v]\!]$. We therefore denote for all $v \in V$, $I_v^- = [\![-d_v, -\lceil\frac{d_v}{2}\rceil]\!]$, $I_v^+ = [\![\lceil\frac{d_v}{2}\rceil, d_v]\!]$ and $I_v = I_v^- \cup I_v^+$.

$$\text{(MIP2)} \begin{cases} Z^\star_{\text{MIP2}} = \frac{1}{2}\max \sum_{v\in V}\sum_{k\in I_v} |k| y_k^v \\ \text{s.t.} \\ \sum_{k\in I_v} y_k^v = 1, \ \forall v \in V, \\ \sum_{k\in I_v} k y_k^v = \sum_{uv\in E}(\sum_{k\in I_v^+} y_k^v - \sum_{k\in I_u^+} y_k^u), \ \forall v \in V, \\ y_k^v \in \{0,1\}, \ \forall v \in V, \ \forall k \in I_v. \end{cases}$$

Formulation (MIP2) involves about $O(|E|)$ variables and $O(|V|)$ constraints. It will be studied in a forthcoming paper. We will rather consider a strengthening of the linear relaxation of (MIP2) through reformulation-linearization techniques. The latter is obtained by multiplying constraints of (MIP2) and then linearizing. Some other constraints follow from the afore mentioned interpretation of the variables in (MIP2). So, let Y_{kl}^{uv}, with $(u,v) \in V^2$, $(k,l) \in I_u \times I_v$, denote a binary variable representing the product $y_k^u y_l^v$. Then, given that the variables y_k^v are binary and satisfy (2) we have $Y_{kl}^{uu} = 0, \forall u \in V, \forall k \neq l$. Considering then the product of the left side of (2) with itself we deduce $\sum_{k\in I_v} Y_{kk}^{vv} = 1$. Using other equations obtained by multiplying variables of the form y_k^v with equations

(2) and others obtained from (3), we can deduce from (MIP2) the following exact formulation for the maximum cardinality cut problem.

$$
\text{(MIP3)}\begin{cases}
Z^\star_{\text{MIP3}} = \dfrac{1}{2} \max \sum_{v \in V} \sum_{k \in I_v} |k| Y^{vv}_{kk} \\[2mm]
\text{s.t.} \\[1mm]
\sum_{k \in I_v} Y^{vv}_{kk} = 1, \ \forall v \in V, & \text{(4a)} \\[3mm]
\sum_{k \in I_v} k Y^{vv}_{kk} = \sum_{uv \in E} \Big(\sum_{k \in I^+_v} Y^{vv}_{kk} - \sum_{k \in I^+_u} Y^{uu}_{kk} \Big), \ \forall v \in V, & \text{(4b)} \\[3mm]
Y^{vv}_{kk} = \sum_{l \in I_u} Y^{vu}_{kl}, \ \forall u, v \in V, \ \forall k \in I_v, & \text{(4c)} \\[3mm]
(d_v - k) Y^{vv}_{kk} = \sum_{uv \in E} \sum_{l \in I^+_u} Y^{vu}_{kl}, \ \forall v \in V, \ \forall k \in I^+_v, & \text{(4d)} \\[3mm]
- k Y^{vv}_{kk} = \sum_{uv \in E} \sum_{l \in I^+_u} Y^{vu}_{kl}, \ \forall v \in V, \ \forall k \in I^-_v, & \text{(4e)} \\[3mm]
\sum_{l \in I_u} l Y^{vu}_{kl} = \sum_{uw \in E} \Big(\sum_{l \in I^+_w} Y^{vw}_{kl} - \sum_{l \in I^+_u} Y^{vu}_{kl} \Big), \ \forall v \neq u \in V, \ \forall k \in I_v, & \text{(4f)} \\[3mm]
Y^{vu}_{kl} = Y^{uv}_{lk}, \ \forall u, v \in V, \ \forall (k,l) \in I_v \times I_u, & \text{(4g)} \\[2mm]
Y^{vu}_{kl} \in \{0,1\}, \ \forall u, v \in V, \ \forall (k,l) \in I_v \times I_u, & \text{(4h)}
\end{cases}
$$

Proposition 2. *The optimal objective value of (MIP3) equals the maximum cardinality of a cut in the graph G: $Z^\star_{MIP3} = w^\star$.* □

Observe that (MIP3) was derived from (MIP2) using lifting. Other formulations could be obtained using some other well-known lifting techniques such as the one of Lassere, the Sherali-Adams technique or the lifting of Lovász-Schrijver [23,26,28]. Since the aim of this paper is not to compare lifting techniques, we do not elaborate more on this topic.

3   A Semidefinite Programming Bound

Let Y denote a symmetric matrix with rows and columns indexed by all pairs (u,k) with $u \in V$ and $k \in I_u$. The entry in the row indexed by (u,k) and column indexed by (v,l) corresponds to the variable Y^{uv}_{kl}. Then, let (SDP3) denote the relaxation obtained from (MIP3) replacing the symmetry (4g) and binary constraints (4h) by the following ones

$$Y - Diag(Y)Diag(Y)^t \succeq 0 \tag{5}$$

$$Y^{vu}_{kl} \geq 0, \forall u, v \in V, \forall k \in I^-_v \cup I^+_v, l \in I^-_u \cup I^+_u, \tag{6}$$

where $Diag(Y)$ denotes the vector corresponding to the diagonal of Y.

We are going to prove that the semidefinite relaxation (SDP3) provides a generally better upper bound for the maximum cardinality cut problem than that from Goemans &Williamson's relaxation. For, consider the following semidefinite relaxation of a 0/1 formulation of the maximum cardinality cut problem whose optimal objective value coincides with $Z^\star_{SDP0}$,

$$(\text{SDP4}) \begin{cases} Z^\star_{SDP4} = \max \frac{1}{2} \sum_{u \in V} \sum_{v:uv \in E} (x_u + x_v - 2X_{uv}) \\ \text{s.t.} \\ Diag(X) = x, \\ X - xx^t \succeq 0, \\ x \in \mathbb{R}^{|V|}, \ X \in \mathbb{S}^{|V|}, \end{cases}$$

where X_{uv} stands for the entry of the matrix X in the row corresponding to node u, and column corresponding to node v.

Proposition 3. *The following inequality holds* $Z^\star_{SDP3} \leq Z^\star_{SDP4}(= Z^\star_{SDP0})$.

Proof. Let Y denote a feasible solution for the formulation (SDP3) and let $(X, x) \in \mathbb{R}^{|V| \times |V|} \times \mathbb{R}^{|V|}$ be defined as follows: $X_{uv} := \sum_{k=-d_u}^{-1} \sum_{l=-d_v}^{-1} Y_{kl}^{uv}$, $\forall u, v \in V$ with $u \neq v$ and $X_{vv} = x_v := \sum_{k=-d_v}^{-1} Y_{kk}^{vv}, \forall v \in V$.

We now show $X - xx^t \succeq 0$. For, let $z \in \mathbb{R}^{|V|}$ and define $\bar{z}$ of convenient dimension as follows: $\bar{z}_k^v = z_v$ if $k < 0$ and 0 otherwise. Then, we have

$$\begin{aligned} \bar{z}^t Y \bar{z} &= \sum_{u \in V} \sum_{k=-d_u}^{-1} \sum_{v \in V} \sum_{l=-d_v}^{-1} \bar{z}_k^u \bar{z}_l^v Y_{kl}^{uv} \\ &= \sum_{u \in V} \sum_{v \in V} z_u z_v (\sum_{k=-d_u}^{-1} \sum_{l=-d_v}^{-1} Y_{kl}^{uv}) \\ &= \sum_{u \in V} \sum_{v \in V} z_u z_v X_{uv} = z^t X z. \end{aligned}$$

Also,

$$\begin{aligned} \bar{z}^t Diag(Y) &= \sum_{v \in V} \sum_{k=-d_v}^{-1} \bar{z}_k^v Y_{kk}^{vv} \\ &= \sum_{v \in V} (z_v \sum_{k=-d_v}^{-1} Y_{kk}^{vv}) \\ &= \sum_{v \in V} z_v x_v = z^t x. \end{aligned}$$

It follows that $z^t(X - xx^t)z = \bar{z}^t(Y - Diag(Y)Diag(Y)^t)\bar{z} \geq 0$, where the last inequality follows from the feasibility of Y w.r.t. (SDP3), and thus $X - xx^t \succeq 0$. So, we have shown (X, x) is a feasible solution for (SDP4).

Since Y is symmetric, the same holds for X. We now show the objective value of (X, x) w.r.t. (SDP4), denoted Z_X equals that of Y w.r.t. (SDP3).

$$\begin{aligned} Z_X &= \frac{1}{2}(\sum_{u \in V} \sum_{v: uv \in E} (x_u + x_v - 2X_{uv})) \\ &= \sum_{v \in V} d_v x_v - \sum_{u \in V} \sum_{uv \in E} X_{uv} \\ &= \sum_{v \in V} d_v \sum_{k=-d_v}^{-1} Y_{kk}^{vv} - \sum_{u \in V} \sum_{uv \in E} \sum_{k=-d_u}^{-1} \sum_{l=-d_v}^{-1} Y_{kl}^{uv} \\ &= \sum_{v \in V} d_v \sum_{k=-d_v}^{-1} Y_{kk}^{vv} - \sum_{u \in V} \sum_{k=-d_u}^{-1} (k + d_u) Y_{kk}^{uu} \\ &= - \sum_{u \in V} \sum_{k=-d_u}^{-1} k Y_{kk}^{uu} \\ &= \sum_{u \in V} \sum_{k=1}^{d_u} k Y_{kk}^{uu}. \end{aligned}$$

$\square$

By Proposition 3, it follows that a randomized algorithm similar to the one by Goemans and Williamson [15] but applied to an optimal solution Y of (SDP3) has the same approximation ratio. To be more precise, let $Z \in \mathbb{R}^{|V| \times |V|}$ denote the matrix with entries $Z_{uv} = 4(X_{uv} - x_u x_v) + (2x_u - 1)(2x_v - 1)$, where $X_{uv} := \sum_{k=-d_u}^{-1} \sum_{l=-d_v}^{-1} Y_{kl}^{uv}, \forall u, v \in V$ with $u \neq v$ and $X_{vv} = x_v := \sum_{k=-d_v}^{-1} Y_{kk}^{vv}, \forall v \in V$. It can be checked that Z is a feasible solution for the formulation (SDP0) with the same objective value as Y. Then, let $H \in \mathbb{R}^{m \times |V|}$ (for some $m \leq |V|$) denote a matrix such that $Z = H^t H$ and let r denote a vector which is randomly generated according to a uniform distribution on the unit sphere in $\mathbb{R}^m$. The cut returned by the algorithm is then $\delta(S)$ with $S := \{v \in V : r^t h_v \geq 0\}$, where h_v stands for the column of H corresponding to node $v \in V$.

The new bound is exact (i.e., equal to the maximum cardinality of a cut) for some graph classes. For space limitation reasons, we will only consider the case of complete graphs. We already know that the bound provided by relaxation (SDP0) is exact for even complete graphs (K_n with n even). This does not hold for odd complete graphs. We prove that the bound given by (SDP3) is exact for all complete graphs.

Proposition 4. *For a complete graph, the optimal objective value of (SDP3) is exact:* $Z_{SDP3}^\star = w^\star$.

Proof. See Appendix A.

In fact the proof of Proposition 4 implies that the linear relaxation of (MIP3) is exact. Details about this linear relaxation will follow in an extended version of the paper.

4 Further Mixed Integer Linear Programming Formulations

In this section, we present three new exact formulations for the unweighted maximum cardinality cut problem with interesting computational performances. The first one stems from (MIP2) using the fact that for all $v \in V$, $\sum_{k \in I_v^-} y_k^v = 1 - \sum_{k \in I_v^+} y_k^v$ deleting all the variables y_k^v where k is negative. We obtain the following exact formulation.

$$(\text{MIP5}) \begin{cases} Z_{\text{MIP5}}^\star = \max \sum_{v \in V} \sum_{k \in I_v^+} k y_k^v \\ \text{s.t.} \\ \sum_{k \in I_v^+} y_k^v \leq 1, \ \forall v \in V, \\ \sum_{uv \in E} (\sum_{k \in I_v^+} y_k^v - \sum_{k \in I_u^+} y_k^u) \leq \sum_{k \in I_v^+} k y_k^v - \lceil \frac{d_v}{2} \rceil (1 - \sum_{k \in I_v^+} y_k^v), \ \forall v \in V, \\ \sum_{uv \in E} (\sum_{k \in I_v^+} y_k^v - \sum_{k \in I_u^+} y_k^u) \geq \sum_{k \in I_v^+} k y_k^v - d_v (1 - \sum_{k \in I_v^+} y_k^v), \ \forall v \in V, \\ y_k^v \in \{0, 1\}, \ \forall v \in V, \forall k \in I_v^+. \end{cases}$$

It involves about half as many variables as (MIP2) and has generally better performance, detailed results can be found in Sect. 5. In order to further reduce the number of variables for an exact formulation, we can aggregate the variables y_k^v with $k \in I_v^+$ for a vertex v to form a variable x^v equal to $\sum_{k \in I_v^+} y_k^v$. For doing so, we need another variable z^v equal to $\sum_{k \in I_v^+} k y_k^v$ in order to keep the information about the difference between the outdegree and indegree of v important for the objective function. We thus obtain the following exact formulation.

$$
\text{(MIP6)} \begin{cases}
Z^\star_{\mathsf{MIP6}} = \max \sum_{v \in V} z^v \\
\text{s.t.} \\
\lceil \frac{d_v}{2} \rceil + \lfloor \frac{d_v}{2} \rfloor x^v - z^v \leq \sum_{uv \in E} x^u \leq d_v - z^v, \ \forall v \in V, \\
\lceil \frac{d_v}{2} \rceil x^v \leq z^v \leq d_v x^v, \ \forall v \in V, \\
x \in \{0,1\}^V, \ z \in \mathbb{R}^V.
\end{cases}
$$

(MIP6) involves $2|V|$ variables, half of which are integer variables and its performance is better than that of (MIP5) for many instances (see Sect. 5). This formulation can also be obtained using the linearization technique of Glover [14] applied to the standard quadratic program modeling the maximum cut problem. One can also propose a third formulation somewhat in between (MIP5), i.e. no aggregation of variables, and (MIP6), i.e. total aggregation of the variables for each vertex. To do so, we partition the interval I_v^+ for each vertex $v \in V$. Let $\alpha > 1$, we parametrize such a partition with α defining the following sequences for each $v \in V$

$$
\begin{cases}
a_1^v = \lceil \frac{d_v}{2} \rceil, \\
a_i^v = 1 + b_{i-1}^v, \text{ for } i > 1, \\
b_i^v = \min(\lfloor \alpha * a_i^v \rfloor, d_v),
\end{cases}
$$

and compute k_v, the smallest integer such that $b_{k_v}^v = d_v$. Then similarly to the formulations (MIP1), (MIP2) and (MIP3), we take a variable y_k^v for each vertex $v \in V$ and each $k \in [\![1, k_v]\!]$ whose interpretation is the following: $y_k^v = 1$ if and only if $z^v \in [\![a_k^v, b_k^v]\!]$. We therefore obtain the following exact formulation for all $\alpha > 1$.

$$
\text{(MIP7[}\alpha\text{])} \begin{cases}
Z^\star_{\mathsf{MIP7}} = \max \sum_{v \in V} z^v \\
\text{s.t.} \\
\lceil \frac{d_v}{2} \rceil + \lfloor \frac{d_v}{2} \rfloor x^v - z^v \leq \sum_{uv \in E} x^u \leq d_v - z^v, \ \forall v \in V, \\
\sum_{k=1}^{k_v} y_k^v = x^v, \ \forall v \in V, \\
\sum_{k=1}^{k_v} a_k^v y_k^v \leq z^v \leq \sum_{k=1}^{k_v} b_k^v y_k^v, \ \forall v \in V, \\
x \in [0,1]^V, \ y^v \in \{0,1\}^{k_v}, \ \forall v \in V, \ z \in \mathbb{R}^V.
\end{cases}
$$

For the purpose of comparing performances, we now give a basic exact formulation for the unweighted maximum cut problem based on the triangle inequalities. It involves one variable $x_{i,j}$ for each unordered pair of vertices $\{i,j\} \subset V$ $(i \neq j)$. Hence $O(n^2)$ variables and $O(n^3)$ constraints.

$$
\text{(MIP8)} \begin{cases}
\max \sum_{ij \in E} x_{i,j} \\
\text{s.t.} \\
x_{i,j} + x_{j,k} + x_{i,k} \leq 2, \ \forall \{i,j,k\} \subset V, \ |\{i,j,k\}| = 3, \\
x_{i,j} + x_{j,k} - x_{i,k} \geq 0, \ \forall (i,j,k) \in V^3, \ |\{i,j,k\}| = 3, \\
x_{i,j} \in \{0,1\}, \ \forall \{i,j\} \subset V, \ i \neq j.
\end{cases}
$$

5   Preliminary Computational Experiments

Some numerical experiments have been conducted to evaluate the quality of the new SDP bound. For each problem instance, we report $w^\star$ (the maximum cardinality of a cut), $Z^\star_{\mathsf{SDP0}}$ (the optimal objective value of (SDP0)) and $Z^\star_{\mathsf{SDP3}}$ (the optimal objective value of (SDP3)). We also mention some results related to the Mixed integer Programming formulations of the previous Section reporting the running time of (MIP2), (MIP5), (MIP6), (MIP7[1.5]), (MIP7[1.3]), (MIP7[1.1]) and (MIP8) for each instance.

The algorithms used for these computations were written in C/C++ calling COIN-OR's CSDP library to solve the semidefinite programs and IBM's ILOG CPLEX optimizer© for the linear and mixed integer programs; all have been performed with a processor $1.9\,\mathrm{GHz} \times 4$, $15.6\,\mathrm{GB}$ RAM. In order to further the relevance of our comparison, we also give for each instance the running time (BC) of the semi-definite based solver BiqCrunch [21] compiled in Python and run on the same machine as the mixed integer programs.

The graph instances used for the computations are denoted as follows:

- K_n: the complete graph with n vertices,
- W_n: the wheel graph with n vertices (i.e. $n-1$ spokes),
- Pe, Co, Oc, Do and Ic: the Petersen graph, the Coxeter graph, the octahedron, the dodecahedron and the icosahedron respectively. (Information about the platonic graphs can be found in [24]),
- C_n: the cycle graph with n vertices,
- $G^{t_2}_{n_1,n_2}$: the 2-dimensional toroidal grid graph, i.e. the cartesian product of two cycles $C_{n_1} \bullet C_{n_2}$,
- $G^{t_k}_n$: the n-dimensional toroidal grid graph of length k, i.e. the cartesian product $\bullet^n_{i=1} C_k$,
- $R_{n,d}$: a randomly generated graph with n vertices and density d: $d = \frac{200|E|}{n(n-1)}$,
- $P_{n,D}$: a randomly generated planar graph with n vertices and proportion of edges with respect to a maximum planar graph D: $D = \frac{100|E|}{3(n-2)}$.

The random graphs were generated using rudy, a machine-independent graph generator by Giovanni Rinaldi.

Let us start with the results related to the new SDP bound. The first set of instances considered consists of two basic graph classes : odd complete graphs and wheel graphs. One can see that $Z^\star_{\mathsf{SDP3}} = w^\star$ for complete graphs as shown in Proposition 4. The bound seems to be exact for wheels (according to numerical experiments). We report in Table 2 results obtained on some well-known graphs: the Petersen graph, the Coxeter graph, the octahedron, the dodecahedron and the icosahedron, along with some toroidal grid graphs. The results reported in Table 3 were computed from randomly generated graphs.

The preliminary computational results from Tables 1, 2, and 3 not only confirm the inequality proved in Proposition 3, but clearly point out that the quality of the new bound presented in the previous Sections is sometimes significatively better than that of Goemans and Williamsons relaxation.

Table 1. Computational results of (SDP3) for complete graphs and wheel graphs

Instance	K_5	K_7	K_{11}	W_5	W_8	W_{10}	W_{12}	W_{15}	W_{17}	W_{20}	W_{22}	W_{25}
$w^\star$	6	12	30	6	10	13	16	21	24	28	31	36
$Z^\star_{SDP3}$	6	12	30	6	10	13	16	21	24	28	31	36
$Z^\star_{SDP0}$	6.25	12.25	30.25	6.25	10.614	13.809	16.979	21.875	25	29.566	32.703	37.5

Table 2. Computational results of (SDP3) for special graph classes

Instance	Pe	Co	Oc	Do	Ic	C_3	C_5	C_7	C_9	C_{11}	C_{13}	C_{15}
$w^\star$	12	36	8	24	20	2	4	6	8	10	12	14
$Z^\star_{SDP3}$	12	36.167	9	25	21	2	4	6.125	8.25	10.383	12.463	14.523
$Z^\star_{SDP0}$	12.5	37.9	9	26.18	21.708	2.25	4.523	6.653	8.729	10.777	12.811	14.836

Instance	C_{17}	C_{19}	C_{21}	C_{23}	C_{25}	$G^{t_2}_{3,3}$	$G^{t_2}_{3,4}$	$G^{t_2}_{3,5}$	$G^{t_2}_{4,5}$	$G^{t_2}_{5,5}$	$G^{t_3}_{3}$
$w^\star$	16	18	20	22	24	12	20	22	36	40	54
$Z^\star_{SDP3}$	16.58	18.621	20.653	22.685	24.709	13.5	20	23.639	36	44.168	60
$Z^\star_{SDP0}$	16.855	18.87	20.883	22.893	24.901	13.5	21	24.818	38.09	45.225	60.75

Table 3. Computational results of (SDP3) for randomly generated graphs

Instance	$R_{5,8}$	$R_{10,9}$	$R_{10,14}$	$R_{10,18}$	$R_{10,23}$	$R_{10,27}$	$R_{10,34}$	$R_{10,36}$	$R_{15,21}$	$R_{15,32}$	$R_{15,42}$
$w^\star$	6	8	12	14	17	19	22	23	17	24	30
$Z^\star_{SDP3}$	6	8	12	14	17	19	22	23	17	24.236	30.381
$Z^\star_{SDP0}$	6.25	8.25	12.585	14.399	17.603	19.962	22.676	23.346	18.006	25.357	31.569

Instance	$R_{15,53}$	$R_{20,19}$	$R_{20,38}$	$R_{20,57}$	$R_{30,44}$	$P_{5,7}$	$P_{5,9}$	$P_{10,10}$	$P_{10,12}$	$P_{10,18}$	$P_{10,24}$
$w^\star$	36	16	29	43	37	5	6	8	10	13	16
$Z^\star_{SDP3}$	36.567	16	29.202	43	37.31	5	6	8	10	13	16
$Z^\star_{SDP0}$	37.39	16.679	30.682	44.757	39.005	5.432	6.25	8.409	10.715	13.932	16.992

Instance	$P_{20,11}$	$P_{20,16}$	$P_{20,27}$	$P_{20,41}$	$P_{20,54}$	$P_{25,35}$	$P_{25,52}$	$P_{25,69}$	$P_{30,8}$	$P_{30,17}$	$P_{30,42}$
$w^\star$	9	15	21	30	36	28	39	46	7	15	33
$Z^\star_{SDP3}$	9	15	21	30	36.207	28.091	39	46.446	7	15	33.037
$Z^\star_{SDP0}$	9.25	15.25	22.495	31.289	38.131	29.705	40.614	48.468	7.25	15.25	34.412

Let us now look at the running times of the Formulations (MIP8), (MIP2), (MIP5), (MIP6), (MIP7[1.5]), (MIP7[1.3]) and (MIP7[1.1]) and of the solver (BC) on several bigger instances found in Table 4. For the entries marked ">900", the

Table 4. Running time for the MIP formulations (in seconds)

Instance	w^*	MIP8	MIP2	MIP5	MIP6	MIP7 [1.5]	MIP7 [1.3]	MIP7 [1.1]	BC	$|V|$	$|E|$
$R_{20,90}$	99	3	1	1	3	1	1	1	0	20	171
$R_{25,50}$	97	7	63	9	1	2	5	5	1	25	150
$R_{25,90}$	152	71	12	3	11	20	16	4	0	25	270
$R_{30,50}$	141	30	341	71	13	21	26	39	1	30	217
$R_{30,90}$	219	>900	44	21	144	111	99	23	0	30	391
$R_{40,25}$	136	29	>900	748	33	49	47	114	1	40	195
$P_{50,50}$	373	2	1	0	0	0	0	0	3	46	72
$P_{50,90}$	598	3	17	10	2	2	1	1	4	50	129
$P_{75,40}$	75	89	0	0	1	0	1	0	5	66	87
$P_{75,50}$	90	77	2	1	1	1	1	0	6	70	109
$P_{75,70}$	120	22	7	5	1	1	1	2	2	75	153
$P_{75,100}$	146	>900	>900	>900	172	>900	62	38	47	75	219
$P_{100,40}$	100	73	0	1	0	0	1	0	8	87	117
$P_{100,90}$	190	168	>900	>900	47	102	35	34	40	100	264
$P_{200,50}$	246	>900	>900	272	75	47	31	26	304	188	297
$P_{300,50}$	376	-	>900	>900	340	203	64	53	>900	282	447
K_{50}	625	>900	2	1	0	0	0	0	0	50	1225
K_{100}	2500	>900	17	7	4	1	1	1	64	100	4950
K_{150}	5625	>900	179	42	29	4	3	3	166	150	11175
K_{175}	7656	>900	322	77	>900	>900	>900	3	5	175	15225
K_{200}	10000	>900	704	145	4	7	4	6	368	200	19900
K_{225}	12656	>900	>900	223	>900	>900	>900	8	11	225	25200
K_{300}	22500	-	>900	875	>900	>900	106	208	177	300	44850
G_4^{t4}	1024	>900	0	1	0	0	0	1	19	256	1024
$G_{10,10}^{t2}$	200	121	0	0	0	0	0	0	2	100	200
$G_{8,15}^{t2}$	232	>900	2	1	1	0	1	0	6	120	240
$G_{11,12}^{t2}$	252	>900	40	8	2	2	1	2	5	132	264
$G_{15,20}^{t2}$	580	>900	>900	>900	86	75	32	32	150	300	600
$G_{20,20}^{t2}$	800	>900	0	0	0	0	1	0	109	400	800
$G_{100,100}^{t2}$	20000	-	7	29	43	45	17	17	>900	10000	20000
W_{75}	111	26	2	3	1	1	0	1	4	75	148
W_{100}	148	58	2	7	5	3	0	1	11	100	198
W_{175}	261	>900	9	333	22	14	1	1	44	175	348
W_{250}	373	>900	61	>900	26	26	2	2	301	250	498
W_{400}	598	>900	62	>900	>900	>900	4	3	>900	400	798
W_{550}	823	-	>900	>900	>900	>900	5	5	>900	550	1098
W_{775}	1161	-	>900	>900	>900	>900	11	10	-	775	1548
W_{925}	1386	-	>900	>900	>900	626	15	15	-	925	1848
W_{1250}	1873	-	>900	>900	>900	>900	34	39	-	1250	2548
C_{100}	100	105	0	0	0	0	0	0	5	100	100
C_{175}	174	889	0	0	0	0	0	0	31	175	175
C_{250}	250	>900	0	0	0	0	0	0	71	250	250
C_{550}	550	-	1	0	0	0	0	0	>900	550	550
C_{1250}	1250	-	0	0	0	0	1	0	-	1250	1250
C_{1750}	1750	-	1	0	1	1	1	0	-	1750	1750
C_{3000}	3000	-	1	1	2	1	1	2	-	3000	3000
C_{5725}	5725	-	13	12	37	13	11	13	-	5725	5725
C_{9000}	9000	-	2	7	10	8	7	7	-	9000	9000

running time exceeded 900 s and the process was therefore interrupted and for the entries marked "-", the memory of the machine was full and the process was therefore interrupted. First, one can see that the new formulations introduced in Sect. 4 perform much better than the classical triangular formulation (MIP8) for all the studied graph families except the general random graphs, and on all of these instances, there is one of our formulation that performs better than BiqCrunch does. More specifically, we can see that (MIP5) has generally better performance than (MIP2) and that for some instances, (MIP5) is drastically better than (MIP6), and for others, it is the other way around. Interestingly, we observe that being (MIP7) somewhat in between (MIP5) and (MIP6), there exists for almost each graph instance a value of α for which (MIP7[α]) has the shortest computing time. Practically, (MIP7[1.1]) seems to be the most robust of them.

6 Conclusion

Starting from graph orientation, we have seen that the maximum cut problem can be modeled in several new ways. By lifting, one can get some bounds that are stronger than the standard semidefinite bound of [15]. The bounds are even exact for some graph classes. Several new mixed integer programming formulations have been obtained using discretization and aggregation techniques. The performance of these formulations compares to and is often better than the performance of the BiqCrunch solver on many graph families, it can even be improved if we strengthen the formulations either by adding valid inequalities or using other lifting techniques. Also the new formulations we introduced here for the unweighted maximum cut problem may lead to similar formulations for the weighted case. This will be part of a forthcoming paper.

Appendix A   Proof of Proposition 4

Since $Z^\star_{\mathsf{SDP0}}$ is exact for even complete graphs, $Z^\star_{\mathsf{SDP3}}$ is also exact by Proposition 3.

Let us now consider odd complete graphs. Let Y be an optimal solution of (SDP3). Let π be any permutation of the set of vertices. By symmetry of the complete graph, the solution defined by $Y_{kl}^{\pi(u)\pi(v)} = Y_{kl}^{uv}$ is obviously still an optimal solution of (SDP3). By considering the set of all permutations Π_n (the symmetric group), and combining all solutions, we still get an optimal solution (by linearity) Z where $Z_{kl}^{uv} = \frac{1}{|\Pi_n|}\sum_{\pi\in\Pi_n} Y_{kl}^{\pi(u)\pi(v)}$. Since we consider all permutations, Z_{kl}^{uv} does not depend on u and v. In other words, there are numbers $f(k,l)$ and $g(k)$ such that $Z_{kl}^{uv} = f(k,l)$ if $u \neq v$ and $Z_{kk}^{uu} = g(k)$, $k, l = \frac{n-1}{2}, \cdots, n-1$.

Let us build another solution Z' of (SDP3) as follows: $Z_{kl}'^{uv} = Z_{(-k)(-l)}^{uv}$. By symmetry of complete graphs, Z' is also an optimal solution of (SDP3). Then, $Z'' = \frac{1}{2}(Z + Z')$ is also optimal. Observe that $Z''_{kl}^{uv} = Z''_{(-k)(-l)}^{uv}$. This implies that we can assume that $g(k) = g(-k)$ and $f(k,l) = f(-k,-l)$.

Moreover, constraints (4a) lead to

$$\sum_{k\in[\![\frac{n-1}{2},n-1]\!]\cup[\![1-n,\frac{1-n}{2}]\!]} g(k) = 1 = 2 \sum_{k\in[\![\frac{n-1}{2},n-1]\!]} g(k). \tag{7}$$

From (4c), we deduce that

$$\sum_{l\in[\![\frac{n-1}{2},n-1]\!]\cup[\![1-n,\frac{1-n}{2}]\!]} f(k,l) = g(k), \quad \forall k \in [\![\frac{n-1}{2},n-1]\!]\cup[\![1-n,\frac{1-n}{2}]\!]. \tag{8}$$

Using equalities (4d), we can also write that

$$kg(k) = (n-1) \sum_{l\in[\![1-n,\frac{1-n}{2}]\!]} f(k,l), \quad \forall k \in [\![\frac{n-1}{2},n-1]\!]. \tag{9}$$

Considering (4f) for a positive k, we obtain $\sum_{l\in[\![\frac{n-1}{2},n-1]\!]\cup[\![1-n,\frac{1-n}{2}]\!]} lf(k,l) = -\sum_{l\in[\![1-n,\frac{1-n}{2}]\!]} f(k,l)$, which is equivalent to

$$\sum_{l\in[\![1-n,\frac{1-n}{2}]\!]} -(l+1)f(k,l) = \sum_{l\in[\![\frac{n-1}{2},n-1]\!]} lf(k,l), \quad \forall k \in [\![\frac{n-1}{2},n-1]\!]. \tag{10}$$

Combining (8) and (9), we deduce that

$$(n-1-k) \sum_{l\in[\![1-n,\frac{1-n}{2}]\!]} f(k,l) = k \sum_{l\in[\![\frac{n-1}{2},n-1]\!]} f(k,l), \quad \forall k \in [\![\frac{n-1}{2},n-1]\!]. \tag{11}$$

Observe that (11) implies that when $k = n-1$ then $f(n-1,l) = 0$ for $l \in [\![\frac{n-1}{2},n-1]\!]$. By (10), we get that $f(n-1,l) = 0$ for $l \in [\![1-n,\frac{1-n}{2}]\!]$. We can then assume in the rest of the proof that $k < n-1$.

Observe that the left side of (10) satisfies

$$\sum_{l\in[\![1-n,\frac{1-n}{2}]\!]} -(l+1)f(k,l) \geq \frac{n-3}{2} \sum_{l\in[\![1-n,\frac{1-n}{2}]\!]} f(k,l)$$
$$= \frac{n-3}{2}\frac{k}{n-1-k} \sum_{l\in[\![\frac{n-1}{2},n-1]\!]} f(k,l)$$

where the last equality is induced by (11). Let k^{max} be the largest k such that $f(k,l) \neq 0$ for some l. We already know that $k^{max} \leq n-2$. The right side of (10) necessarily satisfies $\sum_{l\in[\![\frac{n-1}{2},n-1]\!]} lf(k,l) \leq k^{max} \sum_{l\in[\![\frac{n-1}{2},n-1]\!]} f(k,l)$. Combining the two previous inequalities together with (10), we obtain

$$(k^{max} - \frac{n-3}{2}\frac{k}{n-1-k}) \sum_{l\in[\![\frac{n-1}{2},n-1]\!]} f(k,l) \geq 0, \quad \forall k \in [\![\frac{n-1}{2},k^{max}]\!]. \tag{12}$$

By considering the case $k = k^{max}$ in (12), the sum $\sum_{l\in[\![\frac{n-1}{2},n-1]\!]} f(k^{max},l)$ is strictly positive, leading to $k^{max} - \frac{n-3}{2}\frac{k^{max}}{n-1-k^{max}} \geq 0$. In other words, we necessarily have $k^{max} \leq \frac{n+1}{2}$. This implies that $g(k) = 0$ and $f(k,l) = 0$ if either $k > \frac{n+1}{2}$ or $k < -\frac{n+1}{2}$ (we use here the fact that $g(k) = g(-k)$ and (8)).

Writing (10) and (11) for $k = \frac{n-1}{2}$ and $k = \frac{n+1}{2}$, we get the next 4 equations.

$$(n-3)[f(\frac{n+1}{2}, \frac{-1-n}{2}) + f(\frac{n+1}{2}, \frac{1-n}{2})] = (n+1)[f(\frac{n+1}{2}, \frac{n+1}{2}) + f(\frac{n+1}{2}, \frac{n-1}{2})] \quad (13)$$

$$(n-1)f(\frac{n+1}{2}, \frac{-1-n}{2}) + (n-3)f(\frac{n+1}{2}, \frac{1-n}{2}) = (n+1)f(\frac{n+1}{2}, \frac{n+1}{2}) + (n-1)f(\frac{n+1}{2}, \frac{n-1}{2}) \quad (14)$$

$$f(\frac{n-1}{2}, \frac{-1-n}{2}) + f(\frac{n-1}{2}, \frac{1-n}{2}) = f(\frac{n-1}{2}, \frac{n+1}{2}) + f(\frac{n-1}{2}, \frac{n-1}{2}) \quad (15)$$

$$(n-1)f(\frac{n-1}{2}, \frac{-1-n}{2}) + (n-3)f(\frac{n-1}{2}, \frac{1-n}{2}) = (n+1)f(\frac{n-1}{2}, \frac{n+1}{2}) + (n-1)f(\frac{n-1}{2}, \frac{n-1}{2}) \quad (16)$$

Substracting (13) from (14) leads to $f(\frac{n+1}{2}, \frac{n-1}{2}) = -f(\frac{n+1}{2}, \frac{-1-n}{2})$. By non-negativity of the f values, we deduce that $f(\frac{n+1}{2}, \frac{n-1}{2}) = f(\frac{n+1}{2}, \frac{-1-n}{2}) = 0$ and $f(\frac{n+1}{2}, \frac{1-n}{2}) = \frac{n+1}{n-3}f(\frac{n+1}{2}, \frac{n+1}{2})$. Substracting $(n-1) \times$ (15) from (16) leads in a similar way to $f(\frac{n-1}{2}, \frac{1-n}{2}) = f(\frac{n-1}{2}, \frac{n+1}{2}) = 0$ and $f(\frac{n-1}{2}, \frac{-1-n}{2}) = f(\frac{n-1}{2}, \frac{n-1}{2})$.

Using (9) and the previous observations we get that:

$$g(\frac{n+1}{2}) = 2\frac{n-1}{n+1}f(\frac{n+1}{2}, \frac{1-n}{2}) \text{ and } g(\frac{n-1}{2}) = 2f(\frac{n-1}{2}, \frac{-n-1}{2}).$$

Using the fact that $f(\frac{n+1}{2}, \frac{1-n}{2}) = f(\frac{n-1}{2}, \frac{-n-1}{2})$ and $g(\frac{n+1}{2}) + g(\frac{n-1}{2}) = \frac{1}{2}$, one can deduce that $f(\frac{n-1}{2}, \frac{-n-1}{2}) = \frac{n+1}{8n}$. Consequently, $g(\frac{n+1}{2}) = \frac{n-1}{4n}$ and $g(\frac{n-1}{2}) = \frac{n+1}{4n}$.

Remember that $Z^{\star}_{\mathsf{SDP3}} = \frac{1}{2}\sum_{v \in V}\sum_k |k|Y^{vv}_{kk} = \sum_{v \in V}\sum_{k>0}kY^{vv}_{kk}$, leading to $Z^{\star}_{\mathsf{SDP3}} = n\left(\frac{n+1}{2}g(\frac{n+1}{2}) + \frac{n-1}{2}g(\frac{n-1}{2})\right) = \frac{n^2-1}{4}$, and ending the proof. $\qquad\square$

References

1. Anjos, M., Wolkowicz, H.: Strengthened semidefinite relaxations via a second lifting for the Max-Cut problem. Discrete Appl. Math. **119**, 79–106 (2002)
2. Ashcraft, C.C., Liu, J.W.H.: Using domain decomposition to find graph bisectors. Technical report CS-95-08, York University (1995)
3. Barahona, F., Mahjoub, A.R.: On the cut polytope. Math. Program. **36**, 157–173 (1986)
4. Barahona, F., Grötschel, M., Jünger, M., Reinelt, G.: An application of combinatorial optimization to statistical physics and circuit layout design. Oper. Res. **36**, 493–513 (1998)
5. Ben-Ameur, W., Mahjoub, A.R., Neto, J.: The maximum cut problem. In: Paschos, V. (ed.) Paradigms of Combinatorial Optimization, pp. 131–172. Wiley-ISTE, Hoboken (2010)
6. Ben-Ameur, W., Neto, J.: On a gradient-based randomized heuristic for the maximum cut problem. Int. J. Math. Oper. Res. **4**(3), 276–293 (2012)
7. Ben-Ameur, W., Neto, J.: Spectral bounds for the maximum cut problem. Networks **52**(1), 8–13 (2008)
8. Ben-Ameur, W., Neto, J.: Spectral bounds for unconstrained (-1, 1)-quadratic optimization problems. Eur. J. Oper. Res. **207**(1), 15–24 (2010)

9. Calzarossa, M., Serazzi, G.: Workload characterization: a survey. Proc. IEEE **81**, 1136–1150 (1993)
10. Deza, M., Laurent, M.: Geometry of Cuts and Metrics. Springer, Berlin (1997)
11. Dolezal, O., Hofmeister, T., Lefmann, H.: A comparison of approximation algorithms for the MaxCut problem, Report CI-/99, Universität Dortmund (1999)
12. Festa, P., Pardalos, P., Resende, M., Ribeiro, C.: Randomized heuristics for the MAX-CUT problem. Optim. Methods Softw. **7**, 1033–1058 (2002)
13. Fischer, I., Gruber, G., Rendl, F., Sotirov, R.: Computational experience with a bundle approach for semidefinite cutting plane relaxations of Max-Cut and equipartition. Math. Program. **105**, 451–469 (2006)
14. Glover, F.: Improved linear integer programming formulations of nonlinear integer problems. Manage. Sci. **22**, 455–460 (1975)
15. Goemans, M., Williamson, D.: Improved approximation algorithms for maximum cut and satisfiability problems using semidefinite programming. J. ACM **42**, 1115–1145 (1995)
16. Grötschel, M., Pulleyblank, W.R.: Weakly bipartite graphs and the Max-Cut problem. Oper. Res. Lett. **1**, 23–27 (1981)
17. Hammer, P.: Some network flow problems solved with pseudo-boolean programming. Oper. Res. **32**, 388–399 (1965)
18. Hastad, J.: Some optimal inapproximability results. J. ACM **48**, 798–859 (2001)
19. Helmberg, C.: A cutting plane algorithm for large scale semidefinite relaxations. Technical report ZR-01-26, Konrad-Zuse-Zentrum Berlin (2001)
20. Karp, R.M.: Reducibility among combinatorial problems. In: Miller, R.E., Thatcher, J.W. (eds.) Complexity of Computer Computation, pp. 85–103. Plenum Press, New York (1972)
21. Krislock, N., Malick, J., Roupin, F.: Improved semidefinite bounding procedure for solving Max-Cut problems to optimality. Math. Program. **143**(1–2), 61–86 (2014)
22. Laurent, M.: Tighter linear and semidefinite relaxations for Max-Cut based on the Lovász-Schrijver lift-and-project procedure. SIAM J. Optim. **2**, 345–375 (2001)
23. Laurent, M.: A comparison of the Sherali-Adams, Lovász-Schrijver and Lasserre relaxations for 0–1 programming. Math. Oper. Res. **28**(3), 470–496 (2003)
24. Read, R.C., Wilson, R.J.: An Atlas of Graphs. Clarendon Press, Oxford (1998)
25. Rendl, F., Rinaldi, G., Wiegele, A.: Solving Max-Cut to optimality by intersecting semidefinite and polyhedral relaxations, IASI Research report 08–11 (2008)
26. Rothvoß, T.: The Lasserre hierarchy in approximation algorithms. Lecture Notes for the MAPSP Tutorial Preliminary Version (2013)
27. Sahni, S., Gonzalez, T.: P-complete approximation algorithms. J. Assoc. Comput. Mach. **23**(3), 555–565 (1976)
28. Wiegele, A.: Nonlinear Optimization Techniques Applied to Combinatorial Optimization Problems. Ph.D. thesis, Alpen-Adria-Universitt Klagenfurt (2006)

Maximum Weight Independent Sets in ($S_{1,1,3}$, bull)-free Graphs

T. Karthick[1]([⊠]) and Frédéric Maffray[2]

[1] Indian Statistical Institute, Chennai Centre,
Chennai 600113, India
karthick@isichennai.res.in
[2] CNRS, Laboratoire G-SCOP, University of Grenoble-Alpes,
Grenoble, France
frederic.maffray@grenoble-inp.fr

Abstract. The MAXIMUM WEIGHT INDEPENDENT SET (MWIS) problem on graphs with vertex weights asks for a set of pairwise nonadjacent vertices of maximum total weight. The MWIS problem is well known to be NP-complete in general, even under substantial restrictions. The computational complexity of the MWIS problem for $S_{1,1,3}$-free graphs is unknown. In this note, we give a proof for the solvability of the MWIS problem for ($S_{1,1,3}$, bull)-free graphs in polynomial time. Here, an $S_{1,1,3}$ is the graph with vertices $v_1, v_2, v_3, v_4, v_5, v_6$ and edges $v_1 v_2, v_2 v_3, v_3 v_4, v_4 v_5, v_4 v_6$, and the *bull* is the graph with vertices v_1, v_2, v_3, v_4, v_5 and edges $v_1 v_2$, $v_2 v_3, v_3 v_4, v_2 v_5, v_3 v_5$.

Keywords: Graph algorithms · Weighted independent set · Modular decomposition · Claw-free graph · Fork-free graph · Bull-free graph

1 Introduction

In a graph G, an *independent set* (also called *stable set*) is a subset of pairwise nonadjacent vertices. In the MAXIMUM INDEPENDENT SET PROBLEM (MIS for short), given a graph G, the task is to find an independent set of maximum cardinality. In the MAXIMUM WEIGHT INDEPENDENT SET PROBLEM (MWIS for short), the input is a vertex-weighted graph G, where every vertex has a non-negative integer weight, and the task is to find an independent set of maximum total weight. The M(W)IS problem is a fundamental and extremely well-studied algorithmic graph problem. The MIS problem is known to be NP-complete in general [21], even under various restrictions [14,32], is hard to approximate within a factor of $O(n^\epsilon)$ for $\epsilon < 1$ [3], and is not fixed parameter tractable unless FPT = W[1], see [16].

Here we will focus on studying M(W)IS problem on restricted classes of graphs. If $\mathcal{F}$ is a family of graphs, a graph G is said to be $\mathcal{F}$-*free* if it contains

F. Maffray—Partially supported by ANR project STINT under reference ANR-13-BS02-0007.

T.N. Dinh and M.T. Thai (Eds.): COCOON 2016, LNCS 9797, pp. 385–392, 2016.
DOI: 10.1007/978-3-319-42634-1_31

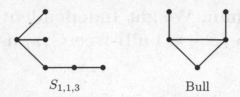

$S_{1,1,3}$ Bull

Fig. 1. Some special graphs.

no induced subgraph isomorphic to any graph in $\mathcal{F}$. Alekseev [1] showed that
the M(W)IS problem remains NP-complete on H-free graphs, whenever H is
connected, but neither a path nor a subdivision of the claw ($K_{1,3}$). On the other
hand, the M(W)IS problem is known to be solvable in polynomial time on many
graph classes, such as: P_4-free graphs [15], chordal graphs [20], perfect graphs
[19], $2K_2$-free graphs [17], some classes of apple-free graphs [9,27], some subclasses of subcubic graphs [11,24,28], some classes of hole-free graphs [4,5,7,10],
and for some classes of planar graphs [2,30].

For integers $i, j, k \geq 1$, let $S_{i,j,k}$ denote a tree with exactly three vertices of
degree one, being at distance i, j and k from the unique vertex of degree three.
The graph $S_{1,1,1}$ is called a *claw* and $S_{1,1,2}$ is called a *chair or fork*. Also, note
that $S_{i,j,k}$ is a subdivision of a claw.

Minty [31] showed that the MWIS problem can be solved in polynomial time
for claw-free graphs. Using modular decomposition techniques, Lozin and Milanič
[26] showed that the MWIS problem can be solved in polynomial time for fork-
free graphs. The complexity of the MWIS problem is unknown for the class of
$S_{1,1,3}$-free graphs. However, the M(W)IS problem can be solved in polynomial
time for some subclasses of $S_{i,j,k}$-free graphs; see [18, Table 1] and [23]. Note that
the class of $S_{1,1,3}$-free graphs extend the class of fork-free graphs and the class
of P_5-free graphs. It is also known that the MWIS problem in P_5-free graphs can
be solved in polynomial time [25].

The *bull* is the graph with vertex-set $\{v_1, v_2, v_3, v_4, v_5\}$ and edge-set $\{v_1v_2,$
$v_2v_3, v_3v_4, v_2v_5, v_3v_5\}$; see Fig. 1. Our main result is the following.

Theorem 1. *The MWIS problem can be solved in time $O(n^9)$ in the class of
$(S_{1,1,3},$ bull)-free graphs.*

Before presenting the proofs, we recall some related results. Brandstädt and
Mosca [10] showed that MWIS can be solved in polynomial time in the class of
(odd-hole, bull)-free graphs. This class does not contain the class of $(S_{1,1,3},$ bull)-
free graphs, since the latter can contain arbitrarily long holes. Thomassé et al.
[34] use the decomposition theorem for bull-free trigraphs, due to Chudnovsky
[12,13], to prove that MWIS is FPT in the class of bull-free graphs. The bottleneck against polymiality is a subclass called $\mathcal{T}_1$. It might be that one can
prove that MWIS is polynomial in the class of $S_{1,1,3}$-free graphs in $\mathcal{T}_1$. However
our algorithm is, we believe, conceptually much simpler.

2 Preliminaries

For missing notation and terminology, we refer to [8]. Let G be a finite, undirected and simple graph with vertex-set $V(G)$ and edge-set $E(G)$. We let $|V(G)| = n$ and $|E(G)| = m$. Let P_n and C_n denote respectively the path, and the cycle on n vertices. In a graph G, the neighborhood of a vertex x is denoted by $N(x)$, and we also use the notation $N[x]$ for $N(x) \cup \{x\}$. For two disjoint subsets $A, B \subset V(G)$, we say that A is *complete* to B if every vertex in A is adjacent to every vertex in B. In case A contains only one vertex a, we may write that a (rather than $\{a\}$) is complete to B. A vertex $z \in V(G)$ *distinguishes* two other vertices $x, y \in V(G)$ if z is adjacent to one of them and nonadjacent to the other. A vertex set $M \subseteq V(G)$ is a *module* in G if no vertex from $V(G) \setminus M$ distinguishes two vertices from M. The *trivial modules* in G are $V(G)$, $\emptyset$, and all one-elementary vertex sets. A graph G is *prime* if it contains only trivial modules. Note that prime graphs with at least three vertices are connected.

A class of graphs $\mathcal{G}$ is *hereditary* if every induced subgraph of a member of $\mathcal{G}$ is also in $\mathcal{G}$. We will use the following theorem by Lozin and Milanič [26].

Theorem 2 ([26]). *Let $\mathcal{G}$ be a hereditary class of graphs. If the MWIS problem can be solved in $O(n^p)$-time for prime graphs in $\mathcal{G}$, where $p \geq 1$ is a constant, then the MWIS problem can be solved for graphs in $\mathcal{G}$ in time $O(n^p + m)$.*   □

A *clique* in G is a subset of pairwise adjacent vertices in G. A *clique separator/clique cutset* in a connected graph G is a subset Q of vertices in G such that Q is a clique and such that the graph induced by $V(G) \setminus Q$ is disconnected. A graph is an *atom* if it does not contain a clique separator.

Let $\mathcal{C}$ be a class of graphs. A graph G is *nearly* $\mathcal{C}$ if for every vertex v in $V(G)$ the graph induced by $V(G) \setminus N[v]$ is in $\mathcal{C}$. Let $\alpha_w(G)$ denote the weighted independence number of G. Obviously, we have:

$$\alpha_w(G) = \max\{w(v) + \alpha_w(G \setminus N[v]) \mid v \in V(G)\}. \tag{1}$$

The following observation follows obviously from (1); see also [6].

Observation 1 ([6]). *If the MWIS problem is solvable in time T on a class $\mathcal{C}$, then it is solvable on nearly $\mathcal{C}$ graphs in time $n \cdot T$.*   □

We will also use the following theorem given in [4]; see also [22]. Though the theorem (Theorem 1 of [4]) is stated only for hereditary class of graphs, the proof also work for any class of graphs, and is given below:

Theorem 3 ([4]). *Let $\mathcal{C}$ be a class of graphs such that MWIS can be solved in time $O(f(n))$ for every graph in $\mathcal{C}$ with n vertices. Then in any class of graphs whose atoms are all nearly $\mathcal{C}$ the MWIS problem can be solved in time $O(n \cdot f(n) + nm)$.*   □

Theorem 4 ([29]). *The MWIS problem can solved in time $O(n^7)$ for (P_6, bull)-free graphs.*   □

3 MWIS on $(S_{1,1,3}, \text{bull})$-free Graphs

This section contains the proof of Theorem 1: we show that the MWIS problem can be solved in time $O(n^9)$ for $(S_{1,1,3}, \text{bull})$-free graphs, by analyzing the structure of this class of graphs.

A 6-*fan* is a graph with vertices $v_1, v_2, \ldots, v_6$ and a such that v_1-v_2-v_3-v_4-v_5-v_6 is a P_6 and a is adjacent to v_i for all $i \in \{1, \ldots, 6\}$.

Lemma 1. *Let G be a prime bull-free graph. Then G does not contain a 6-fan.*

Proof. Suppose that G contains a 6-fan with vertices $v_1, \ldots, v_6$ and u such that v_1-v_2-v_3-v_4-v_5-v_6 is a P_6 and u is adjacent to v_i for all $i \in \{1, \ldots, 6\}$. Let $P = \{v_1, \ldots, v_6\}$ and $A = \{a \in V(G) \setminus P \mid a \text{ is complete to } P\}$. So $A \neq \emptyset$. Let H be the component of $G \setminus A$ that contains P. We claim that:

$$A \text{ is complete to } V(H). \tag{2}$$

Proof of (2): Suppose on the contrary that there exist $a \in A$ and $x \in V(H)$ such that a is not adjacent to x. By the definition of A we have $x \in V(H) \setminus P$. By the definition of H, there is a shortest path x_0-x_1-$\cdots$-x_q such that $x_0 \in P$, $x_1, \ldots, x_q \in V(H) \setminus P$, and $x_q = x$, and $q \geq 1$.

First suppose that $q = 1$. Since x has a neighbor (x_0) in P and $x \notin A$, there are two integers $i, j \in \{1, \ldots, 6\}$ such that $|i - j| = 1$ and x is adjacent to v_i and not to v_j, and we may assume up to symmetry that $\{i, j\} \in \{\{1, 2\}, \{2, 3\}, \{3, 4\}\}$. Suppose that $\{i, j\} = \{1, 2\}$. Then x is adjacent to v_k for each $k \in \{4, 5, 6\}$, for otherwise $\{x, v_i, v_j, a, v_k\}$ induces a bull. If x is not adjacent to v_3, then $\{v_3, v_4, v_5, x, v_1\}$ induces a bull if $i = 1$, while $\{v_1, a, v_3, v_4, x\}$ induces a bull if $i = 2$. So x is adjacent to v_3. If $i = 1$, then $\{v_2, v_3, v_4, x, v_6\}$ induces a bull; if $i = 2$, then $\{v_1, v_2, v_3, x, v_6\}$ induces a bull, a contradiction. Next, suppose that $\{i, j\} = \{2, 3\}$. Then x is adjacent to v_k for each $k \in \{5, 6\}$, for otherwise $\{x, v_i, v_j, a, v_k\}$ induces a bull. If x is not adjacent to v_4, then either $\{v_4, v_5, v_6, x, v_1\}$ induces a bull or $\{x, v_5, v_4, a, v_1\}$ induces a bull. So x is adjacent to v_4. Likewise, if x is not adjacent to v_3, then either $\{v_3, v_4, v_5, x, v_1\}$ induces a bull or $\{x, v_4, v_3, a, v_1\}$ induces a bull. So x is adjacent to v_3. So $i = 3$; but then $\{v_2, v_3, v_4, x, v_6\}$ induces a bull, a contradiction. Finally suppose that $\{i, j\} = \{3, 4\}$. Since we may assume that we are not in the preceding cases, x is complete to $\{v_1, v_2, v_3\}$ and has no neighbor in $\{v_4, v_5, v_6\}$. But then $\{x, v_3, v_4, a, v_6\}$ induces a bull.

Now suppose that $q \geq 2$. By the preceding point, a is adjacent to x_1. Since x_1 has a neighbor and a non-neighbor in P, there are non-adjacent vertices $v, v' \in P$ such that x_1 is adjacent to v and not to v'. Then, by induction on $j = 2, \ldots, q$, we see that a is adjacent to x_j, for otherwise $\{v', a, x_{j-2}, x_{j-1}, x_j\}$ induces a bull. So a is adjacent to x. Thus (2) holds.

By (2) and the definition of H, and the fact that $A \neq \emptyset$, we see that $V(H)$ is a homogeneous set in G, which contradicts the hypothesis that G is prime. $\square$

A *hole* is a chordless cycle C_k, where $k \geq 5$, and a *long hole* is a hole C_k, where $k \geq 7$.

Theorem 5. *Let G be a prime $(S_{1,1,3}$, bull, long hole)-free graph. Then every atom of G is nearly P_6-free.*

Proof. Let G' be an atom of G. Suppose on the contrary that G' has a vertex x whose non-neighborhood contains a P_6, say, v_1-v_2-v_3-v_4-v_5-v_6. Let $P = \{v_1, \ldots, v_6\}$ and $A = N(P)$. Let X be the component of $G' \setminus A$ that contains x. Let $A^+ = \{a \in A \mid a \text{ has a neighbor in } X\}$. So A^+ is a separator that separates X from P. We claim that:

$$\text{Every vertex } a \in A^+ \text{ satisfies } N_P(a) = \{v_1\} \text{ or } N_P(a) = \{v_6\}. \qquad (3)$$

Proof of (3). Consider any $a \in A^+$. By the definition of this set, a has a neighbor $z \in X$. So z has no neighbor in P. Suppose that there is an integer i such that a is adjacent to v_i and v_{i+1} ($i \in \{1, \ldots, 5\}$). If a is complete to P, then $P \cup \{a\}$ induces a 6-fan, which contradicts Lemma 1. So a has a non-neighbor in P. Then, possibly up to reversing the labelling, there is an integer j such that a is adjacent to v_j and v_{j+1} and not to v_{j+2}; but then $\{z, a, v_j, v_{j+1}, v_{j+2}\}$ induces a bull. So there is no such integer i, in other words $N_P(a)$ is a stable set. Suppose that $|N_P(a)| \geq 3$. Then, up to symmetry, $N_P(a)$ is equal to either $\{v_1, v_3, v_5\}$ or $\{v_1, v_3, v_6\}$. If $N_P(a) = \{v_1, v_3, v_5\}$, then $\{v_4, v_5, v_6, a, v_1, v_2\}$ induces an $S_{1,1,3}$. If $N_P(a) = \{v_1, v_3, v_6\}$, then $\{z, v_1, a, v_6, v_5, v_4\}$ induces an $S_{1,1,3}$. Hence $|N_P(a)| \leq 2$.

Suppose that $|N_P(a)| = 2$. Then, up to symmetry, $N_P(a)$ is equal to either $\{v_1, v_3\}$, $\{v_1, v_4\}$, $\{v_1, v_5\}$, $\{v_1, v_6\}$, $\{v_2, v_4\}$, or $\{v_2, v_5\}$.
If $N_P(a) = \{v_1, v_3\}$, then $\{a, v_2, v_3, v_4, v_5, v_6\}$ induces an $S_{1,1,3}$.
If $N_P(a) = \{v_1, v_4\}$, then $\{z, v_1, a, v_4, v_5, v_6\}$ induces an $S_{1,1,3}$.
If $N_P(a) = \{v_1, v_5\}$, then $\{v_6, v_5, v_4, a, v_1, v_2\}$ induces an $S_{1,1,3}$.
If $N_P(a) = \{v_1, v_6\}$, then $P \cup \{a\}$ induces a C_7.
If $N_P(a) = \{v_2, v_4\}$, then $\{z, v_2, a, v_4, v_5, v_6\}$ induces an $S_{1,1,3}$.
If $N_P(a) = \{v_2, v_5\}$, then $\{v_1, v_2, v_3, a, v_5, v_6\}$ induces an $S_{1,1,3}$, a contradiction.

Hence $|N_P(a)| = 1$. Now if $N_P(a) = \{v_2\}$, then $\{a, v_1, v_2, v_3, v_4, v_5\}$ induces an $S_{1,1,3}$, and if $N_P(a) = \{v_3\}$, then $\{a, v_2, v_3, v_4, v_5, v_6\}$ induces an $S_{1,1,3}$; and the cases $N_P(a) = \{v_4\}$ or $\{v_5\}$ are symmetric. Thus (3) holds.

For each $i \in \{1, 6\}$, let $B_i = \{a \in A^+ \mid N_P(a) = \{v_i\}\}$. By (3) we have $A^+ = B_1 \cup B_6$. We observe that:

$$\text{Each of } B_1 \text{ and } B_6 \text{ is a clique.} \qquad (4)$$

Proof of (4). If B_1 contains two non-adjacent vertices a, b, then $\{a, b, v_1, v_2, v_3, v_4\}$ induces an $S_{1,1,3}$. The same holds for B_6. Thus (4) holds.

Since G' is an atom, the set A^+ is not a clique separator of G', so, since it is a separator, it is not a clique. Hence, by (4), there are non-adjacent vertices $a \in B_1$ and $b \in B_6$. By the definition of A^+, there is a shortest path Q from a to b with all its interior vertices in X. Then the union of P and Q is a C_k, for some $k \geq 9$, a contradiction. $\qquad \square$

Theorem 6. *The MWIS problem can be solved in time $O(n^8)$ for $(S_{1,1,3}$, bull, long hole)-free graphs.*

Proof. Let G be an $(S_{1,1,3}$, bull, long hole)-free graph. First suppose that G is prime. By Theorem 5, every atom of G is nearly P_6-free. Since the MWIS problem for $(P_6$, bull)-free graphs can be solved in time $O(n^7)$ (by Theorem 4), MWIS can be solved in time $O(n^8)$ for G, by Theorem 3. Then the time complexity is the same when G is not prime, by Theorem 2.   □

A *k-wheel* is a graph that consists of a k-cycle plus a vertex (called the center) adjacent to all the vertices of the cycle.

Lemma 2 ([33]). *Let G be a prime bull-free graph. Then G does not contain a k-wheel, for any $k \geq 6$.*

Theorem 7. *Prime $(S_{1,1,3}, bull)$-free graphs are nearly long hole-free.*

Proof. Let G be a prime $(S_{1,1,3}$, bull)-free graph. Assume to the contrary that there is a vertex p in G such that $G \setminus N[p]$ contains an induced C_k, where $k \geq 7$, say H with vertex set $V(H) = \{v_1, v_2, \ldots, v_k\}$ and with edge set $E(H) = \{v_1v_2, v_2v_3, \ldots, v_{k-1}v_k, v_kv_1\}$. Let $A = N(H)$. Let X be the component of $G \setminus A$ that contains p. Let $A^+ = \{a \in A \mid a$ has a neighbor in $X\}$. So A^+ is a separator that separates X from H.

Since G is prime, it is connected, so $A^+ \neq \emptyset$. Pick any $x \in A^+$. So there exists a vertex $y \in X$ such that $xy \in E(G)$.

Suppose that x is adjacent to two consecutive vertices, say, v_j, v_{j+1} of H. Then x must be adjacent to v_{j+2}, for otherwise $\{y, x, v_j, v_{j+1}, v_{j+2}\}$ induces a bull. The same argument repeated along H implies that x is complete to H. But then $V(H) \cup \{x\}$ induces a k-wheel, with $k \geq 7$, a contradiction to Lemma 2.

Therefore, assume that the vertex x is not adjacent to two consecutive vertices of H. Since $x \in N(H)$, there exists a vertex $v_i \in V(H)$ $(1 \leq i \leq k)$ such that $xv_i \in E(G)$. By our assumption, $xv_{i-1}, xv_{i+1} \notin E(G)$. Suppose that $xv_{i+2} \notin E(G)$. Then since $\{v_{i+3}, v_{i+2}, v_{i+1}, v_i, v_{i-1}, x\}$ does not induce an $S_{1,1,3}$, we have $xv_{i+3} \in E(G)$. Then since $\{v_{i-2}, v_{i-1}, v_i, x, v_{i+3}, y\}$ does not induce an $S_{1,1,3}$, we have $xv_{i-2} \in E(G)$. But, now $\{v_{i+2}, v_{i+1}, v_i, x, y, v_{i-2}\}$ induces an $S_{1,1,3}$ in G, which is a contradiction. So, suppose that $xv_{i+2}, xv_{i-2} \in E(G)$. By assumption $xv_{i-3}, xv_{i+3} \notin E(G)$. If $k = 7$, then $\{v_{i-3}, v_{i+3}, v_{i+2}, x, y, v_i\}$ induces an $S_{1,1,3}$ in G, and if $k \geq 8$, then $\{v_{i-3}, v_{i-2}, x, v_{i+2}, v_{i+3}, v_{i+1}\}$ induces an $S_{1,1,3}$ in G, a contradiction.   □

Now we can give the proof of Theorem 1. Let G be an $(S_{1,1,3}$, bull)-free graph. First suppose that G is prime. By Theorem 7, G is nearly long-hole-free. Since the MWIS problem for $(S_{1,1,3}$, bull, long hole)-free graphs can be solved in time $O(n^8)$ (by Theorem 6), MWIS can be solved in time $O(n^9)$ for G, by Observation 1. Finally, when G is not prime the time complexity is the same, by Theorem 2.   □

References

1. Alekseev, V.E.: The effect of local constraints on the complexity of determination of the graph independence number. In: Combinatorial-Algebraic Methods in Applied Mathematics, pp. 3–13. Gorkiy University Press, Gorky (1982). (in Russian)
2. Alekseev, V.E., Lozin, V.V., Malyshev, D., Milanič, M.: The maximum independent set problem in planar graphs. In: Ochmański, E., Tyszkiewicz, J. (eds.) MFCS 2008. LNCS, vol. 5162, pp. 96–107. Springer, Heidelberg (2008)
3. Arora, S., Barak, B.: Computational Complexity - A Modern Approach. Cambridge University Press, Cambridge (2009)
4. Basavaraju, M., Chandran, L.S., Karthick, T.: Maximum weight independent sets in hole- and dart-free graphs. Discrete Appl. Math. **160**, 2364–2369 (2012)
5. Brandstädt, A., Giakoumakis, V., Maffray, F.: Clique separator decomposition of hole- and Diamond-free graphs and algorithmic consequences. Discrete Appl. Math. **160**, 471–478 (2012)
6. Brandstädt, A., Hoàng, C.T.: On clique separators, nearly chordal graphs, and the maximum weight stable set problem. Theor. Comput. Sci. **389**, 295–306 (2007)
7. Brandstädt, A., Karthick, T.: Weighted efficient domination in two subclasses of P_6-free graphs. Discrete Appl. Math. **201**, 38–46 (2016)
8. Brandstädt, A., Le, V.B., Spinrad, J.P.: Graph Classes: A Survey. SIAM Monographs on Discrete Mathematics, vol. 3. SIAM, Philadelphia (1999)
9. Brandstädt, A., Lozin, V.V., Mosca, R.: Independent sets of maximum weight in apple-free graphs. SIAM J. Discrete Math. **24**(1), 239–254 (2010)
10. Brandstädt, A., Mosca, R.: Maximum weight independent sets in odd-hole-free graphs without dart or without bull. Graphs Comb. **31**, 1249–1262 (2015)
11. Brause, C., Le, N.C., Schiermeyer, I.: The maximum independent det problem in subclasses of subcubic graphs. Discrete Math. **338**, 1766–1778 (2015)
12. Chudnovsky, M.: The structure of bull-free graphs I: three-edge paths with centers and anticenters. J. Comb. Theor. B **102**, 233–251 (2012)
13. Chudnovsky, M.: The structure of bull-free graphs II and III: a summary. J. Comb. Theor. B **102**, 252–282 (2012)
14. Corneil, D.G.: The complexity of generalized clique packing. Discrete Appl. Math. **12**, 233–240 (1985)
15. Corneil, D.G., Perl, Y., Stewart, L.K.: A linear recognition for cographs. SIAM J. Comput. **14**, 926–934 (1985)
16. Downey, R.G., Fellows, M.R.: Parameterized Complexity. Springer, New York (1999)
17. Farber, M.: On diameters and radii of bridged graphs. Discrete Math. **73**, 249–260 (1989)
18. Gerber, M.U., Hertz, A., Lozin, V.V.: Stable sets in two subclasses of banner-free graphs. Discrete Appl. Math. **132**, 121–136 (2004)
19. Grötschel, M., Lovász, L., Schrijver, A.: The ellipsoid method and its consequences in combinatorial optimization. Combinatorica **1**, 169–197 (1981)
20. Frank, A.: Some polynomial algorithms for certain graphs and hypergraphs. In: Proceedings of the Fifth BCC, Congressus Numerantium, XV, pp. 211–226 (1976)
21. Karp, R.M.: Reducibility among combinatorial problems. In: Miller, R.E., Thatcher, J.W., Bohlinger, J.D. (eds.) Complexity of Computer Computations, pp. 85–103. Springer, New York (1972)
22. Karthick, T.: Weighted independent sets in a subclass of P_6-free graphs. Discrete Math. **339**, 1412–1418 (2016)

23. Karthick, T., Maffray, F.: Maximum weight independent sets inclasses related to claw-free graphs. Discrete Appl. Math. (2015). http://dx.doi.org/10.1016/j.dam.2015.02.012

24. Le, N.C., Brause, C., Schiermeyer, I.: New sufficient conditions for α-redundant vertices. Discrete Math. **338**, 1674–1680 (2015)

25. Lokshtanov, D., Vatshelle, M., Villanger, Y.: Independent set in P_5-free graphs in polynomial time. In: Proceedings of the Twenty-Fifth Annual ACM-SIAM Symposium on Discrete Algorithms, pp. 570–581 (2014)

26. Lozin, V.V., Milanič, M.: A polynomial algorithm to find an independent set of maximum weight in a fork-free graph. J. Discrete Algorithms **6**, 595–604 (2008)

27. Lozin, V.V., Milanič, M., Purcell, C.: Graphs without large apples and the maximum weight independent set problem. Graphs Comb. **30**, 395–410 (2014)

28. Lozin, V.V., Monnot, J., Ries, B.: On the maximum independent set problem in subclasses of subcubic graphs. J. Discrete Algorithms **31**, 104–112 (2015)

29. Maffray, F., Pastor, L.: The maximum weight stable set problem in $(P_6, bull)$-free graphs (2016). arXiv:1602.06817v1

30. Malyshev, D.S.: Classes of subcubic planar graphs for which the indepedent set problem is polynomial-time solvable. J. Appl. Ind. Math. **7**, 537–548 (2013)

31. Minty, G.M.: On maximal independent sets of vertices in claw-free graphs. J. Comb. Theor. Ser. B **28**, 284–304 (1980)

32. Poljak, S.: A note on stable sets and colorings of graphs. Commun. Math. Univ. Carolinae **15**, 307–309 (1974)

33. Reed, B., Sbihi, N.: Recognizing bull-free perfect graphs. Graphs Comb. **11**, 171–178 (1995)

34. Thomassé, S., Trotignon, N., Vušković, K.: A polynomial Turing-kernel for weighted independent set in bull-free graphs. Algorithmica (2015, in press)

Decomposing Cubic Graphs into Connected Subgraphs of Size Three

Laurent Bulteau[1], Guillaume Fertin[2], Anthony Labarre[1(✉)], Romeo Rizzi[3], and Irena Rusu[2]

[1] Université Paris-Est, LIGM (UMR 8049), CNRS, ENPC, ESIEE Paris, UPEM, 77454 Marne-la-Vallée, France
Anthony.Labarre@u-pem.fr
[2] Laboratoire d'Informatique de Nantes-Atlantique, UMR CNRS 6241, Université de Nantes, 2 rue de la Houssinière, 44322 Nantes Cedex 3, France
[3] Department of Computer Science, University of Verona, Verona, Italy

Abstract. Let $S = \{K_{1,3}, K_3, P_4\}$ be the set of connected graphs of size 3. We study the problem of partitioning the edge set of a graph G into graphs taken from any non-empty $S' \subseteq S$. The problem is known to be NP-complete for any possible choice of S' in general graphs. In this paper, we assume that the input graph is cubic, and study the computational complexity of the problem of partitioning its edge set for any choice of S'. We identify all polynomial and NP-complete problems in that setting, and give graph-theoretic characterisations of S'-decomposable cubic graphs in some cases.

1 Introduction

General context. Given a connected graph G and a set S of graphs, the S-DECOMPOSITION problem asks whether G can be represented as an edge-disjoint union of subgraphs, each of which is isomorphic to a graph in S. The problem has a long history that can be traced back to Kirkman [7] and has been intensively studied ever since, both from pure mathematical and algorithmic points of view. One of the most notable results in the area is the proof by Dor and Tarsi [3] of the long-standing "Holyer conjecture" [6], which stated that the S-DECOMPOSITION problem is NP-complete when S contains a single graph with at least three edges.

Many variants of the S-DECOMPOSITION problem have been studied while attempting to prove Holyer's conjecture or to obtain polynomial-time algorithms in restricted cases [11], and applications arise in such diverse fields as traffic grooming [10] and graph drawing [5]. In particular, Dyer and Frieze [4] studied a variant where S is the set of connected graphs with k edges for some natural k, and proved the NP-completeness of the S-DECOMPOSITION problem for any $k \geq 3$, even under the assumption that the input graph is planar and bipartite (see Theorem 3.1 in [4]). They further claimed that the problem remains NP-complete under the additional constraint that all vertices of the input graph have degree 2 or 3. Interestingly, if one looks at the special case where $k = 3$ and G is a bipartite *cubic* graph (i.e., each vertex has degree 3), then G can clearly be decomposed in polynomial time, using $K_{1,3}$'s only, by selecting either part of the bipartition and making each

© Springer International Publishing Switzerland 2016
T.N. Dinh and M.T. Thai (Eds.): COCOON 2016, LNCS 9797, pp. 393–404, 2016.
DOI: 10.1007/978-3-319-42634-1_32

vertex in that set the center of a $K_{1,3}$. This shows that focusing on the case $k = 3$ and on cubic graphs can lead to tractable results — as opposed to general graphs, for which when $k = 3$, and for any non empty $S' \subseteq S$, the S'-DECOMPOSITION problems all turn out to be NP-complete [4,6].

In this paper, we study the S-DECOMPOSITION problem on cubic graphs in the case $k = 3$ — i.e., $S = \{K_{1,3}, K_3, P_4\}$. For any non-empty $S' \subseteq S$, we settle the computational complexity of the S-DECOMPOSITION problem by showing that the problem is NP-complete when $S' = \{K_{1,3}, P_4\}$ and $S' = S$, while all the other cases are in P. Table 1 summarises the state of knowledge regarding the complexity of decomposing cubic and arbitrary graphs using connected subgraphs of size three, and puts our results into perspective.

Table 1. Known complexity results on decomposing graphs using subsets of $\{K_{1,3}, K_3, P_4\}$.

Allowed subgraphs			Complexity according to graph class	
$K_{1,3}$	K_3	P_4	Cubic	Arbitrary
✓			in P (Proposition 3)	NP-complete [4, Theorem 3.5]
	✓		$O(1)$ (impossible)	NP-complete [6]
		✓	in P [8]	NP-complete [4, Theorem 3.4]
✓	✓		in P (Proposition 6)	NP-complete [4, Theorem 3.5]
✓		✓	NP-complete (Theorem 2)	NP-complete [4, Theorem 3.1]
	✓	✓	in P (Proposition 2)	NP-complete [4, Theorem 3.4]
✓	✓	✓	NP-complete (Theorem 3)	NP-complete [4, Theorem 3.1]

Terminology. We follow Brandstädt et al. [2] for notation and terminology. All graphs we consider are simple, connected and nontrivial (i.e. $|V(G)| \geq 2$ and $|E(G)| \geq 1$). Given a set S of graphs, a graph G *admits an S-decomposition*, or *is S-decomposable*, if $E(G)$ can be partitioned into subgraphs, each of which is isomorphic to a graph in S. Throughout the paper, S denotes the set of connected graphs of size 3, i.e. $S = \{K_3, K_{1,3}, P_4\}$. We study the following problem:

S'-DECOMPOSITION

Input: a cubic graph $G = (V, E)$, a non-empty set $S' \subseteq S$.
Question: does G admit a S'-decomposition?

We let $G[U]$ denote the subgraph of G induced by $U \subseteq V(G)$. Given a graph $G = (V, E)$, *removing* a subgraph $H = (V' \subseteq V, E' \subseteq E)$ of G consists in removing edges in E' from G as well as the possibly resulting isolated vertices. Finally, let G and G' be two graphs. Then:

- *subdividing* an edge $\{u, v\} \in E(G)$ consists in inserting a new vertex w into that edge, so that $V(G)$ becomes $V(G) \cup \{w\}$ and $E(G)$ is replaced with $E(G) \setminus \{u, v\} \cup \{u, w\} \cup \{w, v\}$;

- *attaching G' to a vertex $u \in V(G)$* means building a new graph H by identifying u and some $v \in V(G')$;
- *attaching G' to an edge $e \in E(G)$* consists in subdividing e using a new vertex w, then attaching G' to w.

Figure 1 illustrates the process of attaching an edge to an edge of the cube graph, and shows other small graphs that we will occasionally use in this paper.

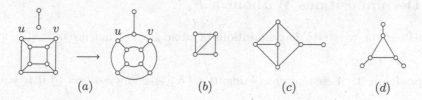

$$(a) \qquad\qquad (b) \qquad\qquad (c) \qquad\qquad (d)$$

Fig. 1. (a) Attaching a new edge to $\{u, v\}$; (b) the diamond graph; (c) the co-fish graph; (d) the net graph.

2 Decompositions Without a $K_{1,3}$

In this section, we study decompositions of cubic graphs that use only P_4's or K_3's. Note that no cubic graph is $\{K_3\}$-decomposable, since all its vertices have odd degree. According to Bouchet and Fouquet [1], Kotzig [8] proved that a cubic graph admits a $\{P_4\}$-decomposition iff it has a perfect matching. However, the proof of the forward direction as presented in [1] is incomplete, as it requires the use of Proposition 1(b) below, which is missing from their paper. Therefore, we provide the following proposition for completeness, together with another result which will also be useful for the case where $S' = \{K_3, P_4\}$.

Proposition 1. *Let G be a cubic graph that admits a $\{K_3, P_4\}$-decomposition D. Then, in D, (a) no K_3 is used, and (b) no three P_4's are incident to the same vertex.*

Proof. Partition $V(G)$ into three sets V_1, V_2 and V_3, where V_1 (resp. V_2, V_3) is the set of vertices that are incident to exactly one P_4 (resp. two, three P_4's) in D. Note that V_1 is exactly the set of vertices involved in K_3's in D. Let $n_i = |V_i|$, $1 \leq i \leq 3$. Our goal is to show that $n_1 = n_3 = 0$, i.e. $V_1 = V_3 = \emptyset$. For this, note that (1) each vertex in V_3 is the extremity of three different P_4's, (2) each vertex in V_2 is simultaneously the extremity of one P_4 and an inner vertex of another P_4, while (3) each vertex in V_1 is the extremity of one P_4. Since each P_4 has two extremities and two inner vertices, if p is the number of P_4's in D, we have:

- $p = \frac{3n_3 + n_2 + n_1}{2}$ (by (1), (2) and (3) above, counting extremities);
- $p = \frac{n_2}{2}$ (by (2) above, counting inner vertices).

Putting together the above two equalities yields $n_1 = n_3 = 0$, which completes the proof.   □

Since K_3's cannot be used in cubic graphs for $\{K_3, P_4\}$-decompositions by Proposition 1 above, we directly obtain the following result, which implies that $\{K_3, P_4\}$-decomposition is in P.

Proposition 2. *A cubic graph admits a* $\{K_3, P_4\}$*-decomposition iff it has a perfect matching.*

3 Decompositions Without a P_4

In this section, we study decompositions of cubic graphs that use only $K_{1,3}$'s or K_3's.

Proposition 3. *A cubic graph* G *admits a* $\{K_{1,3}\}$*-decomposition iff it is bipartite.*

Proof. For the reverse direction, select either set of the bipartition, and make each vertex in that set the center of a $K_{1,3}$. For the forward direction, let D be a $\{K_{1,3}\}$-decomposition of G, and let C and L be the sets of vertices containing, respectively, all the centers and all the leaves of $K_{1,3}$'s in D. We show that this is a bipartition of $V(G)$. First, $C \cup L = V$ since D covers all edges and therefore all vertices. Second, $C \cap L = \emptyset$ since a vertex in $C \cap L$ would have degree at least 4. Finally, each edge in D connects the center of a $K_{1,3}$ and a leaf of another $K_{1,3}$ in D, which belong respectively to C and L. Therefore, G is bipartite. □

We now prove that $\{K_{1,3}, K_3\}$-decompositions can be computed in polynomial time. Recall that a graph is *H-free* if it does not contain an induced subgraph isomorphic to a given graph H. Since bipartite graphs admit a $\{K_{1,3}\}$-decomposition (by Proposition 3), we can restrict our attention to non-bipartite graphs that contain K_3's (indeed, if they were K_3-free, then only $K_{1,3}$'s would be allowed and Proposition 3 would imply that they admit no decomposition). Our strategy consists in iteratively removing subgraphs from G and adding them to an initially empty $\{K_{1,3}, K_3\}$-decomposition until G is empty, in which case we have an actual decomposition, or no further removal operations are possible, in which case no decomposition exists. Our analysis relies on the following notion: a K_3 induced by vertices $\{u, v, w\}$ in a graph G is *isolated* if $V(G)$ contains no vertex x such that $\{u, v, x\}$, $\{u, x, w\}$ or $\{x, v, w\}$ induces a K_3.

Lemma 1. *If a cubic graph* G *admits a* $\{K_{1,3}, K_3\}$*-decomposition* D*, then every isolated* K_3 *in* G *belongs to* D*.*

Proof (Contradiction). If an isolated K_3 were not part of the decomposition, then exactly one vertex of that K_3 would be the center of a $K_{1,3}$, leaving the remaining edge uncovered and uncoverable. □

$\overline{C_6}$ is a minimal example of a cubic non-bipartite graph with K_3's that admits no $\{K_{1,3}, K_3\}$-decomposition: both K_3's in that graph must belong to the decomposition (by Lemma 1), but their removal yields a perfect matching.

Observation 1. *Let G be a connected cubic graph. Then no sequence of at least one edge or vertex removal from G yields a cubic graph.*

Proof (Contradiction). If after applying at least one removal operation on G we obtain a cubic graph G', then the graph that precedes G' in this removal sequence must have had a vertex of degree at least four, since G is connected.

$\square$

Proposition 4. *For any non-bipartite cubic graph G whose K_3's are all isolated, one can decide in polynomial time whether G is $\{K_{1,3}, K_3\}$-decomposable.*

Proof. We build a $\{K_{1,3}, K_3\}$-decomposition by iteratively removing $K_{1,3}$'s and K_3's from G, which we add as we go to an initially empty set D. By Lemma 1, all isolated K_3's must belong to D, so we start by adding them all to D and removing them from G; therefore, G admits a $\{K_{1,3}, K_3\}$-decomposition iff the resulting subcubic graph G' admits a $\{K_{1,3}\}$-decomposition. Observe that G' contains vertices of degree 1 and 2; we note that:

1. each vertex of degree 1 must be the leaf of some $K_{1,3}$ in D;
2. each vertex of degree 2 must be the meeting point of two $K_{1,3}$'s in D.

The only ambiguity arises for vertices of degree 3, which may either be the center of a $K_{1,3}$ in D or the meeting point of three $K_{1,3}$'s in D; however, there will always exist at least one other vertex of degree 1 or 2 until the graph is empty (by Observation 1). Therefore, we can safely remove $K_{1,3}$'s from our graph and add them to D by following the above rules in the stated order; if we succeed in deleting the whole graph in this way, then D is a $\{K_{1,3}, K_3\}$-decomposition of G, otherwise no such decomposition exists. $\square$

We conclude with the case where the graph may contain non-isolated K_3's.

Proposition 5. *If a cubic graph G contains a diamond, then one can decide in polynomial time whether G is $\{K_{1,3}, K_3\}$-decomposable.*

Proof. The only cubic graph on 4 vertices is K_4, which is diamond-free and $\{K_{1,3}, K_3\}$-decomposable, so we assume $|V(G)| \geq 6$. Let $\mathcal{D}$ be a diamond in G induced by vertices $\{u, v, w, x\}$ and such that $\{u, x\} \notin E(G)$, as shown in Fig. 2(a). $\mathcal{D}$ is connected to two other vertices u' and x' of G, which are respectively adjacent to u and x, and there are only two ways to use the edges of $\mathcal{D}$ in a $\{K_{1,3}, K_3\}$-decomposition, as shown in Fig. 2(b) and (c). If $u' = x'$, regardless of the decomposition we choose for $\mathcal{D}$, u' and its neighbourhood induce a P_3 in the graph obtained from G by removing the parts added to D. But then that P_3 cannot be covered, so no $\{K_{1,3}, K_3\}$-decomposition exists for G. Therefore, we assume that $u' \neq x'$.

As Fig. 2(b) and (c) show, either $\{u, v, w\}$ or $\{v, w, x\}$ must form a K_3 in D, thereby forcing either $\{v, w, x, x'\}$ or $\{u', u, v, w\}$ to form a $K_{1,3}$ in D. In both cases, removing the K_3 and the $K_{1,3}$ yields a graph G' which contains vertices of degree 1, 2 or 3. As in the proof of Proposition 4, Observation 1 allows us to obtain the following helpful properties:

1. every leaf in G' must be the leaf of some $K_{1,3}$ in D;
2. every vertex y of degree two in G' must either belong to a K_3 or be a leaf of
   two distinct $K_{1,3}$'s in D, which can be decided as follows:
   (a) if y belongs to a K_3 in G', then it must also belong to a K_3 in D; otherwise,
       it would be the leaf of a $K_{1,3}$ and the graph obtained by removing that
       $K_{1,3}$ would contain a P_3, which we cannot cover;
   (b) otherwise, y must be a leaf of two $K_{1,3}$'s in D.

We therefore iteratively remove subgraphs from our graph and add them to
D according to the above rules, which we follow in the stated order; if we succeed
in deleting the whole graph in this way using either decomposition in Fig. 2(b)
or (c) as a starting point, then D is a $\{K_{1,3}, K_3\}$-decomposition of G, otherwise
no such decomposition exists.   □

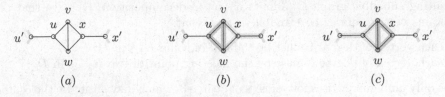

Fig. 2. (a) A diamond in a cubic graph, and (b), (c) the only two ways to decompose
it in a $\{K_{1,3}, K_3\}$-decomposition. (Color figure online)

All the arguments developed in this section lead to the following result.

Proposition 6. *The $\{K_{1,3}, K_3\}$-DECOMPOSITION problem on cubic graphs is
in P.*

4   Decompositions that Use both $K_{1,3}$'s and P_4's

In this section, we show that problems $\{K_{1,3}, P_4\}$-DECOMPOSITION and $\{K_{1,3},
K_3, P_4\}$-DECOMPOSITION are NP-complete. Our hardness proof relies on two
intermediate problems that we define below and is structured as follows:

> CUBIC PLANAR MONOTONE 1-IN-3 SATISFIABILITY
> $\leq_P$ DEGREE-2,3 $\{K_{1,3}, K_3, P_4\}$-DECOMPOSITION WITH MARKED EDGES   (Theorem 1 page 10)
> $\leq_P$ $\{K_{1,3}, K_3, P_4\}$-DECOMPOSITION WITH MARKED EDGES   (Lemma 4 page 9)
> $\leq_P$ $\{K_{1,3}, P_4\}$-DECOMPOSITION   (Lemma 3 page 7)

We start by introducing the following intermediate problem:

$\{K_{1,3}, K_3, P_4\}$-DECOMPOSITION WITH MARKED EDGES

Input: a cubic graph $G = (V, E)$ and a subset $M \subseteq E$ of edges.
Question: does G admit a $\{K_{1,3}, K_3, P_4\}$-decomposition D such that no
           edge in M is the middle edge of a P_4 in D and such that every
           K_3 in D has either one or two edges in M?

The drawings that illustrate our proofs in this section show marked edges as dotted edges. The proof of Lemma 3 uses the following result.

Lemma 2. *Let e be a bridge in a cubic graph G which admits a $\{K_{1,3}, K_3, P_4\}$-decomposition D. Then e must be the middle edge of a P_4 in D.*

Proof (Contradiction). First note that e cannot belong to a K_3 in D. Now suppose e is part of a $K_{1,3}$ in D. The situation is as shown below (without loss of generality):

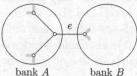

bank A bank B

If we remove from G the $K_{1,3}$ in D that contains e, then summing the terms of the degree sequence of $G[V(B)]$ yields $2 + 3(|V(B)| - 1) = 2|E(B)|$, which means that $2|E(B)| \equiv 2 \pmod 3$, so $|E(B)| \not\equiv 0 \pmod 3$ and therefore B admits no decomposition into components of size three. The very same argument shows that if e belongs to a P_4 in D, then it must be its middle edge, which completes the proof.   □

Lemma 3. *Let (G, M) be an instance of $\{K_{1,3}, K_3, P_4\}$-DECOMPOSITION WITH MARKED EDGES, and G' be the graph obtained by attaching a co-fish to every edge in M. Then G can be decomposed iff G' admits a $\{K_{1,3}, P_4\}$-decomposition.*

Proof. We prove each direction separately.

$\Rightarrow$: we show how to transform a decomposition D of (G, M) into a decomposition D' of G'. The subgraphs in D that have no edge in M are not modified. For the other subgraphs, we distinguish between four cases:

(a) if an edge of M belongs to a $K_{1,3}$ in D, then attaching a co-fish does not prevent us from adapting the decomposition of G in G':

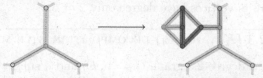

(b) if an edge of M belongs to a P_4 in D, then it is an extremity of that P_4 and attaching a co-fish does not prevent us from adapting that part of the decomposition:

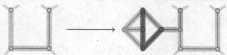

(c) if a K_3 in D has one edge in M, we can adapt the partition as follows:

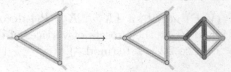

(d) if a K_3 in D has two edges in M, we can adapt the partition as follows:

$\Leftarrow$: we now show how to transform any $\{K_{1,3}, P_4\}$-decomposition D' of G' into a decomposition of (G, M). Again, the only parts of D' that will need adapting are those connected to the co-fishes that we inserted when transforming G into G'. Since the leaf u of the co-fish we inserted has a neighbour x such that $\{u, x\}$ is a bridge in G', $\{u, x\}$ is the middle edge of a P_4 in D' (Lemma 2) and we may therefore assume without loss of generality that our starting point in G' is as follows:

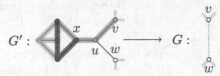

with $\{v, w\} \notin E(G')$ since G is simple; therefore $\{u, w\}$ cannot belong to a K_3 in G', and we have two cases to consider:

(a) if $\{u, w\}$ belongs to a $K_{1,3}$ in D', that $K_{1,3}$ can be mapped onto a $K_{1,3}$ in D by replacing $\{u, w\}$ with $\{v, w\}$;

(b) otherwise, $\{u, w\}$ is an extremal edge of a P_4 in D'; since $\{u, w\} \notin E(G)$, either that edge will remain in a P_4 when removing the co-fish and replacing $\{u, w\}$ with $\{v, w\}$, or it will end up in a K_3 with either one or two marked edges. Either way, the part can be added as such to D. $\square$

We now show that we can restrict our attention to the following variant of $\{K_{1,3}, K_3, P_4\}$-DECOMPOSITION WITH MARKED EDGES. We say a graph is *degree-2,3* if its vertices have degree only 2 or 3.

DEGREE-2,3 $\{K_{1,3}, K_3, P_4\}$-DECOMPOSITION WITH MARKED EDGES

Input: a degree-2,3 graph $G = (V, E)$ and a subset $M \subseteq E$ of edges.
Question: does G admit a $\{K_{1,3}, K_3, P_4\}$-decomposition D such that no edge in M is the middle edge of a P_4 in D and such that every K_3 in D has either one or two edges in M?

The following observation will help.

Observation 2. *Let G be a degree-2,3 graph with $|V_2|$ degree-2 vertices. If G is $\{K_{1,3}, K_3, P_4\}$-decomposable, then $|V_2| \equiv 0 \pmod 3$.*

Proof. If $G = (V, E)$ admits a $\{K_{1,3}, K_3, P_4\}$-decomposition, then $|E| \equiv 0 \pmod 3$. Let V_2 and V_3 be the subsets of vertices of degree 2 and 3 in G. Then $2|V_2| + 3|V_3| = 2|E|$, so $2|V_2| \equiv 0 \pmod 3$. $\square$

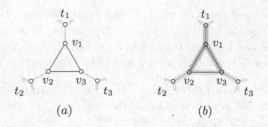

Fig. 3. Adding a net (a) to a graph with degree-2 vertices t_1, t_2, t_3 (dotted edges belong to M'), and (b) its only possible decomposition (up to symmetry). (Color figure online)

We prove that allowing degree-2 vertices does not make the problem substantially more difficult, by adding the following gadgets until all vertices have degree 3. Let (G, M) be an instance of DEGREE-2,3 $\{K_{1,3}, K_3, P_4\}$-DECOMPOSITION WITH MARKED EDGES, where G has at least three degree-2 vertices t_1, t_2, t_3; by *adding a net over* $\{t_1, t_2, t_3\}$, we mean attaching a net by its leaves to v_1, v_2 and v_3 and adding the edges incident to the net's leaves to M (see Fig. 3(a)).

Proposition 7. *Let (G, M) be an instance of* DEGREE-2,3 $\{K_{1,3}, K_3, P_4\}$-DECOMPOSITION WITH MARKED EDGES, *where G has at least three degree-2 vertices t_1, t_2, t_3, and let (G', M') be the instance obtained by adding a net to (G, M). Then (G', M') has three degree-2 vertices less than (G, M), and (G, M) can be decomposed iff (G', M') can be decomposed.*

Proof. By construction, G' has fewer degree-2 vertices, since t_1, t_2, t_3 now have degree 3 instead of 2, other vertices of G are unchanged, and new vertices $\{v_1, v_2, v_3\}$ have degree 3. We now prove the equivalence.

$\Rightarrow$: given a decomposition D for (G, M), we only need to add the $K_{1,3}$ induced by $\{v_1, t_1, v_2, v_3\}$ and the P_4 induced by $\{t_2, v_2, v_3, t_3\}$ to cover the edges of the added net in order to obtain a decomposition D' for (G', M') (see Fig. 3(b)).

$\Leftarrow$: we show that the only valid decompositions must include the choice we made in the proof of the forward direction. Indeed, the marked edges cannot be middle edges in a P_4, and the K_3 induced by v_1, v_2 and v_3 cannot appear as a K_3 in a decomposition. Moreover, no marked edge can be the extremity of a P_4 with two edges lying in the K_3, since this would force another marked edge to be the middle edge of a P_4. Therefore the only possible decomposition of the net is the one defined above (up to symmetry), and we can safely remove the P_4 and the $K_{1,3}$ from D' while preserving the rest of the decomposition.

□

Lemma 4. DEGREE-2,3 $\{K_{1,3}, K_3, P_4\}$-DECOMPOSITION WITH MARKED EDGES $\leq_P \{K_{1,3}, K_3, P_4\}$-DECOMPOSITION WITH MARKED EDGES.

Proof. Given an instance (G, M) of DEGREE-2,3 $\{K_{1,3}, K_3, P_4\}$-DECOMPOSITION WITH MARKED EDGES, create an instance (G', M') by successively adding a net to any triple of degree-2 vertices, until no such triple remains. By Proposition 7, (G, M) is decomposable iff (G', M') is decomposable. Moreover, either G' is cubic (and then (G', M') is an instance of $\{K_{1,3}, K_3, P_4\}$-DECOMPOSITION WITH MARKED EDGES), or G is trivially a no-instance by Observation 2   □

Finally, we show that DEGREE-2,3 $\{K_{1,3}, K_3, P_4\}$ DECOMPOSITION WITH MARKED EDGES is NP-complete. Our reduction relies on the CUBIC PLANAR MONOTONE 1-IN-3 SATISFIABILITY problem [9]:

CUBIC PLANAR MONOTONE 1-IN-3 SATISFIABILITY

Input: a Boolean formula $\phi = C_1 \wedge C_2 \wedge \cdots \wedge C_n$ without negations over a set $\Sigma = \{x_1, x_2, \ldots, x_m\}$, with exactly three distinct variables per clause and where each literal appears in exactly three clauses; moreover, the graph with clauses and variables as vertices and edges joining clauses and the variables they contain is planar.

Question: does there exist an assignment of truth values $f : \Sigma \to \{$TRUE, FALSE$\}$ such that exactly one literal is TRUE in every clause of ϕ?

Theorem 1. DEGREE-2,3 $\{K_{1,3}, K_3, P_4\}$-DECOMPOSITION WITH MARKED EDGES *is NP-complete.*

Proof. We first show how to transform an instance $\phi = C_1 \wedge C_2 \wedge \cdots \wedge C_n$ of CUBIC PLANAR MONOTONE 1-IN-3 SATISFIABILITY into an instance (G, M) of DEGREE-2,3 $\{K_{1,3}, K_3, P_4\}$-DECOMPOSITION WITH MARKED EDGES. The transformation proceeds by:

1. mapping each variable x_i onto a $K_{1,3}$ denoted by $K(x_i)$ and whose edges all belong to M;
2. mapping each clause $C = \{x_i, x_j, x_k\}$ onto a cycle with five vertices in such a way that $K(x_i)$, $K(x_j)$ and $K(x_k)$ each have a leaf that coincides with a vertex of the cycle and exactly two such leaves are adjacent in the cycle.

Figure 4 illustrates the construction, which yields a degree-2,3 graph. We now show that ϕ is satisfiable iff (G, M) admits a decomposition.

⇒: we apply the following rules for transforming a satisfying assignment for ϕ into a decomposition D for (G, M):
  - if variable x_i is set to FALSE, then the corresponding $K(x_i)$ is added as such to D;
  - otherwise, the three edges of $K(x_i)$ will be the meeting points of three different $K_{1,3}$'s in the decomposition, one of which will have two edges in the current clause gadget.

Two cases can be distinguished based on whether or not a leaf of $K(x_i)$ is adjacent to a leaf of $K(x_j)$ or $K(x_k)$, but in both cases the rest of the clause gadget yields a P_4 that we add as such to the decomposition (see Fig. 4(b) and (c)).

$\Leftarrow$: we now show how to convert a decomposition D for (G, M) into a satisfying truth assignment for ϕ. First, we observe that D must satisfy the following crucial structural property:

For each clause $C = (x_i \lor x_j \lor x_k)$, exactly two subgraphs out of $K(x_i)$, $K(x_j)$ and $K(x_i)$ appear as $K_{1,3}$'s in D.

Indeed, G is K_3-free by construction, and:

(a) if all of them appear as $K_{1,3}$'s in D, then the remaining five edges of the clause gadget cannot be decomposed;

(b) if only $K(x_i)$ appears as a $K_{1,3}$ in D, then x_j — without loss of generality — must be a leaf either of a $K_{1,3}$ in D with a center in the clause gadget or of a P_4 in D with two edges in the clause gadget (the P_4 cannot connect x_j and x_k, otherwise the rest of the gadget cannot be decomposed); in both cases, the remaining three edges of the clause gadget must form a P_4, thereby causing $K(x_k)$ to appear as a $K_{1,3}$ in D, a contradiction (a similar argument allows us to handle $K(x_j)$ and $K(x_k)$);

(c) finally, if none of them appear as $K_{1,3}$'s in D, then x_i must be the leaf either of a $K_{1,3}$ in D with a center in the clause gadget, or of a P_4 with two edges in the clause gadget; in both cases, the remaining three edges of the clause gadget must form a P_4 in D, which in turn makes it impossible to decompose the rest of the graph.

Therefore, D yields a satisfying assignment for ϕ in the following simple way: if $K(x_i)$ appears as a $K_{1,3}$ in D, set it to FALSE, otherwise set it to TRUE.   □

Theorem 2. $\{K_{1,3}, P_4\}$-DECOMPOSITION *is NP-complete.*

Proof. Immediate from Lemmas 3 and 4 and Theorem 1.   □

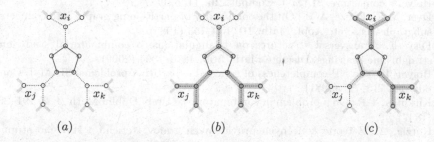

(a)   (b)   (c)

Fig. 4. (a) Connecting clause and variable gadgets in the proof of Theorem 1; dotted edges belong to M. (b), (c) Converting truth assignments into decompositions in the proof of Theorem 1; the only variable set to TRUE is mapped onto a $K_{1,3}$ in the decomposition; (b) shows the case where the only variable set to TRUE — namely, x_i — is such that $K(x_i)$ has no leaf adjacent to a leaf of $K(x_j)$ nor $K(x_k)$; (c) shows the other case, where x_j is set to TRUE and $K(x_i)$ and $K(x_k)$ have leaves made adjacent by the clause gadget. (Color figure online)

A like-minded reduction allows us to prove the hardness of $\{K_{1,3}, K_3, P_4\}$-DECOMPOSITION.

Theorem 3. $\{K_{1,3}, K_3, P_4\}$-DECOMPOSITION *is NP-complete, even on K_3-free graphs.*

5   Conclusions and Future Work

We provided in this paper a complete complexity landscape of $\{K_{1,3}, K_3, P_4\}$-DECOMPOSITION for cubic graphs. A natural generalisation, already studied by other authors, is to study decompositions of k-regular graphs into connected components with k edges for $k > 3$. We would like to determine whether our positive results generalise in any way in that setting. It would also be interesting to identify tractable classes of graphs in the cases where those decomposition problems are hard, and to refine our characterisation of hard instances; for instance, does there exist a planarity-preserving reduction for Theorem 3? Finally, we note that some applications relax the size constraint by allowing the use of graphs with at most k edges in the decomposition [10]; we would like to know how that impacts the complexity of the problems we study in this paper.

References

1. Bouchet, A., Fouquet, J.-L. Trois types de décompositions d'un graphe en chaénes. In: Berge, C., Bresson, D., Camion, P., Maurras, J.F., Sterboul, F. (eds.) Combinatorial Mathematics: Proceedings of the International Colloquium on Graph Theory and Combinatorics. North-Holland Mathematics Studies, vol. 75, pp. 131–141, North-Holland (1983)
2. Brandstädt, A., Lę, V.B., Spinrad, J.P.: Graph classes: a survey. SIAM Monographs on Discrete Mathematics and Applications, Society for Industrial Mathematics (1987)
3. Dor, D., Tarsi, M.: Graph decomposition is NP-complete: a complete proof of Holyer's conjecture. SIAM J. Comput. **26**, 1166–1187 (1997)
4. Dyer, M.E., Frieze, A.M.: On the complexity of partitioning graphs into connected subgraphs. Discrete Appl. Math. **10**, 139–153 (1985)
5. Fusy, É.: Transversal structures on triangulations: a combinatorial study and straight-line drawings. Discrete Math. **309**, 1870–1894 (2009)
6. Holyer, I.: The NP-completeness of some edge-partition problems. SIAM J. Comput. **10**, 713–717 (1981)
7. Kirkman, T.P.: On a problem in combinatorics. Camb. Dublin Math. J. **2**, 191–204 (1847)
8. Kotzig, A.: Z teorie konečných pravidelných grafov tretieho a štvrtého stupňa, Časopis pro pěstování matematiky, pp. 76–92 (1957)
9. Moore, C., Robson, J.M.: Hard tiling problems with simple tiles. Discrete Comput. Geom. **26**, 573–590 (2001)
10. Muñoz, X., Li, Z., Sau, I.: Edge-partitioning regular graphs for ring traffic grooming with a priori placement of the ADMs. SIAM J. Discrete Math. **25**, 1490–1505 (2011)
11. Yuster, R.: Combinatorial and computational aspects of graph packing and graph decomposition. Comput. Sci. Rev. **1**, 12–26 (2007)

Automorphisms of the Cube n^d

Pavel Dvořák[1]($\boxtimes$) and Tomáš Valla[2]

[1] Faculty of Mathematics and Physics, Charles University in Prague,
Prague, Czech Republic
koblich@iuuk.mff.cuni.cz

[2] Faculty of Information Technology, Czech Technical University in Prague,
Prague, Czech Republic
tomas.valla@fit.cvut.cz

Abstract. Consider a hypergraph H_n^d where the vertices are points of the d-dimensional combinatorial cube n^d and the edges are all sets of n points such that they are in one line. We study the structure of the group of automorphisms of H_n^d, i.e., permutations of points of n^d preserving the edges. In this paper we provide a complete characterization. Moreover, we consider the COLORED CUBE ISOMORPHISM problem of deciding whether for two colorings of the vertices of H_n^d there exists an automorphism of H_n^d preserving the colors. We show that this problem is GI-complete.

1 Introduction

Combinatorial cube n^d (or simply a cube n^d) is a set of points $[n]^d$, where $[n] = \{0, \dots, n-1\}$. A *line* ℓ of a cube n^d is a set of n points of n^d which lie in a geometric line in the d-dimensional space where the cube n^d is embedded. We denote the set of all lines of the cube n^d by $\mathbb{L}(n^d)$. Thus, the hypergraph H_n^d is defined as $(n^d, \mathbb{L}(n^d))$.

We denote the group of all permutations on n elements by $\mathbb{S}_n$. A permutation $P \in \mathbb{S}_{n^d}$ is an *automorphism* of the cube n^d if $\ell = \{v_1, \dots, v_n\} \in \mathbb{L}(n^d)$ implies $P(\ell) = \{P(v_1), \dots, P(v_n)\} \in \mathbb{L}(n^d)$. Informally, an automorphism of the cube n^d is a permutation of the cube points which preserves the lines. We denote the set of all automorphisms of n^d by T_n^d. Note that all automorphisms of n^d with a composition $\circ$ form a group $\mathbb{T}_n^d = (T_n^d, \circ, Id)$.

Our main result is the characterization of the generators of the group $\mathbb{T}_n^d$ and computing the order of $\mathbb{T}_n^d$. Surprisingly, the structure of $\mathbb{T}_n^d$ is richer than only the obvious rotations and symmetries. We use three groups of automorphisms for characterization of the group $\mathbb{T}_n^d$ as follows. The first one is a group $\mathbb{R}_d$ of rotations of the d-dimensional hypercube. Generators of $\mathbb{R}_d$ are the rotations

$$R_{ij}([x_1, \dots, x_i, \dots, x_j, \dots, x_d]) = [x_1, \dots, n - x_j - 1, \dots, x_i, \dots, x_d]$$

P. Dvořák—The research leading to these results has received funding from the European Research Council under the European Union's Seventh Framework Programme (FP/2007-2013) / ERC Grant Agreement no. 616787.

T. Valla—Supported by the Centre of Excellence – Inst. for Theor. Comp. Sci. 79 (project P202/12/G061 of GA ČR).

© Springer International Publishing Switzerland 2016
T.N. Dinh and M.T. Thai (Eds.): COCOON 2016, LNCS 9797, pp. 405–416, 2016.
DOI: 10.1007/978-3-319-42634-1_33

for every $i, j \in \{1, \ldots, d\}$. The second group is a group of permutation automorphisms $\mathbb{F}_n$ contains mappings $F_\pi([x_1, \ldots, x_d]) = [\pi(x_1), \ldots, \pi(x_d)]$ where $\pi \in \mathbb{S}_n$ such that it has a *symmetry property*: if $\pi(p) = q$ then $\pi(n - p - 1) = n - q - 1$. The last one is a group of axial symmetry $\mathbb{X}$ which contains the automorphisms Id and $X([x_1, \ldots, x_{d-1}, x_d]) = [x_1, \ldots, x_d, x_{d-1}]$. Our main result is summarized in the following theorem. For the proof we use and generalize some ideas of Silver [11] who proved a same result for the cube 4^3.

Theorem 1. *The group $\mathbb{T}_n^d$ is generated by the elements of $\mathbb{R}_d \cup \mathbb{F}_n \cup \mathbb{X}$. The order of the group $\mathbb{T}_n^d$ is $2^{d-1+k}d!k!$ where $k = \lfloor \frac{n}{2} \rfloor$.*

An *isomorphism* of two hypergraphs $H_1 = (V_1, E_1), H_2 = (V_2, E_2)$ is a bijection $f : V_1 \rightarrow V_2$ such that for each $\{s_1, \ldots, s_r\} \subseteq V_1, \{s_1, \ldots, s_r\} \in E_1 \Leftrightarrow \{f(s_1), \ldots, f(s_r)\} \in E_2$. A *coloring* of a hypergraph $H = (V, E)$ by k colors is a function $s : V \rightarrow [k]$. The following problem is well studied.

PROBLEM: COLORED HYPERGRAPH ISOMORPHISM (CHI)
Instance: Hypergraphs $H_1 = (V_1, E_1), H_2 = (V_2, E_2)$, colorings $s_1 : V_1 \rightarrow [k], s_2 : V_2 \rightarrow [k]$
Question: Is there an isomorphism $f : V_1 \rightarrow V_2$ of H_1 and H_2 such that it preserves the colors? I.e., it holds $s_1(v) = s_2(f(v))$ for every vertex v in V_1.

There are several FPT algorithms[1] for CHI—see Arvind et al. [2,3]. The problem COLORED CUBE ISOMORPHISM is defined as the problem CHI where both $H_1, H_2 = H_n^d$. Since we know the structure of the group $\mathbb{T}_n^d$, it is natural to ask if COLORED CUBE ISOMORPHISM is an easier problem than CHI. We prove that the answer is negative. The class of decisions problems GI contains all problems with a polynomial reduction to the problem GRAPH ISOMORPHISM.

PROBLEM: GRAPH ISOMORPHISM
Instance: Graphs G_1, G_2
Question: Are the graphs G_1 and G_2 isomorphic?

It is well known that CHI is GI-complete, see Booth and Colbourn [6]. We prove the same result for COLORED CUBE ISOMORPHISM.

Theorem 2. *The problem COLORED CUBE ISOMORPHISM is GI-complete even if both input colorings has a form $n^d \rightarrow [2]$.*

The paper is organized as follows. First we count the order of the group $\mathbb{T}_2^d$, whose structure is different from other automorphism groups. Next, for clarity reasons we characterize the generators for $\mathbb{T}_n^3$, and then we generalize the results for the general group $\mathbb{T}_n^d$. In Sect. 5 we count the order of the group $\mathbb{T}_n^d$. In the last section we study the complexity of COLORED CUBE ISOMORPHISM and show some idea of a prove of Theorem 2.

---

[1] The parameter is the maximum number of vertices colored by the same color.

1.1   Motivation

A natural motivation for this problem comes from the game of Tic-Tac-Toe. It is usually played on a 2-dimensional square grid and each player puts his tokens (usually crosses for the first player and rings for the second) at the points on the grid. A player wins if he occupies a line with his token vertically, horizontally or diagonally (with the same length as the grid size) faster than his opponent. Tic-Tac-Toe is a member of a large class of games called strong positional games. For an extraordinary reference see Beck [5]. The size of a basic Tic-Tac-Toe board is 3×3 and it is easy to show by case analysis that the game ends as a draw if both players play optimally. However, the game can be generalized to larger grid and more dimensions. The d-dimensional Tic-Tac-Toe is played on the points of a d-dimensional combinatorial cube and it is often called the game n^d. With larger boards the case analysis becomes unbearable even using computer search and clever algorithms have to be devised.

The only (as far as we know) non-trivial solved 3-dimensional Tic-Tac-Toe is the game 4^3, which is called Qubic. Qubic is a win for the first player, which was shown by Patashnik [10] in 1980. It was one of the first examples of computer-assisted proofs based on a brute-force algorithm, which utilized several clever techniques for pruning the game tree. Another remarkable approach for solving Qubic was made by Allis [1] in 1994, who introduced several new methods. However, one technique is common for both authors: the detection of isomorphisms of game configurations. As the game of Qubic is highly symmetric, this detection substantially reduces the size of the game tree.

For the game n^d, theoretical result are usually achieved for large n or large d. For example, by the famous Hales and Jewett theorem [8], for any n there is (an enormously large) d such that the hypergraph H_n^d is not 2-colorable, that means, the game n^d cannot end in a draw. Using the standard Strategy Stealing argument, n^d is thus a first player's win. In two dimensions, each game $n^2, n > 2$, is a draw (see Beck [5]). Also, several other small n^d are solved.

All automorphisms for Qubic were characterized by Silver [11] in 1967. As in the field of positional games the game n^d is intensively studied and many open problems regarding n^d are posed, the characterization of the automorphism group of n^d is a natural task.

The need to characterize the automorphism group came from our real effort to devise an algorithm and computer program that would be able to solve the game 5^3, which is the smallest unsolved Tic-Tac-Toe game. While our effort of solving 5^3 is currently not yet successful, we were able to come up with the complete characterization of the automorphism group n^d, giving an algorithm for detection of isomorphic positions not only in the game 5^3, but also in n^d in general.

A game configuration can be viewed as a coloring s of n^d by crosses, rings and empty points, i.e., $s : n^d \to [3]$. Since we know the structure of the group $\mathbb{T}_n^d$, this characterization yields an algorithm for detecting isomorphic game positions by simply trying all combinations of the generators (the number of the combinations is given by the order of the group $\mathbb{T}_n^d$). A natural question arises: can one obtain a faster algorithm? Note that the hypergraph H_n^d has polynomially many edges in the number of vertices. Therefore, from a polynomial point of view it does not matter if there are hypergraphs H_d^n with colorings or only colorings on the input. Due to Theorem 2 we conclude that deciding if two game configurations are isomorphic is as hard as deciding if two graphs are isomorphic.

Although our primary motivation came from the game of Tic-Tac-Toe, we believe our result has much broader interest as it presents an analogy of auto-morphism characterization results of hypercubes (see e.g. [7,9]).

2   Preliminaries

Beck [5] gives a different point of view on the lines of n^d. Let $s = (s^1, \ldots, s^n)$ be a sequence of n distinctive points of a cube n^d. Let $s^i = [s_1^i, \ldots, s_d^i]$ for every $1 \le i \le n$. We say that s is *linear* if for every $1 \le j \le d$ a sequence $\tilde{s}_j = (s_j^1, \ldots, s_j^n)$ is strictly increasing, strictly decreasing or constant and at least one sequence $\tilde{s}_j$ has to be nonconstant. A set of points $\{p^1, p^2, \ldots, p^n\} \subseteq n^d$ is a line if it can be ordered into a linear sequence $(q^1, q^2, \ldots, q^n)$. Beck [5] worked with ordered lines (the linear sequences in our case). However, for us it is more convenient to have unordered lines because some automorphisms will change the order of points in the line.

Let ℓ be a line and $q = (q^1, \ldots, q^n)$ be an ordering of ℓ into a linear sequence. Note that every line in $\mathbb{L}(n^d)$ has two such ordering. Another ordering of ℓ into a linear sequence is $(q^n, \ldots, q^1)$. We define a *type* of a sequence $\tilde{q}_j = (q_j^1, \ldots, q_j^n)$ as $+$ if $\tilde{q}_j$ is strictly increasing, $-$ if $\tilde{q}_j$ is strictly decreasing, c if $\tilde{q}_j$ is constant and $q_j^i = c$ for every $1 \le i \le n$. A type of q is $type(q) = (type(\tilde{q}_1), \ldots, type(\tilde{q}_n))$.

Type of a line ℓ is a type of an ordering of ℓ into a linear sequence. Since every line has two such ordering, every line has also two types. How-ever, the second type of ℓ can be obtained by switching $+$ and $-$ in the first type. For example, let $\ell = \{[0,0,3],[0,1,2],[0,2,1],[0,3,0]\} \in \mathbb{L}(4^3)$ then $type(\ell) = \{(0,+,-),(0,-,+)\}$. However, for better readability we write only $type(\ell) = (0,+,-)$. We denote the i-th entry in $type(\ell)$ by $type(\ell)_i$.

Let us now define several terms we use in the rest of the paper. A *dimension* $\dim(\ell)$ of a line $\ell \in \mathbb{L}(n^d)$ is $\dim(\ell) = |\{i \in \{1, \ldots, d\} | type(\ell)_i \in \{+, -\}\}|$. A *degree* $\deg(p)$ of a point $p \in n^d$ is a number of incident lines, formally $\deg(p) = |\{\ell \in \mathbb{L}(n^d) | p \in \ell\}|$. Two points $p_1, p_2 \in n^d$ are *collinear*, if there exists a line $\ell \in \mathbb{L}(n^d)$, such that $p_1 \in \ell$ and $p_2 \in \ell$. A point $p \in n^d$ is called a *corner* if p has coordinates only 0 and $n-1$. A point $p = [x_1, \ldots, x_d] \in n^d$ is an *outer point* if there exists at least one $i \in \{1, \ldots, d\}$ such that $x_i \in \{0, n-1\}$. If a point $p \in n^d$ is not an outer point then p is called an *inner point*.

A line $\ell \in \mathbb{L}(n^d)$ is called an *edge* if $\dim(\ell) = 1$ and ℓ contains two corners. Two corners are *neighbors* if they are connected by an edge. A line $\ell \in \mathbb{L}(n^d)$ with $\dim(\ell) = d$ is called *main diagonal*. We denote the set of all main diagonals by $\mathbb{L}_m(n^d)$. For better understanding the notions see Fig. 1 with some examples in the cube 4^3.

A k-dimensional *face* F of the cube n^d is a maximal set of points of n^d, such that there exist two index sets $I, J \subseteq \{1, \ldots, d\}, I \cap J = \emptyset, |I| + |J| = d - k$ and

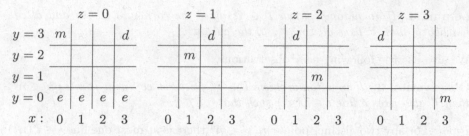

Fig. 1. A cube 4^3 with some examples of lines. An edge e has a type $(+,0,0)$, a line d has a dimension 2 and a type $(+,3,-)$ and a main diagonal m has a type $(+,-,+)$.

for each point $[x_1,\ldots,x_d]$ in F holds that $x_i = 0$ for each $i \in I$ and, $x_j = n-1$ for each $j \in J$. For example, $\{[x,y,0,n-1]|x,y \in [n]\}$ is a 2-dimensional face of the cube n^4.

A point $p \in n^d$ is *fixed* by an automorphism T if $T(p) = p$. A set of points $\{p_1,\ldots,p_k\}$ is fixed by an automorphism T if $\{p_1,\ldots,p_k\} = \{T(p_1),\ldots,T(p_k)\}$. Note that if a set S is fixed it does not necessarily mean every point of S is fixed.

2.1   Order of $\mathbb{T}_2^d$

The cube 2^d is different from other cubes because every two points are collinear. Thus, we have the following proposition.

Proposition 1. *Order of the group $\mathbb{T}_2^d$ is $(2^d)!$.*

Proof. Every permutation of the points of the cube 2^d is an automorphism, as the graph H_2^d is the complete graph on 2^d vertices.   □

We further assume that $n > 2$.

3   Automorphisms of n^3

For better understanding of our technique, we first show the result for the 3-dimensional case of the group $\mathbb{T}_n^3$. Here we state several general lemmas how an arbitrary automorphism maps main diagonals, edges and corners. The proofs are technical and are omitted from this conference paper.

Lemma 1. *Let $F =\{[x,y,0,\ldots,0]|x,y \in [n]\}$ be a face of n^d, and let an automorphism $T \in \mathbb{T}_n^d$ fixes all 4 corners of F, i.e., points $[0,\ldots,0]$, $[n-1,0,\ldots,0]$, $[0,n-1,0,\ldots,0]$ and $[n-1,n-1,0,\ldots,0]$. Then, if T fixes a point $[i,0,\ldots,0], i \in [n]$ it also fixes a point $[n-i-1,0,\ldots,0]$.*

Lemma 2. *Every automorphism $T \in \mathbb{T}_n^d$ maps a main diagonal $m \in \mathbb{L}_m(n^d)$ onto a main diagonal $m' \in \mathbb{L}_m(n^d)$.*

Lemma 3. *Let $T \in \mathbb{T}_n^d$, e be an edge and p be a corner, such that $p \in e$. If the corner p is fixed by T, then $T(e) = e'$ is an edge such that $p \in e'$.*

Lemma 4. *If an automorphism $T \in \mathbb{T}_n^d$ fixes the corner $[0, \ldots, 0]$ and all its neighbors, then T fixes all corners of the cube n^d.*

We also use the following easy observations.

Observation 3. *If an automorphism $T \in \mathbb{T}_n^d$ fixes two collinear points $p, q \in n^d$, then T also fixes a line $\ell \in \mathbb{L}(n^d)$ such that $p, q \in \ell$.*

Proof. For any two distinct points $p_1, p_2 \in n^d$ there is at most one line $\ell \in \mathbb{L}(n^d)$ such that $p_1, p_2 \in \ell$. Therefore, if the points p and q are fixed then the line ℓ has to be fixed as well. $\square$

Observation 4. *If two lines $\ell_1, \ell_2 \in \mathbb{L}(n^d)$ are fixed by $T \in \mathbb{T}_n^d$ then their intersection, a point $p = \ell_1 \cap \ell_2$, is fixed by T.*

Proof. For any two lines ℓ, ℓ' there is at most one point in $\ell \cap \ell'$. Therefore, if the lines ℓ_1 and ℓ_2 are fixed then the point p has to be fixed as well. $\square$

3.1   Generators of $\mathbb{T}_n^3$

In this section we characterize generators of the group $\mathbb{T}_n^3$. We use two basic groups of automorphisms. The group of permutation automorphisms $\mathbb{F}_n$. The group second group is the group of rotations $\mathbb{R}$ of a 3-dimensional cube. The generators of $\mathbb{R}$ are rotations

$$R_x([x, y, z]) = [x, n - z - 1, y],$$
$$R_y([x, y, z]) = [n - z - 1, y, x],$$
$$R_z([x, y, z]) = [n - y - 1, x, z].$$

Definition 1. *Let $\mathbb{A}_n^3$ be a group generated by elements of $\mathbb{R} \cup \mathbb{F}_n$.*

We prove that $\mathbb{A}_n^3 = \mathbb{T}_n^3$. The idea of the proof, that resembles a similar proof of Silver [11], is composed of two steps:

1. For any automorphism $T \in \mathbb{T}_n^3$ we find an automorphism $A \in \mathbb{A}_n^3$ such that $T \circ A$ fixes every point in a certain set S.
2. If an automorphism $T' \in \mathbb{T}_n^3$ fixes every point in S then T' is the identity.

Hence, for every $T \in \mathbb{T}_n^3$ we find an inverse element T' such that T' is composed only by elements of $\mathbb{R} \cup \mathbb{F}_n$, therefore $T \in \mathbb{A}_n^3$. The proof of the second part is very similar to the proof for a general cube n^d. Thus, it is proved only for the general cube in the next section.

Theorem 5. *For every $T \in \mathbb{T}_n^3$ there exists $A \in \mathbb{A}_n^3$, such that $T \circ A$ fixes all corners and every point of the line $\ell = \{[i, 0, 0] | i \in [n]\}$.*

Proof. First we find an automorphism $A' \in \mathbb{A}_n^3$ such that $T \circ A'$ fixes all corners. We start with the point $p_0 = [0, 0, 0]$. A point $T(p_0)$ has to be on a main diagonal (by Lemma 2). Without loss of generality $T(p_0) = [i, i, n-i-1]$. We take $f_\pi \in \mathbb{F}_n$ such that $\pi(i) = 0$, $\pi(0) = i$, $\pi(n - i - 1) = n - 1$, $\pi(n - 1) = n - i - 1$, and $\pi(k) = k$ otherwise. Therefore, $T \circ F_\pi(p_0)$ is a corner. Then we take $R_1 \in \mathbb{R}$ such that the automorphism $T_1 = T \circ F_\pi \circ R_1$ fixes p_0.

By Lemma 3 the line $T_1(\ell)$ must be mapped onto an edge e such that $p_0 \in e$. If the corner $p_1 = [n-1, 0, 0]$ is fixed by T_1, we take $T_2 = T_1$. Otherwise it can be mapped onto $[0, n-1, 0]$ (or $[0, 0, n-1]$). We take a rotation $R_2([x, y, z]) = [y, z, x]$ (or $[z, x, y]$). Thus, the automorphism $T_2 = T_1 \circ R_2$ fixes corners p_1 and p_0. Note that $R_2([0, 0, 0]) = [0, 0, 0]$.

If a corner $p_2 = [0, n - 1, 0]$ is fixed by T_2 we take $T_3 = T_2$. Otherwise it can be mapped only onto $[0, 0, n - 1]$. We take a rotation $R_3([x, y, z]) = [n - x - 1, n - z - 1, n - y - 1]$ and permutation automorphism F_σ, where $\sigma(i) = n - i - 1$. Hence, $T_3 = T_2 \circ R_3 \circ F_\sigma$ fixes the points p_0, p_1, p_2 as follows. For p_0,

$$T_2 \circ R_3 \circ F_\sigma([0, 0, 0]) = R_3 \circ F_\sigma([0, 0, 0]) = F_\sigma([n - 1, n - 1, n - 1]) = [0, 0, 0].$$

For p_1,

$$T_2 \circ R_3 \circ F_\sigma([n-1, 0, 0]) = R_3 \circ F_\sigma([n-1, 0, 0]) = F_\sigma([0, n-1, n-1]) = [n-1, 0, 0].$$

For p_2,

$$T_2 \circ R_3 \circ F_\sigma([0, n-1, 0]) = R_3 \circ f_\sigma([0, 0, n-1]) = f_\sigma([n-1, 0, n-1]) = [0, n-1, 0].$$

A corner $p_3 = [0, 0, n - 1]$ is fixed by T_3 automatically, because it is neighbor of p_0 and all others neighbors are already fixed. All other corners are fixed due to Lemma 4. The automorphism $T_3 = T \circ A'$ for some $A' \in \mathbb{A}_n^3$ fixes all corners of the cube n^3.

Now we find an automorphism A such that $T \circ A$ fixes all corners and all points on the line ℓ. The line ℓ is fixed by T_3 due to Observation 3. Let $k = \lfloor \frac{n}{2} \rfloor - 1$. We construct the automorphism A by induction over $i \in \{0, \dots, k\}$. We show that in a step i an automorphism Y_i fixes all corners and every point in a set

$$Q_i = \{[j, 0, 0], [n - j - 1, 0, 0] | 0 \le j \le i\}.$$

First, let $i = 0$ and $Y_0 = T_3$. The automorphism Y_0 fixes all corners and Q_0 contains only $[0, 0, 0]$ and $[n - 1, 0, 0]$, which are also corners. Suppose that $i > 0$. By induction hypothesis, we have an automorphism Y_{i-1} which fixes all corners and every point in the set Q_{i-1}. If $Y_{i-1}([i, 0, 0]) = [i, 0, 0]$ then $Y_i = Y_{i-1}$. Otherwise $Y_{i-1}([i, 0, 0]) = [j, 0, 0]$. Note that $i < j < n - i - 1$ because points from Q_{i-1} are already fixed. Let us consider $F_\pi^i \in \mathbb{F}_n$ where $\pi(j) = i$, $\pi(i) = j$, $\pi(n - j - 1) = n - i - 1$, $\pi(n - i - 1) = n - j - 1$, and $\pi(k) = k$ otherwise. The automorphism $Y_i = Y_{i-1} \circ F_\pi^i$ fixes the following points:

1. All corners, as the automorphism Y_{i-1} fixes all corners by the induction hypothesis and $\pi(0) = 0$ and $\pi(n - 1) = n - 1$.

2. Set Q_{i-1}, as the automorphism Y_{i-1} fixes the set Q_{i-1} by the induction hypothesis and $\pi(k) = k$ for all $k < i$ and $k > n - i - 1$.
3. Point $[i, 0, 0]$: $Y_{i-1} \circ F_{\pi}^{i}([i, 0, 0]) = F_{\pi}^{i}([j, 0, 0]) = [i, 0, 0]$.
4. Point $[n - i - 1, 0, 0]$ by Lemma 1.

Note that if n is odd a point $[\frac{n-1}{2}, 0, \ldots, 0]$ is fixed as well by an automorphism Y_k. Thus, the automorphism $Y_k = T \circ A$ for some $A \in \mathbb{A}_n^d$ fixes all points of the line ℓ and all corners of the cube.   □

4   Generators of the Group $\mathbb{T}_n^d$

In this section we characterize the generators of the general group $\mathbb{T}_n^d$. As we stated in Sect. 1, we use the groups $\mathbb{R}_d$, $\mathbb{F}_n$ and $\mathbb{X}$.

Definition 2. *Let $\mathbb{A}_n^d$ be a group generated by elements of $\mathbb{R}_d \cup \mathbb{F}_n \cup \mathbb{X}$.*

We prove that $\mathbb{A}_n^d = \mathbb{T}_n^d$ in the same two steps as we proved $\mathbb{A}_n^3 = \mathbb{T}_n^3$.

1. For any automorphism $T \in \mathbb{T}_n^d$ we find an automorphism $A \in \mathbb{A}_n^d$, such that $T \circ A$ fixes all corners of the cube n^d and one edge.
2. If an automorphism $T' \in \mathbb{T}_n^d$ fixes all corners and one edge then T' is identity.

Theorem 6. *For all $T \in \mathbb{T}_n^d$ there exists $A \in \mathbb{A}_n^d$ such that $T \circ A$ fixes every corner of the cube n^d and every point of a line $\ell = \{[i, 0, \ldots, 0] | i \in [n]\}$.*

Proof (Sketch). First we construct an automorphism $A' \in \mathbb{A}_d^n$ such that $T \circ A'$ fixes all corners. We start with the point $p_0 = [0, \ldots, 0]$. By Lemma 2, the point $T(p_0)$ has to be on a main diagonal. We choose $F \in \mathbb{F}_n$ such that $T \circ F(p_0)$ is a corner. Then, we choose $R \in \mathbb{R}_d$ such that $T \circ F \circ R(p_0) = p_0$.

By induction over i we can construct automorphisms Z_i to fix the points p_0 and

$$p_i = [0, \ldots \underset{i}{n - 1}, \ldots 0]$$

for all $i \in \{0, \ldots, d-2\}$. We start with the automorphism $Z_0 = T \circ F \circ R$ and in a step i we compose the automorphism Z_{i-1} with a suitable rotation in $\mathbb{R}^d$. If Z_{d-2} fixes p_{d-1}, then $Z_{d-1} = Z_{d-2}$. Otherwise p_{d-1} is mapped onto p_d and then $Z_{d-1} = Z_{d-2} \circ X$, where $X \in \mathbb{X}$ and $X \neq Id$. Thus, the automorphism Z_{d-1} fixes all points of P_{d-1} and the corner p_d is fixed automatically because there is no other possibility where the corner p_d can be mapped. The automorphism Z_{d-1} fixes the corner $p_0 = [0, \ldots, 0]$ and all its neighbors. Therefore by Lemma 4, the automorphism $Z_{d-1} = T \circ A'$ for some $A' \in \mathbb{A}_d^n$ fixes all corners of the cube.

The automorphism fixing points on the line ℓ is constructed in the same way as in the proof of Theorem 5. We find an automorphism Y fixing all corners and points on the line ℓ by induction. We start with the automorphism Z_{d-1}. In step i of the induction we compose the automorphism from the step $i - 1$ and an automorphism $F_i \in \mathbb{F}_n$ which fixes points $[i, 0, \ldots, 0]$ and $[n - i - 1, 0, \ldots, 0]$. □

It remains to prove that if an automorphism $T \in \mathbb{T}_n^d$ fixes all corners and all points in the line $\ell = \{[i, 0, \ldots, 0] | i \in [n]\}$ then T is the identity. We prove it in two parts. First, we prove that if $d = 2$ then the automorphism T is the identity. Then, we prove it for a general dimension by an induction argument.

Theorem 7. *Let an automorphism $T \in \mathbb{T}_n^2$ fixes all corners of the cube and all points in the line $\ell = \{[i, 0] | i \in [n]\}$. Then, the automorphism T is the identity.*

Proof. Let $d_1, d_2 \in \mathbb{L}(n^2)$. Thus, $type(d_1) = (+, +)$ and $type(d_2) = (+, -)$. Since all corners are fixed, the diagonals d_1 and d_2 are fixed as well due to Observation 3. Let $p \in d_1 \cup d_2$ such that p is not a corner. The point p is collinear with the only one point $q \in \ell$ such that q is not a corner. Therefore, every point on the diagonals d_1 and d_2 is fixed.

Now we prove that every line in $\mathbb{L}(n^2)$ is fixed. Let $\ell_1 \in \mathbb{L}(n^2)$ be a line of a dimension 1. Suppose n is even. The line ℓ_1 intersects the diagonals d_1 and d_2 in distinct points, which are fixed. Therefore, the line ℓ_1 is fixed as well by Observation 3.

Now suppose n is odd. If ℓ_1 does not contain the face center $c_1 = [\frac{n-1}{2}, \frac{n-1}{2}, 0, \ldots, 0]$ then ℓ is fixed by the same argument as in the previous case. Thus, suppose $c_1 \in \ell_1$. There are two lines ℓ_2, ℓ_3 in $\mathbb{L}(n^2)$ of dimension 1 which contains c_1. Their types are $type(\ell_2) = (\frac{n-1}{2}, +)$ and $type(\ell_3) = (+, \frac{n-1}{2})$. The line ℓ_2 also intersects the line ℓ. Therefore, the lines contains two fixed points c_1 and $[\frac{n-1}{2}, 0]$ and thus the line ℓ_2 is fixed. The line ℓ_3 is fixed as well because every other line is fixed. For better understanding of all lines and points used in the proof see Fig. 2 with example of the cube 5^2.

Every point in 5^2 is fixed due to Observation 4 because every point is in an intersection of at least two fixed lines.   □

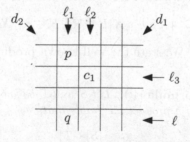

Fig. 2. Points and lines used in the proof of Theorem 7.

Theorem 8. *Let an automorphism $T \in \mathbb{T}_n^d$ fix all corners of the cube n^d and all points of an arbitrary edge e. Then, the automorphism T is the identity.*

Proof. We prove the theorem by induction over dimension d of the cube n^d. The basic case for $d = 2$ is Theorem 7.

Therefore, we can suppose $d > 2$ and the theorem holds for all dimensions smaller then d. Without loss of generality, $e = \{[i, 0, \ldots, 0] | i \in [n]\}$. We consider the face $F = \{[x_1, \ldots, x_{d-1}, 0] | x_1, \ldots, x_{d-1} \in [n]\}$. The face F has a dimension $d-1$ and $e \subset F$. Therefore, all points of F are fixed by the induction hypothesis. Then we take all faces G of dimension $d-1$ such that $F \cap G \neq \emptyset$. Corners $c \in G$ are fixed. There is at least one edge f such that $f \subseteq F \cap G$. Therefore the points of f are also fixed and the points $p \in G$ are fixed by the induction hypothesis. By this argument we show that every outer point is fixed. Every line $\ell \in \mathbb{L}(n^d)$ is fixed due to Observation 3 because every line contains at least two outer points. Therefore by Observation 4, every point $q \in n^d$ is fixed because every point is an intersection of at least two lines. □

5   Order of the Group $\mathbb{T}_n^d$

In the previous section we characterized the generators of the group $\mathbb{T}_n^d$. Now we compute the order of $\mathbb{T}_n^d$. First, we state several technical lemmas whose proofs are omitted in this conference paper.

Lemma 5. *Orders of the basic groups are as follows.*

1. $|\mathbb{R}_d| = 2d|\mathbb{R}_{d-1}| = 2^{d-1}d!$, $|\mathbb{R}_2| = 4$
2. $|\mathbb{F}_n| = \prod_{i=0}^{\lfloor \frac{n}{2} \rfloor - 1} (2\lfloor \frac{n}{2} \rfloor - 2i)$
3. $|\mathbb{X}| = 2$

Lemma 6. *The groups $\mathbb{R}_d$ and $\mathbb{F}_n$ commute, and the groups $\mathbb{X}$ and $\mathbb{F}_n$ commute.*

Lemma 7. *Let $X \in \mathbb{X}$ such that $X \neq Id$. Then, for all $R_1 \in \mathbb{R}_d$ there exists $R_2 \in \mathbb{R}_d$ such that $R_1 \circ X = X \circ R_2$.*

By Lemmas 6 and 7 we can conclude that any automorphism $A \in \mathbb{T}_n^d$ can be written as $A = R \circ F \circ X$ where $R \in \mathbb{R}_d$, $F \in \mathbb{F}_n$ and $X \in \mathbb{X}$. Thus, the product

$$\mathbb{R}_d \mathbb{F}_n \mathbb{X} = \{R \circ F \circ X | R \in \mathbb{R}_d, F \in \mathbb{F}_n, X \in \mathbb{X}\}$$

is exactly the group $\mathbb{T}_n^d$. We state the well-known product formula for a group product.

Lemma 8 (Product formula [4]). *Let S and T be subgroups of a finite group G. Then, for an order of a product ST holds that*

$$|ST| = \frac{|S| \cdot |T|}{|S \cap T|}.$$

Thus, for computing the order of $\mathbb{T}_n^d$ we need to compute the orders of intersections of the basic groups $\mathbb{R}_d$, $\mathbb{F}_n$ and $\mathbb{X}$.

Lemma 9. *If d is odd, then $\mathbb{R}_d \cap \mathbb{F}_n = \{Id\}$. If d is even, then $\mathbb{R}_d \cap \mathbb{F}_n = \{Id, F_\sigma\}$ where $\sigma(i) = n - i - 1$.*

Lemma 10. *The group $\mathbb{X}$ can be generated by elements of the groups $\mathbb{R}_d$ and $\mathbb{F}_n$ if and only if d is odd.*

Theorem 9. *The order of the group* $\mathbb{T}_n^d$ *is* $|\mathbb{R}_d| \cdot |\mathbb{F}_n|$.

Proof. If d is odd $|\mathbb{R}_d \cap \mathbb{F}_n| = 1$ due to Lemma 9. Moreover, the group $\mathbb{X}$ is a subset of $\mathbb{R}_d\mathbb{F}_n$ due to Lemma 10. Therefore, the group $\mathbb{T}_n^d$ is exactly a product $\mathbb{R}_d\mathbb{F}_n$ and the theorem holds by Lemma 8.

Now suppose d is even. By Lemma 9, $|\mathbb{R}_d \cap \mathbb{F}_n| = 2$. Thus by Lemma 8, $|\mathbb{R}_d\mathbb{F}_n| = |\mathbb{R}_d| \cdot |\mathbb{F}_n|/2$. The order of the intersection $|\mathbb{X} \cap \mathbb{R}_d\mathbb{F}_n|$ is 1 by Lemma 10. Hence, $|\mathbb{T}_n^d| = 2|\mathbb{R}_d\mathbb{F}_n| = |\mathbb{R}_d| \cdot |\mathbb{F}_n|$. $\square$

As a corollary of Theorem 9 we get the second part of Theorem 1.

Corollary 1. *Let* $k = \lfloor \frac{n}{2} \rfloor$. *Then,* $|\mathbb{T}_d^n| = 2^{d-1+k}d!k!$.

Proof. By Theorem 9, the order $|\mathbb{T}_d^n|$ is $2^{d-1}d! \prod_{i=0}^{k-1} (2k - 2i)$ for $k = \lfloor \frac{n}{2} \rfloor$. There are k even numbers from 2 to $2k$ in the product $\prod_{i=0}^{k-1} (2k - 2i)$. Therefore, it can be rewritten as $2^k k!$. $\square$

Corollary 2. *The groups* $\mathbb{T}_{2k}^d$ *and* $\mathbb{T}_{2k+1}^d$ *are isomorphic for* $k \geq 2$.

Proof. The rotation group for generating $\mathbb{T}_{2k}^d$ and $\mathbb{T}_{2k+1}^d$ is the same. For every permutation $\pi \in \mathbb{S}_{2k+1}$ with the symmetry property holds that $\pi(k) = k$. Therefore, the group $\mathbb{F}_{2k}$ is isomorphic to the group $\mathbb{F}_{2k+1}$. Whether $\mathbb{X}$ is generated by $\mathbb{F}_n$ and $\mathbb{R}_d$ depends only on the dimension. $\square$

6   The Complexity of Colored Cube Isomorphism

In this section we prove Theorem 2. As we stated before, CHI is in GI. Therefore, COLORED CUBE ISOMORPHISM as a subproblem of CHI is in GI as well. It remains to prove the problem is GI-hard.

First, we describe how we reduce the input of GRAPH ISOMORPHISM to the input of COLORED CUBE ISOMORPHISM. Let $G = (V, E)$ be a graph. Without loss of generality $V = \{0, \ldots, n - 1\}$. We construct the coloring $s^G : [k]^2 \to [2], k = 2n + 4$ as follows. The value of $s^G([i, j])$ is 1 if $[i, j] = [n, n]$ or $[i, j] = [n, n + 1]$ or $i, j \leq n - 1$ and $\{i, j\} \in E$. The value of $s^G(p)$ for all other point p is 0. We can view the coloring s^G as a matrix M^G such that $M_{i,j}^G = s^G([i, j])$. The submatrix of M^G consisting of the first n rows and n columns is exactly the adjacency matrix of the graph G.

The idea of the reduction is as follows. If two colorings s^{G_1}, s^{G_2} are isomorphic via a cube automorphism $A \in \mathbb{T}_n^d$ then A can be composed only of permutation automorphisms in $\mathbb{F}_k$. Moreover, if $A = F_\pi$ for some permutation π then the permutation π maps the numbers in $[n]$ to the numbers in $[n]$ and describes the isomorphism between the graphs G_1 and G_2.

Lemma 11. *Let* G_1, G_2 *be graphs without vertices of degree 0. If colorings* s^{G_1}, s^{G_2} *are isomorphic via a cube automorphism* A *then* $A = F_\pi \in \mathbb{F}_k$. *Moreover,* $\pi(i) \leq n - 1$ *if and only if* $i \leq n - 1$.

Proof (Sketch). Let $A = R \circ X \circ F$ where $R \in \mathbb{R}_2, X \in \mathbb{X}, F \in \mathbb{F}_k$ and m_1, m_2 be main diagonals of $[k]^2$ of type $(+, +)$ and $(+, -)$, respectively. Due to the colors of $[n, n]$ and $[n, n + 1]$ we can show that A has to fix m_1 and m_2 and that $A \in \mathbb{F}_k$. Moreover, if $A = F_\pi$ then $\pi(n) = n$ and $\pi(n + 1) = n + 1$.

For every $i \leq n - 1$ there is at least one point with color 1 on a line of type $(+, i)$ in both colorings s^{G_1}, s^{G_2} because graphs G_1 and G_2 do not contain any vertex of degree 0. On the other hand, for every $i \geq n + 2$ there are only points with color 0 on a line of type $(+, i)$ in both colorings. Therefore, if $i \leq n - 1$ then i has to be mapped on $j \leq n - 1$ by π. $\square$

The proof of the following theorem follows from Lemma 11.

Theorem 10. *Let $G_1 = (V_1, E_1)$ and $G_2 = (V_2, E_2)$ be graphs without vertices of degree 0. Then, the graphs G_1 and G_2 are isomorphic if and only if the colorings s^{G_1} and s^{G_2} are isomorphic.*

We may suppose that inputs graphs G_1 and G_2 have minimum degree at least 1 for the purpose of the polynomial reduction of GRAPH ISOMORPHISM to COLORED CUBE ISOMORPHISM. Thus, Theorem 2 follows from Theorem 10.

References

1. Allis, L.V.: Searching for solutions in games and artificial intelligence. Ph.D. thesis, University of Limburg (1994)
2. Arvind, V., Das, B., Köbler, J., Toda, S.: Colored hypergraph isomorphism is fixed parameter tractable. Algorithmica **71**, 120–138 (2015)
3. Arvind, V., Köbler, J.: On hypergraph and graph isomorphism with bounded color classes. In: Durand, B., Thomas, W. (eds.) STACS 2006. LNCS, vol. 3884, pp. 384–395. Springer, Heidelberg (2006)
4. Ballester-Bolinches, A., Esteba-Romero, R., Asaad, M.: Products of Finite Groups. (Walter) De Gruyter, Berlin (2010)
5. Beck, J.: Tic-Tac-Toe Theory. Cambridge University Press, Cambridge (2006)
6. Booth, K.S., Colbourn, C.J.: Problems polynomially equivalent to graph isomorphism. Technical report CS-77/04, Department of Computer Science, University of Waterloo (1977)
7. Choudum, S.A., Sunitha, V.: Automorphisms of augmented cubes. Int. J. Comput. Math. **85**, 1621–1627 (2008)
8. Hales, A.W., Jewett, R.: Regularity and positional games. Trans. Am. Math. Soc. **106**, 222–229 (1963)
9. Harary, F.: The automorphism group of a hypercube. J. Univers. Comput. Sci. **6**, 136–138 (2000)
10. Patashnik, O.: Qubic: $4 \times 4 \times 4$ tic-tac-toe. Math. Mag. **53**, 202–216 (1980)
11. Silver, R.: The group of automorphisms of the game of 3-dimensional ticktacktoe. Am. Math. Mon. **74**, 247–254 (1967)

Hadwiger's Conjecture and Squares of Chordal Graphs

L. Sunil Chandran[1], Davis Issac[2(✉)], and Sanming Zhou[3]

[1] Indian Institute of Science, Bangalore 560012, India
sunil@csa.iisc.ernet.in
[2] Max Planck Institute for Informatics, Saarbrücken, Germany
dissac@mpi-inf.mpg.de
[3] School of Mathematics and Statistics, The University of Melbourne,
Parkville, VIC 3010, Australia
sanming@unimelb.edu.au

Abstract. Hadwiger's conjecture states that for every graph G, $\chi(G) \leq \eta(G)$, where $\chi(G)$ is the chromatic number and $\eta(G)$ is the size of the largest clique minor in G. In this work, we show that to prove Hadwiger's conjecture in general, it is sufficient to prove Hadwiger's conjecture for the class of graphs $\mathcal{F}$ defined as follows: $\mathcal{F}$ is the set of all graphs that can be expressed as the square graph of a split graph. Since split graphs are a subclass of chordal graphs, it is interesting to study Hadwiger's Conjecture in the square graphs of subclasses of chordal graphs. Here, we study a simple subclass of chordal graphs, namely 2-trees and prove Hadwiger's Conjecture for the squares of the same. In fact, we show the following stronger result: If G is the square of a 2-tree, then G has a clique minor of size $\chi(G)$, where each branch set is a path.

Keywords: Hadwiger's conjecture · 2-trees · Square graphs · Minors

1 Introduction

The Four Color Theorem is perhaps the most famous theorem in graph theory. It states that the chromatic number of a planar graph is at most 4. The history of the development of graph theory itself is intimately linked with the attempts to solve the Four Color Conjecture. Wagner [Wag37] showed in 1937 that the four color conjecture is equivalent to the following statement: If a graph is K_5-minor free, then it is 4-colorable. In 1943, Hadwiger [Had43] proposed the following conjecture.

Conjecture 1 (Hadwiger's Conjecture). For $t \geq 1$, every graph G without a K_{t+1} minor is t-colorable.

L. Sunil Chandran—Part of the work was done when this author was visiting *Max Planck Institute for Informatics, Saarbruecken, Germany* supported by *Alexander von Humboldt Fellowship*.

© Springer International Publishing Switzerland 2016
T.N. Dinh and M.T. Thai (Eds.): COCOON 2016, LNCS 9797, pp. 417–428, 2016.
DOI: 10.1007/978-3-319-42634-1_34

This conjecture if proved, would give a far reaching generalization of the 4-color theorem. Hadwiger [Had43] proved the conjecture for $t \leq 3$. The Four color theorem was proved by Appel et al. [AH+77, AHK+77] in 1977. In view of Wagner's theorem [Wag37], this implies that Hadwiger's conjecture is true for $t = 4$. In 1993, Robertson et al. [RST93] proved Hadwiger's conjecture for $t = 5$. It remains unsolved for $t > 5$. Kawarabayashi and Toft [KT05] showed that any graph that is K_7-minor free and $K_{4,4}$-minor free is 6-colorable.

For perfect graphs Hadwiger's conjecture is trivially true. Reed and Seymour [RS04] proved Hadwiger's conjecture for line graphs. Belkale and Sunil Chandran [BC09] proved the conjecture for proper circular arc graphs. In 2008, Chudnovsky and Fradkin [CF08] published a work which generalizes both the above results: They proved Hadwiger's conjecture for a class of graphs called quasi-line graphs, which properly contains both proper circular arc graphs and line graphs.

As far as we know, not many results are known regarding Hadwiger's conjecture with respect to square graphs, even for the squares of well known special classes of graphs. This is surprising considering the fact that the chromatic number is well studied with respect to the squares of several special classes of graphs. For e.g., see the extensive work on Wegner's conjecture [Weg77]. We believe that this may be due to the difficulty level involved in dealing with this problem, as we show in Theorem 1 that proving Hadwiger's conjecture for squares of chordal graphs will also prove the conjecture for general graphs.

Hadwiger's conjecture for powers of cycles and their complements was proved by Li and Liu [LL07]; Sunil Chandran et al. [CKR08] studied the Hadwiger number with respect to the Cartesian product of the graphs.

1.1 Our Contributions

Hadwiger's conjecture is well-known to be a tough problem. Bollobás et al. [BCE80] describe it as "one of the deepest unsolved problems in graph theory." It could be useful if we can show that it is sufficient to concentrate on certain class of graphs. **Chordal graphs** are those graphs that have no induced cycle of length 4 or more. **Split graphs** are those graphs whose vertices can be partitioned into two sets such that one induces an independent set and the other induces a clique. Split graphs form a subclass of chordal graphs. In Sect. 2, *we show that, in order to prove the Hadwiger's conjecture in general, it is sufficient to prove it for the class of graphs that are squares of split graphs.* (See Theorem 1.)

We understand that our reduction may not really help in making Hadwiger's conjecture easier. But it does show that the squares of split graphs captures the complexity of the general problem. For an optimistic researcher, it opens up the question of studying the Hadwiger's conjecture on the squares of various special classes of graphs in the hope of getting some new insights about the problem.

In light of Theorem 1, it is interesting to study Hadwiger's conjecture for squares of subclasses of chordal graphs. It can be shown that chordal graphs are exactly the class of graphs that can be constructed by starting with a clique and doing the following operation, a finite number of times: Pick a clique C in the current graph, introduce a new vertex v, and make v adjacent to all the vertices in C.

k-*trees* are a special case of chordal graphs, where we start with a k-clique and at each step we pick a k-clique. Hence it is interesting to prove Hadwiger's conjecture for squares of k-trees. As a first step, *we prove Hadwiger's conjecture for squares of 2-trees* in Sect. 3 of this paper. (See Theorem 2.) A slightly more general class than 2-trees allows one to join a fresh vertex to a clique of size at most 2 instead of exactly 2. We remark that it is easy to extend Theorem 2 to this class of graphs.

Structure of Branch Sets[1]: Although proving the Hadwiger's conjecture requires only to show a clique minor of size at least $\chi(G)$, it is also interesting to study the structure of branch sets forming such a clique minor. For example, in the case of graphs with independence number at most 2, Seymour proposed the following stronger conjecture [Bla05]: If G has no stable set of size 3, then G has a clique minor of size at least $|V(G)|/2$ using only edges or single vertices as branch sets. *For squares of 2-trees we show that there exists a clique minor of size $\chi(G)$ where the branch sets forming the clique minor are paths.*(See Theorem 2).

Towards Generalizing the Result of Chudnovsky and Fradkin: Chudnovsky and Fradkin [CF08] proved that Hadwiger's conjecture is true for quasi-line graphs. A graph G is a quasi-line graph if for every vertex $v \in V(G)$, the set of neighbors of v in G can be expressed as the union of two cliques. A natural way to generalize concept of quasi-line graph is the following:

Definition 1 (Generalized Quasi-Line Graphs[2]**).** *A graph G is a generalized quasi-line graph if for any subset $S \subseteq V(G)$, there exists a vertex $u \in S$ such that the neighbors of u induce union of two cliques in $G[S]$ (the induced subgraph on S).*

It is natural to consider the problem of generalizing the result of Chudnovsky and Fradkin to generalized quasi-line graphs.

Open Problem 1. *Prove Hadwiger's conjecture for generalized quasi-line graphs.*

Taking into account the difficulty level of [CF08], the above question might turn out to be difficult. Therefore it is natural to try to prove the conjecture for non-trivial subclasses of generalized quasi-line graphs.

Observation 1. *The squares of 2-trees form a subclass of generalized quasi-line graphs.*

Remark 1. It is interesting to note that squares of 2-degenerate graphs do not form a subclass of generalized quasi-line graphs.

---

[1] See Subsect. 1.3 for the definition.

[2] This generalization is in the same spirit as the generalization of graphs of maximum degree k to k-degenerate graphs. A graph G is a maximum degree k graph, if every vertex has at most k neighbors. A graph G is a k-degenerate graph is for any subset $S \subseteq V(G)$, there exists a vertex $u \in S$, such that u has at most k neighbors in $G[S]$. Graph classes which can be considered to be generalizations of quasi-line graphs can also be found in [KT14], for e.g. k-perfectly groupable graphs, k-simplicial graphs, k-perfectly orientable graphs etc.

Note that squares of 2-trees do not form a subclass of quasi-line graphs. Hence, by Theorem 2, we prove Hadwiger's conjecture for a special class of generalized quasi-line graphs that is not contained in quasi-line graphs.

1.2   Future Directions

An obvious next step will be to prove Hadwiger's Conjecture for squares of k-trees for fixed $k \geq 3$. It is also interesting to try to prove Hadwiger's conjecture for squares of other special classes of graphs such as planar graphs. Another direction may be to work towards solving the Open Problem 1. It will be interesting to look at other non-trivial subclasses of generalized quasi-line graphs with respect to Hadwiger's conjecture.

1.3   Preliminaries

For any graph G, we denote the vertices of G by $V(G)$ and the edges of G by $E(G)$. When we are talking about a singleton set $\{x\}$, we may abuse the notation and use x for the sake of conciseness. We say disjoint vertex sets $V_1, V_2 \subseteq V(G)$ are adjacent in G, if there exist $v_1 \in V_1$ and $v_2 \in V_2$ such that $\{v_1, v_2\} \in E(G)$.

Definition 2 (Square of a Graph). *For any graph G, the square of G, denoted by G^2, is the graph on the same vertex set as G, such that there is an edge between a pair of vertices u and v if and only if they are adjacent in G or are adjacent to a common vertex in G.*

Definition 3 (Clique). *Any $C \subseteq V(G)$ is called a **clique** of G if there is an edge in G between every pair of vertices in C. We use $\omega(G)$ to denote the size of the largest clique in G.*

Definition 4 (Coloring, Proper Coloring, Optimal Coloring, Chromatic Number). *A **coloring** of graph G is defined as a mapping from $V(G)$ to a set of colors. A coloring of graph G is called a **proper coloring** if no two adjacent vertices have the same color in it. An **optimal coloring** of graph G is any proper coloring of G that minimizes the number of colors used. The **Chromatic number** of G (denoted by $\chi(G)$) is defined as the number of colors used by an optimal coloring of G.*

For any coloring μ of G and any $S \subseteq V(G)$, we use $\mu(S)$ to denote $\{\mu(v) : v \in S\}$.

Definition 5 (2-Tree). *A **2-tree** is a graph that can be constructed by starting with an edge and doing the following operation a finite number of times: Pick an edge $e = \{u, v\}$ in the current graph, introduce a new vertex w and add edges $\{u, w\}$ and $\{v, w\}$.*

Definition 6 (Edge Contraction). *The operation of contraction of an edge $e = \{u, v\}$ is defined as follows: the vertices u and v are deleted and a new vertex v_e is added to the graph. Edges are added between v_e and all the vertices that were adjacent to at least one of u and v.*

Definition 7 (Minor). *A graph H is called a* **minor** *of a graph G if H can be obtained from G using any sequence of the following operations:*

1. *Deleting a vertex;*
2. *Deleting an edge;*
3. *Contracting an edge.*

An equivalent definition of minors is as follows.

Definition 8 (Minor, Branch Sets). *A graph H with $V(H) = \{h_1, h_2 \ldots, h_n\}$ is said to be a minor of G if there exists $S_1, S_2, \ldots, S_n \subseteq V(G)$ such that*

1. *for all $1 \leq i \leq n$, $G[S_i]$ is connected,*
2. *for all $i \neq j$, $S_i \cap S_j = \emptyset$ and*
3. *S_i is adjacent to S_j in G if $\{h_i, h_j\} \in E(H)$.*

The sets $S_1, S_2 \ldots S_n$ are called the **branch sets** *of the minor H of G.*

Definition 9 (Clique Minor, Hadwiger Number). *A* **clique minor** *of G is defined as a minor of G that is a clique. A* **clique minor of size** k *of G is defined as a minor of G that is a k-clique. The* **Hadwiger Number** *of G is the largest k such that G has a clique minor of size k. We denote the Hadwiger number of G by $\eta(G)$.*

Note that the necessary and sufficient conditions for $S_1, S_2, \ldots S_n \subseteq V(G)$ to be the branch sets of a clique minor of G are that $G[S_i]$ is connected for all i, $S_i \cap S_j = \emptyset$ for all $i \neq j$, and S_i is adjacent to S_j for all $i \neq j$.

2   Reduction to Square Graphs of Split Graphs

Theorem 1. *If Hadwiger's conjecture is shown to be true for the class of graphs that can be represented as the square of some split graph, then Hadwiger's conjecture is true for the general case.*

Proof. Let G be an arbitrary graph. We assume that G has no isolated vertices since they do not affect chromatic number or Hadwiger's number. We will construct a split graph H from G, such that if Hadwiger's conjecture is true for the square of H, then it is also true for G. By definition, the vertex set of the split graph H can be split into two classes, say C and S, where C induces a clique and S induces an independent set. We will make C correspond to $E(G)$, in the sense that for each edge $e \in E(G)$ we have a vertex in C: say for $e \in E(G)$, $v_e \in C$. We make $S = V(G)$ and each vertex $x \in S$ is made adjacent to all the vertices of C that correspond to the edges incident on x; i.e., for $v \in S$, $N_H(v) = \{v_e \in C : e$ is an edge in G incident on $v\}$. Here, $N_H(v)$ denotes the neighborhood of v in H.

In the square of H, the induced subgraph on S is exactly the graph G. The reason is this: Let x, y be two vertices in S (i.e. $V(G)$). They are adjacent in the square of H if and only if there exists a common neighbor for x and y in H.

Clearly, this common neighbor has to be from C, and therefore has to correspond to a common incident edge on x and y. This is possible only if there is an edge between x and y in G. This means that in the square of H, x and y are adjacent if and only if x and y are adjacent in G.

Also, the vertex $x \in S$ is connected to all the vertices of C in the square of H, since C is a clique in H.

The chromatic number of the square of H equals $\chi(G) + |C|$. To see that $\chi(H^2) \leq \chi(G) + |C|$, use the following coloring for H^2: color the vertices of C with $|C|$ different colors and then color S using an optimal coloring of G. This is possible since the subgraph induced in H^2 on S is the same as G. Now, suppose $\chi(H^2) < \chi(G) + |C|$. Then, since C requires $|C|$ different colors and they should all be different from any color in S, we get that S was colored with fewer colors than $\chi(G)$. But then we could color G using this coloring of S to get a proper coloring of G with less than $\chi(G)$ colors, which is a contradiction.

Finally, the biggest clique minor in the square of H has exactly $|C| + \eta(G)$ vertices. It is easy to see that a clique minor of that size exists: Consider the branch sets of G corresponding to the biggest clique minor of G in the induced subgraph on S; then consider each vertex of C as a separate branch set. Clearly these branch sets produce a clique minor of size $\eta(G) + |C|$ in the square of H. Now if a larger clique minor exists, then let $B_1, B_2, \ldots, B_k$ be the corresponding branch sets. Define $B_i' = B_i$, if $B_i \cap C = \emptyset$, else let $B_i' = B_i \cap C$. It is easy to see that $B_1', B_2', \ldots, B_k'$ also can produce a clique minor of size k in the square of H. Thus, if $k > |C| + \eta(G)$, then there should be more than $\eta(G)$ branch sets that does not intersect with C; which means that G has a clique minor of size greater than $\eta(G)$, contradicting the definition of $\eta(G)$.

Therefore, if Hadwiger's conjecture is true for the square of H, we have $\eta(G) + |C| \geq \chi(G) + |C|$, which implies that $\eta(G) \geq \chi(G)$. In other words, Hadwiger's conjecture will be true for G also.   □

3   Hadwiger's Conjecture for Square Graphs of 2-Trees

By definition, any 2-tree can be constructed by starting with an edge and doing the following operation a finite number of times: Pick an edge $e = \{u, v\}$, introduce a new vertex w and add edges $\{u, w\}$ and $\{v, w\}$. We call each of these operations, a **step** in the construction. If $e = \{u, v\}$ is the edge picked and w is the newly introduced vertex in a step, then we say that e is getting **processed** in that step and that w is a **vertex-child** of e. We also say that each of $\{u, w\}$ and $\{v, w\}$ is an **edge-child** of e and that e is the **parent** of each of w, $\{u, w\}$ and $\{v, w\}$. And also, $\{u, w\}$ and $\{v, w\}$ are called **siblings** of each other. Note that an edge e can be processed in more than one step. But, without loss of generality, we can assume that all the steps in which e is processed occur contiguously. Now, for each edge and vertex we define a **level** inductively. We define the level of the first edge and its end points to be 0. Any vertex-child or edge-child of an edge of level k is said to have level $k + 1$. Observe that two edges that are siblings of each other have the same *level*. Without loss of generality, we also assume that the order of processing of edges follows a **breadth-first ordering**, i.e., an edge of level i will be processed before an edge of *level* j, if $i < j$.

Theorem 2. *For any 2-tree T, $\chi(T^2) \leq \eta(T^2)$. Moreover, T^2 has a clique minor of size $\chi(T^2)$ where all the branch sets are paths.*

In the rest of this section, we prove Theorem 2.[3] We will prove by induction on $\chi(T^2)$. We know that $\chi(T^2) \geq 2$. So, we take the base case as when $\chi(T^2) = 2$. Since T^2 has an edge, we get that $\eta(T^2) \geq 2 = \chi(T^2)$. And, since all the branch sets are singletons, they are paths. So the base case is done.

Now, consider a 2-tree T with $\chi(T^2) > 2$. In the construction of T as described above, let T_i be the 2-tree resulting after i^{th} step. Consider the step j such that $\chi(T^2) = \chi(T_j^2) = \chi(T_{j-1}^2) + 1$. In the rest of the proof, we will prove that $\chi(G^2) \leq \eta(G^2)$ where $G = T_j$, and also that G^2 has a clique minor of size $\chi(G^2)$ where each branch set is a path.

Let l_{max} be the *level* of the edge with the largest *level* in G. Note that the largest *level* of any vertex in G is also l_{max}. Also, observe that the *level* of the last edge processed is $l_{max} - 1$. None of the edges that have *level* l_{max} have been processed due to the breadth-first ordering of the processing of edges. If $l_{max} \leq 1$, then G^2 is a clique and hence $\chi(G^2) = \omega(G^2)$. Moreover, we have a clique minor of size $\chi(G^2)$ where each branch set is a singleton set. Hence, we can assume that $l_{max} > 1$ for the rest of the proof.

For any vertex a, we let $N(a)$ denote the neighbors of a in G. $N(a)$ does not include a. $N[a]$ denotes $N(a) \cup a$. For a vertex set X, we define $N[X]$ as $\bigcup_{x \in X} N[x]$. We also define $N^2[a] = N[N[a]]$ for vertex a and $N^2[X] = N[N[X]]$ for vertex set X. $N^2(a)$ is defined as $N^2[a] \setminus a$ for a vertex a and for set X, $N^2(X) = N^2[X] \setminus X$.

Lemma 1. *There exists an optimal coloring μ of G^2 and a vertex p such that $level(p) = l_{max}$ and p is the only vertex with color $\mu(p)$.*

Proof. Let v be the vertex introduced in step j. There exists a coloring μ' of T_{j-1}^2 using $\chi(G^2) - 1$ colors from the definition of G and T_{j-1}. μ' together with a new color for v gives the required coloring of G^2.   □

We fix a coloring μ and a vertex p as given in Lemma 1 such that the *level* of the vertex with smallest *level* in $N(p)$ is as large as possible. We call p as the **pivot** vertex and μ as the **pivotal coloring**. From now on, when we say the color of a vertex, we mean the color of the vertex under the coloring μ, unless stated otherwise.

Lemma 2. *All colors of μ are present in $N^2[p]$ where p is the pivot vertex.*

Let $\{u, w\}$ be the *parent* of *pivot* p. Since $l_{max} > 1$, we can assume without loss of generality that there is a vertex t such that w is a *child* of $\{u, t\}$. This

---

[3] We omit proofs of some lemmas here due to space constraint. They can be found in the full version of the paper at http://arxiv.org/abs/1603.03205.

implies that $\{u, w\}$ and $\{w, t\}$ are siblings and have *level* equal to $l_{max} - 1$. Also, *level* $(\{u, t\}) = l_{max} - 2$. Let B be the set of all *children* of $\{w, t\}$ and C be that of $\{u, w\}$.

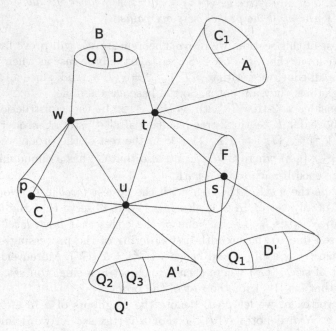

Fig. 1. The figure shows different vertex sets of 2-tree G that we use in the proof.

Lemma 3. *For any vertex* $b \in B$, $N(b) = \{w, t\}$. *And for any vertex* $c \in C$, *we have* $N(c) = \{u, w\}$.

Let F be $(N(u) \cap N(t)) \setminus w$. Let C_1 be defined as $\{v \in N(t) \mid \mu(v) \in \mu(C)\}$ and A be defined as $N(t) \setminus (B \cup F \cup C_1 \cup \{u, w\})$.

Lemma 4. $\mu(A) \subseteq \mu(N(u) \setminus (C \cup F \cup \{w, t\}))$.

By Lemma 4, for each color $c \in \mu(A)$, there is a c-colored vertex in $N(u)$. Note that there cannot be more than one c-colored vertex in $N(u)$. Let $A' \subseteq N(u)$ be such that $\mu(A') = \mu(A)$. For each $a' \in A'$, let **couple** (a') be defined as the vertex $a \in A$ with $\mu(a) = \mu(a')$. Similarly, for each $a \in A$, let **couple** (a) be defined as the vertex $a' \in A'$ with $\mu(a') = \mu(a)$. Note that since A and A' are disjoint, a vertex and its **couple** are always distinct. Let D be defined as $\{x \in B \mid \mu(x) \notin \mu(N(u))\}$ and let $Q = B \setminus D$. We also define $Q' = N(u) \setminus (A' \cup C \cup F \cup \{w, t\})$. Note that $A', A, F, Q', Q, D, C, C_1$ and $\{u, w, t\}$ are all disjoint with each other.

Lemma 5. *If* $D = \emptyset$, *then* $\chi(G^2) \leq \eta(G^2)$. *Moreover,* $\chi(G^2) = \omega(G^2)$ *and hence* G^2 *has a clique minor of size* $\chi(G^2)$ *where each branch set is a singleton set.*

Due to Lemma 5, for the rest of the proof we can assume that $D \neq \emptyset$.

Lemma 6. *For any $d \in D$, there is no other vertex in $N^2[p]$ with color $\mu(d)$.*

Lemma 7. $\mu(Q') = \mu(Q)$.

Lemma 8. *If $A = \emptyset$, then $\chi(G^2) \leq \eta(G^2)$. Moreover, $\chi(G^2) = \omega(G^2)$ and hence G^2 has a clique minor of size $\chi(G^2)$ where each branch set is a singleton set.*

Due to Lemma 8 we assume that A is not empty for the rest of the proof. Note that this also implies that A' is not empty.

Lemma 9. $l_{max} > 2$. *(Note that we assume here that A' and D are not empty.)*

Lemma 10. $level(u) = l_{max} - 2$.

Observation 2. *Due to Lemmas 10 and 9, we can assume that there is a vertex $s \in F$ such that $\{u, t\}$ is the child of $\{s, t\}$. Then, level of $\{s, t\}$ is $l_{max} - 3$. Also, $\{s, u\}$ is the sibling of $\{u, t\}$ and hence has level $l_{max} - 2$. (see Fig. 1.)*

Definition 10. *For any coloring ϕ of G^2 and any two colors r and g, we define a (ϕ, r, g)-**bicolored path** as any path in G^2 such that all the vertices in the path are colored either r or g under ϕ.*

Lemma 11. *Consider vertices $a' \in A'$ and $d \in D$. There exists a $(\mu, \mu(a'), \mu(d))$-bicolored path from a' to **couple**(a) in G^2.*

Proof. Let $\mu(a') = r$ and $\mu(d) = g$. Let $a = \mathbf{couple}(a')$. Now, consider the induced (bicolored) subgraph H of G^2 on all the vertices with colors r and g in μ. In particular, consider the connected component H' of H containing a'. Suppose H' does not contain a for the sake of contradiction. We show that then we can construct a proper coloring μ' of G^2, with fewer colors than μ which will be a contradiction. For all $v \in V(G) \setminus (V(H') \cup p)$, we set $\mu'(v) = \mu(v)$. For all $v \in V(H')$, we set $\mu'(v) = g$ if $\mu(v) = r$ and $\mu'(v) = g$ otherwise. In other words, for the vertices in H', we exchange the colors. In particular, $\mu'(a') = g$. Finally, we set $\mu'(p) = r$. Clearly, μ' has fewer colors than μ. It remains to prove that μ' is a proper coloring of G^2. It is easy to see that exchanging colors within H' does not violate the properness of the coloring. So, we only have to prove that there is no $v \in N^2(p)$ with $\mu'(v) = \mu'(p) = r$. Suppose there was such a v for the sake of contradiction. We consider two separate cases, namely, when $v \in V(H')$ and when $v \notin V(H')$. When $v \in V(H')$, $\mu(v) = g$. But then $v = d$ because by Lemma 6, d is the only vertex in $N^2[p]$ with color g under μ. But, $d \notin V(H')$ because a is the only vertex in $N^2[d]$ with color r under μ and $a \notin V(H')$. Hence, we get $v \notin V(H')$ which is a contradiction in this case. Now, let us consider the case when $v \notin V(H')$. In this case $\mu(v) = r$. Observe that in the coloring μ, a' is the only vertex in $N^2[p]$ with color r because $N^2[p] \subseteq N^2[a'] \cup N^2(a)$. But then, $v = a' \in V(H')$ which is a contradiction in this case.   □

An edge e_2 is said to be the **edge-descendant** of edge e_1 if $e_2 = e_1$ or if the parent of e_2 is an *edge-descendant* of e_1. A vertex v is said to be a **vertex-descendant** of edge e if v is the *vertex-child* of an *edge-descendant* of e.

Lemma 12. *For an edge $e = \{v_1, v_2\}$ such that $level(e) = l_{max} - 2$, if x is a vertex-descendant of e, then $N^2(x) \subseteq N[\{v_1, v_2\}]$.*

Lemma 13. $N^2(A' \cup Q') \subseteq N[\{u, t, s\}]$.

Lemma 14. *If $v \in N^2(A' \cup Q')$ and $\mu(v) \in \mu(B)$, then $v \in N(\{u, s\})$.*

Lemma 15. *If $v \in N^2(A' \cup Q')$ and $\mu(v) \in \mu(D)$, then $v \in N(s)$.*

Proof. Consider such a v. By Lemma 14, we have that $v \in N[\{u, s\}]$. But $v \notin N[u]$ by definition of D. $\square$

Let D' be defined as $\{x \in N(s) \mid \mu(x) \in \mu(D)\}$.

Lemma 16. *1. $\mu(D') = \mu(D)$ and*
2. For each $d' \in D'$ and for each $a' \in A'$, there exists a $(\mu, \mu(a'), \mu(d'))$-bicolored path from a' to $couple(a')$ in G^2 such that d' is adjacent to a' in this path.

We now extend the definition of *couple* for the set D'. For $d' \in D'$, we define $couple(d')$ as the vertex in D with color the same as d' in μ.

Corollary 1. *Each $d' \in D'$ is adjacent in G^2 to each $a' \in A'$.*

Corollary 2. *For all $d' \in D'$ and $a' \in A'$, there exists a $(\mu, \mu(d'), \mu(a'))$-bicolored path from d' to $couple(d')$ in G^2.*

Definition 11. Bridging-set. *For any $k \geq 0$, $\{q_1, q_2, \ldots, q_k\} \subseteq N(s) \setminus D'$ is called a bridging-set if for each $1 \leq i \leq k$, there exists a vertex $q_i' \in Q'$ such that $\mu(q_i') = \mu(q_i)$ and q_i' is non-adjacent in G^2 to at least one vertex in $D' \cup \{q_1, q_2 \ldots, q_{i-1}\}$. The vertex q_i' is called the **bridging-partner** of q_i denoted by $bp(q_i)$. Also, we designate one vertex in $D' \cup \{q_1, q_2 \ldots, q_{i-1}\}$ to which q_i' is non-adjacent in G^2 as the **bridging-non-neighbor** of q_i' denoted by $bn(q_i')$. If there is more than candidate, we fix one of them arbitrarily as the bridging-non-neighbor.*

Note that an empty set is a bridging-set. Also, note that for any q in the bridging-set, $bp(q) \neq q$ because otherwise $bp(q)$ is adjacent in G^2 to all vertices in $N(s)$ and hence there is no possible candidate for the bridging-non-neighbor of $bp(q)$. This contradicts Definition 11.
Let Q_1 be a bridging-set with maximum cardinality.

Definition 12. Bridging-sequence. *For each $v \in Q_1 \cup D'$, the bridging-sequence of v is defined as a sequence of distinct vertices $s_1, s_2 \ldots, s_j$ where $s_1 = v$, $s_j \in D'$ and for all $2 \leq i \leq j$, s_i is the bridging-non-neighbor of bridging-partner of s_{i-1}. (From Definition 11, it is easy to see that such a sequence should exist for all $v \in Q_1 \cup D'$. Note that for a vertex $d \in D'$, the bridging-sequence consist of only one vertex, that is d.)*

Lemma 17. *Let $q \in Q_1$, $x = bp(q)$ and $y = bn(x)$. If there exists $v \in N^2(x)$ such that $\mu(v) = \mu(y)$, then $y \in Q_1$ and $v = bp(y)$.*

Definition 13. Bridging-re-coloring. *Given any $z \in Q_1 \cup D'$, we define the bridging-re-coloring of μ with respect to z (denoted by ψ_z) by the following construction:*

1. *For all $x \in V(G)$, initialize $\psi_z(x) = \mu(x)$.*
2. *Suppose $s_1, s_2, \ldots, s_j$ is the bridging-sequence of z. For all $1 \le i < j$, set $\psi_z(bp(s_i)) = \mu(s_{i+1})$. (Observe that $\forall i \ne j$, $\mu(s_i) \ne \mu(s_j)$ since $s_i, s_j \in N(s)$ and hence each color is used for recoloring at most once in this step.)*

Lemma 18. *For all $z \in Q_1 \cup D'$, ψ_z is an optimal coloring of G^2.*

Lemma 19. *Let $a' \in A'$, $q \in Q_1 \cup D'$, $r = \mu(a')$, $g = \mu(q)$, $V_1 = \{x | \psi_q(x) \in \{r, g\}\}$ and $V_2 = \{x | \mu(x) \in \{r, g\}\}$. Then, $V_1 \subseteq V_2$. (In fact, $V_2 \setminus V_1 = bp(q)$).*

Lemma 20. *For all $q \in Q_1$ and $a' \in A'$, there exists a $(\mu, \mu(a'), \mu(q))$-bicolored path from a' to $\mathbf{couple}(a')$ in G^2 such that q is adjacent to a' in the path.*

Corollary 3. *Each $q \in Q_1$ is adjacent in G^2 to each $a' \in A'$.*

We now extend the definition of *couple* to set Q_1. For $q \in Q_1$, let *couple*(q) be defined as the vertex in Q with the same color as q in the coloring μ. Then, we have the following corollary to Lemma 20.

Corollary 4. *For each $q \in Q_1$ and $a' \in A'$, there is a $(\mu, \mu(a'), \mu(q))$-bicolored path from q to couple(q).*

Proof. This follows because *couple*(a') is adjacent in G^2 to *couple*(q). $\qquad\square$

Let $Q_2 = \{bp(q) \mid q \in Q_1\}$. Note that $Q_2 \subseteq Q'$ and $\mu(Q_2) = \mu(Q_1)$. Let $Q_3 = Q' \setminus Q_2$. Note that $\mu(Q_3 \cup Q_1) = \mu(Q') = \mu(Q)$.

Lemma 21. *For all $q' \in Q_3$, $Q_1 \cup D' \subseteq N^2[q']$.*

Let $a'_1, a'_2 \ldots a'_{n_a}$ be the vertices in A' and $z_1, z_2, \ldots, z_{n_z}$ be the vertices in $D' \cup Q_1$. If $n_a \le n_z$, then we define vertex disjoint paths $P_1, P_2, \ldots, P_{n_a}$ such that each P_i is a $(\mu, \mu(a'_i), \mu(z_i))$-bicolored path from a_i to couple(a_i). Such a path exists for each i due to Lemmas 16 and 20. If $n_z < n_a$, then we define vertex disjoint paths $P_1, P_2, \ldots, P_{n_z}$ such that each P_i is a $(\mu, \mu(a_i), \mu(z_i))$-bicolored path from z_i to couple(z_i). These paths exist due to Corollaries 2 and 4. Note that in both cases, the paths are vertex disjoint with each other because the color of the vertices in P_i and P_j are disjoint for $i \ne j$.

In the case when $n_a \le n_z$, we define $\mathcal{B}$ as the set of following branch sets: each vertex in $N[w]$ as a singleton branch set, each vertex in F as a singleton branch set, and path branch sets $V(P_i)$ for $1 \le i \le n_a$. In the case when $n_z < n_a$, we define $\mathcal{B}$ as the set of following branch sets: each vertex in $N[u] \setminus Q_2$ as a singleton branch set and the path branch sets $V(P_i)$ for $1 \le i \le n_z$. In both cases, it is easy to see that all the branch sets are connected to each other in G^2 and hence forms a clique minor of G^2. We prove this in Lemma 22. It is also easy to see that the number of branch sets is at least $\chi(G^2)$ in each of the two cases. We prove this in Lemma 23. Since, the paths P_i are vertex disjoint with each other, all branch sets in $\mathcal{B}$ are disjoint with each other. This completes the proof of Theorem 2.

Lemma 22. *Each pair of branch sets in $\mathcal{B}$ are adjacent to each other in G^2.*

Lemma 23. $|\mathcal{B}| \geq \chi(G^2)$

References

[AH+77]   Appel, K., Haken, W., et al.: Every planar map is four colorable. part i: discharging. Ill. J. Math. **21**(3), 429–490 (1977)

[AHK+77]   Appel, K., Haken, W., Koch, J., et al.: Every planar map is four colorable. part ii: reducibility. Ill. J. Math. **21**(3), 491–567 (1977)

[BC09]   Belkale, N., Sunil Chandran, L.: Hadwiger's conjecture for proper circular arc graphs. Eur. J. Comb. **30**(4), 946–956 (2009)

[BCE80]   Bollobs, B., Catlin, P.A., Erdős, P.: Hadwiger's conjecture is true for almost every graph. Eur. J. Comb. **1**(3), 195–199 (1980)

[Bla05]   Blasiak, J.: A special case of Hadwiger's conjecture. arXiv preprint arXiv:math/0501073 (2005)

[CF08]   Chudnovsky, M., Fradkin, A.O.: Hadwiger's conjecture for quasi-line graphs. J. Graph Theory **59**(1), 17–33 (2008)

[CKR08]   Sunil Chandran, L., Kostochka, A., Krishnam Raju, J.: Hadwiger number and the cartesian product of graphs. Graphs Comb. **24**(4), 291–301 (2008)

[Had43]   Hadwiger, H.: Über eine klassifikation der streckenkomplexe. Vierteljschr. Naturforsch. Ges. Zürich **88**, 133–142 (1943)

[KT05]   Kawarabayashi, K., Toft, B.: Any 7-chromatic graphs has k 7 or k 4, 4 as a minor. Combinatorica **25**(3), 327–353 (2005)

[KT14]   Kammer, F., Tholey, T.: Approximation algorithms for intersection graphs. Algorithmica **68**(2), 312–336 (2014)

[LL07]   Li, D., Liu, M.: Hadwigers conjecture for powers of cycles and their complements. Eur. J. Combinatorics **28**(4), 1152–1155 (2007)

[RS04]   Reed, B., Seymour, P.: Hadwiger's conjecture for line graphs. Eur. J. Comb. **25**(6), 873–876 (2004)

[RST93]   Robertson, N., Seymour, P., Thomas, R.: Hadwiger's conjecture fork 6-free graphs. Combinatorica **13**(3), 279–361 (1993)

[Wag37]   Wagner, K.: Über eine Eigenschaft der ebenen Komplexe. Math. Ann. **114**(1), 570–590 (1937)

[Weg77]   Wegner, G.: Graphs with given diameter and a coloring problem (1977)

Computational Geometry

Minimum Width Color Spanning Annulus

Ankush Acharyya[✉], Subhas C. Nandy, and Sasanka Roy

Indian Statistical Institute, Kolkata 700108, India
{ankush_r,nandysc,roysasanka}@isical.ac.in

Abstract. Given a set P of n points in $\mathbb{R}^2$, each assigned with one of the k distinct colors, we study the problem of finding the minimum width color spanning annulus of different shapes. Specifically, we consider the circular annulus ($CSCA$) and axis-parallel square annulus ($CSSA$). The time and space complexities of the proposed algorithms for both the problems are $O(n^4 \log n)$ and $O(n)$, respectively.

Keywords: Circular annulus · Color-spanning · Minimum width · Arrangement

1 Introduction

Given a set $P = \{p_1, p_2, \ldots, p_n\}$ of points in $\mathbb{R}^2$, each colored with one of the colors $\{1, 2, \ldots, k\}$, $k \leq n$, a region is said to be *color-spanning* if it contains at least one point of each color in that region. An *annulus* is a region bounded by two closed concentric geometric curves of same type, named as *inner boundary* and *outer boundary* respectively [3]. A color-spanning annulus is an annulus that contains at least one point of each color. In this paper we study the problem of identifying the minimum width color spanning circular annulus ($CSCA$) and axis-parallel square annulus ($CSSA$).

The motivation for studying color-spanning objects comes from the facility location problem. Here the input consists of different types of facilities, each having multiple copies, spread over a region. The output is a region of desired shape with at least one copy of each facility. The first natural variation of the problem is the minimum radius color-spanning circle. The best known result is by Abellanas et al. [1], which points out that the smallest color spanning circle can be solved in $O(kn \log n)$ time by using the technique of computing the upper envelope of Voronoi surfaces [7,8]. Abellanas et al. [2] also showed that the narrowest color-spanning strip and smallest axis-parallel color-spanning rectangle can be found in $O(n^2 \alpha(k) \log k)$ and $O(n(n-k) \log^2 n)$ time respectively. Later Das et al. [5] improved the time complexity of narrowest color spanning corridor problem to $O(n^2 \log n)$, and smallest color-spanning axis-parallel rectangle problem to $O(n(n-k) \log k)$. They provided a solution for the arbitrary oriented color-spanning rectangle problem in $O(n^3 \log k)$ time using $O(n)$ space. The problem of computing the minimum width annulus is also studied in the literature. Given a set of n monochromatic points ($k = 1$), the minimum width

© Springer International Publishing Switzerland 2016
T.N. Dinh and M.T. Thai (Eds.): COCOON 2016, LNCS 9797, pp. 431–442, 2016.
DOI: 10.1007/978-3-319-42634-1_35

circular annulus containing all the points can be computed in $O(n^2)$ time and $O(n)$ space [3].

In this paper, we first show that if the annulus-center is given, then the $CSCA$ problem can be solved in $\Theta(n \log n)$ time. Next, we solve the $CSCA$ problem, with the annulus-center constrained to lie on a given line, in $O(n^2 \log n)$ time. Finally we show that the unconstrained version of the $CSCA$ problem can be solved in $O(n^4 \log n)$ time. Next, we show that similar logic works for formulating an algorithm for the $CSSA$ problem in $O(n^4 \log n)$ time. Both of the algorithms use $O(n)$ space. Similar methods work for (i) computing the minimum width color-spanning circular annulus in L_1 norm [6], and (ii) computing the minimum width circular and square annulii containing at least k monochromatic points.

Note that the $CSCA$ problem and $CSSA$ problem are both circular annulii problem in L_2 and L_∞ norm. If we change the norm of measuring the distance then the geometric characterizations for the annulus changes drastically. We could solve the problem in L_1, L_2 and L_∞ norm. It remains an interesting open question to solve this problem in other norms.

2 Preliminaries

The *inner boundary* and *outer boundary* of a minimum width color spanning annulus $\mathcal{A}$ are defined by C_{in} and C_{out} respectively. The common center of C_{in} and C_{out} is referred to as the *annulus-center*. The width of an annulus is the difference of radii between two circles C_{in} and C_{out} in the corresponding norm. Interior of an annulus $\mathcal{A}$, defined by $INT(\mathcal{A})$, is the region inside $\mathcal{A}$ excluding C_{in} and C_{out}. The points lying on C_{in} and C_{out} of an annulus $\mathcal{A}$ are said to define the annulus $\mathcal{A}$.

Observation 1. *(a) The points defining a mimimal width annulus $\mathcal{A}$ are of distinct colors, and (b) $INT(\mathcal{A})$ does not contain any point of color that define the annulus $\mathcal{A}$.*

Throughout the paper, we use $d(p_i, p_j)$ to denote the Euclidean distance between the pair of points $p_i, p_j \in P$ in our specified norm. The closest distance of a line segment ℓ from a point p_i will be denoted by $d(\ell, p_i)$.

2.1 Tight Bounds for a Constrained Version

Theorem 1. *The time complexity of computing the minimum width color spanning circular annulus around a given annulus-center π, in any norm of computing the distances, is $\Theta(n \log n)$.*

Proof. We first propose an $O(n \log n)$ time algorithm; next we show that the lower bound of the problem is $\Omega(n \log n)$. Let o be the origin of the coordinate system, and consider a half-line ℓ from o along the positive part of the x-axis. For each point $p \in P$, we put a point p' on ℓ such that $d(\pi, p) = d(o, p')$. Each element is attached with its corresponding color. Next, we compute the smallest color-spanning interval[1] on the x-axis in $O(n \log n)$ time [4]. Let $I = [p'_i, p'_j]$ be

---

[1] A color-spanning interval on a line is an interval containing points of each color.

the smallest color-spanning interval. The minimum width $CSCA$ with π as the annulus-center has $p_i \in C_{in}$ and $p_j \in C_{out}$.

We prove the lower bound of this problem using a reduction from the well-known *set-disjointness problem*, where two sets $X = \{x_1, x_2, \ldots, x_n\}$ and $Y = \{y_1, y_2, \ldots, y_n\}$ are given, and the objective is to test whether $X \cap Y = \emptyset$. The lower bound of the time complexity for the *set-disjointness problem* is $\Omega(n \log n)$ [9]. We can reduce this problem to the color spanning interval problem with two colors in $O(n)$ time. The members of X and Y are put on a real line $\mathbb{R}$, and assigned colors red to all the elements in X and color blue to all the elements in Y. Now, if the minimum color spanning interval is found to be of length zero, then $X \cap Y \neq \emptyset$. Thus, the time complexity of the color-spanning interval problem is $\Omega(n \log n)$. To show that the smallest color-spanning annulus problem in $\mathbb{R}^2$ is in $\Omega(n \log n)$, we put the annulus-center $c \in \mathbb{R}$ which is a real number strictly smaller than the minimum among the values of all the elements in $X \cup Y$. Thus, a smallest color-spanning annulus with annulus-centered at c corresponds to the minimum length color-spanning interval on $\mathbb{R}$ with the points $X \cup Y$.   □

3   Circular Annulus

For the given set $P = \{p_1, p_2 \ldots p_n\}$ of points, each one is assigned with a color in the set $\{1, 2, \ldots, k\}$, our objective is to determine the minimum width color-spanning circular annulus ($CSCA$).

3.1 $CSCA$ with Annulus-Center on a Line

We now consider the problem of computing of minimum width $CSCA$ with annulus-center on a query line L. Let $\mathcal{A}^*$ be a minimum width $CSCA$ with annulus-center constrained on a given line L. Without loss of generality, we may assume that the line L is the x-axis. Consider a point $p_i = (\alpha, \beta) \in P$. The squared distance from point p_i to each point $(x, 0) \in L$ is a parabola $y = f_i = (x - \alpha)^2 + \beta^2$ with its vertex at the point (α, β^2) and x-axis as the directrix.

Lemma 1. *For any pair of points p_i and p_j, the curves f_i and f_j intersect exactly at one point, say (a, b), where $(a, 0)$ is the point of intersection of the perpendicular bisector of the line segment $[p_i, p_j]$ with the x-axis, and b is the distance of both p_i and p_j from the point $(a, 0)$.*

Definition 1. *The mirror-image of a point p with respect to a line L is a point p' on the other side of p with respect to the line L such that $d(p', L) = d(p, L)$. The point p' will be referred to as $image_L(p)$.*

Observation 2. *Consider an annulus U with its annulus-center lying on a line L. For any point p, either both p and $image_L(p)$ lie inside (including the boundary) of U or both of them lie outside U.*

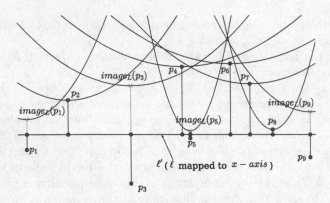

Fig. 1. Transformation of distance information of the points from the line L

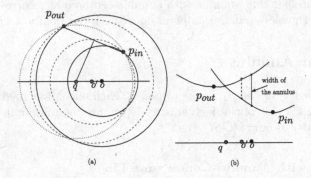

Fig. 2. Demonstration of Lemma 3

We consider the line L as the x-axis, and transform the points in P as follows. Consider $image_L(p)$ for all the points with negative y-coordinate; the points of P with positive y-coordinates are considered as it is. Let the transformed point set be $P' = \{p'_1, p'_2, \ldots, p'_n\}$, where all the points in P' have positive y-coordinates. Consider the arrangement $\mathcal{Z}$ of curves $\{f_i | p'_i \in P'\}$ (see Fig. 1). Any vertical line segment $\mathcal{I} = [(a, \phi), (a, \psi)]$ in the arrangement $\mathcal{Z}$ corresponds to a circular annulus with annulus-center $(a, 0) \in L$ and width $(\psi - \phi) > 0^2$. If $\mathcal{I}$ represents a $CSCA$ then $\mathcal{I}$ must intersect at least one curve of each color[3]. Such a vertical line segment $\mathcal{I}$ will be referred to as a *color-spanning stick*.

Lemma 2. *The width of an annulus-centered on a line L with its C_{in} and C_{out} passing through $p_{in} = (\alpha_{in}, \beta_{in})$ and $p_{out} = (\alpha_{out}, \beta_{out})$ respectively is a convex function of x on L.*

---

[2] We are considering color spanning annulus of non-zero width, so we only need to consider the non-zero length sticks.

[3] The reason is that, for a $CSCA$ there exists at least one point of each color j at distance d_j from $(a, 0)$ satisfying $\phi \leq d_j \leq \psi$ for all $j \in [1, 2, \ldots, k]$.

Proof. For the sake of simplicity, let L be the x-axis. The width of the annulus with annulus-center $x \in L$ is $W(x) = |((x - \alpha_{out})^2 + \beta_{out}^2) - ((x - \alpha_{in})^2 + \beta_{in}^2)|$ $= |2x(\alpha_{in} - \alpha_{out}) + (\beta_{out}^2 - \beta_{in}^2 + \alpha_{out}^2 - \alpha_{in}^2)|$. Since any function of the form $|cx + d|$ is convex for constants c and d, $W(x)$ is a convex function. $\qquad\square$

The width $W(x)$ centered at a point $x \in L$ is a convex function. Thus it attains minima exactly at one point on L which is the point of intersection of the perpendicular bisector of $[p_{in}, p_{out}]$ with the line L. For a demonstration, let us consider Fig. 2(a), where the change in the annulus for different annulus-centers, namely o, o' and q, on the line L are shown using solid, dashed and dotted circles. The Fig. 2(b) shows the plots of the distances of p_{out} and p_{in} from different points on the line L. These are parabolas, namely Π_{out} and Π_{in} respectively. The width of the annulus centered at a point is the length of the line segment between Π_{out} and Π_{in} on the line perpendicular to L at the corresponding point. These are shown using solid, dashed and dotted vertical line segments in Fig. 2(b).

Lemma 3. *(a) The necessary and sufficient condition for computing the $\mathcal{A}^*$ is to find minimum among all the minimal width CSCA's constrained on line L defined by three distinct colors, and (b) $INT(\mathcal{A}^*)$ does not contain any point of those three distinct colors that define $\mathcal{A}^*$.*

Proof. (a) Note that, both C_{in} and C_{out} of $\mathcal{A}^*$ must be defined by at least one point; otherwise, its width can be reduced keeping the annulus-center same.

For a contradiction, let $\mathcal{A}^*$ be defined by exactly two points $p_{out} \in C_{out}$ and $p_{in} \in C_{in}$ (see Fig. 2). Let $o \in L$ be the annulus-center, and there exists a small positive real number ϵ such that if the annulus-center is moved in either direction from o on L by an amount ϵ, the circles C'_{out} and C'_{in} of the new annulus $\mathcal{R}'$ still pass through only p_{out} and p_{in} respectively and the points that were inside previous annulus $\mathcal{R}$ still remain in $\mathcal{R}'$. Let the perpendicular bisector of p_{in} and p_{out} intersects L at a point q. As stated in Lemma 2, the width of the annulus $\mathcal{A}^*$ decreases as the annulus-center o moves toward the point q along L.

To prove that three points are enough to define a $CSCA$, let us consider that $\mathcal{A}^*$ is defined by at least four points. So, one of the circles must be defined by at least two points, say p_i and p_j. The perpendicular bisector of these two points defines the unique annulus-center on the line L. Again, since the annulus-center and one of the boundaries, say C_{in}, is fixed, only one point is enough to define the other boundary C_{out} satisfying the color spanning property of $\mathcal{A}^*$. Thus, we have a contradiction. It needs to mention that, more than three points may appear on the boundary of $\mathcal{A}^*$. Here, all possible three points will define the same $CSCA$, i.e., $\mathcal{A}^*$.

(b) The distinctness of colors on the boundary follows from Observation 1(a). The fact that $INT(\mathcal{A}^*)$ does not contain any point from those three distinct colors follows from Observation 1(b). $\qquad\square$

Lemma 3 leads to the following facts:

- Each pair of curves in $\mathcal{Z}$ intersect only once. Thus, the number of vertices in $\mathcal{Z}$ is $O(n^2)$. Each intersection point (a, b) (of f_i and f_j) in $\mathcal{Z}$ represents a circle with center at $(a, 0)$ and passing through the points p_i and p_j.

- One of the end-points of the color-spanning stick (vertical line segment) defining the minimum width $CSCA$ will coincide with a vertex of $\mathcal{Z}$ and its other end-point lies on an edge[4] (or a vertex in the degenerate case) of $\mathcal{Z}$.

Thus, at every vertex v of $\mathcal{Z}$, we need to consider two color-spanning sticks of minimum length above and below v with their one end anchored at the vertex v. We execute a line sweep to process the vertices of $\mathcal{Z}$ for computing a color-spanning stick of minimum length as in [5]. Thus, we have the following result.

Lemma 4. *The minimum width $CSCA$ with annulus-center on a given line can be computed in $O(n^2 \log n)$ time.*

3.2 Unconstrained Circular Annulus

We now consider the unconstrained version of $CSCA$, where annulus-center can be anywhere on the plane. Let $\mathcal{A}^+$ be a unconstrained minimum width $CSCA$.

Lemma 5. *(a) The necessary and sufficient condition to find $\mathcal{A}^+$ is to find minimum among all minimal width $CSCA$'s defined by four distinct colors, and (b) $INT(\mathcal{A}^+)$ does not contain any point of those four colors that defines $\mathcal{A}^+$.*

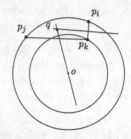

Fig. 3. Circles with three points cannot define the optimum annulus

Proof. (a) We first prove that $\mathcal{A}^+$ can not be defined by three points of distinct colors. Similar argument holds to show that $\mathcal{A}^+$ can not be defined with less than three points.

The width of $\mathcal{A}^+$ with three points on C_{out} (resp. C_{in}) and no point on C_{in} (resp. C_{out}) can be reduced by increasing the radius of C_{in} (resp. decreasing the radius of C_{out}) maintaining the annulus-center unchanged.

For a contradiction, let us assume that $\mathcal{A}^+$ be defined by two points p_i and p_j on C_{out}, and one point p_k on C_{in} (see Fig. 3). The annulus-center o of $\mathcal{A}^+$ lies on the perpendicular bisector L of $[p_i, p_j]$. Let q be the point of intersection of the perpendicular bisector of $[p_i, p_k]$ with L. As in Lemma 3(a), the width of the annulus reduces as we move from the point o towards q along L. Using the similar arguments of Lemma 3 we conclude that $\mathcal{A}^+$ is defined by four points.

The sufficiency of four points for defining an unconstrained $CSCA$ can be proved exactly as in Lemma 3.

Here we need to admit that more than four points may appear on the boundary of $\mathcal{A}^+$. Any four of these points will define the same $\mathcal{A}^+$. The distinctness of the colors on the boundary of $\mathcal{A}^+$ can be proved as in Observation 1(a). Part (b) also can be proved as in Observation 1(b). ☐

---

[4] here, the edges of $\mathcal{Z}$ are curve segments.

When C_{in} and C_{out} have four points, then we have the following configurations.

Type A: Both C_{in} and C_{out} have two points.
Type B: C_{in} has three points and C_{out} has one point.
Type C: C_{out} has three points and C_{in} has one point.

For computing $\mathcal{A}^+$ of *Type A*, we consider all possible pairs of bi-colored points $p_i, p_j \in P$. Let L_{ij} be the perpendicular bisector of (p_i, p_j). We execute the algorithm of Sect. 3.1 to compute $\mathcal{A}^*$ with one of its C_{in} or C_{out} passing through (p_i, p_j).

Observe that $image_{L_{ij}}(p_i) = p_j$ or vice versa. Thus, the parabolas corresponding to p_i and p_j with line L_{ij} as the x-axis are the same. Let us name this curve as h_{ij}. The one end of all the vertical color-spanning sticks representing the annulii with one circle passing through (p_i, p_j) must touch h_{ij}. In order to be a $CSCA$, the other circle of such an annulus must pass through two points (by Observation 5). Thus, the vertical line segment corresponding to such a $CSCA$ must also incident to the point of intersection of two parabolas. Thus, in order to get these $CSCA$s', we scan all the vertices of $\mathcal{Z}$. At each vertex, we draw a vertical line segment up to h_{ij}. If this vertical line segment intersects the parabola of all the colors, then it corresponds to a $CSCA$, and we note down its length. Finally, the one with minimum length corresponds to the $\mathcal{A}^+$ of *Type A*.

During this process, we also recognize the $\mathcal{A}^+$ of *Type B* and *Type C*. These correspond to the color-spanning sticks at all the vertices on the curve h_{ij}. From each of these vertices, we generate two $CSCA$s (above and below) as mentioned in Sect. 3.1 and return the optimum one. Thus, we have the following result.

Theorem 2. *Given a set of n points, each point is assigned with one of the k possible colors, the minimum width $CSCA$ can be computed in $O(n^4 \log n)$ time using $O(n)$ space.*

4 Axis Parallel Square Annulus

For the given set $P = \{p_1, p_2 \ldots p_n\}$ of n points, each one is assigned with a color in the set $\{1, 2, \ldots, k\}$, our objective is to determine the smallest width color-spanning axis-parallel square annulus ($CSSA$). For the sake of simplicity, we assume that the point set P are in general position, that is no two points have same x- or y-coordinate. A *corner point* is a common point of two adjacent sides of C_{in} (resp. C_{out}). Similarly, a *non-corner point* is any point $p \in C_{in}$ (resp. $p \in C_{out}$) other than a corner point. The points of P that appear on C_{in} and C_{out} are said to *define* the corresponding $CSSA$. Before computing the unconstrained version of the $CSSA$, here also we shall consider the cases when the annulus-center is *(i) on a given horizontal or vertical line* and *(ii) on a given line making $45°$ or $135°$ angle with the x-axis*.

4.1 $CSSA$ Centered on a Horizontal or a Vertical Line

We now explain the algorithm when the annulus-center is on a given a horizontal line L and the objective is to compute the minimum width $CSSA$ with annulus-center lying on L. A similar method works when L is vertical. We use S^* to denote a minimum width $CSSA$, with its annulus-center constrained on the horizontal or vertical line L. As in Sect. 3.1, here also we consider $image_L(p)$ for each point $p \in P$ that lie below the line L. Thus, the transformed points P' are all above the line L. For each point $p_i' \in P'$, we plot the distance function f_i for every point on the line L in the L_∞ norm (see Fig. 4(a)). It consists of three parts, namely $left$, mid and $right$, where $left$ (resp. $right$) is a half-line making angle 135° (resp. 45°) with L and mid is a horizontal line segment connecting the end-points of $left$ and $right$. Both the end-points of the horizontal edge of a curve is referred as corner of the curve. Again, arguing as in Lemma 1, we can show that f_i and f_j intersect at a single point o^5. The x-coordinate of the point o is same as the point where the Voronoi partitioning line of p_i and p_j in L_∞ norm intersects the line L (see Fig. 4(b)).

The arrangement $\mathcal{Z}^+$ of f_i's (see Fig. 6) consists of $O(n^2)$ vertices. A $CSSA$ with annulus-center at a point α on the line L corresponds to a stick I which is perpendicular on L at the point α, and intersects at least one curve of each color in $\{f_i, i = 1, 2, \ldots, n\}$. The length of the stick defines the width of the annulus. As in Lemma 2, here also the width of a $CSSA$ centered on L and defined by a set of points is a convex function of the position of its center x on L.

Lemma 6. *(a) The necessary and sufficient condition to find S^* is to find minimum among all the minimal width $CSSA$'s defined by three distinct colored points, and (b) $INT(S^*)$ does not contain any point of those three colors that define S^*.*

Proof: In S^*, both C_{in} and C_{out} must be defined by at least one point; otherwise, its width can be reduced keeping the annulus-center unchanged. For a contradiction, let S^* be defined by exactly two points; $p_{out} \in C_{out}$ and $p_{in} \in C_{in}$ (see Fig. 5). Let $o \in L$ be the annulus-center. Note that, there exists a small positive real number ϵ such that if the annulus-center is moved ϵ distance in either direction along L, the new annulus R' is still color spanning, C_{out} and C_{in} pass through only p_{out} and p_{in} respectively. Let the voronoi partitioning line of p_{in} and p_{out} intersects L at a point q. Now, if we move the annulus-center o towards q the width of the annulus remains same up to the corner point $c \in f_{in}$ (see Fig. 5 (b)). If we further continue to move o towards q the width of the annulus decreases and finally becomes zero at point q. Thus we have a contradiction. This supports the fact that, S^* must be defined by at least three points.

To prove the sufficiency of three points, let us assume four points on the boundary of S^*. Two cases may arise. *(i) Both the boundaries of C_{in} and C_{out} have two points each.* and *(ii) One of the boundaries of C_{in} and C_{out} has three*

---

[5] The curves f_i and f_j may intersect in an interval if the corresponding points p_i' and p_j' are in same horizontal line and $d(p_i', p_j')$ is less than their distance from L.

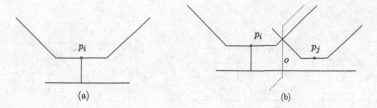

Fig. 4. (a) The function f_i, and (b) point of intersection of two curves f_i and f_j

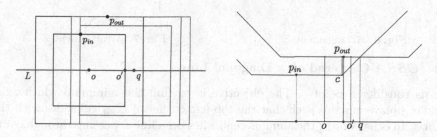

Fig. 5. Demonstration of Lemma 6

points and the other boundary has one point. In *Case(i)* considering two points of C_{in} that defines the diagonal or vertical line that contains the center intersects the line L at a point which is the annulus-center. As the center is defined, C_{out} can be defined by one point only. In *Case (ii)* let C_{in} has three points. As argued in earlier case, two points are enough to define C_{in} uniquely and the third point of C_{in} is redundant to define S^*.

The distinctness of colors on the boundary of S^* and of $INT(S^*)$ follows from Observation 1(a) and (b) respectively. □

Degenerate case arises when a point p appears at a corner of C_{in} or C_{out}. We draw a diagonal line through p which intersects L at a point defining the annulus-center. One more point is enough to define the other square. Hence in this case two points are enough to define S^*. To stand by Lemma 6, we consider a corner point as a point on its two adjacent sides. If p_i and p_j defines such a $CSSA$ with p_i at a corner point, then in the arrangement of $\mathcal{Z}^+$ one of the end-points of the vertical stick I defining that $CSSA$ lies on a corner point of a curve f_i.

As in Sect. 3.1, we compute the minimum width $CSSA$ centered on a horizontal line by sweeping a vertical line to process all the event points in the arrangement of $\mathcal{Z}^+$, where the event points are the vertices (intersection points of the curves) of the arrangement and the corner points of the curves. Thus, we have the following result:

Lemma 7. *The minimum width $CSSA$ centered on a given horizontal or vertical line can be computed in $O(n^2 \log n)$ time.*

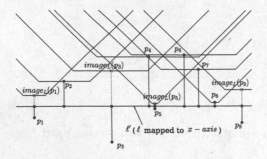

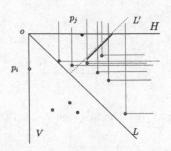

Fig. 6. Arrangement $\mathcal{Z}^+$ **Fig. 7.** Arrangement $\mathcal{Z}^{++}$

4.2 $CSSA$ Centered on a Diagonal Line

Let us consider a point o. The objective is to find the minimum width axis-parallel square annulus such that the top-left corner of C_{out} coincides with the point o. In other words, the annulus-center lies on a line L passing through o and making an angle 135° with the x-axis. The other cases of defining the diagonals which contain the annulus-center can be handled symmetrically. We use $\mathcal{S}^{**}$ to denote a minimum width $CSSA$, with its annulus-center constrained on a diagonal line L.

Draw a horizontal line H and a vertical line V through the point o. The points in the bottom-right quadrant $\mathcal{R}$ formed by the lines V and H play the role for describing this annulus. Let $\hat{\mathcal{R}}$ be the region above the line L in $\mathcal{R}$. We consider $image_L(p)$ of all points $p \in \mathcal{R} - \hat{\mathcal{R}}$. The points originally lying in the region $\hat{\mathcal{R}}$ remains as it is. At each point $p_i \in \hat{\mathcal{R}}$, define a curve f_i by drawing a vertical half-line upwards and a horizontal half-line towards right (see Fig. 7). Let $\mathcal{Z}^{++}$ be the arrangement of these curves. Each curve remains active in the quadrant $\hat{\mathcal{R}}$. Also each pair of curve intersects at most once. Thus the number of vertices in the arrangement $\mathcal{Z}^{++}$ is $O(n^2)$ in the worst case.

A $\mathcal{S}^{**}$ with top-left corner at o corresponds to a line-segment (stick) orthogonal to L and starting from H towards bottom-left that intersects at least one curve of each color in the arrangement $\mathcal{Z}^{++}$ (see the thick line segment in Fig. 7). We sweep a half-line L' orthogonal to L with its one end anchored on the line L.

The sweep in $\mathcal{Z}^{++}$ starts from the point o. The event points of this sweep are (i) points of intersection of the members of $\mathcal{Z}^{++}$ with H, (ii) the corner points in each curve, and (iii) points of intersection of each pair of curves. Initially, the stick is of length zero and is not color-spanning. As the sweep progresses, new curves appear on the sweep line by processing event points of type (i). At some point of time, the stick becomes color-spanning. During the further sweep, the event points are handled as in [5]. The sweep line status is maintained as a heap of size k whose entries correspond to the k colors, Each entry of the heap is attached with a linked list storing the presence of all curves of that color which are intersected by the stick. Processing of each type (i) event for inserting it in the sweep line status needs $O(\log n)$ time. As earlier, the processing of each type (ii) and type (iii) events also need $O(\log n)$ time for updating the event queue and sweep line status. While processing an event-point if the color of the curve appeared at the end of the stick is same as that of the point defining the lines H or V or the corner point o, then that curve is deleted from the stick. Thus, at each instance the length of the stick is also determined. Since the number of event points is $O(n^2)$ in the worst case, the Lemma 8 follows.

Lemma 8. *The minimum width CSSA centered on a given diagonal line can be computed in $O(n^2 \log n)$ time.*

4.3   Unconstrained Axis Parallel Square Annulus

Now, we are going to consider the unconstrained version of the $CSSA$ problem where the annulus-center can be anywhere in the given plane. We use $\mathcal{S}^+$ to denote a unconstrained minimum width $CSSA$. To find the unconstrained $\mathcal{S}^+$, we divide our analysis in the following two cases depending on the positions of the points in P that define $\mathcal{S}^+$:

Non-Corner $CSSA$**:** The points of P that define the $\mathcal{S}^+$ appear on the edge of the boundary of C_{in} and C_{out}.

Corner $CSSA$**:** At least one of the point of P that define the $\mathcal{S}^+$ is at a corner point of the boundary of $\mathcal{S}^+$.

We now show that computation of unconstrained $\mathcal{S}^+$ can be formulated using the two subproblems described in Sects. 4.1 and 4.2.

Lemma 9. *Either C_{in} or C_{out} of the minimum width non-corner $\mathcal{S}^+$ has two different color points on two different boundaries.*

Proof. As in Lemma 6, here also we can show that C_{in} or C_{out} has more than one point. We have assumed that no two points in P have same x- or y- coordinate. Thus, the points will appear on different boundaries of C_{in} or C_{out}.   □

As a consequence of Lemma 9, we consider each pair of points as non-corner points of some $CSSA$. For each pair we define the four types of lines which may contain the optimum annulus-center, these are the horizontal line, the vertical line, the diagonal lines making 45° and 135° angle with the x-axis. For each these lines the optimum annulus can be computed in $O(n^2 \log n)$ time (see Sect. 4.1 or 4.2). Considering $O(n^2)$ pairs of points, the non-corner $CSSA$ can be found in $O(n^4 \log n)$ time.

Corner $\mathcal{S}^+$ has a corner point on its boundary. We consider each point in P as a corner point and define two diagonal lines making 45° and 135° respectively with the x-axis. For every possible diagonal lines, we can find the $\mathcal{S}^+$ centered on that line in $O(n^2 \log n)$ time (see Sect. 4.2). Thus, a corner $\mathcal{S}^+$ can be solved in $O(n^3 \log n)$ time.

Combining the solutions for the *non-corner* and *corner* cases we have the following result:

Theorem 3. *Given a set of n points, each point is assigned with one of the k possible colors, the minimum width CSSA can be computed in $O(n^4 \log n)$ time using $O(n)$ space.*

References

1. Abellanas, M., Hurtado, F., Icking, C., Klein, R., Langetepe, E., Ma, L., Palop, B., Sacristán, V.: Smallest color-spanning objects. In: Meyer auf der Heide, F. (ed.) ESA 2001. LNCS, vol. 2161, pp. 278–289. Springer, Heidelberg (2001)
2. Abellanas, M., Hurtado, F., Icking, C., Klein, R., Langetepe, E., Ma, L., Palop, B., Sacristán, V.: The farthest color Voronoi diagram and related problems. In: 17th EuroCG, pp. 113–116 (2001)
3. de Berg, M., Cheong, O., van Kreveld, M., Overmars, M.: Computational Geometry: Algorithms and Applications, 3rd edn. Springer, Heidelberg (2008)
4. Chen, D.Z., Misiolek, E.: Algorithms for interval structures with applications. Theor. Comput. Sci. 508, 41–53 (2013)
5. Das, S., Goswami, P.P., Nandy, S.C.: Smallest color-spanning object revisited. Int. J. Comput. Geom. Appl. 19, 457–478 (2009)
6. Gluchshenko, O.N., Hamacher, H.W., Tamir, A.: An optimal O(nlogn) algorithm for finding an enclosing planar rectilinear annulus of minimum width. Oper. Res. Lett. 37, 168–170 (2009)
7. Huttenlocher, D.P., Kedem, K., Sharir, M.: The upper envelope of Voronoi surfaces and its applications. Discret. Comput. Geom. 9, 267–291 (1993)
8. Sharir, M., Agarwal, P.K.: Davenport-Schinzel Sequences and Their Geometric Applications. Cambridge University Press, Cambridge (1995)
9. Ben-Or, M.: Lower bounds for algebraic computation trees. In: Proceedings of the Fifteenth Annual ACM Symposium on Theory of Computing, pp. 80–86 (1983)

Computing a Minimum-Width Square or Rectangular Annulus with Outliers

[Extended Abstract]

Sang Won Bae[✉]

Department of Computer Science, Kyonggi University, Suwon, South Korea
swbae@kgu.ac.kr

Abstract. A square or rectangular annulus is the closed region between a square or rectangle and its offset. In this paper, we address the problem of computing a minimum-width square or rectangular annulus that contains at least $n - k$ points out of n given points in the plane. The k excluded points are considered as outliers of the n input points. We present several first algorithms to the problem.

1 Introduction

Covering a given point set by a certain geometric shape, such as a circle, a square and a rectangle, is a fundamental problem in computational geometry. Problems of this kind often appear in the form of an optimization problem with various criteria and constraints. Among these covering problems, finding a minimum shape covering all but a few input points is of another great interest in view of outlier removal. In such a problem, more precisely, given n input points and an integer k, we are asked to find a smallest shape enclosing at least $n - k$ out of the n input points. From the viewpoint of optimization, excluding the k points reduces the objective value the most; in other words, including them would cause a relatively high increase in cost. In this sense, such excluded points are considered to be *outliers* of the given point set.

There has been quite a lot of attention to the problem of covering n given points, but excluding k outliers, in particular, with an axis-parallel square or a rectangle. Aggarwal et al. [2] achieved a running time of $O((n - k)^2 n \log n)$ both for the square and rectangle cases. For the rectangle case, Segal and Kedem [15] presented an $O(n + k^2(n - k))$-time algorithm and Atanassov et al. [4] and Ahn et al. [3] came up with an improvement to $O(n + k^3)$ time. For the square case, a randomized algorithm that runs in $O(n \log n)$ time was presented by Chan [7], and later Ahn et al. [3] improved it to $O(n + k \log k)$ time. In particular, Ahn et al. [3] extended their idea to the problem of covering $n - k$ points by p disjoint squares or rectangles, resulting in efficient algorithms for small $p \leq 3$.

This research was supported by Basic Science Research Program through the National Research Foundation of Koresa (NRF) funded by the Ministry of Education (2015R1D1A1A01057220).

T.N. Dinh and M.T. Thai (Eds.): COCOON 2016, LNCS 9797, pp. 443–454, 2016.
DOI: 10.1007/978-3-319-42634-1_36

Atanassov et al. [4] also considered other variants of the problem for the purpose of outlier removal, including the problem of finding a minimum convex set containing $n - k$ points. More general frameworks to handle outliers as violations of constraints for LP-type optimization problems have also been introduced by Matoušek [13]. Note that computing a minimum enclosing circle, square, or rectangle falls into the class of LP-type problems.

Besides squares and rectangles, *square annuli* or *rectangular annuli* are also of much interest in this stem of research, in particular, when the input points are assumed to be sampled from the boundary of a square or rectangle with noise or error. A square/rectangular annulus is defined to be a closed region between a square/rectangle and its offset, and the width of such an annulus means the distance between the two squares/rectangles defining it.

It was relatively recent that the problem of computing a *minimum-width square/rectangular annulus* containing input points has been introduced and studied in a theoretical point of view. Abellanas et al. [1] presented an $O(n)$-time algorithm for the rectangular annulus problem and considered several variations of the problem. Gluchshenko et al. [11] gave an $O(n \log n)$-time algorithm for the square annulus, and proved that this is optimal. Also, it is known that a minimum-width annulus over all orientations can be found in $O(n^3 \log n)$ time for the square annulus [5] and in $O(n^2 \log n)$ time for the rectangular annulus [14].

In this paper, we study the minimum-width square/rectangular annulus problems with k outliers. More precisely, given n points in the plane and an integer $k \geq 0$, our problems, the k-SQUAREANNULUS problem and the k-RECTANNULUS problem, asks to cover at least $n - k$ points by a minimum-width square and rectangular annulus, respectively. We present several algorithms for the k-SQUAREANNULUS and k-RECTANNULUS problems. Among them, we show that the k-SQUAREANNULUS problem can be solved in $O(k^2 n \log n + k^3 n)$ time, and the k-RECTANNULUS problem can be solved in $O(nk^2 \log k + k^4 \log^3 k)$ time. It is worth mentioning that our algorithms are optimal when k is a constant, matching time bounds for the case of no outlier allowance. To our best knowledge, no nontrivial algorithm for the problems for $k \geq 1$, except that the general framework for LP-type problems by Matoušek [13]. Note that the minimum-width rectangular annulus problem is an LP-type problem with combinatorial dimension five, while the square annulus problem is not. The framework of Matoušek [13] thus automatically results in an $O(nk^5)$-time algorithm for the problem with k outliers.

Due to page limit, all proofs are removed but will be found in a full version.

2   Preliminaries

In this paper, we only handle axis-parallel squares and rectangles. Henceforth, any square or rectangle we discuss is assumed to be axis-parallel, unless stated otherwise. Consider a rectangle, or possibly a square, R in the plane $\mathbb{R}^2$. We call the intersection point of its two diagonals the *center* of R. The *height* and the *width* of R are the lengths of its vertical side and horizontal side, respectively. An *(inward) offset* of R by $\delta > 0$ is a rectangle obtained by sliding the four

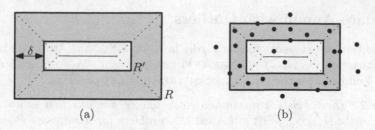

Fig. 1. (a) A rectangle R and its offset by δ, defining a rectangular annulus (shaded area). The base of R is depicted as a solid line segment in the middle. (b) A minimum-width rectangular containing given points with 2 inside outliers and 5 outside outliers.

sides of R inwards by δ. If R is of height h and width w, then the offset of R by $\delta = \frac{1}{2}\min\{h,w\}$ is degenerated to a line segment or a point, called the *base* of R. Note that the base of R is either vertical if $h > w$ or horizontal if $h < w$, and the center of R is the midpoint of the base of R.

For any positive $\delta \leq \frac{1}{2}\min\{h,w\}$, consider an inward offset R' of R by δ. Then, the closed region A between R and R', including its boundary, is called a *rectangular annulus* with the *outer rectangle* R and the *inner rectangle* R'. See Fig. 1(a). When R is a square and so is R', the annulus A is called a *square annulus*. The distance δ between the sides of R and R' is called the *width* of the annulus. The complement $\mathbb{R}^2 \setminus A$ of the annulus A is separated into two connected components. We shall call the outside of R the *outside* of A and the inside of R' the *inside* of A.

Given a set P of n point in $\mathbb{R}^2$ and a nonnegative integer k, our problems, the k-SQUAREANNULUS and the k-RECTANNULUS problems, ask a minimum-width square or rectangular annulus that contains at least $n-k$ points of P. See Fig. 1(b). The at most k points that are not covered by the resulting annulus are called *outliers*. Since the complement of any annulus are separated into its inside and outside, such an outlier may lie either in the inside or in the outside. We call an outlier an *outside outlier* if it lies in the outside of the resulting annulus, or an *inside outlier*, otherwise. In some applications, no inside outlier would be allowed while outside outliers are allowed, or vice versa, or even the numbers of inside and outside outliers are prescribed. This motivates to a variation of the problems, called the $(k_{\mathrm{in}}, k_{\mathrm{out}})$-SQUAREANNULUS and the $(k_{\mathrm{in}}, k_{\mathrm{out}})$-RECTANNULUS problems for nonnegative integers k_{in} and k_{out}, in which at most k_{in} inside outliers and at most k_{out} outside outliers are allowed.

When no outlier is allowed, i.e., $k = 0$, the 0-SQUAREANNULUS and 0-RECTANNULUS problems are solved in $O(n \log n)$ and $O(n)$ time, respectively, by Abellanas et al. [1] and Gluchshenko et al. [11]. Both algorithms are based on the following observation, which will be useful also for our further discussions.

Lemma 1 (Abellanas et al. [1] and Gluchshenko et al. [11]). *There exists a minimum-width square/rectangular annulus A containing P such that the outer square/rectangle of A is a smallest square/rectangle enclosing P.* ⊡

3   Square Annuli with Outliers

In this section, we present efficient algorithms for k-SQUAREANNULUS problem and its variation (k_{in}, k_{out})-SQUAREANNULUS problem. We start with an analogue to Lemma 1 for the (k_{in}, k_{out})-SQUAREANNULUS problem.

Lemma 2. *There exists a minimum-width square annulus that is an optimal solution to the (k_{in}, k_{out})-SQUAREANNULUS problem for a point set P such that there are two points in P lying in the opposite sides of its outer square.*

Note that this also implies the existence of an optimal square annulus to the k-SQUAREANNULUS problem with the same property, as is an optimal solution to an instance of the (k_{in}, k_{out})-SQUAREANNULUS problem with $k = k_{in} + k_{out}$.

We first consider special cases of the (k_{in}, k_{out})-SQUAREANNULUS problem where $k_{in} = 0$ or $k_{out} = 0$, and then proceed to the general case.

3.1   Case of No Outside Outlier

Here, we consider the (k_{in}, k_{out})-SQUAREANNULUS problem with $k_{in} \geq 1$ and $k_{out} = 0$, so at most k_{in} outliers must lie in the inside of the resulting annulus. In this case, there exists an optimal annulus whose outer square is a smallest square enclosing all the input points P by Lemma 2.

Let R be the smallest enclosing rectangle of P, and h and w be the height and width of R. We assume without loss of generality that $h \geq w$. Then, the trace of the centers of all smallest squares enclosing P forms a horizontal line segment C. Note that the length of C is exactly equal to $h - w$ and its midpoint coincides with the center of R. Lemma 2 implies that the center of an optimal square annulus to the $(k_{in}, 0)$-SQUAREANNULUS problem lies in C. Let $S(c)$ be the smallest square enclosing P with center c. For a fixed center $c \in C$, the width of the corresponding optimal annulus is determined by the $(k_{in} + 1)$-th nearest point in P from c with respect to the L_∞ metric since the inner square with center c now can contain at most k_{in} outliers in its interior.

We thus consider the L_∞ distance between $p \in P$ and $c \in C$, denoted by $f_p(c)$ as a function of c over C. These functions f_p indeed play a central role in previous results on minimum-width square annuli [5,11], based on the following properties.

Lemma 3 (Gluchshenko [11] and Bae [5]). *The functions f_p are continuous and piecewise linear such that its graph consists of at most three linear pieces whose slopes are -1, 0, and 1 in this order. In addition, any two of them properly cross at most once. Therefore, the graphs of the functions f_p form a family of n pseudo-lines.* ⬜

Now, we define

$$L_{k_{in}+1}(c) := \text{the } (k_{in} + 1)\text{-th smallest value among } f_p(c) \text{ for } p \in P,$$

for every $c \in C$. Since the side length of the outer square is fixed as h, the width of the annulus at center $c \in C$ is determined as $\frac{1}{2}h - L_{k_{in}+1}(c)$. Thus, our problem is equivalent to maximizing $L_{k_{in}+1}(c)$ over $c \in C$.

The function $L_{k_{\text{in}}+1}$ is known as the $(k_{\text{in}}+1)$-*level* of the functions f_p for $p \in P$. By Lemma 3, it is known that the graph of $L_{k_{\text{in}}+1}$ consists of $O(nk_{\text{in}}^{\frac{1}{3}})$ line segments [9] and can be computed in $O(n \log n + nk_{\text{in}}^{\frac{1}{3}})$ time [8]. Hence, we conclude the following.

Lemma 4. *Given n points and an integer $k_{\text{in}} \geq 0$, the $(k_{\text{in}}, 0)$-* SQUAREANNULUS *problem can be solved in $O(n \log n + nk_{\text{in}}^{\frac{1}{3}})$ time.*   ☐

3.2 Case of No Inside Outlier

Now, we consider the case where $k_{\text{in}} = 0$ and $k_{\text{out}} \geq 1$.

By Lemma 2, we can restrict ourselves to those annuli whose outer squares are determined by two points in P lying on its opposite sides. Without loss of generality, we assume that the outer square of an optimal annulus has one point on its top side and another on its bottom side. The other case can be handled in a symmetric way. In the following, we fix two points $t, b \in P$ and find a best possible annulus whose outer square is determined by these two points on its top and bottom sides. Our overall algorithm will try all possible combinations of t and b, and finally compute an optimal square annulus.

For i and j with $0 \leq i, j \leq k_{\text{out}}$, let $t \in P$ be the $(i+1)$-th highest point in P and $b \in P$ be the $(j+1)$-th lowest point in P. We only consider the points in P below t and above b, including t and b, denoted by P_{ij}, and square annuli containing at least $n - k_{\text{out}}$ points whose outer square is defined by t and b. Note that $|P_{ij}| \geq n - (i+j)$. Let h be the vertical distance between t and b and ℓ be the horizontal line such that the vertical distance from t and b to ℓ is $h/2$. For each $c \in \ell$, define $S(c)$ be the square of side length h and center c. Also, define $n_{ij}(c) := |P_{ij} \cap S(c)|$ be the number of points in P lying in the square $S(c)$.

For a fixed center $c \in \ell$, now we have a fixed outer square $S(c)$ and the inner square is uniquely determined by the nearest point in P_{ij} from c with respect to the L_∞ distance. As done above, we define the function f_p for each $p \in P_{ij}$ over the line ℓ such that $f_p(c)$ is the L_∞ distance from c to p for each $c \in \ell$. Let $L(c) := \min_{p \in P_{ij}} f_p(c)$ be the lower envelope of the functions f_p. The width of the unique square annulus $A(c)$ with center $c \in \ell$ is determined as $\frac{1}{2}h - L(c)$. The last thing to check is whether or not $A(c)$ contains at least $n - k_{\text{out}}$ points in P_{ij}. Thus, our problem is equivalent to the following:

$$\text{maximize } L(c) = \min_{p \in P_{ij}} f_p(c) \text{ over } c \in \ell \text{ subject to } n_{ij}(c) \geq n - k_{\text{out}}.$$

Note that $n_{ij}(c)$, as a function of c over ℓ, is piecewise constant whose break points correspond to those $c \in \ell$ at which a point in P_{ij} lies on the left or right side of $S(c)$. Thus, the function n_{ij} induces at most n intervals on ℓ. Among the intervals, interesting to us are those with $n_{ij}(c) \geq n - k_{\text{out}}$. We call them *feasible intervals*.

Hence, we compute the lower envelope of f_p over ℓ, and find the highest point on the lower envelope in the feasible intervals. This can be done in $O(n \log n)$ time [12] by Lemma 3. We obtain the overall optimal solution by running the indices i and j from 0 to k_{out}.

Lemma 5. *Given n points and $k_{out} \geq 1$, the $(0, k_{out})$-SQUAREANNULUS problem can be solved in $O(k_{out}^2 n \log n)$ time.* ⊡

3.3 The General Case

In order to handle the general (k_{in}, k_{out})-SQUAREANNULUS problem for $k_{in}, k_{out} > 0$, we combine the ideas of the above two special cases of $k_{in} = 0$ or $k_{out} = 0$. Again by Lemma 2, we only consider annuli whose outer squares are determined by two opposite extremes. Without loss of generality, we assume that the outer square of an optimal annulus has one point on its top side and another on its bottom side.

For i and j with $0 \leq i, j \leq k_{out}$, let $t \in P$ be the $(i + 1)$-th highest point in P and $b \in P$ be the $(j + 1)$-th lowest point in P. Define P_{ij}, ℓ, h, $S(c)$ and $n_{ij}(c)$ for $c \in \ell$, as declared above. Also, for each $p \in P_{ij}$, define $f_p(c)$ for $c \in \ell$ to be the L_∞ distance from c to p. In this case, we can exclude at most k_{in} inside outliers. So, we are interested in the $(k_{in} + 1)$-level $L_{k_{in}+1}$ of the functions f_p over ℓ. Recall that for each $c \in \ell$, $L_{k_{in}+1}(c)$ is defined to be the $(k_{in} + 1)$-th smallest number out of $f_p(c)$ for $p \in P_{ij}$. Then, our (k_{in}, k_{out})-SQUAREANNULUS problem in this case is equivalent to

$$\text{maximize } L_{k_{in}+1}(c) \text{ over } c \in \ell \text{ subject to } n_{ij}(c) \geq n - k_{out}.$$

This can be solved by computing the $(k_{in} + 1)$-level $L_{k_{in}+1}$ of the functions f_p, specifying the feasible intervals on ℓ, and finding the highest point of $L_{k_{in}+1}$ over the feasible intervals. By Lemma 3 and the aforementioned discussions, this can be done in $O(n \log n + k_{in}^{\frac{1}{3}} n)$ time. By repeating this for all $0 \leq i, j, \leq k_{out}$, we obtain the following result.

Theorem 1. *Given n points and two positive integers k_{in} and k_{out}, the (k_{in}, k_{out})-SQUAREANNULUS problem can be solved in $O(k_{out}^2 n \log n + k_{out}^2 k_{in}^{\frac{1}{3}} n)$ time.* ⊡

Now, we turn to the k-SQUAREANNULUS problem for $k \geq 0$. Note that we can solve it by handling $k + 1$ instances of the (k_{in}, k_{out})-SQUAREANNULUS problem with $k_{out} = k - k_{in}$. This gives us an algorithm of running time $O(k^3 n \log n + k^{\frac{10}{3}} n)$. Fortunately, we can reduce the time complexity by almost a factor of k by computing the 1-level, the 2-level, ..., the $(k + 1)$-level of the functions f_p at the same time using the algorithm by Everett et al. [10].

Theorem 2. *Given n points and an integer $k \geq 1$, the k-SQUAREANNULUS problem can be solved in $O(k^2 n \log n + k^3 n)$ time.*

4 Rectangular Annuli with Outliers

In this section, we consider the k-RECTANNULUS problem and the (k_{in}, k_{out})-RECTANNULUS problem. We start with an analog to Lemma 2.

Lemma 6. *There exists a minimum-width rectangular annulus that is an optimal solution to the (k_{in}, k_{out})-RECTANNULUS problem for a point set P such that each side of its outer rectangle contains at least one point in P.*

4.1 The Offset Rectangle Problem

First, we consider the *offset rectangle problem* in which we are given a set P of n points, a rectangle R, and an integer $k_{in} \geq 0$, and want to find a largest offset of R that contains at most k_{in} points of P in its interior. This problem can be easily solved in $O(n)$ time, regardless of k_{in}, as follows. First, compute the L_∞ distance d_p from each $p \in P$ to the base of R, and then find the $(k_{in} + 1)$-th nearest point in P from the base of R by running any linear-time selection algorithm.

We can devise algorithms for the (k_{in}, k_{out})-RECTANNULUS problem by calling the above algorithm for the offset rectangle problem as a subroutine. Observe that the resulting offset of R, together with R, forms a minimum-width rectangular annulus with at most k_{in} inside outliers. In the case where no outside outlier is allowed, i.e., $k_{out} = 0$, there exists an optimal rectangular annulus whose outer rectangle is equal to the smallest rectangle enclosing P by Lemma 6. Thus, this special case can be handled by solving the offset rectangle problem with the smallest rectangle enclosing P in $O(n)$ time.

Now, consider the general case of $k_{out} > 0$. Again, by Lemma 6, it suffices to try all rectangles defined by four points $t, b, l, r \in P$ as outer rectangles R such that t is the $(i + 1)$-th highest point in P, b is the $(j + 1)$-th lowest point in P, l is the $(i' + 1)$-th leftmost point in P, and r is the $(j' + 1)$-th rightmost point in P, for $0 \leq i, j, i', j' \leq k_{out}$. This implies that the (k_{in}, k_{out})-RECTANNULUS problem is reduced to $O(k_{out}^4)$ instances of the case of no outside outlier, yielding an $O(k_{out}^4 n)$-time algorithm. One can even drop a factor of k_{out} by a simple observation.

Lemma 7. *Given n points and integers $k_{in}, k_{out} > 0$, the (k_{in}, k_{out})-RECTANNULUS problem can be solved in $O(k_{out}^3 n)$ time, regardless of k_{in}. Therefore, the k-RECTANNULUS problem can be solved in $O(k^4 n)$ time for any $k > 0$.*

4.2 Offset Rectangle Queries

As discussed above, the offset rectangle problem is a central subproblem for our purpose, which needs to be solved a heavy number of times. Though an instance of the offset rectangle problem can be solved in optimal linear time, if we can do it more efficiently by paying some preprocessing cost, then we will be able to obtain more efficient algorithms for our main problems.

Here, we present a data structure that answers the *offset rectangle query* in poly-logarithmic time for any query (R, k_{in}) of rectangle R and integer k_{in}. As done above, let d_p be the L_∞ distance from the base of R to p, for each $p \in P$. In order to answer the offset rectangle query in a desired time, we need to select the $(k_{in} + 1)$-th smallest value among d_p for all $p \in P$ in poly-logarithmic time.

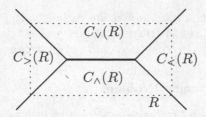

Fig. 2. The four sections induced by a rectangle R (dotted).

For the purpose, we consider the four *sections* induced by the given rectangle R. Specifically, shoot two rays from each endpoint of the base of R to its two nearest corners of R. These four rays, together with the base of R, result in a partition of the plane $\mathbb{R}^2$ into the four sections, each of which contains a side of R. Let $C_\vee(R)$, $C_\wedge(R)$, $C_>(R)$, and $C_<(R)$ denote the sections corresponding to the top, bottom, left and right sides of R, respectively. If the width of R is longer than the width, then each of $C_>(R)$ and $C_<(R)$ forms a right-angled cone heading for the x or $-x$ directions, respectively, while each of $C_\vee(R)$ and $C_\wedge(R)$ forms a right-angled cone cut by the base of R. See Fig. 2 for an illustration. Then, the L_∞ distance $d_p(R)$ from each point $p \in P$ to the base of R is just the vertical distance if $p \in C_\vee(R) \cup C_\wedge(R)$ or the horizontal distance if $p \in C_>(R) \cup C_<(R)$.

Thus, we consider range queries on P with ranges representing one of these sections. Our ranges are thus bounded by at most three lines in directions $\{\pi/4, 3\pi/4, \pi\}$ or $\{\pi/4, 3\pi/4, 0\}$. Range queries of this kind can be answered by using a three-dimensional range tree. More precisely, we build four standard range trees $\mathcal{T}_\vee$, $\mathcal{T}_\wedge$, $\mathcal{T}_>$, and $\mathcal{T}_<$ with fractional cascading for the following point sets, respectively, in three dimensional space:

$$P_\vee := \{(x+y, -x+y, y) \in \mathbb{R}^3 \mid (x, y) \in P\},$$
$$P_\wedge := \{(-x-y, x-y, -y) \in \mathbb{R}^3 \mid (x, y) \in P\},$$
$$P_> := \{(-x+y, -x-y, -x) \in \mathbb{R}^3 \mid (x, y) \in P\}, \text{ and}$$
$$P_< := \{(x-y, x+y, x) \in \mathbb{R}^3 \mid (x, y) \in P\}.$$

Observe that each of the sections $C_\vee(R)$, $C_\wedge(R)$, $C_>(R)$, and $C_<(R)$ for P exactly corresponds to a three-dimensional orthogonal range for $P_\vee$, $P_\wedge$, $P_>$, and $P_<$, respectively. See the textbook [6] for more details about range trees.

Our data structure $\mathcal{D}$ consists of these four range trees $\mathcal{T}_\vee$, $\mathcal{T}_\wedge$, $\mathcal{T}_>$, and $\mathcal{T}_<$, and is shown to efficiently answer the offset rectangle query.

Lemma 8. *There exists a data structure $\mathcal{D}$ of size $O(n \log n)$ that can be built in $O(n \log^2 n)$ time for n input points such that the offset rectangle query for any given rectangle R and integer k_{in} can be processed in $O(\log^2 n \log k_{in})$ time.*

In order to solve the (k_{in}, k_{out})-RECTANNULUS problem and k-RECTANNULUS problem, we build the structure $\mathcal{D}$ and exploit it to solve each necessary instance of the offset rectangle problem. This results in the following theorem.

Theorem 3. *Given n points and two integers $k_{\text{in}}, k_{\text{out}} > 0$, the $(k_{\text{in}}, k_{\text{out}})$-RECTANNULUS problem can be solved in $O(n \log^2 n + k_{\text{out}}^3 \log k_{\text{in}} \log^2 n)$ time. Therefore, the k-RECTANNULUS problem can be solved in $O(n \log^2 n + k^4 \log k \log^2 n)$ time for any integer $k > 0$.*

4.3   Kernelization

The time bounds in Theorem 3 have a term of $n \log^2 n$, which does not match the case of no outlier allowance. A further improvement can be made by a kernelization that extracts a small number of points K from P.

Lemma 6 implies that the outer rectangle of an optimal annulus is determined by at most four points of P and then the inner rectangle is determined by one point as observed above. The idea of our kernelization is thus filtering out those in P that have no chance to appear either on the outer or inner boundary of any optimal annulus.

For the outer rectangle, let $K_{\text{out}} \subseteq P$ be the set of points p such that p is either the $(i+1)$-th highest, the $(j+1)$-th lowest, $(i'+1)$-th leftmost, or $(j'+1)$-th rightmost point in P for $0 \le i, j, i', j' \le k_{\text{out}}$. Then, it is obvious that the outer rectangle of any optimal annulus for P is always determined by four points in K_{out} by Lemma 6; otherwise, we would have more than k_{out} outside outliers. We specify K_{out} in $O(n)$ time by selecting the $(k_{\text{out}} + 1)$-th point in each direction.

It is more involved to compute such a kernel K_{in} for the inner rectangle. Our approach is to extract a sufficient number of points in P from each section of every possible outer rectangle R that are nearest from the base of R. The following observation is now quite clear from our discussions so far.

Lemma 9. *Let R be a rectangle and k_{in} be a nonnegative integer. The largest offset of R containing at most k_{in} points of P in its interior is determined by a point included in the set consisting of the following points: the $k_{\text{in}} + 1$ lowest points from $P \cap C_{\vee}(R)$, the $k_{\text{in}} + 1$ highest points from $P \cap C_{\wedge}(R)$, the $k_{\text{in}} + 1$ rightmost points from $P \cap C_{>}(R)$, and the $k_{\text{in}} + 1$ leftmost points from $P \cap C_{<}(R)$.*

On the other hand, by Lemma 6, our outer rectangle should be $R = R(i, j; i', j')$ determined by the four indices (i, j, i', j'): for each $0 \le i, j, i', j' \le k_{\text{out}}$, let $\ell_t(i)$ be the horizontal line through the $(i + 1)$-th highest point in P, $\ell_b(j)$ be the horizontal line through the $(j + 1)$-th lowest point in P, $\ell_l(i')$ be the vertical line through the $(i' + 1)$-th leftmost point in P, and $\ell_r(j')$ be the vertical line through the $(j' + 1)$-th rightmost point in P. Then, $R(i, j; i', j')$ is the rectangle bounded by these four lines. As above, we consider the four sections of $R(i, j; i', j')$, denoted by $C_{\vee}(i, j; i', j')$, $C_{\wedge}(i, j; i', j')$, $C_{>}(i, j; i', j')$, and $C_{<}(i, j; i', j')$.

Our goal is thus to find a subset $K_{\text{in}} \subset P$ that includes at least the $k_{\text{in}} + 1$ lowest points from $P \cap C_{\vee}(i, j; i', j')$, the $k_{\text{in}} + 1$ highest points from $P \cap C_{\wedge}(i, j; i', j')$, the $k_{\text{in}} + 1$ rightmost points from $P \cap C_{>}(i, j; i', j')$, the $k_{\text{in}} + 1$ leftmost points from $P \cap C_{<}(i, j; i', j')$, for every $0 \le i, j, i', j' \le k_{\text{out}}$. Since some $R(i, j; i', j')$ is the outer rectangle of an optimal annulus by Lemma 6, its inner counterpart must be determined by one point in such a subset K_{in} by Lemma 9.

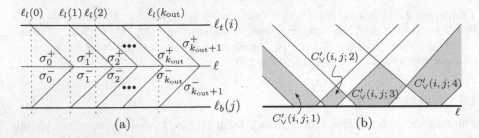

Fig. 3. (a) Decomposition for the first phase. (b) Decomposition for the second phase.

In the following, we show how to build K_{in}. Our algorithm consists of two phases, collecting points from P that are candidates to define the inner rectangle of an optimal annulus.

Phase 1. We fix two opposite indices i and j, giving us a horizontal slab between $\ell_t(i)$ and $\ell_b(j)$. Let h be the vertical distance between $\ell_t(i)$ and $\ell_b(j)$, and ℓ be the middle horizontal line such that the distance between $\ell_t(i)$ and ℓ is $h/2$. Then, every $R(i, j; i', j')$ is determined between $\ell_t(i)$ and $\ell_b(i)$ with height h and center on ℓ. Recall that the left section $C_>(i, j; i', j')$ forms a cone if the width of $R(i, j; i', j')$ is at least its height, or a cone cut by the base of $R(i, j; i', j')$, otherwise. In the former case, observe that $C_>(i, j; i', j')$ is independent of choice of j'. In the first phase of our algorithm, we collect candidate points from P that may define the offset rectangle of $R(i, j; i', j')$ as the $(k_{\text{in}} + 1)$-th nearest point in a cone, not being cut by the base of R.

For each $0 \le i' \le k_{\text{out}}$, consider the point $z \in \ell$ that is $h/2$ distant from $\ell_l(i') \cap \ell$ to the right. Let $w_{i'}^+$ be the segment between z and the intersection point $\ell_t(i) \cap \ell_l(i')$, and $w_{i'}^-$ be the segment between z and $\ell_b(j) \cap \ell_l(i')$. Note that $w_{i'}^+$ is of slope -1 and $w_{i'}^-$ of slope 1. Now, we consider the decomposition of the slab between $\ell_t(i)$ and $\ell_b(j)$ by the line ℓ and the segments $w_{i'}^+$ and $w_{i'}^-$ for all $0 \le i' \le k_{\text{out}}$. This decomposition of the slab consists of $2(k_{\text{out}} + 2)$ cells: $\sigma_0^+, \sigma_1^+, \ldots, \sigma_{k_{\text{out}}+1}^+$ from left to right above ℓ and $\sigma_0^-, \sigma_1^-, \ldots, \sigma_{k_{\text{out}}+1}^-$ from left to right below ℓ. See Fig. 3(a). For each $0 \le i' \le k_{\text{out}} + 1$, we collect the $k_{\text{in}} + 1$ lowest points and the $k_{\text{in}}+1$ rightmost points from $P \cap \sigma_{i'}^+$, and the $k_{\text{in}}+1$ highest points and the $k_{\text{in}}+1$ rightmost points from $P \cap \sigma_{i'}^-$. If $P \cap \sigma_{i'}^+$ or $P \cap \sigma_{i'}^-$ consists of less than $k_{\text{in}}+1$ points, then we collect all of them. Let $K_>(i, j; i') \subset P$ denote the set of these collected points for each $0 \le i' \le k_{\text{out}} + 1$.

We perform the same procedure for $\ell_r(j')$ for each $0 \le j' \le k_{\text{out}}$ in the symmetrical way to obtain the subsets $K_<(i, j'; j')$. Also, repeat the above procedure for every choice of i and j, with $0 \le i, j \le k_{\text{out}}$. We then observe the following.

Lemma 10. *For $0 \le i, j, i', j' \le k_{\text{out}}$, suppose that $R(i, j; i', j')$ has the longer width than the height. Then, the $k_{\text{in}} + 1$ rightmost points in $P \cap C_>(i, j; i', j')$ are contained in $\bigcup_{0 \le i'' \le i'} K_>(i, j; i'')$ and the $k_{\text{in}} + 1$ leftmost points in $P \cap C_<(i, j; i', j')$ are contained in $\bigcup_{0 \le j'' \le j'} K_<(i, j; j'')$.*

We do the same procedure for vertical lines $\ell_l(i')$ and $\ell_r(j')$ for every choice of (i', j'), resulting in the analogous subsets $K_\vee(i', j'; i)$ and $K_\wedge(i', j'; j)$. Let $K_1 \subset P$ be the union of all these subsets of points collected in the first phase. We then proceed to the second phase.

Phase 2. Again, fix i and j and consider the slab between $\ell_t(i)$ and $\ell_b(j)$. We call a pair (i', j') of indices *separated* if the width of $R(i, j; i', j')$ is at least its height. Also, a separated pair (i', j') is called *minimally separated* if either $i' = k_{out}$ or $j' = k_{out}$, or both $(i'+1, j')$ and $(i', j'+1)$ are not separated. We find all minimally separated pairs in order $(i'_1, j'_1), \ldots, (i'_m, j'_m)$ such that $i'_1 < i'_2 < \cdots < i'_m$. Then, by definition we also have $j'_1 > j'_2 > \cdots > j'_m$, so $m \leq k_{out} + 1$. Now consider the upward sections $C_\vee(i, j; i'_a, j'_a)$ for $1 \leq a \leq m$. Let $C'_\vee(i, j; a) := C_\vee(i, j; i'_a, j'_a) \setminus (C_\vee(i, j; i'_{a-1}, j'_{a-1}) \cup C_\vee(i, j; i'_{a+1}, j'_{a+1}))$. See Fig. 3(b). Then for each $1 \leq a \leq m$, we define $K'_\vee(i, j; a) \subset P$ to be the set of the $k_{in} + 1$ lowest points in $P \cap C'_\vee(i, j; a)$. Apply the same to the downwards sections $C_\wedge(i, j; i'_a, j'_a)$ to obtain the sets $K'_\wedge(i, j; a)$.

We repeat the above for every (i, j) with $0 \leq i, j \leq k_{out}$. Then, we observe:

Lemma 11. *For $0 \leq i, j, i', j' \leq k_{out}$, suppose that $R(i, j; i', j')$ has the longer width than the height, and let (i'_a, j'_a) be the minimally separated pair such that $i'_a \geq i'$ and $j'_a \geq j'$. Then, the $k_{in} + 1$ lowest points in $P \cap C_\vee(i, j; i', j')$ are contained in $K'_\vee(i, j; a) \cup K_1$, and the $k_{in} + 1$ highest points in $P \cap C_\wedge(i, j; i', j')$ are contained in $K'_\wedge(i, j; a) \cup K_1$.*

We apply the same procedure for vertical lines $\ell_l(i')$ and $\ell_r(j')$ for every choice of (i', j'), resulting in the analogous subsets $K'_>(i', j'; a)$ and $K'_<(i', j'; a)$. Let K_2 be the union of all these subsets obtained in the second phase. Then, we let $K_{in} := K_1 \cup K_2$. Our resulting kernel K of P is just $K_{out} \cup K_{in}$.

Finally, we conclude the following.

Lemma 12. *Let P be a set of n points and k_{in}, k_{out} be two positive integers. There is a subset $K \subset P$ with $|K| = O(k_{out}^3 k_{in})$ such that for any $0 \leq k'_{in} \leq k_{in}$ and $0 \leq k'_{out} \leq k_{out}$, the (k'_{in}, k'_{out})-RECTANNULUS problem for P is equivalent to that for K. Such a set K can be computed in $O(nk_{out}^2 \log k_{out} + k_{out}^3 k_{in})$ time.*

4.4 Putting It All Together

In order to solve the k-RECTANNULUS problem, we first compute the kernel K of P by Lemma 12 for $k_{in} = k_{out} = k$. Second, build the data structure $\mathcal{D}$ of Lemma 8. As observed in Theorem 3 and its proof, the problem can be solved by $O(k^4)$ offset rectangle queries. Therefore, we conclude the following.

Theorem 4. *Given n points and an integer $k \geq 1$, the k-RECTANNULUS problem can be solved in $O(nk^2 \log k + k^4 \log^3 k)$ time.*

For the (k_{in}, k_{out})-RECTANNULUS problem, if we apply the same approach as above, then the kernel K has size $O(k_{out}^3 k_{in})$ and thus it takes

$O(k_{out}^3 k_{in} \log^2(k_{out} + k_{in}))$ time for building the data structure for offset rectangle queries. However, this would be too much since the total query time takes only $O(k_{out}^3 \log^3(k_{out} + k_{in}))$ time as it suffices to perform $O(k_{out}^3)$ queries. In this case, indeed, while computing the kernel K of Lemma 12, we obtain enough information to directly solve the problem without the structure of Lemma 8, concluding the following.

Theorem 5. *Given n points and two integers $k_{in}, k_{out} \geq 0$, the (k_{in}, k_{out})-* RECTANNULUS *problem can be solved in $O(nk_{out}^2 \log k_{out} + k_{out}^3 k_{in})$ time.*

References

1. Abellanas, M., Hurtado, F., Icking, C., Ma, L., Palop, B., Ramos, P.: Best fitting rectangles. In: Proceedings of the European Workshop on Computational Geometry (EuroCG 2003) (2003)
2. Aggarwal, A., Imai, H., Katoh, N., Suri, S.: Finding k points with minimum diameter and related problems. J. Algorithms **12**, 38–56 (1991)
3. Ahn, H.K., Bae, S.W., Demaine, E.D., Demaine, M.L., Kim, S.S., Korman, M., Reinbacher, I., Son, W.: Covering points by disjoint boxes with outliers. Comput. Geom. Theor. Appl. **44**(3), 178–190 (2011)
4. Atanassov, R., Bose, P., Couture, M., Maheshwari, A., Morin, P., Paquette, M., Smid, M., Wuhrer, S.: Algorithms for optimal outlier removal. J. Discrete Alg. **7**(2), 239–248 (2009)
5. Bae, S.W.: Computing a minimum-width square annulus in arbitrary orientation [extended abstract]. In: Proceedings of the 10th International Workshop on Algorithms and Computation (WALCOM 2016), vol. 9627, pp. 131–142 (2016)
6. de Berg, M., van Kreveld, M., Overmars, M., Schwarzkopf, O.: Computationsl Geometry: Alogorithms and Applications, 2nd edn. Springer, Heidelberg (2000)
7. Chan, T.M.: Geometric applications of a randomized optimization technique. Discrete Comput. Geom. **22**(4), 547–567 (1999)
8. Chan, T.M.: Remarks on k-level algorithms in the plane (1999). Manuscript
9. Dey, T.K.: Improved bounds on planar k-sets and related problems. Discrete Comput. Geom. **19**, 373–382 (1998)
10. Everett, H., Robert, J.M., van Kreveld, M.: An optimal algorithm for computing $(\leq k)$-levels, with applications. Int. J. Comput. Geom. Appl. **6**(3), 247–261 (1996)
11. Gluchshenko, O.N., Hamacher, H.W., Tamir, A.: An optimal $O(n \log n)$ algorithm for finding an enclosing planar rectilinear annulus of minimum width. Oper. Res. Lett. **37**(3), 168–170 (2009)
12. Hershberger, J.: Finding the upper envelope of n line segments in $O(n \log n)$ time. Inform. Proc. Lett. **33**, 169–174 (1989)
13. Matoušek, J.: On geometric optimization with few violated constraints. Discrete Comput. Geom. **14**, 365–384 (1995)
14. Mukherjee, J., Mahapatra, P., Karmakar, A., Das, S.: Minimum-width rectangular annulus. Theor. Comput. Sci. **508**, 74–80 (2013)
15. Segal, M., Kedem, K.: Enclosing k points in the smallest axis parallel rectangle. Inform. Process. Lett. **65**, 95–99 (1998)

Approximating the Maximum Rectilinear Crossing Number

Samuel Bald[1]([⊠]), Matthew P. Johnson[1,2], and Ou Liu[1]

[1] The Graduate Center of the City University of New York, New York, USA
sbald@gradcenter.cuny.edu
[2] Lehman College, City University of New York, New York, USA

Abstract. Drawing a graph in a way that minimizes the number of edge-crossings is a well-studied problem. Recently there has been work characterizing both the minimum and maximum number of edge-crossings possible in various graph classes, assuming rectilinear (straight-line) edges. In this paper, we investigate the algorithmic problem of maximizing the number of edge-crossings over all rectilinear drawings a graph. We show that this problem is NP-hard and lies in $\exists \mathbb{R}$. We give a non-trivial derandomization of the natural randomized 1/3-approximation algorithm, which generalizes to a weighted setting as well as to an *ordering constraint satisfaction problem*. We evaluate these algorithms and other heuristics in simulation.

1 Introduction

The problem of drawing a graph in the plane with a minimum number of edge-crossings—called the graph's *crossing number*—is a well-studied, NP-hard [11] problem dating from the first half of the twentieth century [7,22]. The problem is quite difficult: the best approximation result known is an $O(n^{9/10} \cdot \text{poly}(d \log n))$-approximation ($d$ is the maximum degree) [6], and the *rectilinear* version—where edges must be drawn straight-line—is $\exists \mathbb{R}$-complete [20], and hence NP-hard.

Recently there has been interest in characterizing both the minimum and *maximum* number of edge-crossings possible in different graph classes for various edge-crossing variants [15,16,23], including the *maximum rectilinear crossing number* (MRCN) of a graph G, denoted by $\overline{\text{CR}}(G)$ [2]. (We denote the *minimum rectilinear crossing number* (mRCN) of a graph G by $\overline{\text{cr}}(G)$).

In this paper we investigate the *algorithmic* problem corresponding to MRCN, where, given a graph G, we seek a rectilinear drawing maximizing the number of edge-crossings. We prove that this problem is NP-hard and lies in $\exists \mathbb{R}$, but quite different in character from mRCN. We present an efficient derandomization of the natural randomized 1/3-approximation algorithm [23], which we extend to an edge-weighted generalization, as well as to an *ordering constraint satisfaction problem* (OCSP) [12], which subsumes, e.g., problem settings where we have two graphs G_1, G_2, and we only care about "inter-graph" crossings, i.e., crossings of edge pairs $(e_1, e_2) \in E(G_1) \times E(G_2)$. We also show that the same

T.N. Dinh and M.T. Thai (Eds.): COCOON 2016, LNCS 9797, pp. 455–467, 2016.
DOI: 10.1007/978-3-319-42634-1_37

approximation guarantee holds for a "continuous" variant of the randomized algorithm. Our experimental results suggest that these algorithms, as well as other heuristics, perform well in practice.

The maximum crossing number taken only over *convex* rectilinear drawings, i.e., rectilinear drawings where the vertices form a convex set, is denoted by $\overline{CR}^{\circ}(G)$, and we refer to the corresponding optimization problem as cMRCN. Since all convex drawings preserve relative vertex-ordering, and hence, crossings, we can restrict our attention here to simple convex drawings called *circle drawings*, where the vertices of G are placed along the circumference of a unit circle. Many specific graphs and graph families are known to satisfy $\overline{CR}(G) = \overline{CR}^{\circ}(G)$, and it has been conjectured that $\overline{CR}(G) = \overline{CR}^{\circ}(G)$ holds for *all* simple graphs [2]. Under this *convex conjecture* (CC), MRCN is transformed from a geometric problem into a purely combinatorial one, specifically, into an OCSP, albeit one with added structure. Our algorithms produce drawings that are convex; we emphasize, however, that their approximation guarantees hold regardless of the status of CC, i.e., the approximation is with respect to the *unrestricted* optimum. Our hardness results also hold regardless of CC, but in fact also extend to cMRCN.

2   Related Work

Minimizing the number of crossings is perhaps the best known in a large class of problems varying both in methods of counting edge-crossings and in types of drawings permitted [17,18,21]. The problem is NP-hard even when restricted to cubic graphs [14] and near-planar graphs [4]; the best known approximation is a randomized $O(n^{9/10} \cdot \text{poly}(d \log n))$-approximation ($d$ is the maximum degree) [6]. mRCN is $\exists \mathbb{R}$-complete [20], and hence NP-hard. Better approximation results are known for MRCN when G is k-colorable, for $k = 3, 4$, and when G is a triangulation [15]. The value of $\overline{CR}(G)$ is known for many specific graphs G, including the Peterson graph (49 [9]), the complete graph K_n ($\binom{n}{4}$, achieved by a circle drawing [19]), the cycle graph C_n ($\overline{CR}(C_n) = (n(n-4))/2 + 1$ when $n = 2k$, and $\overline{CR}(C_n) = \binom{n}{2} - n$ when $n = 2k+1$, achieved by drawing the graph as a *star polygon* $\{2k+1, k\}$ [10,13,24]), and the path graph P_n ($\binom{n-1}{2} - (n-2)$, achieved by a drawing similar to a star polygon).

The 1/3-approximation guarantee (in expectation) for the natural randomized algorithm for MCRN was given in [23], where a second optimization objective for graph drawings was also considered, motivated by the "Planarity Game": given a drawing of a planar graph with many edge-crossings, the player tries to rearrange the vertices in order to eliminate all edge-crossings. It was shown that the problem of finding a drawing that is maximally difficult for the game player—in the sense of requiring the largest number of single-vertex moves to remove all edge-crossings, called the "shift complexity"—is NP-hard.

3 Preliminaries

3.1 Problem Statements and Bounds

Let $G = (V, E)$ be an unweighted simple graph with $n = |V|$. Let $\mathcal{D}(G)$ denote the set of all possible rectilinear drawings of G. Let $D \in \mathcal{D}(G)$. Then $V(D)$ and $E(D)$ denote the sets of vertices and edges of G respectively, taken in their relative positions in D. We let $cross(D)$ denote the number of crossings in D. The goal in the MRCN problem is to find $\overline{CR}(G) := \max_{D \in \mathcal{D}(G)} cross(D)$ as well as the corresponding D which maximizes this quantity.

We also define the weighted version of MRCN, which we denote wMRCN. Let $w : E \to \mathbb{R}$ be an assignment of weights to each edge in G. For a drawing $D \in \mathcal{D}(G)$ and $e, f \in E(D)$, let $e \oslash f$ denote that e and f cross. We define the $crossing\text{-}cost$ function $c : E(D) \times E(D) \to \mathbb{R}$ as

$$c(e, f) := \begin{cases} w(e) \cdot w(f) & \text{if } e \oslash f \\ 0 & \text{otherwise} \end{cases}$$

For $D \in \mathcal{D}(G)$, define $cost(D) := \sum_{e, f \in E(D)} c(e, f)$. Then we want to find $\overline{CR}_w(G) := \max_{D \in \mathcal{D}(G)} cost(D)$ as well as the maximizing D. Note that wMRCN reduces to MRCN by setting all edge-weights uniformly to 1.

We now define some important upper bounds on $\overline{CR}(G)$ and $\overline{CR}_w(G)$. Let $B(G) = \{\{e, f\} \subseteq E \mid e \cap f = \emptyset\}$ be the set of disjoint edge pairs of G, so that $|B(G)|$ upper-bounds $\overline{CR}(G)$. Let $B'(G) = \sum_{u,v,w,x \in V} P(\{u, v, w, x\})$ where

$$P(\{u, v, w, x\}) = \begin{cases} 1 & \text{some partition of } \{u, v, w, x\} \text{ is in } B(G) \\ 0 & \text{otherwise} \end{cases}$$

so that $B'(G)$ counts at most one disjoint edge pair for every distinct set of 4 vertices. We also define $B_w(G) = \sum_{e, f \in B(G)} w(e) \cdot w(f)$ for weight-assignment w to E, so that $B_w(G) \geq \overline{CR}_w(G)$. We write B, B', B_w when there is no ambiguity.

3.2 Convex Position

We will examine the relationship of MRCN to its convex counterpart, cMRCN. Here we wish to compute $\overline{CR}^\circ(G) := \max_{D \in \mathcal{D}^\circ(G)} cross(D)$, where $\mathcal{D}^\circ(G)$ denotes the set of all possible $convex$ rectilinear drawings of G.

The validity of CC would imply a $\Theta(n!)$ algorithm for MRCN, namely, the brute-force procedure that computes the number of crossings for the drawings induced by all $\frac{1}{2}(n - 1)!$ nonisomorphic circular permutations (i.e., those that are inequivalent under flipping the circle). Under the assumption that CC holds (or alternatively, in the cMRCN problem variant), MRCN is transformed from a geometric problem to a purely combinatorial one, specifically, to a problem within the class of $ordering\ constraint\ satisfaction\ problems$ (OCSPs) [12].

A *k-arity* Π-OCSP P is specified by a subset $\Pi \subseteq S_k$, where S_k is the set of all permutations on $\{1, ..., k\}$. Such a problem P is specified by a pair (V, C), where V is a set of variables and C is a set of constraints in the form of ordered k-tuples of V. The objective is to find a permutation σ of V maximizing the number of *satisfied* constraints, where constraint $c \in C$ is satisfied with respect to σ if the ordering $\sigma_{|c}$ of c induced by σ is in Π. (Prominent examples of OCSPs include the Maximum Acyclic Subgraph problem $(k = 2)$, and the Betweenness problem $(k = 3)$.) To construct the 4-arity OCSP for a cMRCN instance G, let $V = V(G)$, and $C = \{(u, v, w, x) \mid \{(u, v), (w, x)\} \in B\}$. Then we may define $\Pi = \{1324, 2314, 1423, 2413, 3142, 4132, 3241, 4231\}$, i.e., Π is the set of all crossing-potential permutations on the constraints in a convex drawing of G. We note that the constructed ordering problem is a special case of an OCSP since there is additional structure on the set of 4-tuples to which the constraints apply, due to the underlying graph (i.e., we only take disjoint edge pairs). We refer to the general 4-arity Π-OCSP (i.e., no structure on C) as CROCS.

4   Complexity

The decision problem for MRCN asks whether there exists a rectilinear drawing $D \in \mathcal{D}(G)$ with $cross(D) \geq k$ for some nonnegative integer k. It is clear that in the setting restricted to convex drawings, i.e., for $\overline{\mathrm{CR}}^{\circ}(G)$, the problem lies in NP, since here we can restrict our attention to circle drawings.

In the general setting where non-convex drawings are permitted, it is not known whether the problem is in NP (expressing the vertex-coordinates of an optimal drawing may require arbitrary precision)—also true of mRCN. This issue would be dissolved if CC were proven. In any event, MRCN is similar to mRCN in the sense that:

Corollary 1 [20]. *MRCN* $\in \exists \mathbb{R}$.

The proof (omitted) is a straightforward modification of the proof for mRCN [20]. There are two interesting consequences of Corollary 1: First, since $\exists \mathbb{R}$ is solvable in exponential time and $\exists \mathbb{R} \subseteq$ PSPACE [5], we have that:

Corollary 2. *MRCN is solvable in exponential-time with polynomial space.*

Second, since MRCN is expressible in $\exists \mathbb{R}$ using only *strict* inequalities (so too for mRCN), an optimal drawing can be constructed where vertices lie at rational—and hence, integral—coordinates. Of course, exponential precision may still be required. We note that mRCN is in fact $\exists \mathbb{R}$-hard [3], but we do not know how to adapt the proof for MRCN.

We now show that MRCN is NP-hard (implying the NP-hardness of wMRCN). All drawings constructed are convex, implying that cMRCN is also NP-hard.

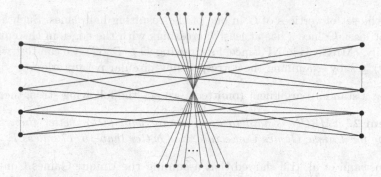

Fig. 1. A convex drawing of graph G' achieving $tk + \binom{t}{2}$ crossings. The t disjoint separators (blue) all cross a cut of size k in the underlying graph G (black).

Theorem 1. *MRCN is NP-hard, and cMRCN is NP-complete.*

Proof. We reduce from the MAX-CUT problem. We are given a MAX-CUT instance, consisting of a graph G together with a positive integer k. We'd like to determine if G admits a cut of size at least k. We construct an MRCN instance as follows: we consider the graph $G' = G \cup t \cdot P_2$ where $t \cdot P_2$ denotes t disjoint copies of P_2 (i.e., t disjoint edges) for some $t > \binom{n}{4} \geq \overline{CR}(G)$ (but $t = \text{poly}(n)$). We reduce the problem of determining if G admits a cut of size at least k to determining whether $\overline{CR}(G') \geq tk + \binom{t}{2}$.

Suppose that G admits a cut $(C, V - C)$ of size at least k. We describe how to draw G' to achieve at least $tk + \binom{t}{2}$ crossings. Fix any unit square in the plane. Arbitrarily arrange the vertices in C along the left edge of the unit square at unique positions, and the vertices of $V - C$ along the right edge in the same manner. Now arrange the t auxiliary edges (or *separators*) so that for each separator, one vertex lies on the top edge of the unit square, and the other on the bottom edge, and all of the separators cross one another (see Fig. 1). Note that the constructed drawing is convex. Since the cut is of size at least k, every one of the t separators will cross at least k edges of G. Moreover, the t separators generate $\binom{t}{2}$ crossings amongst themselves, so we have at least $tk + \binom{t}{2}$ crossings.

Now suppose we have a drawing $D \in \mathcal{D}(G')$ such that $cross(D) \geq tk + \binom{t}{2}$. One of the t separators must have at least k crossings with the edges of G. For suppose not. Then there would be at most $tk - t$ crossings between the t separators and the edges of G. Since the t separators can generate at most $\binom{t}{2}$ crossings amongst themselves, this would imply that there must be at least t crossings involving only edges of G. But this is impossible since $\overline{CR}(G) < t$. Thus there exists at least one separator e that crosses at least k edges of G. We move the other $t - 1$ separators so that they lie nearly-vertical with e (as in Fig. 1) if this is not the case already; note that we still have at least as many crossings as we had before, since each separator will now have at least k crossings with the edges of G. Note we also have a convex drawing of G'. Consider the cut $(C, V - C)$ induced by extending any separator s infinitely in both directions, with C the set of vertices of G in one of the resulting halfplanes. Such a cut is of size at least k since s has at least k crossings with the edges in the cut.

Finally, cMRCN is in NP, since, for drawing $D \in \mathcal{D}^\circ(G)$, we can test whether $cross(D) \geq k$ by examining the corresponding circular permutation.   □

Using a standard argument (omitted), we get the following strengthening:

Theorem 2. *wMRCN is hard to approximate with factor better than ≈ 0.878 assuming the Unique Games Conjecture, and better than $16/17$ unless $P = NP$.*

Guruswami et al. [12] showed that assuming the Unique Games Conjecture (UGC), *every* constant-arity OCSP P is *approximation-resistant*, i.e., if ρ is the expected fraction of satisfied constraints in a random ordering, then it is UG-hard to approximate P with a factor $\rho' > \rho$. This result extends to CROCS. In contrast, cMRCN corresponds to a special case of an OCSP, so that the result of [12] does not imply approximation resistance in this case. We do not know how to adapt the proof from [12] for cMRCN.

5   Randomized and Derandomized Algorithms

Due to the NP-hardness of MRCN, we turn to approximation. The natural randomized algorithm of choosing a random circular permutation (which we refer to simply as RAND) appeared in [23], where it was shown to provide a 1/3-approximation in expectation. The proof comes from observing that for any pair of disjoint edges, the edges will cross under 8 of 24 possible permutations of the four vertices, and from linearity of expectation. Implicit in this argument is the same 1/3-approximation in expectation for edge crossings weighted based on edge weights, or indeed, edge crossings with arbitrary weights.

Corollary 3 [23]. *RAND is a 1/3-approximation for wMRCN in expectation.*

We remark that the guarantee of 1/3 is tight, as there exists an infinite family of graphs on which the approximation factor 1/3 is achieved. (Consider, e.g., graphs consisting of n vertices, for any $n \geq 4$, and only two disjoint edges.) We also note that even though the solution constructed is convex, the approximation guarantee is with respect to the global (not necessarily convex) optimum.

In [23] it was stated without details that RAND can be derandomized via the method of conditional expectations (see, e.g., [1]). Explicitly constructing the derandomized algorithm, however, is nontrivial. We now show how to accomplish the derandomization in the unweighted setting (which can be straightforwardly adapted to the weighted setting).

Theorem 3. *There exists a deterministic 1/3-approximation for MRCN.*

Proof. Let $G = (V, E)$ be the graph under consideration, with $n = |V|$. We derandomize by the method of conditional expectations. We sequentially place the vertices into specified "slots" on the unit circle centered at the origin, in each round choosing a location maximizing the conditional expectation of the solution value. Let $\theta_i = \frac{2\pi(i-1)}{n}$ for $i \in \{1, ..., n\}$, where the angles are formed with respect to the standard x-axis. The i-th vertex v_i to be placed will have its chosen location on the unit circle specified by an angle θ_j not occupied by any other vertices. For convenience, we say that v_i is placed in *slot* $T_i = j$, where we refer to v_i's chosen angle by its *index*. The coordinates of v_i's slot are recovered as $(\cos(\theta_{T_i}), \sin(\theta_{T_i}))$. For a vertex u, let $s(u)$ denote the index of u's slot when applicable, so that if u is in slot θ_j, then $s(u) = j$.

Now, suppose we have placed $i - 1$ vertices in slots $T_1, ..., T_{i-1}$. Let R_i be the set of unoccupied slots in round i. Then we want to choose a slot $T_i \in R_i$ for the vertex v_i that maximizes the expected value of the complete solution value, where the expectation is taken over the future locations $T_{i+1}, \ldots, T_n$ of the vertices yet to be placed. Formally, let W be a random variable for the number of crossings generated by RAND, and let $W(r_1, \ldots, r_i) = E[W \mid T_1 = r_1, \ldots, T_i = r_i]$. Notice that we have $W(\emptyset) = E[W] \geq \frac{1}{3}\overline{\mathrm{CR}}(G)$. In round i, we wish to choose a slot $r^* \in R_i$ with $r^* = \arg\max_{r \in R_i} W(r_1, \ldots, r_{i-1}, r)$. Let V^- be the set of already-placed vertices, and V^+ as those yet to be placed. Partition E into five classes E_{ab}: $E_{--}, E_{-i}, E_{+i}, E_{++}$, and E_{-+}, where the subscripts $a, b \in \{-, +, i\}$ refer to three types of vertices in a corresponding edge: already-placed vertices $(-)$, vertices not yet considered $(+)$, and vertex v_i currently being placed (i).

Let W_{--}^{-i} be the number (or more generally the weight[1]) of crossings between edges in E_{--} and E_{-i}, and similarly for all other edge pairing classes, so that

$$W = \sum_{ab, cd \in \{--, -i, +i, ++, -+\}} W_{ab}^{cd}$$

Now, the placement of v_i will only potentially affect the crossings of a disjoint edge pair $\{e, f\} \in B$ if two conditions are met: (1) one of the two edges e, f is in E_{--} or E_{-i}, and (2) one of the two edges e, f is in one of E_{-i}, E_{+i}, E_{-+} or E_{++}. This is so because if no edges are present, then placing v_i places no restrictions on whether e, f will cross or not. On the other hand if both edges are present, then whether they cross or not is already determined. This leaves eight classes W_{ab}^{cd} to consider. However, notice that clearly $W_{-i}^{-i} = W_{-i}^{+i} = 0$, since any of the corresponding edges are both incident on vertex v_i. Thus we are left to consider the crossings of six classes when placing v_i: $W_{--}^{-i}, W_{--}^{+i}, W_{++}^{-i}, W_{-+}^{-i}, W_{--}^{++}$, and W_{--}^{-+}. We analyze each of these in turn, fixing a slot $r \in R_i$ and computing the expected number of crossings for each such class if v_i is placed at slot r. That is, we show how to compute $w_{ab}^{cd}(r) = E[W_{ab}^{cd} \mid T_1 = r_1, \ldots, T_i = r]$, ranging over the relevant values of (ab, cd). A deterministic selection is effected by placing v_i in a slot r^* maximizing the total expected number of crossings over all six classes. Note computing each quantity W_{ab}^{cd} for a given slot can be done in time

---

[1] For ease of exposition, we refer simply to the *number* of crossings for the remainder of the proof, although exactly the same analysis applies in the weighted setting, whether the weights of crossings are based on underlying edge weights or are permitted to be completely arbitrary, as in an OCSP.

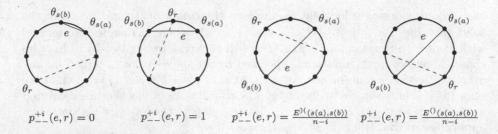

$$p^{+i}_{--}(e,r) = 0 \qquad p^{+i}_{--}(e,r) = 1 \qquad p^{+i}_{--}(e,r) = \frac{E^{)(}(s(a),s(b))}{n-i} \qquad p^{+i}_{--}(e,r) = \frac{E^{()}(s(a),s(b))}{n-i}$$

Fig. 2. The four possible scenarios in the case of W^{+i}_{--}, and the corresponding values of $p^{+i}_{--}(e,r)$

$O(|E_{ab}| \cdot |E_{cd}|)$, by checking every pair of edges in $E_{ab} \times E_{cd}$. Hence the algorithm (see Algorithm 1) is polynomial-time.

W^{-i}_{--}: For a fixed slot $r \in R_i$, the value $w^{-i}_{--}(r)$ is deterministic since all relevant edges are present when v_i is placed; thus, we simply compute the number of crossings present for this class for each unoccupied slot $r \in R_i$.

W^{+i}_{--}: For each unoccupied slot $r \in R_i$ we compute the expected number of crossings for this class. First, for each edge $e = (a,b) \in E_{--}$ (with $s(a) < s(b)$), we compute the probability $p^{+i}_{--}(e,r)$ that an edge in E_{+i} crosses e if $T_i = r$. The degenerate case where a and b are in adjacent slots clearly satisfies $p^{+i}_{--}(e,r) = 0$. Another triviality occurs when a and b are separated by one slot, and r is the intermediate slot, in which case $p^{+i}_{--}(e,r) = 1$. In general we have

$$p^{+i}_{--}(e,r) = \begin{cases} \dfrac{E^{)(}(s(a),s(b))}{n-i} & s(a) < r < s(b) \\[2mm] \dfrac{E^{()}(s(a),s(b))}{n-i} & \text{otherwise} \end{cases}$$

where, counting the tentative placement of v_i at slot r, $E^{)(}(s(a),s(b))$ denotes the number of empty slots with indices less than $s(a)$ or greater than $s(b)$, and $E^{()}(s(a),s(b))$ denotes the number of empty slots between $s(a)$ and $s(b)$. In total we have $n - i$ empty slots, hence the denominators. (See Fig. 2 for illustrations of the possible cases for $p^{+i}_{--}(e,r)$.) Thus for a given slot r, we can compute the expected number of crossings in this class *involving a given edge* in E_{--} as $|E_{+i}| \cdot p^{+i}_{--}(e,r)$, and we compute the expected number of *all* crossings in this class for the given slot r as

$$w^{+i}_{--}(r) = |E_{+i}| \cdot \sum_{e \in E_{--}} p^{+i}_{--}(e,r) \tag{1}$$

W^{-i}_{++}: Fix $r \in R_i$. For each edge $e = (i,j) \in E_{-i}$ (where, without loss of generality, assume $r < s(j)$), the expected number of edges of E_{++} it crosses is $|E_{++}| \cdot p^{-i}_{++}(e,r)$ where $p^{-i}_{++}(e,r)$ is the probability that an edge in E_{++} crosses e if $T_i = r$. As in the case of W^{+i}_{--}, we have the degenerate case where

$p_{++}^{-i}(e, r) = 0$; however, the trivial case $p_{++}^{-i}(e, r) = 1$ does not occur since no vertices from the edges in E_{++} are fixed. In general we have, by symmetry,

$$p_{++}^{-i}(e, r) = 2 \left(\frac{E^{()}(r, s(j))}{n - i} \cdot \frac{E^{)(}(r, s(j))}{n - i} \right)$$

Taking the expectation over all edges in E_{-i}, we get

$$w_{++}^{-i}(r) = |E_{++}| \cdot \sum_{e \in E_{-i}} p_{++}^{-i}(e, r) \tag{2}$$

as the expected number of crossings over this class for slot r.

W_{-+}^{-i}: Fix $r \in R_i$. For each edge $e = (i, j) \in E_{-i}$ (and again assume $r < s(j)$), the expected number of edges of E_{-+} that e crosses is $\sum_{f \in E_{-+}} p_{-+}^{-i}(e, f, r)$ where $p_{-+}^{-i}(e, f, r)$ is the probability that $f \in E_{-+}$ crosses e if $T_i = r$. Note that the summation cannot be reduced, since in general, $p_{-+}^{-i}(e, f, r)$ may differ for each $f \in E_{-+}$. Now, let $f = (a, b) \in E_{-+}$ where $a \in V^-$. We have in general that

$$p_{-+}^{-i}(e, f, r) = \begin{cases} \frac{E^{)(}(r, s(j))}{n - i} & r < s(a) < s(j) \\ \frac{E^{()}(r, s(j))}{n - i} & \text{otherwise} \end{cases}$$

and again we have the degenerate cases where $p_{-+}^{-i}(e, f, r) = 0, 1$. Now taking the expectation over all edges in E_{-i} we get

$$w_{-+}^{-i}(r) = \sum_{e \in E_{-i}} \sum_{f \in E_{-+}} p_{-+}^{-i}(e, f, r) \tag{3}$$

as the expected number of crossings over this class for slot r. The cases of W_{--}^{++} and W_{--}^{-+} (omitted) are similar to those of W_{++}^{-i} and W_{-+}^{-i}, respectively. Finally, for each empty slot $r \in R_i$, we sum the expected number of crossings over all classes, greedily placing v_i in the best slot. We call the corresponding deterministic algorithm DERAND (shown in Algorithm 1).   □

Algorithm 1. DERAND

1: $R \leftarrow \{1, ..., n\}$
2: **for** $i = 1$ to n
3:     **for-each** r in R
4:        compute $w_{--}^{-i}(r)$ by counting the number of relevant crossings present
5:        compute $w_{--}^{+i}(r)$ according to Eq. 1
6:        compute $w_{++}^{-i}(r)$ according to Eq. 2
7:        compute $w_{--}^{-i}(r)$ (Equation omitted)
8:        compute $w_{--}^{++}(r)$ according to Eq. 3
9:        compute $w_{--}^{-+}(r)$ (Equation omitted)
10:    $r^* \leftarrow \arg\max_{r \in R} w_{--}^{-i}(r) + w_{--}^{+i}(r) + w_{++}^{-i}(r) + w_{-+}^{-i}(r) + w_{--}^{++}(r) + w_{--}^{-+}(r)$
11:    $T_i \leftarrow r^*$
12:    $R \leftarrow R - \{r^*\}$

A different approach to a randomized algorithm would be to randomly select angles about the unit circle when placing vertices, as opposed to selecting discrete slots. The behavior of this *continuous* randomized algorithm is not identical to RAND, but it nonetheless provides the same approximation guarantee.

For, consider two non-adjacent edges (a, b) and (c, d), where without loss of generality the algorithm places c and d prior to a and b, with $\theta_c < \theta_d$. Let $\theta = \theta_d - \theta_c$. An edge-crossing occurs precisely when $\theta_c < \theta_a < \theta_d$ and $\theta_b > \theta_d$ or $\theta_b < \theta_c$, or vice versa. Thus for *fixed* θ, the probability that (a, b) crosses (c, d) is, by symmetry, $2\left(\frac{\theta}{2\pi} \cdot \frac{2\pi - \theta}{2\pi}\right)$. Let p be the probability that (a, b) crosses (c, d). Integrating over all angles θ and normalizing the probability to 1, we get

$$p = \frac{1}{2\pi} \int_0^{2\pi} 2\left(\frac{\theta}{2\pi} \cdot \frac{2\pi - \theta}{2\pi}\right) d\theta = \frac{1}{3}$$

The 1/3-approximation guarantee again follows by linearity of expectation. This algorithm can also be derandomized by the method of conditional expectations, but the analysis is much more cumbersome, and requires calculus.

6   Experimental Results

We now present the results of experiments measuring the performance of several natural heuristics for MRCN. In order to measure average performance, we generated Erdős-Rényi random graphs $G(n, p)$ [8]. We performed two types of experiments: in the first type, we fixed p (at, e.g., $p = 1/2$) and varied n in the range $4 \leq n \leq 45$ (n-type); in the second type, we fixed $n = 30$ and varied p from 0 to 1 in increments of 0.05 (p-type). In addition to testing RAND and DERAND, the other two heuristics tested were:

– *greedy*: vertices are placed sequentially along the unit circle; in each round, a location maximizing the number of crossings in the subgraph constructed thus far is selected
– *local search*: first, the vertices are placed about the unit circle in arbitrary locations; then, we repeatedly relocate a vertex whose relocation increases the number of crossings; the process stops when no move offers a gain

We examined performance in three ways: average solution value, average ratio of solution value to $|B|$, and average ratio of solution value to B'. For brevity, we present only the results using the latter two performance measures. We did not compare to $\overline{\mathrm{CR}}(G)$ and $\overline{\mathrm{CR}}^\circ(G)$ for tractability reasons.

As an illustration of how values were plotted, we describe one of the n-type experiments: For each input size n, 100 graphs were generated from the $G(n, p)$ distribution. For each graph G_i, $r_i = h_i/|B|$ was computed, where h_i is the result of running heuristic H on G_i. Then, the average of the ratios $\frac{1}{100}\sum r_i$ was plotted as the performance of heuristic H for input size n.

Note that all experiments indicate local search to be the best heuristic, although, the plots indicate similar *asymptotic* performance for greedy and

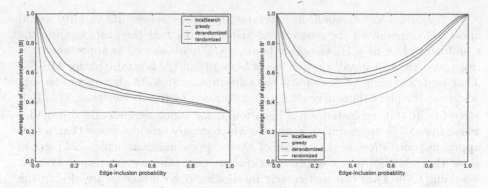

Fig. 3. p-type experiments on $G(30, p)$ graphs. Average ratio of solution value to $|B|$ (left); to B' (right).

DERAND. In contrast, RAND exhibited consistently worse performance due to it being a $1/3$-approximation in expectation.

In the p-type experiments (Fig. 3), we find that unique trends are displayed when taking ratios relative to $|B|$ and B'. In the case of $|B|$ (Fig. 3, left), we see that the average ratios of all heuristics excluding RAND vary smoothly from 1 to $1/3$. Intuitively, the heuristics are very likely to be close to OPT for sparse graphs, and OPT will be near $|B|$; in contrast, *any* circle drawing of the complete graph K_n is optimal, and we have $\overline{\mathrm{CR}}(K_n)/B(K_n) = \binom{n}{4}/\left(\binom{\binom{n}{2}}{2} - (n-2)\binom{n}{2}\right) = 1/3$. In the case of B' (Fig. 3, right), we see that the average ratios of all heuristics excluding RAND form smooth "u-shaped" curves, reaching 1 at both extrema. Intuitively, OPT is near B' for very sparse and very dense graphs (e.g., $\overline{\mathrm{CR}}(K_n) = B'(K_n)$), and the heuristics will be close to OPT in these cases. Whereas for intermediate p-values, there is little correlation between B' and OPT. RAND exhibits a more dramatic "u-curve" in the B'-plot; the average ratios initially drop, and then approach 1 as OPT and B' grow closer (as $p \to 1$), albeit always underperforming the other heuristics.

In the n-type experiments (figures omitted), we found similar trends for $|B|, B'$, where for B', the average ratios were slightly higher since $B' \leq |B|$. The experimental results showed a decrease in heuristic performance as $n \to \infty$, starting ($n = 4$) at an average ratio ≈ 1, and apparently leveling off at ≈ 0.5 ($n = 45$). It is unclear if these trends would also hold true with respect to OPT.

7   Open Problems

As mentioned in Sect. 4, assuming UGC, *every* constant-arity OCSP P is *approximation-resistant* [12]. Since cMRCN corresponds to a special case of an OCSP, approximation-resistance is not immediate. Therefore an interesting question is to determine whether the OCSP corresponding to cMRCN is approximation-resistant assuming UGC. An answer in the affirmative would imply the optimality of the $1/3$-approximation for cMRCN.

A resolution to CC would have interesting consequences. Its validity would allow us to circumvent the geometry of MRCN and attack the problem from the standpoint of OCSPs. On the other hand, if CC holds as well as approximation-resistance, then no approximation factor better than $1/3$ is attainable for MRCN. That said, assuming CC's validity, the experimental results above seem to suggest that the algorithms may offer approximation factors better than $1/3$. Relative to $\overline{CR}^{\circ}(G)$, no instance has been found for which derandomized provides worse than a $3/8$-approximation, and for which greedy provides worse than a $1/2$-approximation. Proving that either of these approximation factors hold would show that cMRCN is not approximation-resistant (and by extension, MRCN, assuming CC). Thus two particularly interesting open problems are closing the gap between $1/3$ and $3/8$ for derandomized, and proving an approximation guarantee—conceivably $1/2$—for greedy.

Acknowledgements. This work was supported in part by PSC-CUNY Research Award 67665-00 45, CUNY Collaborative Incentive Research Grant (CIRG 21) 2153, and a Research in the Classroom Idea Grant.

References

1. Alon, N., Spencer, J.: The Probabilistic Method, chap. 15, pp. 249–258. Wiley, Hoboken (1992)
2. Alpert, M., Feder, E., Harborth, H.: The maximum of the maximum rectilinear crossing numbers of d-regular graphs of order n. Electr. J. Comb. **16**(1) (2009)
3. Bienstock, D.: Some provably hard crossing number problems. Discrete Comput. Geom. **6**(5), 443–459 (1991)
4. Cabello, S., Mohar, B.: Adding one edge to planar graphs makes crossing number and 1-planarity hard. SIAM J. Comput. **42**(5), 1803–1829 (2013)
5. Canny, J.: Some algebraic and geometric computations in pspace. In: Proceedings of the Twentieth Annual ACM Symposium on Theory of Computing, STOC 1988, pp. 460–467. ACM, New York (1988)
6. Chuzhoy, J.: An algorithm for the graph crossing number problem. CoRR, abs/1012.0255 (2010)
7. Erdős, P., Guy, R.K.: Crossing number problems. Am. Math. Mon. **80**, 52–58 (1973)
8. Erdős, P., Rényi, A.: On random graphs. i. Publicationes Math. **6**, 290–297 (1959)
9. Feder, E., Harborth, H., Herzberg, S., Klein, S.: The maximum rectilinear crossing number of the Petersen graph. Congr. Numerantium **206**, 31–40 (2010)
10. Furry, W., Kleitman, D.: Maximal rectilinear crossings of cycles. Stud. Appl. Math. **56**, 159–167 (1977)
11. Garey, M.R., Johnson, D.S.: Crossing number is NP-complete. SIAM J. Algebraic Discrete Methods **4**(3), 312–316 (1983)
12. Guruswami, V., Hstad, J., Manokaran, R., Raghavendra, P., Charikar, M.: Beating the random ordering is hard: every ordering CSP is approximation resistant. SIAM J. Comput. **40**(3), 878–914 (2011)
13. Harborth, H.: Drawing of the cycle graph. Congr. Numer. **66**, 15–22 (1988)
14. Hlinený, P.: Crossing number is hard for cubic graphs. J. Comb. Theory Ser. B **96**(4), 455–471 (2006)

15. Kang, M., Pikhurko, O., Ravsky, A., Schacht, M., Verbitsky, O.: Obfuscated drawings of planar graphs. CoRR, abs/0803.0858 (2008)
16. Pach, J., Ábrego, B., Fernández-Merchant, S., Salazar, G.: The rectilinear crossing number of K_n: closing in (or are we?). In: Pach, J. (ed.) Thirty Essays on Geometric Graph Theory, pp. 5–18. Springer, New York (2013)
17. Pach, J., Tóth, G.: Thirteen problems on crossing numbers. Geombinatorics **9**, 194–207 (2000)
18. Pach, J., Tóth, G.: Which crossing number is it anyway? J. Comb. Theory, Ser. B **80**(2), 225–246 (2000)
19. Ringel, G.: Extremal problems in the theory of graphs. In: Fiedler, M. (ed.) Theory of Graphs and Its Applications, Proceedings of Symposium Smolenice 1963, Prague, pp. 85–90 (1964)
20. Schaefer, M.: Complexity of some geometric and topological problems. In: Eppstein, D., Gansner, E.R. (eds.) GD 2009. LNCS, vol. 5849, pp. 334–344. Springer, Heidelberg (2010)
21. Schaefer, M.: The graph crossing number, its variants: a survey. Electron. J. Combin. Dyn. Surv. **21** (2014)
22. Turán, P.: A note of welcome. J. Graph Theory **1**(1), 7–9 (1977)
23. Verbitsky, O.: On the obfuscation complexity of planar graphs. Theor. Comput. Sci. **396**(1–3), 294–300 (2008)
24. Weisstein, E.W.: Star polygon (2010)

An Improved Approximation Algorithm for rSPR Distance

Zhi-Zhong Chen[1($\boxtimes$)], Eita Machida[1], and Lusheng Wang[2]

[1] Department of Information System Design, Tokyo Denki University,
Hatoyama, Saitama 350-0394, Japan
zzchen@mail.dendai.ac.jp
[2] Department of Computer Science, City University of Hong Kong,
Tat Chee Avenue, Kowloon, Hong Kong SAR
lwang@cs.cityu.edu.hk

Abstract. The problem of computing the rSPR distance of two given trees has many applications but is unfortunately NP-hard. The previously best approximation algorithm for rSPR distance achieves a ratio of 2.5 and it was open whether a better approximation algorithm for rSPR distance exists. In this paper, we answer this question in the affirmative by presenting an approximation algorithm for rSPR distance that achieves a ratio of $\frac{7}{3}$. Our algorithm is based on the new notion of *key* and several new structural lemmas.

1 Introduction

When studying the evolutionary history of a set X of existing species, one can obtain a phylogenetic tree T_1 with leaf set X with high confidence by looking at a segment of sequences or a set of genes [10]. When looking at another segment of sequences, a different phylogenetic tree T_2 with leaf set X can be obtained with high confidence, too. In this case, we want to measure the dissimilarity of T_1 and T_2. The rooted subtree prune and regraft (rSPR) distance between T_1 and T_2 has been used for this purpose [9]. It can be defined as the minimum number of edges that should be deleted from each of T_1 and T_2 in order to transform them into *essentially identical* rooted forests F_1 and F_2. Roughly speaking, F_1 and F_2 are *essentially identical* if they become identical forests (called *agreement forests* of T_1 and T_2) after repeatedly contracting an edge (p, c) in each of them such that c is the unique child of p (until no such edge exists).

The rSPR distance is an important metric that often helps us discover reticulation events. In particular, it provides a lower bound on the number of reticulation events [1,2], and has been regularly used to model reticulate evolution [11,12]. Unfortunately, it is NP-hard to compute the rSPR distance of two given phylogenetic trees [5,9]. This has motivated researchers to design approximation algorithms for the problem [3,4,9,13]. Hein *et al.* [9] were the first to come up with an approximation algorithm. They also introduced the important notion of maximum agreement forest (MAF) of two phylogenetic trees. Their algorithm was correctly analyzed by Bonet *et al.* [3]. Rodrigues *et al.* [13] modified

© Springer International Publishing Switzerland 2016
T.N. Dinh and M.T. Thai (Eds.): COCOON 2016, LNCS 9797, pp. 468–479, 2016.
DOI: 10.1007/978-3-319-42634-1_38

Hein *et al.*'s algorithm so that it achieves an approximation ratio of 3 and runs in quadratic time. Whidden and Zeh [16] came up with a very simple approximation algorithm that runs in linear time and achieves an approximation ratio of 3. Although the ratio 3 is achieved by a very simple algorithm in [16], no polynomial-time approximation algorithm had been designed to achieve a better ratio than 3 before Shi *et al.* [8] presented a polynomial-time approximation algorithm that achieves a ratio of 2.5.

In certain real applications, the rSPR distance between two given phylogenetic trees is small enough to be computed exactly within reasonable amount of time. This has motivated researchers to take the rSPR distance as a parameter and design fixed-parameter algorithms for computing the rSPR distance of two given phylogenetic trees [5,7,14–16]. These algorithms are basically based on the branch-and-bound approach and use the output of an approximation algorithm (for rSPR distance) to decide if a branch of the search tree should be cut. Thus, better approximation algorithms for rSPR distance also lead to faster exact algorithms for rSPR distance. It is worth noting that approximation algorithms for rSPR distance can also be used to speed up the computation of hybridization number and the construction of minimum hybridization networks [6].

Let T_1 and T_2 be two phylogenetic trees on the same set X of leaves. In a nutshell, the simple ratio-3 approximation algorithm in [16] proceeds in stages until T_1 and T_2 become identical forests. Roughly speaking, in each stage, the algorithm chooses two arbitrary sibling leaves in T_1 and uses them to find and remove at most three edges from T_2 such that the removal decreases the rSPR distance of T_1 and T_2 by at least 1. Since at most three edges are removed from T_2 in each stage and at least one of the removed edges is also removed from T_2 by an optimal solution, the algorithm achieves an approximation ratio of 3. Shi *et al.* [8] improve Whidden *et al.*'s algorithm by refining each stage. In each stage, Shi *et al.*'s algorithm carefully chooses a dangling *subtree* S of T_1 with at most 4 leaves and uses S to carefully choose and remove a set B of edges from T_2. On the positive side, B has a crucial property that the removal of the edges of B decreases the rSPR distance of T_1 and T_2 by at least $\frac{2}{5}|B|$. Because of this property, their algorithm achieves a ratio of 2.5. On the negative side, S and B have a drawback that after removing the edges of B from T_2, each leaf of S becomes isolated in T_2. Moreover, their search for B is by complicated case analysis and the complicatedness makes it hard to refine their algorithm so that it achieves a better ratio. Indeed, Shi *et al.* [8] ask whether a ratio better than 2.5 can be achieved by a polynomial-time approximation algorithm, and point out that new ideas are necessary.

In this paper, we answer Shi *et al.*'s question in the affirmative by presenting a quadratic-time approximation algorithm for rSPR distance that achieves a ratio of $\frac{7}{3}$. Our algorithm also proceeds in stages until the input trees T_1 and T_2 become identical forests. Roughly speaking, in each stage, our algorithm carefully chooses a dangling *subforest* (rather than a dangling subtree) S of T_1 with at most 6 leaves and uses S to carefully choose and remove a set B of edges from T_2. Similar to but better than Shi *et al.*'s algorithm [8], B has a crucial property that the removal of the edges of B decreases the rSPR distance of T_1 and T_2 by at least $\frac{3}{7}|B|$. Because of this property, our algorithm achieves a ratio of $\frac{7}{3}$. The search of S and B in our algorithm is based on the new notion of *key*. Using a key (as a tool), it is possible for our algorithm to find S and B such that even after removing the edges of B from T_2, some leaves of S remain connected in T_2. In other words, keys enable us to overcome the drawback in Shi *et al.*'s algorithm [8]. At first glance, since the subforest S in our algorithm can often be larger than the subtree S in Shi *et al.*'s algorithm, the search for B in our algorithm can become even more complicated. Fortunately, we can prove four structural lemmas which enable us to construct B recursively and hence significantly simplify the case analysis. We believe that the lemmas can be used to obtain better approximation algorithms for rSPR distance. Indeed, we conjecture that rSPR distance can be approximated within a ratio of $2 + \epsilon$ for any $\epsilon > 0$.

The remainder of this paper is organized as follows. In Sect. 2, we give the basic definitions that will be used throughout the paper. In Sects. 3 and 4, we define the structures for which we will search the input phylogenetic trees (or forests). In Sect. 5, we describe the approximation algorithm and analyze its performance. Section 6 concludes the paper. Due to space limit, the four structural lemmas and their proofs are omitted and so are the proofs of Lemmas 1 and 2 and Theorem 1. We also omit the case analysis for searching a desired set B. The full version of the paper is available upon request to the first author.

2   Preliminaries

Throughout this paper, a *rooted forest* always means a directed acyclic graph in which every node has in-degree at most 1 and out-degree at most 2.

Let F be a rooted forest. F is *binary* if the out-degree of every node in F is either 0 or 2. The *roots* (respectively, *leaves*) of F are those nodes whose in-degrees (respectively, out-degrees) are 0. We use $L(F)$ to denote the set of leaves in F. A $v \in L(F)$ is a *root leaf* if it is also a root of F. For a non-root v of F, $e_F(v)$ denotes the edge entering v in F. Each edge e leaving a node p and entering another node c in F is denoted by (p, c); moreover, p is the *tail* of e and the *parent* of c in F, while c is the *head* of e and a *child* of p in F. A node v of F is *bifurcate* (respectively, *unifurcate*) if it has two children (respectively, only one child) in F. If a root v of F is unifurcate, then *contracting v in F* is the operation that modifies F by deleting v. If a non-root v of F is unifurcate, then *contracting v in F* is the operation that modifies F by first adding an edge from the parent of v to the child of v and then deleting v. *Binarizing F* is the operation that modifies F by repeatedly contracting a unifurcate node until no node is unifurcate. Note that binarizing F yields a rooted binary forest. For example, binarizing the tree in Fig. 1(4) yields the tree in Fig. 1(5).

For convenience, we view each node u of F as an ancestor and descendant of itself. A *proper* ancestor (respectively, descendant) of u in F is an ancestor (respectively, descendant) of u different from u. For a node v of F, the *subtree*

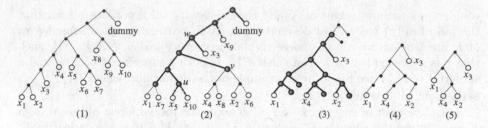

Fig. 1. (1) A tree T, (2) another tree F, (3) $F|^+_{\{x_1,\ldots,x_4\}}$, (4) $F\uparrow_{\{x_1,\ldots,x_4\}}$, and (5) $F\updownarrow_{\{x_1,\ldots,x_4\}}$.

F_v *of* F *rooted at* v is the tree obtained from F by deleting all nodes that are not descendants of v in F. For each node v of F, F_v is called a *dangling subtree* of F. In particular, if v is a root of F, then F_v is a *component* of F. We use $|F|$ to denote the number of components in F. A node u is *lower* (respectively, *higher*) than another node v in F if u is a proper descendant (respectively, ancestor) of v in F. For a set or sequence S of leaves in the same component of F, $\ell_F(S)$ denotes the lowest common ancestor (LCA) of the leaves in S.

Let u and v be two nodes in the same component of F. If u and v have the same parent in F, then they are *siblings* in F. We use $u \sim_F v$ to denote the (undirected) path between u and v in F. The *length* of $u \sim_F v$ is the number of edges in $u \sim_F v$. Each node of $u \sim_F v$ other than u and v is an *internal node* of $u \sim_F v$. If u is a root and v is a leaf in F, then $u \sim_F v$ is a *root-leaf path*. The *height* of F is the maximum length of a root-leaf path in F. If neither $\ell_F(u,v) = u$ nor $\ell_F(u,v) = v$, then u and v are *incomparable* in F; otherwise, they are *comparable*. A *dangling edge between* u *and* v in F is an edge e in F such that the tail of e is an internal node of $u \sim_F v$ but the head of e is not a node of $u \sim_F v$. $D_F(u,v)$ denotes the set of dangling edges between u and v in F. For example, in Fig. 1, $D_F(x_1, x_9) = \{e_F(x_7), e_F(u), e_F(v), e_F(x_3)\}$, while $D_F(x_1, w) = \{e_F(x_7), e_F(u), e_F(v)\}$. A *dangling subtree between* u *and* v in F is the subtree rooted at the head of an edge in $D_F(u,v)$. A *hovering subforest* of F is a subforest F' of F such that the in- and out-degrees of each non-leaf v in F' are the same as those of v in F, respectively. Note that F' can be obtained from F by deleting zero or more components and deleting the proper descendants of zero or more incomparable nodes. Of course, if F is a tree, then each nonempty hovering subforest of F is indeed a tree and is hence called a *hovering subtree* of F. For example, in Fig. 1(2), the bold (dashed or solid) edges show a hovering subtree of F.

Let X be a subset of $L(F)$. $F\uparrow_X$ denotes the rooted forest obtained from F by first removing all nodes with no leaf descendant in X and then repeatedly contracting a unifurcate root until no root is unifurcate. $F\updownarrow_X$ denotes the rooted forest obtained by binarizing $F\uparrow_X$. $F\uparrow^+_X$ denotes the rooted forest obtained from F by deleting those nodes u such that no node of $F\uparrow_X$ is the parent or a child of u in F. $F|^+_X$ denotes the rooted forest obtained from F by first removing all components without leaves in X and then removing all non-roots v such that the tail of $e_F(v)$ has no leaf descendant in X. In particular, when some leaves of F are labeled, we always delete the label of every leaf $y \notin X$ in $F|_X^+$ such that y is a labeled leaf of F. Note that $F\uparrow_X^+$ is both a superforest of $F\uparrow_X$ and a subforest of $F|_X^+$. See Fig. 1 for an example of the definitions, where $F\uparrow_{\{x_1,\ldots,x_4\}}^+$ is the bold subtree in Fig. 1(3).

Let C be a set of edges in F. $F - C$ denotes the rooted forest obtained from F by deleting the edges in C. A leaf in $F - C$ is *old* if it is also a leaf in F; otherwise, it is *new*. $F \ominus C$ denotes the rooted binary forest obtained from $F - C$ by first deleting all nodes with no old leaf descendants and then binarizing it. C_F^+ denotes $C \cup D_1 \cup D_2$, where D_1 is the set of edges (p, c) in $F - C$ such that every leaf descendant of c in $F - C$ is new, while D_2 is the set of edges (p, c) in $F - (C \cup D_1)$ such that each ancestor of p in $F - (C \cup D_1)$ is unifurcate. C is a *cut* of F if every component of $F - C$ has an old leaf. C is a *canonical cut* of F if all leaves of $F - C$ are old. For example, if C is the set of 7 dashed edges in Fig. 1(2), then C is a cut of F but is not canonical (because of the new leaf u), $F \ominus C$ is the tree in Fig. 1(5) together with the 7 root leaves $x_5, \ldots, x_{10}$, *dummy*, while $C_F^+ = C \cup \{e_F(u), e_F(w), e_F(dummy)\}$. It is known that if C is a set of edges in F, then F has a canonical cut C' such that $F \ominus C = F \ominus C'$ [4]. For this reason, canonical cuts C have been frequently used in the literature when we are concerned about only $F \ominus C$.

For a node v with parent p and sibling u in F, *detaching the dangling subtree with root v* is the operation that modifies F by first deleting the edge (p, v) and then contracting p. A *detaching operation* on F is the operation of detaching a dangling subtree of F.

Suppose that F and $\hat{F}$ are two rooted forests such that some leaves of $\hat{F}$ may be unlabeled but the labeled leaves of $\hat{F}$ one-to-one correspond to the leaves of F (i.e., each pair of corresponding leaves have the same label). We can extend the correspondence between the labeled leaves of F and $\hat{F}$ to their non-leaf vertices recursively as follows. If a non-root vertex v of F corresponds to a non-root vertex $\hat{v}$ of $\hat{F}$ and the parents of v and $\hat{v}$ are both unifurcate, then their parents correspond to each other. Similarly, if v and v' are siblings in F and they respectively correspond to $\hat{v}$ and $\hat{v}'$ in $\hat{F}$ such that $\hat{v}$ and $\hat{v}'$ are siblings, then their parents correspond to each other. For convenience, whenever a vertex v of F corresponds to a vertex $\hat{v}$ of $\hat{F}$, we always view F_v and $\hat{F}_{\hat{v}}$ as the same tree (although they are in different forests).

3 Configurations and Search Trees

In general, in order to find a difference between a rooted tree T and a rooted forest F with the same set of labeled leaves, it suffices to look at only a portion of T and a portion of F (rather than the whole T and F). This motivates us to define a *configuration* to be a pair (T, F), where T is a rooted binary tree whose leaves have distinct labels and F is a rooted binary forest satisfying the following conditions:

- Some leaves of F may be unlabeled but the labeled leaves one-to-one correspond to the leaves of T (i.e., each pair of corresponding leaves have the same label).
- No component of F consists of a single node and no two unlabeled leaves are siblings in F.
- If two labeled leaves are siblings in F, then they are not siblings in T.

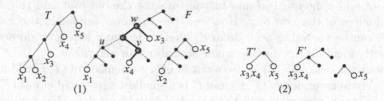

Fig. 2. (1) A configuration (T, F) and (2) an induced subconfiguration (T', F') of (T, F).

Figure 2(1) shows an example of configuration. For convenience, we view $(\perp, \perp)$ as a configuration, where $\perp$ stands for the *empty tree* (i.e., a tree without nodes). The *size* of (T, F) is the number of leaves in T, while the *height* of (T, F) is that of T. (T, F) is a *final* configuration if all leaves of F are labeled. Note that if (T, F) is final and v is a node in T, then $(T_v, F|_{L(T_v)}^+)$ is a configuration. If (T, F) is not final, then an *extension* of (T, F) is a final configuration (E_1, E_2) such that T is a dangling subtree of E_1 and F is a hovering subforest of E_2.

The *subconfiguration of* (T, F) *induced by a nonempty cut* C *of* F is the configuration (T', F') obtained as follows.

1. Initially, $T' = T$, and F' is obtained from $F \ominus C$ by deleting those components with no labeled leaves.
2. While F' has a component K that is also a dangling subtree of T', delete K from F' and modify T' as follows: if $K = T'$, delete K; otherwise, detach K and further delete it.
3. While T' and F' have a common pair (x, y) of sibling leaves, merge x, y, and their parent into a single labeled leaf whose label is xy (i.e., the concatenation of the labels of x and y).
4. For each non-leaf v of F' such that v has no labeled leaf descendant in F' but the parent of v has one, merge the dangling subtree with root v into a single unlabeled leaf.

For example, if C consists of the 4 dashed edges in Fig. 2(1), then the subconfiguration induced by C is as in Fig. 2(2). Note that if (T, F) is final, so is (T', F'). This holds no matter whether C is canonical or not, because the leaves of $F \ominus C$ are labeled when so are the leaves of F. If (T, F) is final and $(T', F') = (\perp, \perp)$, then C is called an *agreement cut* of (T, F) and $F \ominus C$ is called an *agreement forest* of (T, F). If in addition, C is a canonical cut of F, then C is a *canonical agreement cut* of (T, F). The smallest size of an agreement cut of a final (T, F) is called the *rSPR distance* between T and F, and is denoted by $d(T, F)$. It is widely known that $d(T, F)$ is also the smallest number of detaching operations we can perform on each of T and F so that they become the same rooted binary forest.

It is worth pointing out that to compute the rSPR distance for a final (T, F), it is required in the literature that we preprocess each of T and F by first adding a new root and a *dummy* leaf and further making the old root and the *dummy* be the children of the new root. However, the common *dummy* in the modified T and F can be viewed as an ordinary (labeled) leaf and hence we do not have to explicitly mention the *dummy* when describing an algorithm.

Obviously, a cut of (T, F) is also a cut of every extension of (T, F). Moreover, if (E_1, E_2) is an extension of (T, F) and C is a smallest agreement cut of (E_1, E_2), then $E_2 \ominus C$ is an agreement forest of (E_1, E_2) with $|E_2| + d(E_1, E_2)$ components. So, if C is a cut of (T, F), then for every extension (E_1, E_2) of (T, F), C together with a smallest agreement cut S of the subconfiguration (E_1', E_2') of (E_1, E_2) induced by C yields an agreement cut $C \cup S$ of (E_1, E_2) such that $E_2 \ominus (C \cup S)$ is an agreement forest of (E_1, E_2) with $(|E_2| + |C|) + d(E_1', E_2')$ components. The *local ratio* achieved by a cut C of (T, F) is

$$r(C) = \max_{(E_1, E_2)} \frac{(|C| + d(E_1', E_2')) - d(E_1, E_2)}{d(E_1, E_2) - d(E_1', E_2')},$$

where (E_1, E_2) ranges over all extensions of (T, F) and (E_1', E_2') is the subconfiguration of (E_1, E_2) induced by C.

If for some small constant $\delta \geq 0$, we can always find a cut C of a given configuration (T, F) with $r(C) \leq \delta$, then we can design a recursive approximation algorithm for rSPR distance that achieves a ratio of $1 + \delta$. This can be easily shown but we omit the details here because finding such a C seems to be very difficult in general. In some special cases such as the following G1 and G2, we do know how to find a cut C such that $d(E_1, E_2) - d(E_1', E_2') = 1$ and $r(C) = 0$:

G1. T has a pair (x, y) of sibling leaves with $|D_F(x, y)| = 1$.
G2. F has a pair (x, y) of sibling leaves such that x and y are labeled leaves, $|D_T(x, y)| = 1$, and the head of the unique edge in $D_T(x, y)$ is a leaf z of T.

G1 was observed in [16]. G2 follows from G1 and was first used in [7]. Since in the two cases we can achieve the best $r(C)$, we will always assume that *neither G1 nor G2 occurs in a configuration* (except when we describe an algorithm).

Since calculating $d(E_1, E_2) - d(E_1', E_2')$ is difficult, an alternative way is to calculate an upper bound on $(d(E_1', E_2') + |C|) - d(E_1, E_2)$ and a lower bound on $d(E_1, E_2) - d(E_1', E_2')$. This motivates us to consider *search trees* below.

If (T, F) is final, a simple way to compute $d(T, F)$ is to build a *search tree* Γ as follows. The root of Γ is $(\emptyset, \emptyset)$. In general, each node of Γ is a pair (C_T, C_F) satisfying the following conditions:

- C_T and C_F are canonical cuts of T and F, respectively.
- All but one of the roots of $T - C_T$ are *agreed*, where a node u of $T - C_T$ is *agreed* if $F - C_F$ has a node v such that binarizing the dangling subtree of $T - C_T$ with root u and binarizing the dangling subtree of $F - C_F$ with root v yield the same tree.
- If (C_T, C_F) is left as a leaf in Γ, then all roots of $T - C_T$ must be agreed.

Now, suppose that a node (C_T, C_F) of Γ has been constructed but should not be left as a leaf in Γ. To construct the children of (C_T, C_F) in Γ, we first select a bifurcate node u in $T - C_T$ such that u is still not agreed but its children u_1 and u_2 in $T - C_T$ are already agreed. We can find such a u because modifying $T - C_T$ by deleting the proper descendants of all agreed nodes yields a binary forest in which all but one components are root leaves. For each $i \in \{1, 2\}$, let v_i be the highest node in $F - C_F$ agreed with u_i. For convenience, let $F' = F - C_F$. The children of (C_T, C_F) are constructed by distinguishing three cases as follows:

Case 1: v_1 or v_2 is a root of F'. If v_1 is a root of F', then $(C_T \cup \{(u, u_1)\}, C_F)$ is the only child of (C_T, C_F) in Γ; otherwise, $(C_T \cup \{(u, u_2)\}, C_F)$ is the only child of (C_T, C_F) in Γ.

Case 2: v_1 and v_2 fall into different components of F' but Case 1 does not occur. In this case, (C_T, C_F) has two children in Γ, where for each $i \in \{1, 2\}$, the i-th child of (C_T, C_F) in Γ is $(C_T \cup \{(u, u_i)\}, C_F \cup \{e_{F'}(v_i)\})$.

Case 3: v_1 and v_2 fall into the same component of F'. In this case, (T', F') has three children in Γ. The first two are constructed as in *Case 2*. The third child is $(C_T, C_F \cup D_{F'}(v_1, v_2))$.

Note that the children of (C_T, C_F) in Γ are ordered. This finishes the construction of Γ.

Let P be a root-leaf path in Γ. We use $C(P)$ to denote the canonical cut of F contained in the leaf of P. Obviously, $C(P)$ is a canonical agreement cut of (T, F). For a positive integer $k \leq 3$, P *picks a cherry* (Y_1, Y_2) *in the k-th way* if P has a non-leaf (C_T, C_F) satisfying the following conditions:

- The k-th child of (C_T, C_F) exists in Γ and also appears in P.
- For the bifurcate node u in $T - C_T$ selected to construct the children of (C_T, C_F) in Γ, Y_1 (respectively, Y_2) is the set of leaf descendants of u_1 (respectively, u_2) in $T - C_T$, where u_1 and u_2 are the children of u in $T - C_T$.

P picks a cherry (Y_1, Y_2) if it does so in the k-th way for some positive $k \leq 3$. For a set Y of leaves in T, P *isolates* Y if it picks a cherry (Y, Z) in the first way or a cherry (Z, Y) in the second way, while P *creates* Y if it picks a cherry (Y_1, Y_2) with $Y = Y_1 \cup Y_2$ in the third way.

(T, F) may have multiple search trees (depending on the order of cherries picked by root-leaf paths in a search tree). Nonetheless, it is widely known that if (T, F) is final, then for each search tree Γ of (T, F), $d(T, F) = \min_P |C(P)|$, where P ranges over all root-leaf paths in Γ [15].

Even if (T, F) is not final, we can still construct a search tree Γ of (T, F) as above. For each root-leaf path P in Γ, $C(P)$ is still a canonical cut of F but is not an agreement cut of (T, F) because unlabeled leaves exist in F. Moreover, if (E_1, E_2) is an extension of (T, F), then each search tree of (T, F) is a hovering subtree of some search tree of (E_1, E_2) because T is a dangling subtree of E_1 and F is a hovering subforest of E_2.

Let X be a set of leaves in T such that each component of $T{\uparrow}_X$ is a dangling subtree of T. An X-search tree Γ_X of (T, F) is constructed almost in the same way as a search tree of (T, F); the only differences are as follows:

- For each leaf (C_T, C_F) in Γ_X, it is impossible to find a bifurcate node u in $T - C_T$ such that all leaf descendants of u in $T - C_T$ belong to X and u is still not agreed but its children in $T - C_T$ are.
- To construct the children of a non-leaf (C_T, C_F) in Γ_X, we always select a bifurcate node u in $T - C_T$ such that all leaf descendants of u in $T - C_T$ belong to X and u is still not agreed but its children in $T - C_T$ are.

Obviously, an X-search tree Γ_X of (T, F) is a hovering subtree of some search tree of (T, F). Moreover, for each X-search tree Γ_X of (T, F) and for each extension (E_1, E_2) of (T, F), the following hold (because Γ_X is a hovering subtree of some search tree of (E_1, E_2)):

- For each root-leaf path P in Γ_X, $C(P)$ is a canonical cut of F and can be extended to a canonical agreement cut of (E_1, E_2).
- Γ_X has at least one path P such that $C(P)$ can be extended to a smallest canonical agreement cut of (E_1, E_2).

4   Keys and Local Ratios

Throughout this section, let (T, F) be a configuration. Instead of cuts, we consider a more useful notion of *key*. A *key* of (T, F) is a triple $\kappa = (X, B, R)$ satisfying the following conditions:

1. X is a set of leaves in T such that each component of $T{\uparrow}_X$ is a dangling subtree of T.
2. B is a set of edges in $F{\uparrow}_X^+$ such that each component of $F - B$ contains at least one (labeled or unlabeled) leaf of F.
3. R is a set of edges in $F{\uparrow}_X$ such that $R \cap B_F^+ = \emptyset$.
4. If $R = \emptyset$, then $e_F(x) \in B_F^+$ for every $x \in X$. Otherwise, for the set $Y = \{x \in X \mid e_F(x) \notin B_F^+\}$, we have that $T{\downarrow}_Y = F{\downarrow}_Y$, R is the edge set of $F{\uparrow}_Y$, and B contains every edge (p, c) of F such that (p, c) is not an edge of $F{\uparrow}_Y$ but p is a node of $F{\uparrow}_Y$.

For example, if (T, F) is as in Fig. 2(1), then $\kappa_e = (X, B, R)$ is a key of (T, F), where $X = \{x_1, \ldots, x_4\}$, B consists of the 4 dashed edges in F, and R is the edge set of $x_3 \sim_F x_4$.

The intuition behind the definition of key is as follows. Consider an X-search tree Γ_X of (T, F) and an extension (E_1, E_2) of (T, F). Recall that for each leaf (C_T, C_F) of Γ_X, C_T and C_F are respectively canonical cuts of T and F, and each vertex u of $T\!\uparrow_X - C_T$ is agreed with a vertex v of $F - C_F$. So, if we let $B = C_F$ and R be the edge set of $F\!\uparrow_Y$ with $Y = \{x \in X \mid e_F(x) \notin B_F^+\}$, then (X, B, R) is a key of (T, F). Moreover, at least one leaf (C_T, C_F) of Γ_X satisfies that C_F can be extended to a smallest canonical agreement cut of (E_1, E_2). However, by only looking at Γ_X, we don't know which leaf of Γ_X has this property. So, instead of finding such a C_F, we compromise by finding a key of (T, F).

In most cases, $X = L(T)$ holds and we simply write $\kappa = (B, R)$ instead of $\kappa = (X, B, R)$. If $R = \emptyset$, then κ is *normal* and we simply write $\kappa = (X, B)$ instead of $\kappa = (X, B, R)$; otherwise, it is *abnormal*. In case $X = L(T)$ and $R = \emptyset$, we further simply write $\kappa = B$ instead of $\kappa = (X, B, R)$ and $\kappa = (X, B)$. In essence, only normal keys were considered in [8].

In the remainder of this section, let $\kappa = (X, B, R)$ be a key of (T, F). The *size* of κ is $s(\kappa) = |B|$. The *subconfiguration of (T, F) induced by* κ is the subconfiguration (T', F') of (T, F) induced by B. Let P be a root-leaf path in an X-search tree of (T, F), and $S = C(P) \setminus R$. An edge $e \in B$ is *free* with respect to (w.r.t.) P if $e \in S$ or the leaf descendants of the head of e in $F - (S \cup B)$ are all new. Let $f_e(\kappa, P)$ denote the set of edges in B that are free w.r.t. P. A component K in $F - (S \cup (B \setminus f_e(\kappa, P)))$ is *free* if the leaves of K are new and there is at least one edge $e \in B \setminus f_e(\kappa, P)$ whose tail is a leaf of K. Let $f_c(\kappa, P)$ denote the set of free components in $F - (S \cup (B \setminus f_e(\kappa, P)))$. For example (cf., Fig. 2, if κ_e is the above example key and P isolates x_2 and further creates $\{x_1, x_3\}$ and isolates it, then $C(P) \cap R = \{e_F(v)\}$, $f_e(\kappa_e, P) = \{e_F(x_2), e_F(u)\}$ but $f_c(\kappa_e, P) = \emptyset$. On the other hand, if we modify the example κ_e by setting $R = \emptyset$ and $B = \{e_F(x_1), e_F(x_2), e_F(x_3)\}$, then for the same P, $C(P) \cap R = \emptyset$, $f_e(\kappa_e, P) = \{e_F(x_2)\}$, and $f_c(\kappa_e, P)$ contains a unique component (which is just a path from w to the parent of x_1 in F). The *lower bound* achieved by κ w.r.t. P is $b(\kappa, P) = |f_e(\kappa, P)| + |f_c(\kappa, P)| + |C(P) \cap R|$.

Lemma 1. *Let P be a root-leaf path in an X-search tree of (T, F). Then, for every extension (E_1, E_2) of (T, F), if $C(P)$ is a subset of some smallest canonical agreement cut of (E_1, E_2), then $d(E_1, E_2) - d(E_1', E_2') \geq b(\kappa, P)$, where (E_1', E_2') is the subconfiguration of (E_1, E_2) induced by κ.*

The *lower bound* achieved by κ is $b(\kappa) = \max_{\Gamma_X} \min_P b(\kappa, P)$, where Γ_X ranges over all X-search trees of (T, F) and P ranges over all root-leaf paths in Γ_X. The *local ratio* achieved by κ is $r(\kappa) = \frac{s(\kappa) - b(\kappa)}{b(\kappa)}$. If $r(\kappa) \leq \frac{4}{3}$, then we say that κ is *good*. The next lemma may help the reader understand the definitions (especially for abnormal keys).

Lemma 2. *If T has a sibling-leaf pair (x_1, x_2) such that $\ell_F(x_1, x_2)$ exists and $|D_F(x_1, x_2)| = 2$, then $\kappa = (\{x_1, x_2\}, D_F(x_1, x_2), R)$ is an abnormal key of (T, F) with $s(\kappa) = 2$ and $b(\kappa) \geq 1$, where R is the edge set of $x_1 \sim_F x_2$.*

Sometimes, finding a good key of (T, F) may be difficult. In such cases, we look for a *combined key* of (T, F) which is an (ordered) pair $\Psi = (\kappa_1, \kappa_2)$ such that κ_1 is a key of (T, F) and κ_2 is a key of the subconfiguration (T', F') of (T, F) induced by κ_1. Ψ is *normal* if both κ_1 and κ_2 are normal; otherwise, Ψ is *abnormal*. The *subconfiguration of (T, F) induced by Ψ* is the subconfiguration of (T', F') induced by κ_2. The *size* of Ψ is $s(\Psi) = s(\kappa_1) + s(\kappa_2)$. The *lower bound* achieved by Ψ is $b(\Psi) = b(\kappa_1) + b(\kappa_2)$. The *local ratio* achieved by Ψ is $r(\Psi) = \frac{s(\Psi) - b(\Psi)}{b(\Psi)}$. If $r(\Psi) \leq \frac{4}{3}$, then we say that Ψ is *good*.

5 The Algorithm and Its Performance

To approximate $d(T, F)$ for a final configuration (T, F), we proceed as follows.

1. If T has at most six leaves, then compute $d(T, F)$ exactly (say, by brute force) and return it.
2. If T has a sibling-leaf pair (x_1, x_2) with $|D_F(x_1, x_2)| = 1$, then perform the following steps:
   (a) Compute the subconfiguration (T', F') of (T, F) induced by $D_F(x_1, x_2)$.
   (b) Recursively compute an approximation d_1 of $d(T', F')$.
   (c) Return $d_1 + 1$ (as the approximation of $d(T, F)$).
3. If F has a sibling-leaf pair (x_1, x_2) such that both x_1 and x_2 are labeled, $|D_T(x_1, x_2)| = 1$, and the head of the unique edge in $D_T(x_1, x_2)$ is a leaf x_3, then perform the following steps:
   (a) Compute the subconfiguration (T', F') of (T, F) induced by $\{e_F(x_3)\}$.
   (b) Recursively compute an approximation d_2 of $d(T', F')$.
   (c) Return $d_2 + 1$ (as the approximation of $d(T, F)$).
4. Find a sibling-leaf pair (x_1, x_2) in T such that the distance from the root to the parent v of x_1 and x_2 in T is maximized over all sibling-leaf pairs in T.
5. Try to find a good key κ or a good combined key $\Psi = (\kappa_1, \kappa_2)$ of $(T_v, F|_{L(T_v)}^+)$. (The search for κ and Ψ is done via case analysis. In particular, we can show that κ or Ψ exists if T_v has at least four leaves. Note that each unlabeled leaf in $F|_{L(T_v)}^+$ is the root of a dangling subtree of F.)
6. If neither κ nor Ψ is found in Step 5, then replace v by its parent in T, and go back to Step 5.
7. If κ is found in Step 5, then perform the following steps:
   (a) Compute the subconfiguration (T', F') of (T, F) induced by κ.
   (b) Recursively compute an approximation d_3 of $d(T', F')$.
   (c) Return $d_3 + s(\kappa)$ (as the approximation of $d(T, F)$).
8. If Ψ is found in Step 5, then perform the following steps:
   (a) Compute the subconfiguration (T', F') of (T, F) induced by κ_1, and then compute the subconfiguration (T'', F'') of (T', F') induced by κ_2.
   (b) Recursively compute an approximation d_4 of $d(T'', F'')$.
   (c) Return $d_4 + s(\Psi)$ (as the approximation of $d(T, F)$).

Theorem 1. *The algorithm achieves an approximation ratio of $\frac{7}{3}$ and runs in quadratic time.*

6 Concluding Remarks

We conjecture that rSPR distance can be approximated within a ratio of $2 + \epsilon$ for any constant $\epsilon > 0$. Basically, we have shown that for each configuration (T, F) such that T has at most 6 leaves, we can find a key κ whose local ratio is at most $\frac{4}{3}$. To design an approximation algorithm with a better ratio than $\frac{7}{3}$, it seems promising to let the algorithm search for larger configurations (T, F). The main difficulty is how to avoid complicated case analysis.

References

1. Baroni, M., Grunewald, S., Moulton, V., Semple, C.: Bounding the number of hybridisation events for a consistent evolutionary history. J. Math. Biol. **51**, 171–182 (2005)
2. Beiko, R.G., Hamilton, N.: Phylogenetic identification of lateral genetic transfer events. BMC Evol. Biol. **6**, 159–169 (2006)
3. Bonet, M.L., John, K.S., Mahindru, R., Amenta, N.: Approximating subtree distances between phylogenies. J. Comput. Biol. **13**, 1419–1434 (2006)
4. Bordewich, M., McCartin, C., Semple, C.: A 3-approximation algorithm for the subtree distance between phylogenies. J. Discrete Algorithms **6**, 458–471 (2008)
5. Bordewich, M., Semple, C.: On the computational complexity of the rooted subtree prune and regraft distance. Ann. Comb. **8**, 409–423 (2005)
6. Chen, Z.-Z., Wang, L.: FastHN: a fast tool for minimum hybridization networks. BMC Bioinformatics **13**, 155 (2012)
7. Chen, Z.-Z., Fan, Y., Wang, L.: Faster exact computation of rSPR distance. J. Comb. Optim. **29**(3), 605–635 (2015)
8. Shi, F., Feng, Q., You, J., Wang, J.: Improved approximation algorithm for maximum agreement forest of two rooted binary phylogenetic trees. J. Comb. Optim (to appear). doi:10.1007/s10878-015-9921-7
9. Hein, J., Jing, T., Wang, L., Zhang, K.: On the complexity of comparing evolutionary trees. Discrete Appl. Math. **71**, 153–169 (1996)
10. Ma, B., Wang, L., Zhang, L.: Fitting distances by tree metrics with increment error. J. Comb. Optim. **3**, 213–225 (1999)
11. Maddison, W.P.: Gene trees in species trees. Syst. Biol. **46**, 523–536 (1997)
12. Nakhleh, L., Warnow, T., Lindner, C.R., John, L.S.: Reconstructing reticulate evolution in species - theory and practice. J. Comput. Biol. **12**, 796–811 (2005)
13. Rodrigues, E.M., Sagot, M.-F., Wakabayashi, Y.: The maximum agreement forest problem: approximation algorithms and computational experiments. Theoret. Comput. Sci. **374**, 91–110 (2007)
14. Wu, Y.: A practical method for exact computation of subtree prune and regraft distance. Bioinformatics **25**(2), 190–196 (2009)
15. Whidden, C., Beiko, R.G., Zeh, N.: Fast FPT algorithms for computing rooted agreement forests: theory and experiments. In: Festa, P. (ed.) SEA 2010. LNCS, vol. 6049, pp. 141–153. Springer, Heidelberg (2010)
16. Whidden, C., Zeh, N.: A unifying view on approximation and FPT of agreement forests. In: Salzberg, S.L., Warnow, T. (eds.) WABI 2009. LNCS, vol. 5724, pp. 390–402. Springer, Heidelberg (2009)

Scheduling Algorithms
and Circuit Complexity

Online Non-preemptive Scheduling to Optimize Max Stretch on a Single Machine

Pierre-Francois Dutot[1], Erik Saule[2], Abhinav Srivastav[1(✉)],
and Denis Trystram[1]

[1] University of Grenoble Alpes, Saint-Martin-d'Hères, France
{pfdutot,srivasta,trystram}@imag.fr
[2] Department of Computer Science, University of North Carolina at Charlotte,
Charlotte, USA
esaule@uncc.edu

Abstract. We consider in this work a classical online scheduling problem with release times on a single machine. The quality of service of a job is measured by its *stretch*, which is defined as the ratio of its response time over its processing time. Our objective is to schedule the jobs non-preemptively in order to optimize the maximum stretch. We present both positive and negative theoretical results. First, we provide an online algorithm based on a waiting strategy which is $(1 + \frac{\sqrt{5}-1}{2}\Delta)$-competitive where Δ is the upper bound on the ratio of processing times of any two jobs. Then, we show that no online algorithm has a competitive ratio better than $\frac{\sqrt{5}-1}{2}\Delta$. The proposed algorithm is asymptotically the best algorithm for optimizing the maximum stretch on a single machine.

1 Introduction

Scheduling independent jobs that arrive over time is a fundamental problem that arises in many applications. Often, the aim of a scheduler is to optimize some function(s) that measure the performance or quality of service delivered to the jobs. The most popular and relevant metrics include throughput maximization, minimization of maximum or average completion times and optimizing the flow time [1]. These metrics have received a lot of attention over the last years in various scenarios: on single or multiple machines, in online or offline settings, in weighted or unweighted settings, etc. One of the most relevant performance measures in job scheduling is the *fair* amount of time that the jobs spend in the system. This includes the waiting time due to processing some other jobs as well as the actual processing time of the job itself. Such scheduling problems arise for instance while scheduling jobs in parallel computing platforms. The *stretch* is the factor by which a job is slowed down with respect to the time it takes on an unloaded system [2].

Here, we are interested in scheduling a stream of jobs to minimize the maximum stretch (*max-stretch*) on a single machine. This problem is denoted as $1|r_i, online|S_{max}$ in the classical 3-fields notation of scheduling problems [3]. While this problem admits no constant approximation algorithm in the offline

© Springer International Publishing Switzerland 2016
T.N. Dinh and M.T. Thai (Eds.): COCOON 2016, LNCS 9797, pp. 483–495, 2016.
DOI: 10.1007/978-3-319-42634-1_39

case [2], interesting results can be derived by introducing an instance-dependent parameter Δ: the ratio between the largest and the smallest processing time in the instance.

We show using an adversary technique, that no online algorithm can achieve a competitive ratio better than $\alpha\Delta$ where $\alpha = \frac{\sqrt{5}-1}{2}$ (the golden ratio). This improves upon the previously best known lower bound of $\frac{1+\Delta}{2}$ by Saule *et al.* [4].

Based on the observation that no greedy algorithm can reach this lower bound, we designed Wait-Deadline Algorithm (WDA) which enforces some amount of waiting time for large jobs, before they can be scheduled. We prove that WDA has a competitive ratio of $1 + \alpha\Delta$, which improves upon the best known competitive ratio of Δ achieved by First-Come First-Served and presented by Legrand *et al.* [5].

The competitive ratio of WDA $(1+\alpha\Delta)$ and the lower bound on best achievable competitive ratio $(\alpha\Delta)$ are asymptotically equal when Δ goes to infinity. In other words, this paper essentially closes the problem of minimizing max-stretch on a single machine.

This paper is organized as follows. Section 2 defines the problem formally and summarizes the main positive and negative results that relate to optimizing the maximum stretch objective. Section 3 provides lower bounds on the competitive ratio of deterministic algorithms for both objectives and it indicates that algorithms with good competitive ratios have to wait before executing large jobs. Section 4 presents the wait-deadline algorithm (WDA). Then we provide the corresponding detailed analysis for the competitive ratio of max-stretch in Sect. 5. Finally, we provide concluding remarks in Sect. 6 and discuss future issues for the continuation of this work.

2   Problem Definition and Related Works

We study the problem of scheduling on a single machine n independent jobs that arrive over time. A scheduling instance is specified by the set of jobs J. The objective is to execute the continuously arriving stream of jobs. We consider the clairvoyant version of the problem where the processing time p_i of each job i is only known at its release time r_i. Without loss of generality, we assume that the smallest and largest processing times are equal to 1 and Δ, respectively. We also assume that the scheduler knows the value of Δ which is a common assumption in online schedulers.

In a given schedule σ_i, C_i and S_i denote the start time, completion time and stretch of job i, respectively where $S_i = \frac{C_i - r_i}{p_i}$. We are interested in minimizing $S_{max} = \max_{j \in J} S_j$.

An online algorithm is said to be ρ-competitive if the worst case ratio (over all possible instances) of the objective value of the schedule generated by the algorithm is no more that ρ times the performance of the optimal (offline clairvoyant) algorithm [6].

Bender *et al.* introduced the stretch performance objective to study the fairness for HTTP requests arriving at web servers [2]. They showed that the problem of optimizing *max-stretch* in a non-preemptive offline setting cannot be approximated within a factor of $\Omega(n^{1-\epsilon})$, unless $P = NP$. They also showed that any online algorithm has a competitive ratio in $\Omega(\Delta^{\frac{1}{3}})$. Finally, they provided an online preemptive algorithm using the classical EDF strategy (*earliest deadline first*) and showed that it is $O(\sqrt{\Delta})$ competitive.

Later, Legrand *et al.* showed that the First-Come First-Served algorithm (FCFS) is Δ-competitive for the *max-stretch* problem on a single machine [5]. Since preemption is not used in FCFS, the above bound is also valid in the non-preemptive case. They also showed that the problem of optimizing *max-stretch* on a single machine with preemption cannot be approximated within a factor of $\frac{1}{2}\Delta^{\sqrt{2}-1}$. Saule *et al.* showed that all approximation algorithms for the single machine problem and m parallel machine of optimizing max-stretch cannot have a competitive ratio better than $\frac{1+\Delta}{2}$ and $(1 + \frac{\Delta}{m+1})/2$, respectively [4]. Bansal and Pruhs [7], Golovin *et al.* [8], Im and Moseley [9] and Anand *et al.* [10] studied similar problems with resource augmentation.

3   Lower Bounds on Competitive Ratios for Max-Stretch

Observation 1. *Any greedy algorithm for scheduling jobs on a single machine has a competitive ratio of at least Δ for max-stretch.*

For non-preemptive schedules, it is easy to prove that any greedy algorithm is at least Δ-competitive using the following adversary technique. At time 0 a large job of processing time Δ arrives. Any greedy algorithm schedules it immediately. At time ϵ, a small job of processing time 1 is released. Since preemption is not allowed, the greedy algorithm can only schedule the small job at time $t = \Delta$ and thus $S_{max} \approx \Delta$. The optimal algorithm finishes the small job first and hence has a stretch close to 1; more precisely of $S^* = \frac{\Delta+\epsilon}{\Delta}$.

Hence, for an improved bound, the algorithm should incorporate some waiting time strategies. We show below a lower bound on the competitive ratio of such algorithms using a similar adversary technique.

Theorem 2. *There is no ρ-competitive non-preemptive algorithm for optimizing max-stretch for any fixed $\rho < \frac{\sqrt{5}-1}{2}\Delta$.*

Proof. Let ALG be any scheduling algorithm. Consider the following behaviour of the adversary. At time 0 a job of size Δ is released. On the first hand, if ALG schedules this job of size Δ at time t such that $0 \le t \le \frac{\sqrt{5}-1}{2}\Delta$, then the adversary sends a job of size 1 at time $t + \epsilon$ where $0 < \epsilon \ll 1$. In which case, ALG achieves a max stretch of $S_{max} = \Delta + 1$ while the optimal schedule has a max stretch of $S^*_{max} = \frac{t+1}{\Delta} + 1$. Therefore, the competitive ratio of ALG is greater than (or equal to) $\frac{\sqrt{5}-1}{2}\Delta$, for sufficiently large values of Δ. On the other hand if $\Delta > t > \frac{\sqrt{5}-1}{2}\Delta$, then the adversary sends a job of size 1 at time Δ.

ALG reaches a max-stretch of $S_{max} = t + 1$ while the optimal solution has a max-stretch of $S^*_{max} = 1$. Hence, ALG has a competitive ratio greater than (or equal to) $\frac{\sqrt{5}-1}{2}\Delta$. Lastly, if ALG schedules the job at time t such that $t \geq \Delta$, then the adversary releases a job of size 1 at time $t + \epsilon$, where $0 < \epsilon \ll 1$. The competitive ratio of ALG is greater than $\frac{\sqrt{5}-1}{2}\Delta$ times the optimal schedule, since ALG achieves a max-stretch of $S_{max} = \Delta + 1$ while the optimal schedule has a max-stretch of $S^*_{max} = 1$. ◻

4   The Wait-Deadline Algorithm (WDA) for Streams of Jobs

We design an online non-preemptive algorithm for optimizing max-stretch on a single machine. To develop the intuition, we briefly consider the case where all the jobs have been released. The feasibility of scheduling the set of jobs within a given maximum stretch S can be easily determined since the stretch formula sets a deadline for each job. Knowing these deadlines, the best order of execution for the jobs is determined by the Earliest Deadline First (EDF) algorithm which schedules the jobs as soon as possible in the order of non-decreasing deadlines. EDF is known to schedule all released jobs before their deadlines on a single machine if such a schedule exists [1].

In the online setting, these deadlines cannot be computed in advance. Our algorithm emulates these deadlines in two ways: firstly by holding the large jobs for a fixed amount of time to avoid small worst cases as explained below, secondly by computing a feasible deadline for the currently available jobs and using it to select the next one to start.

Observation 1 indicates that any algorithm with a competitive ratio better than Δ for max-stretch must wait for some time before it starts scheduling large jobs due to the non-clairvoyant nature of arrival times of the jobs. Waiting strategies have been studied for the problem of minimizing weighted completion time [11,12]. To best our knowledge, this is the first work which studies waiting time strategies in the context of flow time. As stated before, our algorithm also needs to maintain an estimate of the *max-stretch* and adjust this estimate whenever EDF can not produce a feasible schedule.

We now describe the Wait-Deadline algorithm (WDA). We classify the jobs into two sets, namely *large set* and *small set* (denoted by J_{large} and J_{small}, respectively), based on their processing time. More specifically, $J_{small} = \{i \in J : 1 \leq p_i \leq 1 + \alpha\Delta\}$ and $J_{large} = \{i \in J : 1 + \alpha\Delta < p_i \leq \Delta\}$

We maintain two separate queues: the *Ready queue* (denoted by Q_R) and the *Wait queue* (denoted by Q_W). Whenever a job $i \in J_{small}$ is released, it is placed directly into the *Ready queue*. On the other hand, when a job $i \in J_{large}$ is released, it is initially placed in the *Wait queue* for αp_i units of time and then moved to the *Ready queue*.

Our algorithm is based on three kinds of events: (i) a job is released, (ii) a waiting period ends and (iii) a job ends. Whenever an event occurs the queues

Data: Ready queue Q_R at time t
Result: Job to be scheduled at time t
Perform binary search on *max-stretch* to find the appropriate deadline to
schedule all the jobs of Q_R;
Store the *max-stretch* estimate as a lower bound for the next binary search;
Return the job of Q_R with the earliest deadline where ties are broken according
to the processing time of the job (the shortest job is returned);

Algorithm 1. Job selection in WDA

Data: Q_R and Q_W are initially empty sets
Result: An online schedule
Wait for events to occur.
Let t be the time at which events occurred.
while *At least one event occurring at time t has not been processed* **do**
    switch *Event* **do**
        case *Job i has been released*
            if *the new job is in J_{small}* **then**
                | Update Q_R.
            else
                | Create a new event at time $t + \alpha p_i$ and update Q_W.
        case *Job i finished its waiting period*
            | Remove i from Q_W and add it to Q_R.
        case *Job i finished its execution*
            | Nothing special to do in this case for Q_R and Q_W.
if $Q_R \neq \emptyset$ *and the machine is idle* **then**
    | Select a new job to execute using Algorithm 1 and remove it from Q_R.
Return to the first line to wait for the next time instant when events occur.

Algorithm 2. Wait-Deadline algorithm

are updated, then if the *Ready queue* is not empty and the machine is idle, a job
is selected as depicted in the job selection pseudo-code in Algorithm 1.

Intuitively, we modify the release time of every job $i \in J_{large}$ to a new value
$r_i + \alpha p_i$. Let t be the time at which the machine becomes idle. Then the algorithm
sets the deadline $d_i(t)$ for each job $i \in Q_R$ where $d_i(t) = r_i + S(t)p_i$ and $S(t)$ is
the estimated *max-stretch* such that all the jobs in Q_R can be completed. Note
that the deadline $d_i(t)$ uses the original release time r_i rather than the modified
release date. For already released jobs, $S(t)$ can be computed in polynomial time
using a binary search similarly to the technique used in [2]. The upper bound for
the binary search can be derived from the FCFS schedule, while 1 is a natural
lower bound at time $t = 0$. At any later time $t > 0$, whenever a job has to
be selected for execution, WDA uses the previous stretch estimate as a lower
bound for the new binary search. As indicated in Algorithm 1, the job with
the earliest deadline is scheduled. Note that $S(t)$ is increasing with respect to
time t. We also assume that Δ is already known to WDA, which is a common
hypothesis for online scheduling algorithms. The entire procedure is summarized
in Algorithm 2.

Before we start with the competitive analysis, remember that $\alpha = \frac{\sqrt{5}-1}{2}$. Indeed Theorem 2 suggests that for an instance of two jobs with size 1 and Δ, it is optimal to wait for $\alpha\Delta$ time units before the job of size Δ is scheduled. When the size of the jobs can take any values between 1 and Δ, the partitioning of jobs in J_{small} and J_{large} ensures that small jobs can be scheduled as soon as they arrive while large jobs wait a fraction α of their processing time before they can be scheduled.

5   WDA Is $(1 + \alpha\Delta)$-Competitive for Max-Stretch

5.1   General Framework

We denote WDA the schedule produced by our algorithm and OPT some fixed optimal schedule. For the rest of this analysis, a superscript of $*$ indicates that the quantities in question refer OPT. We use r'_i to denote the modified released time of job i, that is $r'_i = r_i$ if job $i \in J_{small}$, otherwise $r'_i = r_i + \alpha p_i$. Moreover $d_i(t)$ denotes the estimated deadline of job i at time t i.e., $d_i(t) = r_i + S(t)p_i$.

Let z be the job in WDA that attains the *max-stretch* among the jobs in J. We remove all jobs from the instance J that are released after the start of job z without changing the S_z and without increasing the optimal stretch. Similarly, we also remove the set of jobs that are scheduled after the job z in WDA, without changing S_z and without increasing the optimal stretch. Therefore, we assume, without loss of generality, that z is the latest job in J that is processed in WDA.

Definition 3. *We define the set of jobs* Before z, *denoted by J_B, as the set of jobs that are scheduled during the interval $[r'_z, \sigma_z)$, that is: $J_B = \{i \in J : r'_z \leq \sigma_i < \sigma_z\}$.*

Property 4. *For all jobs in set* Before z, *at their start times, the deadlines of jobs are at most the deadline of job z. More formally, $d_i(\sigma_i) \leq d_z(\sigma_i) : \forall i \in J_B$.*

This simply stems from the fact that the job i starting at time $t = \sigma_i$ is selected because its deadline is the earliest.

Property 5. *The schedule WDA ensures that $\forall i \in J$ the machine is busy for during time interval $[r'_i, C_i)$.*

As soon as a job is completed, an event will be generated and a new job is selected to run if Q_R is not empty. Job i is in Q_R from its modified release date r'_i until its starting time σ_i.

Our general approach is to relate the stretch of job z with the stretch of another job in the optimal schedule. The completion time of job z in WDA can be written as $C_z = r_z + S_z p_z$.

In the optimal schedule OPT, there is a job which completes at or after time $C_z - \alpha\Delta$. This is due to the fact that $\alpha\Delta$ is the maximum difference between the makespan of schedules WDA and OPT. In the rest of this analysis, we denote

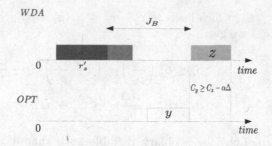

Fig. 1. Representation of jobs z and y in WDA and OPT schedule, respectively

such a job by y (refer to Fig. 1). Hence, the completion time of job y can be written as $C_y^* = r_y + S_y^* p_y \geq r_z + S_z p_z - \alpha\Delta$. Isolating S_z in the previous equation, we get:

$$S_z \leq S_y^* \left(\frac{p_y}{p_z} \right) + \frac{r_y - r_z}{p_z} + \frac{\alpha\Delta}{p_z} \qquad (1)$$

Theorem 6. *WDA is* $(1+\alpha\Delta)$-*competitive for the problem of minimizing max-stretch non-preemptively.*

The proof is constructed mainly in three separate parts: Lemmas 7, 11 and 17. Each part mostly relies on refining Eq. 1 in different cases. They are devised based on ratio of processing time of job z and job y, as defined earlier. We further divided them into few sub cases depending upon the execution time of job y in WDA. In most of the sub cases, the lower bound on *max-stretch* are different and are derived using tricky mathematical arguments. To elaborate the proof more specifically, Lemma 7 considers the case when $p_y \leq p_z$; Lemma 11 consider the case when $p_z < p_y \leq (1+\alpha\Delta)p_z$; Lastly, Lemma 17 considers the case when $(1+\alpha\Delta)p_z < p_y$.

Frequently, we refer to the intermediate stretch at time t. As aforementioned, we use the notation $S(t)$ to refer to the intermediate maximum stretch at time t such that all jobs in the *Ready queue* can be scheduled within their respective deadlines. Note that $S(\sigma_i) \geq S_i$ for all job $i \in J$.

5.2   Proving the Bound When $p_y \leq p_z$

Lemma 7. *If* $p_y \leq p_z$, *then* $S_z \leq S_y^* + \alpha\Delta$.

Proof. We consider two cases:

1. *Suppose* $y \in J_B$. Then Property 4 implies that $r_y + S(\sigma_y)p_y \leq r_z + S(\sigma_y)p_z$. Since the stretch of job z is greater than the intermediate stretch at any time,

we have $S(\sigma_y) \le S_z$, which leads to $r_y - r_z \le S_z(p_z - p_y)$. Substituting this inequality in Eq. 1 we get,

$$S_z \le S_y^* \left(\frac{p_y}{p_z}\right) + \left(1 - \frac{p_y}{p_z}\right) S_z + \frac{\alpha\Delta}{p_z}$$

$$S_z \le S_y^* + \frac{\alpha\Delta}{p_y} \quad \le S_y^* + \alpha\Delta$$

2. *Suppose $y \notin J_B$.* Let δ be a binary variable such that it is 0 when job z belongs to class J_{small}, otherwise it is 1. Then the modified release time of job z can we re-written as $r_z' = r_z + \delta\alpha p_z$. The start time of job y is earlier than the modified released time of job z, that is $r_y \le \sigma_y < r_z'$. This implies that $r_y < r_z + \delta\alpha p_z$. Substituting this inequality in Eq. 1 we get,

$$S_z \le S_y^* \left(\frac{p_y}{p_z}\right) + \frac{\alpha\Delta}{p_z} + \delta\alpha \quad \le S_y^* + \frac{\alpha\Delta}{1 + \delta\alpha\Delta} + \delta\alpha \quad \le S_y^* + \alpha\Delta$$

When $\delta = 1$, the last inequality follows from that fact that $\frac{\alpha\Delta}{1+\alpha\Delta} + \alpha < \alpha\Delta$ when $\Delta \ge 2$. $\square$

5.3   Proving the Bound When $p_z < p_y \le (1 + \alpha\Delta)p_z$

Observation 8. *In WDA, there does not exist a job i such that job z is released no later than job i and the processing time of job i is more than that of job z. More formally, $\nexists i \in J : r_i \ge r_z$ and $p_i > p_z$.*

The proof of Observation 8 is omitted to accommodate space constraints. For the remaining cases, it follows that job z is processed before job y in OPT, $p_z < p_y$ and $r_y < r_z$. Before moving on to analysis of such cases, we define the notion of *limiting* jobs which play a crucial role in the analysis to follow.

Definition 9. *We say that job i limits job j if the following statements are true.*

- *processing time of job i is more than that of job j, $p_i > p_j$*
- *job i is scheduled at or after the modified released time of job j, both in WDA and OPT*
- *job i is processed earlier than job j in WDA, $\sigma_i < \sigma_j$*
- *job j is processed earlier than job i in OPT, $\sigma_j^* < \sigma_i^*$*

Property 10. *If i limits j then the stretch of job i in WDA is at least $1 + \frac{p_i}{p_j} - \frac{p_j}{p_i}$.*

The proof of this property is omitted to accommodate space constraints.

Now we have all the tools to show the bound for *max-stretch* in the case where $p_y \le (1 + \alpha\Delta)p_z$.

Lemma 11. *If $p_z < p_y$ and $p_y \leq (1 + \alpha\Delta)p_z$ then $S_z \leq S^*(1 + \alpha\Delta)$.*

Proof. Suppose that the completion time of job z in schedule WDA is no later than the completion time of job y in OPT, that is $C_y^* \geq C_z$. Similar to Eq. 1, the relationship between the stretch of job z in WDA and the stretch of job y in OPT can be written as $S_z \leq S_y^* \frac{p_y}{p_z} + \frac{r_y - r_z}{p_z}$. From Observation 8, it follows that job y is released earlier than job z, *i.e.* $r_y - r_z \leq 0$. Thus combining both inequalities, we have $S_z \leq S_y^* \frac{p_y}{p_z} \leq S_y^*(1 + \alpha\Delta) \leq S^*(1 + \alpha\Delta)$. Therefore, we assume $C_y^* < C_z$ for the rest of this proof. We further split the analysis in three separate cases.

Case A: *Job $y \in J_B$.* Observe that the start time of job y is at or after the modified release time of job z *i.e.* $\sigma_y \geq r_z'$. Applying Property 4, we have $r_y + S(\sigma_y)p_y \leq r_z + S(\sigma_y)p_z$. Since $p_y > p_z$ and the stretch of any job is at least 1, we can re-write the above inequality as $r_y - r_z \leq p_z - p_y$. Using this inequality in Eq. 1 along with the fact that $p_y \leq (1 + \alpha\Delta)p_z$ proves that the bound holds in this case.

Case B: *Job $y \notin J_B$ and $C_y \leq r_z'$.* The assumption $C_y \leq r_z'$ implies that $r_y + S_y p_y \leq r_z + \delta\alpha p_z$ where $\delta = 0$ if $z \in J_{small}$ or 1 otherwise. Since the stretch of job y is greater than 1 or $1 + \alpha$, depending upon class of job y, job y is released at least p_y time units earlier than job z, that is $r_z - r_y \geq p_y$. Using this inequality with Eq. 1 proves that the bound holds in this case.

Case C: *Job $y \notin J_B$ and $C_y > r_z'$.* Since $C_y^* < C_z$, there exists a job k such that $[\sigma_k, C_k) \subseteq [\sigma_y, C_z)$ and $[\sigma_k^*, C_k^*) \nsubseteq [\sigma_y, C_z)$.

Case C.1: *Consider $r_k \geq \sigma_y$.* Since job k is released after the start time of job y, the completion time of job k in OPT is strictly larger than the completion time of job z in WDA, *i.e.* $C_k^* > C_z$. Suppose that $p_k \leq p_z$, then Lemma 7 implies that bound is true. On the contrary if $p_k > p_z$, then Observation 8 implies that job k is released earlier than job z. Moreover, the difference in the release time of job z and job k is at most p_y. Hence $r_z - r_k \leq (1 + \alpha\Delta)p_z$. Using Property 4, we have $r_k + S(\sigma_k)p_k \leq r_z + S(\sigma_k)p_z$ and $S(\sigma_k) \geq \frac{p_k + p_z}{p_z}$. Consequently, we get that the difference in release time of job z and job k is at least $\frac{p_k^2 - p_z^2}{p_z}$. Equating this lower bound with upper bound on $r_k - r_z$, we get $p_k \leq p_z(\sqrt{2 + \alpha\Delta})$. As $C_k^* > C_z$ and $p_k \leq p_z(\sqrt{2 + \alpha\Delta})$, we get $S_z \leq S_k^*(\sqrt{2 + \alpha\Delta}) \leq S^*(1 + \alpha\Delta)$.

Case C.2: *Consider $r_k < \sigma_y$.* If $p_k \leq p_z$ then by Property 10, we have $S_y > 1 + \frac{p_y}{p_k} - \frac{p_k}{p_y} > 1 + \frac{p_y}{p_k} - \frac{p_z}{p_y}$. Since $r_z' \leq C_y$ and $y \notin J_B$, we have $r_y + S_y p_y - p_y < r_z$. Using both inequalities in Eq. 1 proves that our bound holds in this case. Conversely suppose that $p_k > p_z$. Since $k \in J_B$, using Property 4 we have $r_k + S(\sigma_k)p_k \leq r_z + S(\sigma_k)p_z$. As intermediate stretch estimate is a non-decreasing function of time, $p_k > p_z$ and $\sigma_y \leq \sigma_k$, we have $r_k + S(\sigma_y)p_k < r_z + S(\sigma_y)p_z$. Hence $r_y + S(\sigma_y)p_y < r_k + S(\sigma_y)p_k < r_z + S(\sigma_y)p_z$. The above facts imply that $r_y - r_z < S(\sigma_y)(p_z - p_y) < p_z - p_y$ since $p_z - p_y < 0$. Substituting this inequality in Eq. 1 gives $S_z \leq S_y^* \frac{p_y}{p_z} + 1 - \frac{p_y}{p_z} + \frac{\alpha\Delta}{p_z} \leq (S_y^* - 1)\frac{p_y}{p_z} + 1 + \alpha\Delta \leq S_y^*(1 + \alpha\Delta)$. $\square$

5.4   Proving the Bound When $(1 + \alpha\Delta)p_z < p_y$

Now we build up the tools for the last major Lemma 17 which shows that $S_z \leq S^*(1 + \alpha\Delta)$ when $p_z(1 + \alpha\Delta) \leq p_y$. Observe that for this particular case job z and job y belongs to class J_{small} and J_{large}, respectively. To simplify the notations, from here on we will refer to r'_z as r_z.

Definition 12. *At any time t, we define $J_U(t)$ as set of jobs that are unfinished at time t, i.e. $J_U(t) = \{i \in J : r_i \leq t < C_i\}$*

Then the following lemma relates the stretch estimates $S(t)$ shortly after r_z with the jobs in $J_U(r_z)$.

Lemma 13. *Denote by j the first job started in WDA after r_z. For $t \geq \sigma_j$, $S(t)$ is at least $\dfrac{\sum_{i \in J_U(r_z)} p_i + \sigma_j - r_z}{p_z}$.*

The proof is omitted here to accommodate space constraints.

Before we proceed onto last case analysis in Lemma 17, we define two sets of jobs that are useful for the further analysis. Our aim is to relate the set of jobs in WDA and OPT that are executed after r_z. Informally, we first define a set consisting of jobs that were processed during the interval $[r_z, C_y^*)$, in OPT, such that for each job, their processing time is at most the processing time of job z.

Definition 14. *We define J_S as the set of all jobs in OPT for which the following conditions are met:*

– *job i starts no earlier than r_z, i.e. $\sigma_i^* \geq r_z$.*
– *$p_i \leq p_z$ or the deadline of job i is at most the deadline of job z, according to the optimal stretch S^*, i.e. $r_i + S^* p_i \leq r_z + S^* p_z$.*
– *Job i completes before job y, i.e. $C_i^* < C_y^*$.*

Observe that job z belongs to J_S. Hence J_S is a non-empty set. Now we define the set of big jobs that were processed *consecutively*[1] just before job y (see Fig. 2).

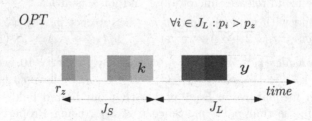

$$OPT \qquad\qquad \forall i \in J_L : p_i > p_z$$

Fig. 2. Representing set of jobs in J_S and J_L

---

[1] Here we assume that the optimal schedule is non-lazy, that is all jobs are scheduled at the earliest time and there is no unnecessary idle time.

Definition 15. *We define J_L as the set of jobs in schedule OPT that are executed between the completion time of latest job in set J_S and completion time of job y (refer to Fig. 2). Formally, $J_L = \{i \in J : \sigma_i^* \in [C_k^*, C_y^*)\}$ where $k \in J_S$ and $\sigma_k^* \geq \sigma_i^*, \forall i \in J_S$. Moreover, λ and $|J_L|$ denote the length of time interval $[C_k^*, C_y^*)$ and the number of jobs in J_L, respectively.*

Note that job y belongs to J_L(hence $\lambda \geq p_y$) and $\forall i \in J_L$, we have $p_i > p_z$ and $r_z + S^* p_z < r_i + S^* p_i$.

Property 16. *If $p_z(1 + \alpha\Delta) < p_y \leq \Delta$, then the total processing time of the jobs in $J_U(r_z)$ is at least $\lambda - p_y + \alpha\Delta$.*

Now we have all the tools necessary to prove the Lemma 17.

Lemma 17. *If $p_z(1 + \alpha\Delta) < p_y \leq \Delta$, then $S_z < S^*(1 + \alpha\Delta)$, where S^* is the maximum stretch of some job in OPT.*

Proof. Let k be the latest job in set J_S (see Fig. 2). More formally, $k \in J_S$ and $\forall i \in J_S : \sigma_i^* \leq \sigma_k^*$. From Definition 15, we have $C_k^* = C_y^* - \lambda$. We can re-write this equality in terms of the stretch of job y and k as $p_y S_y^* = p_k S_k^* + \lambda + r_k - r_y$. Substituting this expression in Eq. 1, we get:

$$S_z \leq S_k^* \frac{p_k}{p_z} + \frac{r_k - r_z}{p_z} + \frac{\alpha\Delta + \lambda}{p_z} \tag{2}$$

Remember that in this subsection we denote by j the first job that starts its execution after time r_z, that is $\sigma_j \leq \sigma_i : \forall i \in J_B$. Now we organize this proof into two parts.

Case A : *Suppose $\sigma_y \geq r_z$.* From Property 4 we have $r_y + S(\sigma_y)p_y < r_z + S(\sigma_y)p_z < r_z + S_z p_z$. Using this inequality in Eq. 1, we get $S^* \geq S(\sigma_y) - 1$. Since $\sigma_y \geq r_z$, it follows that job $y \in J_B$ and $S(\sigma_j) \leq S(\sigma_y)$. Also note that job y *limits* job z. Therefore using Property 16 and Lemma 13, we have $S(\sigma_z) \geq S(\sigma_j) \geq 1 + \frac{\lambda - p_y + \alpha\Delta}{p_z}$. Therefore, we have $S^* > \frac{\lambda - p_y + \alpha\Delta}{p_z}$.

Case A.1: *Assume $r_k \leq r_z$.* Plugging $r_k - r_z \leq 0$ and the above lower bound on S^* in Eq. 2 we have the desired results.

Case A.2: *Assume $r_k > r_z$.* From Observation 8, we have $p_k < p_z$. Observe that job k belongs to J_B. From Property 4, we have the $r_k - r_z \leq S(\sigma_k)(p_z - p_k) \leq S_z(p_z - p_k)$. Combining this with above lower bound on S^* and using in Eq. 2, we obtain bounded competitive ratio.

Case B: *Suppose that $\sigma_y < r_z$.* Again by Properties 16 and 13, it follows that $S(\sigma_j) \geq 1 + \frac{\lambda - p_y + \alpha\Delta}{p_z}$.

Case B.1: Suppose that there exists some job l such that $l \in J_L$ and $l \in J_B$.[2] Then replace job y with job l in *Case A* and the proof follows.

---

[2] Note that job l starts processing after time r_z in both schedule OPT and WDA.

Case B.2: Now assume that there does not exist any job l such that $l \in J_L$ and $l \in J_B$. Recall that $|J_L| \geq 2$ as stated in case hypothesis B.1. Let v be the smallest job in J_L. Observe that v starts before time r_z in schedule WDA since $v \notin J_B$. Therefore there must be a job $w \in J_B$ such that $\sigma_w^* < r_z$. Now we split the proof into two sections based on processing times of such jobs.

Assume that there exists at least one such job w with $p_v \leq p_w$. Job v is scheduled before job w in the WDA, this implies that $r_v + S(\sigma_v)p_v \leq r_w + S(\sigma_v)p_w$. Since $\sigma_v < r_z \leq \sigma_j$ and $p_v \leq p_w$, we have $r_v + S(\sigma_j)p_v \leq r_w + S(\sigma_j)p_w$. Also z is the last job to be scheduled, which states that $r_w + S(\sigma_j)p_w \leq r_z + S(\sigma_j)p_z$. Hence, we have $r_v + S(\sigma_j)p_v \leq r_w + S(\sigma_j)p_w \leq r_z + S(\sigma_j)p_z$. Since job $v \in J_L$, we also have $r_z + S^*p_z \leq r_v + S^*p_v$. This implies that $S(\sigma_j) \leq S^*$. Using this lower bound in Eq. 2, our competitive ratio holds.

On the contrary, we assume that there exists no job w such that $p_v \leq p_w$. Then it implies that there are at least $|J_L|$ are jobs in J_B such that they are started before time r_z in OPT (call such jobs J_M). Moreover $\forall i \in J_M$, $p_i \leq p_v$. Since all jobs belonging to set J_L starts execution before r_z in OPT, there exist a job (denoted by x) in J_M that is delayed at least by λ time units before its start time in WDA. Hence $S^* > S(\sigma_v) \geq \frac{\lambda + p_x}{p_x}$. Now we look at two cases together. First, as we assume that $p_x < 2p_z$. This implies that $S^* \geq \frac{\lambda + 2p_z}{2p_z}$. Second, if $S^* \geq \frac{\lambda + p_x}{p_x} \geq (2|J_L| + 1)$. Using last terms as lower bounds on S^* in Eq. 2, our bound holds.

It remains to prove the case where $\frac{\lambda + p_x}{p_x} < (2|J_L| + 1)$ and $p_x \geq 2p_z$. Then we have $p_x > \frac{\lambda}{2|J_L|} \geq \frac{p_v}{2}$. Since job x belongs to set J_B, we have $r_x + S(\sigma_j)p_x \leq r_z + S(\sigma_j)p_z$. Note that at time σ_v, we have $r_v + S(\sigma_v)p_v < r_x + S(\sigma_v)p_x$. Since $p_x < p_v$, we have $r_v < r_x$. Moreover as $v \in J_L$, we also have $r_z + S^*p_z \leq r_v + S^*p_v$. This implies that $r_z + S^*p_z \leq r_v + S^*p_v \leq r_x + 2S^*p_x$. Combining this with $r_x + S(\sigma_j)p_x \leq r_z + S(\sigma_j)p_z$, we get $S^* \geq \frac{S(\sigma_j)(p_x - p_z)}{(2p_x - p_z)}$. Using this as lower bound in Eq. 2, we have our desired results.   □

6   Concluding Remarks

We investigated the online non-preemptive problem scheduling of a set of jobs on a single machine that are released over time so as to optimize the maximum stretch of the jobs. We showed that no algorithm can achieve a competitive ratio better than $\frac{\sqrt{5}-1}{2}\Delta$ for the maximum stretch objective. We proposed a new algorithm which delays the execution of large jobs and achieves a competitive ratio $1 + \frac{\sqrt{5}-1}{2}\Delta$. This paper essentially closes the problem of optimizing the maximum stretch on a single machine. Indeed, when Δ goes to infinity, these upper and lower bounds are both equal to $\frac{\sqrt{5}-1}{2}\Delta$.

The following questions will receive our attention next. Is WDA competitive for the average stretch? Can the waiting strategy of WDA be extended to the more general weighted flow time objectives? Can we design an algorithm better than Δ competitive for max-stretch when multiple machines are available?

Acknowledgments. This work has been partially supported by the LabEx PERSYVAL-Lab (ANR-11-LABX-0025-01) funded by the French program Investissement d'avenir. Erik Saule is a 2015 Data Fellow of the National Consortium for Data Science (NCDS) and acknowledges the NCDS for funding parts of the presented research.

References

1. Brucker, P.: Scheduling Algorithms, 3rd edn. Springer, New York (2001)
2. Bender, M., Chakrabarti, S., Muthukrishnan, S.: Flow and stretch metrics for scheduling continuous job streams. In: Proceedings of the SODA, pp. 270–279 (1998)
3. Lawler, E.L., Lenstra, J.K., Rinnooy Kan, A.H.G., Shmoys, D.B.: Sequencing and scheduling: algorithms and complexity. Handbooks Oper. Res. Manag. Sci. **4**, 445–522 (1993)
4. Saule, E., Bozdağ, D., Çatalyürek, İ.V.: Optimizing the stretch of independent tasks on a cluster: from sequential tasks to moldable tasks. J. Parallel Distrib. Comput. **72**(4), 489–503 (2012)
5. Legrand, A., Su, A., Vivien, F.: Minimizing the stretch when scheduling flows of divisible requests. J. Sched. **11**(5), 381–404 (2008)
6. Hochbaum, D.S.: Approximation Algorithms for NP-Hard Problems. PWS, Boston (1997)
7. Bansal, N., Pruhs, K.: Server scheduling in the Lp norm: a rising tide lifts all boat. In: Proceedings of the ACM STOC, pp. 242–250 (2003)
8. Golovin, D., Gupta, A., Kumar, A., Tangwongsan, K.: All-norms and all-Lp-norms approximation algorithms. In: Proceedings of the FSTTCS, pp. 199–210 (2008)
9. Im, S., Moseley, B.: An online scalable algorithm for minimizing l_k-norms of weighted flow time on unrelated machines. In: Proceedings of the ACM-SIAM SODA, pp. 98–108 (2011)
10. Anand, S., Garg, N., Kumar, A.: Resource augmentation for weighted flow-time explained by dual fitting. In: Proceedings of the ACM-SIAM SODA, pp. 1228–1241 (2012)
11. Lu, X., Sitters, R.A., Stougie, L.: A class of on-line scheduling algorithms to minimize total completion time. Oper. Res. Lett. **31**, 232–236 (2003)
12. Megow, N., Schulz, A.S.: On-line scheduling to minimize average completion time revisited. Oper. Res. Lett. **32**, 485–490 (2003)

Complex-Demand Scheduling Problem
with Application in Smart Grid

Majid Khonji[1,2]([⊠]), Areg Karapetyan[1], Khaled Elbassioni[1],
and Chi-Kin Chau[1]

[1] Department of Electrical Engineering and Computer Science,
Masdar Institute of Science and Technology, Abu Dhabi, UAE
{mkhonji,akarapetyan,kelbassioni,ckchau}@masdar.ac.ae
[2] Department of Research and Development,
Dubai Electricity and Water Authority (DEWA), Dubai, UAE
majid.khonji@dewa.gov.ae

Abstract. We consider the problem of scheduling complex-valued
demands over a discretized time horizon. Given a set of users, each user is
associated with a set of demands representing different user's preferences.
A demand is represented by a complex number, a time interval, and a
utility value obtained if the demand is satisfied. At each time slot, the
magnitude of the total selected demands should not exceed a given capacity. This naturally captures the supply constraints in alternating current
(AC) electric systems. In this paper, we consider maximizing the aggregate user utility subject to power supply limits over a time horizon. We
present approximation algorithms characterized by the maximum angle
ϕ between any two complex-valued demands. More precisely, a PTAS is
presented for the case $\phi \in [0, \frac{\pi}{2}]$, a bi-criteria FPTAS for $\phi \in [0, \pi\text{-}\delta]$ for
any polynomially small δ, assuming the number of time slots in the dis-
cretized time horizon is a constant. Furthermore, if the number of time
slots is polynomial, we present a reduction to the real-valued unsplittable flow on a path problem with only a constant approximation ratio.
Finally, we present a practical greedy algorithm for the single time slot
case with an approximation ratio of $\frac{1}{2} \cos \frac{\phi}{2}$, while the running time is
$O(n \log n)$, which can be implemented efficiently in practice.

Keywords: Algorithms · Scheduling · Smart grid · Unsplittable flow ·
Knapsack

1 Introduction

One of the most important worldwide developments is to revolutionize the legacy
electricity grid infrastructure by upgrading to computationally smarter grid,
which will be capable of managing a diversity of distributed generations and
renewable energy sources. A key aspect of the emerging *smart grid* is to optimize
power supply to match consumers' demands. *Microgrids* are typically medium-
to-low voltage networks with integrated *distributed generation*. A microgrid could

© Springer International Publishing Switzerland 2016
T.N. Dinh and M.T. Thai (Eds.): COCOON 2016, LNCS 9797, pp. 496–509, 2016.
DOI: 10.1007/978-3-319-42634-1_40

run short of power supply due to emergency conditions, high purchase price in the bulk market, or renewable fluctuation over time. Thus, consumers with deferrable loads such as electric vehicles or dishwashers can be scheduled efficiently to match the available supply. This, in fact, models the day-ahead electric market at the distribution network whereby customers provide their deferrable demand preferences along with the amount they are willing to pay, and the grid operator decides the best allocation.

Although resource allocation and scheduling mechanisms have been well-studied in many systems from transportation to communication networks, the rise of the smart grid presents a new range of algorithmic problems, which are a departure from these systems. One of the main differences is the presence of periodic time-varying entities (e.g., current, power, voltage) in AC electric systems, which are often expressed in terms of non-positive real, or even complex numbers. In power terminology [13], the real component of the complex number is called the *active* power, the imaginary is known as *reactive* power, and the magnitude as *apparent* power. For example, purely resistive appliances have positive active power and zero reactive power. Appliances and instruments with capacitive or inductive components have non-zero reactive power, depending on the phase lag with the input power. Machinery, such as in factories, has large inductors, and hence has positive power demand. On the contrary, shunt-capacitor equipped electric vehicle charging stations can generate reactive power.

We consider a variable power generation capacity over a discrete time horizon. Every user of the smart grid is associated with a set of demand preferences, wherein a demand is represented by a complex-valued number, a time interval at which the demand should be supplied, and a utility value obtained if the demand is satisfied. Some demands are inelastic (i.e., unsplittable) that are either fully satisfied, or completely dropped. At each time slot, the magnitude of the total satisfied demands among all different preferences should not exceed the generation capacity of the power grid represented by the magnitude of the aggregate complex-valued demand. This, in fact, captures the variation in supply constraints over time in alternating current (AC) electric systems. This problem captures the demand-response management in power systems.

Conventionally, demands in AC systems are represented by complex numbers in the first and fourth quadrants of the complex plane. We note that our problem is invariant when the arguments of all demands are shifted by the same angle. For convenience, we assume the demands are rotated such that one of the demands is aligned along the positive real axis. In realistic setting of power systems, the active power demand is positive, but the power factor (i.e., the cosine of the demand's argument) is bounded from below by a certain threshold, which is equivalent to restricting the argument of complex-valued demands.

We present approximation algorithms characterized by the maximum angle ϕ between any two complex-valued demands. More precisely, we present a PTAS for the case $\phi \in [0, \frac{\pi}{2}]$, a bi-criteria FPTAS for $\phi \in [0, \pi\text{-}\delta]$ for any polynomially small δ, assuming the number of time slots in the discretized time horizon is constant. Furthermore, if the number of time slots is polynomial, we present a reduction to the *unsplittable flow on a path problem* [5] that adds only a constant factor to the approximation ratio. We remark that the unsplittable flow problem considers only real-valued demands which is indeed simpler than our setting. Finally, we present a practical greedy algorithm for the single time slot case with an approximation ratio of $\frac{1}{2}\cos\frac{\phi}{2}$, and the running time is an order of $O(n\log n)$, which can be implemented in real world power systems.

The paper is structured as followed. In Sect. 2, we briefly present the related works. In Sect. 3, we provide the problem definitions and notations needed. Then we present algorithms for the case of a constant number of time slots in Sect. 4, namely, a PTAS for $\phi \in [0, \frac{\pi}{2}]$ and an FPTAS for $\phi \in [0, \pi\text{-}\delta]$. In Sect. 5 we present the reduction to the unsplittable flow problem for the case of a polynomial number of time slots. Our greedy algorithm is provided in Sect. 6, followed by the conclusion in Sect. 7. Due to the lack of space, all proofs are omitted and provided in the technical report [17].

2   Related Work

Several recent studies consider resource allocation with inelastic demands. For a single time slot case, our problem resembles the *complex-demand knapsack problem* (CKP) [23]. Let ϕ be the maximum angle between any complex valued demands. Yu and Chau [23] obtained a $\frac{1}{2}$-approximation for the case where $0 \leq \phi \leq \frac{\pi}{2}$. Woeginger [22] (also [23]) proved that no fully polynomial-time approximation scheme (FPTAS) exists for CKP. Recently, Chau et al. [6][1] provided a polynomial-time approximation scheme (PTAS), and a bi-criteria FPTAS (allowing constraint violation) for $\frac{\pi}{2} < \phi < \pi - \delta$, which closes the approximation gap. It is shown that when $\phi \in (\frac{\pi}{2}, \pi]$, there is no α-approximation to CKP for any α with polynomial number of bits [19]. Additionally, when δ is arbitrarily close to zero (i.e., $\phi \to \pi$) there is no (α, β)-approximation in general for any α, β with polynomial number of bits. Therefore, the PTAS and the bi-criteria FPTAS [6] are the best approximation possible for CKP.

Elbassioni and Nguyen [11] extended CKP to handle a constant number of quadratic (and linear) constraints. A fast greedy algorithm is provided in [15] for solving CKP (with a single quadratic constraint) that runs in $O(n\log n)$, and achieves a constant approximation ratio. A recent work [18] extends the greedy algorithm to solve the optimal power flow problem (OPF), a generalization of CKP to a networked setting including voltage constraints.

When the demands are real-valued, our problem considering multiple time slots is related to the unsplittable flow problem on a path (UFP). In UFP, each demand is associated with a unique path from a source to a sink. UFP is strongly NP-hard [9]. A Quasi-PTAS is obtained by Bansal et al. [2]. Anagnostopoulos et al. [1] obtained a $1/(2 + \epsilon)$-approximation (where $\epsilon > 0$ is a constant). This matched the previously known approximation with the *no bottleneck assumption* (NBA) [8], i.e., the largest demand is at most the smallest capacity. The UFP with *bag constraints* (BAG-UFP) is the generalization of UFP where each users

---

[1] The complete work is provided in a technical report [7].

has a set of demands among which at most one is selected [4]. This problem is APX-hard even in the case of unit demands and capacities [21]. Under the NBA assumption, Elbassioni et al. [10] obtained a $\frac{1}{65}$-approximation which was later improved by Chakrabarthi et al. [4] to $\frac{1}{17}$. Recently, Grandoni et al. [14] obtained an $O(\log n / \log \log n)^{-1}$-approximation without NBA. A constant factor approximation to BAG-UFP remains an interesting open question.

When the number of time slots is constant, our problem generalizes CKP (see [6]) to multiple time slots, and also extends that of [11] by considering multiple demands per user, adding n extra constraints. We also include elastic demands, i.e., demands that can be partially satisfied, along with inelastic demands in the problem formulation. Furthermore, for the case of a polynomial number of time slots, our problem is a generalization of the unsplittable flow problem on paths to accommodate complex-valued demands. Finally, we extend the greedy algorithm in [15] (for the single time slot case) to handle multiple demands per user keeping the same approximation ratio and running time.

3 Problem Definitions and Notations

In this section we formally define the complex-demand scheduling problem. Throughout this paper, we denote ν^R as the real part and ν^I as the imaginary part of a given complex number ν. The magnitude of ν is denoted by $|\nu|$. Unless we state otherwise, we denote μ_t (and sometimes $\mu(t)$ whenever we use subscripts for other purposes) as the t-th component of the sequence μ.

3.1 Complex-Demand Scheduling Problem

We consider a discrete time horizon denoted by $\mathcal{T} \triangleq \{1, ..., m\}$. At each time slot $t \in \mathcal{T}$, the generation capacity of the power grid is denoted by $C_t \in \mathbb{R}_+$ (where $0 \in \mathbb{R}_+$). Denote $\mathcal{N} \triangleq \{1, ..., n\}$ by the set of all users. Each user $k \in \mathcal{N}$ declares a set of demand preferences indexed by the set D_k, among which at most one inelastic demand is selected (see Cons. 3 below). Each demand $j \in D_k$ is defined over a time interval $T_j \subseteq \mathcal{T}$, that is, $T_j = \{t_1, t_1 + 1, ..., t_2\}$ where $t_1, t_2 \in \mathcal{T}$ and $t_1 \leq t_2$. Demand j is also associated with a set of complex numbers $\{s_{k,j}(t)\}_{t \in T_j}$ where $s_{k,j}(t) \triangleq s_{k,j}^R(t) + i s_{k,j}^I(t) \in \mathbb{C}$ is a complex power demand at time t. A positive utility $u_{k,j}$ is associated with each user demand (k, j) if satisfied.

Some user demands are inelastic, denoted by $\mathcal{I} \subseteq \mathcal{N} \times \bigcup_k D_k$, which are required to be either fully satisfied or fully dropped. The rest of demands, denoted by $\mathcal{F} \subseteq \mathcal{N} \times \bigcup_k D_k$ such that $\mathcal{F} \cap \mathcal{I} = \varnothing$, are elastic demands, which can be partially satisfied. The goal is to decide a solution of control variables $\left((x_{k,j})_{(k,j) \in \mathcal{I}}, (x_{k,j})_{(k,j) \in \mathcal{F}}\right) \in \{0, 1\}^{|\mathcal{I}|} \times [0, 1]^{|\mathcal{F}|}$ that maximizes the total utility of satisfiable users subject to the generation capacity over time. We define the complex-demand scheduling problem over m discrete time slots (m-CSP) by the following mixed integer programming problem.

$$(m\text{-CSP}) \quad \max \sum_{k \in \mathcal{N}} \sum_{j \in D_k} u_{k,j} x_{k,j} \tag{1}$$

subject to $\left| \sum_{k \in \mathcal{N}} \sum_{j \in D_k : T_j \ni t} s_{k,j}(t) \cdot x_{k,j} \right| \leq C_t,$ $\forall t \in \mathcal{T}$ (2)

$$\sum_{j \in D_k} x_{k,j} \leq 1,$$ $\forall k \in \mathcal{N}$ (3)

$x_{k,j} \in \{0,1\}$ $\forall (k,j) \in \mathcal{I}$ and $x_{k,j} \in [0,1]$ $\forall (k,j) \in \mathcal{F}.$ (4)

Cons. 2 captures the capacity constraint, and Cons. 3 forces at most one inelastic demand for every user to be selected (since each inelastic demand (k,j) is associated with a discrete variable $x_{k,j} \in \{0,1\}$).

1-CSP (i.e., $|\mathcal{T}| = 1$) is called the complex-demand knapsack, denoted by CKP. (We drop subscripts t and j when $|\mathcal{T}| = 1$ and $|D_k| = 1$ for all $k \in \mathcal{N}$.) Evidently, m-CSP is NP-complete, because the knapsack problem is a special case when we set all $s_{k,j}^{\mathrm{I}}(1) = 0$, $\mathcal{T} = \{1\}$, and $|D_k| = 1$. We will write m-CSP$[\phi_1, \phi_2]$ for the restriction of problem m-CSP subject to $\phi_1 \leq \max_{k \in \mathcal{N}} \arg(s_k) \leq \phi_2$, where $\arg(s_k) \geq 0$ for all $k \in \mathcal{N}$.

3.2 Approximation Algorithms

Given a solution $x \triangleq (x_{k,j})_{k \in \mathcal{N}, j \in D_k}$, we denote the total utility by $u(x) \triangleq \sum_{k \in \mathcal{N}} \sum_{j \in D_k} u_{k,j} x_{k,j}$. We denote an optimal solution to m-CSP by x^* and $\mathrm{OPT} \triangleq u(x^*)$.

Definition 1. *For $\alpha \in (0,1]$ and $\beta \geq 1$, we define a bi-criteria (α, β)-approximation to m-CSP as a solution $\hat{x} = \left((\hat{x}_{k,j})_{(k,j) \in \mathcal{I}}, (\hat{x}_{k,j})_{(k,j) \in \mathcal{F}} \right) \in \{0,1\}^{|\mathcal{I}|} \times [0,1]^{|\mathcal{F}|}$ satisfying Cons. 3-4, and*

$$\left| \sum_{k \in \mathcal{N}} \sum_{j \in D_k : T_j \ni t} s_{k,j}(t) \hat{x}_{k,j} \right| \leq \beta \cdot C_t   \text{for all } t \in \mathcal{T}$$ (5)

such that $u(\hat{x}) \geq \alpha \mathrm{OPT}$.

In the above definition, α characterizes the approximation ratio between an approximate solution and the optimal solution, whereas β characterizes the violation bound of constraints. In particular, a *polynomial-time approximation scheme* (PTAS) is a $(1 - \epsilon, 1)$- approximation algorithm for any $\epsilon > 0$. The running time of a PTAS is polynomial in the input size for every fixed ϵ, but the exponent of the polynomial might depend on $1/\epsilon$. An even stronger notion is a *fully polynomial-time approximation scheme* (FPTAS), which requires the running time to be polynomial in both input size and $1/\epsilon$. In this paper, we are interested in bi-criteria FPTAS, which is a $(1, 1 + \epsilon)$-approximation algorithm for any $\epsilon > 0$, with the running time to be polynomial in the input size and $1/\epsilon$. When $\beta = 1$, we call an (α, β)-approximation an α-approximation.

4 m-CSP with a Constant Number of Time Slots

In this section we assume the number of time slots $|\mathcal{T}|$ is a constant. This assumption is practical in the realistic setting, where users declare their demands on hourly basis one day ahead (i.e., $|\mathcal{T}| = 24$) [12]. We remark that the results in this section do not require T_j to be a continuous interval in $\mathcal{T}$.

4.1   PTAS for m-CSP$[0, \frac{\pi}{2}]$

Define a convex relaxation of m-CSP (denoted by RLXCSP), such that Cons. 4 is relaxed by $x_{k,j} \in [0,1]$ for all $(k,j) \in \mathcal{I} \cup \mathcal{F}$. Given two subsets of inelastic demands $S_1, S_0 \subseteq \mathcal{I}$, we define another convex relaxation that will be used in the PTAS denoted by RLXCSP$[S_1, S_0]$. This is equivalent to RLXCSP subject to partial substitution such that $x_{k,j} = 1$, for all $(k,j) \in S_1$ and $x_{k,j} = 0$, for all $(k,j) \in S_0$, where $S_1 \cap S_0 = \varnothing$:

$$(\text{RLXCSP}[S_1, S_0]) \qquad \max_{x_{k,j}} \sum_{k \in \mathcal{N}} \sum_{k \in D_k} u_{k,j} x_{k,j}, \qquad \text{such that} \qquad (6)$$

$$\Big(\sum_{k \in \mathcal{N}} \sum_{j \in D_k : t \in T_j} s^{\text{R}}_{k,j}(t)\, x_{k,j} \Big)^2 + \Big(\sum_{k \in \mathcal{N}} \sum_{j \in D_k : t \in T_j} s^{\text{I}}_{k,j}(t) \quad x_{k,j} \Big)^2 \leq C_t^2, \; \forall t \in \mathcal{T} \qquad (7)$$

$$\sum_{j \in D_k} x_{k,j} \leq 1, \qquad\qquad\qquad \forall k \in \mathcal{N} \qquad (8)$$

$$x_{k,j} = 1 \qquad\qquad\qquad \forall (k,j) \in S_1 \qquad (9)$$

$$x_{k,j} = 0, \qquad\qquad\qquad \forall (k,j) \in S_0 \qquad (10)$$

$$x_{k,j} \in [0,1] \qquad\qquad\qquad \forall (k,j) \in (\mathcal{F} \cup \mathcal{I}). \qquad (11)$$

The above relaxation can be solved approximately in polynomial time using standard convex optimization algorithms (see, e.g., [20]). In fact, such algorithms can find a feasible solution x^{cx} to the convex relaxation such that $u(x^{\text{cx}}) \geq \text{OPT}^* - \gamma$, where γ is a constant and OPT^* is the optimal objective value of RLXCSP$[S_1, S_0]$, in time polynomial in the input size (including the bit complexity) and $\log \frac{1}{\gamma}$. Notice that we can obtain $(1 - \frac{\epsilon}{2})$-approximation (i.e., $u(x^{\text{cx}}) \geq (1 - \frac{\epsilon}{2}) \cdot \text{OPT}^*$) using such algorithms, by setting γ to $\frac{\epsilon}{2}$.

We provide a $(1 - \epsilon)$-approximation for m-CSP$[0, \frac{\pi}{2}]$ in Algorithm 1, denoted by m-CSP-PTAS. The idea of m-CSP-PTAS is based on that proposed in [11] with two extensions. First, we consider multiple demands per user. This in fact adds n extra constraints to that in [11], and thus the rounding procedure requires further analysis. The second extension is the addition of elastic demands $\mathcal{F}$. We remark that [6] considers multiple inelastic demands per user for the single time slot case (denoted by CKP); however, their algorithm is based on a completely different geometric approach that is more complicated than that in [11].

Given a feasible solution x^* to RLXCSP$[S_1, S_0]$, a restricted set of demands $R \subseteq \mathcal{I} \cup \mathcal{F}$, and vectors $C^{\text{R}}, C^{\text{I}} \in \mathbb{R}_+^m$, we define the following relaxation, denoted by $LP[C^R, C^I, x^*, R]$:

$$(LP[C^R, C^I, x^*, R]) \quad \max_{x_{k,j} \in [0,1]} \sum_{k \in \mathcal{N}} \sum_{j \in D_k} u_{k,j} x_{k,j} \tag{12}$$

$$\text{subject to} \quad \sum_{k \in \mathcal{N}} \sum_{j \in D_k : t \in T_j} s_{k,j}^R(t) \cdot x_{k,j} \leq C_t^R, \qquad \forall t \in \mathcal{T} \tag{13}$$

$$\sum_{k \in \mathcal{N}} \sum_{j \in D_k : t \in T_j} s_{k,j}^I(t) \cdot x_{k,j} \leq C_t^I, \qquad \forall t \in \mathcal{T} \tag{14}$$

$$\sum_{j \in D_k} x_{k,j} \leq 1, \qquad \forall k \in \mathcal{N} \tag{15}$$

$$x_{k,j} = x_{k,j}^* \qquad \forall (k,j) \in R. \tag{16}$$

Algorithm 1 proceeds as follows. We guess $S_1 \subseteq \mathcal{I}$ to be the $\frac{8m}{\epsilon}$ largest inelastic utility demands in the optimal solution; this defines an excluded set of demands $S_0 \subseteq \mathcal{I} \setminus S_1$ whose utilities exceed one of the utilities in S_1 (Line 4). For each such S_1 and S_0, we solve the convex program $\text{RLXCSP}[S_1, S_0]$ and obtain a $(1 - \frac{\epsilon}{2})$-approximation x^{cx} (note that the feasibility of the convex program is guaranteed by the conditions in Line 3). We remark that $\text{RLXCSP}[S_1, S_0]$ is convex only when $s_{k,j}^R(t), s_{k,j}^I(t) \geq 0$, i.e., all demands lie in the first quadrant of the complex plane. The real and imaginary projections over all time slots of solution x^{cx}, denoted by $C^R \in \mathbb{R}_+^m$ and $C^I \in \mathbb{R}_+^m$, are used to define the linear program $LP[C^R, C^I, x^{cx}, \mathcal{F} \cup S_1 \cup S_0]$ over the restricted set of demands $\mathcal{F} \cup S_1 \cup S_0$. We solve the linear program in Line 8, and then round down the solution corresponding to demands $(k,j) \in \mathcal{I}$ in Line 9. Finally, we return a solution $\hat{x}$ that attains maximum utility among all obtained solutions.

Algorithm 1. $m\text{-CSP-PTAS}[\{u_{k,j}, \{s_{k,j}(t)\}_{t \in T_j}\}_{k \in \mathcal{N}, j \in D_k}, (C_t)_{t \in \mathcal{T}}, \epsilon]$

Require: Users' utilities and demands $\{u_{k,j}, \{s_{k,j}(t)\}_{t \in T_j}\}_{k \in \mathcal{N}, j \in D_k}$; capacity over time C_t; accuracy parameter ϵ

Ensure: $(1 - \epsilon)$-solution $\hat{x}$ to $m\text{-CSP}[0, \frac{\pi}{2}]$

1: $\hat{x} \leftarrow \mathbf{0}$

2: **for** each set $S_1 \subseteq \mathcal{I}$ such that $|S_1| \leq \frac{8m}{\epsilon}$ **do**

3:   **if** $\left| \sum_{(k,j) \in S_1 : t \in T_j} s_{k,j}(t) \right| \leq C_t$ and $\sum_{j \in D_k} x_{k,j} \leq 1, \quad \forall t \in \mathcal{T}, \forall k \in \mathcal{N}$ **then**

4:     $S_0 \leftarrow \{(k,j) \in \mathcal{I} \setminus S_1 \mid u_{k,j} > \min_{(k',j') \in S_1} u_{k',j'}\}$

5:     $x^{cx} \leftarrow$ Solution of $\text{RLXCSP}[S_1, S_0]$ ▷ Obtain a $(1 - \frac{\epsilon}{2})$-approximation

6:     **for** all $t \in \mathcal{T}$ **do**

7:       $C_t^R \leftarrow \sum_{k \in \mathcal{N}} \sum_{j \in D_k : t \in T_j} s_{k,j}^R(t) \cdot x_{k,j}^{cx}; C_t^I \leftarrow \sum_{k \in \mathcal{N}} \sum_{j \in D_k : t \in T_j} s_{k,j}^I(t) \cdot x_{k,j}^{cx}$

8:     $x^{lp} \leftarrow$ Solution of $LP[C^R, C^I, x^{cx}, \mathcal{F} \cup S_1 \cup S_0]$

                                    ▷ Round the LP solution

9:     $\bar{x} \leftarrow \{(\bar{x}_{k,j})_{k \in \mathcal{N}, j \in D_k} \mid \bar{x}_{k,j} = \lfloor x_{k,j}^{lp} \rfloor$ for $(k,j) \in \mathcal{I}$, and $\bar{x}_{k,j} = x_{k,j}^{lp}$ for $(k,j) \in \mathcal{F}\}$

10:     **if** $u(\bar{x}) > u(\hat{x})$ **then** $\hat{x} \leftarrow \bar{x}$

11: **return** $\hat{x}$

Theorem 1. *For any fixed ϵ, Algorithm 1 obtains a $(1 - \epsilon, 1)$-approximation in polynomial time.*

We remark that the PTAS is the best approximation one can hope for, since it is shown in [22,23] that it is NP-Hard to obtain an FPTAS for the single time slot version (1-CSP$[0, \frac{\pi}{2}]$).

4.2 Bi-Criteria FPTAS for m-CSP$[0, \pi\text{-}\delta]$

In the previous section, we have restricted our attention to the setting where all demands lie in the positive quadrant of the complex plane (i.e., m-CSP$[0, \frac{\pi}{2}]$). In this section, we extend our study to the second quadrant (m-CSP$[0, \pi\text{-}\delta]$) for any arbitrary small constant $\delta > 0$, that is, we assume $\arg(s_{k,j}(t)) \leq \pi - \delta$ for all $k \in \mathcal{N}, j \in D_k, t \in T_j$. It is shown in [19] for the case $|\mathcal{T}| = 1$ that m-CSP$[0, \pi]$ is inapproximable and there is no α-approximation for m-CSP$[0, \pi\text{-}\delta]$. Therefore, a bi-criteria $(1, 1+\epsilon)$ is the best approximation one can hope for. Additionally, it is shown that if δ is arbitrarily close to zero, then there is no (α, β)-approximation in general for any α, β with polynomial number of bits. Furthermore the running time should depend on the maximum angle $\phi \triangleq \max_{k \in \mathcal{N}, j \in D_k, t \in T_j} \arg(s_{k,j}(t))$. This algorithm is an extension of that presented by [6] to multiple time slots and also incorporates elastic demands. We consider the following technical assumptions: for any user k,

(i) if $(k, j) \in \mathcal{F}$, then user k has a unique demand (i.e., $|D_k| = 1$)[2]; and
(ii) all demands $s_{k,j}(t), j \in D_k$ reside in one quadrant of the complex plane.

For convenience, we let $\theta = \max\{\phi - \frac{\pi}{2}, 0\}$ (see Fig. 1 for an illustration). We present a $(1, 1 + \epsilon)$-approximation for m-CSP$[0, \pi\text{-}\delta]$ in Algorithm 2, denoted by m-CSP-BIFPTAS, that is polynomial in both $\frac{1}{\epsilon}$ and n (i.e., FPTAS). We assume that $\tan\theta$ is bounded by a polynomial in n; without this assumption, a bi-criteria FPTAS is unlikely to exist (see [19]).

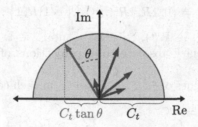

Fig. 1. We measure $\theta = \phi - \frac{\pi}{2}$ from the imaginary axis.

---

[2] This assumption is mainly needed for the dynamic program that is invoked by Algorithm 2.

Let $\mathcal{N}_+ \triangleq \{k \in \mathcal{N} \mid s_{k,j}^{\mathrm{R}}(t) \geq 0, \forall j \in D_k, \ t \in T_j\}$ and $\mathcal{N}_- \triangleq \{k \in \mathcal{N} \mid s_{k,j}^{\mathrm{R}}(t) < 0, \forall j \in D_k, \ t \in T_j\}$ be the subsets of users with demands in the first and second quadrants, respectively. Note that $\mathcal{N}_+$ and $\mathcal{N}_-$ partition the set of users $\mathcal{N}$ by assumption (ii) above.

Consider any solution $\widehat{x}$ to m-CSP$[0, \pi\text{-}\delta]$. The basic idea of Algorithm m-CSP-BiFPTAS is to enumerate the guessed total projections on real and imaginary axes of all time slots for $\sum_{k\in\mathcal{N}_+} \sum_{j\in D_k : t\in T_j} \widehat{x}_{k,j} s_{k,j}(t)$ and $\sum_{k\in\mathcal{N}_-} \sum_{j\in D_k : t\in T_j} \widehat{x}_{k,j} s_{k,j}(t)$ respectively. We can use $\tan\theta$ to upper bound the total projections for any feasible solution $\widehat{x}$ as follows, for all $t \in \mathcal{T}$:

$$\sum_{k\in\mathcal{N}} \sum_{j\in D_k : t\in T_j} s_{k,j}^{\mathrm{I}}(t) \cdot \widehat{x}_{k,j} \leq C_t, \qquad \sum_{k\in\mathcal{N}_-} \sum_{j\in D_k : t\in T_j} -s_{k,j}^{\mathrm{R}}(t) \cdot \widehat{x}_{k,j} \leq C_t \tan\theta,$$

$$\sum_{k\in\mathcal{N}_+} \sum_{j\in D_k : t\in T_j} s_{k,j}^{\mathrm{R}}(t) \cdot \widehat{x}_{k,j} \leq C_t(1 + \tan\theta), \qquad (17)$$

which is illustrated in Fig. 1. We then solve two separate multi-dimensional knapsack problems of dimension $2m$ (denoted by $2m$DKP), to find subsets of demands that satisfy the individual guessed total projections. But since $2m$DKP is generally NP-hard, we need to round-up the demands to get a problem that can be solved efficiently by dynamic programming. We show that the violation of the optimal solution to the rounded problem w.r.t. the original problem is small in ϵ.

Next, we describe the rounding in detail. First, we define $L_t \triangleq \frac{\epsilon C_t}{n(\tan\theta+1)}$, for all $t \in \mathcal{T}$ such that the new rounded-up demands $\widehat{s}_{k,j}(t)$ are defined by:

$$\widehat{s}_{k,j}(t) = \widehat{s}_{k,j}^{\mathrm{R}}(t) + \mathbf{i}\widehat{s}_{k,j}^{\mathrm{I}}(t) \triangleq \begin{cases} \left\lceil \frac{s_{k,j}^{\mathrm{R}}(t)}{L_t} \right\rceil \cdot L_t + \mathbf{i} \left\lceil \frac{s_{k,j}^{\mathrm{I}}(t)}{L_t} \right\rceil \cdot L_t, & \text{if } s_{k,j}^{\mathrm{R}}(t) \geq 0, \\ \left\lfloor \frac{s_{k,j}^{\mathrm{R}}(t)}{L_t} \right\rfloor \cdot L_t + \mathbf{i} \left\lceil \frac{s_{k,j}^{\mathrm{I}}(t)}{L_t} \right\rceil \cdot L_t, & \text{otherwise.} \end{cases}$$

$$(18)$$

We also define $R \triangleq \frac{\epsilon}{3n(\tan\theta+1)^2}$, such that the values of any elastic $x_{k,j}, (k,j) \in \mathcal{F}$ are selected from the discrete set $\mathcal{R}$ of integer multiples of R defined by

$$\mathcal{R} \triangleq \left\{0, 1R, 2R, ..., (\lceil \tfrac{1}{R} \rceil - 1)R, 1\right\}.$$

Let $\xi_+ \in \mathbb{R}_+^m$ (and $\xi_- \in \mathbb{R}_+^m$), $\zeta_+ \in \mathbb{R}_+^m$ (and $\zeta_- \in \mathbb{R}_+^m$) be respectively the guessed real and imaginary absolute total projections of the rounded optimal solution.

Then the possible values of ξ_+, ξ_-, ζ_+ and ζ_- in each component t are integer mutiples of $(R \cdot L_t)$:

$$\xi_+(t) \in \mathcal{A}_+(t) \triangleq \left\{0, (RL_t), 2(RL_t), ..., \left\lceil \frac{C_t(1 + \tan\theta)}{RL_t} \right\rceil \cdot (RL_t)\right\},$$

$$\xi_-(t) \in \mathcal{A}_-(t) \triangleq \left\{0, (RL_t), 2(RL_t), ..., \left\lceil \frac{C_t \cdot \tan\theta}{RL_t} \right\rceil \cdot (RL_t)\right\},$$

$$\zeta_+(t), \zeta_-(t) \in \mathcal{B}(t) \triangleq \left\{0, (RL_t), 2(RL_t), ..., \left\lceil \frac{C_t}{RL_t} \right\rceil \cdot (RL_t)\right\}. \qquad (19)$$

The next step is to solve the rounded instance exactly. Assume an arbitrary order on $\mathcal{N} = \{1, ..., n\}$. We use recursion to define a table, with each entry $U(k, C^{\mathrm{R}}, C^{\mathrm{I}})$, $C^{\mathrm{R}}, C^{\mathrm{I}} \in \mathbb{R}^m_+$, as the maximum utility obtained from a subset of users $\{1, 2, \ldots, K\} \subseteq \mathcal{N}$ with demands $\{\widehat{s}_{k,j}(t)\}_{k \in \{1,...,K\}, j \in D_k, t \in T_j}$ that can fit exactly within capacities $\{C^{\mathrm{R}}_t\}_{t \in \mathcal{T}}$ on the real axis and $\{C^{\mathrm{I}}_t\}_{t \in \mathcal{T}}$ on the imaginary axis. We denote by $2m\mathrm{DKP\text{-}EXACT}[\cdot]$ the algorithm for solving exactly the rounded $2m\mathrm{DKP}$ by dynamic programming. We provide the detailed description of $2m\mathrm{DKP\text{-}EXACT}[\cdot]$ in the appendix.

Algorithm 2. $m\text{-CSP-BIFPTAS}[\{u_{k,j}, \{s_{k,j}(t)\}_{t \in T_j}\}_{k \in \mathcal{N}, j \in D_k}, (C_t)_{t \in \mathcal{T}}, \epsilon]$

Require: Users' utilities and demands $\{u_{k,j}, \{s_{k,j}(t)\}_{t \in T_j}\}_{k \in \mathcal{N}, j \in D_k}$; capacity over time C_t; accuracy parameter ϵ

Ensure: $(1, 1 + 4\epsilon)$-solution $\widehat{x}$ to $m\text{-CSP}[0, \pi\text{-}\delta]$

1: $\widehat{x} \leftarrow 0$
2: **for all** $s_{k,j}(t)$, $k \in \mathcal{N}$, $j \in D_k$, and $t \in T_j$ **do**
3:     Set $\widehat{s}_{k,j}(t) \leftarrow \widehat{s}^{\mathrm{R}}_{k,j}(t) + i\widehat{s}^{\mathrm{I}}_{k,j}(t)$ as defined by 18   ▷ Round up
4: **for all** $\xi_+ \in \prod_{t \in \mathcal{T}} \mathcal{A}_+(t), \xi_- \in \prod_{t \in \mathcal{T}} \mathcal{A}_-(t), \zeta_+, \zeta_- \in \prod_{t \in \mathcal{T}} \mathcal{B}(t)$ **do**
5:     **if** $(\xi_+(t) - \xi_-(t))^2 + (\zeta_+(t) + \zeta_-(t))^2 \leq (1 + 2\epsilon)^2 C_t^2$ for all $t \in \mathcal{T}$ **then**
                                         ▷ Solve $2m\mathrm{DKP}$
6:       $y_+ \leftarrow 2m\mathrm{DKP} - Exact\Big(\{u_{k,j}, (\widehat{s}_{k,j}(t)/L_t)_t\}_{k \in \mathcal{N}_+, j \in D_k}, (\xi_+(t)/L_t)_t, (\zeta_+(t)/L_t)_t\Big)$
7:       $y_- \leftarrow 2m\mathrm{DKP} - Exact\Big(\{u_{k,j}, (-\widehat{s}_{k,j}(t)/L_t)_t\}_{k \in \mathcal{N}_-, j \in D_k}, (\xi_-(t)/L_t)_t, (\zeta_-(t)/L_t)_t\Big)$
8:       **if** $u(y_+ + y_-) > u(\widehat{x})$ **then**
9:          $\widehat{x} \leftarrow y_+ + y_-$
10: **return** $\widehat{x}$

Theorem 2. *Algorithm* $m\text{-CSP-BIFPTAS}$ *is a* $(1, 1 + 4\epsilon)$-*approximation for* $m\text{-}CSP[0, \pi\text{-}\delta]$ *and its running time is polynomial in both* n *and* $\frac{1}{\epsilon}$.

5 $m\text{-CSP}[0, \frac{\pi}{2}]$ with Polynomial number of Time Slots

In this section, we extend our results to polynomial number of time slots $|\mathcal{T}|$. We assume in this section that all demands lie in the first quadrant of the complex plane (i.e., $\phi \triangleq \max_k \arg(s_k) \leq \frac{\pi}{2}$). We provide a reduction to the unsplittable flow problem on a path with bag constraints (BAG-UFP) for which recent approximation algorithms are developed in the literature [3,10,14]. We remake that BAG-UFP considers only real-valued demand, wherein $m\text{-CSP}$ demands are complex-valued. We will show that such a reduction will increase the approximation ratio of BAG-UFP by a constant factor of $\cos\frac{\phi}{2}$, where $\phi \leq \frac{\pi}{2}$ is the maximum argument of any demand. To this extent, we will restrict our setting by the following assumptions to accommodate the setting of BAG-UFP:

(i) We assume constant demands over time: $s_{k,j}(t) = s_{k,j}(t')$ for any $t, t' \in T_j$. To simplify notations, we will write $s_{k,j}$ to denote the unique demand over all time steps T_j.

(ii) All demands are inelastic, i.e., $\mathcal{F} = \varnothing$.

For convenience, we will call our problem m-CSP$'$ when restricted to the above assumptions. When all demands in m-CSP$'$ are real-valued, the problem is equal to BAG-UFP. We denote m-CSP* (resp., BAG-UFP*) as the linear relaxation of m-CSP$'$ (resp., BAG-UFP), that is, $x_{k,j} \in [0,1]$ for all $k \in \mathcal{N}$, $j \in D_k$. Let OPT and $\overline{\text{OPT}}$ be the optimal objective value of m-CSP$'$ and BAG-UFP, respectively. Also denote OPT* and $\overline{\text{OPT}}^*$ by the optimal objective value of m-CSP* and BAG-UFP* respectively.

We will show in Lemma 1 and Theorem 3 below that one can use the algorithms developed for BAG-UFP with bounded integrality gap to obtain approximate solutions to m-CSP$'[0, \frac{\pi}{2}]$.

Lemma 1. *Given a solution $\bar{x} \in \{0,1\}^n$ to BAG-UFP such that $u(\bar{x}) \geq \psi \cdot \overline{\text{OPT}}^*$, $\psi \in [0,1]$ then $\bar{x}$ is feasible for m-CSP$'[0, \frac{\pi}{2}]$ and $u(\bar{x}) \geq \psi \cos \frac{\phi}{2} \cdot \text{OPT}$.*

We can apply Lemma 1 using the recent LP-based algorithm by Grandoni et al. [14] to obtain the following result.

Theorem 3. *Assume $\phi \leq \frac{\pi}{2}$, there exists an $\Omega(\log n / \log \log n)$-approximation for m-CSP$'[0, \frac{\pi}{2}]$. Additionally, if all demands have the same utility, we obtain $\Omega(1)$-approximation.*

Prior work has addressed an important restriction of UFP (also BAG-UFP) called the no bottleneck assumption (NBA), namely, $\max_{k \in \mathcal{N}, j \in D_k} |s_{k,j}| \leq C_{\min} \triangleq \min_t C_t$ states that the largest demand is at most the smallest capacity over all time slots. Define the bottleneck time of demand (k, j) by $b_{k,j} \triangleq \arg\min_{t \in T_j} C_t$. Given a constant $\delta \in [0,1]$, we call a demand (k, j) δ-small if $|s_{k,j}| \leq \delta C_{b_{k,j}}$, otherwise we call it δ-large. We remark that NBA naturally holds in smart grids since individual demands are typically much smaller than the generation capacity over all time slots. In the following, we show that there exists an $\Omega(1)$-approximation for m-CSP$'[0, \frac{\pi}{2}]$. This is achieved by splitting demands to δ-small and δ-large and solving each instance separately then taking the maximum utility solution. The next lemma is an extension to an earlier work by Chakrabarti et al. [5] (to accommodate complex-valued demands) used to derive a dynamic program that approximates δ-large demands.

Lemma 2. *The number of δ-large demands that cross an edge in any feasible solution is at most $2\lfloor \frac{1}{\delta^2} \cdot \sec \frac{\phi}{2} \rfloor$.*

Theorem 4. *Under the NBA assumption, there exists an $\Omega(1)$-approximation for m-CSP$'[0, \frac{\pi}{2}]$. The running time is $O(\bar{n}^2)$, where $\bar{n} \triangleq \sum_{k \in \mathcal{N}} |D_k|$.*

6   Practical Greedy Approximation for 1-CSP$[0, \frac{\pi}{2}]$

In this section we give a practical greedy constant-factor approximation algorithm, presented in Algorithm 3, and denoted by 1-CSP-GREEDY, for the single time slot case (1-CSP$[0, \frac{\pi}{2}]$) where $|\mathcal{T}| = 1$. Our algorithm is an extension of MCKP-GREEDY algorithm introduced in [16] (in chapter 11). Despite the theoretical value of the PTAS and FPTAS presented in Sect. 4 (particularly, $|\mathcal{T}| = 1$), the running time is quite large and hence impractical for real world applications. On the other hand, Algorithm achieves $\left(\frac{1}{2}\cos\frac{\phi}{2}\right)$-approximation only in $O(\bar{n}\log\bar{n})$ time for 1-CSP, where $\bar{n} \triangleq \sum_{k\in\mathcal{N}}|D_k|$.

Algorithm 3 starts by sorting demands in each set D_k in a non-decreasing order of their utilities and successively testing for *LP-dominance*. The concepts of *dominance* and *LP-dominance* are covered in detail in chapter 11 of [16]. Once LP-dominated items are determined and eliminated, Algorithm 3 computes the incremental utilities of the reduced user set and orders users in a non-decreasing order of their incremental efficiencies (i.e., utility to demand ratio). Algorithm 3 then selects users sequentially in that order whenever feasible. Lastly, 1-CSP-GREEDY returns the best out of two candidate solutions: the greedily obtained solution or the highest utility user. We remark that such a simple greedy algorithm can be also used as a fast heuristic for multiple time slots. For instance, in the setting where users arrive online, 1-CSP-GREEDY could be applied to each time slot, after reducing the capacity by the magnitude of demands consumed in previous time slots.

Algorithm 3. 1-CSP-GREEDY$[\{u_{k,j}, s_{k,j}\}_{k\in\mathcal{N}, j\in D_k}, C]$

Require: Users' utilities and demands $\{u_{k,j}, s_{k,j}\}_{k\in\mathcal{N}, j\in D_k}$; capacity C
Ensure: $(\frac{1}{2}\cos\frac{\phi}{2})$-solution $\bar{x}$ to 1-CSP
Initialization:
- Add a dummy demand with zero utility and zero demand to each set D_k, $k \in \mathcal{N}$
- Sort demands in each set D_k, $k \in \mathcal{N}$ by their magnitude in a non-decreasing order
- For each $k \in \mathcal{N}$, define a new set $R_k \subseteq D_k$ of *LP-dominating* demands (see [16]).
- $E \leftarrow \varnothing$, $\tilde{x} \leftarrow \mathbf{0}$; $\tilde{x}_{k,1} \leftarrow 1$ for all $k \in \mathcal{N}$; $\tau \leftarrow 0$; $\hat{x} \leftarrow \mathbf{0}$

1: **for** $k \in \mathcal{N}$, $j = 2, ..., |R_k|$ **do**
2:     $\tilde{u}_{k,j} \leftarrow u_{k,j} - u_{k,j-1}$; $\tilde{s}_{k,j} \leftarrow s_{k,j} - s_{k,j-1}$; $E \leftarrow E \cup \{(k,j)\}$
3: Sort items in E by their efficiency $\left(\frac{\tilde{u}_{k,j}}{|\tilde{s}_{k,j}|}\right)$ in a non-increasing order
4: **for** $(k,j) \in E$ **do**
5:     **if** $|\tau + \tilde{s}_{k,j}| \leq C$ **then**
6:         $\tilde{x}_{k,j} \leftarrow 1$; $\tilde{x}_{k,j-1} \leftarrow 0$; $\tau \leftarrow \tau + \tilde{s}_{k,j}$
7:     **else if** $(k,j) \in \mathcal{F}$ **then**
8:         $\tilde{x}_{k,j} \leftarrow \arg\max\{0, \frac{C-|\tau|}{|\tilde{s}_{k,j}|}\}$; **break**
9: Set $\hat{x}_{k',j'} \leftarrow 1$ for $(k',j') \triangleq \arg\max_{j\in R_k, k\in\mathcal{N}}\{u_{k,j}\}$
10: Set $\bar{x} \leftarrow \arg\max_{x\in\{\hat{x}, \tilde{x}\}} u(x)$
11: **return** $\bar{x}$

Theorem 5. *Algorithm* 1-CSP-GREEDY *is* $\left(\frac{1}{2}\cos\frac{\phi}{2}\right)$*-approximation for* 1-CSP$[0, \frac{\pi}{2}]$. *The running time is* $O(\bar{n}\log\bar{n})$, *where* $\bar{n} \triangleq \sum_{k\in\mathcal{N}}|D_k|$.

7   Conclusion

This paper extends the previous results known for the single time slot case (CKP) to a more general scheduling setting. When the number of time slots m is constant, both the previously known PTAS and FPTAS are extended to handle multiple-time slots, multiple user preferences, and handle mixed elastic and inelastic demands. For polynomial m, a reduction is presented from CSP$[0, \frac{\pi}{2}]$ to the real-valued BAG-UFP, which can be used to obtain algorithms for CSP$[0, \frac{\pi}{2}]$ based on BAG-UFP algorithms that have bounded integrability gap for their LP-relaxation. We further presented a practical greedy algorithm with efficient running time that can be implement in real systems. As a future work, we shall improve the second case (polynomial m) to a constant-factor approximation. Additionally, we may consider different objective functions such as minimizing the maximum peak consumption at any time slot.

References

1. Anagnostopoulos, A., Grandoni, F., Leonardi, S., Wiese, A.: Amazing 2+ ε approximation for unsplittable flow on a path. In: Proceedings of SODA, pp. 26–41. SIAM (2014)
2. Bansal, N., Chakrabarti, A., Epstein, A., Schieber, B.: A quasi-PTAS for unsplittable flow on line graphs. In: Proceedings of STOC, STOC 2006, pp. 721–729. ACM (2006)
3. Chakaravarthy, V.T., Choudhury, A.R., Gupta, S., Roy, S., Sabharwal, Y.: Improved algorithms for resource allocation under varying capacity. In: Schulz, A.S., Wagner, D. (eds.) ESA 2014. LNCS, vol. 8737, pp. 222–234. Springer, Heidelberg (2014)
4. Chakaravarthy, V.T., Pandit, V., Sabharwal, Y., Seetharam, D.P.: Varying bandwidth resource allocation problem with bag constraints. In: Parallel & Distributed Processing (IPDPS), pp. 1–10. IEEE (2010)
5. Chakrabarti, A., Chekuri, C., Gupta, A., Kumar, A.: Approximation algorithms for the unsplittable flow problem. Algorithmica **47**(1), 53–78 (2007)
6. Chau, C.K., Elbassioni, K., Khonji, M.: Truthful mechanisms for combinatorial AC electric power allocation. In: Proceedings of AAMAS (2014). http://arxiv.org/abs/1403.3907
7. Chau, C.K., Elbassioni, K., Khonji, M.: Truthful mechanisms for combinatorial allocation of electric power in alternating current electric systems for smart grid. ACM Trans. Econ. Comput. (2016). http://arxiv.org/abs/1507.01762
8. Chekuri, C., Mydlarz, M., Shepherd, F.B.: Multicommodity demand flow in a tree and packing integer programs. ACM Trans. Algorithms (TALG) **3**(3), 27 (2007)
9. Darmann, A., Pferschy, U., Schauer, J.: Resource allocation with time intervals. Theor. Comput. Sci. **411**(49), 4217–4234 (2010)
10. Elbassioni, K., Garg, N., Gupta, D., Kumar, A., Narula, V., Pal, A.: Approximation algorithms for the unsplittable flow problem on paths and trees. In: LIPIcs-Leibniz International Proceedings in Informatics, vol. 18 (2012)

11. Elbassioni, K., Nguyen, T.T.: Approximation schemes for multi-objective optimization with quadratic constraints of fixed CP-rank. In: Walsh, T. (ed.) ADT 2015. LNCS, vol. 9346, pp. 273–287. Springer, Heidelberg (2015)
12. Fang, X., Misra, S., Xue, G., Yang, D.: Smart grid the new and improved power grid: a survey. IEEE Commun. Surv. Tutorials 14(4), 944–980 (2012)
13. Grainger, J., Stevenson, W.: Power System Analysis. McGraw-Hill, New York City (1994)
14. Grandoni, F., Ingala, S., Uniyal, S.: Improved approximation algorithms for unsplittable flow on a path with time windows. In: Sanità, L., et al. (eds.) WAOA 2015. LNCS, vol. 9499, pp. 13–24. Springer, Heidelberg (2015). doi:10.1007/978-3-319-28684-6_2
15. Karapetyan, A., Khonji, M., Chau, C.K., Elbassioni, K., Zeineldin, H.: Efficient algorithm for scalable event-based demand response management in microgrids. Technical report, Masdar Institute (2015)
16. Kellerer, H., Pferschy, U., Pisinger, D.: Knapsack Problems. Springer, Heidelberg (2010)
17. Khonji, M., Karapetyan, A., Elbassioni, K., Chau, C.K.: Complex-demand scheduling problem with application in smart grid. Technical report, Masdar Institute (2016). http://arxiv.org/abs/1603.01786
18. Khonji, M., Chau, C.K., Elbassioni, K.: Optimal power flow with inelastic demands for demand response in radial distribution networks. Technical report, Masdar Institute (2016). http://arxiv.org/abs/1507.01762
19. Khonji, M., Chau, C.K., Elbassioni, K.M.: Inapproximability of power allocation with inelastic demands in AC electric systems and networks. In: ICCCN, pp. 1–6 (2014)
20. Nemirovski, A.S., Todd, M.J.: Interior-point methods for optimization. Acta Numerica 17(1), 191–234 (2008)
21. Spieksma, F.C.: On the approximability of an interval scheduling problem. J. Sched. 2(5), 215–227 (1999)
22. Woeginger, G.J.: When does a dynamic programming formulation guarantee the existence of a fully polynomial time approximation scheme (fptas)? INFORMS J. Comput. 12(1), 57–74 (2000)
23. Yu, L., Chau, C.K.: Complex-demand knapsack problems and incentives in AC power systems. In: Proceedings of AAMAS, pp. 973–980. Richland, SC (2013)

From Preemptive to Non-preemptive Scheduling Using Rejections

Giorgio Lucarelli[1]($\boxtimes$), Abhinav Srivastav[1,2], and Denis Trystram[1]

[1] LIG, Université Grenoble-Alpes, Grenoble, France
{giorgio.lucarelli,abhinav.srivastav,denis.trystram}@imag.fr
[2] Verimag, Université Grenoble-Alpes, Grenoble, France

Abstract. We study the classical problem of scheduling a set of independent jobs with release dates on a single machine. There exists a huge literature on the preemptive version of the problem, where the jobs can be interrupted at any moment. However, we focus here on the non-preemptive case, which is harder, but more relevant in practice. For instance, the jobs submitted to actual high performance platforms cannot be interrupted or migrated once they start their execution (due to prohibitive management overhead). We target on the minimization of the total stretch objective, defined as the ratio of the total time a job stays in the system (waiting time plus execution time), normalized by its processing time. Stretch captures the quality of service of a job and the minimum total stretch reflects the fairness between the jobs. So far, there have been only few studies about this problem, especially for the non-preemptive case. Our approach is based to the usage of the classical and efficient for the preemptive case shortest remaining processing time (SRPT) policy as a lower bound. We investigate the (offline) transformation of the SRPT schedule to a non-preemptive schedule subject to a recently introduced resource augmentation model, namely the rejection model according to which we are allowed to reject a small fraction of jobs. Specifically, we propose a $\frac{2}{\epsilon}$-approximation algorithm for the total stretch minimization problem if we allow to reject an ϵ-fraction of the jobs, for any $\epsilon > 0$. This result shows that the rejection model is more powerful than the other resource augmentations models studied in the literature, like speed augmentation or machine augmentation, for which non-polynomial or non-scalable results are known. As a byproduct, we present a $\frac{1}{\epsilon}$-approximation algorithm for the total flow-time minimization problem which also rejects at most an ϵ-fraction of jobs.

1 Introduction

In this work we are interested in the analysis of an efficient algorithm for scheduling jobs non-preemptively under the objective of minimizing the total (or average)

G. Lucarelli and D. Trystram—This work has been partially supported by the projet Moebus (ANR-13-INFR-0001) funded by ANR.

A. Srivastav—This work has been partially supported by the LabEx PERSYVAL-Lab (ANR-11-LABX-0025-01) funded by the French program "Investissement d'avenir".

© Springer International Publishing Switzerland 2016
T.N. Dinh and M.T. Thai (Eds.): COCOON 2016, LNCS 9797, pp. 510–519, 2016.
DOI: 10.1007/978-3-319-42634-1_41

stretch of the jobs. Stretch is the most relevant metric used in the context of resource management in large scale parallel computing platforms. Informally, the stretch of a job is the total time it spends in the system normalized by its processing time. Thus, the average stretch over all the jobs represents a quality of service measure in terms of fairness among the jobs. The jobs whose execution requires more time are more appropriate to wait longer than short ones. Non-preemptive scheduling policies are usually considered in computing platforms since practically, interrupting jobs during their execution is not allowed. This is due to significant communication overhead and extra memory costs that are induced by such interruptions. However, from the combinatorial side, scheduling non-preemptively is harder and as a consequence, a much less studied problem.

More formally, we consider the offline problem of scheduling a set $\mathcal{J}$ of n independent jobs on a single machine. Each job $j \in \mathcal{J}$ is characterized by a *processing time* p_j and a *release date* r_j. Given a schedule $\mathcal{S}$, we denote by $\sigma_j^{\mathcal{S}}$ and $C_j^{\mathcal{S}}$ the *starting time* and *completion time*, respectively, of the job j. Then, its *flow time* is defined as $F_j^{\mathcal{S}} = C_j^{\mathcal{S}} - r_j$, that is the total time that j remains to the system. The *stretch* of j in a schedule $\mathcal{S}$ is defined as $s_j^{\mathcal{S}} = \frac{F_j^{\mathcal{S}}}{p_j}$, that is the flow time of j is normalized with respect to its processing time. When there is no ambiguity, we will simplify the above notation by dropping $\mathcal{S}$. Our objective is to create a *non-preemptive* schedule that minimizes the total stretch of all jobs in $\mathcal{J}$, i.e., $\sum_{j \in \mathcal{J}} s_j$. This problem is known as the *total (or average) stretch minimization* problem.

The total stretch minimization problem is a special case of the *total weighted flow-time minimization* problem where each job $j \in \mathcal{J}$ is additionally characterized by a weight w_j and the objective is to minimize $\sum_{j \in \mathcal{J}} w_j F_j$. The above problem reduces to the total stretch minimization problem if we consider that $w_j = \frac{1}{p_j}$ for each $j \in \mathcal{J}$. Another closely related problem which is also a special case of the total weighted flow-time minimization problem is the *total flow-time minimization* problem in which the weights of all jobs are equal. Although total flow-time and total stretch objectives do not have an immediate relation, the latter is generally considered to be a more difficult problem since w_j depends on the job's processing time, while in the former all the jobs have the same weight.

Based on the inapproximability results for different variants of the above problems (see for example [2,16] and the related work below), Kalyanasundaram and Pruhs [14] and Phillips *et al.* [18] proposed to study the effect of resource augmentation, in which the algorithm is applied to a more powerful environment than the optimal one. For instance, in the *machine augmentation model* the algorithm can use more machines than the optimal solution, while in the *speed augmentation model* the algorithm can execute the jobs on faster machines comparing to the machines of the optimal schedule. More specifically, given some optimization objective (e.g., total weighted flow-time), an algorithm is said to be ℓ-machine ρ-approximation if it uses ℓm machines and it is a ρ-approximation with respect to an optimal scheduling algorithm using m machines, for some $\ell > 1$; similarly, we can define a v-speed ρ-approximation algorithm. Recently, Choudhury *et al.* [10] proposed the *rejection model*, in which the algorithm can reject a bounded fraction of the jobs (or a set of jobs whose total weight is a bounded fraction of the total weight of all jobs), while the optimal solution should execute all the jobs of the instance. In this paper, we study the total stretch minimization problem with respect to the rejection model.

Related Work. When preemptions are allowed, the well-known online Shortest Remaining Processing Time (SRPT) strategy returns the optimal solution for the total flow-time minimization problem [1] and a 2-competitive solution for the total stretch minimization problem [17]. A polynomial time approximation scheme has been also presented in [7] for the total stretch objective. On the other hand, for the total weighted flow-time minimization problem, the best known guarantee is given by a randomized algorithm which achieves an approximation ratio of $O(\log \log \Delta)$ [5], where Δ is the ratio of the largest processing time over the smallest processing time in the input instance. Furthermore, algorithms of competitive ratios $O(\log W)$ [4] and $O(\log^2 \Delta)$ [9] are known, while any algorithm should have a competitive ratio $\Omega(\min\{\sqrt{\frac{\log W}{\log \log W}}, \sqrt{\frac{\log \log \Delta}{\log \log \log \Delta}}\})$ [2], where W is the ratio of the largest weight over the smallest weight in the input instance.

In the non-preemptive context, even for the objectives of total flow-time and total stretch, the problem becomes much harder to approximate. Specifically, there is no approximation algorithm for the total flow-time minimization problem with ratio $O(n^{\frac{1}{2}-\epsilon})$, for any $\epsilon > 0$, unless $P = NP$ [15]. On the other hand, an algorithm that matches this ratio has been presented in the same paper. In the online setting, in [9] it is mentioned that any algorithm should have a competitive ratio $\Omega(n)$ even for the total flow-time objective. In [8], the greedy online Shortest Processing Time (SPT) strategy is proven to be $\frac{\Delta+1}{2}$-competitive for the total flow-time minimization problem and this ratio is the best possible for this problem. Similarly, the weighted generalization of SPT is $(\Delta+1)$-competitive for the total weighted flow-time objective and this ratio is optimal [20].

In the resource augmentation framework, an $(1 + \epsilon)$-speed $O(\frac{1}{\epsilon})$-competitive algorithm is known for the total weighted flow-time minimization problem when preemptions are allowed [6]. In [11], an $O(\frac{1}{\epsilon^{12}})$-competitive algorithm has been presented for the total weighted flow-time objective which rejects an ϵ-fraction of jobs; this result holds also for parallel machines.

If preemptions are not allowed, a 12-speed $(2+\epsilon)$-approximation algorithm for the total flow-time objective and a 12-speed 4-approximation algorithm for the total weighted flow-time objective have been presented in [3]. In [13], a dynamic programming framework has been presented that runs in quasi-polynomial time. This framework works also for the parallel machine setting and leads to a $(1+\epsilon)$-speed and $(1 + \epsilon)$-approximate solution for the total weighted flow-time minimization problem and to a $(1+\epsilon)$-speed and 1-approximate solution for the total flow-time minimization problem. In [18], an $O(\log \Delta)$-machine 1-competitive algorithm has been proposed for the total weighted flow-time objective even for parallel processors. For the unweighted version, an $O(\log n)$-machine $(1 + o(1))$-competitive algorithm and an $O(\log n)$-machine $(1 + o(1))$-speed 1-competitive algorithm have been proposed in the same paper. Note that the algorithms in [18] work in the online setting but they need to know the minimum and the maximum processing times in advance. Moreover, an ℓ-machine $(1+\Delta^{1/\ell})$-competitive algorithm designed in [12] for the total flow-time minimization problem, if Δ is known a priori to the algorithm. They also provided a lower bound which shows that their algorithm is optimal up to a constant factor for any constant ℓ.

No results for the total stretch minimization problem without preemptions in the resource augmentation context are known, except from the results that derive from the more general problem of minimizing the weighted flow-time.

Contribution and Organization of the Paper. In this paper, we explore the relation between preemptive and non-preemptive schedules with respect to the total stretch objective subject to the rejection model. More specifically, we consider the SRPT policy for creating a preemptive schedule. In Sect. 2 we describe several structural properties of this schedule. Next, we show how to transform the preemptive schedule created by the SRPT policy to a non-preemptive schedule with given worst-case guarantees.

In Sect. 3, we use the rejection model and we give an $\frac{2}{\epsilon}$-approximation algorithm if we are permitted to delete a subset of jobs such that their total weight is an ϵ-fraction of the total weight of all jobs. Note that, the relation among the rejection model and other resource augmentation models is not clear. For example, in Fig. 1 we give an instance for which the best possible solution using rejections is worse than the best possible solution using speed-augmentation, when the same constant ϵ is selected for both models. However, our result shows the strength of the rejection model, particularly in the non-preemptive context, since the known results subject to other resource augmentation models either need quasi-polynomial time [13] or they cannot arrive arbitrarily close to the

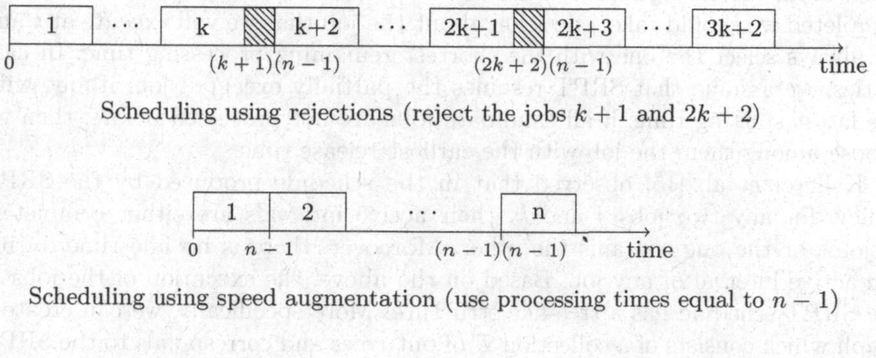

Scheduling using rejections (reject the jobs $k+1$ and $2k+2$)

Scheduling using speed augmentation (use processing times equal to $n-1$)

Fig. 1. An instance of $n = 3k + 2$ jobs with equal processing times $p_j = n$, equal weights $w_j = 1$, and release dates $r_j = (j-1)(n-1)$, where $1 \leq j \leq n$. By setting $\epsilon = \frac{1}{n-1}$, in the rejection model we are allowed to reject at most $\epsilon n \leq 2$ jobs, while in the speed augmentation model the processing time of each job becomes $\frac{p_j}{1+\epsilon} = n-1$. The total flow time using rejections is $3\sum_{j=1}^{k}(n+j-1) = \frac{21}{2}k^2 + \frac{9}{2}k$, while the total flow time using speed augmentation is $n(n-1) = 9k^2 + 9k + 2$ which is better for large enough k.

classical model without resource augmentation [3,18] even for the total flow-time objective. Contrarily, our result is the best possible we can expect in the rejection model.

Finally, using the same rejection strategy and analysis, we obtain an $\frac{1}{\epsilon}$-approximation algorithm if we are allowed to delete an ϵ-fraction of jobs. We conclude in Sect. 4. Before continuing, we give some additional notation which we use throughout the paper.

Notations. In what follows, for each job $j \in \mathcal{J}$ and schedule $\mathcal{S}$, we define the interval $[\sigma_j^\mathcal{S}, C_j^\mathcal{S}]$ to be the *active interval* of j in $\mathcal{S}$. In the case where preemptions are allowed, the active interval of j may have a length bigger than p_j. A job j is *available* at a time t if it is released but it is not yet completed, i.e., $r_j \le t < C_j$. We call a schedule *compact* if it does not leave any idle time whenever there is a job available for execution.

2   Structure and Properties of SRPT and an Intermediate Schedule

In this section we deal with the structure of a preemptive schedule created by the Shortest Remaining Processing Time (SRPT) policy and we give some useful properties that we will use in the following sections. According to the SRPT policy, at any time, we select to execute the available job with the shortest remaining processing time. Since the remaining processing time of the executed job $j \in \mathcal{J}$ decreases over time, its execution may be interrupted only in the case where a new job $k \in \mathcal{J}$ is released and the processing time of k is smaller than the remaining processing time of j at r_k. Hence, the SRPT policy can be seen as an event-driven algorithm in which at each time t where a job is released or completed we should take a decision about the job that we will execute at t and we always select the one with the shortest remaining processing time. In case of ties, we assume that SRPT resumes the partially executed job, if any, with the latest starting time; if all candidate jobs are not processed before, then we choose among them the job with the earliest release time.

Kellerer et al. [15] observed that in the schedule produced by the SRPT policy, for any two jobs j and k, their active intervals are either completely disjoint or the one contains the other. Moreover, there is no idle time during the active interval of any job. Based on the above, the execution of the jobs in the SRPT schedule has a tree-like structure. More specifically, we can create a graph which consists of a collection $\mathcal{T}$ of out-trees and corresponds to the SRPT schedule as follows (see Fig. 2): for each job $j \in \mathcal{J}$, we create a vertex u_j. For each pair of jobs $j, k \in \mathcal{J}$, we add an arc (u_j, u_k) if and only if $[\sigma_k, C_k] \subset [\sigma_j, C_j]$ and there is no other job $i \in \mathcal{J}$ so that $[\sigma_k, C_k] \subset [\sigma_i, C_i] \subset [\sigma_j, C_j]$.

In what follows, we denote by $root(T)$ the root of each out-tree $T \in \mathcal{T}$. Intuitively, each vertex $root(T)$ corresponds to a job for which at any time t during its execution there is no other job which has been partially executed at t. We denote also by $a(j)$ the parent of the vertex that corresponds to the job

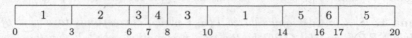

jobs	1	2	3	4	5	6
r_j	0	3	6	7	14	16
p_j	7	3	3	1	5	1

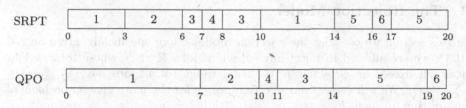

Shortest Remaining Processing Time (SRPT) schedule

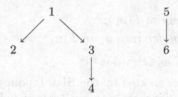

Collection of out-trees

Fig. 2. A schedule created by the SRPT policy and its corresponding collection of out-trees.

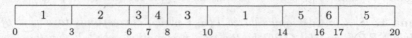

Fig. 3. Transformation from SRPT to QPO schedule

$j \in \mathcal{J}$ in T. Moreover, let $T(u_j)$ be the subtree of $T \in \mathcal{T}$ rooted at a vertex u_j in T. Note that, we may refer to a job j by its corresponding vertex u_j and vice versa.

In this paper, we use the schedule created by the SRPT policy for the preemptive variant of our problem as a lower bound to the non-preemptive variant. The SRPT policy is known to be optimal [16,19] for the problem of minimizing the sum $\sum_{j \in \mathcal{J}} F_j$ when preemptions of jobs are allowed. However, for the preemptive variant of the total stretch minimization problem, SRPT is a 2-approximation algorithm [17].

Consider now the collection of out-trees $\mathcal{T}$ obtained by an SRPT schedule and let $T(u_j)$ be the subtree rooted at any vertex u_j. We construct a non-preemptive schedule for the jobs in $T(u_j)$ as follows: during the interval $[\sigma_j, C_j]$, we run the jobs in $T(u_j)$ starting with j and then running the remaining jobs in order of increasing SRPT completion time as shown in Fig. 3. This policy has been proposed in [8] for the problem of minimizing the sum $\sum_{j \in \mathcal{J}} F_j$ and corresponds to a post order transversal of the subtree $T(u_j)$ excluding its root which is scheduled in the first position. We call the above policy as *Quasi Post Order* (QPO) and we will use it for the problem of minimizing the sum $\sum_{j \in \mathcal{J}} s_j$. The following lemma presents several observations for the QPO policy.

Lemma 1. [8] *Consider any subtree $T(u_k)$ which corresponds to a part of the schedule $\mathcal{SRPT}$ and let $\mathcal{QPO}$ be the non-preemptive schedule for the jobs on $T(u_k)$ created by applying the Quasi Post Order policy. Then,*

(i) *all jobs in $\mathcal{QPO}$ are executed during the interval $[\sigma_k^{\mathcal{SRPT}}, C_k^{\mathcal{SRPT}}]$ without any idle period,*

(ii) $\sigma_j^{\mathcal{QPO}} \geq r_j$ *for each u_k in $T(u_k)$,*

(iii) $C_j^{\mathcal{QPO}} \leq C_j^{\mathcal{SRPT}} + p_k$ *for each u_j in $T(u_k)$ with $j \neq k$, and*

(iv) $C_k^{\mathcal{QPO}} = C_k^{\mathcal{SRPT}} - \sum_{u_j \in T(u_k): j \neq k} p_j.$

Note that, the schedule created by the SRPT policy is a compact schedule, since it always execute a job if there is an available one. Therefore, by Lemma 1.i, the following directly holds.

Corollary 1. *The schedule created by the QPO policy is compact.*

3   The Rejection Model

In this section we consider the rejection model. More specifically, given an $\epsilon \in (0,1)$, we are allowed to reject any subset of jobs $\mathcal{R} \subset \mathcal{J}$ whose total weight does not exceed an ϵ-fraction of the total weight of all jobs, i.e., $\sum_{j \in \mathcal{R}} w_j \leq \epsilon \sum_{j \in \mathcal{J}} w_j$. We will present our rejection policy for the more general problem of minimizing $\sum_{j \in \mathcal{J}} w_j F_j$.

Our algorithm is based on the tree-like structure of the SRPT schedule. Let us focus first on a single out-tree $T \in \mathcal{T}$. The main idea is to reject the jobs that appear in the higher levels of T (starting with its root) and run the remaining jobs using the QPO policy. The rejected jobs are, in general, long jobs which are preempted several times in the SRPT schedule and their flow time can be used as an upper bound for the flow time of the smaller jobs that are released and completed during the life interval of the longest jobs. In order to formalize this, for each job $j \in \mathcal{J}$ we introduce a charging variable x_j. In this variable we accumulate the weight of jobs whose flow time will be upper bounded by the flow time of job j in the SRPT schedule. At the end of the algorithm, this variable will be exactly equal to $\frac{1}{\epsilon} w_j$ for each rejected job $j \in \mathcal{R}$, while $x_j < \frac{1}{\epsilon} w_j$ for each non-rejected job $j \in \mathcal{J} \setminus \mathcal{R}$. In fact, for most of the non-rejected jobs this variable will be equal to zero at the end of the algorithm. Our algorithm considers the jobs in a bottom-up way and charges the weight of the current job to its ancestors in T which are closer to the root and their charging variable is not yet full; that is the vertices to be charged are selected in a top-down way. Note that, we may charge parts of the weight of a job to more than one of its ancestors.

Algorithm 1 describes formally the above procedure. For notational convenience, we consider a fictive vertex u_0 which corresponds to a fictive job with $w_0 = 0$. We connect u_0 with the vertex $root(T)$ of each out-tree $T \in \mathcal{T}$ in such a way that u_0 becomes the parent of all of them. Let T^* be the created tree with root u_0.

Algorithm 1.

1: Create a preemptive schedule $\mathcal{SRPT}$ and the corresponding out-tree T^*
2: Initialization: $\mathcal{R} \leftarrow \emptyset$, $x_j \leftarrow w_j$ for each $j \in \mathcal{J}$, $x_0 \leftarrow 0$
3: **for** each vertex u_j of T^* with $x_j = w_j$ in post-order traversal **do**
4:     **while** $x_j \neq 0$ and $x_{a(j)} < \frac{1}{\epsilon} w_{a(j)}$ **do**
5:         Let u_k be a vertex in the path between u_0 and u_j such that
            $x_{a(k)} = \frac{1}{\epsilon} w_{a(k)}$ and $x_k < \frac{1}{\epsilon} w_k$
6:         Let $y \leftarrow \min\{x_j, \frac{1}{\epsilon} w_k - x_k\}$
7:         $x_j \leftarrow x_j - y$ and $x_k \leftarrow x_k + y$
8: **for** each job $j \in \mathcal{J}$ **do**
9:     **if** $x_j = \frac{1}{\epsilon} w_j$ **then**
10:         Reject j, i.e., $\mathcal{R} \leftarrow \mathcal{R} \cup \{j\}$
11: **return** S: the non-preemptive schedule for the jobs in $\mathcal{J} \setminus \mathcal{R}$ using QPO

Note that, the for-loop in Lines 3–7 of Algorithm 1 is not executed for all jobs. In fact, it is not applied to the jobs that will be rejected as well as to some children of them for which at the end of the algorithm it holds that $w_j < x_j < \frac{1}{\epsilon} w_j$. The weight of these jobs is charged to themselves. Moreover, the while-loop in Lines 4–7 of Algorithm 1 terminates either if the whole weight of j is charged to its ancestors or if the parent of u_j is already fully charged, i.e., $x_{a(j)} = \frac{1}{\epsilon} w_{a(j)}$.

Theorem 1. *For the schedule S created by Algorithm 1 it holds that*

$$(i) \sum_{j \in \mathcal{J} \setminus \mathcal{R}} w_j F_j^{\mathcal{S}} \leq \frac{1}{\epsilon} \sum_{j \in \mathcal{J}} w_j F_j^{\mathcal{SRPT}}, \text{ and}$$

$$(ii) \sum_{j \in \mathcal{R}} w_j \leq \epsilon \sum_{j \in \mathcal{J}} w_j.$$

Proof. Consider first any vertex u_k such that $k \in \mathcal{J} \setminus \mathcal{R}$ and $a(k) \in \mathcal{R}$. By the execution of the algorithm, all the jobs corresponding to vertices in the path from u_0 to $a(k)$ do not appear in S. Hence, k starts in S at the same time as in $\mathcal{SRPT}$, i.e., $\sigma_k^{\mathcal{S}} = \sigma_k^{\mathcal{SRPT}}$. Thus, by Lemma 1, the jobs that correspond to the vertices of the subtree $T^*(u_k)$ are scheduled in S during the interval $[\sigma_k^{\mathcal{SRPT}}, C_k^{\mathcal{SRPT}}]$. In other words, for any job j in $T^*(u_k)$ it holds that $C_j^{\mathcal{S}} \leq C_k^{\mathcal{SRPT}}$, while by the construction of T^* we have that $\sigma_k^{\mathcal{SRPT}} < r_j$. Assume now that the weight of j is charged by Algorithm 1 to the jobs $j_1, j_2, \ldots, j_{q_j}$, where q_j is the number of these jobs. Let w_j^i be the weight of j charged to $j_i \in \{j_1, j_2, \ldots, j_{q_j}\}$; note that $w_j = \sum_{i=1}^{q_j} w_j^i$. By the definition of the algorithm, each $j_i \in \{j_1, j_2, \ldots, j_{q_j}\}$ is an ancestor of both k and j in T^* (one of them may coincides with k). Therefore, by the definition of T^*, it holds that $\sigma_{j_i}^{SRPT} < r_j < C_j^{\mathcal{S}} \leq C_{j_i}^{SRPT}$, for each $j_i \in \{j_1, j_2, \ldots, j_{q_j}\}$. Then, we have

$$\sum_{j \in \mathcal{J} \setminus \mathcal{R}} w_j F_j^{\mathcal{S}} \leq \sum_{j \in \mathcal{J} \setminus \mathcal{R}} \sum_{i=1}^{q_j} w_j^i F_{j_i}^{SRPT} \leq \sum_{j \in \mathcal{J}} x_j F_j^{SRPT} \leq \sum_{j \in \mathcal{J}} \frac{1}{\epsilon} w_j F_j^{SRPT}$$

where the second inequality holds by regrouping the flow time of all appearances of the same job, and the last one by the fact that Algorithm 1 charges at each job j at most $(1 + \frac{1}{\epsilon})w_j$. Finally, since the weight of each job is charged exactly once (probably to more than one other jobs) we have $\sum_{j \in \mathcal{J}} w_j \geq \frac{1}{\epsilon} \sum_{j \in \mathcal{R}} w_j$ and the theorem holds. □

Since SRPT creates an optimal preemptive schedule for the problem of minimizing $\sum_{j \in \mathcal{J}} F_j$ on a single machine and an optimal preemptive schedule is a lower bound for a non-preemptive one the following theorem holds.

Theorem 2. *Algorithm 1 is a $\frac{1}{\epsilon}$-approximation algorithm for the single-machine total flow-time minimization problem without preemptions if we are allowed to reject an ϵ-fraction of the jobs.*

By combining Theorem 1 and the fact that SRPT is a 2-approximation algorithm for the preemptive variant of the total stretch minimization problem [17], the following theorem holds.

Theorem 3. *Algorithm 1 is a $\frac{2}{\epsilon}$-approximation algorithm for the single-machine total stretch minimization problem without preemptions if we are allowed to reject a set of jobs whose total weight is no more than an ϵ-fraction of the total weight of all jobs.*

4   Concluding Remarks

We studied the effects of applying resource augmentation in the transformation of a preemptive schedule to a non-preemptive one for the problem of minimizing total stretch on a single machine. Specifically, we show the power of the rejection model for scheduling without preemptions comparing with other resource augmentation models, by presenting an algorithm which has a performance arbitrarily close to optimal. Note that, SRPT is a 14-competitive algorithm for minimizing total stretch on parallel machines when preemptions and migrations are allowed [17]. So, an interesting question is to explore the general idea of this paper about transforming preemptive to non-preemptive schedules subject to the rejection model on parallel machines based on the above result.

References

1. Baker, K.R.: Introduction to Sequencing and Scheduling. Wiley, New York (1974)
2. Bansal, N., Chan, H.-L.: Weighted flow time does not admit $o(1)$-competitive algorithms. In: SODA, pp. 1238–1244 (2009)
3. Bansal, N., Chan, H.-L., Khandekar, R., Pruhs, K., Stein, C., Schieber, B.: Non-preemptive min-sum scheduling with resource augmentation. In: FOCS, pp. 614–624 (2007)
4. Bansal, N., Dhamdhere, K.: Minimizing weighted flow time. ACM Trans. Algorithms **3**(4), 39 (2007)
5. Bansal, N., Pruhs, K.: The geometry of scheduling. SIAM J. Comput. **43**, 1684–1698 (2014)
6. Becchetti, L., Leonardi, S., Marchetti-Spaccamela, A., Pruhs, K.: Online weighted flow time and deadline scheduling. J. Discrete Algorithms **4**, 339–352 (2006)
7. Bender, M.A., Muthukrishnan, S., Rajaraman, R.: Approximation algorithms for average stretch scheduling. J. Sched. **7**, 195–222 (2004)
8. Bunde, D.P.: SPT is optimally competitive for uniprocessor flow. Inf. Process. Lett. **90**, 233–238 (2004)
9. Chekuri, C., Khanna, S., Zhu, A.: Algorithms for minimizing weighted flow time. In: STOC, pp. 84–93 (2001)
10. Choudhury, A.R., Das, S., Garg, N., Kumar, A.: Rejecting jobs to minimize load and maximum flow-time. In: SODA, pp. 1114–1133 (2015)
11. Choudhury, A.R., Das, S., Kumar, A.: Minimizing weighted lp-norm of flow-time in the rejection model. In: FSTTCS, pp. 25–37 (2015)
12. Epstein, L., van Stee, R.: Optimal on-line flow time with resource augmentation. Discrete Appl. Math. **154**, 611–621 (2006)
13. Im, S., Li, S., Moseley, B., Torng, E.: A dynamic programming framework for non-preemptive scheduling problems on multiple machines [extended abstract]. In: SODA, pp. 1070–1086 (2015)
14. Kalyanasundaram, B., Pruhs, K.: Speed is as powerful as clairvoyance. J. ACM **47**, 617–643 (2000)
15. Kellerer, H., Tautenhahn, T., Woeginger, G.J.: Approximability and nonapproximability results for minimizing total flow time on a single machine. SIAM J. Comput. **28**, 1155–1166 (1999)
16. Leonardi, S., Raz, D.: Approximating total flow time on parallel machines. J. Comput. Syst. Sci. **73**, 875–891 (2007)
17. Muthukrishnan, S., Rajaraman, R., Shaheen, A., Gehrke, J.E.: Online scheduling to minimize average stretch. SIAM J. Comput. **34**, 433–452 (2005)
18. Phillips, C.A., Stein, C., Torng, E., Wein, J.: Optimal time-critical scheduling via resource augmentation. Algorithmica **32**, 163–200 (2002)
19. Schrage, L.: A proof of the optimality of the shortest remaining processing time discipline. Oper. Res. **16**, 687–690 (1968)
20. Tao, J., Liu, T.: WSPT's competitive performance for minimizing the total weighted flow time: from single to parallel machines. Math. Probl. Eng. (2013). doi:10.1155/2013/343287

Flow Shop for Dual CPUs in Dynamic Voltage Scaling

Vincent Chau[1,2], Ken C. K. Fong[2], Minming Li[2(✉)], and Kai Wang[2]

[1] Department of Computer Science, Hong Kong Baptist University,
Kowloon Tong, Hong Kong
vincchau@comp.hkbu.edu.hk
[2] Department of Computer Science, City University of Hong Kong,
Kowloon Tong, Hong Kong
{ken.fong,kai.wang}@my.cityu.edu.hk, minming.li@cityu.edu.hk

Abstract. We study the following flow shop scheduling problem on two processors. We are given n jobs with a common deadline D, where each job j has workload $p_{i,j}$ on processor i and a set of processors which can vary their speed dynamically. Job j can be executed on the second processor if the execution of job j is completed on the first processor. Our objective is to find a feasible schedule such that all jobs are completed by the common deadline D with minimized energy consumption. For this model, we present a linear program for the discrete speed case, where the processor can only run at specific speeds in $S = \{s_1, s_2, \cdots, s_q\}$ and the job execution order is fixed. We also provide a $m^{\alpha-1}$-approximation algorithm for the arbitrary order case and for continuous speed model where m is the number of processors and α is a parameter of the processor.

We then introduce a new variant of flow shop scheduling problem called sense-and-aggregate model motivated by data aggregation in wireless sensor networks where the base station needs to receive data from sensors and then compute a single aggregate result. In this model, the first processor will receive unit size data from sensors and the second processor is responsible for calculating the aggregate result. The second processor can decide when to aggregate and the workload that needs to be done to aggregate x data will be $f(x)$ and another unit size data will be generated as the result of the partial aggregation which will then be used in the next round aggregation. Our objective is to find a schedule such that all data are received and aggregated by the deadline with minimum energy consumption. We present an $O(n^5)$ dynamic programming algorithm when $f(x) = x$ and a greedy algorithm when $f(x) = x - 1$.

1 Introduction

Energy efficiency has become a major concern nowadays, especially when more and more data centers are used. One of the technologies used to save energy is speed-scaling where the processors are capable of varying its speed dynamically. The faster the processors run, the more energy they consume. The idea is to complete all the given jobs before their deadlines with the slowest possible speed to minimize the energy usage.

© Springer International Publishing Switzerland 2016
T.N. Dinh and M.T. Thai (Eds.): COCOON 2016, LNCS 9797, pp. 520–531, 2016.
DOI: 10.1007/978-3-319-42634-1_42

The energy minimization problem of scheduling n jobs with release times and deadlines on a single processor that can vary its speed dynamically where preemption is allowed has been first studied in the seminal paper by Yao et al. [12]. The time complexity has been improved in [5,8,9]. More works along this line can be found in the surveys by Albers [1,2].

On the other hand, researchers have also studied various shop scheduling problems on multiple processors. For example, in the classical flow shop scheduling on m processors, there exists dependencies on execution order between processors for the same job, such that for each job j, it has to be completed on processor i before being executed on processor $i + 1$ where $1 \leq i \leq m - 1$. This problem is known to turn NP-hard by adding constraints [3]. One of the polynomial case is when there are two processors and one aims to minimize the makespan, i.e. the completion time of the last job, which can be solved by a greedy algorithm and is proposed by Johnson [6].

Recently, Mu and Li [10] combined the classical flow shop scheduling problem with the speed-scaling model. They proposed a polynomial time algorithm for the flow shop speed scaling problem for two processors when the order of jobs is fixed. They also showed that the order returned by Johnson's algorithm is not optimal for the speed-scaling environment. Fang et al. [4] studied the flow shop scheduling problem and aim to find an optimal order of jobs with a restriction on peak power consumption, i.e. the energy consumption is bounded by a constant for a fixed size interval at any time in the schedule.

Our contributions. In this paper, we first extend the study of combining the classical flow shop scheduling problem and the speed-scaling model proposed by Mu and Li [10] to the discrete speed setting where the processor can only run at specific speeds in $S = \{s_1, s_2, \cdots, s_q\}$ with predefined job execution order. We present a linear program for this case and also give a simple $m^{\alpha-1}$-approximation algorithm for the general case of flow shop scheduling problem where job execution order is not given.

We then introduce a new variant of flow shop scheduling problem called sense-and-aggregate model motivated by data aggregation in wireless sensor networks where the base station needs to receive data from sensors and then compute a single aggregate result like minimum, maximum or average value of all the input. When dealing with these kinds of aggregation functions, we can either wait for all the input data to be ready, or instead of waiting for all the input data, we can collect part of the data, and compute the sub results. Eventually, we can compute the final aggregation result based on the sub results that we computed previously. The second approach will play a significant role in energy efficient computation.

In the sense-and-aggregate model, we simulate the aggregation calculation in the setting of two processors, where the first processor performs the data collection task, and the second processor is responsible for the calculations. Note that both processors can change speeds when executing workloads. We are given a set of n unit jobs in the first processor and the workload-consideration-function $f(x)$ to indicate the amount of workload that the second processor will process to aggregate x units of data and a deadline D. After each aggregation, one unit of data will be produced which will be used in the next partial aggregation. There is a trade off between waiting and calculating. The more we wait, the more we have to schedule faster (thus consuming more energy), but the less we have to schedule for some function f. Based on this trade off, our objective is to find a schedule with minimum energy consumption to aggregate all the data by the deadline D. For this model, we present an $O(n^5)$ dynamic programming algorithm when $f(x) = x$ and a greedy algorithm when $f(x) = x - 1$.

The remainder of this paper is organized as follows. In Sect. 2, we present the formulation of the standard flow shop speed scaling problem. In Sect. 3, we propose the solution for the flow shop problem with discrete speed scaling model where the job order is predefined. In Sect. 4, we present a lower bound and approximation algorithm for the general setting of the flow shop speed scaling problem. In Sect. 5, we define the sense-and-aggregate model and present a greedy algorithm to solve the aggregation model when $f(x) = x - 1$, together with a dynamic programming approach when $f(x) = x$.

2   Formulation of Standard Flow Shop with Speed Scaling

We are given m processors and n jobs with a common deadline D, where each job j has workload $p_{i,j}$ on processor i. Each job has to be completed on processor i before it can be performed on the next processor $i + 1$. The energy consumption depends on the processor speed. If $s(t)$ is the speed of the processor at time t, then the energy consumption of the processor in the interval $[u, u')$ is $\int_u^{u'} P(s(t))dt$ where $P(s(t)) = s(t)^\alpha$ is the power function of the processor and α is a parameter of the processor, e.g. 1.11 for Intel PXA 270, 1.62 for Pentium M770 and 1.66 for a TCP offload engine [11]. We only assume that $\alpha > 1$ in order to keep the convexity of the power function P. The objective is to find a schedule (when to execute what job at which speed) to finish all the jobs by the deadline D conforming to the flow shop requirement to minimize the total energy consumption of all the processors.

Given a schedule, we define *critical jobs* to be the jobs that are executed on the second processor as soon as they are completed on the first processor. Given a list of *critical jobs* $\{c_1, c_2, \ldots, c_k\}$ the total cost can be computed as follows [10]:
$$\frac{\left(p_{1,1} + \sum_{i=1}^{k-1} \sqrt[\alpha]{\left(\sum_{j=1+c_i}^{c_{i+1}} p_{1,j}\right)^\alpha + \left(\sum_{j=c_i}^{-1+c_{i+1}} p_{2,j}\right)^\alpha} + p_{2,n}\right)^\alpha}{D}.$$
Note that jobs between two critical jobs $\{c_i + 1, c_i + 2, \ldots, c_{i+1}\}$ are executed at the same speed on the same machine.

3   Discrete Speed Setting with Given Job Order

In this section, we have a fixed order of jobs that have to be executed on m processors.

Moreover, processors can only run at specific speeds in $S = \{s_1, s_2, \ldots, s_q\}$. The goal is to find a schedule with minimum energy to finish all jobs by the deadline D. The discrete setting of the classical speed scaling problem was studied

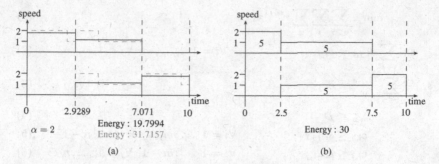

Fig. 1. Example of instance where the discrete optimal schedule cannot be obtained by first computing the continuous optimal schedule.

in [7]. They solved the problem by first computing an optimal schedule in the continuous setting, and then perform the transformation to the solution in the discrete setting by adjusting the speeds. However, in the flow shop speed scaling problem, such a transformation technique cannot be used to obtain an optimal schedule. The example below with two processors demonstrates that computing the optimal schedule in the continuous setting by using the algorithm proposed by Mu and Li [10] cannot obtain the optimal schedule in the discrete speed setting.

The instance contains two jobs with $p_{i,j} = 5 \; \forall i, j$. Moreover, we set $\alpha = 2$ and $D = 10$. In the discrete setting, we have the set of speeds $S = \{1, 2\}$. According to Fig. 1(a), we use the algorithm proposed by Mu and Li [10] to compute the result schedule with continuous speed, and convert the schedule to discrete speed setting with energy consumption of 31.7157. However, the optimal schedule only consumes 30 as illustrated in Fig. 1(b). This shows that adjusting the speeds obtained from the algorithm proposed by [10] to the adjacent available discrete speeds cannot obtain an optimal solution in the discrete speed setting.

Therefore, it is necessary to design a new method to calculate the optimal solution for the discrete speed setting. We propose the following linear program in Fig. 2. Without loss of generality, we assume that jobs are scheduled in the following order $1, \ldots, n$.

Let $x_{i,j,v}$ be the workload done for job j on processor i at speed v. Let $s_{i,j}$ (resp. $c_{i,j}$) be the starting time (resp. completion time) of job j on processor i. The constraints (2) are to ensure that all jobs are fully executed while the constraints (3), (4), (5), (6) ensure that deadlines are not violated. Especially, since the execution order is fixed, for each processor, jobs need to be scheduled in the order $1, \ldots, n$. Constraints (5) ensure this order. Job j needs to be completed before that job $j + 1$ starts. Similarly, constraints (6) ensure that a job j must to be finished on processor i before starting on the next processor $i + 1$. Constraints (3) ensure that the completion time of job j on processor i is exactly after its processing time $\sum_{v \in S} \frac{x_{i,j,v}}{v}$ from its starting time $s_{i,j}$. This linear program is correct as we can obtain a feasible schedule in the given order of jobs with deadlines satisfied and jobs scheduled entirely. Given the solution of the linear program, we know exactly the processing time of each job on each processor, i.e. the processing time of job j on processor i is $\sum_{v \in S} x_{i,j,v}$. Then

$$\min \ \sum_{v \in S} \sum_i \sum_j v^{\alpha-1} x_{i,j,v} \tag{1}$$

$$\text{s.t.} \sum_{v \in S} x_{i,j,v} = p_{i,j} \qquad\qquad \forall i = 1, 2, \ldots, m, \ \forall j = 1, \ldots, n \tag{2}$$

$$s_{i,j} + \sum_{v \in S} \frac{x_{i,j,v}}{v} = c_{i,j} \qquad\qquad\qquad\qquad \forall i, j \tag{3}$$

$$c_{m,n} \le D \tag{4}$$

$$c_{i,j} \le s_{i,j+1} \qquad\qquad \forall i = 1, \ldots, m, \ \forall j = 1, \ldots, n-1 \tag{5}$$

$$c_{i,j} \le s_{i+1,j} \qquad\qquad \forall i = 1, \ldots, m-1, \ \forall j = 1, \ldots, n \tag{6}$$

$$x_{i,j,v}, \ s_{i,j}, \ c_{i,j} \ge 0 \qquad\qquad\qquad\qquad \forall i, j, v \tag{7}$$

Fig. 2. Linear program for discrete speed setting flow shop

we simply schedule jobs on the first processor with the corresponding processing time. Each job j on the second processor is scheduled either immediately after the previous job $j-1$, or exactly at the time it is completed on the first processor. The constraint (4) guarantees that the returned scheduled has a makespan at most D.

4 Continuous Speed Setting with Arbitrary Order

In this section, we investigate the flow shop scheduling problem where the execution order of jobs is not predefined.

Theorem 1 [10]. *The problem of flow shop with arbitrary speeds given the order of execution of jobs on two processors is polynomial.*

We propose a linear time approximation algorithm for the flowshop scheduling problem for an arbitrary number of processors.

Let $V_i = \sum_{j=1}^n p_{i,j}$ be the processing volume on the i-th processor.

Proposition 1. $\sum_i (V_i)^\alpha D^{1-\alpha}$ *is a valid lower bound.*

Proof. By omitting the precedence constraints, we can schedule jobs with uniform speed. Since the workload is fixed on each processor, the minimum energy for processor i is $V_i^\alpha D^{1-\alpha}$. □

Proposition 2. $\sum_i (V_i)^\alpha D^{1-\alpha} \ge m(\sum_i V_i/m)^\alpha D^{1-\alpha}$.

Proof. Since the power function is convex, the minimum energy is given by executing the same workload on all processors. □

The main idea of the algorithm is to schedule jobs at speed $\frac{\sum_i V_i}{D}$ in any order. In fact, it is sufficient to schedule jobs on the first processor, then schedule jobs on the second processor once the last job ends on the previous processor, and so on. In other words, we schedule jobs in arbitrary order on each processor i in $[\sum_{j=0}^{i-1} V_j, V_i + \sum_{j=0}^{i-1} V_j]$. In this way, all precedence constraints are satisfied and no job misses its deadline. Therefore it is a feasible schedule. The energy consumption of this schedule is exactly $(\sum_i V_i)^\alpha D^{1-\alpha}$.

Theorem 2. *There exists a linear time algorithm which is an $m^{\alpha-1}$-approximation.*

Proof. We analyze the approximation ratio by dividing the energy cost of the returned schedule by the lower bound.

$$\frac{(\sum_i V_i)^\alpha D^{1-\alpha}}{m(\sum_i V_i/m)^\alpha D^{1-\alpha}} = \frac{(\sum_i V_i)^\alpha}{m(\sum_i V_i/m)^\alpha} = \frac{(\sum_i V_i)^\alpha}{(\sum_i V_i)^\alpha m(1/m)^\alpha} = m^{\alpha-1}$$

With the discussion above, we can get a feasible schedule in linear time.   □

Proposition 3. *The bound of the algorithm is tight for m processors.*

Proof. We create an instance with m jobs and m processors. The first job has workload $(1, \ldots, 1, K)$, the second job has workload $(1, \ldots, 1, K, 1)$ and so on. The last job has workload $(K, 1, \ldots, 1)$, with $K > 0$.

Let S be a schedule such that jobs are scheduled in the following order $1, 2, \ldots, n$. We formulate the cost of this schedule such that each job is a critical job. Note that this cost may be larger than the optimal cost. Since we have $m^{\alpha-1} \geq \frac{ALG}{OPT} \geq \frac{ALG}{S}$. We only need to show that $\frac{ALG}{S}$ is at least $m^{\alpha-1}$.

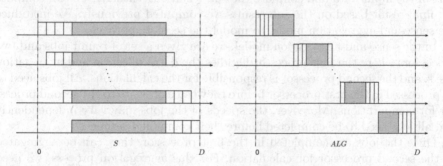

Fig. 3. Tight example: first job has workload $(1, \ldots, 1, K)$, and the last job has workload $(K, 1, \ldots, 1)$. The gray block represents a workload of K of each job.

The energy consumption of the schedule S (shown in Fig. 3) is

$$S = (\sqrt[\alpha]{1} + \sqrt[\alpha]{2} + \ldots + \sqrt[\alpha]{m-1} + \sqrt[\alpha]{mK^\alpha} + \sqrt[\alpha]{m-1} + \ldots + \sqrt[\alpha]{1})^\alpha D^{-1}$$

$$= \left(2\sum_{i=1}^{m-1} \sqrt[\alpha]{i} + \sqrt[\alpha]{mK^\alpha}\right)^\alpha D^{-1}$$

while the energy usage in our algorithm is $ALG = (m(m-1+K))^\alpha D^{-1}$.

$$\frac{ALG}{S} = \frac{(m(m-1+K))^\alpha D^{-1}}{\left(2\sum_{i=1}^{m-1} \sqrt[\alpha]{i} + \sqrt[\alpha]{mK^\alpha}\right)^\alpha D^{-1}} = \left(\frac{m(m-1+K)}{2\sum_{i=1}^{m-1} \sqrt[\alpha]{i} + \sqrt[\alpha]{mK^\alpha}}\right)^\alpha$$

Then

$$\lim_{K \to +\infty} \left(\frac{m(m-1+K)}{2\sum_{i=1}^{m-1} \sqrt[\alpha]{i} + \sqrt[\alpha]{mK^\alpha}} \right)^\alpha = \lim_{K \to +\infty} \left(\frac{m(\frac{m-1}{K}+1)}{\frac{2\sum_{i=1}^{m-1} \sqrt[\alpha]{i}}{K} + \sqrt[\alpha]{m}} \right)^\alpha = m^{\alpha-1}$$

Finally, we have $m^{\alpha-1} \geq \frac{ALG}{OPT} \geq \frac{ALG}{S} \geq m^{\alpha-1} - \varepsilon$ for small $\varepsilon > 0$ since the limit is asymptotic. Thus, ALG is a $m^{\alpha-1}$-approximation algorithm and the analysis is tight.   □

5   Sense-and-Aggregate Model

In this section, we present a new variant of the flow shop speed scaling problem denoted as sense-and-aggregation model which is motivated by data aggregation and computation tasks in the wireless sensor network.

In the wireless sensor network, the station needs to receive data from the sensors and compute the single aggregation result like minimum, maximum, or average of all inputs. For these kinds of aggregation functions, we have two ways to deal with it. We can either wait for all inputs to be ready or we can collect part of the input data and compute the sub results. Eventually, we can compute the final result based on the sub results we computed previously. We introduce the sense-and-aggregation model to model the second approach.

In the sense-and-aggregation model, we are given a set of n unit jobs and two processors where the first processor handles the data collection and aggregation tasks, and the second processor is responsible for the calculations. All jobs need to be processed in the first processor before they can aggregate to the second processor for computations. Moreover, the speeds of the jobs may vary independently and all jobs need to be completed before their common deadline D.

Once the jobs are completed in the first processor, they can be aggregated to the second processor for calculation. For this aggregation process, we need to make use of the workload-consideration-function $f(x)$ to indicate the amount of workload that the second processor will process to aggregate x units of data. Then based on the workload-consideration function, we can compute how fast we need to schedule the jobs in order to complete all jobs before the deadline D. The main idea of performing partial aggregation is to compute the sub results in order to speed up the computation process. After each partial aggregation, one unit of data will be produced for the sub result which will be used in the next partial aggregation.

Observe that there is a trade off between waiting and calculating. The more we wait, the more we have to schedule faster (thus consuming more energy), but the less we have to schedule for some function f. Our objective is to find a feasible schedule such that all jobs are completed before their common deadline D in both processors with the minimum usage of energy.

Without loss of generality, we assume that each aggregation on the second processor has at least two units of workload.

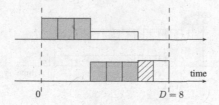

Fig. 4. Optimal schedule of 4 jobs for sense-and-aggregate model with $f(x) = x$.

In the example showed in Fig. 4, we are going to find out the optimal schedule of 4 jobs with $f(x) = x$. We decide to collect 3 unit jobs on the first processor before scheduling them on the second processor (in gray). Then we schedule the last job on the first processor and the second processor aggregates it with the output data from the previous aggregation, a total workload of 2 for the second aggregation. The output of the first aggregation is shown hatched.

We first analyze when $f(x) = x - 1$ and propose a greedy algorithm to schedule the jobs with minimum energy consumption. Then we present a dynamic programming approach to solve the problem when $f(x) = x$.

5.1 $f(x) = x - 1$

For aggregation functions such as sum, min and max, there holds that $f(x) = x - 1$. We first show that the total workload on the second processor is not affected by the decision of the second processor.

Proposition 4. *The workload on the second processor is $n - 1$.*

Proof. Suppose we have k aggregations with sizes $B_1, B_2, \ldots, B_k$. Then $\sum_{i=1}^{k} B_i = n + k - 1$, since the first $k - 1$ aggregations will generate one unit of data each to be the input of subsequent aggregations. Therefore, $\sum_{i=1}^{k} f(B_i) = \sum_{i=1}^{k} (B_i - 1) = n - 1$.   □

The idea is to schedule jobs as soon as possible on the second processor since the total workload on the second processor remains unchanged. We obtain the minimum energy consumption by scheduling jobs in the following way.

We first schedule two unit jobs on the first processor. Then we aggregate these jobs on the second processor and generate one unit workload ($f(2) = 1$). At the same time, we continue to schedule jobs on the first processor for the next aggregation (see Fig. 5). It is easy to see that the precedence constraints are guaranteed and it is a feasible schedule. Note that it is not possible to get another schedule with lower energy consumption. Indeed, it is not possible to schedule on the second processor when the first two jobs are scheduled. Similarly, we cannot schedule any job on the first processor when we schedule the last aggregation. Finally, the minimum energy consumption is $\dfrac{\left(2 + \sqrt[\alpha]{(n-2)^\alpha + (n-2)^\alpha + 1}\right)^\alpha}{D}$ according to [10].

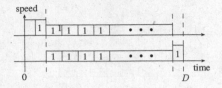

Fig. 5. Optimal schedule when the workload-consideration-function is $f(x) = x - 1$.

5.2 $f(x) = x$

The idea of the algorithm for this case is based on [10]. Indeed, the calculation of the energy consumption depends on critical jobs. A job is critical when the completion time of a job on the first processor is equal to the starting time of the same job on the second processor. Moreover, we only need to focus on schedules in which jobs are executed without preemption (each job is scheduled in one piece) and the execution order is the same on both processors. Recall that given a list of critical jobs $\{c_1, c_2, \ldots, c_k\}$, the total cost can be computed as follows [10]: $\dfrac{\left(p_{1,1} + \sum_{i=1}^{k-1} \sqrt[\alpha]{\left(\sum_{j=c_i+1}^{c_{i+1}} p_{1,j}\right)^{\alpha} + \left(\sum_{j=c_i}^{c_{i+1}-1} p_{2,j}\right)^{\alpha}} + p_{2,n}\right)^{\alpha}}{D}$.

We adapt the algorithm proposed in [10] for our case. We have to guess the total workload on the second processor since the number of the aggregations is not known in advance. In the following, we suppose that the total workload on the second processor is W. Finally, we show that W is polynomial.

Definition 1. *Let $F(s, w, g)$ be the minimum cost of the jobs $s + 1, \ldots, n$ with a workload of w on the second processor and a pending workload of g on the first processor (before job $s + 1$).*

Note that the objective function is $\min_{\substack{1 \leq j \leq n \\ 1 \leq w \leq W}} \dfrac{(F(j+1,w,j))^{\alpha}}{D}$.

Algorithm 1. Flowshop with accumulation

1: Procedure $F(s, w, g)$
2: $Min \leftarrow +\infty$
3: **for** $i = s + 1, \ldots, n$ **do**
4:     **for** $B = 1, \ldots, w$ **do**
5:         //test whether s to i can be a segment in the optimal schedule with the minimum pending workload on the first processor
6:         $k \leftarrow A(i - s, g, B, B)$
7:         $Min = \min\{Min, \sqrt[\alpha]{(i - s)^{\alpha} + B^{\alpha}} + F(i + 1, w - B, (i - s) - k)\}$
8:     **end for**
9:     **return** Min
10: **end for**

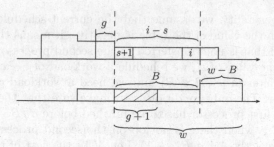

Fig. 6. Illustration of Algorithm 1 and Definition 1.

Definition 2. $A(i, g, B, e)$ *is the maximum workload on the first processor that can be aggregated such that:*

- *there is at most a workload of i on the first processor*
- *there is at most a workload of B on the second processor*
- *There is a pending workload of g (already scheduled on the first processor on a previous critical interval)*
- *the second processor has already scheduled a workload of e.*

With the definitions above, we can construct the dynamic program for computing function A.

Proposition 5 (Fig. 7).

$$A(i, g, B, e) = \max \begin{cases} A(i, g, B, e-1) \\ \max_{\substack{0 \le e' < e \\ (A(i,g,B,e')+(e-e'-1))/i \le e'/B}} A(i, g, B, e') + (e - e' - 1) \end{cases}$$

$$A(i, g, B, 0) = -g$$

Proof. The conditions under the minimization are to ensure that we get a feasible schedule. The initialization means that we have g pending units of workload.

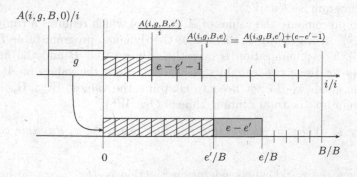

Fig. 7. Illustration of Proposition 5.

Without loss of generality, we assume that the current schedule starts at time 0. Note that when the value of the table is negative, it means that there is still pending workload that is not transferred to the second processor.

By choosing the value of e', we schedule a workload of $e - e'$ on the second processor in the interval $[e'/B, e/B)$ and we need a workload of $e - e' - 1$ on the first processor. The only condition is to make sure that the workload that we gather on the first processor has to be finished before e'/B.

The first case is when there is no job on the second processor in $[\frac{e-1}{B}, \frac{e}{B})$. Then the cost is exactly $A(i, g, B, e - 1)$. Let A' be the cost of the second case.

We first prove that $A(i, g, B, e) \leq A'$. Let $\mathcal{S}$ be a schedule (on both processors) that realizes $A(i, g, B, e')$. Let e be a value such that $(A(i, g, B, e') + (e - e' - 1))/i \leq e'/B \leq 1$. We build a schedule from time 0 to $\frac{A(i,g,B,e')+(e-e'-1)}{i}$ on the first processor (note that the last value may be negative and is still feasible) from the schedule $\mathcal{S}$ on the same time interval. On the second processor, we construct from 0 to e'/B with the schedule $\mathcal{S}$ within the same time interval and we schedule the gathered jobs (in gray in Fig. 7) from time e'/B to time e/B. This is a feasible schedule and its cost is exactly $A(i, g, B, e') + (e - e' - 1)$. Thus we have $A(i, g, B, e) \leq A'$.

We then prove that $A(i, g, B, e) \geq A'$. Let $\mathcal{O}$ (resp. $\mathcal{S}$) be a schedule that realizes $A(i, g, B, e)$ (resp. $A(i, g, B, e')$) in which the starting time of the last job is maximum on the second processor, i.e. the time e'/B is maximum. Then the restriction $\mathcal{S}$ of $\mathcal{O}$ in $[e'/B, e/B)$ is a schedule that meets all constraints related to $A(i, g, B, e')$. Hence its cost is greater than $A(i, g, B, e')$. Thus we have $A(i, g, B, e) \geq A'$. □

Proposition 6. *The flow shop problem with aggregation can be solved in* $O(n^3W^2)$ *time.*

Proof. We first show that the time complexity of the dynamic program in Proposition 5 is $O(n^3W^2)$. The size of the dynamic program table is $O(n^2W^2)$. The maximization is over the values of e'. Actually, e' can be expressed as $e - f(y)$ for $y = 1, \ldots, n+1$. Otherwise, there is at least $1/B$ unit time which is idle and we can consider $A(i, j, B, e'')$ instead of $A(i, j, B, e')$ with $e'' > e$. Therefore the running time of the maximization is $O(n)$. Thus, the total running time of the dynamic program is $O(n^3W^2)$.

We can precompute the values of $A(i, g, B, e)$ which requires a running time of $O(n^3W^2)$. In Algorithm 1, the size of the dynamic program table $F(s, w, g)$ is $O(n^2W)$. The minimization is over the values i and B and the number of possibilities for these two variables is $O(nW)$. Since the values of $A(i, g, B, e)$ are precomputed, we do not need to compute the values $A(i, g, B, e)$ at each time. To sum up, the total running time is $O(n^3W^2)$. □

Corollary 1. *When the aggregation function is* $f(x) = x$, *the time complexity is* $O(n^5)$.

Proof. When the workload-consideration-function is $f(x) = x$, we have $W \leq 2n - 1$. It is easy to see that there is at least a workload of n to schedule on the second processor. The maximum workload that we can create after groupings is $n - 1$. Indeed, each aggregation should contain at least one unit of workload from the first processor and at most one unit of workload from the previous aggregation. Finally, we have a maximum workload of $2n - 1$ on the second processor.   □

With the previous proof, we can deduce that W is strongly related to the workload-consideration-function $f(x)$. If there are k aggregations and x_i is the workload of the i-th aggretation, the total workload on the second processor is $\max_k \sum_{i=1}^{k} f(x_i + 1)$ such that $\sum_i x_i = n$. We can roughly bound this with $n(1 + \max f(x))$.

Acknowledgments. This work was fully supported by a grant from the Research Grants Council of the Hong Kong Special Administrative Region, China [Project No. CityU 117913].

References

1. Albers, S.: Energy-efficient algorithms. Commun. ACM **53**(5), 86–96 (2010)
2. Albers, S.: Algorithms for dynamic speed scaling. In: Proceedings of 28th STACS 2011, vol. 9, pp. 1–11. LIPIcs (2011)
3. Brucker, P.: Scheduling Algorithms, vol. 3. Springer, Heidelberg (2007)
4. Fang, K., Uhan, N.A., Zhao, F., Sutherland, J.W.: Flow shop scheduling with peak power consumption constraints. Ann. OR **206**(1), 115–145 (2013)
5. Gaujal, B., Navet, N.: Navet.: Dynamic voltage scaling under EDF revisited. Real-Time Syst. **37**(1), 77–97 (2007)
6. Johnson, S.M.: Optimal two-and three-stage production schedules with setup times included. Nav. Res. Logist. Q. **1**(1), 61–68 (1954)
7. Kwon, W.-C., Kim, T.: Optimal voltage allocation techniques for dynamically variable voltage processors. ACM TECS **4**(1), 211–230 (2005)
8. Li, M., Yao, A.C., Yao, F.F.: Discrete and continuous min-energy schedules for variable voltage processors. Proc. Nat. Acad. Sci. U.S.A. **103**(11), 3983–3987 (2006)
9. Li, M., Yao, F.F., Yuan, H.: An $O(n^2)$ algorithm for computing optimal continuous voltage schedules. CoRR abs/1408.5995 (2014)
10. Mu, Z., Li, M.: DVS scheduling in a line or a star network of processors. J. Comb. Optim. **29**(1), 16–35 (2015)
11. Wierman, A., Andrew, L.L.H., Tang, A.: Power-aware speed scaling in processor sharing systems: optimality and robustness. Perform. Eval. **69**(12), 601–622 (2012)
12. Yao, F.F., Demers, A.J., Shenker, S.: A scheduling model for reduced CPU energy. In: 36th FOCS, pp. 374–382. IEEE Computer Society (1995)

Computational Geometry
and Computational Biology

Algorithms for k-median Clustering over Distributed Streams

Sutanu Gayen[✉] and N.V. Vinodchandran

University of Nebraska-Lincoln, Lincoln, NE 68588, USA
{sgayen,vinod}@cse.unl.edu

Abstract. We consider the k-median clustering problem over distributed streams. In the distributed streaming setting there are multiple computational nodes where each node receives a data stream and the goal is to maintain an approximation of a function of interest at all time over the union of the local data at all the nodes. The approximation is maintained at a coordinator node which has bidirectional communication channels to all the nodes. This model is also known as the distributed functional monitoring model. A natural variant of this model is the distributed sliding window model where we are interested only in maintaining approximation over a recent period of time.

This paper gives new algorithms for the k-median clustering problem in the distributed streaming model and its sliding-window counter part.

1 Introduction

In the distributed streaming model, a set of computational nodes gets a stream of data items. They have a two-way communication channel to a coordinator node which keeps track of an approximate value of a function of interest computed over the union of all the local streams. Such distributed scenario arise in many natural data management situations including monitoring over LAN devices and over sensor networks. A natural variant of this distributed streaming model is the 'sliding window' version where the interest is only on a window of most recent data items. The main algorithm design goal is to minimize the resources: space, time, and total communication, required to compute/approximate the function, to the extent possible. There has been significant body of research in computing many natural functions. In particular efficient algorithms for maintaining statistics such as a random sample, sum, join sizes, k-center clustering, heavy hitters, and quantiles has been already proposed in these models [3,6–9,13,15]. We refer the reader to a survey by Cormode [5] for recent work and references in this area.

In this paper, we consider the k-median clustering problem in distributed streaming models. We design algorithms in both the infinite window and the sliding window models. For both, we achieve constant factor approximations for the k-median clustering cost with efficient communication. To the best of our knowledge, this important clustering problem has not been studied in the distributed streaming scenario and this paper presents first results in this direction.

© Springer International Publishing Switzerland 2016
T.N. Dinh and M.T. Thai (Eds.): COCOON 2016, LNCS 9797, pp. 535–546, 2016.
DOI: 10.1007/978-3-319-42634-1_43

1.1   Related Work and Contribution of This Paper

We briefly outline prior work related to k-median clustering in streaming models and the new results in this paper. Clustering is an important computational task and has applications in all area of data analytics. The k-median variant of this general problem has received much attention in traditional streaming model. Refer to Sect. 2 for definitions.

The following table summarizes the known results for k-median clustering problem in the streaming models. Here n is the number of points, W is the window size, and OPT is the optimum value. It is standard to measure space in terms of weighted points (Table 1).

Table 1. k-median previous work

Paper	Model	Input space	Space complexity in weighted points	Approx
[2]	Insertion-only stream	Metric	$O(k \log n)$	$O(1)$
[11]	Insertion-only stream	Euclidean	$O(k\epsilon^{-d} \log^{2d+2} n)$	$1 + \epsilon$
[10]	Insertion-deletion stream	Geometric	$\text{poly}(k, \frac{1}{\epsilon^d}, \log n)$	$1 + \epsilon$
[1]	Insertion-only sliding window	Metric	$O(k^3 \log^4 W \log^2 \text{OPT})$	$O(1)$
[1]	Insertion-only sliding window	Euclidean	$\text{poly}(k, \frac{1}{\epsilon^d}, \log W, \log \text{OPT})$	$1 + \epsilon$

To the best of our knowledge k-median clustering has not been considered in distributed streaming models. Cormode et al. [8] have studied the related k-center problem. Also, Zhang et al. [16] investigated Euclidean k-median clustering on a related but different distributed model. We summarize our results for k-median clustering in the following table. In the table we only give the communication cost. Please refer to the main section for space complexity bounds. Here n is the number of points, m is the number of distributed nodes, OPT is the optimum value, and W is the window size. All the results are in the insertion-only model (Table 2).

Table 2. Our results

Model	Input space	Communication	Approx
Distributed streaming	Metric	$O(m \log n \log^2 \text{OPT})$ bits $+$ $O(km \log^2 n \log \text{OPT})$ weighted points	$O(1)$
Distributed sliding window	Euclidean	$O(m \log W \log^2 \text{OPT} + s^2 m^3 \log \text{OPT})$ bits $+$ $O(km \log^2 W \log \text{OPT})$ weighted points	$O(1)$

2   Distributed Streaming Models and High Level Ideas of This Paper

2.1   The Distributed Streaming Models

First we briefly describe the well known data streaming model. In this model a set of data items (say of size n) arrives at a computational node one after another.

The goal is to compute a function of interest over the set. The computational node may not have enough memory to store the entire set of data and hence an exact computation may not be possible. Instead it stores a sketch of the data of size much smaller than n, typically, poly($\log n$) or $O(n^\epsilon)$ for some $\epsilon < 1$. After the entire data is observed, it can compute on the sketch to output an approximation of the function of interest over the data. This basic model in known as insertion-only streaming model. There are other variations of this basic model. One might be interested to produce output at all times on the set of data items seen so far. This is known as the online streaming model. If the data items are coming from some finite domain, the source might delete some of the data elements that arrived previously. This variant is known as insertion-deletion stream. Instead of the entire stream, one might be interested to compute over the most recent W items that arrived at all times. Then, it must use space much smaller than W. This variation is known as the sliding window model. The data streaming and above-mentioned variants have received considerable attention because it nicely models many real-life computations over large data sets.

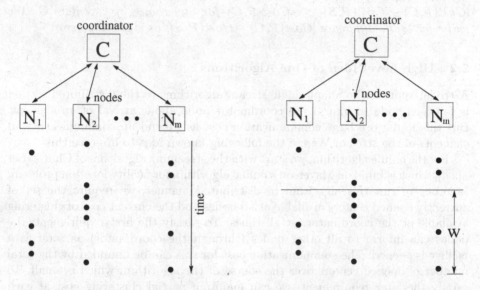

Fig. 1. distributed streaming   **Fig. 2.** distributed sliding window

In the distributed streaming model we have m computational nodes. At any point in time, an element from a global data stream may be sent to one of the nodes (thus the data stream is distributed over m nodes). The goal is to compute a function of interest over the entire data stream. The function computation is done at a special node called the *coordinator* node which has a bi-directional communication channel to each of the m computational nodes. Thus in addition to time and space used by the nodes, total communication is also an important resource and a main design objective is to use as little communication as possible to maintain an approximate value of the function. Similar to the single stream case, we also have the *online* and the *sliding window* variation for distributed streams. In the sliding-window model, it is assumed that all the data elements also comes with a time-stamp. A schematic diagram of both the distributional streaming and sliding window models is given above.

Before we describe high level ideas used in this paper we give two important definitions of k-median clustering and the notion of ϵ-coreset.

Definition 1 (k-median Clustering). *Given a set of points P from a metric space X, identify a subset $C^* \subseteq P$ of k centers such that*

$$C^* = \arg \min_{C \subseteq P, |C|=k} Cost(P, C)$$

where $Cost(P, C) = \sum_{p \in P} \min_{c \in C} d(p, c)$ is the clustering *cost and d is the distance function of X.*

Definition 2 (ϵ-Coreset for k-median Clustering). *For k-median clustering over a set of points P, a weighted set $S \subseteq P$ is called an ϵ-coreset if $|Cost(P, C) - Cost(P, S)| \le \epsilon Cost(P, C)$ for any choice of k centers C. We will refer to the difference $|Cost(P, C) - Cost(P, S)|$ as clustering error.*

2.2   High Level Idea of Our Algorithms

A trivial approach to adapt single stream algorithms to the distributed setting is for each node to inform the coordinator node about arrival of an element. But this leads to a large communication cost as we end up communicating all elements of the stream. We use the following known idea to improve this:

For the online algorithm, we start with the streaming algorithm of Charikar et al. [4], which is build on Mayerson's online algorithm for facility location problem. In order to run this algorithm in distributed manner, we require the set of currently opened centers available at all nodes and the current cost of clustering available at the coordinator, at all times. To satisfy the first requirement, the nodes can inform to all other nodes (through the coordinator) as soon as a facility is opened. The communication cost for this can be bounded by the total number of opened centers over the course of the algorithm, which is small. To satisfy the later requirement, we can maintain partial clustering cost at each node restricted to their part of the stream. As soon as this cost doubles to say c, they communicate to the coordinator $\lceil \log_2 c \rceil$. This way the coordinator will always have at most 2-approximation of the current clustering cost. The number of such communications will be logarithmic in the clustering cost. This algorithm requires the knowledge of n, the number of points, beforehand. We get around this by making binary search for n.

For the sliding window algorithm, we use the forward-backward technique from [9]. For this we first assume that the the stream starts at $t = 0$. We can divide the entire time into disjoint but consecutive windows of the form $[aW, (a + 1)W)$ where W is the size of the window. Then, at any moment, the current stream window intersect at most two of the consecutive time windows say

$[aW, (a+1)W)$ and $[(a+1)W, (a+2)W)$. Let us call the *forward part* the part of the stream in $[(a+1)W, (a+2)W)$ and the *backward part* that in $[aW, (a+1)W)$. Computation over the forward part can be done by running an online algorithm starting from times $(a + 1)W$. Computation over the backward part is more difficult as elements expire one by one with time. We notice that an algorithm from Braverman et al. [1] can be used here. This algorithm maintains instances of an online algorithm starting from a set of indices from the current window. This online algorithm builds a succinct representation of the input points called a *coreset*. In the distributed setting, each node can maintain such indices and notify the coordinator about the new coreset whenever any such index is crossed by the current window. To notice this crossing, the nodes need to know the gap between two indices in the global stream. Therefore as in [9], we will make an assumption that the data elements are augmented with a unique global time-stamp. In fact, it can be seen such a time-stamp modulo W suffices for our purpose. This is achievable in practice with a little more total cost of transmission. One down side is that for many nodes, the error of the coresets may multiply so that we need to rescale the error for each online algorithm. This is where we use the fact that arbitrary approximation schemes are known for Euclidean k- median clustering. Hence while infinite window algorithm works for any metric space, our finite sliding window algorithm only works for the k-median problem over Euclidean space.

3   Algorithms for k-median Clustering

3.1   Online Algorithm for Distributed Streams

Our starting point is the algorithm of Charikar et al. [4] for k-median clustering on insertion-only streams rooted on Meyerson's algorithm [14] for computing online facility location problem. The later problem is closely related to k-median clustering. Given a set of points P, the online facility problem asks to find a set $F \subseteq P$ of facilities to open such that $(f.|F| + Cost(P, F))$ is as small as possible. Here, $Cost$ is the clustering cost function from Definition 1; f is the cost of opening a single facility. The first part of the sum is known as the facility cost and the second part of the sum is known as the service cost.

Let us first go over Meyerson's algorithm briefly. It keeps in memory the set of already opened facilities F. As the new point p arrives, it computes the distance d from p to the closest facility from F. We open facility at p with probability $\frac{d}{f}$, where f is the cost of opening a single facility. It can be shown if we set f to be $\frac{L}{k(1+\log n)}$, the expected service cost becomes at most $(L + 4\text{OPT})$, where OPT is the optimal k-median clustering cost. The expected number of opened facilities becomes $k(1 + \log n)(1 + \frac{4\text{OPT}}{L})$. Moreover, it was shown by Braverman et al. [2] that these values are close to their expectations with high probability[1]. For metric space, we restate this result in the following lemma.

---

[1] Throughout this paper with high probability means with probability at least $(1 - \frac{1}{n})$.

Lemma 1 (Theorem 3.1 of [2]**).** *If we run the online facility location algorithm of [14] with* $f = \frac{L}{k(1+\log n)}$ *for some* $L \leq \mathrm{OPT}$*, the service cost is at most* $(3 + \frac{2e}{e-1})\mathrm{OPT}$ *and the number of opened facilities is at most* $7k(1 + \log n)\frac{\mathrm{OPT}}{L}$ *with probability at least* $1 - \frac{1}{n}$*. Here* OPT *is the optimum* k*-clustering cost.*

Given these facts, one can do binary search to guess $L = \Theta(\mathrm{OPT})$ so that we get a set of $O(k \log n)$ weighted points which represents the original points with additional clustering cost of $O(\mathrm{OPT})$. This idea was used in both the algorithms of [2,4]. We start with a small value of L and check that the clustering cost is at most αL and the number of opened facilities is at most $\beta k(1 + \log n)$ for some constants α, β from [2]. As soon as this condition is violated we increase the current value of L by a factor of γ for some $\gamma > 1$. With this new value of L, we start a new round and continue seeing all the elements of the stream again. But now the already seen elements will be replaced by the current set of at most $\beta k(1 + \log n)$ weighted centers produced. In this way after all elements of the stream are seen, we can run an offline k-median algorithm on these weighted centers to get the final k-clusters. This result is summarized in the theorem below.

Theorem 1 (Theorem 3.2 of [2] **Restated).** *With high probability, there is an algorithm for* k*-median clustering that achieves a constant approximation. This uses exactly* k *facilities and stores* $O(k \log n)$ *weighted points in memory.*

To translate the above algorithm in the distributed setting we address the following questions. Firstly, we need to maintain the set of facilities in distributed manner. This is not difficult. Since there are $O(k \log n)$ many weighted centers for each round, we can announce to all other sites as soon as a center is created. Moreover, since after each round, the value of L is doubled, there are at most $O(\log \mathrm{OPT})$ many rounds. Secondly, we need to keep track of the current service cost efficiently. For this, each of the nodes keep track of partial cost for their part of the stream. As soon as this partial cost gets within 2^l to 2^{l+1}, they communicate to the coordinator $(l + 1)$. So, the coordinator always has at most 2-factor approximation of the true service cost. Then the coordinator can do the aforementioned two checks and announce the change of round. At the change of round, coordinator can collect all the weighted centers from previous rounds so that it can replay the seen part of the stream. Opened centers from this replay are broadcast to all the nodes.

The total communication cost can be broken up into 4 pieces. For announcing an opened facility $O(k \log n \log \mathrm{OPT})$ many points need to be communicated over all rounds. A factor m more points with weights need to be communicated during change of rounds. For keeping track of the number of arrived points, communication requires $O(\log n \log \log n)$ bits and that for the cost requires $O(m \log^2 \mathrm{OPT})$ bits. Finally, we need the knowledge of n beforehand to set up the initial facility cost and to check one of the conditions for changing rounds. This is addressed by the coordinator making $O(\log n)$ guesses for n starting from 1 to some upper bound with factor of 2. The streaming algorithms are run for all these guesses in parallel. The nodes inform about the arrival of $(1 + \delta)^l$ elements for some fixed $\delta > 0$ to the coordinator so that the later always has an $(1 + \delta)$-factor approximate value for n. This is used to recognize the correct guess and to produce the correct output. Due to the guessing, all costs increase by $O(\log n)$-factor. Since we will be using constant factor approximation for n and the total service cost, the constants α and β need to be rescaled. Finally, we have the following theorem.

Theorem 2. *There is an online algorithm $\mathcal{A}$ for k-median clustering of a set of n points from metric space, appearing in a stream distributed across m nodes upto $O(1)$ approximation ratio with high probability. The total communication cost of $\mathcal{A}$ is $O(m \log n \log^2 \mathrm{OPT} + km \log^2 n \log \mathrm{OPT}.S)$ bits where S is the space required to store a single point from the metric space along with an integral weight of at most n and OPT is the optimum clustering cost for the current set of points.*

3.2   Sliding Window Algorithm for Distributed Streams

We next move on to clustering problems on sliding windows. As we said previously, we will use the forward-backward technique from [9]. The entire stream can be broken up into disjoint windows of length W starting at times $(aW + 1)$ and ending at times $(a + 1)W$ for some integer a. Then, at any time $(aW + t)$, current window will intersect at most 2 of such windows. The time interval for the former part will be $(aW - (W - t - 1))$ to aW and that for the later part will be $(aW + 1)$ to $(aW + t)$. In the later part elements arrive one after another while in the former part they expire. So, assuming we have a communication-efficient distributed online algorithm $\mathcal{A}$, the later part can be computed by running an instance of $\mathcal{A}$ starting from the times aW. This is known as the forward algorithm.

In the k-median clustering scenario, online algorithms with at most $(1 + \epsilon)$-approximation ratio are known for Euclidean space based on merge and reduce technique of coresets. From Definition 2, coresets are a small subset of the input points with weights which approximately preserves the clustering cost. Adapting the online algorithm over distributed stream in straightforward manner requires communication of all the points. Also, the online algorithm works hierarchically in bottom-up manner by including each input point into the coreset in the beginning. This way, after opening a coreset point at some time, it may be closed and merged with some other coreset point at a later time. So, it is not straightforward to compose the coresets for the local streams in online manner. We feel it is a difficult barrier to overcome. So, for the forward algorithm we will use the algorithm of Theorem 2. This will work because Euclidean space is a specific case of metric space.

For the backward algorithm, one idea is to combine sketches corresponding to the local streams. But then, for the global stream, for m such nodes, the clustering error increases by a factor m. So, the online algorithm of Theorem 2 cannot be used since it achieves $O(1)$ approximation ratio. In general, it is hard to achieve approximation ratio better than $(1 + \frac{2}{e})$ for metric k-median clustering [12]. However, for Euclidean space, approximation ratio of $(1 + \epsilon)$ is possible for any $\epsilon > 0$. This is precisely the reason that our algorithm is limited to the Euclidean case for sliding windows.

The online algorithm is based on the *merge and reduce* technique. It keeps a pair of buckets of size at most s at different levels. Here s is the size of a single offline coreset as in [11]. The first level of the buckets are filled with input points as they arrive. Once both the buckets are full, we merge these buckets and reduce them to a single bucket of size at most s using the offline coreset construction algorithm. These s points are placed in a third bucket at the next level. If this next level also becomes full, we keep on merging and reducing until at some level we end up having exactly 1 bucket. It can be shown that $O(\log n)$ many levels suffice for n points. Since each level adds a clustering error by at most ϵ fraction, ϵ must be rescaled to $\frac{\epsilon}{O(\log n)}$ for the offline algorithm.

The offline algorithm runs by defining a set R of *regions* from the Euclidean space and by keeping a weighted sample of the input points from region. We restate a lemma of [1] that shows the online coreset construction approximates density of each of the regions well.

Lemma 2 (Lemma 4.3 of [1] Restated). *Let B_i be a bucket at level i of the merge and reduce algorithm. P_i be the set of input points which B_i represents then for all region R of the Euclidean space, the error $||P_i \cap R| - |B_i \cap R||$ is small and can be ignored for k-median clustering.*

It was also shown that if we add or remove ϵ-fraction of points from each region of the coreset points, the resulting coreset is also a 2ϵ-coreset of the original points.

Lemma 3 (Lemma 4.1 of [1]). *Let K be an ϵ-coreset for Euclidean clustering of a set of points P and Λ_K the corresponding partition. Suppose for each region $R \in \Lambda_K$ we add or remove ϵ fraction of points from the coreset. The resulting coreset K' is an ϵ-coreset of K.*

Braverman et al. [1] provided the first sliding window algorithm for Euclidean k-median clustering. We briefly go over it below. Keep the online algorithm running for a number of indices such that just after traversing each index the number of points in some region of the coreset reduces by ϵ factor. Then since by Lemma 2 density of the coresets approximates the true density of each region and since by Lemma 3 it is permissible to drop at most ϵ fraction of points from each region, any subset of points exactly in between two indices can be approximated by the coreset for the former index. Moreover, since, just after traversing each index the number of points in some region of the coreset reduces by ϵ factor, there can be at most $O(\frac{s}{\epsilon^2} \log \text{OPT})$ many such indices. This sliding window algorithm is summarized in the following theorem.

Theorem 3 (Theorem 4.1 of [1] Restated). *There is a sliding window algorithm for Euclidean k-median clustering upto approximation ratio $(1+\epsilon)$ for any $\epsilon > 0$ in $(s^2\epsilon^{-2}\log OPT)$ space where OPT is the maximum cost of clustering over all windows and s is the size of a single online coreset based on merge and reduce.*

This algorithm can be adapted in the distributed setting in the following manner. Each node maintains the sliding window algorithm for their part of the stream for time interval $(2kW, 2(k+1)W]$. Online algorithms are maintained at the indices just after which the cardinality of some region reduces by ϵ fraction. At time $2(k+1)W$, the coordinator collects all these indices and sorts them in increasing order of time. Within each consecutive pair of indices, the number of points in each region of the coreset for each node reduces by at most ϵ fraction. So, the union of the individual coresets is also a coreset for the entire bucket but with error at most $\epsilon(OPT_1 + OPT_2 + \cdots + OPT_m)$ where OPT_i is the optimum clustering cost for the part of the window corresponding to the ith node. We can upper bound each OPT_i by OPT for the entire bucket and choose $\epsilon = \frac{\epsilon'}{m}$ to make the error at most $\epsilon'OPT$. The space complexity at each node becomes $O(\frac{s^2m^2}{\epsilon'^2}\log OPT)$ and at the coordinator it will be a factor m more. The communication cost will be $O(\frac{s^2m^3}{\epsilon'^2}\log OPT)$ for time interval W. All the resources may be further improved by letting the nodes communicate only when the current window crosses one of their indices locally.

Using time-stamps, the coordinator can recognize appropriate buckets to combine and output. The nodes also need to maintain a forward algorithm. As we said before, this is exactly the algorithm of Lemma 2. It will require additional communication of $O(mk\log n\log^2 OPT)$ bits and additional space usage of $O(k\log n\log OPT)$. The error increases upto $(\alpha + \epsilon)$-factor where α is that for the metric counterpart. Therefore, we have the following theorem.

Theorem 4. *There is an algorithm $\mathcal{B}$ for k-median clustering over a sliding window of W points from a Euclidean space, appearing in a stream distributed across m nodes upto $O(1)$-approximation factor with high probability. $\mathcal{B}$ requires communication of $O(s^2m^3\log OPT + m\log W\log^2 OPT + km\log^2 W\log OPT.S)$ bits. Here S is the space required to store a single point from the metric space along with an integral weight of at most W, s is the size of a single online coreset based on merge and reduce for Euclidean space and OPT is the maximum of the optimum clustering costs over any window of size W.*

The algorithm is formally presented below in Algorithms 1 and 2.

Algorithm 1. Sliding window algorithm at the ith node

Data: A set of points $\{p_1, p_2, \cdots, p_n, \cdots, p_N\}$ from metric space presented in a stream distributed across m nodes

Result: Approximate k-median for $\{p_{n-W+1}, p_2, \cdots, p_n\} \forall n \leq N$ where p_n is the most recently arrived element

Upon arrival of new point p:

1   Update the sliding window algorithm;
2   **if** p has timestamp aW for some a **then**
3     $|$ Send change of round to the coordinator;
4   **else**
5     $|$ Update the online algorithm;
6   **end**

Upon change of round:

7   Send the current set of coresets $C_i = \{C_{i,1}, C_{i,2}, \cdots, C_{i,l}\}$ from the sliding window algorithm to the coordinator;
8   Reset the online algorithm of Theorem 2 for subsequent points;
9   Reset the sliding window algorithm of Theorem 3 for subsequent points;

Algorithm 2. Sliding window algorithm at the coordinator

Data: A set of points $\{p_1, p_2, \cdots, p_n, \cdots, p_N\}$ from metric space presented in a stream distributed across m nodes

Result: Approximate k-median for $\{p_{n-W+1}, p_2, \cdots, p_n\} \forall n \leq N$ where p_n is the most recently arrived element

Upon receipt of change of round:

1   Broadcast change of round;
2   Collect sets of coresets C_i from all nodes $i = 1$ to m;
3   Reset the online algorithm of Theorem 2 ;

At all other times:

4   Update the online algorithm;

Output at current time t:

5   **for** $i=1$ to m **do**
6     $|$ Find consecutive coresets C_{i,j_i} and C_{i,j_i+1} such that $(t - W + 1)$ lies between them;
7   **end**
8   Let C be the set of weighted centers from the online algorithm;
9   Output $\cup_i C_{i,j_i} \cup C$;

---

4   Open Problems

In this paper we designed communication-efficient algorithms for metric k-median problem in distributed streaming model and for Euclidean k-median problem in the distributed sliding window model. We get constant approximation guarantee. For the metric case, it still remains open to design communication-efficient algorithm for distributed sliding windows. There is an online algorithm for the Euclidean case that achieves $(1 + \epsilon)$ approximation for any ϵ. It is open whether we can adapt this algorithm in the distributed setting. If successful, such an algorithm could be directly plugged into our forward algorithm to achieve arbitrary approximation for the distributed sliding windows setting as well.

References

1. Braverman, V., Lang, H., Levin, K., Monemizadeh, M.: Clustering problems on sliding windows. In: Krauthgamer, R. (ed.) Proceedings of the Twenty-Seventh Annual ACM-SIAM Symposium on Discrete Algorithms, SODA 2016, Arlington, VA, USA, 10–12 January 2016, pp. 1374–1390. SIAM (2016)
2. Braverman, V., Meyerson, A., Ostrovsky, R., Roytman, A., Shindler, M., Tagiku, B.: Streaming k-means on well-clusterable data. In: Randall, D. (ed.) Proceedings of the Twenty-Second Annual ACM-SIAM Symposium on Discrete Algorithms, SODA 2011, San Francisco, California, USA, 23–25 January 2011, pp. 26–40. SIAM (2011)
3. Chan, H.-L., Lam, T.W., Lee, L.-K., Ting, H.-F.: Continuous monitoring of distributed data streams over a time-based sliding window. In: Marion, J.-Y., Schwentick, T. (eds.) 27th International Symposium on Theoretical Aspects of Computer Science, STACS 2010, Nancy, France, 4–6 March 2010. LIPIcs, vol. 5, pp. 179–190. Schloss Dagstuhl - Leibniz-Zentrum fuer Informatik (2010)
4. Charikar, M., O'Callaghan, L., Panigrahy, R.: Better streaming algorithms for clustering problems. In: Larmore, L.L., Goemans, M.X. (eds.) Proceedings of the 35th Annual ACM Symposium on Theory of Computing, 9–11 June 2003, San Diego, CA, USA, pp. 30–39. ACM (2003)
5. Cormode, G.: Algorithms for continuous distributing monitoring: a survey. In: Laura, L., Querzoni, L. (eds.) First International Workshop on Algorithms and Models for Distributed Event Processing 2011, Proceedings, Rome, Italy, 19 September 2011. ACM International Conference Proceeding Series, vol. 585, pp. 1–10. ACM (2011)
6. Cormode, G., Muthukrishnan, S., Yi, K.: Algorithms for distributed functional monitoring. In: Teng, S.-H. (ed.) Proceedings of the Nineteenth Annual ACM-SIAM Symposium on Discrete Algorithms, SODA 2008, San Francisco, California, USA, 20–22 January 2008, pp. 1076–1085. SIAM (2008)
7. Cormode, G., Muthukrishnan, S., Yi, K., Zhang, Q.: Optimal sampling from distributed streams. In: Paredaens, J., Van Gucht, D. (eds.) Proceedings of the Twenty-Ninth ACM SIGMOD-SIGACT-SIGART Symposium on Principles of Database Systems, PODS 2010, Indianapolis, Indiana, USA, 6-11 June 2010, pp. 77–86. ACM (2010)

8. Cormode, G., Muthukrishnan, S., Zhuang, W.: Conquering the divide: continuous clustering of distributed data streams. In: Chirkova, R., Dogac, A., Tamer Özsu, M., Sellis, T.K. (eds.) Proceedings of the 23rd International Conference on Data Engineering, ICDE 2007, The Marmara Hotel, Istanbul, Turkey, 15-20 April 2007, pp. 1036–1045. IEEE (2007)

9. Cormode, G., Yi, K.: Tracking distributed aggregates over time-based sliding windows. In: Ailamaki, A., Bowers, S. (eds.) SSDBM 2012. LNCS, vol. 7338, pp. 416–430. Springer, Heidelberg (2012)

10. Frahling, G., Sohler, C.: Coresets in dynamic geometric data streams. In: Gabow, H.N., Fagin, R. (eds.) Proceedings of the 37th Annual ACM Symposium on Theory of Computing, Baltimore, MD, USA, 22–24 May 2005, pp. 209–217. ACM (2005)

11. Har-Peled, S., Mazumdar, S.: On coresets for k-means and k-median clustering. In: Babai, L. (ed.) Proceedings of the 36th Annual ACM Symposium on Theory of Computing, Chicago, IL, USA, 13–16 June 2004, pp. 291–300. ACM (2004)

12. Jain, K., Mahdian, M., Saberi, A.: A new greedy approach for facility location problems. In: Reif, J.H. (ed.) Proceedings on 34th Annual ACM Symposium on Theory of Computing, Montréal, Québec, Canada, 19–21 May 2002, pp. 731–740. ACM (2002)

13. Keralapura, R., Cormode, G., Ramamirtham, J.: Communication-efficient distributed monitoring of thresholded counts. In: Chaudhuri, S., Hristidis, V., Polyzotis, N. (eds.) Proceedings of the ACM SIGMOD International Conference on Management of Data, Chicago, Illinois, USA, 27–29 June 2006, pp. 289–300. ACM (2006)

14. Meyerson, A.: Online facility location. In: 42nd Annual Symposium on Foundations of Computer Science, FOCS 2001, Las Vegas, Nevada, USA, 14–17 October 2001, pp. 426–431. IEEE Computer Society (2001)

15. Woodruff, D.P., Zhang, Q.: Tight bounds for distributed functional monitoring. In: Karloff, H.J., Pitassi, T. (eds.) Proceedings of the 44th Symposium on Theory of Computing Conference, STOC 2012, New York, NY, USA, 19–22 May 2012, pp. 941–960. ACM (2012)

16. Zhang, Q., Liu, J., Wang, W.: Approximate clustering on distributed data streams. In: Alonso, G., Blakeley, J.A., Chen, A.L.P. (eds.) Proceedings of the 24th International Conference on Data Engineering, ICDE 2008, Cancún, México, 7–12 April 2008, pp. 1131–1139. IEEE (2008)

Polygon Simplification by Minimizing Convex Corners

Yeganeh Bahoo[1], Stephane Durocher[1(✉)], J. Mark Keil[2], Saeed Mehrabi[3], Sahar Mehrpour[1], and Debajyoti Mondal[1]

[1] Department of Computer Science, University of Manitoba, Winnipeg, Canada
{bahoo,durocher,mehrpour,jyoti}@cs.umanitoba.ca
[2] Department of Computer Science, University of Saskatchewan, Saskatoon, Canada
keil@cs.usask.ca
[3] Cheriton School of Computer Science, University of Waterloo, Waterloo, Canada
smehrabi@uwaterloo.ca

Abstract. Let P be a polygon with $r > 0$ reflex vertices and possibly with holes. A subsuming polygon of P is a polygon P' such that $P \subseteq P'$, each connected component R' of P' subsumes a distinct component R of P, i.e., $R \subseteq R'$, and the reflex corners of R coincide with the reflex corners of R'. A subsuming chain of P' is a minimal path on the boundary of P' whose two end edges coincide with two edges of P. Aichholzer et al. proved that every polygon P has a subsuming polygon with $O(r)$ vertices. Let $\mathcal{A}_e(P)$ (resp., $\mathcal{A}_v(P)$) be the arrangement of lines determined by the edges (resp., pairs of vertices) of P. Aichholzer et al. observed that a challenge of computing an optimal subsuming polygon P'_{min}, i.e., a subsuming polygon with minimum number of convex vertices, is that it may not always lie on $\mathcal{A}_e(P)$. We prove that in some settings, one can find an optimal subsuming polygon for a given simple polygon in polynomial time, i.e., when $\mathcal{A}_e(P'_{min}) = \mathcal{A}_e(P)$ and the subsuming chains are of constant length. In contrast, we prove the problem to be NP-hard for polygons with holes, even if there exists some P'_{min} with $\mathcal{A}_e(P'_{min}) = \mathcal{A}_e(P)$ and subsuming chains are of length three. Both results extend to the scenario when $\mathcal{A}_v(P'_{min}) = \mathcal{A}_v(P)$.

1 Introduction

Polygon simplification is well studied in computational geometry, with numerous applications in cartographic visualization, computer graphics and data compression [8,9]. Techniques for simplifying polygons and polylines have appeared in the literature in various forms. Common goals of these simplification algorithms include to preserve the shape of the polygon, to reduce the number of vertices, to reduce the space requirements, and to remove noise (extraneous bends) from the polygon boundary (e.g., [2,4,5]). In this paper we consider a specific version of polygon simplification introduced by Aichholzer et al. [1], which keeps

S. Durocher—Work of the author is supported in part by the Natural Sciences and Engineering Research Council of Canada (NSERC).

T.N. Dinh and M.T. Thai (Eds.): COCOON 2016, LNCS 9797, pp. 547–559, 2016.
DOI: 10.1007/978-3-319-42634-1_44

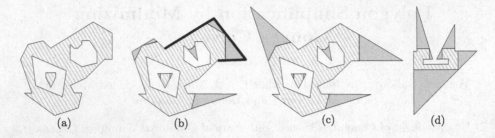

(a)   (b)   (c)   (d)

Fig. 1. (a) A polygon P, where the polygon is filled and the holes are empty regions. (b) A subsuming polygon P', where P' is the union of the filled regions. A subsuming chain is shown in bold. (c) A min-convex subsuming polygon P'_{min}, where $\mathcal{A}_e(P'_{min}) = \mathcal{A}_e(P)$. (d) A polygon P such that for any min-convex subsuming polygon P'_{min}, $\mathcal{A}_e(P) \neq \mathcal{A}_e(P'_{min})$.

reflex corners intact, but minimizes the number of convex corners. Aichholzer et al. showed that such a simplification can help achieve faster solutions for many geometric problems such as answering shortest path queries, computing Voronoi diagrams, and so on.

Let P be a polygon with r reflex vertices and possibly with holes. A *reflex corner* of P consists of three consecutive vertices u, v, w on the boundary of P such that the angle $\angle uvw$ inside P is more than $180°$. We refer the vertex v as a *reflex* vertex of P. The vertices of P that are not reflex are called *convex* vertices. By a *component* of P, we refer to a connected region of P. A polygon P' *subsumes* P if $P \subseteq P'$, each component R' of P' subsumes a distinct component R of P, i.e., $R \subseteq R'$, and the reflex corners of R coincide with the reflex corners of R'. A *k-convex subsuming polygon* P' contains at most k convex vertices. A *min-convex subsuming polygon* is a subsuming polygon that minimizes the number of convex vertices. Figure 1(a) illustrates a polygon P, and Figs. 1(b) and (c) illustrate a subsuming polygon and a min-convex subsuming polygon of P, respectively. A *subsuming chain* of P' is a minimal path on the boundary of P' whose end edges coincide with a pair of edges of P, as shown in Fig. 1(b).

Aichholzer et al. [1] showed that for every polygon P with n vertices, $r > 0$ of which are reflex, one can compute in linear time a subsuming polygon P' with at most $O(r)$ vertices. Note that although a subsuming polygon with $O(r)$ vertices always exists, no polynomial-time algorithm is known for computing a min-convex subsuming polygon. Finding an optimal subsuming polygon seems challenging since it does not always lie on the arrangement of lines $\mathcal{A}_e(P)$ (resp., $\mathcal{A}_v(P)$) determined by the edges (resp., pairs of vertices) of the input polygon. Figure 1(c) illustrates an optimal polygon P'_{min} for the polygon P of Fig. 1(a), where $\mathcal{A}_e(P'_{min}) = \mathcal{A}_e(P)$. On the other hand, Fig. 1(d) shows that a min-convex subsuming polygon may not always lie on $\mathcal{A}_e(P)$ or $\mathcal{A}_v(P)$. Note that the input polygon of Fig. 1(d) is a *simple polygon*, i.e., it does not contain any hole. Hence determining min-convex subsuming polygons seems challenging even for simple polygons. In fact, Aichholzer et al. [1] posed an open question that asks to determine the complexity of computing min-convex subsuming polygons, where the input is restricted to simple polygons.

Let P be a simple polygon. In this paper we show that if there exists a min-convex subsuming polygon P'_{min} such that $\mathcal{A}_e(P'_{min}) = \mathcal{A}_e(P)$ and the subsuming chains of P'_{min} are of constant length, then one can compute such an optimal subsuming polygon in polynomial time. In contrast, if P contains holes, then we prove the problem to be NP-hard. The hardness result holds even when the min-convex subsuming polygon P'_{min} lies on the arrangement $\mathcal{A}_e(P)$, and the length of every subsuming chain of P'_{min} is three. Both results extend to the scenario when $\mathcal{A}_v(P'_{min}) = \mathcal{A}_v(P)$.

The rest of the paper is organised as follows. In Sect. 2 we describe the techniques for computing subsuming polygons. Section 3 includes the NP-hardness result. Finally, Sect. 4 concludes the paper discussing directions to future research.

2   Computing Subsuming Polygons

In this section we show that for any simple polygon P, if there exists a min-convex subsuming polygon P_{min} such that $\mathcal{A}_e(P) = \mathcal{A}_e(P'_{min})$ and the subsuming chains are of length at most t, then one can compute an optimal polygon in $O(t^{O(1)}n^{f(t)})$ time. Therefore, if $t = O(1)$, then the time complexity of our algorithm is polynomial in n. We first present definitions and preliminary results on outerstring graphs, which will be an important tool for computing subsuming polygons.

2.1   Independent Set in Outerstring Graphs

A graph G is a *string graph* if it is an intersection graph of a set of simple curves in the plane, i.e., each vertex of G is a mapped to a curve (string), and two vertices are adjacent in G if and only if the corresponding curves intersect. G is an *outerstring graph* if the underlying curves lie interior to a simple cycle C, where each curve intersects C at one of its endpoints. Figure 2(a) illustrates an outerstring graph and the corresponding arrangement of curves. Later in our algorithm, the polygon will correspond to the cycle of an outerstring graph, and some polygonal chains attached to the boundary of the polygon will correspond to the strings of that outerstring graph.

A set of strings is called *independent* if no two strings in the set intersect, the corresponding vertices in G are called an independent set of vertices. Let G be a weighted outerstring graph with a set $\mathcal{T}$ of weighted strings. A *maximum weight independent set* $\texttt{MWIS}(\mathcal{T})$ (resp., $\texttt{MWIS}(G)$) is a set of independent strings $T \subseteq \mathcal{T}$ (resp., vertices) that maximizes the sum of the weights of the strings in T. Observe that $\texttt{MWIS}(\mathcal{T})$ is also a maximum weight independent set $\texttt{MWIS}(G)$ of G. By $|\texttt{MWIS}(G)|$ we denote the weight of $\texttt{MWIS}(G)$.

Let $\Gamma(G)$ be the arrangement of curves that corresponds to G, e.g., see Fig. 2(a). Let R be a geometric representation of $\Gamma(G)$, where C is represented as a simple polygon P, and each curve is represented as a simple polygonal chain inside P such that one of its endpoints coincides with a distinct vertex of

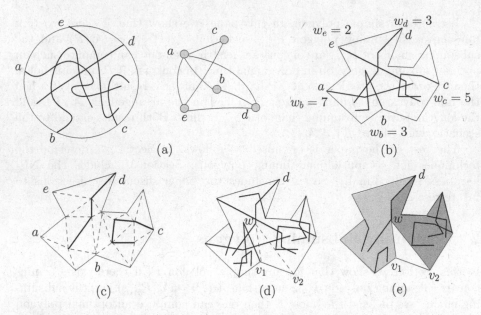

Fig. 2. (a) Illustration for G and $\Gamma(G)$. (b) A geometric representation R of G. (c) A triangulated polygon obtained from an independent set of G. (d)–(e) Dynamic programming to find maximum weight independent set.

P. Keil et al. [6] showed that given a geometric representation R of G, one can compute a maximum weight independent set of G in $O(s^3)$ time, where s is the number of line segments in R.

Theorem 1 (Keil et al. [6]**).** *Given the geometric representation R of a weighted outerstring graph G, there exists a dynamic programming algorithm that computes a maximum weight independent set of G in $O(s^3)$ time, where s is the number of straight line segments in R.*

Figure 2(b) illustrates a geometric representation R of some G, where each string is represented with at most 4 segments. Keil et al. [6] observed that any maximum weight independent set of strings can be triangulated to create a triangulation P_t of P, as shown in Fig. 2(c). Let $\mathcal{T}$ be the strings in R. Then the problem of finding $\mathtt{MWIS}(\mathcal{T})$ can be solved by dividing the problem into subproblems, each described using only two points of R. We illustrate how the subproblems are computed very briefly using Fig. 2(d). Let $P(v_1, v_2)$ be the problem of finding $\mathtt{MWIS}(\mathcal{T}_{v_1,v_2})$, where $\mathcal{T}_{v_1,v_2}$ consists of the strings that lie to the left of $v_1 v_2$. Let $w v_1 v_2$ be a triangle in P_t, where w is a point on some string d inside $P(v_1, v_2)$; see Fig. 2(d). Since P_t is a triangulation of the maximum weight string set, d must be a string in the optimal solution. Hence $P(v_1, v_2)$ can be computed from the solution to the subproblems $P(v_1, w)$ and $P(w, v_2)$, as shown in Fig. 2(e). Keil et al. [6] showed that there are only a few different cases depending on whether the points describing the subproblems belong to the polygon or the strings. We will use this idea of computing $\mathtt{MWIS}(\mathcal{T})$ to compute subsuming polygons.

2.2   Subsuming Polygons via Outerstring Graphs

Let $P = (v_0, v_1, \ldots, v_{n-1})$ be a simple polygon with n vertices, $r > 0$ of which are reflex vertices. A *convex chain* of P is a path $C_{ij} = (v_i, v_{i+1}, \ldots, v_{j-1}, v_j)$ of strictly convex vertices, where the indices are considered modulo n.

Let $P' = (w_0, w_1, \ldots, w_{m-1})$ be a subsuming polygon of P, where $\mathcal{A}_e(P') = \mathcal{A}_e(P)$, and the subsuming chains are of length at most t. Let $C'_{qr} = (w_q, \ldots, w_r)$ be a subsuming chain of P'. Then by definition, there is a corresponding convex chain C_{ij} in P such that the edges (v_i, v_{i+1}) and (v_{j-1}, v_j) coincide with the edges (w_q, w_{q+1}) and (w_{r-1}, w_r). We call the vertex v_i the *left support* of C'_{qr}. Since $\mathcal{A}_e(P') = \mathcal{A}_e(P)$, the chain C'_{qr} must lie on $\mathcal{A}_e(P)$. Moreover, since P' is a min-convex subsuming polygon, the number of vertices in C'_{qr} would be at most the number of vertices in C_{ij}.

We claim that the number of paths in $\mathcal{A}_e(P)$ from v_i to v_j is $O(n^t)$. Since t is an upper bound on the length of the subsuming chains, any subsuming chain can have at most $(t-1)$ line segments. Since there are only $O(n)$ straight lines in the arrangement $\mathcal{A}_e(P)$, there can be at most n^j paths of j edges, where $1 \leq j \leq t-1$. Consequently, the number of candidate chains that can subsume C_{ij} is $O(n^t)$.

Lemma 1. *Given a simple polygon P with n vertices, every convex chain C of P has at most $O(n^t)$ candidate subsuming chains in $\mathcal{A}_e(P)$, each of length at most t.*

In the following we construct an outerstring graph using these candidate subsuming chains. We first compute a simple polygon Q interior to P such that for each edge e in P, there exists a corresponding edge e' in Q which is parallel to e and the perpendicular distance between e and e' is ϵ, as shown in dashed line in Fig. 3(a). We choose ϵ sufficiently small[1] such that for each component w of P, Q contains exactly one component inside w. We now construct the strings. Let v_j be a convex corner of P. Let S_j be the set of candidate subsuming chains such that for each chain in S_j, the left support of the chain appears before v_j while traversing the unbounded face of P in clockwise order. For example, the subsuming chains that correspond to v_j are (v_{j-2}, z_1, v_{j+1}), $(v_{j-3}, z_{13}, z_2, v_{j+1})$, $(v_{j-3}, z_{14}, z_3, v_{j+1})$, $(v_{j-3}, z_{11}, z_4, v_{j+1})$, $(v_{j-3}, z_{15}, z_5, v_{j+1})$, $(v_{j-3}, z_8, z_5, v_{j+1})$, (v_{j-3}, z_7, v_{j+1}), as shown in Fig. 3(b). For each of these chains, we create a unique endpoint on the edge e' of Q, where e' corresponds to the edge $v_j v_{j+1}$ in P, as shown in Fig. 3(c). We then attach these chains to Q by adding a segment from v_j to its unique endpoint on Q.

We attach the chains for all the convex vertices of P to Q. Later we will use these chains as the strings of an outerstring graph. We then assign each chain a weight, which is the number of convex vertices of P it can reduce. For example in Fig. 3(b), the weight of the chain $(v_{j-3}, z_8, z_5, v_{j+1})$ is one.

Although the strings are outside of the simple cycle, it is straightforward to construct a representation with all the strings inside a simple cycle Q: Consider placing a dummy vertex at the intersection points of the arrangement, and

---

[1] Choose $\epsilon = \delta/3$, where δ is the distance between the closest visible pair of boundary points.

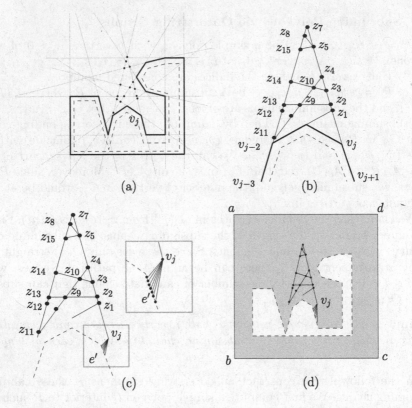

Fig. 3. (a) Illustration for the polygon P (in bold), $\mathcal{A}_e(P)$ (in gray), and Q (in dashed lines). (b) Chains of v_j. (c) Attaching the strings to Q. (d) Dynamic programming inside the gray region.

then find a straight-line embedding of the resulting planar graph such that the boundary of Q corresponds to the outerface of the embedding. Consequently, Q and its associated strings correspond to an outerstring graph representation R. Let G be the underlying outerstring graph. We now claim that any $\mathtt{MWIS}(G)$ corresponds to a min-convex subsuming polygon of P.

Lemma 2. *Let P be a simple polygon, where there exists a min-convex subsuming polygon that lies on $\mathcal{A}_e(P)$, and let G be the corresponding outerstring graph. Any maximum weight independent set of G yields a min-convex subsuming polygon of P.*

Proof. Let T be a set of strings that correspond to a maximum weight independent set of G. Since T is an independent set, the corresponding subsuming chains do not create edge crossings. Moreover, since each subsuming chain is weighted by the number of convex corners it can remove, the subsuming chains corresponding to T can remove $|\mathtt{MWIS}(G)|$ convex corners in total.

Assume now that there exists a min-convex subsuming polygon that can remove at least k convex corners. The corresponding subsuming chains would correspond to an independent set T' of strings in G. Since each string is weighted by the number of convex corners the corresponding subsuming chain can remove, the weight of T' would be at least k.   □

2.3 Time Complexity

To construct G, we first placed a dummy vertex at the intersection points of the chains, and then computed a straight-line embedding of the resulting planar graph such that all the vertices of Q are on the outerface. Therefore, the geometric representation used at most nt edges to represent each string. Since each convex vertex of P is associated with at most $O(n^t)$ strings, there are at most $n \times O(n^t)$ strings in G. Consequently, the total number of segments used in the geometric representation is $O(tn^{2+t})$. A subtle point here is that the strings in our representation may partially overlap, and more than three strings may intersect at one point. Removing such degeneracy does not increase the asymptotic size of the representation. Finally, by Theorem 1, one can compute the optimal subsuming polygon in $O(t^3 n^{6+3t})$ time.

The complexity can be improved further to as follows. Let $abcd$ be a rectangle that contains all the intersection points of $\mathcal{A}_e(P)$. Then every optimal solution can be extended to a triangulation of the closed region between $abcd$ and Q. Figure 3(d) illustrates this region in gray. We now can apply a dynamic programming similar to Sect. 2.1 to compute the maximum weight independent string set, where each subproblem finds a maximum weight set inside some subpolygon. Each such subpolygon can be described using two points v_1, v_2, each lying either on Q or on some string, and a subset of $\{a, b, c, d\}$ that helps enclosing the subpolygon.

Since there are $n \times O(n^t)$ strings, each containing at most t points, the number of vertices in the geometric representation is $O(tn^{1+t})$. Therefore, the size of the dynamic programming table is $O(tn^{1+t}) \times O(tn^{1+t}) \times O(1)$. Since there can be at most $O(tn^{1+t})$ candidate triangles $v_1 v_2 w$, we take $O(tn^{1+t})$ time to fill an entry of the table. Hence the dynamic program takes at most $O(t^3 n^{3+3t})$ time in total.

Theorem 2. *Given a simple polygon P with n vertices such that there exists a min-convex subsuming polygon that lie on $\mathcal{A}_e(P)$ and the subsuming chains are of length at most t, one can compute such a min-convex subsuming polygon in $O(t^3 n^{3+3t})$ time.*

2.4 Generalizations

We can generalize the results for any given line arrangements. However, such a generalization may increase the time complexity. For example, consider the case when the given line arrangement is $\mathcal{A}_v(P)$, which is determined by the pairs of vertices of P. Since we now have $O(n^2)$ lines in the arrangement $\mathcal{A}_v(P)$, the time complexity increases to $O(t^3 (n^2)^{3+3t})$, i.e., $O(t^3 n^{6+6t})$.

3 NP-Hardness of Min-Convex Subsuming Polygon

In this section we prove that it is NP-hard to find a subsuming polygon with minimum number of convex vertices. We denote the problem by MIN-CONVEX-SUBSUMING-POLYGON. We reduce the NP-complete problem monotone planar 3-SAT [3], which is a variation of the 3-SAT problem as follows: Every clause in a monotone planar 3-SAT consists of either three negated variables (*negative clause*) or three non-negated variables (*positive clause*). Furthermore, the bipartite graph constructed from the variable-clause incidences, admits a planar drawing such that all the vertices corresponding to the variables lie along a horizontal straight line l, and all the vertices corresponding to the positive (respectively, negative) clauses lie above (respectively, below) l. The problem remains NP-hard even when each variable appears in at most four clauses [7].

The idea of the reduction is as follows. Given an instance of a monotone planar 3-SAT I with variable set X and clause set C, we create a corresponding instance $\mathcal{P}_I$ of MIN-CONVEX-SUBSUMING-POLYGON. Let λ be the number of convex vertices in $\mathcal{P}_I$. The reduction ensures that if there exists a satisfying truth assignment of I, then $\mathcal{P}_I$ can be subsumed by a polygon with at most $\lambda - |X||C|^2 - 3|C|$ convex vertices, and vice versa.

Given an instance I of monotone planar 3-SAT, we first construct an orthogonal polygon P_o with holes. We denote each clause and variable using a distinct axis-aligned rectangle, which we refer to as the *c-rectangle* and *v-rectangle*, respectively. Each edge connecting a clause and a variable is represented as a thin vertical strip, which we call an *edge tunnel*. Figures 4(a) and (b) illustrate an instance of monotone planar 3-SAT and the corresponding orthogonal polygon, respectively. While adding the edge tunnels, we ensure for each v-rectangle that the tunnels coming from top lie to the left of all the tunnels coming from the bottom. Figure 4(b) marks the top and bottom edge tunnels by upward and downward rays, respectively. The v-rectangles, c-rectangles and the edge tunnels may form one or more holes, whereas the polygon is shown in diagonal line pattern. We now transform P_o to an instance $\mathcal{P}_I$ of MIN-CONVEX-SUBSUMING-POLYGON.

We first introduce a few notations. Let $abcd$ be a convex quadrangle and let l_{ab} be an infinite line that passes through a and b. Assume also that l_{bc} and l_{ad} intersect at some point e, and c, d, e all lie on the same side of l_{ab}, as shown in Figs. 4(c) and (d). Then we call the quadrangle $abcd$ a *tip* on l, and the triangle cde a *cap* of $abcd$.

3.1 Variable Gadget

We construct variable gadgets from the v-rectangles. We add some top-right (and the same number of top-left) tips at the bottom side of the v-rectangle, as show in Fig. 4(e). There are three top-right and top-left tips in the figure. For convenience we show only one top-left and one top-right tip in the schematic representation, as shown in Fig. 4(f). However, we assign weight to these tips to denote how many tips there should be in the exact construction. We will ensure

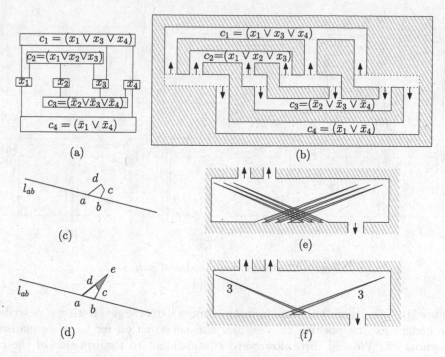

Fig. 4. (a) An instance I of monotone planar 3-SAT. (b) The orthogonal polygon P_o corresponding to I. (c)–(f) Illustration for the variable gadget.

a few more properties: (I) The caps do not intersect the boundary of the v-rectangle, (II) no two top-left caps (or, top-right caps) intersect, and (III) every top-left (resp., top-right) cap intersects all the top-right (resp., top-left) caps.

Observe that each top-left tip contributes to two convex vertices such that covering them with a cap reduces the number of convex vertices by 1. The peak of the cap reaches very close to the top-left corner of the v-rectangle, which will later interfere with the clause gadget. Specifically, this cap will intersect any downward cap of the clause gadget coming through the top edge tunnels. Similarly, each top-right tip contributes to two convex vertices, and the corresponding cap intersects any upward cap coming through the bottom edge tunnels.

Note that the optimal subsuming polygon P cannot contain the caps from both the top-left and top-right tips. We assign the tips with a weight of $|C|^2$. In the hardness proof this will ensure that either the caps of top-right tips or the caps of top-left tips must exist in P, which will correspond to the true and false configurations, respectively.

3.2 Clause Gadget

Without loss of generality assume that each clause is incident to three edge tunnels, otherwise, we can create necessary multi-edges to satisfy this constraint.

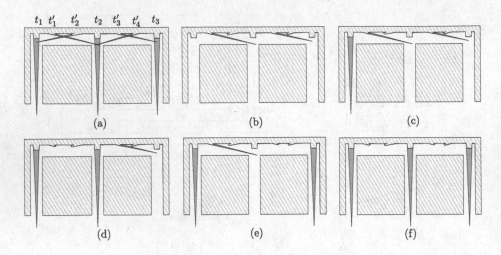

t_1 t_1' t_2' t_2 t_3' t_4' t_3

(a)   (b)   (c)

(d)   (e)   (f)

Fig. 5. Illustration for the clause gadget.

Figure 5(a) illustrates the transformation for a c-rectangle. Here we describe the gadget for the positive clauses, and the construction for negative clauses is symmetric. We add three downward tips incident to the top side of the c-rectangle, along its three edge tunnels. Each of these downward tip contributes to two convex vertices such that covering the tip with a cap reduces the number of convex vertices by 1. Besides, the corresponding caps reach almost to the bottom side of the v-rectangles, i.e., they would intersect the top-left caps of the v-rectangles. Let these tips be t_1, t_2, t_3 from left to right, and let $\gamma_1, \gamma_2, \gamma_3$ be the corresponding caps.

We then add a down-left and a down-right tip at the top side of the c-rectangle between t_i and t_{i+1}, where $1 \leq i \leq 2$, as shown in Fig. 5(a). Let the tips be $t_1', \ldots, t_4'$ from left to right, and let the corresponding caps be $\gamma_1', \ldots, \gamma_4'$. Note that the caps corresponding to t_j' and t_{j+1}', where $1 \leq j \leq 4$, intersect each other. Therefore, at most two of these four caps can exist at the same time in the solution polygon. Observe also that the caps corresponding to t_1, t_2, t_3 intersect the caps corresponding to $\{t_2'\}, \{t_1', t_4'\}, \{t_3'\}$, respectively. Consequently, any optimal solution polygon containing none of $\{\gamma_1, \gamma_2, \gamma_3\}$ have at least 12 convex vertices along the top boundary of the c-rectangle, as shown in Fig. 5(b).

We now show that any optimal solution polygon P containing at least $\alpha > 0$ caps from $\Gamma = \{\gamma_1, \gamma_2, \gamma_3\}$ have exactly 11 convex vertices along the top boundary of the c-rectangle. We consider the following three cases:

Case 1 ($\alpha = 1$): If γ_1 (resp., γ_3) is in P, then P must contain $\{\gamma_1', \gamma_3'\}$ (resp., $\{\gamma_2', \gamma_4'\}$). Figure 5(c) illustrates the case when P contains γ_1. If γ_2 is in P, then P must contain $\{\gamma_2', \gamma_3'\}$. In all the above scenario the number of convex vertices along the top boundary of the c-rectangle is 11.

Case 2 ($\alpha = 2$): If P contains $\{\gamma_1, \gamma_3\}$, then either γ_1' or γ_4' must be in P. Otherwise, P contains either $\{\gamma_1, \gamma_2\}$ or $\{\gamma_2, \gamma_3\}$. If that P contains $\{\gamma_1, \gamma_2\}$, as in Fig. 5(d), then γ_3' must lie in P. In the remaining case, γ_2' must lie in P. Therefore, also in this case the number of convex vertices along the top boundary of the c-rectangle is 11.

Case 3 ($\alpha = 3$): In this scenario P cannot contain any of $\gamma_1', \ldots, \gamma_4'$. Therefore, as shown in Fig. 5(e), the number of convex vertices along the top boundary of the c-rectangle is 11.

As a consequence we obtain the following lemma.

Lemma 3. *If a clause is satisfied, then any optimal subsuming polygon reduces exactly three convex vertex from the corresponding c-rectangle.*

3.3 Reduction

Although we have already described the variable and clause gadgets, the optimal subsuming polygon still may come up with some unexpected optimization that interferes with the convex corner count in our hardness proof. Figure 6(left) illustrates one such example. Therefore, we replace each convex corner that does not correspond to the tips by a small polyline with alternating convex and reflex corners, as shown Fig. 6(right).

We now prove the NP-hardness of computing optimal subsuming polygon.

Theorem 3. *Finding an optimal subsuming polygon is NP-hard.*

Proof. Let $I = (X, C)$ be an instance of the monotone planar 3-SAT and let $\mathcal{P}_I$ be the corresponding instance of MIN-CONVEX-SUBSUMING-POLYGON. Let λ be the number of convex vertices in $\mathcal{P}_I$. We now show that I admits a satisfying truth assignment if and only if $\mathcal{P}_I$ can be subsumed using a polygon having at most $\lambda - |X||C|^2 - 3|C|$ convex vertices.

First assume that I admits a satisfying truth assignment. For each variable x, we choose either the top-right caps or the top-left caps depending on whether x is assigned true or false. Consequently, we save at least $|X||C|^2$ convex vertices. Consider any clause $c \in C$. Since c is satisfied, one or more of its variables are assigned true. Therefore, for each positive (resp., negative) clause, we can have one or more downward (resp., upward) caps that enter into the v-rectangles.

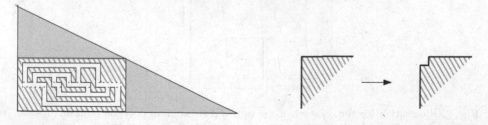

Fig. 6. Refinement of $\mathcal{P}_I$.

By Lemma 3, we can save at least three convex vertices from each c-rectangle. Therefore, we can find a subsuming polygon with at most $\lambda - |X||C|^2 - 3|C|$ convex vertices.

Assume now that some polygon P with at most $\lambda - |X||C|^2 - 3|C|$ convex vertices can subsume $\mathcal{P}_I$. We now find a satisfying truth assignment for I. Note that the maximum number of convex vertices that can be reduced from the c-rectangles is at most $3|C|$. Therefore, P must reduce at least $|C|^2$ convex vertices from each v-rectangle. Recall that in each v-rectangle, either the top-right or the top-left caps can be chosen in the solution, but not both. Therefore, the v-rectangles cannot help reducing more than $|X||C|^2$ convex vertices. If P contains the top-right caps of the v-rectangle, then we set the corresponding variable to true, otherwise, we set it to false. Since P has at most $\lambda - |X||C|^2 - 3|C|$ convex vertices, and each c-rectangle can help to reduce at most 3 convex vertices (Lemma 3), P must have at least one cap from $\gamma_1, \gamma_2, \gamma_3$ at each c-rectangle. Therefore, each clause must be satisfied. Recall that the downward (resp., upward) caps coming from edge tunnels are designed carefully to have conflict with the top-left (resp., top-right) caps of v-variables. Since top-left and top-right caps of v-variables are conflicting, the truth assignment of each variable is consistent in all the clauses that contains it. □

4   Conclusion

In this paper we have developed a polynomial-time algorithm that can compute optimal subsuming polygons for a given simple polygon in restricted settings. On the other hand, if the polygon contains holes, then we show the problem of computing an optimal subsuming polygon is NP-hard. Therefore, the question of whether the problem is polynomial-time solvable for simple polygons, remains open.

Our algorithm can find an optimal solution if the optimal subsuming polygon lies on some prescribed arrangement of lines, e.g., $\mathcal{A}_e(P)$ or $\mathcal{A}_v(P)$. The running time of our algorithm depends on the length of the subsuming chains, i.e., the

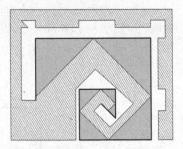

Fig. 7. Illustration for the case when the optimal subsuming polygon contains a subsuming chain of length $\Omega(n)$. The subsuming chain is shown in bold.

running time is polynomial if the subsuming chains are of constant length. However, there exist polygons whose optimal subsuming polygons contain subsuming chains of length $\Omega(n)$. Figure 7 illustrates such an example optimal solution that is lying on $\mathcal{A}_e(P)$. Therefore, it would be interesting to find algorithms whose running time is polynomial in the size of $\mathcal{A}_e(P)$ or $\mathcal{A}_v(P)$.

Another interesting research direction would be to examine whether there exists a good approximation algorithm for the problem.

References

1. Aichholzer, O., Hackl, T., Korman, M., Pilz, A., Vogtenhuber, B.: Geodesic-preserving polygon simplification. Int. J. Comput. Geom. Appl. **24**(4), 307–324 (2014)
2. Arge, L., Deleuran, L., Mølhave, T., Revsbæk, M., Truelsen, J.: Simplifying massive contour maps. In: Epstein, L., Ferragina, P. (eds.) ESA 2012. LNCS, vol. 7501, pp. 96–107. Springer, Heidelberg (2012)
3. de Berg, M., Khosravi, A.: Optimal binary space partitions for segments in the plane. Int. J. Comput. Geom. Appl. **22**(3), 187–206 (2012)
4. Douglas, D.H., Peucker, T.K.: Algorithm for the reduction of the number of points required to represent a line or its caricature. Can. Cartographer **10**(2), 112–122 (1973)
5. Guibas, L.J., Hershberger, J., Mitchell, J.S.B., Snoeyink, J.: Approximating polygons and subdivisions with minimum link paths. Int. J. Comput. Geom. Appl. **3**(4), 383–415 (1993)
6. Keil, J.M., Mitchell, J.S.B., Pradhan, D., Vatshelle, M.: An algorithm for the maximum weight independent set problem on outerstring graphs. In: Proceedings of CCCG, pp. 2–7 (2015)
7. Kempe, D.: On the complexity of the "reflections" game (2003). http://www-bcf.usc.edu/dkempe/publications/reflections.pdf
8. Mackaness, W.A., Ruas, A., Sarjakoski, L.T.: Generalisation of Geographic Information: Cartographic Modelling and Applications. Elsevier, Amsterdam (2011)
9. Ratschek, H., Rokne, J.: Geometric Computations with Interval and New Robust Methods: Applications in Computer Graphics, GIS and Computational Geometry. Horwood Publishing, Chichester (2003)

Combinatorial Scoring of Phylogenetic Networks

Nikita Alexeev and Max A. Alekseyev[✉]

The George Washington University, Washington, D.C., USA
maxal@gwu.edu

Abstract. Construction of phylogenetic trees and networks for extant species from their characters represents one of the key problems in phylogenomics. While solution to this problem is not always uniquely defined and there exist multiple methods for tree/network construction, it becomes important to measure how well the constructed networks capture the given character relationship across the species.

In the current study, we propose a novel method for measuring the *specificity* of a given phylogenetic network in terms of the total number of distributions of character states at the leaves that the network may impose. While for binary phylogenetic trees, this number has an exact formula and depends only on the number of leaves and character states but not on the tree topology, the situation is much more complicated for non-binary trees or networks. Nevertheless, we develop an algorithm for combinatorial enumeration of such distributions, which is applicable for arbitrary trees and networks under some reasonable assumptions.

1 Introduction

The evolutionary history of a set of species is often described with a rooted phylogenetic tree with the species at the leaves and their common ancestor at the root. Each internal vertex and its outgoing edges in such a tree represent a speciation event followed by independent descents with modifications, which outlines the traditional view of evolution. Phylogenetic trees do not however account for *reticulate events* (i.e., partial merging of ancestor lineages), which may also play a noticeable role in evolution through hybridization, horizontal gene transfer, or recombination [1,2]. Phylogenetic networks represent a natural generalization of phylogenetic trees to include reticulate events. In particular, phylogenetic networks can often more accurately describe the evolution of characters (e.g., phenotypic traits) observed in the extant species. Since there exists a number of methods for construction of such phylogenetic networks [3,4], it becomes important to measure how well the constructed networks capture the given character relationship across the species.

In the current study, we propose a novel method for measuring the *specificity* of a given phylogenetic network in terms of the total number of distributions of character states at the leaves that the network may impose. While for binary

---

The work is supported by the National Science Foundation under the grant No. IIS-1462107.

T.N. Dinh and M.T. Thai (Eds.): COCOON 2016, LNCS 9797, pp. 560–572, 2016.
DOI: 10.1007/978-3-319-42634-1_45

phylogenetic trees, this number has an exact formula and depends only on the number of leaves and character states but not on the tree topology [5,6], the situation is much more complicated for non-binary trees or network. Nevertheless, we propose an algorithm for combinatorial enumeration of such distributions, which is applicable for arbitrary trees and networks under the assumption that reticulate events do not much interfere with each other as explained below.

We view a *phylogenetic network* $\mathcal{N}$ as a rooted directed acyclic graph (DAG). Let $\mathcal{N}^\star$ be an undirected version of $\mathcal{N}$, where the edge directions are ignored. We remark that if there are no reticulate events in the evolution, then $\mathcal{N}$ represents a tree and thus $\mathcal{N}^\star$ contains no cycles. On the other hand, reticulate events in the evolution result in appearance of parallel directed paths in $\mathcal{N}$ and cycles in $\mathcal{N}^\star$. In our study, we restrict our attention to *cactus networks* $\mathcal{N}$, for which $\mathcal{N}^\star$ represents a cactus graph, i.e., every edge in $\mathcal{N}^\star$ belongs to at most one simple cycle. In other words, we require that the simple cycles in $\mathcal{N}^\star$ (resulting from reticulate events) are all pairwise edge-disjoint. This restriction can be interpreted as a requirement for reticulate events to appear in "distant" parts of the network. Trivially, trees represent a particular case of cactus networks. We remark that some problems, which are NP-hard for general graphs, are polynomial for cactus graphs [7], and some phylogenetic algorithms are also efficient for cactus networks [8]. We also remark that cactus networks generalize galled trees (where cycles are vertex-disjoint) and represent a particular case of galled networks (where cycles may share edges) [9].

We assume that the *leaves* (i.e., vertices of outdegree 0) of a given phylogenetic network $\mathcal{N}$ represent extant species. A *k-state character* is a partition of the species (i.e., leaves of $\mathcal{N}$) into k nonempty sets. In this paper, we consider only *homoplasy-free* multi-state characters (see [10]), and enumerate the possible number of such k-state characters for any particular $\mathcal{N}$.

2 Methods

Let $\mathcal{N}$ be a cactus network. For vertices u, v in $\mathcal{N}$, we say that u is an *ancestor* of v and v is a *descendant* of u, denoted $u \succcurlyeq v$, if there exists a path from u to v (possibly of length 0 when $u = v$). Similarly, we say that v is *lower than* u, denoted $u \succ v$, if $u \succcurlyeq v$ and $u \neq v$. For a set of vertices V of $\mathcal{N}$, we define a *lowest common ancestor* as a vertex u in $\mathcal{N}$ such that $u \succcurlyeq v$ for all $v \in V$ and there is no vertex u' in $\mathcal{N}$ such that $u \succ u' \succcurlyeq v$ for all $v \in V$. While for trees a lowest common ancestor is unique for any set of vertices, this may be not the case for networks in general. However, in Sect. 2.2, we will show that a lowest common ancestor in a cactus network is also unique for any set of vertices.

A k-state character on $\mathcal{N}$ can be viewed as a *k-coloring* on the leaves of $\mathcal{N}$, i.e., a partition of the leaves into k nonempty subsets, each colored with a unique color numbered from 1 to k (in an arbitrary order). A k-state character is homoplasy-free if the corresponding k-coloring $\mathcal{C}$ of the leaves of $\mathcal{N}$ is *convex*, i.e., the coloring $\mathcal{C}$ can be expanded to some internal vertices of $\mathcal{N}$ such that the subgraphs induced by the vertices of each color are rooted and connected.

Our goal is to compute the number of homoplasy-free multi-state characters on the leaves of $\mathcal{N}$, which is the same as the total number of convex colorings $p(\mathcal{N}) = \sum_{k=1}^{\infty} p_k(\mathcal{N})$, where $p_k(\mathcal{N})$ is the number of convex k-colorings on the leaves of $\mathcal{N}$.

2.1 Trees

In this section, we describe an algorithm for computing $p(\mathcal{T})$, where $\mathcal{T}$ is a rooted phylogenetic tree.

We uniquely expand each convex k-coloring of the leaves of $\mathcal{T}$ to a *partial k-coloring* of its internal vertices as follows. For the set L_i of the leaves of color i, we color to the same color i their lowest common ancestor r_i and all vertices on the paths from r_i to the leaves in L_i (since the k-coloring of leaves is convex, this coloring procedure is well-defined). We call such partial k-coloring of (the vertices of) $\mathcal{T}$ *minimal*. Alternatively, a partial k-coloring on $\mathcal{T}$ is minimal if and only if for each $i = 1, 2, \ldots, k$, the induced subgraph $\mathcal{T}_i$ of color i is rooted and connected (i.e., forms a subtree of $\mathcal{T}$), and removal of any vertex of $\mathcal{T}_i$ that is not a leaf of $\mathcal{T}$ breaks the connectivity of $\mathcal{T}_i$.

By construction, $p_k(\mathcal{T})$ equals the number of minimal k-colorings of $\mathcal{T}$.

The number $p_k(\mathcal{T})$ in the case of *binary* trees is known [5,6] and depends only on the number of leaves in a binary tree $\mathcal{T}$, but not on its topology.

Theorem 1 [5, **Proposition** 1]. *Let $\mathcal{T}$ be a rooted binary tree with n leaves. Then the number of convex k-colorings of $\mathcal{T}$ is $\binom{2n-k-1}{k-1}$. Correspondingly, $p(\mathcal{T})$ equals the Fibonacci number F_{2n-1}.*

The case of arbitrary (non-binary) trees is more sophisticated.

Let $\mathcal{T}$ be a rooted tree. For a vertex v in $\mathcal{T}$, we define $\mathcal{T}_v$ as the *full subtree* of $\mathcal{T}$ rooted at v and containing all descendants of v.

Let $\mathcal{T}'$ be any rooted tree larger than $\mathcal{T}$ such that $\mathcal{T}$ is a full subtree of $\mathcal{T}'$. We call a k-coloring of $\mathcal{T}$ *semiminimal* if this coloring is induced by some minimal coloring on $\mathcal{T}'$ (which may use more than k colors). Clearly, all minimal colorings are semiminimal, but not all semiminimal colorings are minimal. We remark that a semiminimal k-coloring of $\mathcal{T}$, in fact, does not depend on the topology of $\mathcal{T}'$ outside $\mathcal{T}$ and thus is well-defined for $\mathcal{T}$.

Lemma 1. *A semiminimal k-coloring of $\mathcal{T}$ is well-defined.*

Proof. Let $\mathcal{C}'$ be a minimal coloring of $\mathcal{T}'$ and $\mathcal{C}$ be its induced coloring on $\mathcal{T}$.

If $\mathcal{C}'$ is such that $\mathcal{T}$ and $\mathcal{T}' \setminus \mathcal{T}$ have no common colors, then $\mathcal{C}$ is minimal, and this property does not depend on the topology of $\mathcal{T}' \setminus \mathcal{T}$.

If $\mathcal{C}'$ is such that $\mathcal{T}$ and $\mathcal{T}' \setminus \mathcal{T}$ have some common color i, then the root r of $\mathcal{T}$ and its parent in $\mathcal{T}'$ are colored into i (hence, the shared color i is unique). Then the coloring on $\mathcal{T} \cup \{(r, l)\}$, where $\mathcal{T}$ inherits its coloring from $\mathcal{T}'$ and l is a new leaf colored into i, is minimal. This property does not depend on the topology of $\mathcal{T}' \setminus \mathcal{T}$ either. $\square$

For a semiminimal k-coloring $\mathcal{C}$ on $\mathcal{T}_v$, there exist three possibilities:

- Vertex v is colored and shares its color with at least two of its children. In this case, coloring $\mathcal{C}$ is minimal.
- Vertex v is colored and shares its color with exactly one of its children, and coloring $\mathcal{C}$ is not minimal.
- Vertex v is not colored. In this case, coloring $\mathcal{C}$ represents a minimal coloring of $\mathcal{T}_v$.

Correspondingly, for each vertex v, we define

- $f_k(v)$ is the number of minimal k-colorings of $\mathcal{T}_v$ such that at least two children of v have the same color (the vertex v must also have this color);
- $g_k(v)$ is the number of semiminimal k-colorings of $\mathcal{T}_v$ such that the vertex v shares its color with exactly one of its children (i.e., semiminimal but not minimal k-colorings);
- $h_k(v)$ is the number of minimal k-colorings of $\mathcal{T}_v$ such that the vertex v is not colored.

We remark that the number of minimal k-colorings of $\mathcal{T}$ equals $f_k(r) + h_k(r)$, where r is the root of the tree $\mathcal{T}$.

We define the following generating functions:

$$F_v(x) = \sum_{k=1}^{\infty} f_k(v) \cdot x^k; \qquad G_v(x) = \sum_{k=1}^{\infty} g_k(v) \cdot x^k; \qquad H_v(x) = \sum_{k=1}^{\infty} h_k(v) \cdot x^k.$$

$$(1)$$

For a leaf v of $\mathcal{T}$, we assume $f_k(v) = \delta_{k,1}$ (Kronecker's delta) and $g_k(v) = h_k(v) = 0$ for any $k \geq 1$. Correspondingly, we have $F_v(x) = x$ and $G_v(x) = H_v(x) = 0$.

If a vertex v has d children $u_1, u_2, \ldots, u_d$, then one can compute $F_v(x)$, $G_v(x)$, and $H_v(x)$ using the generating functions at the children of v as follows.

Theorem 2. *For any internal vertex v of $\mathcal{T}$, we have*

$$H_v(x) = \prod_{i=1}^{d} (F_{u_i}(x) + H_{u_i}(x)); \tag{2}$$

$$G_v(x) = \sum_{i=1}^{d} (F_{u_i}(x) + G_{u_i}(x)) \prod_{\substack{j=1 \\ j \neq i}}^{d} (F_{u_j}(x) + H_{u_j}(x)) = H_v(x) \cdot \sum_{i=1}^{d} \frac{F_{u_i}(x) + G_{u_i}(x)}{F_{u_i}(x) + H_{u_i}(x)}; \tag{3}$$

$$F_v(x) = x \prod_{i=1}^{d} \left(F_{u_i}(x) + H_{u_i}(x) + \frac{F_{u_i}(x) + G_{u_i}(x)}{x} \right) - x \cdot H_v(x) - G_v(x); \tag{4}$$

where $u_1, u_2, \ldots, u_d$ are the children of v.

Proof. Suppose that vertex v is not colored in a minimal k-coloring of $\mathcal{T}$. Then each its child is either not colored or has a color different from those of the other children. Furthermore, if a child of v is colored, its color must appear at least twice among its own children. Thus, the number of semiminimal colorings of $\mathcal{T}_{u_i}$ in this case is $f_{k_i}(u_i) + h_{k_i}(u_i)$, where k_i is the number of colors in $\mathcal{T}_{u_i}$ $(i = 1, 2, \ldots, d)$. Also, the subtrees $\mathcal{T}_{u_i}$ cannot share any colors with each other. Hence, the number of minimal k-colorings of $\mathcal{T}_v$ with non-colored v equals

$$\sum_{k_1 + \cdots + k_d = k} \prod_{i=1}^{d} (f_{k_i}(u_i) + h_{k_i}(u_i)),$$

implying formula (2).

Now suppose that vertex v is colored in a minimal k-coloring of $\mathcal{T}$. Then v must share its color with at least one of its children. Consider two cases.

Case 1. Vertex v shares its color with exactly one child, say u_i. Then there are $f_{k_i}(u_i) + g_{k_i}(u_i)$ semiminimal k_i-colorings for $\mathcal{T}_{u_i}$. For any other child u_j $(j \neq i)$, similarly to the above, we have that the number of semiminimal k_j-colorings equals $f_{k_j}(u_j) + h_{k_j}(u_j)$. Hence, the number of semiminimal k-colorings of $\mathcal{T}_v$ in this case equals

$$\sum_{k_1 + \cdots + k_d = k} (f_{k_i}(u_i) + g_{k_i}(u_i)) \prod_{\substack{j=1 \\ j \neq i}}^{d} (f_{k_j}(u_j) + h_{k_j}(u_j)),$$

implying formula (3).

Case 2. Vertex v shares its color with children $u_i, i \in I, |I| \geq 2$, but not with u_j for $j \notin I$. Since the color of v is the only color shared by $\mathcal{T}_{u_i}$, we have $k_1 + \cdots + k_d = k + |I| - 1$. Similarly to Case 1, we get that the number of minimal k-colorings of $\mathcal{T}_v$ is a coefficient at $x^{k+|I|-1}$ in

$$\prod_{i \in I} (F_{u_i}(x) + G_{u_i}(x)) \prod_{j \notin I} \left(F_{u_j}(x) + H_{u_j}(x) \right),$$

which is the same as the coefficient of x^k in

$$x \prod_{i \in I} \frac{(F_{u_i}(x) + G_{u_i}(x))}{x} \prod_{j \notin I} \left(F_{u_j}(x) + H_{u_j}(x) \right).$$

Summation of this expression over all subsets $I \subset \{1, 2, \ldots, d\}$ gives us the first term of (4), from where we subtract the sum over I with $|I| = 0$ (the term $x \cdot H_v(x)$) and with $|I| = 1$ (the term $G_v(x)$) to prove (4).   □

2.2 Cactus Networks

In this section, we show how to compute $p(\mathcal{N})$, where $\mathcal{N}$ is a cactus network. The following lemma states an important property of cactus networks.

Lemma 2. *Let $\mathcal{N}$ be a cactus network. Then for any set of vertices of $\mathcal{N}$, their lowest common ancestor is unique.*

Proof. For any set of vertices, there exists at least one common ancestor, which is the root of $\mathcal{N}$.

It is enough to prove the statement for 2-element sets of vertices. Indeed, if for any pair of vertices their lowest common ancestor is unique, then in any set of vertices we can replace any pair of vertices with their lowest common ancestor without affecting the lowest common ancestors of the whole set. After a number of such replacements, the set reduces to a single vertex, which represents the unique lowest common ancestor of the original set.

Suppose that for vertices u_1 and u_2 in $\mathcal{N}$, there exist two lowest common ancestors r_1 and r_2. Let r' be a lowest common ancestor of r_1 and r_2 (clearly, $r' \neq r_1$ and $r' \neq r_2$), and P_1 and P_2 be paths from r' to r_1 and r_2, respectively. Then P_1 and P_2 are edge-disjoint. Let $Q_{i,j}$ be paths from r_i to $u_j (i, j = 1, 2)$. It easy to see that for each $i = 1, 2$, the paths $Q_{i,1}$ and $Q_{i,2}$ are edge-disjoint. Then the paths $P_1, Q_{1,1}, P_2, Q_{2,1}$ form a simple cycle in $\mathcal{N}^*$; similarly, the paths $P_1, Q_{1,2}, P_2, Q_{2,2}$ form a simple cycle in $\mathcal{N}^*$. These simple cycles share the path P_1 (and P_2), a contradiction to $\mathcal{N}$ being a cactus network.   □

In contrast to trees, cactus networks may contain branching paths, which end at vertices of indegree 2 called *sinks* (also known as *reticulate vertices* [2,9]). Clearly, there are no vertices of indegree greater than 2 in a cactus network (if there are three incoming edges to some vertex then each of them belongs to two simple cycles in $\mathcal{N}^*$). We will need the following lemma.

Lemma 3. *Let p_l and p_r be the parents of some sink in a cactus network $\mathcal{N}$, and s be their lowest common ancestor. Let P_l and P_r be paths from s to p_l and p_r, respectively. Then P_l and P_r (i) are edge-disjoint; (ii) do not contain sinks, except possibly vertex s; and (iii) are unique.*

Proof. Since s is the lowest common ancestor of p_l and p_r, the paths P_l and P_r are edge-disjoint.

Suppose that there is an edge (u, t') on a path from s to p_l such that t' is a sink. Then this edge belongs to two different simple cycles in $\mathcal{N}^*$, a contradiction to $\mathcal{N}$ being a cactus network.

It is easy to see that if there exists a path P_l' from s to p_l different from P_l, then the paths P_l and P_l' would share a sink (different from s), which does not exist on P_l. Hence, the path P_l is unique and so is P_r.   □

Let t be a sink in $\mathcal{N}$ and p_l and p_r be its parents. Let s be the lowest common ancestor of p_l and p_r (which exists by Lemma 2). Lemma 3 implies that the paths P_1 and P_2 from s to t that visit vertices p_l and p_r, respectively, are unique and edge-disjoint. We call such a vertex s *source* and refer to the unordered pair of paths $\{P_1, P_2\}$ as a *simple branching path* (denoted $s \rightrightarrows t$) and to each of these paths as a *branch* of $s \rightrightarrows t$. Notice that one source may correspond to two or more sinks in $\mathcal{N}$.

For any two vertices u and v connected with a unique path in $\mathcal{N}$, we denote this path by $u \rightarrow v$ (which is a null path if $v = u$). A *branching path* between vertices p and q in $\mathcal{N}$ is an alternating sequence of unique and simple branching paths

$$p \rightarrow s_1 \rightrightarrows t_1 \rightarrow \cdots \rightarrow s_m \rightrightarrows t_m \rightarrow q,$$

where some of the unique paths may be null.

Lemma 4. *For any vertices $u \succcurlyeq v$ in a cactus network $\mathcal{N}$, the union of all paths between them forms a branching path.*

Proof. Suppose $\mathcal{N}$ is a cactus network and $u \succcurlyeq v$ in $\mathcal{N}$. Let $\mathcal{N}'$ be a subnetwork of $\mathcal{N}$ formed by all paths from u to v. Clearly, $\mathcal{N}'$ is a rooted (at u) cactus. We will prove that $\mathcal{N}'$ is a branching path by induction on the number of sinks in $\mathcal{N}'$. If there are no sinks in $\mathcal{N}'$, then a path from u to v is unique, then the statement holds. Otherwise, there exists a sink t in $\mathcal{N}'$ such that $u \succ t \succcurlyeq v$ and a path from t to v is unique, while there exist multiple paths from u to t. Let s be the source in $\mathcal{N}'$ corresponding to t. Then the branching path from s to v has the form $s \rightrightarrows t \to v$. Since every path from u to v visits t, it also must visit s (by Lemma 3 a path cannot enter into a simple branching path $s \rightrightarrows t$ other than through vertex s). Let $\mathcal{N}''$ be the subnetwork of $\mathcal{N}'$ consisting of all paths from u to s. Since the number of sinks in $\mathcal{N}''$ is one less than in $\mathcal{N}'$, by induction it is a branching path. Then $\mathcal{N}'$ is a branching path obtained from $\mathcal{N}''$ by concatenating it with the path $s \rightrightarrows t \to v$.   □

For vertices $u \succcurlyeq v$ in $\mathcal{N}$, the union of all paths from u to v is called the *maximal branching path*.

We generalize the notion of the minimal k-coloring to the case of cactus networks by expanding any convex k-coloring of the leaves of $\mathcal{N}$ to a partial coloring of the internal vertices of $\mathcal{N}$ as follows.

Network Coloring Procedure. For a given convex k-coloring of the leaves of $\mathcal{N}$, let L_i be the set of the leaves of color i. We consider maximal branching paths from the lowest common ancestor $r^{(i)}$ of L_i to all $l \in L_i$, which by Lemma 4 have the form:

$$r^{(i)} \to s_1^{(i)} \rightrightarrows t_1^{(i)} \to \cdots \to s_{m_i}^{(i)} \rightrightarrows t_{m_i}^{(i)} \to l.$$

At the first step, we color all vertices in the unique subpaths of such maximal branching paths into color i (Fig. 1a, b). Lemma 5 below shows that this coloring procedure (performed for all $i = 1, 2, \ldots, k$) is well-defined. At the second step, we color some branches of simple branching subpaths of the maximal branching paths from $r^{(i)}$ to the leaves in L_i. Namely, for each branch between $s_j^{(i)}$ and $t_j^{(i)}$ we check if its vertices are colored (at the first step) in any color other than i; if no other color besides i is present in the branch, we color all its vertices into i (Fig. 1b, c). Lemma 5 below shows that at least one branch of each simple branching subpath is colored this way, implying that the induced subgraphs of each color in the resulting partial coloring are connected. We refer to the resulting partial coloring as a *minimal k-coloring* of $\mathcal{N}$.

Lemma 5. *For any convex k-coloring on the leaves of a cactus network $\mathcal{N}$, the corresponding minimal k-coloring of $\mathcal{N}$ is well-defined. Moreover, the induced subgraph of $\mathcal{N}$ of each color is connected.*

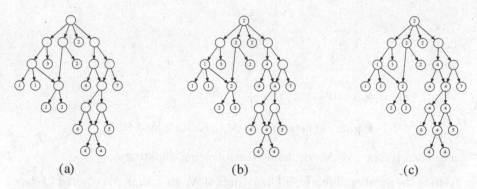

Fig. 1. (a) A convex coloring of the leaves of a cactus network $\mathcal{N}$, where the colors are denoted by labels. (b) The partial coloring of $\mathcal{N}$ constructed at the first step of the coloring procedure. (c) The minimal coloring of $\mathcal{N}$.

Proof. By the definition of convexity, a given convex k-coloring of the leaves of $\mathcal{N}$ can be expanded to a partial k-coloring $\mathcal{C}$ of $\mathcal{N}$ such that the induced subgraphs of each color are connected. The partial coloring of $\mathcal{N}$ that we obtain at the first step is a subcoloring of $\mathcal{C}$. Indeed, since the induced subgraphs of each color in $\mathcal{C}$ are connected, the unique subpaths of the maximal branching paths are colored (at the first step) into the same color as in $\mathcal{C}$. Hence, no conflicting colors can be imposed at the first step.

On the second step, we color a branch in some color i only if corresponding source and sink are colored in color i, so there are no conflicts on the second step. By Lemma 3, each non-source vertex belongs to at most one simple branching path, and thus the second step and the whole coloring procedure are well-defined.

For any simple branching subpath $s_j^{(i)} \rightrightarrows t_j^{(i)}$, at least one branch, say b, is colored into i in $\mathcal{C}$. In the subcoloring of $\mathcal{C}$ obtained at the first step, the branch b cannot contain any colors besides i. Hence, we will color all vertices of b into i at the second step. That is, at least one branch of every simple branching subpath will be colored, implying that the induced subgraph of each color is connected. $\square$

Similarly to the case of trees, we compute $p(\mathcal{N})$ as the number of minimal colorings of a cactus network $\mathcal{N}$.

Let $\mathcal{N}'$ be any rooted network larger than $\mathcal{N}$ such that $\mathcal{N}$ is a rooted subnetwork of $\mathcal{N}'$ and all edges from $\mathcal{N}' \setminus \mathcal{N}$ to $\mathcal{N}$ end at the root of $\mathcal{N}$. We call a k-coloring of $\mathcal{N}$ *semiminimal* if this coloring is induced by some minimal coloring on $\mathcal{N}'$. Similarly, to the case of trees (Lemma 1), a semiminimal k-coloring of $\mathcal{N}$ does not depend on the topology of $\mathcal{N}' \setminus \mathcal{N}$ and thus is well-defined.

For each vertex v in $\mathcal{N}$, we define a subnetwork $\mathcal{N}_v$ of $\mathcal{N}$ rooted at v and containing all descendants of v. An internal vertex in $\mathcal{N}$ is *regular* if the subnetworks rooted at its children are pairwise vertex-disjoint. It is easy to see that sources in $\mathcal{N}$ are not regular.

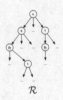 

$\mathcal{N}_s$ $\mathcal{N}_{s\setminus t}$ and $\mathcal{N}_t$ $\mathcal{L}$ $\mathcal{R}$ $\mathcal{B}$

Fig. 2. Subnetworks $\mathcal{N}_s, \mathcal{N}_{s\setminus t}, \mathcal{N}_t, \mathcal{L}, \mathcal{R}$, and $\mathcal{B}$.

For each vertex v in $\mathcal{N}$, we define the following quantities:

- $f_k(v)$ is the number of minimal k-colorings of $\mathcal{N}_v$ such that v is colored (f-type coloring);
- $g_k(v)$ is the number of semiminimal but not minimal k-colorings of $\mathcal{N}_v$ (g-type coloring);
- $h_k(v)$ is the number of minimal k-colorings of $\mathcal{N}_v$ such that the vertex v is not colored (h-type coloring).

Our goal is to compute $p_k(\mathcal{N}) = f_k(r) + h_k(r)$ for each positive integer k, where r is the root of $\mathcal{N}$. As before, for a leaf v of $\mathcal{N}$, we have $f_k(v) = \delta_{k,1}$ and $g_k(v) = h_k(v) = 0$ for any $k \geq 1$. We define the generating function $F_v(x), G_v(x)$, and $H_v(x)$ as in (1). Whenever we compute these functions in a subnetwork $\mathcal{M}$ of $\mathcal{N}$, we refer to them as $F_v^{\mathcal{M}}(x), G_v^{\mathcal{M}}(x)$, and $H_v^{\mathcal{M}}(x)$.

It is easy to see that Theorem 2 holds for all regular vertices v of $\mathcal{N}$ and therefore gives us a way to compute $F_v(x), G_v(x)$, and $H_v(x)$, provided that these functions are already computed at the children of v. So it remains to describe how to compute these functions at the sources in $\mathcal{N}$.

Let s be a source in $\mathcal{N}$ and t be any sink corresponding to s. We define p_l and p_r be the parents of t. To obtain formulas for $F_s(x), G_s(x)$, and $H_s(x)$, we consider the auxiliary subnetworks $\mathcal{N}_s, \mathcal{N}_t, \mathcal{N}_{s\setminus t} = \mathcal{N}_s \setminus \mathcal{N}_t, \mathcal{L}, \mathcal{R}$, and $\mathcal{B}$ (Fig. 2), where

- $\mathcal{L}$ is the subnetwork obtained from $\mathcal{N}_s$ by removing the edge (p_l, t);
- $\mathcal{R}$ is the subnetwork obtained from $\mathcal{N}_s$ by removing the edge (p_r, t);
- $\mathcal{B}$ is the subnetwork obtained from $\mathcal{N}_s$ by removing all the edges in the simple branching path $s \rightrightarrows t$.

It is easy to see that the vertex sets of the subnetworks $\mathcal{N}_s, \mathcal{L}, \mathcal{R}, \mathcal{N}_t \cup \mathcal{N}_{s\setminus t}$, and $\mathcal{B}$ coincide, and therefore a partial coloring of one subnetwork translates to the others.

Lemma 6. *Let t be a sink in a cactus network $\mathcal{N}$ and s be the corresponding source. If $\mathcal{C}$ is a partial coloring of $\mathcal{N}_s$ of f-type, g-type, or h-type, then $\mathcal{C}$ contains a partial subcoloring of $\mathcal{L}$ or $\mathcal{R}$ of the same type.*

Proof. We say that a partial coloring of $\mathcal{N}$ *uses* an edge (u, v) if vertices u and v are colored into the same color. Let p_l and p_r be the parents of t. Note that if $\mathcal{C}$ does not use the edge (p_l, t) then it is a partial coloring of the same type on $\mathcal{L}$. Similarly, if $\mathcal{C}$ does not use the edge (p_r, t) then it is a partial coloring of the same type on $\mathcal{R}$. So, it remains to consider the case when $\mathcal{C}$ uses both edges (p_l, t) and (p_r, t).

Suppose that $\mathcal{C}$ uses both edges (p_l, t) and (p_r, t). Notice that such $\mathcal{C}$ cannot be of h-type (since p_r and p_l share the same color, their lowest common ancestor s has to be colored as well). So, $\mathcal{C}$ has f-type or g-type. Let i be the color of t. From the second step of the coloring procedure, it follows that each vertex v such that $s \succcurlyeq v \succ t$ is also colored into i. Hence, removal of one of the edges (p_l, t) and (p_r, t) does not break the connectivity of the induced subgraph of color i. Thus, if $\mathcal{C}$ has f-type, then it contains a partial subcoloring of both $\mathcal{L}$ and $\mathcal{R}$ of f-type. Now suppose that $\mathcal{C}$ has g-type and $\mathcal{C}'$ is a subcoloring of $\mathcal{C}$ constructed at the first step of the network coloring procedure. Then at least one branch in $s \rightrightarrows t$ does not contain vertices of color i in $\mathcal{C}'$ (otherwise $\mathcal{C}$ would have f-type). If this branch contains p_l then $\mathcal{C}$ contains a partial subcoloring of $\mathcal{L}$ of g-type; otherwise $\mathcal{C}$ contains a partial subcoloring of $\mathcal{R}$ of g-type. $\qquad\square$

Theorem 3. *Let s be a source in $\mathcal{N}$ and t be any sink corresponding to s. Then*

$$H_s(x) = H_s^{\mathcal{L}}(x) + H_s^{\mathcal{R}}(x) - H_s^{\mathcal{N}_{s\setminus t}}(x) \cdot (F_t(x) + H_t(x)); \tag{5}$$

$$\begin{aligned} G_s(x) = {} & G_s^{\mathcal{L}}(x) + G_s^{\mathcal{R}}(x) \\ & - G_s^{\mathcal{N}_{s\setminus t}}(x) \cdot (F_t(x) + H_t(x)) - (F_t(x) + G_t(x)) \cdot \prod_{v:\, s\succcurlyeq v\succ t} H_v^{\mathcal{B}}(x); \end{aligned} \tag{6}$$

$$\begin{aligned} F_s(x) = {} & F_s^{\mathcal{L}}(x) + F_s^{\mathcal{R}}(x) - F_s^{\mathcal{N}_{s\setminus t}}(x) \cdot (F_t(x) + H_t(x)) \\ & - (F_t(x) + G_t(x)) \cdot \left(\prod_{v:\, s\succcurlyeq v\succ t} \left(\frac{F_v^{\mathcal{B}}(x) + G_v^{\mathcal{B}}(x)}{x} + H_v^{\mathcal{B}}(x) \right) - \prod_{v:\, s\succcurlyeq v\succ t} H_v^{\mathcal{B}}(x) \right) \end{aligned} \tag{7}$$

under the following convention: if a non-leaf vertex v in $\mathcal{N}$ turns into a leaf in a network $\mathcal{N}' \in \{\mathcal{N}_{s\setminus t}, \mathcal{L}, \mathcal{R}, \mathcal{B}\}$, then we re-define $F_v^{\mathcal{N}'}(x) = G_v^{\mathcal{N}'}(x) = 0$ and $H_v^{\mathcal{N}'}(x) = 1$.

Proof. We say that a partial coloring of $\mathcal{N}$ *uses* an edge (u, v) if vertices u and v are colored into the same color. Let p_l and p_r be the parents of t.

Let us enumerate h-type colorings of $\mathcal{N}_s$ first. We remark that such coloring cannot use both edges (p_l, t) and (p_r, t) (if it uses both these edges, the source s would be colored by the definition of minimal coloring). That is, any h-type coloring of $\mathcal{N}_s$ represents an h-type coloring of $\mathcal{L}$ or $\mathcal{R}$, or both these networks. The number of h-type k-colorings of $\mathcal{L}$ and $\mathcal{R}$ is the coefficient of x^k in $H_s^{\mathcal{L}}(x)$ and $H_s^{\mathcal{R}}(x)$, respectively. By the inclusion-exclusion principle, the number of h-type k-colorings of $\mathcal{N}_s$ equals the sum of those of $\mathcal{L}$ or $\mathcal{R}$ minus the number of h-type k-colorings of both $\mathcal{L}$ and $\mathcal{R}$. A coloring of the last kind does not use either of the edges (p_l, t) and (p_r, t), and thus is formed by an h-type coloring of $\mathcal{N}_{s\setminus t}$ and a minimal coloring of $\mathcal{N}_t$ (the colors of the two colorings are disjoint). The number of such coloring pairs equals the coefficient of x^k in $H_s^{\mathcal{N}_{s\setminus t}}(x)(F_t(x) + H_t(x))$, which completes the proof of (5).

We use similar reasoning to prove (6) and (7). The first two terms in these formulas are similar to those in (5) that correspond to same-type colorings on

$\mathcal{L}$ or $\mathcal{R}$ (at least one of which always exists by Lemma 6). The case of colorings of g-type or f-type on both $\mathcal{L}$ and $\mathcal{R}$ is more complicated and is split into two subcases depending on whether none or both of the edges (p_l, t) and (p_r, t) are used (if exactly one of the edges is used, it cannot be removed without making the induced subgraph of this color disconnected). The subcase of using none of the edges is similar to h-type colorings and gives us the third term in formulas (6) and (7). So it remains to enumerate colorings on both $\mathcal{L}$ and $\mathcal{R}$ that use both edges (p_l, t) and (p_r, t).

Let $\mathcal{C}$ be a g-type k-coloring of $\mathcal{N}_s$ that is a coloring on both $\mathcal{L}$ and $\mathcal{R}$ and uses both edges (p_l, t) and (p_r, t). Let $\mathcal{C}'$ be a subcoloring of $\mathcal{C}$ constructed at the first step of the network coloring procedure. Vertices s and t have the same color in both $\mathcal{C}$ and $\mathcal{C}'$, but any vertex v with $s \succ v \succ t$ is colored in $\mathcal{C}$ but not in $\mathcal{C}'$. So $\mathcal{C}$ corresponds to a semiminimal coloring on $\mathcal{N}_t$ with a colored root t (i.e., of f-type or g-type) and a coloring on $\mathcal{B} \setminus \mathcal{N}_t$ such that vertices v with $s \succcurlyeq v \succ t$ (in $\mathcal{N}_s$) are not colored. Since $\mathcal{B}$ is the union of vertex-disjoint subnetworks $\mathcal{N}_v$ with $s \succcurlyeq v \succcurlyeq t$, the number of such coloring pairs equals the coefficient of x^k in

$$(F_t(x) + G_t(x)) \cdot \prod_{v:\ s \succcurlyeq v \succ t} H_v^{\mathcal{B}}(x).$$

Now, let $\mathcal{C}$ be an f-type k-coloring of $\mathcal{N}_s$ that is a coloring on both $\mathcal{L}$ and $\mathcal{R}$ and uses both edges (p_l, t) and (p_r, t). Let $\mathcal{C}'$ be a subcoloring of $\mathcal{C}$ constructed at the first step of the network coloring procedure. Vertices s and t have the same color in both $\mathcal{C}$ and $\mathcal{C}'$, and any vertex v with $s \succ v \succ t$ is either colored into the same color or not colored in $\mathcal{C}'$. So $\mathcal{C}$ corresponds to a semiminimal coloring on $\mathcal{N}_t$ with the colored root t (i.e., of f-type or g-type) and a coloring on $\mathcal{B} \setminus \mathcal{N}_t$ such that at least one vertex v with $s \succcurlyeq v \succ t$ (in $\mathcal{N}_s$) is colored into the same color. The number of such colorings is the coefficient of x^k in

$$(F_t(x) + G_t(x)) \cdot \left(\prod_{v:\ s \succcurlyeq v \succ t} \left(\frac{F_v^{\mathcal{B}}(x) + G_v^{\mathcal{B}}(x)}{x} + H_v^{\mathcal{B}}(x) \right) - \prod_{v:\ s \succcurlyeq v \succ t} H_v^{\mathcal{B}}(x) \right).$$

The first product in the parentheses represents the generating function for the number of colorings of $\mathcal{B} \setminus \mathcal{N}_t$, where each vertex v with $s \succcurlyeq v \succ t$ (in $\mathcal{N}_s$) is either colored in a reserved color (accounted by the term $\frac{F_v^{\mathcal{B}}(x) + G_v^{\mathcal{B}}(x)}{x}$) or not colored (accounted by the term $H_v^{\mathcal{B}}(x)$). Subtraction of the second product eliminates the case where no vertex v with $s \succcurlyeq v \succ t$ (in $\mathcal{N}_s$) is colored.   □

3   Algorithm for Computing $p(\mathcal{N})$

Theorem 2 (for regular vertices) and Theorem 3 (for sources) allow us to compute the generating functions F, G, H at the root r of a cactus network $\mathcal{N}$ recursively. Namely, to compute $F_v(x)$, $G_v(x)$, and $H_v(x)$ for a vertex v (starting at $v = r$), we proceed as follows:[1]

---

[1] An implementation of the present algorithm in the SageMath mathematical software is available at http://cblab.org/projects/.

- if v is a leaf, then $F_v(x) = x, G_v(x) = H_v(x) = 0$ (except for the special case of a newly formed leaf described in Theorem 3, when $F_v(x) = G_v(x) = 0$ and $H_v(x) = 1$);
- if v is regular, we recursively proceed with computing $F_u(x), G_u(x)$, and $H_u(x)$ for every child u of v, and then combine the results with formulae (2), (3), (4);
- if v is a source, we select any sink t corresponding to s, and apply Theorem 3 to compute $F_v(x), G_v(x)$, and $H_v(x)$ from the generating functions computed in smaller subnetworks. We remark that while s may be a source for more than one sink and thus may still remain a source in the subnetworks, the number of sinks in each of the subnetworks decreases as compared to $\mathcal{N}$, implying that our recursion sooner or later will turn s into a regular vertex and then recursively proceed down to its children.

From the generating functions at the root r of $\mathcal{N}$, we can easily obtain the number $p_k(\mathcal{N})$ of convex k-colorings of $\mathcal{N}$ as the coefficient of x^k in $F_r(x)+H_r(x)$. This further implies that $p(\mathcal{N})$ can be computed as

$$p(\mathcal{N}) = \sum_{k=0}^{\infty} p_k(\mathcal{N}) = F_r(1) + H_r(1).$$

4   Applications

Network Specificity. We propose to measure the specificity of a cactus network $\mathcal{N}$ with n leaves as a decreasing function of $p(\mathcal{N})$. Notice that the value of $p(\mathcal{N})$ can be as small as $2^n - n$ (for a tree with n leaves all being children of the root) and as large as the Bell number B_n (enumerating set partitions of the leaves). We therefore find it convenient to define the *specificity score* of $\mathcal{N}$ as

$$\tau(\mathcal{N}) = \frac{n}{\log_2(p(\mathcal{N}) + n)}.$$

In particular, we always have $0 < \tau(\mathcal{N}) \leq 1$, where the upper bound is achievable. The asymptotic of B_n further implies that $\tau(\mathcal{N})$ can be asymptotically as low as $\frac{\log 2}{\log n}$, which vanishes as n grows.

From Theorem 1, it can be easily seen that for a binary tree $\mathcal{T}$ with n leaves, we have $\tau(\mathcal{T}) \approx \frac{n}{\log_2 \phi^{2n-1}} \approx 0.72$ when n is large, where $\phi = \frac{1+\sqrt{5}}{2}$ is the golden ratio.

Network Comparison. Existing methods for construction of phylogenetic networks (e.g., hybridization networks from a given set of gene trees [3,4]) often rely on the parsimony assumption and attempt to minimize the number of reticulate events. Such methods may generate multiple equally parsimonious networks, which will then need to be evaluated and compared from a different perspective. It is equally important to compare phylogenetic networks constructed by different methods. If the number of reticulate events in a constructed network is small, it is quite likely that this network represents a cactus network. Furthermore, there exist methods that explicitly construct phylogenetic cactus networks [8]. This makes our method well applicable for evaluation and comparison of such networks in terms of their specificity as defined above.

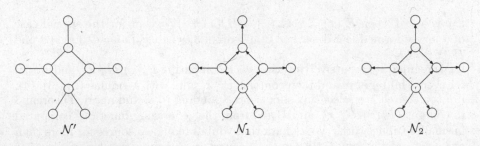

$$\mathcal{N}' \qquad\qquad \mathcal{N}_1 \qquad\qquad \mathcal{N}_2$$

Fig. 3. Networks $\mathcal{N}_1$ and $\mathcal{N}_2$ represent different orientations of the same undirected network $\mathcal{N}'$, i.e., $\mathcal{N}_1^\star = \mathcal{N}_2^\star = \mathcal{N}'$.

Orientation of Undirected Networks. Some researchers consider undirected phylogenetic networks (called "abstract" in the survey [2]) that describe evolutionary relationship of multiple species but do not correlate their evolution with time. For a given undirected cactus network $\mathcal{N}'$, our method allows one to find a root and an orientation of $\mathcal{N}'$, i.e., a directed rooted network $\mathcal{N}$ with $\mathcal{N}^\star = \mathcal{N}'$, that maximizes the specificity score $\tau(\mathcal{N})$. Indeed, different orientations of the same undirected network may result in different scores even if they are rooted at the same vertex. For example, in Fig. 3 the network $\mathcal{N}_1$ has $p(\mathcal{N}_1) = 35$ convex colorings and the score $\tau(\mathcal{N}_1) \approx 0.94$, while the network $\mathcal{N}_2$ has $p(\mathcal{N}_2) = 37$ convex colorings and the score $\tau(\mathcal{N}_2) \approx 0.927$.

References

1. Holland, B.R., Huber, K.T., Moulton, V., Lockhart, P.J.: Using consensus networks to visualize contradictory evidence for species phylogeny. Mol. Biol. Evol. **21**(7), 1459–1461 (2004)
2. Huson, D.H., Scornavacca, C.: A survey of combinatorial methods for phylogenetic networks. Genome Biol. Evol. **3**, 23–35 (2011)
3. Ulyantsev, V., Melnik, M.: Constructing parsimonious hybridization networks from multiple phylogenetic trees using a SAT-solver. In: Dediu, A.-H., Hernández-Quiroz, F., Martín-Vide, C., Rosenblueth, D.A. (eds.) AlCoB 2015. LNCS, vol. 9199, pp. 141–153. Springer, Heidelberg (2015)
4. Wu, Y.: An algorithm for constructing parsimonious hybridization networks with multiple phylogenetic trees. J. Comput. Biol. **20**(10), 792–804 (2013)
5. Steel, M.: The complexity of reconstructing trees from qualitative characters and subtrees. J. Classif. **9**(1), 91–116 (1992)
6. Kelk, S.: A note on convex characters and Fibonacci numbers (2015). arXiv eprint: arXiv:1508.02598
7. Korneyenko, N.: Combinatorial algorithms on a class of graphs. Discret. Appl. Math. **54**(23), 215–217 (1994)
8. Brandes, U., Cornelsen, S.: Phylogenetic graph models beyond trees. Discret. Appl. Math. **157**(10), 2361–2369 (2009)
9. Huson, D.H., Rupp, R., Berry, V., Gambette, P., Paul, C.: Computing galled networks from real data. Bioinformatics **25**(12), i85–i93 (2009)
10. Semple, C., Steel, M.: Tree reconstruction from multi-state characters. Adv. Appl. Math. **28**(2), 169–184 (2002)

Highly Bi-Connected Subgraphs for Computational Protein Function Annotation

Jucheol Moon[1], Iddo Friedberg[2], and Oliver Eulenstein[1(✉)]

[1] Department of Computer Science, Iowa State University, Ames, IA 50011, USA
{moon,oeulenst}@iastate.edu
[2] Department of Veterinary Microbiology and Preventive Medicine,
Iowa State University, Ames, IA 50011, USA
idoerg@iastate.edu

Abstract. Identifying highly connected subgraphs in biological networks has become a powerful tool in computational biology. By definition a highly connected graph with n vertices can only be disconnected by removing more than $\frac{n}{2}$ of its edges. This definition, however, is not suitable for bipartite graphs, which have various applications in biology, since such graphs cannot contain highly connected subgraphs. Here, we introduce a natural modification of highly connected graphs for bipartite graphs, and prove that the problem of finding such subgraphs with the maximum number of vertices in bipartite graphs is NP-hard. To address this problem, we provide an integer linear programming solution, as well as a local search heuristic. Finally, we demonstrate the applicability of our heuristic to predict protein function by identifying highly connected subgraphs in bipartite networks that connect proteins with their experimentally established functionality.

Keywords: Highly bi-connected · Highly connected · Bipartite graph · Computational protein function annotation

1 Introduction

Protein function comprises the biochemical, cellular, and phenotypic aspects of the molecular events that are executed by proteins. Ultimately, functions of a protein are made known by various types of biological experiments [15]. However, the experimental function annotation of proteins is expensive and can rarely be applied to large amounts of sequence data. Therefore, computational methods have emerged as a powerful tool to predict and elucidate the function of proteins [7]. One way to predict function is by analyzing protein interaction networks, and inferring the function from various types of associations between proteins [16,17]. However, network-based functional annotation is frequently challenged by biological networks that are incomplete and error-prone [4]. To confront these challenges, highly connected subgraphs have been identified in such networks, which are then analyzed for functional protein annotation [9].

© Springer International Publishing Switzerland 2016
T.N. Dinh and M.T. Thai (Eds.): COCOON 2016, LNCS 9797, pp. 573–584, 2016.
DOI: 10.1007/978-3-319-42634-1_46

An undirected graph with n vertices is *highly connected* if it can only be disconnected by removing more than $\frac{n}{2}$ of its edges. While this approach has produced several credible results [17], it cannot be applied to the large class of biological networks that is represented by using bipartite graphs. This is due to the fact that highly connected subgraphs do not exist in bipartite graphs.

Here we overcome this stringent limitation by proposing a natural adaptation of the definition for highly connected subgraphs to bipartite graphs. A bipartite graph $G = (U, V, E)$ is *highly bi-connected* if more than $\frac{1}{2}\min(|U|, |V|)$ of its edges are required to disconnect it. To identify useful highly connected subgraphs in bipartite biological networks, we analyze the *highly bi-connected (HBC)* problem that given a bipartite graph and a natural number k, decides whether this graph contains a highly bi-connected subgraph with k vertices. We show that this problem, like its related problem for identifying highly connected subgraphs [12], is NP-Hard. Consequently, to address the HBC problem, we describe an *integer linear programming (ILP)* formulation and a heuristic algorithm that can handle large-scale instances. Finally, we demonstrate the performance of our heuristic through a comparative study using exact ILP solutions and an applicability study for to protein function annotation.

Related Work. Many computational methods are used to predict protein function (for comprehensive reviews, see [7, 16]). Most commonly, methods based on the *amino acid sequence similarity* propose that if two sequences have a high degree of similarity, then they have evolved from a common ancestor and may have similar functions. *Phylogenomics* based methods state that the evolutionary history of putative homologs must be considered when annotating protein functions so that the protein function annotation should be taken from the closest ortholog, rather than from the most similar sequence. In *genomic context* based methods, protein function is inferred by matching the inter-genomic profiles of the unknown protein to those that are known. Among the non sequence-based methods, using *protein protein interaction (PPI)* networks is quite promising for predicting the protein function in the context of the biological process in which it participates.

A key idea of *graph clustering* is to identify densely connected subgraphs as clusters that have many interactions within themselves and few interactions outside of themselves in the graph [11]. A study by Przulj *et al.* [14] determines clusters, which could indicate protein functions, by using the HCS algorithm in PPI networks. A highly connected subgraph is defined as a subgraph with n vertices such that more than $\frac{n}{2}$ of its edges must be removed in order to disconnect the subgraph. The concept of a highly connected graph is very similar to that of a *quasi-clique* (i.e., a graph where every vertex has a degree at least $\frac{n-1}{2}$ [11]). Hartuv and Shamir [10] proved that the HCS algorithm, which is based on the $\frac{n}{2}$ connectivity requirement, produces clusters with good homogeneity and separation properties [14]. However, the concept of highly connected subgraphs is not applicable to bipartite graphs, since they do not contain such subgraphs.

Bipartite graphs are frequently used to represent biological networks. Bicliques in bipartite PPI networks play an important role in identifying functional protein groups [18]. A study by Andreopoulos *et al.* [1] identifies locally significant proteins, that mediate the function of proteins, by exploring bicliques in PPI networks. However, a biclique is too stringent for identifying the functional groups [8]. A *quasi-biclique* allows a specified number of missing edges in a biclique [18]. Bu *et al.* [3] show that quasi-bicliques consist of relevant protein functions and also propose a method to predict protein functions based on the classification of known proteins within the quasi-bicliques. Although a quasi-biclique is less stringent, it allows for the inclusion of proteins that interact with few other proteins in the quasi-biclique [4].

Our Contribution. We propose the HBC problem to identify highly bi-connected subgraphs in bipartite networks. Essential for this work is Theorem 3 that describes a highly bi-connected graph $G = (U, V, E)$ equivalently as a graph where the minimum degree is larger than $\frac{1}{2}$ of the minimum cardinality of the vertex sets U and V. Using this theorem we show the NP-hardness of the HBC problem by a polynomial time reduction from the exact 3-sets cover problem. Further, Theorem 3 is also used to describe an initial IQP formulation for the HBC problem that contains quadratic constraints. This initial IP is then transformed into an ILP by replacing the quadratic constraints with linear ones using simplified variables by adapting implication rules. Our heuristic follows a seed based approach, where seeds are expanded to highly bi-connected subgraphs with the maximum number of vertices, and resulting subgraphs with the largest number of vertices are returned. Finally, we demonstrate the performance of our heuristic algorithm by comparing its results with exact ILP solutions for small-scale instances of the HBC problem, and through an experimental study that annotates protein function by analyzing a bipartite protein-function network built from data provided by the *UniProt-GOA* human database [5].

2   Preliminaries and Basic Definitions

A *graph* is a pair $G = (V, E)$ of sets satisfying $E \subseteq V^2$. The elements of V are the *vertices* of the graph G, the elements of E are its *edges*. The vertex set of a graph G is referred to as $V(G)$, and its edge set as $E(G)$. A vertex v is *incident* with an edge e if $v \in e$; then e is an edge at v. The two vertices incident with an edge are its *endvertices* or *ends*, and an edge *joins* its ends. The set of all edges in E at a vertex $v \in V$ is denoted by $E(v)$. Two vertices u, v of G are *adjacent*, or *neighbors*, if (u, v) is an edge of G. If $G' \subseteq G$ and G' contains all the edges $(u, v) \in E$ with $u, v \in V'$, then G' is an induced subgraph of G and denoted $G' := G[V']$. The set of neighbors of a vertex v in G is denoted by $N_G(v)$, or briefly $N(v)$. More generally for $U \subseteq V$, the neighbors in $V \setminus U$ of vertices in U are called *neighbors* of U; their set is denoted by $N(U)$. The *degree* $d_G(v) = d(v)$ of a vertex v is the number $|E(v)|$. The *degree of v in V'* is the number of vertices in V' that are adjacent to v, denoted by $d(v, V') = |\{u | u \in V' \text{ and } (v, u) \in E\}|$. The number $\delta(G) := \min\{d(v) | v \in V\}$ is the *minimum degree* of G. A *path* is a non-empty graph $P = (V, E)$ of the form $V = \{x_0, x_1, ..., x_k\}$, $E = \{x_0 x_1, x_1 x_2, ..., x_{k-1} x_k\}$ where the x_i are all distinct. The number of edges of a path is its *length*. The *distance* $dst_G(x, y)$ in G of two vertices x, y is the length of a shortest $x - y$ path in G; if no such path exists, we set $dst(x, y) := \infty$. If $|V| > 1$ and $G' = (V, E \setminus F)$ is connected for every set $F \subseteq E$ of fewer than l edges, then G is called l-edge-connected. The greatest integer l such that G is the l-edge-connected is edge-connectivity $\lambda(G)$ of G. For every non-trivial graph G, we have $\delta(G) \geq \lambda(G)$. A graph $G = (X \cup Y, E)$ is called bipartite if $V = X \cup Y$ admits a partition into two disjoint subsets X and Y such that every edge connects a vertex in X to one in Y.

Definition 1 (Highly Connected Subgraph). *A graph G is called* highly connected *if $\lambda(G) > \frac{|V|}{2}$. An induced subgraph $G[V']$ (where $V' \subseteq V$) that is highly connected is called a* highly connected subgraph*(HCS).*

Theorem 1. *There is no highly connected bipartite graph.*

Proof. Let $G = (X \cup Y, E)$ be a bipartite graph and highly connected. Then $\frac{|X|+|Y|}{2} \geq \min(|X|, |Y|) \geq \delta(G) \geq \lambda(G) > \frac{|X|+|Y|}{2}$, which is a contradiction.

3   Highly Bi-Connected Subgraph Problem

Definition 2 (Highly Bi-Connected Subgraph). *Let $G = (X \cup Y, E)$ be a bipartite graph. A bipartite graph G is called* highly bi-connected *if $\lambda(G) > \frac{1}{2}\min(|X|, |Y|)$. An induced bipartite subgraph $G[X' \cup Y'], (X' \subseteq Y, Y' \subseteq Y)$ is a* highly bi-connected subgraph*(HBCS) if $G[X' \cup Y']$ is highly bi-connected.*

Theorem 2. *Let $G = (X \cup Y, E)$ be a bipartite graph. If $\lambda(G) > \frac{1}{2}\min(|X|, |Y|)$ and $|X| \geq |Y|$, then $dst(u, v) = 2$ for all distinct $u, v \in X$.*

Proof. $d(u), d(v) \geq \delta(G) \geq \lambda(G) > \frac{1}{2}\min(|X|, |Y|) = \frac{|Y|}{2}$. There exists at least one vertex $z \in Y$ such that $P = uzv$ is a path because $|N_G(u) \cap N_G(v)| \geq 1$.

Corollary 1. *(Dankelmann and Volkmann [6, page 273]) Let $G = (X \cup Y, E)$ be a bipartite graph. If $dst(u, v) = 2$ for all distinct $u, v \in X$, then $\lambda(G) = \delta(G)$.*

Theorem 3. *Let $G = (X \cup Y, E)$ be a bipartite graph. If $\delta(G) > \frac{1}{2}\min(|X|, |Y|)$, then $\delta(G) = \lambda(G)$.*

Proof. Supposed that $|X| \geq |Y|$, and $u, v \in X$. $d(u), d(v) \geq \delta(G) > \frac{|Y|}{2}$. There exists at least one vertex $z \in Y$ such that $P = uzv$ is a path because $|N_G(u) \cap N_G(v)| \geq 1$. Hence, $d(u, v) = 2$ and $\delta(G) = \lambda(G)$ by Theorem 1.

Corollary 2. *A bipartite graph G is highly bi-connected if $\delta(G) > \frac{1}{2}\min(|X|, |Y|)$, $\delta(G) \geq \lceil\frac{1}{2}\min(|X|, |Y|)\rceil$, or $2\delta(G) \geq \min(|X|, |Y|) + 1$.*

Problem 1 (HBCS Problem).
Instance: A undirected bipartite graph $G = (X \cup Y, E)$ and positive integer k.
Question: Is there a vertex set $X' \cup Y'$ such that $|X'| + |Y'| = k$ and $G' = G[X' \cup Y']$ is highly bi-connected?

Theorem 4. *HBCS problem is NP-hard.*

Proof. The exact cover by 3-sets (X3C) problem *is known to be NP-hard* [13]. *The reduction algorithm takes an instance* $\langle S, T \rangle$ *of the X3C problem where* S *is a finite set of* $3k$ *elements and* T *is a collection of* l *triples (three-element subsets of* S). *Without loss of generality, we assume that* $k < l < 2k$.

Step 1. A bipartite graph $G_A = (X_A \cup Y_A, E_A)$ *is created by linking an element* $s_i \in S$ *with* $x_i \in X_A$ *and* $t_j \in T$ *with* $y_j \in Y_A$. *An edge* $(x_i, y_j) \in E_A$ *is established iff* $s_i \notin t_j$. *Note that* $|X_A| = 3k$, $|Y_A| = l$, *and* $d(y_j, X_A) = 3k - 3$.

Step 2. A bipartite graph $G_B = (X_B \cup Y_B, E_B)$ *is constructed as* $X_B = X_a \cup X_b \cup X_c$ *where* $|X_a| = |X_b| = 3k$, $|X_c| = 6$ *and* $Y_B = Y_a \cup Y_b \cup Y_c$ *where* $|Y_a| = |Y_b| = l$, $|Y_c| = 3$. *We set edges to make that* $G[X_a \cup Y_a]$ *and* $G[X_b \cup Y_b]$ *are equivalent with* G_A. *Each vertex in* X_c *is adjacent to all vertices in* $Y_a \cup Y_b$, *and similarly, each vertex in* Y_c *is adjacent to all vertices in* $X_a \cup X_b$. *Note that* $|X_B| = 6k + 6$ *and* $|Y_B| = 2l + 3$.

Step 3. A bipartite graph $G = (X \cup Y, E)$ *is built as* $X = X_1 \cup X_2 \cup \cdots \cup X_{3k}$ *and* $Y = Y_B$ *where* $|X_1| = |X_2| = \cdots = |X_{3k}| = 6k + 6$. *We connect edges to achieve that* $G[X_i \cup Y]$ $(1 \le i \le 3k)$ *is identical to* G_B. *Note that* $|X| = 3k(6k + 6)$ *and* $|Y| = 2l + 3$.

These steps can be done in polynomial time. The output of the reduction algorithm is an instance $\langle G, 18k^2 + 20k + 3 \rangle$ *of the HBCS problem. Suppose that* $\langle S, T \rangle$ *has a perfect cover* $T' \subseteq T$ *where* $|T'| = k$. *We claim that* G *has a HBCS* $G' = G[X' \cup Y']$ *such that* $X' = X$, $Y' = Y'_a \cup Y'_b \cup Y_c$ *where* Y'_a *and* Y'_b *contain* k *vertices associated with* k *triples in* T'. *In the induced bipartite* G', $d(x, Y') \ge (k - 1) + 3 = k + 2 > \frac{1}{2} \min(|X'|, |Y'|) = \frac{1}{2}|Y'| = \frac{1}{2}(2k + 3) = k + \frac{3}{2}$ $(x \in X')$ *and* $d(y, X') \ge 3k(3k + 3) > k + \frac{3}{2}$ $(y \in Y')$. *Thus,* $\delta(G') > \frac{1}{2} \min(|X'|, |Y'|)$ *and* $|X'| + |Y'| = 3k(6k + 6) + (2k + 3) = 18k^2 + 20k + 3$.

Conversely, Suppose that G *has a HBCS* $G' = G[X' \cup Y']$ *where* $|X'| + |Y'| = 18k^2 + 20k + 3$. *We claim that* $\langle S, T \rangle$ *has a perfect cover* $T' \subseteq T$ *such that* $X' = X$, $Y' = Y'_a \cup Y'_b \cup Y_c$ *where* Y'_a *and* Y'_b *contain* k *vertices associated with* k *triples in* T'.

First, we prove that $|Y'_a| + |Y'_b| = 2k$. *If* $|Y'_a| + |Y'_b| < 2k$, *then* $|X| > 3k(6k+6)$. *This is a contradiction. If* $|Y'_a| + |Y'_b| > 2k$, *then there is a positive integer* p *such that* $2k + (2p - 1) \le |Y'_a| + |Y'_b| \le 2k + 2p$. *We assume that* $|Y'_a| \le |Y'_b|$, *hence* $|Y'_a| \le k + p$. *Now, we consider* $X'_{i,a} = X_{i,a} \cap X'$ *and prove* $X'_{i,a} \subsetneq X_{i,a}$. *Suppose that* $X'_{i,a} = X_{i,a}$. *For a vertex* $x \in X'_{i,a}$, $d(x, Y') > \frac{1}{2}|Y'| \ge \frac{1}{2}(2k + 2p - 1 + 3) = k + p + 1$. *By the construction,* $d(x, Y'_a) > k + p - 2$ *because* $d(x, Y'_b) = 0$ *and* $d(x, Y'_c) = 3$. *The inequality can be written* $d(x, Y'_a) \ge k + p - 1$ *since a degree is an integer. The number of edges between all* $X'_{i,a}$ *and* Y'_a *is at least* $3k(k + p - 1) = 3k^2 + 3pk - 3k$. *For a vertex* $y \in Y_a$, $d(y, X'_{i,a}) = 3k - 3$. *The number of edges between all* $X'_{i,a}$ *and* Y'_a *is at most* $(k + p)(3k - 3) = 3k^2 + 3pk - 3k - 3p$. *There is no integer* e *such that* $3k^2 + 3pk - 3k \le e \le 3k^2 + 3pk - 3k - 3p$ *for a positive integer* p. *Thus,* $X'_{i,a} \subset X_{i,a}$ *and there are at least* $3k$ *vertices in* X *not in* X' *since there is at least one vertex in* $X_{i,a}$ *not in* $X'_{i,a}$. $|X'| \le 3k(6k + 6) - 3k = 18k^2 + 15k$,

$|Y'| < 4k+3$ $(\because l < 2k)$, and $|X'|+|Y'| < 18k^2+19k+3$. This is a contradiction, hence $|Y_a'| + |Y_b'| = 2k$.

Second, we prove that $|Y_a'| = |Y_b'| = k$. We assume that $|Y_a'| < k$ and $X_{i,a}' = X_{i,a}$. For a vertex $x \in X_{i,a}'$, $d(x,Y') > \frac{1}{2}|Y'| = \frac{1}{2}(2k+3) = k + \frac{3}{2}$. Hence, $d(x,Y_a') \geq k-1$ $(\because d(x,Y_a') > k+\frac{3}{2}-3)$ and the number of edges between all $X_{i,a}'$ and Y_a' is at least $3k(k-1) = 3k^2 - 3k$. For a vertex $y \in Y_a$, $d(y,X_{i,a}') = 3k-3$. The number of edges between all $X_{i,a}'$ and Y_a' is less than $k(3k-3) = 3k^2 - 3k$. This is a contradiction. Therefore, $X_{i,a}' \subset X_{i,a}$ and there are at least $3k$ vertices in X not in X'. $|X'| \leq 3k(6k+6) - 3k = 18k^2 + 15k$, $|Y'| = 2k+3$, and $|X'| + |Y'| \leq 18k^2 + 17k + 3 < 18k^2 + 20k + 3$. Consequently, $|Y_a'| = |Y_b'| = k$.

Finally, we prove the original claim. For each vertex $x \in X_{i,a}'$, $d(x,Y_a') \geq k-1$ because $d(x,Y') > k+\frac{3}{2}$, $d(x,Y_b') = 0$, and $d(x,Y_c') = 3$. Suppose that there exists a vertex x such that $d(x,Y_a') > k-1$. The number of edges between all $X_{i,a}'$ and Y_a' is greater than $3k(k-1)$. For each vertex $y \in Y_a'$, $d(y,X_{i,a}') = 3k-3$. The number of edges between all $X_{i,a}'$ and Y_a' is $k(3k-3) = 3k(k-1)$. This is a contradiction, and hence $d(x,Y_a') = k-1$ and $d(y,X_{i,a}') = 3k-3$ $(\forall x, y \in X_{i,a}', Y_a')$. This means that the corresponding subset T' is an exact cover of S.

4   Integer Linear Programming

The first IQP formulation requires quadratic constraints, which are then replaced by linear constraints such that it can be solved by various optimization software packages [4]. Furthermore, the second ILP formulation is improved by using the implication rule to simplify variables involved.

4.1   Quadratic Programming for Maximum HBCS

Let $G = (X \cup Y, E)$ be a bipartite graph. For each $x \in X$ $(y \in Y)$, a binary variable v_x (v_y) is introduced. The variable v_x (v_y) is 1 if and only if the vertex v_x (v_y) is in X' (Y'). The integer programing is formulated as follows.

$$\text{maximize} \quad \sum_{x \in X} v_x + \sum_{y \in Y} v_y$$

$$\text{subject to} \quad 2\sum_{y \in Y} e_{xy} v_y v_x \geq v_x(W+1) \quad \forall x \in X \quad (1)$$

$$2\sum_{x \in X} e_{xy} v_x v_y \geq v_y(W+1) \quad \forall y \in Y \quad (2)$$

$$\sum_{x \in X} v_x \geq \sum_{y \in Y} v_y \text{ or } \sum_{x \in X} v_x \leq \sum_{y \in Y} v_y \quad (3)$$

$$W = \sum_{y \in Y} v_y \text{ or } W = \sum_{x \in X} v_x \quad (4)$$

$$v_x \in \{0,1\} \quad \forall x \in X \cup Y$$

where
$$e_{xy} = \begin{cases} 0 & xy \notin E \\ 1 & xy \in E \end{cases}$$

The quadratic terms in constraints are necessary because the constraints apply only to vertices in $X' \cup Y'$.

4.2 Linear Programming for Maximum HBCS

v_x (v_y) has two possible values such as 0 or 1. The constraint (1) is turned into $\sum_{y \in Y} e_{uv} v_y \geq W + 1$ in case of $v_x = 1$ and it becomes trivial when $v_x = 0$. Thus, the constraints (1) and (2) are reestablished as follow.

$$2 \sum_{y \in Y} e_{xy} v_y - W - 1 \geq (|X| + |Y|)(v_x - 1) \qquad \forall x \in X$$

$$2 \sum_{x \in X} e_{xy} v_x - W - 1 \geq (|X| + |Y|)(v_y - 1) \qquad \forall y \in Y$$

In order to obtain an optimal solution, we solve the ILP problem twice by setting the constraints (3) and (4) separately each time (i.e., $W = \sum_{x \in X} v_x$ if $\sum_{x \in X} v_x \leq \sum_{y \in Y} v_y$ or $W = \sum_{y \in Y} v_y$ if $\sum_{x \in X} v_x \geq \sum_{y \in Y} v_y$). In summary, this formulation uses variables and constraints linear to the size of input vertices. i.e., $\mathcal{O}(|X| + |Y|)$.

5 Heuristic Algorithm

For a given bipartite graph and a subset of a vertex partition of this graph, Algorithm 1 identifies a subgraph that satisfies the following four conditions: (i) the subgraph is highly bi-connected; (ii) one vertex partition of the subgraph is identical with the given subset; (iii) the number of vertices in the other vertex partition is greater than or equal to the number of vertices in the given subset; and iv) the number of vertices in the subgraph is maximized. Let n be the number of vertices in the given bipartite graph. The time complexity of Algorithm 1 is

---

Algorithm 1. MaxVertex-HBCS(G, Y')

Input: A bipartite graph $G = (X \cup Y, E)$ and a vertex set $Y' \subseteq Y$.
Output: A maximum vertex HBCS $G' = (X' \cup Y', E')$ such that $|X'| \geq |Y'|$.
 1: **for** all $v \in N(Y')$ **do**
 2:     **if** $|N(v) \cap Y'| > \frac{1}{2}|Y'|$ **then**
 3:         $X' = X' \cup \{v\}$
 4:     **end if**
 5: **end for**
 6: **if** $|X'| \geq |Y'|$ AND $G[X' \cup Y']$ is HBCS **then**
 7:     return $G[X' \cup Y']$
 8: **end if**

$\mathcal{O}(n^2)$. This follows directly from $\mathcal{O}(n)$ executions of the for-loop (Steps 1–5), where the time complexity of executing the body of this loop is asymptotically bound by Step 2 requiring $\mathcal{O}(n)$ time.

Algorithm 2 enumerates maximum vertex HBCSs for a given bipartite graph that uses a greedy approach to identify seed vertex sets. The while loop (Steps 3–11) identifies maximum vertex HBCS until no more maximum vertex HBCS can be found from the seed vertices. Algorithm 2 maintains the list of seed vertex sets to avoid repeating the process on the seed vertex sets that are already examined. Let n be the number of vertices in the given bipartite. The time complexity of Step 5 is $\mathcal{O}(n^2)$, and this step is repeated $\mathcal{O}(n^2)$ times through nested for and while loop. Hence, the overall time complexity is $\mathcal{O}(n^4)$.

Algorithm 2. GreedyEnum-MaxVertexHBCS(G)

Input: A bipartite graph $G = (X \cup Y, E)$.
Output: A set of maximum vertex HBCS $G' = (X' \cup Y', E')$ such that $|X'| \geq |Y'|$.
1: **for** $u \in Y$ **do**
2:     $Y' = \{u\}$
3:     **while** $Y' \neq Y$ AND Q does not contain Y' **do**
4:         $Q = Q \cup \{Y'\}$
5:         $G' =$ MaxVertex-HBCS(G, Y')
6:         **if** $G' \neq NULL$ **then**
7:             OUTPUT G'
8:             Find a vertex $v \in N_G(X') \setminus Y'$ that maximize $d_G(v, X')$.
9:             $Y' = Y' \cup \{v\}$
10:        **end if**
11:    **end while**
12: **end for**

---

6   Performance Evaluation of the Heurisitc Algorithm

We analyze the performance of the heuristic algorithm by comparing its results with exact ILP solutions for small-scale instances of the HBC problem, and through an experimental study.

6.1   Comparative Study

We compare heuristic estimates with the exact ILP results for $1,000$ random graphs as input. The random graphs were selected with equal probability from graphs with the following: an overall number of vertices ranging between 10 and 26 vertices and edge densities ranging between 0.6 and 0.8. The resulting differences between exact solutions and heuristic estimates are depicted in Fig. 1.

6.2   Experimental Study

In this experimental study, we present the results of the protein function prediction by using our heuristic algorithm. The *Gene Ontology (GO)* [2] is currently

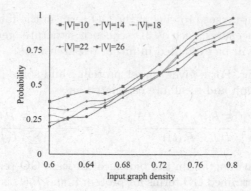

Fig. 1. The performance of the maximum vertex heuristic algorithm is evaluated by comparing its results with exact ILP solutions for small-scale instances.

the dominant approach for machine-legible protein function annotations [7]. GO is a controlled vocabulary that describes three aspects of protein functions: molecular function, biological process, and cellular location. Each aspect is described by a directed acyclic graph of terms and relationships that captures functional information in a standardized fashion that is both computationally amenable and interpretable by humans. We use the *Biological Process* classification scheme in this study. The main task of the experimental study is to predict sets of GO terms for the target proteins with confidence scores.

Target Proteins. We obtained annotated proteins from the January versions of the 2012, 2013, 2014, 2015, and 2016 UniProt-GOA human database [5]. Proteins are considered to be experimentally annotated if they are associated with GO terms having EXP, IDA, IPI, IMP, IGI, IEP, TAS, or IC evidence codes. The set of target proteins is selected by using the following scheme with two distinct time frames t_0 and t $(t_0 < t)$

$$Targets(t) = \text{Set of proteins at least one experimental annotation exist at}(t)$$
$$\cap \text{ Set of proteins only non-experimental annotation exist at}(t_0)$$

The predictive model is trained with non-experimental annotations of target proteins and experimental annotations of non-target proteins at $t_0 = 2012$. The performance of the model is evaluated by comparing the predicted annotations made by us for 2012 to existing experimental annotations in 2013–2016.

Experimental Design. Our predictive model uses the maximum vertex HBCS. For a given set of annotations between proteins and GO terms, we create a bipartite graph that has one vertex partition representing the set of proteins, the other vertex partition representing the set of GO terms, and edges representing the set of annotations. After that, Algorithm 2 finds a list of maximum vertex HBCS from the created bipartite graph. Every pair of a protein and a GO term in each HBCS of the found list is considered as a predictive annotation. The confidence score of the predictive annotation, which indicates the strength of the prediction, is the maximum sequence identity between the target protein and any neighboring non-target proteins of the GO term in the found HBCS. Other sequence identity measures, such as 3D sequence structure, genomic context, or interaction based, will be evaluated in future research work.

Evaluation Metric. For a given target protein i and some decision threshold $t \in [0, 1]$, the precision and recall are calculated as

$$pr_i(t) = \frac{\sum_f I(f \in P_i(t) \wedge f \in T_i)}{\sum_f I(f \in P_i(t))}, \ rc_i(t) = \frac{\sum_f I(f \in P_i(t) \wedge f \in T_i)}{\sum_f I(f \in T_i)}$$

where f is a protein function in the biological process GO terms, T_i is a set of experimentally determined GO terms for protein i, and $P_i(t)$ is a set of predicted protein functions of i with score greater than or equal to t. f ranges over all protein functions and $I(\cdot)$ stands for the indicator function. For a fixed decision threshold t, a point in the precision-recall space is created by averaging precision and recall across targets. Precision and recall at threshold t is calculated as

$$pr(t) = \frac{1}{m(t)} \cdot \sum_{i=1}^{m(t)} pr_i(t), \ rc(t) = \frac{1}{n} \cdot \sum_{i=1}^{n} rc_i(t)$$

where $m(t)$ is the number of proteins on which at least one prediction is made on threshold t and n is the number of proteins in a target set. It should be noted that unlike [15], we did not consider the GO DAG topology, but simply ran our assessment on GO terms as a "flat" vocabulary.

Results and Discussion. The quality of protein function prediction can be measured in different ways that reflect differing motivations for understanding protein functions. For this study, we show the precision-recall curves with all proteins having non-experimental annotations 2012 as the basis for predictions. We used the proteins that gained experimental annotations in 2013–2015 to test our method. The results are shown in Fig. 2.

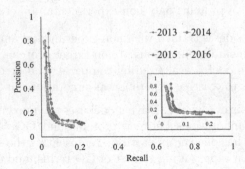

Fig. 2. Precision-recall curves for our method. The model is trained with non-experimental annotations of target proteins and experimental annotations of non-target proteins at $t_0 = 2012$. The performance of the model is evaluated by comparing the predicted annotations made by us for 2012 to existing experimental annotations in 2013–2016.

While our method has an overall low recall rate, it does have a high precision rate at low recall values. For some niche biological applications, such a method may be useful, as biomedical researchers may prefer generating protein function predictions with a high precision rate while trading off recall to minimize false positives for the results they do use. To estimate performance at different recall values, we used the $F_{max(\beta)}$ for the different years defined as

$$F_{max(\beta)} = \max_t \left\{ (1 + \beta^2) \frac{pr(t) \cdot rc(t)}{\beta^2 \cdot pr(t) + rc(t)} \right\}$$

where values for β are 0.1, 0.2, 0.5 and 1.0. F_β is a weighted harmonic mean of the precision and recall. We find the maximal value for each year using different values of β as weight. The lower β, the more weight is given to precision over recall. The results are shown in Table 1.

Table 1. Results of $F_{max(\beta)}$ analysis. See text for details on how $F_{max(beta)}$ is calculated. The lower the value of β, the more precision is weighted over recall. Our method performs best overall with $\beta = 0.1$.

	$F_{max(\beta)}$			
Year	0.1	0.2	0.5	1.0
2013	0.54	0.36	0.14	0.16
2014	0.71	0.47	0.17	0.14
2015	0.58	0.34	0.12	0.11
2016	0.54	0.31	0.12	0.12

7   Conclusion

Our proposed HBC approach sets a way for the functional annotation of proteins based on identifying highly bi-connected subgraphs in bipartite protein-function networks. While we show that the HBC problem is NP-hard, and we describe an ILP formulation and an effective heuristic. The comparative study displays accuracy of our heuristic by comparing its results with exact ILP solutions. Furthermore, the experimental study demonstrates the applicability of the heuristic for functionally annotating proteins. Future research will investigate other maximization objectives for identifying highly bi-connected subgraphs and partitioning problems of bipartite graphs based on highly bi-connected subgraphs.

Acknowledgments. IF and OE acknowledge support from the National Science Foundation award # DBI 1458359 and #GS 133814.

References

1. Andreopoulos, B., An, A., Wang, X., Faloutsos, M., Schroeder, M.: Clustering by common friends finds locally significant proteins mediating modules. Bioinformatics **23**(9), 1124–1131 (2007)
2. Ashburner, M., Ball, C.A., Blake, J.A., Botstein, D., Butler, H., Cherry, J.M., Davis, A.P., Dolinski, K., et al.: Gene ontology: tool for the unification of biology. Nat. Genet. **25**(1), 25–29 (2000)
3. Dongbo, B., Zhao, Y., Cai, L., Xue, H., Zhu, X., Hongchao, L., Zhang, J., et al.: Topological structure analysis of the protein-protein interaction network in budding yeast. Nucleic Acids Res. **31**(9), 2443–2450 (2003)
4. Chang, W.-C., Vakati, S., Krause, R., Eulenstein, O.: Exploring biological interaction networks with tailored weighted quasi-bicliques. BMC Bioinform. **13**(10), S16 (2012)
5. UniProt Consortium, et al.: Uniprot: a hub for protein information. Nucleic Acids Res. **43**, 989 (2014)
6. Dankelmann, P., Volkmann, L.: New sufficient conditions for equality of minimum degree and edge-connectivity. Ars Combinatoria **40**, 270–278 (1995)
7. Friedberg, I.: Automated protein function prediction the genomic challenge. Briefings Bioinform. **7**(3), 225–242 (2006)
8. Geva, G., Sharan, R.: Identification of protein complexes from co-immunoprecipitation data. Bioinformatics **27**(1), 111–117 (2011)
9. Hartuv, E., Schmitt, A.O., Lange, J., Meier-Ewert, S., Lehrach, H., Shamir, R.: An algorithm for clustering cDNA fingerprints. Genomics **66**(3), 249–256 (2000)
10. Hartuv, E., Shamir, R.: A clustering algorithm based on graph connectivity. Inf. Process. Lett. **76**(4), 175–181 (2000)
11. Hüffner, F., Komusiewicz, C., Liebtrau, A., Niedermeier, R.: Partitioning biological networks into highly connected clusters with maximum edge coverage. IEEE/ACM Trans. Comput. Biol. Bioinf. (TCBB) **11**(3), 455–467 (2014)
12. Hüffner, F., Komusiewicz, C., Sorge, M.: Finding highly connected subgraphs. In: Italiano, G.F., Margaria-Steffen, T., Pokorný, J., Quisquater, J.-J., Wattenhofer, R. (eds.) SOFSEM 2015-Testing. LNCS, vol. 8939, pp. 254–265. Springer, Heidelberg (2015)
13. Karp, R.M.: Reducibility among combinatorial problems. In: Miller, R.E., Thatcher, J.W., Bohlinger, J.D. (eds.) Complexity of Computer Computations, pp. 85–103. Springer, New York (1972)
14. Pržulj, N., Wigle, D.A., Jurisica, I.: Functional topology in a network of protein interactions. Bioinformatics **20**(3), 340–348 (2004)
15. Radivojac, P., Clark, W.T., Oron, T.R., Schnoes, A.M., Wittkop, T., Sokolov, A., Graim, K., Funk, C., Verspoor, K., Ben-Hur, A., et al.: A large-scale evaluation of computational protein function prediction. Nat. Methods **10**(3), 221–227 (2013)
16. Rentzsch, R., Orengo, C.A.: Protein function prediction-the power of multiplicity. Trends Biotechnol. **27**(4), 210–219 (2009)
17. Sharan, R., Ulitsky, I., Shamir, R.: Network-based prediction of protein function. Mol. Syst. Biol. **3**(1), 88 (2007)
18. Wang, L.: Near optimal solutions for maximum quasi-bicliques. J. Comb. Optim. **25**(3), 481–497 (2013)

Logic, Algebra and Automata

Cost Register Automata for Nested Words

Andreas Krebs[1], Nutan Limaye[2], and Michael Ludwig[1(✉)]

[1] University of Tübingen, Tübingen, Germany
{krebs,ludwigm}@informatik.uni-tuebingen.de
[2] Indian Institute of Technology, Bombay, India
nutan@cse.iitb.ac.in

Abstract. In the last two decades visibly pushdown languages (VPLs) have found many applications in diverse areas such as formal verification and processing of XML documents. Recently, there has been a significant interest in studying quantitative versions of finite-state systems as well as visibly pushdown systems. In this work, we take forward this study for visibly pushdown systems by considering a functional version of visibly pushdown automata. Our version is formally a generalization of cost register automata (CRA) defined by [Alur et al., 2013]. We observe that our model continues to have all the *good* properties of the CRAs in spite of being a generalization.

Apart from studying the functional properties of the model, we also study the complexity theoretic aspects. Recently such a study was conducted by [Allender and Mertz, 2014] with respect to CRAs. Here we show that CRAs when appended with a visible stack (i.e. in the model defined here), continue to have the same complexity theoretic upper bounds as are known for CRAs. Moreover, we observe that one of the upper bounds shown by Allender et al. which was not tight for CRAs becomes tight for our model. Hence, it answers one of the questions raised in their work.

1 Introduction

Language theory and complexity theory are two major branches of study in theoretical computer science. The study of the interplay between the two has a rich history of more than three decades. The study has not only increased our understanding of many complexity classes, but also has led to very beautiful theory. For example, a well-studied complexity class $\mathbf{NC}^1$ and the finer structure inside the class was revealed through the study of subclasses of regular languages [7,8,19]. Similarly, the class LogCFL and its equivalence with the complexity class $\mathbf{SAC}^1$ is another celebrated example of such a connection [10,15,18,21]. The study of language classes has been relevant in understanding Boolean complexity classes; and similarly, it is helpful to relate arithmetic complexity classes such as $\#\mathbf{NC}^1$, $\#\mathbf{SAC}^1$, $\#\mathbf{P}$ to formal power series (see for example [2]) for improving our understanding of these arithmetic classes.

In the literature of language theory, the study of formal power series has played a central role. One of the most well-studied models, namely weighted

© Springer International Publishing Switzerland 2016
T.N. Dinh and M.T. Thai (Eds.): COCOON 2016, LNCS 9797, pp. 587–598, 2016.
DOI: 10.1007/978-3-319-42634-1_47

automata, are an example of automata which compute an interesting class of formal power series.

More recently, Alur et al. [4,5] defined a generalization of weighted automata, namely *Cost Regsiter Automata*, CRA. In a subsequent work of Allender and Mertz [3] the complexity theoretic study of the formal power series defined by CRAs was performed.

Inspired by the works of [3–5], we extend the complexity theoretical study of formal power series in this work. CRAs defined and studied by [4,5] are models which compute functions from Σ^* to $\mathbb{D}$, where Σ is an input alphabet and $\mathbb{D}$ is a domain such as $\mathbb{Q}, \mathbb{Z}$ or $\mathbb{N}$. A CRA is a finite state automaton which is equipped with a set of registers and an algebra $\mathcal{A}$. The automaton reads the input from left to right; as in the case of a usual finite state automaton, but additionally various registers can get updated by using rules prescribed by the underlying algebra $\mathcal{A}$. It turns out, augmenting a finite state automaton with registers working over an algebra provides a way for performing a quantitative study of regular languages.

Here, we perform a similar quantitative analysis of visibly pushdown languages (VPLs) by augmenting the visibly pushdown automata (VPA) with registers working over an algebra. In this sense, we define a new model of computation, which is a natural extension of two well-known models of computation, namely VPAs and CRAs. We call our model *Cost Register Visibly Pushdown Automata* (CVPA) to signify this combination.

VPLs in their current incarnation were first conceptualized by Alur and Madhusudan [6]. However they have been introduced by Mehlhorn under the name input-driven pushdown languages [20] earlier. Due to their power to generalize regular languages while staying extremely structured, they have found many applications in diverse areas such as formal verification and processing of XML documents. Just like the study of regular systems through weighted automata and CRAs has helped in understanding their quantitative properties, we believe that our work is a step towards understanding quantitative properties of visibly pushdown systems.

Recently, Allender and Mertz [3] developed the complexity theoretic understanding of the arithmetic functions computed by CRAs. They proved many complexity theoretic upper bounds for functions computed by CRAs over many different algebras. In this work, we extend their results to CVPAs. We prove that CVPAs have computational complexity bounds similar to CRAs. This presents a gratifying picture, because CVPAs formally generalize CRAs, however, still stay *computationally efficient*.

Contributions. We start by proving basic language theoretic results about our model. In particular, we first show basic closure properties of CVPAs. Here we get that CVPAs are closed with respect to many natural properties (as shown by [4,5] for CRAs.) We then show that CVPAs formally generalize VPAs and CRAs. It is known that there is a version of CRAs, namely copyless CRAs over $(\mathbb{N}, +)$ which can be simulated by CRAs over $(\mathbb{N}, +c)$[1]. However, we observe

---

[1] Notations explained later.

that this is not true for CVPAs. That is, we give an explicit function which is computable by copyless CVPAs over $(\mathbb{N}, +)$ but cannot be computed by $(\mathbb{N}, +c)$.

The arguments we present here are combinatorial and may be useful in proving similar things for different algebras in the case of CRAs. We also show that CRAs over $(\mathbb{N}, +)$ compute the same class of functions as #NFA. Moreover, we prove that CVPAs generalize the weighted and the counting variant of VPLs as defined in [13,17].

We then perform a complexity theoretical study of CVPAs along the lines of Allender and Mertz [3]. Here our main theorem states that CVPA over the algebra $(\mathbb{Z}, +)$ are $\mathbf{GapNC^1}$-complete. Our $\mathbf{GapNC^1}$ hardness result also holds for a restricted version of CVPAs, namely copyless CVPAs (CCVPAs). No such result is known for the copyless variant of CRAs.

Our $\mathbf{GapNC^1}$ upper bound for CVPA over $(\mathbb{Z}, +)$ builds on several different results. It combines ideas from Buss [11] and Dymond [16] who showed that Boolean formula evaluation (and membership in regular languages) and membership in VPLs are in $\mathbf{NC^1}$ (resp.) with some crucial concepts from [12] which showed that arithmetic formula evaluation is in $\#\mathbf{NC^1}$. This also proves the hardness. Recently, [17] showed an $\#\mathbf{NC^1}$ upper bound (and hardness) for #VPA. In our paper we take this even further and show $\mathbf{GapNC^1}$ upper bound for CVPA over $(\mathbb{Z}, +)$ by appropriately combining all these ideas. We also prove that the class CVPA$(\mathbb{Z}, +c)$ is contained in $\mathbf{NC^1}$, i.e. for any function in CVPA$(\mathbb{Z}, +c)$, the ith bit of the function is computable by an $\mathbf{NC^1}$ circuit.

2 Preliminaries

In this section we establish notation and recall some known definitions and results from circuit complexity and language theory.

2.1 Notation

By Σ and Γ we denote finite *alphabets* and (finite) *words* are finite sequences of letters of the alphabet. The set of all words over Σ is the free monoid Σ^* with concatenation $\circ$ as operation. A *language* is a subset of Σ^*. For $w \in \Sigma^*$, we address the ith letter of w by w_i and by $|w|$ we denote the *length* of w. We use $|w|_a$ to denote the number of letters a appearing in w. The *empty word* is denoted by ϵ.

An *algebra* $\mathcal{A} = (\mathbb{D}, \oplus_1, \ldots, \oplus_k)$ is a *domain* $\mathbb{D}$ which can be infinite, together with a number of *operators* $\oplus_i \colon \mathbb{D}^{\alpha_i} \to \mathbb{D}$ where α_i is the arity of $\oplus_i$. We usually have binary operators and sometimes unary ones. For binary operator $\oplus$ and $c \in \mathbb{D}$, we use $\oplus c$ to indicate that one of the operands of $\oplus$ is a domain constant c. For $k = 1$ and depending on the properties of $\oplus_1$, we get semigroups, monoids, and groups. For $k = 2$, depending on the properties of $\oplus_1$ and $\oplus_2$ we get different kinds of rings. Examples of algebras we use in this paper include $\mathbb{B} = (\{0, 1\}, \wedge, \vee, \neg)$, $(\mathbb{N}, +)$, $(\mathbb{Z}, +)$, $(\mathbb{Z}, +, \times)$, $(\mathbb{Z}, +, \min)$, and $(\mathbb{Z}, +c)$ where $+c$ means addition with a constant.

By $E(\mathcal{A})$ we denote the set of *expressions* (or formulas) over $\mathcal{A}$: All $d \in \mathbb{D}$ are (atomic) expressions and if a and b are expressions, then $(a \oplus_i b)$ is an expression for all $1 \leq i \leq k$. Let $X = \{x_1, \dots, x_l\}$ be a set of variable names (which we later use as *register* names). We consider these variables as atomic expressions, from which we get a set we denote by $E(\mathcal{A}, X)$. An expression $E(\mathcal{A})$ evaluates to an element in $\mathbb{D}$ and an expression in $E(\mathcal{A}, X)$ evaluates to a function $\mathbb{D}^{|X|} \to \mathbb{D}$. With this we are now ready to define the model we wish to study in this paper.

2.2 Circuit Complexity

For the basic circuit complexity we refer e.g. to [22].

A *circuit* $C_n = (V, E, \Sigma, \mathcal{A}, l, v_{\text{out}})$ is a directed acyclic graph (V, E) in which the vertices are called *gates*, the in-degree is called *fan-in*. There are n gates of fan-in 0, which are called *input gates*, which are marked with inputs to the circuit. For a circuit computing functions over an algebra $\mathcal{A} = (\mathbb{D}, \oplus_1, \dots, \oplus_k)$, the map $l \colon V \to \{\oplus_1, \dots, \oplus_k\}$ assigns each gate one of the functions.

If a gate g has fan-in t with gates $g_1, g_2, \dots, g_t$ feeding into it, then it computes the function $l(g)(g_1, g_2, \dots, g_t)$. Fan-in greater than two of a gate g requires associativity of $(\mathbb{D}, l(g))$. The gates naturally compute functions of their fan-ins. The gate $v_{\text{out}} \in V$ is the output gate, which computes a function $\mathbb{D}^n \to \mathbb{D}$. *Formulas* are circuits, in which the underlying DAG is a tree. To treat inputs of arbitrary length, we consider *families of circuits* $C = (C_n)_{n \in \mathbb{N}}$, where there is one circuit for each input length.

For $\mathcal{A} = (\{0, 1\}, \wedge, \vee, \neg)$ we get Boolean circuits, similarly, for $\mathcal{A} = (\mathbb{Z}, +, \times)$ we get arithmetic circuits.

If we want to emphasize a different algebra, we write e.g. $\mathcal{C}(\mathcal{A})$, where $\mathcal{C}$ is a circuit class. The special case of $\mathcal{C}(\mathbb{N}, +, \times)$ restricted to $\{0, 1\}$-inputs is known as $\#\mathcal{C}$ and $\mathcal{C}(\mathbb{Z}, +, \times)$ restricted to $\{-1, 0, 1\}$-inputs is known as **Gap$\mathcal{C}$**. For basics in circuit classes and arithmetic circuit complexity see e.g. [1,22].

2.3 Visibly Pushdown Automata

In our paper we define a new model which computes functions from Σ^* to some domain $\mathbb{D}$. We view our model as a combination of two well-studied models of computations, namely Visibly pushdown automata (VPA) [6] and Cost register automata (CRA) [4,5]. In order to define our model, we first start by recalling the definition of VPA. In the context of VPA we always have to specify a *visible alphabet* $\hat{\Sigma} = (\Sigma_{\text{call}}, \Sigma_{\text{ret}}, \Sigma_{\text{int}})$ such that Σ is the disjoint union of Σ_{call}, Σ_{ret} and Σ_{int}, where Σ_{call} are *call* or *push* letters, Σ_{ret} *return* or *pop* letters and Σ_{int} *internal* letters. For the rest of the paper we assume a fixed visible alphabet unless stated otherwise.

Definition 1 (Visibly Pushdown Automaton (VPA) [6]**).** *A deterministic visibly pushdown automaton is a tuple* $M = (\hat{\Sigma}, Q, q_0, \delta, \Gamma, \perp, F)$ *where* $\hat{\Sigma}$ *is the visible alphabet,* Q *is the finite set of states,* $q_0 \in Q$ *is the initial state,* δ *is the transition function with* $\delta \colon Q \times \Sigma_{\text{int}} \to Q$, $\delta \colon Q \times \Sigma_{\text{call}} \to Q \times \Gamma$, *and*

$\delta \colon Q \times \Sigma_{\mathrm{ret}} \times \Gamma \to Q$. Also, Γ is the finite stack alphabet with bottom-of-stack symbol $\perp \in \Gamma$, and $F \subseteq Q$ is the set of final states.

A configuration of M is an element (q, γ), where q is a state and γ is a word over Γ. A run $(q_0, \gamma_0) \ldots (q_n, \gamma_n)$ of M on a word $w \in \Sigma^n$ is defined as follows:

- If $w_i \in \Sigma_{\mathrm{int}}$, then $\gamma_i = \gamma_{i-1}$ and $q_i = \delta(q_{i-1}, w_i)$.
- If $w_i \in \Sigma_{\mathrm{call}}$, then $\gamma_i = \gamma_{i-1}\gamma$ and $(q_i, \gamma) = \delta(q_{i-1}, w_i)$.
- If $w_i \in \Sigma_{\mathrm{ret}}$, then $\gamma_i \gamma = \gamma_{i-1}$ and $q_i = \delta(q_{i-1}, w_i, \gamma)$. If $\gamma = \perp$, the word is rejected.

A run is said to be accepting if $q_n \in F$. The language accepted by M, denoted by $L(M)$, is defined to be the set of words $w \in \Sigma^*$ such that M has an accepting run on w starting from $(q_0, \perp)$.

As defined above, a VPA M can be thought of as computing a function $f_M \colon \Sigma^* \to \{0, 1\}$, where for all $w \in \Sigma^*$, $f_M(w) = 1$ if and only if $w \in L(M)$. We will append this model with *registers* and interpret its computation over an algebra $\mathcal{A}$ so as to design more general functions $\Sigma^* \to \mathbb{D}$, where $\mathbb{D}$ is a general domain such as $\mathbb{Z}, \mathbb{N}$ etc.

In what follows we will assume that all the input words are *well-matched*, i.e. all our input words have the following two properties.

- any prefix $w_i = w_1 \ldots w_i$ of w the number of call letters in w_i is greater than or equal to the number of return letters in w_i.
- the number of call letters in w equals the number of return letters in w.

This assumption is for the ease of exposition and all our results hold even in the general case, unless stated otherwise. Given some word, we call two positions i, j *matching* if $w_i \ldots w_j$ is well-matched, there is no prefix of $w_i \ldots w_j$ which is well-matched and $w_i \in \Sigma_{\mathrm{call}}$ and $w_j \in \Sigma_{\mathrm{ret}}$.

2.4   Cost Register Automata (CRA)

Here we recall the definition of Cost Register automata, CRA. from [4,5] with a small modification resulting in an equivalent model. In [4,5], the register updates are performed using a tree grammar, but here we consider a version wherein the rules of update are governed by the underlying algebra. It is this version which will help us to define the cost register VPA, the model which we will introduce in the next section.

Definition 2 (Cost Register Automaton (CRA) [4,5]). *A CRA is a tuple* $M = (\Sigma, Q, q_0, \delta, X, \mathcal{A}, v_0, \rho, \mu)$ *where Q is the finite set of states, q_0 is the initial state, $\delta \colon Q \times \Sigma \to Q$ is the transition function, $X = \{x_1, x_2 \ldots, x_k\}$ is the finite set of registers, $\mathcal{A} = (\mathbb{D}, \oplus_1, \ldots, \oplus_m)$ is an algebra, $v_0 \colon X \to \mathbb{D}$ are the initial register values, $\rho \colon Q \times \Sigma \times X \to E(\mathcal{A}, X)$ is the partial register update function, and $\mu \colon Q \to E(\mathcal{A}, X)$ is the partial final cost function.*

A *configuration* of M is an element (q, v), where q is a state and $v\colon X \to E(\mathcal{A}, X)$ assigns each register an expression over the algebra $\mathcal{A}$. A *run* $(q_0, v_0) \ldots (q_n, v_n)$ of M on some input $w \in \Sigma^*$ starts in (q_0, v_0), $q_i = \delta(q_{i-1}, w_i)$ for each $1 \le i \le n$ and for each $x \in X$, $v_i(x)$ is the expression obtained by substituting $v_{i-1}(y)$ for every occurrence of y in $\rho(q_i, w_i, x) \in E(\mathcal{A}, X)$.

The semantics of a CRA M is the function $F_M\colon \Sigma^* \to \mathbb{D}$ which is defined as the evaluation of the expression $\mu(q_n)$ by substituting the expression $v_n(y)$ for every occurrence of $y \in X$ in the expression $\mu(q_n)$. It is undefined if $\mu(q_n)$ is undefined.

Claim ([4, 5]). Given a DFA M, which accepts language $L(M)$, there is a CRA M' such that $F_{M'}(w) = 1$ if $w \in L(M)$ and $F_{M'}(w) = 0$ if $w \notin L(M)$, i.e. M' computes the characteristic function $\Sigma^* \to \{0, 1\}$ of the language $L(M)$.

Note that cost registers are always driven by the states while on the other hand the registers have no impact on the states.

3   Cost Register VPA

Here we define a version of a visibly pushdown automaton by augmenting it with cost registers. A *cost register visibly pushdown automaton*, which we denote as CVPA, is a VPA which along with its visible stack has a finite set of registers, say X and an underlying algebra, say $\mathcal{A}$, which influences the computations performed by the registers. The automaton reads the input from left to right and depending on whether the letter read is a push/pop/internal letter, it updates its stack. This part of the working is exactly as in a VPA.

Additionally, there is a finite number of registers $X = \{x_1, \ldots, x_k\}$. While reading an internal letter the registers are updated as in the case of CRAs. While reading a call letter, the current instance of the register values is allowed to be saved on the stack as a vector $(x_1, \ldots, x_k)$, and the next set of registers are obtained from these. Here, we assume that registers are stored as formal expressions and their evaluations are performed as in the case of CRAs. Finally, while reading the return letter, the registers are updated using the registers available from the previous instance as well as by using the vector of registers on the top of the stack.

Definition 3 (Cost Register VPA (CVPA)). *A cost register visibly pushdown automaton is a tuple* $M = (\hat{\Sigma}, Q, q_0, \delta, \Gamma, \bot, \delta, X, \mathcal{A}, v_0, \rho, \mu)$, *where,* $\hat{\Sigma}$ *is the visible alphabet,* Q *is the finite set of states,* $q_0 \in Q$ *is the initial state,* Γ *is the finite stack alphabet with bottom-of-stack symbol* $\bot \in \Gamma$, $X = \{x_1, \ldots, x_k\}$ *is the finite set of registers,* $\mathcal{A} = (\mathbb{D}, \oplus_1, \ldots, \oplus_m)$ *is an algebra over the domain* $\mathbb{D}$, *and* $v_0\colon X \to \mathbb{D}$ *are the initial register values. Further,* δ *is the (deterministic) transition function similar to the transition function of a VPA,* $\delta\colon Q \times \Sigma_{\mathrm{int}} \to Q$, $\delta\colon Q \times \Sigma_{\mathrm{call}} \to Q \times \Gamma$, $\delta\colon Q \times \Sigma_{\mathrm{ret}} \times \Gamma \to Q$. *The partial register update function* ρ *is as follows:*

$- \rho\colon Q \times \Sigma_{\mathrm{int}} \times X \to E(\mathcal{A}, X_{prev})$

- $\rho\colon Q \times \Sigma_{\text{call}} \times X \to E(\mathcal{A}, X_{prev})$
- $\rho\colon Q \times \Sigma_{\text{ret}} \times \Gamma \times X \to E(\mathcal{A}, X_{prev}, X_{match})$

Here the place holders X_{prev} and X_{match} are both copies of X. $E(\mathcal{A}, X_{prev})$ is the set of expressions over $\mathcal{A}$ using the registers X_{prev} and $E(\mathcal{A}, X_{prev}, X_{match})$ is the set of expressions using registers in X_{prev} and X_{match}. Finally, $\mu\colon Q \to E(\mathcal{A}, X)$ is the partial final cost function.

A configuration of M is an element (q, γ, v), where q is a state, $\gamma \in \Gamma^*$ is the stack and $v\colon X \to E(\mathcal{A}, X)$ assigns each register as expression over $\mathcal{A}$. A run $(q_0, \gamma_0, v_0) \dots (q_n, \gamma_n, v_n)$ of M on some input $w \in \Sigma$ starts in $(q_0, \bot, v_0)$. The part of a configuration without the register (q_i, γ_i) is defined as in a VPA. The register values v_i are updated in the following way:

- If $w_i \in \Sigma_{\text{int}}$ or if $w_i \in \Sigma_{\text{call}}$ then for every $x \in X$, $v_i(x)$ is an expression obtained by substituting $v_{i-1}(y)$ in every occurrence of $y \in X$ in $\rho(q_i, w_i, x)$.
- If $w_i \in \Sigma_{\text{ret}}$ then for every $x \in X$, $v_i(x)$ is the expression obtained by substituting $v_{i-1}(y)$ (resp. $v_j(y)$) for every occurrence of $y \in X_{prev}$ (resp. $y \in X_{match}$) in $\rho(q_i, w_i, g, x)$, where $j < i$ is the matching position of the current index i and $g \in \Gamma$ is the top symbol of the stack content γ_i.

The semantics of a CVPA M is a function $F_M\colon \Sigma^* \to \mathbb{D}$ which is defined on words from $\hat{\Sigma}^*$ as the evaluation of the expression $\mu(q_n)$ by substituting the expression $v_n(y)$ for every occurrence of $y \in X$ in the expression $\mu(q_n)$. It is undefined if $\mu(q_n)$ is undefined. The set of all functions F_M computable by some CVPA M over an algebra $\mathcal{A}$ is denoted by CVPA($\mathcal{A}$). This completes the formal definition of CVPAs.

Remark 4. When specifying register updates we follow the following convention. If e.g. q is a state, a a letter and x a register, then $\rho(q, a, x) = x + x_{match}$ could be a register update rule. If on the right side a plain variable name appears it is meant to be part of X_{prev}. Variable names of the form x_{match} are part of X_{match}.

In the case of CRA interesting and natural subclasses have been defined by imposing a *copyless* restriction (CCRA). For CVPA we can also define a notion of copyless. The simple idea is that each register value may only be used once to calculate a new register value.

Definition 5 (Copyless CVPA (CCVPA)). *A CVPA is called copyless if the following conditions hold:*

- *In each register update expression each variable may only occur at most once.*
- *For any $a \in \Sigma_{\text{call}} \cup \Sigma_{\text{int}}$, variable $x \in X$ and state $q \in Q$, there exists at most one $y \in X$ such that $\rho(q, a, y)$ is an expression containing x.*
- *For any $b \in \Sigma_{\text{ret}}$, variable $x \in X$, state $q \in Q$ and $\gamma \in \Gamma$, there exists at most one $y \in X$ such that $\rho(q, b, \gamma, y)$ is an expression containing x.*
- *For any word $uavb \in \Sigma^*$ where avb is well-matched and $a \in \Sigma_{\text{call}}$ and $b \in \Sigma_{\text{ret}}$ let q_u (resp. q_{ua}, resp. q_{uav}) be the state after reading u (resp. ua, resp. uav), then if a variable x does not occur in $\rho(q_u, a, y)$ for all y there may exist one variable z such that $\rho(q_{uav}, b, \gamma, z)$ contains x_{match}, where $\delta(q_u, a) = (q_{ua}, \gamma)$.*

3.1 Examples

Example 6. Let $\hat{\Sigma} = \{a, b\}$, where $\Sigma_{\text{call}} = \{a\}, \Sigma_{\text{ret}} = \{b\}, \Sigma_{\text{int}} = \emptyset$. Consider the function $f\colon \{a, b\}^* \to \mathbb{N}$ which assigns each prefix of a well-matched word its maximal stack height of VPAs reading the word, e.g. $f(aababababaa) = 3$. This function can be implemented by a CVPA over $(\mathbb{Z}, +c, \max)$, where $+c$ is an operation only allowing adding a constant to a register. The underlying automaton is just a one-state automaton accepting everything. Besides that we need two registers, so $X = \{r, s\}$; further $v_0(r) = v_0(s) = 0$. The register update function is $\rho(q, a, r) = r + 1$, $\rho(q, b, r) = r - 1$ and $\rho(q, a, s) = \rho(q, b, s) = \max(s, r)$. The final output is $\mu(q) = s$.

Example 7. Arithmetic formulas over $(\mathbb{Z}, +, \times)$ can be evaluated by a $\text{CCVPA}(\mathbb{Z}, +, \times)$ machine. An arithmetic formula is e.g. $(-1 \times ((1 + 1) \times (1 + 1 + 1)))$ and it evaluates to -6. For the rest of the example we assume for convenience that the formulas are maximally parenthesized, i.e. we have $(1 + (1 + 1))$ instead of $(1 + 1 + 1)$. We can understand strings that are formulas as visibly words over the alphabet $\{-1, 1, (,), +, \times\}$, where (is a push letter,) a pop letter and $-1, 1, +$, and $\times$ are internal letters. However for the rest we write [and] instead of (and) for better readability.

We now give a formal description of the CVPA $M = (\hat{\Sigma}, Q, q_0, \Gamma, \delta, X, \mathcal{A}, v_0, \rho, \mu)$ evaluating formulas. There are states $Q = \{q_0, q_+, q_\times\}$ and as stack alphabet we introduce the symbols P, T and I (used to indicate that the push happened on the $q_+, q_\times, q_0$, resp.). The transition function is as follows:

- transitions from q_0: $\delta(q_0, [) = (q_0, I)$, $\delta(q_0, 1) = \delta(q_0, -1) = \delta(q_0,], I) = \delta(q_0,], P) = \delta(q_0,], T) = q_0$, $\delta(q_0, +) = q_+$, $\delta(q_0, \times) = q_\times$,
- transitions from q_+ on call and internal letters: $\delta(q_+, [) = (q_0, P)$, $\delta(q_+, 1) = \delta(q_+, -1) = q_0$,
- transitions from $q_\times$ on call and internal letters: $\delta(q_\times, [) = (q_0, T)$, $\delta(q_\times, 1) = \delta(q_\times, -1) = q_0$.

The initial value for x is 0, so $v_0(x) = 0$. The register update function can be described as follows: $\rho(q_0, [, x) = 0$, $\rho(q_0, 1, x) = 1$, $\rho(q_0, -1, x) = -1$, $\rho(q_0, +, x) = \rho(q_0, \times, x) = x$ and $\rho(q_0,], I, x) = x$, $\rho(q_0,], P, x) = x + x_{\text{match}}$ and $\rho(q_0,], T, x) = x \times x_{\text{match}}$. The other states: $\rho(q_+, [, x) = \rho(q_\times, [, x) = 0$, $\rho(q_+, 1, x) = x + 1$, $\rho(q_+, -1, x) = x + (-1)$, $\rho(q_\times, 1, x) = x \times 1$, $\rho(q_\times, -1, x) = x \times (-1)$. Finally we let $\mu(q_0) = x$.

Remark 8. Note that the automaton is copyless. In particular, while pushing we do not use x on the right side for the immediate update; it is only used when the matching return letter is read. Note also that this construction can be easily generalized from $\mathcal{A} = (\mathbb{Z}, +, \times)$ to arbitrary $\mathcal{A}$.

3.2 Language Theoretic Properties of CVPAs

In this section we study some language theoretic properties of CVPAs. The first simple observation is that by definition CVPAs (copyless CVPAs) generalize CRAs (copyless CRAs, resp.), just by not using the stack. We note this as an observation explicitly.

Observation 9. *Let* $f : \Sigma^* \rightarrow \mathbb{D}$ *be a function computed by a CRA over an algebra* $\mathcal{A}$, *then it can also be computed by a CVPA over the same algebra.*

To see that CVPAs generalize VPAs, we first prove a closure property of CVPAs. Given f and g in CVPA($\mathcal{A}$) and a visibly pushdown language L, then we define the function if L then f else g as mapping w to $f(w)$ if $w \in L$ and to $g(w)$ otherwise.

Lemma 10. *If* $f, g \in$ CVPA($\mathcal{A}$) *then* if L then f else g *is in* CVPA($\mathcal{A}$).

A similar result was proved for CRAs by [4]. Our result can be thought of a as a generalization to CVPAs. We can use the previous closure property to attain the characteristic function: if L then 1 else 0.

This gives that the characteristic function of L can be computed by a CVPA over the algebra $\mathbb{B}$. This therefore gives us the following corollary.

Corollary 11. *CVPA generalize VPA, i.e. for a given visibly pushdown language* L, *its characteristic function is in* CVPA($\mathbb{B}$).

Lemma 12. *Let* f_1, f_2 *be two functions over the same visible alphabet computable in* CVPA($\mathcal{A}$) *and* $\oplus$ *be a binary operator from* $\mathcal{A}$, *then* $f_1 \oplus f_2 \in$ CVPA($\mathcal{A}$).

3.3 Relationship Between CVPA($\mathbb{Z}, +c$) and CCVPA($\mathbb{Z}, +$)

In [4] the relationship between the copyless restriction and restrictions on the multiplication was investigated. In the case of CRA it holds that CRA($\mathbb{Z}, +c$) = CCRA($\mathbb{Z}, +$). Unfortunately this is not true for CVPA which can be seen by the following separating example. Let Σ consist of the letters $(,),], a$, where (is a call letter,),] are return letters and a is an internal letter. Consider all well-matched expressions over Σ. Let f be the function counting the number of a's which are immediately enclosed by (and). E.g. on $((aa(a)a)aa]$ the function f evaluates to 4 and on $(a(aa(aaa))]$ it evaluates to 5.

Lemma 13. *The function* f *defined above is in* CCVPA($\mathbb{Z}, +$) *but is not in* CVPA($\mathbb{Z} + c$).

It is easy to see that f has a CCVPA($\mathbb{Z}, +$). To prove that it does not have a CVPA($\mathbb{Z}, +c$), we make use of the fact that if the values can only be incremented or decremented by a constant, then the different configurations that all the registers can reach is limited, thereby proving the lower bound.

4 Comparing the Power of #VPA and CVPA

There are counting variants of nondeterministic VPA and NFA, called #VPA and #NFA. For a fixed VPA (resp. an NFA), the semantics then is a function $\Sigma^* \rightarrow \mathbb{N}$ which assigns each input the number of accepting runs. In [17] it was shown that #VPA is #$\mathbf{NC}^1$-complete, whereas #NFA is only complete for counting paths in bounded-width branching programs, which may be a weaker class [14]. This is interesting since their Boolean counterparts are known to have equal power. We first prove that the class of functions $\mathrm{CRA}(\mathbb{N}, +)$ and #NFA are equal. One direction of this is similar to [4]. We also prove the other direction. To the best of our knowledge this result has not been formally proved in any other work.

Lemma 14. *The set of functions* #NFA *equals* $\mathrm{CRA}(\mathbb{N}, +)$.

In the case of #VPA we need a CVPA with multiplication. Formally,

Lemma 15. *The functions in* #VPA *can be expressed as CVPA over* $(\mathbb{N}, +, \times)$ *which only use multiplication between a just computed register value and a register value from the stack, i.e. between* X_{prev} *and* X_{match} *variables.*

It remains open whether #VPA equals $\mathrm{CVPA}(\mathbb{N}, +)$.

A major motivation for considering cost register models are weighted automata. Weighted automata are NFA where each transition is assigned a weight. Weights are part of an algebra $(\mathbb{D}, \otimes, \odot)$ and in a run all weights are multiplied via $\otimes$. Then the result of the computation is the $\odot$-product of all runs. We consider now the example $(\mathbb{Z}, +, \min)$, which serves a good intuition for the usage of weighted automata as devices quantitatively searching for a "cheapest" run.

Consider a natural weighted variant of VPA over $(\mathbb{Z}, +, \min)$, in which each transition of δ is labeled with a weight and at the end of a computation the weight of the path with the smallest weight is the output.

Lemma 16. *Each weighted VPA function can be expressed as a CVPA over* $(\mathbb{Z}, +c, \min)$. *The resulting automaton does not store register values on the stack.*

5   Complexity Theoretic Results

In this section we present complexity theoretic results about CVPAs and CCVPAs over different algebras. It is known that the membership problem of regular and the visibly pushdown languages is $\mathbf{NC}^1$-complete [9,16]. It is also known that computing a certain bit of the image of a function in $\mathrm{CCRA}(\mathbb{Z}, +)$ or $\mathrm{CCRA}(\Gamma^*, \circ)$ is $\mathbf{NC}^1$-complete [3]. $\mathrm{CRA}(\mathbb{Z}, +)$ is $\mathrm{Gap}\mathbf{NC}^1$-complete [3]. $\mathrm{CCRA}(\mathbb{Z}, +, \times)$ is contained in $\mathrm{Gap}\mathbf{NC}^1$ [3]. #NFA is #BWBP-complete [14] and #VPA is #$\mathbf{NC}^1$-complete [17]. We add to this landscape a few more complexity results by studying CVPAs using the circuit complexity lense.

Theorem 17. *The set* $\mathrm{CVPA}(\mathbb{Z}, +)$ *is* $\mathrm{Gap}\mathbf{NC}^1$*-complete.*

We are given a CVPA M over $(\mathbb{Z}, +)$ and will show that F_M is in $\mathrm{Gap}\mathbf{NC}^1$. The completeness then follows because of the fact that CRA over $(\mathbb{Z}, +)$ are already $\mathrm{Gap}\mathbf{NC}^1$-complete.

Theorem 18. *The set* CCVPA$(\mathbb{Z}, +, \times)$ *is hard for* GapNC^1.

Proof. To prove the above statement, we need to show that any maximally bracketed arithmetic formula over $(\mathbb{Z}, +, \times)$ with inputs from the set $\{-1, 0, 1\}$ can be evaluated by CCVPA$(\mathbb{Z}, +, \times)$, as this is a GapNC^1 hard problem. Recall that the automaton described in Example 7 exactly performed this.   □

Note that a similar GapNC^1 hardness is not known for CCRA$(\mathbb{Z}, +, \times)$.

Theorem 19. *Let f be a function computed by a* CVPA$(\mathbb{Z}, +c)$ *then the ith bit of the output of f can be computed in* **NC**^{1}.

Proof (Sketch). Let $M = (\hat{\Sigma}, Q, q_0, \Gamma, \delta, X, \mathcal{A}, v_0, \rho, \mu)$ be a CVPA over $(\mathbb{Z}, +c)$. First thing to note is that computing the height of the stack at any step can be done in TC^0. (See for instance [17].) Let us denote the circuit that computes the height of the stack (more precisely, it decides whether the jth bit of the height reached after reading i bits of the input is 1 or not) as Height$_n$. Also, as membership testing for VPLs is in **NC**^{1} checking whether after having read $w_1 \ldots w_i$ the state reached is q or not can be done in **NC**^{1}. Let us call the circuit which does this for length n inputs the circuit State$_n$. Therefore, an entire run of the underlying VPA of M can be computed in **NC**^{1} (by creating $|w|$ many copies of State$_{|w|}$, one for each $1 \leq i \leq |w|$).

On $w = w_1 \ldots w_n$ suppose the run of the machine has the following states $q_0, q_1, \ldots, q_n$ and say $\mu(q_n) = x$ then we say that x is the *relevant register* at the nth step. In general, for $i < n$ if x was relevant at $i + 1$th step and at step i $\rho(q_{i-1}, w_i, x) = y + c$ was the register update function, then we say that y is the relevant register at step $i - 1$ and c is the relevant constant at step $i - 1$. Our proof proceeds in the following two steps: We first observe that for a given input the relevant registers and relevant constants can be computed in **NC**^{1}. We then show that once we have this information, computing the final value of the register can be done in **NC**^{1}.   □

References

1. Allender, E.: Arithmetic circuits and counting complexity classes. In: Krajek, J. (ed.) Complexity of Computations and Proofs, Quaderni di Matematica (2004)
2. Allender, E., Arvind, V., Mahajan, M.: Arithmetic complexity, kleene closure, and formal power series. Technical report (1997)
3. Allender, E., Mertz, I.: Complexity of regular functions. In: Dediu, A.-H., Formenti, E., Martín-Vide, C., Truthe, B. (eds.) LATA 2015. LNCS, vol. 8977, pp. 449–460. Springer, Heidelberg (2015)
4. Alur, R., D'Antoni, L., Deshmukh, J.V., Raghothaman, M., Yuan,Y.: Regular functions, cost register automata, and generalized min-cost problems. CoRR, abs/1111.0670 (2011)
5. Alur, R., D'Antoni, L., Deshmukh, J.V., Raghothaman, M., Yuan, Y.: Regular functions and cost register automata. In: 28th Annual ACM/IEEE Symposium on Logic in Computer Science, LICS 2013, New Orleans, LA, USA, 25–28 June 2013, pp. 13–22 (2013)

6. Alur, R., Madhusudan, P.: Visibly pushdown languages. In: Proceedings of the 36th Annual ACM Symposium on Theory of Computing, Chicago, IL, USA, 13–16 June 2004, pp. 202–211 (2004)

7. Barrington, D., Therien, D.: Finite monoids and the fine structure of NC1. In: Proceedings of the Nineteenth Annual ACM Symposium on Theory of Computing, STOC 1987, pp. 101–109, New York, NY, USA. ACM (1987)

8. Barrington, D.A.: Bounded-width polynomial-size branching programs recognize exactly those languages in NC1. In: Proceedings of the Eighteenth Annual ACM Symposium on Theory of Computing, STOC 1986, pp. 1–5, New York, NY, USA. ACM (1986)

9. Mix Barrington, D.A., Compton, K.J., Straubing, H., Thérien, D.: Regular Languages in NC^1. J. Comput. Syst. Sci. **44**(3), 478–499 (1992)

10. Bédard, F., Lemieux, F., McKenzie, P.: Extensions to Barrington's M-program model. Theor. Comput. Sci. **107**(1), 31–61 (1993)

11. Buss, S.R.: The boolean formula value problem is in ALOGTIME. In: Proceedings of the 19th Annual ACM Symposium on Theory of Computing, 1987, New York, New York, USA, pp. 123–131 (1987)

12. Buss, S.R., Cook, S.A., Gupta, A., Ramachandran, V.: An optimal parallel algorithm for formula evaluation. SIAM J. Comput. **21**(4), 755–780 (1992)

13. Caralp, M., Reynier, P.-A., Talbot, J.-M.: Visibly pushdown automata with multiplicities: finiteness and k-boundedness. In: Yen, H.-C., Ibarra, O.H. (eds.) DLT 2012. LNCS, vol. 7410, pp. 226–238. Springer, Heidelberg (2012)

14. Caussinus, H., McKenzie, P., Thérien, D., Vollmer, H.: Nondeterministic NC^1 computation. J. Comput. Syst. Sci. **57**(2), 200–212 (1998)

15. Cook, S.A.: A taxonomy of problems with fast parallel algorithms. Inf. Control **64**(13), 2–22 (1985). International Conference on Foundations of Computation Theory

16. Dymond, P.W.: Input-driven languages are in log n depth. Inf. Process. Lett. **26**(5), 247–250 (1988)

17. Krebs, A., Limaye, N., Mahajan, M.: Counting paths in VPA is complete for $\#NC^1$. Algorithmica **64**(2), 279–294 (2012)

18. Lange, K.-J.: Complexity and structure in formal language theory. Fundam. Inf. **25**(3,4), 327–352 (1996)

19. McKenzie, P., Péladeau, P., Therien, D.: NC1: the automata-theoretic viewpoint. Comput. Complex. **1**(4), 330–359 (1991)

20. Mehlhorn, K.: Pebbling moutain ranges and its application of dcfl-recognition. In: Automata, Languages and Programming, 7th Colloquium, Noordweijkerhout, The Netherland, 14–18 July 1980, Proceedings, pp. 422–435 (1980)

21. Venkateswaran, H.: Properties that characterize LOGCFL. J. Comput. Syst. Sci. **43**(2), 380–404 (1991)

22. Vollmer, H.: Introduction to Circuit Complexity - A Uniform Approach. Texts in Theoretical Computer Science. An EATCS Series. Springer, Heidelberg (1999)

Extending MSVL with Semaphore

Xinfeng Shu[1] and Zhenhua Duan[2]([✉])

[1] School of Computer Science and Technology,
Xi'an University of Posts and Communications, Xi'an 710061, China
shuxf@xupt.edu.cn
[2] Institute of Computing Theory and Technology, Xidian University,
Xi'an 710071, China
zhhduan@mail.xidian.edu.cn

Abstract. Modeling, Simulation and Verification Language (MSVL) is a useful formalism for specification and verification of concurrent systems. To make it more practical and easy to use, we extend MSVL with the technique of semaphore. To do so, the mechanism of MSVL function calls is deeply analyzed. Further, the semaphore type is defined. Moreover, operations over semaphore are formalized. Finally, an example is given to illustrate how to use semaphore to solve the mutual exclusion problem.

Keywords: Temporal logic programming · Projection · Semaphore · Mutual exclusion · Concurrency

1 Introduction

Modeling, Simulation and Verification Language (MSVL) [1], an executable subset of Projection Temporal Logic (PTL) [2,3] with framing technique, is a useful formalism for specification and verification of concurrent and distributed systems [4–6]. It provides a rich set of data types (e.g., char, integer, pointer, string), data structures (e.g., array, list), as well as boolean and arithmetic expressions. Besides, MSVL supports not only the commonly used statements such as assignment, sequential, branch and loop, but also parallel and concurrent statements such as conjunct (S_1 and S_2), parallel ($S_1 \| S_2$) and projection (($S_1, \ldots, S_m$) prj S). Further, Propositional Projection Temporal Logic (PPTL), the propositional subset of PTL, has the expressiveness power of the full regular expressions [7], which enables us to model, simulate and verify the concurrent and reactive systems within a same logical system [8].

In verification of concurrent and distributed systems, an essential problem that must be dealt with is the synchronization and communication between concurrent processes. To solve the problem, some formalisms involve synchronous

This research is supported by Natural Science Foundation of Education Bureau of Shaanxi Province (No. 11JK1037), NSFC Grant Nos. 61133001, 61420106004, 91418201 and 61322202.

T.N. Dinh and M.T. Thai (Eds.): COCOON 2016, LNCS 9797, pp. 599–610, 2016.
DOI: 10.1007/978-3-319-42634-1_48

message passing (e.g., CCS [9] and CSP [10]), and some involve asynchronous channels (e.g., PROMELA [11]). As for MSVL, the communication between parallel components is based on shared variables. Furthermore, MSVL provides a synchronization construct, $await(c)$, to synchronize communication between parallel processes. The meaning of $await(c)$ is simple: it changes no variables, but keeps on waiting until the condition c becomes $true$, at which point it terminates. With this statement, the synchronization between two parallel processes can be easily achieved since another process can cause c to become $true$.

However, the mutual exclusively accessing critical resource for many concurrent processes has not solved in MSVL so far. Therefore, we are motivated to introduce the technique of semaphore [12] to MSVL. To this end, the mechanism of the function of MSVL is deeply analyzed, based on which the semaphore type is defined and the functions to initialize a semaphore variable, allocate as well as release a critical resource are also formalized. Besides, an example is given to illustrate how to use the semaphore of MSVL to solve the synchronization and mutual exclusion problem between currently processes.

The rest of paper is organized as follows. In the next section, PTL and MSVL are briefly introduced. In Sect. 3, the mechanism of MSVL functions calls is analyzed. In Sect. 4, the semaphore is introduced to MSVL. In Sect. 4.1, an example is given to illustrate how to program with semaphore. Finally, conclusions are drawn in Sect. 5.

2   Preliminaries

2.1   Projection Temporal Logic

In this subsection, the syntax and semantics of Projection Temporal Logic (PTL) are briefly introduced. More details can be found in paper [2].

Syntax. Let $Prop$ be a countable set of atomic propositions and V a countable set of typed variables. $B = \{true, false\}$ represents the boolean domain. D denotes the data domain of the underlying logic. The terms e and formulas P of PTL are inductively defined as follows:

$e ::= d \mid a \mid x \mid \bigcirc e \mid f(e_1, \ldots, e_m)$

$P ::= p \mid e_1 = e_2 \mid \rho(e_1, ..., e_m) \mid \neg P \mid P_1 \wedge P_2 \mid \exists v P \mid \bigcirc P \mid (P_1, \ldots, P_m) \, prj \, P$

where $d \in D$ is a constant, $a \in V$ is a static variable, $x \in V$ is a dynamic variable, $v \in V$ is either a static variable or a dynamic one; $p \in Prop$ is an atomic proposition; f is a function and ρ is a predicate both defined over D.

Abbreviation. The conventional constructs $true$, $false$, $\wedge$, $\rightarrow$ as well as $\leftrightarrow$ are defined as usual. Furthermore, we use the following abbreviations:

$$\varepsilon \stackrel{def}{=} \neg\bigcirc true \qquad\qquad \overline{\varepsilon} \stackrel{def}{=} \neg \varepsilon$$
$$\odot P \stackrel{def}{=} \neg\bigcirc\neg P \qquad\qquad P;Q \stackrel{def}{=} (P, Q) \, prj \, \varepsilon$$
$$\Diamond P \stackrel{def}{=} true; P \qquad\qquad len(n) \stackrel{def}{=} \bigcirc^n \varepsilon$$
$$\Box P \stackrel{def}{=} \neg\Diamond\neg P \qquad\qquad keep(P) \stackrel{def}{=} \Box(\overline{\varepsilon} \rightarrow P)$$
$$skip \stackrel{def}{=} \bigcirc \varepsilon \qquad\qquad halt(P) \stackrel{def}{=} \Box(\varepsilon \leftrightarrow P)$$

$$\forall v P \stackrel{\text{def}}{=} \neg \exists v \neg P \qquad\qquad fin(P) \stackrel{\text{def}}{=} \Box(\varepsilon \to P)$$

$$P \| Q \stackrel{\text{def}}{=} ((P;true) \wedge Q) \vee (P \wedge (Q;true)) \vee (P \wedge Q)$$

Semantics. A state s is a pair of assignments (I_p, I_v), which I_p assigns each atomic proposition $p \in Prop$ a truth value in B, whereas I_v assigns each variable $v \in V$ a value in D. An interval (i.e., model) σ is a non-empty sequence of states. The length of σ, denoted by $|\sigma|$, is ω if σ is infinite, or the number of states minus one if σ is finite. We use notation $\sigma_{(i..j)}$ to mean that a subinterval $<s_i, \ldots, s_j>$ of σ with $0 \leq i \preceq j \leq |\sigma|$. The *concatenation* of a finite interval $\sigma = <s_0, \ldots, s_{|\sigma|}>$ with another interval $\sigma' = <s_0', \ldots, s_{|\sigma'|}'>$ (may be infinite) is denoted by $\sigma \bullet \sigma'$ and $\sigma \bullet \sigma' = <s_0, \ldots, s_{|\sigma|}, s_0', \ldots, s_{|\sigma'|}'>$. Let $\sigma = <s_0, s_1, \ldots, s_{|\sigma|}>$ be an interval and $r_1, \ldots, r_h$ be integers $(h \geq 1)$ such that $0 \leq r_1 \leq r_2 \leq \ldots \leq r_h \preceq |\sigma|$. The projection of σ onto $r_1, \ldots, r_h$ is the interval (called projected interval) $\sigma \downarrow (r_1, \ldots, r_h) = <s_{t_1}, \ldots, s_{t_l}>$, $(t_1 < t_2 < \ldots < t_l)$, where $t_1, \ldots, t_l$ is obtained from $r_1, \ldots, r_h$ by deleting all duplicates. For example, $<s_0, s_1, s_2, s_3, s_4, s_5> \downarrow (0, 2, 2, 2, 4, 4, 5) = <s_0, s_2, s_4, s_5>$.

An interpretation, as for PTL, is a triple $\mathcal{I} = (\sigma, i, j)$, where σ is an interval, $i \in N_0$ and $j \in N_\omega$, and $0 \leq i \preceq j \leq |\sigma|$. We use notation (σ, i, j) to mean that a term or a formula is interpreted over a subinterval $<s_i, \ldots, s_j>$ of σ with the current state being s_i. Then, for every term e, the evaluation of e relative to $\mathcal{I}$, denoted by $\mathcal{I}[e]$, is defined by induction on the structure of the term as follows:

$$
\begin{aligned}
\mathcal{I}[d] &= d, \text{ if } d \in D \text{ is a constant value} \\
\mathcal{I}[a] &= I_v^i[a] = I_v^0[a], \text{ if } a \text{ is typed static variable} \\
\mathcal{I}[x] &= I_v^i[x], \text{ if } x \text{ is typed dynamic variable} \\
\mathcal{I}[\bigcirc e] &= \begin{cases} (\sigma, i+1, j)[e], \text{ if } i < j \\ nil, \qquad\qquad \text{ otherwise} \end{cases} \\
\mathcal{I}[f(e_1, \ldots, e_m)] &= \begin{cases} nil, \text{ if } \mathcal{I}[e_h] = nil \text{ for some } h(1 \leq h \leq m) \\ f(\mathcal{I}[e_1], \ldots, \mathcal{I}[e_m]), \text{ otherwise} \end{cases}
\end{aligned}
$$

The satisfaction relation $(\models)$ for PTL formulas is inductively defined as follows:

$\mathcal{I} \models p$ iff $I_p^i[p] = true$, for any given atomic proposition p

$\mathcal{I} \models \rho(e_1, \ldots, e_m)$ iff ρ is a primitive predicate other than $=$ and, for all $h(1 \leq h \leq m)$, $\mathcal{I}[e_h] \neq nil$ and $\rho(\mathcal{I}[e_1], \ldots, \mathcal{I}[e_m]) = true$

$\mathcal{I} \models e_1 = e_2$ iff $\mathcal{I}[e_1] = \mathcal{I}[e_2]$

$\mathcal{I} \models \neg P$ iff $\mathcal{I} \not\models P$

$\mathcal{I} \models P \wedge Q$ iff $\mathcal{I} \models P$ and $\mathcal{I} \models Q$

$\mathcal{I} \models \exists v P$ iff $(\sigma', i, j) \models P$ for some interval σ', $\sigma_{(i..j)} \stackrel{v}{=} \sigma'_{(i..j)}$

$\mathcal{I} \models \bigcirc P$ iff $i < j$ and $(\sigma, i+1, j) \models P$

$\mathcal{I} \models (P_1, \ldots, P_m) \, prj \, Q$ iff there exist integers $i = r_0 \leq \ldots \leq r_{m-1} \leq r_m \preceq j$ such that $(\sigma, r_{l-1}, r_l) \models P_l$ for all $1 \leq l \leq m$, and $(\sigma', 0, |\sigma'|) \models Q$ for one of the following σ':
    (1) $r_m < j$ and $\sigma' = \sigma \downarrow (r_0, \ldots, r_m) \bullet \sigma_{(r_m+1..j)}$
    (2) $r_m = j$ and $\sigma' = \sigma \downarrow (r_0, \ldots, r_h)$ for some $0 \leq h \leq m$

2.2   Modeling, Simulation and Verification Language

Modeling, Simulation and Verification Language (MSVL) is an executable subset of PTL. In the following, we briefly introduce the kernel of MSVL. For more deals, please refer to literature [1].

Expression. The arithmetic expressions e and boolean expressions b of MSVL are inductively defined as follows:

$$e ::= d \mid x \mid \bigcirc e \mid \ominus e \mid e_1 + e_2 \mid e_1 - e_2 \mid e_1 * e_2 \mid e_1/e_2 \mid e_1 \% e_2$$
$$b ::= true \mid false \mid \neg b \mid b_1 \wedge b_2 \mid e_1 = e_2 \mid e_1 \leq e_2$$

where d is is an integer or a floating point number; $x \in V$ is a static or dynamic variable; $\bigcirc e$ ($\ominus e$) refers to the value of expression e at the next (previous) state.

Statement. The elementary statements in MSVL are defined as follows:

(1) Immediate Assign $x \Leftarrow e \overset{\text{def}}{=} x = e \wedge p_x$

(2) Unit Assignment $x := e \overset{\text{def}}{=} \bigcirc x = e \wedge \bigcirc p_x \wedge skip$

(3) Conjunction $S_1 \ and \ S_2 \overset{\text{def}}{=} S_1 \wedge S_2$

(4) Selection $S_1 \ or \ S_2 \overset{\text{def}}{=} S_1 \wedge S_2$

(5) Next $next \ S \overset{\text{def}}{=} \bigcirc S$

(6) Always $always \ S \overset{\text{def}}{=} \Box S$

(7) Termination $empty \overset{\text{def}}{=} \neg \bigcirc true$

(8) Skip $skip \overset{\text{def}}{=} \bigcirc \varepsilon$

(9) Sequential $S_1 ; S_2 \overset{\text{def}}{=} (S_1, S_2) \ prj \ \varepsilon$

(10) Local $exist \ x : S \overset{\text{def}}{=} \exists x : S$

(11) State Frame $lbf(x) \overset{\text{def}}{=} \neg af(x) \rightarrow \exists b : (\ominus x = b \ \wedge x = b)$

(12) Interval Frame $frame(x) \overset{\text{def}}{=} \Box(\overline{\varepsilon} \rightarrow \bigcirc(lbf(x)))$

(13) Projection $(S_1, \ldots, S_m) \ prj \ S$

(14) Condition $if \ b \ then \ S_1 \ else \ S_2 \overset{\text{def}}{=} (b \rightarrow S_1) \wedge (\neg b \rightarrow S_2)$

(15) While $while \ b \ do \ S \overset{\text{def}}{=} (b \wedge S)^\star \wedge \Box(\varepsilon \rightarrow \neg b)$

(16) Await $await(b) \overset{\text{def}}{=} \bigwedge_{x \in V_b} frame(x) \wedge \Box(\varepsilon \leftrightarrow b)$

(17) Parallel $S_1 \| S_2 \overset{\text{def}}{=} ((S_1 ; true) \wedge S_2) \vee (S_1 \wedge (S_2 ; true))$
$\vee (S_1 \wedge S_2)$

where the immediate assignment $x \Leftarrow e$, unit assignment $x \Leftarrow e$, $empty$, $lbf(x)$ and $frame(x)$ are basic statements, and the left are composite ones.

3   Mechanism of MSVL Function

For convenience of modeling for complex software and hardware systems, MSVL takes the divide-and-conquer strategy and employees functions as the basic components like C programming language does. The general grammar of MSVL function is as follows:

$$function\ funcName(in\_type_1\ x_1, \ldots, in\_type_m\ x_m,$$
$$out\_type_1\ y_1, \ldots, out\_type_n\ y_m, return\_type\ RValue)$$
$\{\ S\ \}$ //Function body

where *function* is the keyword to declare a function; funcName is the identifier by which the function can be called; in_type_i x_i (out_type_i y_i) (as many as needed) specifies the i-th input (output) parameter consisting of a type followed by a variable identifier; *return_type* specifies the return type of the function; S, usually a compound MSVL statement, defines the operations inside the function. A function with no input (output) parameters or return value is allowed.

Parameter passing in MSVL is similar to that in C, i.e. all function arguments are passed by values (call-by-value). With call-by-value, the actual argument expression is evaluated, and the resulting value is bound to the corresponding formal parameter in the function. Even if the function may assign a new value to its formal parameter, only its local copy is assigned and anything passed into a function call is unchanged in the callers scope when the function returns. Furthermore, the pointer type is also supported by MSVL, which allows both caller and callee to access and modify a same variable.

To make MSVL more practical and useful, MSVL provides two kinds of function calls, namely internal call and external call [13]. The grammar of the internal call is the *default one*, i.e. $funcName(v_1, \ldots, v_n)$, and the grammar of external call is the general function call statement with the prefixed constraint *ext*, i.e. *ext* $funcName(v_1, \ldots, v_n)$.

For instance, Example 1 is an MSVL program to compute $(1+2+3)*2$. The program consists of two functions *main* and *GetSum1*, which function *main* is the entry of the program and function *GetSum1* is to compute the sum of $1 + .. + n$. Within function *main*, the function call statement marked with (1) is an internal call, whereas the one marked with (2) is an external call.

Example 1. Program to compute $(1+2+3)*2$

```
function GetSum(int n, int *rst) {
    frame(i) and (
        int i and i<== 1 and empty;
        *rst<== 0 and empty;
        while(i<= 3){ *x:= *x+i and i:=i+1 }
    )
};
function main(){
    frame(sum) and (
        int sum and sum<== 0 and skip;
        GetSum(3, &sum);        //(1) Internal function call
        ext GetSum(3, &sum);    //(2) External function call
        sum:=sum*2
    )
};
main()
```

In the following, we make a deep analysis on the difference between the two function calls with Example 1.

Internal Function Call. Internal call means the execution of the called function is transparent to the calling function. The whole model of the calling function is the concatenation of the sub-model of the statements before the function call, the sub-model of the called function and the sub-model of statements after the function call.

For instance, if we remove the statement (2) from the program in Example 1, the execution model of the left program is given in Fig. 1, where the states of the model is marked as the labels on the edges between the nodes. It is the concatenation of the sub-models of statement "int sum and sum<== 0 and skip", function $GetSum1$, and statement "sum:=sum*2".

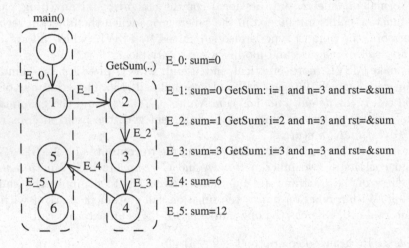

Fig. 1. The execution model of internal function call

Internal function call has a better support for modeling concurrent systems. The basic idea is to describe each concurrent component by an MSVL function and execute them in parallel. The interaction among concurrent components can be realized by shared variables among functions. However, internal call also has its negative side. The parallel execution of functions may lead to the functions affecting each other and getting wrong result. So, we must be very careful to estimate and avoid the error interactions among parallel functions in system modeling.

External Function Call. External call means the execution of the called function is unseen to the calling function. From the view point of the function caller, it focuses on obtaining the computing result and ignores the model over which the called function is executed. The execution of an external called function is completely isolated from the function caller.

For instance, if we remove the statement (1) from the program in Example 1, the execution model of the left program is given in Fig. 2. Although function $GetSum()$ takes 4 states to compute $(1 + 2 + 3)$, the model of $GetSum()$ is abandoned in case of the computing result is obtained. Thus, the execution of external call nearly has no affection on the model of the function caller except for assigning the result 6 to the variable sum.

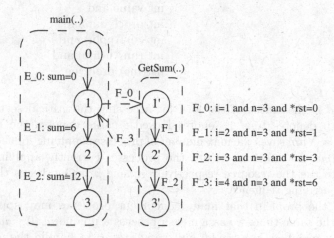

Fig. 2. The execution model of external function call

The external function call completely encapsulates the inner variables (data) and the program logic of the called function, which provides a new system modeling way at a higher level of abstraction. Compared with the internal function call, the external call is used to model the sub-system which has no interaction with function caller during its execution. In most software/hardware system, such sub-systems is the major part. The external function call greatly decreases the complexity of system modeling and helps to ensure the correctness of the system model.

4 Introduction of Semaphore to MSVL

The semaphore [12] is a key technique widely used in concurrent system development, e.g. Operating System and Web Application, to provide mutual exclusion accessing critical resource for many concurrent processes. Intuitively, a semaphore is the entity representing a kind of critical resources. Further, two atomic operations (i.e., a sequence of instructions providing a specific function and its running cannot be interrupted.) *wait* and *signal* are defined over semaphore to allocate and release a unit of resource respectively. In compute system, the realization of atomic operation must be supported by hardware. As for MSVL, we have no such support and hence can only search for an alternative approach to realize semaphore. In the following, we firstly give the definition of semaphore, and then formalize the operations over it.

Semaphore. Let $n \in N_0$ be the number of processes using the semaphore. The *semaphore* is a struct type defined in MSVL as follows:

$$
\text{semaphore}(n) \overset{\text{def}}{=} \text{struct \{}
$$

$$
\begin{aligned}
&\text{int locked and} \\
&\text{int value and} \\
&\text{int procNum and} \\
&\text{int curApp[n] and} \\
&\text{int runAuth[n] and} \\
&\text{list(int) procQue}
\end{aligned}
$$

$$
\text{\}}
$$

where *locked* is the status of the lock for accessing the critical section of the semaphore (0 denotes free, 1 denotes locked); *value* denotes the number of resources; *procNum* saves the max number of processes applying for the resource; arrays *curApp* and *runAuth* record the processes currently applying for and authorized to use the resource respectively; queue *procQue* keeps the blocked processes in arriving sequence.

To solve the problem that many concurrent processes may apply for the resource at the same time, we assign each process a unique *id* $(0 \leq id < n)$ and employ array members *curApp[id]* and *runAuth[id]* to handle the application and authorisation of process *id* respectively.

Semaphore is a parameterized type. To define a semaphore variable, the number of processes to use the resource must be estimated previously. For example, if the process number is 10, then the grammar to define a variable *sem* is: semaphore (10) *sem*.

Before using a semaphore variable, we need call function *sem_init* to initialize it. Function *sem_init* has three parameters, among which *sem* is the semaphore to be initialized; *value* is the initial resource count; *procNum* is the total number of processes using the resource. The function is defined as follows.

```
function sem_init( semaphore(n) *sem, int value, int procNum){
    frame(i) and (
        int i and i<==0 and empty;
        sem→value<==value and empty  //Init resource number
            sem→locked<== 0 and          //Init lock status
            sem→procNum<==procNum ; //Init process number
        while(i < n) {   //Init applying array and authorized array
            sem→curApp[i]:=0 and sem→runAuth[i]:=0;
            i:=i+1
        }
    )
};
```

Function *sem_acquire* corresponds to *wait* operation and allocate one unit of resource to the applier. The function has two parameters: *sem* is the resource related semaphore; *id* is the identifier of the process applying for the resource. It firstly sets the applying flag and running authorisation flag of process *id* to 1 and 0 respectively, and then calls the function *_sem_lock* to lock the semaphore and mutual exclusively enters the critical section. Subsequently, the function minuses $sem \rightarrow value$ by 1 and checks the result. If $sem \rightarrow value < 0$, which means there is no resource left, then it adds the current process *id* to the tail of the blocked process queue *procQue*, frees the semaphore lock(i.e. $sem \rightarrow locked = 0$) and waits for other process to wake up process *id*(i.e. $await(sem \rightarrow runAuth[id] = 1)$. Otherwise, it frees the semaphore lock and uses the resource directly. The definition of function *sem_acquire* is as follows.

```
function sem_acquire( semaphore(n) *sem, int id) {
    sem→curApp[id]:= 1 and        //Set applying flag
        sem→runAuth[id]:= 0;      //Set running authorisation flag
    _sem_lock(sem, id);           //Lock the semaphore
    sem→value:=sem→value-1;
    if(sem→value< 0) then{
        sem→procQue.addtail(id);  //Add to waiting queue
        sem→locked=0;             //Free the semaphore lock
        await(sem→runAuth[id]= 1)
    }else {
        sem→locked=0              //Free the semaphore lock
    }
};
```

Function *sem_release* corresponds to *signal* operation and releases a resource. The function firstly sets the applying flag of process *id* to 1, and then calls the function *_sem_lock* to mutual exclusively enters the critical section. Subsequently, the function increases $sem \rightarrow value$ by 1 and checks the result. If $sem \rightarrow value <= 0$, which means there exists some processes blocked in the queue *procQue*, then it removes the first one from the head of the queue (i.e. $sem \rightarrow procQue.removehead(\&idWake)$) and wakes up it (i.e. $sem \rightarrow runAuth[idWake] := 1$). Finally, it frees the semaphore lock. The definition of function *sem_release* is as follows.

```
function sem_release( semaphore(n) *sem, int id){
    frame(idWake) and (          //ID of the process to be wake up
        int idWake and idWake<== 0 and empty;
        sem→curApp[id]:= 1;      //Set the applying flag
        _sem_lock(sem, id);
        sem→value:=sem→value+1;
        if(sem→value<= 0) then{
            sem→procQue.removehead(&idWake);
            sem→runAuth[idWake]:= 1;
        };
        sem→lock=0               //Free the semaphore lock
    )
};
```

Function _sem_lock is used to identify which process can enter the critical section of the semaphore. The function takes the FCFS (First Come, First Service) strategy to select a process. To this end, it calls function _sem_select with the **external call** to select a process. If current process id is selected, it locks the semaphore (i.e. $sem \rightarrow locked <== 1$) and removes the applying flag of process id (i.e. $sem \rightarrow curApp[id] := 0$). Otherwise, the function increases the $sem \rightarrow curApp[id]$ by 1 to promote its priority to enter the critical section. The definition of function _sem_lock is as follows.

```
function _sem_lock(semaphore(n) *sem, int id){
    frame(idSel) and (
        int idSel and idsel<== −1 and empty;
        while(idSel ! = id) {
            ext _sem_select(sem, &idSel);
            if(idSel=id) then{
                sem→locked<== 1 and empty; //Lock the semaphore
                sem→curApp[id]:= 0           //Remove applying flag
            }else{
                sem→curApp[id]:= sem→curApp[id]+1
            }
        }
    )
};
```

Function _sem_select is used to select the process with the maximum value in the semaphore's applying array curApp. If the semaphore is locked, then no process is selected (i.e. $*idSel := -1$). Otherwise, the function traverses array curApp and finds the maximum one. The definition of _sem_wait is as follows.

```
function _sem_select( semaphore(n) *sem, int *idSel){
    frame(i, max) and (
        int i and i<== 0 and empty;
        int max and max<== 0 and skip;
        if(sem→locked=1) then {
            *idSel := −1
        } else {
            while(i<sem→procNum ) {
                if( sem→curApp[i]> sem→curApp[max]) then{
                    max := i
                };
                i := i+1
            };
            *idSel := max
        }
    )
};
```

4.1   Application of Semaphore

In following, we give an example to illustrate how to employ the technique of semaphore to solve the complex Producer-consumer problem [14]. Without loss of generality, we assume the size of the buffer is 10, and the number of producers and consumers both be 2. In such a problem, there exist 3 kinds of critical resources, i.e., the buffer space, the product and the buffer itself. For each critical resource, we define a semaphore variable, namely $semSpace$, $semProd$ and $semBuf$, and their initial resource numbers are $10, 0, 1$ respectively. The full MSVL program is given in Example 2.

Example 2. Solve Producer-consumer problem using semaphore

```
function Producer( semaphore(4) *sSpace, semaphore(4) *sProd,
                   semaphore(4) *sBuf, list(int) *buf, int id ){
    while(true) {
        sem_acquire(sSpace, id);    //Acquire a buffer space
        sem_acquire(sBuf, id);      //Acquire buffer
        buf→addtail(100);           //Imitate putting product into the buffer
        sem_release(sBuf, id);      //Release buffer
        sem_release(sProd, id);     //Release a product
    }
};
function Consumer( semaphore(4) *sSpace, semaphore(4) *sProd,
                   semaphore(4) *sBuf, list(int) *buf, int id ) {
    while(true) {
        sem_acquire(sProd, id);     //Acquire a product
        sem_acquire(sBuf, id);      //Acquire buffer
        buf→removehead();           //Imitate getting product from the buffer
        sem_release(sBuf, id);      //Release buffer
        sem_release(sSpace, id);    //Release a buffer space
    }
};
function main(){
    frame(semSpace, semProd, semBuf, buffer) and (
        semaphore(4) semSpace and sem_init(&semSpace, 10, 4);
        semaphore(4) semProd and sem_init(&semProd, 0, 4);
        semaphore(4) semBuf and sem_init(&semBuf, 1, 4);
        list(int) buffer;
        Producer(&semSpace, &semProd, &semBuf, &buffer, 0 )
            || Producer(&semSpace, &semProd, &semBuf, &buffer, 1 )
            || Consumer(&semSpace, &semProd, &semBuf, &buffer, 2 )
            || Consumer(&semSpace, &semProd, &semBuf, &buffer, 3 )
    )
};
main()
```

5   Conclusion

In this paper, we extend MSVL by introducing the technique of semaphore. The new semaphore type is defined in MSVL, and three operators *sem_init*, *sem_acquire* and *sem_release* are formalized to initialize a semaphore variable, acquire a resource and release a resource respectively. With the support of semaphore, MSVL can easily solve the synchronization, communication, and mutual exclusion problems between currently processes. In the future, we will apply MSVL to model, simulate and verify more complex concurrent and distributed systems, e.g. Operating System and Service Oriented System.

References

1. Duan, Z., Yang, X., Koutny, M.: Framed temporal logic programming. Sci. Comput. Program. **70**(1), 31–61 (2008)
2. Duan, Z.: Temporal Logic and Temporal Logic Programming. Science Press, Beijing (2005)
3. Duan, Z., Tian, C., Zhang, L.: A decision procedure for propositional projection temporal logic with infinite models. Acta Inf. **45**(1), 43–78 (2008)
4. Wang, M., Duan, Z., Tian, C.: Simulation and verification of the virtual memory management system with MSVL. In: Proceedings of the 2014 IEEE 18th International Conference on Computer Supported Cooperative Work in Design (CSCWD), pp. 360–365, May 2014
5. Yu, Y., Duan, Z., Tian, C., Yang, M.: Model checking C programs with MSVL. In: Liu, S. (ed.) SOFL 2012. LNCS, vol. 7787, pp. 87–103. Springer, Heidelberg (2013)
6. Ma, Q., Duan, Z., Zhang, N., Wang, X.: Verification of distributed systems with the axiomatic system of MSVL. Formal Asp. Comput. **27**(1), 103–131 (2015)
7. Tian, C., Duan, Z.: Expressiveness of propositional projection temporal logic with star. Theor. Comput. Sci. **412**(18), 1729–1744 (2011)
8. Duan, Z., Tian, C.: A unified model checking approach with projection temporal logic. In: Liu, S., Araki, K. (eds.) ICFEM 2008. LNCS, vol. 5256, pp. 167–186. Springer, Heidelberg (2008)
9. Milner, R.: Communication and Concurrency. Prentice Hall, London (1989)
10. Hoare, C.A.R.: Communicating Sequential Processes. Prentice Hall, London (1985)
11. Holzmann, G.J.: The model checker SPIN. IEEE Trans. Softw. Eng. **23**(5), 279–295 (1997)
12. Dijkstra, E.W.: Over de sequentialiteit van procesbeschrijvingen (EWD-35). E.W. Dijkstra Archive. Center for American History, University of Texas at Austin
13. Zhang, N., Duan, Z., Tian, C.: Extending MSVL with function calls. In: Merz, S., Pang, J. (eds.) ICFEM 2014. LNCS, vol. 8829, pp. 446–458. Springer, Heidelberg (2014)
14. Arpaci-Dusseau, R.H.: Operating Systems: Three Easy Pieces [Chapter: Condition Variables]. Arpaci-Dusseau Books (2014)

Satisfiability of Linear Time Mu-Calculus on Finite Traces

Yao Liu[1], Zhenhua Duan[1(✉)], Cong Tian[1(✉)], and Bin Cui[2]

[1] ICTT and ISN Laboratory, Xidian University,
Xi'an 710071, People's Republic of China
yao_liu@stu.xidian.edu.cn, {zhhduan,ctian}@mail.xidian.edu.cn
[2] Key Lab of High Confidence Software Technologies (MOE), School of EECS,
Peking University, Beijing 100871, People's Republic of China
bin.cui@pku.edu.cn

Abstract. In this paper, we study linear time μ-calculus interpreted over finite traces, namely $\nu\mathrm{TL}_f$. We define Present Future form (PF form) for $\nu\mathrm{TL}_f$ formulas and prove that every closed $\nu\mathrm{TL}_f$ formula can be converted into this form. PF form decomposes a formula into two parts: what to be satisfied at the current state and what to be satisfied at the next one. Based on PF form, we provide an algorithm for constructing Present Future form Graph (PFG) that can be employed to depict models of a formula. In addition, a decision procedure for checking satisfiability of $\nu\mathrm{TL}_f$ formulas based on PFG is proposed.

Keywords: Linear time μ-calculus · Finite traces · Present future form graph · Decision procedure · Satisfiability

1 Introduction

Linear Temporal Logic (LTL) is a well-known formalism that can be employed to specify and verify various properties of concurrent systems [21]. Besides, it has been widely used for representing temporally extended goals in planning [1,6,9,20,24] due to its simplicity. Compared with standard LTL interpreted over infinite traces, we are typically interested in finite traces when planning for temporally extended goals. For example, regarding temporally extended goals as finite desirable traces of states, a plan is called correct if its execution succeeds in generating one of these desirable traces. LTL over finite traces, namely LTL_f, is presented in [10] for describing temporal goals. While the expressive power of LTL is (ω-)star-free regular, the expressive power of LTL_f is star-free regular. Due to the restriction of expressive power, many properties are not expressible in LTL_f, e.g. an atomic proposition p holds on every even position. To this end, we investigate linear time μ-calculus ($\nu\mathrm{TL}$) [3] interpreted over finite traces, namely $\nu\mathrm{TL}_f$.

This research is supported by the NSFC Grant Nos. 61133001, 61322202, 91418201, and 61420106004.

T.N. Dinh and M.T. Thai (Eds.): COCOON 2016, LNCS 9797, pp. 611–622, 2016.
DOI: 10.1007/978-3-319-42634-1_49

νTL extends LTL with least and greatest fixpoint operators. Its expressive power is ω-regular [15] with the syntax remaining simple. We focus here on the satisfiability problem of νTL$_f$ formulas. By satisfiability we mean the problem to find a decision procedure for checking if a formula is satisfiable. The satisfiability problem is a fundamental issue in temporal logic and decision procedures for checking satisfiability always play a crucial role in yielding model checking [7] algorithms.

In this paper, we define Present Future form (PF form) for νTL$_f$ formulas and show that every closed νTL$_f$ formula can be converted into this form. Also, we give an algorithm for transforming νTL$_f$ formulas into PF form. PF form decomposes a formula into the present and future parts. The present part is the conjunction of atomic propositions or their negations, stating what should be satisfied at the current state. The future part is the conjunction of elements in the closure of a given formula, stating what should be satisfied at the next state. In addition, based on PF form, we present the notion of Present Future form Graph (PFG), which can be employed to depict models of a formula, and an algorithm for constructing PFGs. Further, we reduce the satisfiability problem of a νTL$_f$ formula to a path searching problem from its PFG. The complexity analysis result shows that the PFG-based decision procedure for checking satisfiability of νTL$_f$ formulas can be accomplished in $2^{O(|\phi|)}$.

In our previous work [19], we have presented Goal Progression Form (GPF) for a fragment $\mathcal{G}_\mu$ of νTL formulas which breaks a $\mathcal{G}_\mu$ formula into the present and future parts. Compared with [19], we consider the full νTL interpreted over finite traces in this paper. The idea of this paper is originally inspired by the normal form of Propositional Projection Temporal Logic (PPTL) [11,12]. Normal form and normal form graph have played a vital role in obtaining a decision procedure for checking satisfiability of PPTL formulas [13,14].

The rest of this paper is organized as follows. Related work is discussed in Sect. 2. The syntax and semantics of νTL$_f$ and some basic notions are introduced in Sect. 3. The PF form of νTL$_f$ formulas is presented in Sect. 4. Section 5 describes an algorithm for constructing PFGs and the decision procedure for checking satisfiability of νTL$_f$ formulas based on PFG is given in Sect. 6. Conclusions are drawn in Sect. 7.

2 Related Work

A lot of work has been done for the decision problems of νTL and the most representative one is given in [25] by Vardi that first applies the automata-theoretic decision procedure for modal μ-calculus [23] to νTL with past operators. Two-way automata are used to deal with the past operators in his work and an algorithm running in $2^{O(|\phi|^4)}$ is obtained eventually. Afterwards, Banieqbal and Barringer [2] prove that if a formula is satisfiable, it is able to generate a good Hintikka structure, which is further reducible to a path searching problem from a graph. Their method has the same time complexity as Vardi's but requires exponential space. In [22], Stirling and Walker first present a tableau method for the decision problems of νTL without discussing the complexity issues. Later, by simplifying the success conditions for a tableau, an improved tableau system [4] is obtained which reuses some notions in [17] and runs in $2^{O(|\phi|^2 \log |\phi|)}$. In [8], Dax, Hofmann, and Lange propose a simple proof system for checking validity of νTL formulas which runs in $2^{O(|\phi|^2 \log |\phi|)}$ and has been implemented in the Objective Categorical Abstract Machine Language (OCAML).

However, all the above-mentioned decision methods consider νTL interpreted over infinite traces. To the best of our knowledge, our work is the first attempt to deal with νTL interpreted over finite traces. The PFG-based decision procedure for νTL$_f$ formulas we propose in this paper is easier to understand and more efficient compared with those for νTL formulas since it avoids the detection of the infinite unfolding problem for least fixpoints. Our decision procedure has two main advantages: (1) it does not involve any process of determinization used in the automata-theoretic methods by considering PFGs and is easy to automate; (2) it gives good insight into why and how a given formula is satisfiable through its PFG.

3   Preliminaries

Let $\mathcal{P}$ be a set of atomic propositions, and $\mathcal{V}$ a set of variables. νTL$_f$ formulas are built based on the following syntax:

$$\phi ::= p \mid \neg p \mid \varepsilon \mid X \mid \phi \vee \phi \mid \phi \wedge \phi \mid \bigcirc \phi \mid \odot \phi \mid \mu X.\phi \mid \nu X.\phi$$

where p ranges over $\mathcal{P}$ and X over $\mathcal{V}$. In particular, $\odot$ is the weak next operator.

We employ σ to indicate μ or ν, and $\dot{p}$ to indicate p or $\neg p$. An occurrence of a variable X in a formula is *bound* if it is in the scope of σX and *free* otherwise. A formula is *closed* if there exists no free variable in it. $\phi[\varphi/Y]$ represents the result of simultaneously replacing all the free occurrences of Y in ϕ by φ. Further, we assume that each variable in a formula is bound at most once. As a result, all the formulas built by the syntax above are in *positive normal form* [18].

νTL$_f$ formulas are interpreted over finite linear structures. A *finite linear structure* over $\mathcal{P}$ is a finite sequence of states, $\rho = s_0, s_1, \ldots, s_n$, with each s_i being a member of $2^{\mathcal{P}}$. We use $s_i \not\rightarrow$ to denote s_i is the last state of ρ and $\mathcal{N}$ the set of states of ρ. The semantics of νTL$_f$ formulas, relative to ρ and an environment $e : \mathcal{V} \to 2^{\mathcal{N}}$, is defined as follows:

$$\|p\|_e^\rho := \{s_i \mid p \in s_i\}$$
$$\|\neg p\|_e^\rho := \{s_i \mid p \notin s_i\}$$
$$\|\varepsilon\|_e^\rho := \{s_i \mid s_i \not\rightarrow\}$$
$$\|X\|_e^\rho := e(X)$$
$$\|\varphi \vee \psi\|_e^\rho := \|\varphi\|_e^\rho \cup \|\psi\|_e^\rho$$
$$\|\varphi \wedge \psi\|_e^\rho := \|\varphi\|_e^\rho \cap \|\psi\|_e^\rho$$
$$\|\bigcirc \varphi\|_e^\rho := \{s_i \mid s_{i+1} \in \|\varphi\|_e^\rho\}$$
$$\|\odot \varphi\|_e^\rho := \{s_i \mid s_{i+1} \in \|\varphi\|_e^\rho\} \cup \{s_i \mid s_i \not\rightarrow\}$$
$$\|\mu X.\varphi\|_e^\rho := \bigcap \{W \subseteq \mathcal{N} \mid \|\varphi\|_{e[X \mapsto W]}^\rho \subseteq W\}$$
$$\|\nu X.\varphi\|_e^\rho := \bigcup \{W \subseteq \mathcal{N} \mid W \subseteq \|\varphi\|_{e[X \mapsto W]}^\rho\}$$

where $e[X \mapsto W]$ is the environment e' that agrees with e except for $e'(X) = W$. e is employed to evaluate free variables and can be discarded for a closed formula.

Given a formula ψ, we say ψ is *true* at the state s_i of a finite linear structure ρ, denoted by $s_i \models \psi$, iff $s_i \in \|\psi\|_e^\rho$. We say ψ is *valid*, denoted by $\models \psi$, iff $s_i \models \psi$ for all the finite linear structures ρ and all the states s_i of ρ; ψ is *satisfiable* iff there exists a finite linear structure ρ and a state s_j of ρ such that $s_j \models \psi$.

A formula is called *guarded* if every occurrence of a bound variable in it is in the scope of a next ($\bigcirc$) or weak next ($\odot$) operator. Every formula can be converted into an equivalent one in guarded form with an exponential increase in the size of the formula in the worst case [5]. Further, for any formula with weak next ($\odot$) operators, we can transform it into an equivalent one containing merely next ($\bigcirc$) operators and ε using the equivalence $\odot\psi \equiv \bigcirc\psi \vee \varepsilon$.

The *closure* $CL(\phi)$ of a $\nu\mathrm{TL}_f$ formula ϕ, based on [16], is the least set of formulas such that

- $\phi, true, \varepsilon \in CL(\phi)$;
- if $\varphi \vee \psi$ or $\varphi \wedge \psi \in CL(\phi)$, then $\varphi, \psi \in CL(\phi)$;
- if $\bigcirc\varphi$ or $\odot\varphi \in CL(\phi)$, then $\varphi \in CL(\phi)$;
- if $\sigma X.\varphi \in CL(\phi)$, then $\varphi[\sigma X.\varphi/X] \in CL(\phi)$.

It has been shown that the size of $CL(\phi)$ is linear in the size of ϕ (denoted by $|\phi|$) [16].

4   PF Form of $\nu\mathrm{TL}_f$ Formulas

In this section, we first define PF form of $\nu\mathrm{TL}_f$ formulas and then demonstrate that every closed $\nu\mathrm{TL}_f$ formula can be converted into this form. For technical reasons, from now on we confine ourselves to guarded formulas where all weak next ($\odot$) operators are replaced by next ($\bigcirc$) operators and ε, and no $\vee$ appears as the main operator under each next ($\bigcirc$) operator. This can be readily achieved using the equivalences $\odot\phi \equiv \bigcirc\phi \vee \varepsilon$ and $\bigcirc(\phi_1 \vee \phi_2) \equiv \bigcirc\phi_1 \vee \bigcirc\phi_2$.

4.1   PF Form

Definition 1 (PF Form). *Let ϕ be a closed νTL_f formula, $\mathcal{P}_\phi$ the set of atomic propositions occurring in ϕ. PF form of ϕ is defined by:*

$$\phi \equiv \bigvee_{j=1}^{m}(\phi_{t_j} \wedge \varepsilon) \vee \bigvee_{i=1}^{n}(\phi_{p_i} \wedge \bigcirc\phi_{f_i})$$

where $\phi_{t_j} \equiv \bigwedge_{k=1}^{m_1} \dot{q}_{jk}$, $\phi_{p_i} \equiv \bigwedge_{h=1}^{n_1} \dot{q}_{ih}$, $q_{jk}, q_{ih} \in \mathcal{P}_\phi$ for each k and h; $\phi_{f_i} \equiv \bigwedge_{l=1}^{n_2} \phi_{il}$, $\phi_{il} \in CL(\phi)$ for each l.

Regarding PF form, we have the following theorem.

Theorem 1. *Every closed νTL_f formula φ can be transformed into PF form.*

Proof. We prove this theorem by induction on the structure of φ. Note that we use $Conj(\psi)$ to denote the set of all the conjuncts in ψ.

- **Base Case:**
  - φ is p (or $\neg p$): it can be written as: $p \equiv p \wedge \varepsilon \vee p \wedge \bigcirc true$ (or $\neg p \equiv \neg p \wedge \varepsilon \vee \neg p \wedge \bigcirc true$). Hence, φ can be converted into PF form in these two cases.
  - φ is ε: it can be written as: $\varepsilon \equiv true \wedge \varepsilon$. Therefore, φ can be transformed into PF form in this case.
- **Induction:**
  - φ is $\bigcirc \phi$: it can be written as $\bigcirc \phi \equiv \bigvee_i (true \wedge \bigcirc \phi_i)$. We can obtain, for each $\phi_c \in Conj(\phi_i)$, that $\phi_c \in CL(\varphi)$ since $\phi \in CL(\varphi)$. Therefore, φ can be transformed into PF form in this case.
  - φ is $\phi_1 \vee \phi_2$: by induction hypothesis, both ϕ_1 and ϕ_2 can be converted into PF form:

$$\phi_1 \equiv \bigvee_{j=1}^m (\phi_{1t_j} \wedge \varepsilon) \vee \bigvee_{i=1}^n (\phi_{1p_i} \wedge \bigcirc \phi_{1f_i}),$$
$$\phi_2 \equiv \bigvee_{j'=1}^{m'} (\phi_{2t_{j'}} \wedge \varepsilon) \vee \bigvee_{i'=1}^{n'} (\phi_{2p_{i'}} \wedge \bigcirc \phi_{2f_{i'}}),$$

where for each $\phi_{1c} \in Conj(\phi_{1f_i})$, $\phi_{1c} \in CL(\phi_1)$; for each $\phi_{2c} \in Conj(\phi_{2f_{i'}})$, $\phi_{2c} \in CL(\phi_2)$. Further, we can obtain that

$$\varphi \equiv \phi_1 \vee \phi_2 \equiv \bigvee_{j=1}^m (\phi_{1t_j} \wedge \varepsilon) \vee \bigvee_{j'=1}^{m'} (\phi_{2t_{j'}} \wedge \varepsilon)$$
$$\vee \bigvee_{i=1}^n (\phi_{1p_i} \wedge \bigcirc \phi_{1f_i}) \vee \bigvee_{i'=1}^{n'} (\phi_{2p_{i'}} \wedge \bigcirc \phi_{2f_{i'}}).$$

We have $\phi_1, \phi_2 \in CL(\varphi)$ since $\phi_1 \vee \phi_2 \in CL(\varphi)$. Next, for each $\phi_{1c} \in CL(\phi_1)$ and $\phi_{2c} \in CL(\phi_2)$, we have $\phi_{1c}, \phi_{2c} \in CL(\varphi)$. Hence, φ can be converted into PF form in this case.
  - φ is $\phi_1 \wedge \phi_2$: by induction hypothesis, both ϕ_1 and ϕ_2 can be converted into PF form:

$$\phi_1 \equiv \bigvee_{j=1}^m (\phi_{1t_j} \wedge \varepsilon) \vee \bigvee_{i=1}^n (\phi_{1p_i} \wedge \bigcirc \phi_{1f_i}),$$
$$\phi_2 \equiv \bigvee_{j'=1}^{m'} (\phi_{2t_{j'}} \wedge \varepsilon) \vee \bigvee_{i'=1}^{n'} (\phi_{2p_{i'}} \wedge \bigcirc \phi_{2f_{i'}}),$$

where for each $\phi_{1c} \in Conj(\phi_{1f_i})$, $\phi_{1c} \in CL(\phi_1)$; for each $\phi_{2c} \in Conj(\phi_{2f_{i'}})$, $\phi_{2c} \in CL(\phi_2)$. Further, we can obtain that

$$\varphi \equiv \phi_1 \wedge \phi_2 \equiv \bigvee_{j=1}^m \bigvee_{j'=1}^{m'} (\phi_{1t_j} \wedge \phi_{2t_{j'}} \wedge \varepsilon)$$
$$\vee \bigvee_{i=1}^n \bigvee_{i'=1}^{n'} (\phi_{1p_i} \wedge \phi_{2p_{i'}} \wedge \bigcirc (\phi_{1f_i} \wedge \phi_{2f_{i'}})).$$

We have $\phi_1, \phi_2 \in CL(\varphi)$ since $\phi_1 \wedge \phi_2 \in CL(\varphi)$. Next, since each $\phi_{1c} \in CL(\phi_1)$ and each $\phi_{2c} \in CL(\phi_2)$, we have $\phi_{1c}, \phi_{2c} \in CL(\varphi)$. Thus, all the conjuncts appearing behind the next ($\bigcirc$) operators in φ are contained in $CL(\varphi)$ and φ can be converted into PF form in this case.
  - φ is $\mu X.\phi$: in this case, we have to unfold $\mu X.\phi$ first using the equivalence $\mu X.\phi \equiv \phi[\mu X.\phi/X]$ in order to convert it into PF form. In other words, we can regard the free variable X appearing in ϕ as an atomic proposition when converting ϕ into PF form since $\mu X.\phi$ will be substituted for X eventually. Consequently, by induction hypothesis, ϕ can be converted into PF form:

$$\phi \equiv \bigvee_{j=1}^m (\phi_{t_j} \wedge \varepsilon) \vee \bigvee_{i=1}^n (\phi_{p_i} \wedge \bigcirc \phi_{f_i}).$$

For each $\phi_c \in Conj(\phi_{f_i})$, we can obtain, by induction hypothesis, that $\phi_c \in CL(\phi)$. Next, by replacing all the free occurrences of X in ϕ by $\mu X.\phi$, we can obtain that

$$\varphi \equiv \phi[\mu X.\phi/X] \equiv \bigvee_{j=1}^m (\phi_{t_j} \wedge \varepsilon) \vee \bigvee_{i=1}^n (\phi_{p_i} \wedge \bigcirc \phi_{f_i}[\mu X.\phi/X]).$$

For each ϕ_c, we have $\phi_c[\mu X.\phi/X] \in CL(\phi[\mu X.\phi/X])$ after the replacement since $\phi_c \in CL(\phi)$. Further, we can obtain that $\phi_c[\mu X.\phi/X] \in CL(\varphi)$ since $\phi[\mu X.\phi/X] \in CL(\varphi)$. Hence, φ can be converted into PF form in this case.
– φ is $\nu X.\phi$: this case can be proved similarly to the case when φ is $\mu X.\phi$.

It follows that every closed $\nu \mathrm{TL}_f$ formula can be transformed into PF form. $\square$

4.2   Algorithm for PF Form Transformation

In this section, we present algorithm $PFTran$ for transforming a closed $\nu \mathrm{TL}_f$ formula into PF form. The basic idea of the algorithm follows directly from the proof of Theorem 1. Therefore, its correctness can be guaranteed.

Algorithm 1. PFTran(ϕ)

1: **case**
2:     ϕ is $true$: **return** $true \wedge \varepsilon \vee true \wedge \bigcirc true$
3:     ϕ is $false$: **return** $false$
4:     ϕ is ε: **return** $true \wedge \varepsilon$
5:     ϕ is ϕ_{prop} where $\phi_{prop} \equiv \bigwedge_{k=1}^n \dot{q}_k$: **return** $\phi_{prop} \wedge \varepsilon \vee \phi_{prop} \wedge \bigcirc true$
6:     ϕ is $\phi_{prop} \wedge \varepsilon$: **return** ϕ
7:     ϕ is $\phi_{prop} \wedge \bigcirc \varphi$: **return** $\bigvee_i (\phi_{prop} \wedge \bigcirc \varphi_i)$
8:     ϕ is $\bigcirc \varphi$: **return** $\bigvee_i (true \wedge \bigcirc \varphi_i)$
9:     ϕ is $\psi \vee \varphi$: **return** PFTran(ψ) $\vee$ PFTran(φ)
10:     ϕ is $\psi \wedge \varphi$: **return** AND(PFTran(ψ), PFTran(φ))
11:     ϕ is $\sigma X.\varphi$: **return** PFTran($\varphi[\sigma X.\varphi/X]$)
12: **end case**

Note that algorithm $PFTran$ utilizes the function AND to handle the boolean connective $\wedge$. It is obvious that the inputs, ψ and φ, for AND are both in PF form. Suppose ψ is of the form $\bigvee_i (\psi_i \wedge \varepsilon) \vee \bigvee_j (\psi_j \wedge \bigcirc \psi_j')$ while φ of the form $\bigvee_k (\varphi_k \wedge \varepsilon) \vee \bigvee_l (\varphi_l \wedge \bigcirc \varphi_l')$. AND returns $\bigvee_i \bigvee_k (\psi_i \wedge \varphi_k \wedge \varepsilon) \vee \bigvee_j \bigvee_l (\psi_j \wedge \varphi_l \wedge \bigcirc(\psi_j' \wedge \varphi_l'))$ eventually.

Theorem 2. *Converting a closed νTL_f formula ϕ into PF form by algorithm PFTran can be done in $2^{O(|\phi|)}$.*

Proof. Intuitively, the running time of $PFTran$ depends mainly on the number of recursive calls for itself as well as the running time of the function AND.

We prove this theorem by induction on the structure of ϕ.

- **Base Case:**
  - ϕ is $true$, $false$, ε, ϕ_{prop}, $\phi_{prop} \wedge \varepsilon$, $\phi_{prop} \wedge \bigcirc\varphi$ (where ϕ_{prop} is of the form $\bigwedge_{k=1}^{n} \dot{q}_k$), or $\bigcirc\varphi$: the theorem holds apparently in these cases.
- **Induction:**
  - ϕ is $\psi \vee \varphi$: we can obtain, by induction hypothesis, that $PFTran(\psi)$ and $PFTran(\varphi)$ can be finished in $2^{O(|\psi|)}$ and $2^{O(|\varphi|)}$, respectively. Therefore, it can be seen that $PFTran(\phi)$ can be done in $2^{O(|\psi|)} + 2^{O(|\varphi|)}$, namely $2^{O(|\phi|)}$.
  - ϕ is $\psi \wedge \varphi$: we have, by induction hypothesis, that $PFTran(\psi)$ and $PFTran(\varphi)$ can be finished in $2^{O(|\psi|)}$ and $2^{O(|\varphi|)}$, respectively. Next, we can easily obtain that the number of disjuncts in ψ (resp. φ) is bounded by $2^{O(|\psi|)}$ (resp. $2^{O(|\varphi|)}$) after the PF form transformation. Therefore, the function AND can be completed in $2^{O(|\psi|+|\varphi|)}$. It follows that $PFTran(\phi)$ can be finished in $2^{O(|\psi|)} + 2^{O(|\varphi|)} + 2^{O(|\psi|+|\varphi|)}$, namely $2^{O(|\phi|)}$.
  - ϕ is $\sigma X.\varphi$: we can convert φ into PF form using algorithm $PFTran$ by treating X as an atomic proposition and it is easy to see, by induction hypothesis, that this process can be done in $2^{O(|\varphi|)}$. Further, by replacing all the free occurrences of X in φ by $\sigma X.\varphi$, we can obtain that $\varphi[\sigma X.\varphi/X]$ can be converted into PF form by algorithm $PFTran$ in $2^{O(|\varphi|)}$. Hence, $PFTran(\sigma X.\varphi)$ can be completed in $2^{O(|\phi|)}$. □

5 Present Future Form Graph

5.1 PFG Definition

For a closed νTL_f formula ϕ, its PFG, denoted by G_ϕ, is a triple (N_ϕ, E_ϕ, n_0) where N_ϕ is a set of nodes, E_ϕ is a set of directed edges, and n_0 is the root node. Each node in N_ϕ is specified by a conjunction of elements in $CL(\phi)$ while each edge in E_ϕ is identified by a triple $(\phi_i, \phi_{ei}, \phi_j)$ where $\phi_i, \phi_j \in N_\phi$ and ϕ_{ei} denotes the label of the edge from ϕ_i to ϕ_j.

Definition 2 (PFG). *For a closed νTL_f formula ϕ, N_ϕ and E_ϕ are inductively defined by: (1) $n_0 = \phi \in N_\phi$; (2) for all $\varphi \in N_\phi \setminus \{\varepsilon, false\}$, if $\varphi \equiv \bigvee_{j=1}^{h}(\varphi_{t_j} \wedge \varepsilon) \vee \bigvee_{i=1}^{k}(\varphi_{p_i} \wedge \bigcirc\varphi_{f_i})$, then $\varepsilon \in N_\phi$, $(\varphi, \varphi_{t_j}, \varepsilon) \in E_\phi$ for each j; $\varphi_{f_i} \in N_\phi$, $(\varphi, \varphi_{p_i}, \varphi_{f_i}) \in E_\phi$ for each i.*

In a PFG, we use a double circle, a black dot, and a single circle to denote the root node, the ε node, and each of other nodes, respectively. Each edge is represented as a directed arc connecting two nodes. Note that, to simplify notations, we use variables to indicate the corresponding fixpoint formulas appearing in each node. An example of PFG for the formula $\mu X.(p \vee \bigcirc X) \vee \nu Y.(q \wedge \bigcirc Y)$ is given in Fig. 1. There are five nodes (including the ε node) in the PFG where n_0 is the root node. (n_0, p, ε) is an edge with the label being p while (n_3, q, n_3) is an edge with the label being q.

A *path* Π in a PFG is a finitely alternate sequence of nodes and edges starting from the root node, while an *ε-path* is a path that ends with the ε node. Let $Atom(\bigwedge_{i=1}^{m} \dot{q}_i)$ represent the set of atomic propositions or their negations in

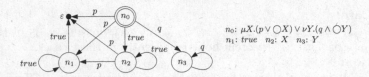

Fig. 1. An example of PFG

$\bigwedge_{i=1}^{m} \dot{q}_i$. As a result, for a given path $\Pi = \phi_0, \phi_{e0}, \phi_1, \phi_{e1}, \dots, \phi_{em}, \phi_{m+1}$, we can obtain its corresponding finite linear structure $Atom(\phi_{e0}), Atom(\phi_{e1}), \dots, Atom(\phi_{em})$. For instance, the path $n_0, true, n_2, p, n_1, true, \varepsilon$ in Fig. 1 corresponds to the finite linear structure $\{true\}\{p\}\{true\}$.

5.2   Algorithm for Constructing PFGs

For a given closed νTL_f formula ϕ, we use algorithm $PFGCon$ to construct its PFG G_ϕ.

Algorithm 2. PFGCon(ϕ)

1: $n_0 = \phi$, $N_\phi = \{n_0\}$, $E_\phi = \emptyset$, H$[n_0] = 0$, AddE = AddN = 0
2: **while** there exists $\varphi \in N_\phi \setminus \{\varepsilon, false\}$ with H$[\varphi] = 0$ **do**
3:     H$[\varphi] = 1$
4:     $\varphi = $ PFTran(φ)
5:     **case**
6:         φ is $\bigvee_{j=1}^{h}(\varphi_{t_j} \wedge \varepsilon)$: AddE = 1
7:         φ is $\bigvee_{i=1}^{k}(\varphi_{p_i} \wedge \bigcirc \varphi_{f_i})$: AddN = 1
8:         φ is $\bigvee_{j=1}^{h}(\varphi_{t_j} \wedge \varepsilon) \vee \bigvee_{i=1}^{k}(\varphi_{p_i} \wedge \bigcirc \varphi_{f_i})$: AddE = AddN = 1
9:     **end case**
10:    **if** AddE = 1 **then**
11:        add the corresponding nodes and edges
12:        AddE = 0
13:    **end if**
14:    **if** AddN = 1 **then**
15:        add the corresponding nodes and edges
16:        AddN = 0
17:    **end if**
18: **end while**
19: **for all** $\varphi \in N_\phi \setminus \{\varepsilon\}$ with no outgoing edge **do**
20:    $N_\phi = N_\phi \setminus \{\varphi\}$
21:    $E_\phi = E_\phi \setminus \bigcup_i \{(\varphi_i, \varphi_e, \varphi)\}$
22: **end for**
23: **return** G_ϕ

Algorithm $PFGCon$ takes ϕ as input and finally returns G_ϕ. It uses $H[\varphi] = 1$ (or 0) to indicate that a formula φ has (or has not) been handled. Moreover, it employs $AddE$ (resp. $AddN$) $= 1$ (or 0) to indicate that the ε (resp. $\bigcirc$) part occurs (or does not occur) in the PF form of the formula currently being handled. At the beginning of $PFGCon$, n_0 is assigned to ϕ; N_ϕ and E_ϕ are initialized to $\{n_0\}$ and empty respectively; $H[n_0]$, $AddE$, and $AddN$ are all assigned to 0. Further, the algorithm repeatedly transforms an unhandled formula $\varphi \in N_\phi$ into PF form by algorithm $PFTran$ and then adds the corresponding nodes and edges to N_ϕ and E_ϕ according to the values of $AddE$ and $AddN$, respectively, until all nodes in N_ϕ have been handled. In addition, it is noteworthy that, throughout the construction of G_ϕ, a *false* node (e.g. $p \wedge \neg p$) may be produced which corresponds to an inconsistent subset of $CL(\phi)$. We use the *for loop* in Line 19 to eliminate these nodes as well as the relative edges.

For a closed νTL_f formula ϕ, since each node in N_ϕ is the conjunction of elements in $CL(\phi)$, we can obtain the following corollary.

Corollary 3. *For any closed νTL_f formula ϕ, both the number of nodes and the number of edges in G_ϕ are bounded by $2^{O(|\phi|)}$.*

Theorem 4. *Constructing the PFG of a closed νTL_f formula ϕ by algorithm $PFGCon$ can be completed in $2^{O(|\phi|)}$.*

Proof. First, we have that the number of iterations of the *while loop* is bounded by $2^{O(|\phi|)}$ according to Corollary 3. Next, algorithm $PFTran$ is called first in each iteration, which can be accomplished in $2^{O(|\phi|)}$ according to Theorem 2. Further, adding nodes and edges for the ε part and the $\bigcirc$ part can be finished in $2^{O(n_p)}$ and $2^{O(|\phi|)}$, respectively, where n_p denotes the number of atomic propositions in ϕ. Hence, the *while loop* can be done in $2^{O(|\phi|)}$. Finally, we can easily obtain that removing inconsistent nodes as well as the relative edges can be completed in $2^{O(|\phi|)}$. It follows that $PFGCon$ can be finished in $2^{O(|\phi|)}$. $\qquad\square$

6   A Decision Procedure Based on PFG

In this section, we demonstrate how to find a model for a given closed νTL_f formula ϕ from G_ϕ. First, we give the following theorem.

Theorem 5. *A closed νTL_f formula ϕ is satisfiable iff an ε-path can be found in G_ϕ.*

Proof. Suppose the PF form of ϕ is:

$$\phi \equiv \bigvee_{j=1}^{m}(\phi_{t_j} \wedge \varepsilon) \vee \bigvee_{i=1}^{n}(\phi_{p_i} \wedge \bigcirc\phi_{f_i}).$$

Given a path $\Pi = \phi, \phi_{e0}, \phi_1, \phi_{e1}, \phi_2, \ldots, \phi_{ek}, \phi_{k+1}$ in G_ϕ, we write Π^i for the prefix, $\phi, \phi_{e0}, \phi_1, \phi_{e1}, \phi_2, \ldots, \phi_{ei}, \phi_{i+1}$, of Π. Given a finite linear structure $\rho = s_0, s_1, \ldots, s_k$, we write ρ^i for the prefix, $s_0, s_1, \ldots, s_i$, of ρ. For each state s_i of ρ, we use $s_i[\dot{r}] = true$ to indicate $r \in s_i$ if $\dot{r} \equiv r$, or $r \notin s_i$ if $\dot{r} \equiv \neg r$.

($\Rightarrow$) Let $\rho_1 = s_0, s_1, \ldots, s_k$ be a model of ϕ (namely $s_0 \models \phi$), $\Pi_1 = \phi, \phi_{e0}, \phi_1, \phi_{e1}, \phi_2, \ldots, \phi_{ek}, \varepsilon$ the ε-path w.r.t. ρ_1. That is, each ϕ_{ei} is the conjunction of each $\dot{p}$ where $s_i[\dot{p}] = true$. We prove that Π_1 can be found in G_ϕ.

(1) When $k = 0$, we have $\rho_1 = s_0$ and $\Pi_1 = \phi, \phi_{e0}, \varepsilon$. By the construction of ϕ_{e0}, we have $s_0 \models \phi_{e0} \wedge \varepsilon$. According to the PF form of ϕ, there exists an edge from ϕ to ε labeled by $\phi_{e0} \equiv \phi_{t_j}$ in G_ϕ $(1 \leq j \leq m)$. Therefore, Π_1 can be found in G_ϕ.

(2) When $k > 0$, the proof proceeds by induction on the length of the prefix Π_1^i of Π_1.

Base Case: $\Pi_1^0 = \phi, \phi_{e0}, \phi_1$. By the construction of G_ϕ, there exists an edge from ϕ to $\phi_1 \equiv \phi_{f_i}$ labeled by $\phi_{e0} \equiv \phi_{p_i} \equiv \bigwedge_{h=1}^{n_1} \dot{p}_h$ $(1 \leq i \leq n)$ such that $s_0 \models \phi_{p_i} \wedge \bigcirc \phi_{f_i}$. Here $s_0[\dot{p}_h] = true$ for each $\dot{p}_h$. Therefore, the prefix Π_1^0 of Π_1 is found in G_ϕ.

Induction: suppose we have found a prefix $\Pi_1^{k-1} = \phi, \phi_{e0}, \phi_1, \phi_{e1}, \phi_2, \ldots, \phi_{e(k-1)}, \phi_k$ of Π_1 in G_ϕ w.r.t. ρ_1^{k-1}. At this point, we can transform ϕ_k into PF form:

$$\phi_k \equiv \bigvee_{j'=1}^{m'}(\phi_{t_{j'}} \wedge \varepsilon) \vee \bigvee_{i'=1}^{n'}(\phi_{p_{i'}} \wedge \bigcirc \phi_{f_{i'}}),$$

such that there exists an edge from ϕ_k to ε labeled by $\phi_{ek} \equiv \phi_{t_{j'}} \equiv \bigwedge_{l=1}^{n_2} \dot{p}_l$ in G_ϕ $(1 \leq j' \leq m')$ and $s_k \models \phi_{t_{j'}} \wedge \varepsilon$. That is, $s_k[\dot{p}_l] = true$ for each $\dot{p}_l$. It follows that Π_1 can be found in G_ϕ.

($\Leftarrow$) Let $\Pi_2 = \phi, \phi_{e0}, \phi_1, \phi_{e1}, \ldots, \phi_{ek}, \varepsilon$ be an ε-path in G_ϕ, $\rho_2 = s_0, s_1, \ldots, s_k$ the corresponding finite linear structure of Π_2. Therefore, for each $\phi_{ei} \equiv \bigwedge_{h=1}^{n_1} \dot{p}_h$, we have $s_i[\dot{p}_h] = true$ for each $\dot{p}_h$. We prove that ρ_2 is a model of ϕ.

(1) When $k = 0$, we have $\Pi_2 = \phi, \phi_{e0}, \varepsilon$ and $\rho_2 = s_0$. Since ρ_2 is the corresponding finite linear structure of Π_2, we have $s_0 \models \phi_{e0}$. We can further obtain that $s_0 \models \phi_{e0} \wedge \varepsilon$. Therefore, ρ_2 is a model of ϕ.

(2) When $k > 0$, the proof proceeds by induction on the length of the prefix ρ_2^i of ρ_2.

Base Case: $\rho_2^0 = s_0$. Since ρ_2 is the corresponding finite linear structure of Π_2, we have $\rho_2^0 \models \phi_{e0}$. That is, ρ_2^0 is a prefix of a model.

Induction: suppose for all $k_1 < k$, $\rho_2^{k_1}$ is a prefix of a model. We prove that ρ_2^k is a prefix of a model of ϕ. By inductive hypothesis, $\rho_2^{k-1} = s_0, s_1, \ldots, s_{k-1}$ is a prefix of the model. According to algorithm $PFGCon$, there must be $\phi_k \equiv \phi_{ek} \wedge \varepsilon$ such that a new node ε and a new edge $(\phi_k, \phi_{ek}, \varepsilon)$ are added to G_ϕ. Since ρ_2 is the corresponding finite linear structure of Π_2, we have $s_k \models \phi_{ek} \wedge \varepsilon$. Since ρ_2^{k-1} is a prefix of the model, $\rho_2^k = \rho_2$ is a model of ϕ.   □

As a result, we reduce the satisfiability problem of a νTL_f formula ϕ to an ε-path searching problem from G_ϕ. In fact, to decide the satisfiability of ϕ, we only need to check whether the ε node exists in G_ϕ. The existence of the ε node indicates that at least one ε-path can be found in G_ϕ.

Consequently, the PFG-based decision procedure for checking satisfiability of νTL_f formulas becomes straightforward, as presented in Algorithm 3.

According to Theorem 4, we obtain the following theorem.

Theorem 6. *For a given closed νTL_f formula ϕ, the decision procedure PFGSAT can be done in $2^{O(|\phi|)}$.*

Algorithm 3. PFGSAT(ϕ)

1: $G_\phi = $ PFGCon(ϕ)
2: **if** $\varepsilon \in N_\phi$ **then**
3:     **return** satisfiable
4: **else**
5:     **return** unsatisfiable
6: **end if**

7   Conclusion

In this paper, we studied νTL interpreted over finite traces, namely νTL$_f$. We presented PF form and PFG of νTL$_f$ formulas as well as the algorithms for PF form transformation and PFG construction, respectively. Further, based on PFG, a decision procedure for checking satisfiability of νTL$_f$ formulas has been proposed. In the near future, we plan to implement the proposed decision procedure. Also, we are going to investigate the PFG-based model checking approach for νTL$_f$ based on the proposed decision procedure and develop a practical model checker for νTL$_f$.

References

1. Bacchus, F., Kabanza, F.: Planning for temporally extended goals. Ann. Math. Artif. Intell. **22**(1–2), 5–27 (1998)
2. Banieqbal, B., Barringer, H.: Temporal logic with fixed points. In: Banieqbal, B., Barringer, H., Pnueli, A. (eds.) Temporal Logic in Specification. LNCS, vol. 398, pp. 62–74. Springer, Heidelberg (1989)
3. Barringer, H., Kuiper, R., Pnueli, A.: A really abstract concurrent model and its temporal logic. In: POPL 1986, pp. 173–183. ACM Press, New York (1986)
4. Bradfield, J.C., Esparza, J., Mader, A.: An effective tableau system for the linear time μ-calculus. In: Meyer auf der Heide, F., Monien, B. (eds.) ICALP 1996. LNCS, vol. 1099, pp. 98–109. Springer, Heidelberg (1996)
5. Bruse, F., Friedmann, O., Lange, M.: On guarded transformation in the modal μ-calculus. Logic J. IGPL **23**(2), 194–216 (2015)
6. Calvanese, D., De Giacomo, G., Vardi, M.Y.: Reasoning about actions and planning in LTL action theories. In: Fensel, D., Giunchiglia, F., McGuinness, D.L., Williams, M.A. (eds.) KR 2002, pp. 593–602. Morgan Kaufmann, San Francisco (2002)
7. Clarke, E.M., Grumberg, O., Peled, D.: Model Checking. MIT Press, Cambridge (1999)
8. Dax, C., Hofmann, M., Lange, M.: A proof system for the linear time μ-calculus. In: Arun-Kumar, S., Garg, N. (eds.) FSTTCS 2006. LNCS, vol. 4337, pp. 273–284. Springer, Heidelberg (2006)
9. De Giacomo, G., Vardi, M.Y.: Automata-theoretic approach to planning for temporally extended goals. In: Biundo, S., Fox, M. (eds.) ECP 1999. LNCS, vol. 1809, pp. 226–238. Springer, Heidelberg (2000)
10. De Giacomo, G., Vardi, M.Y.: Linear temporal logic and linear dynamic logic on finite traces. In: Rossi, F. (ed.) IJCAI 2013, pp. 854–860. AAAI Press, Palo Alto (2013)

11. Duan, Z.: An extended interval temporal logic and a framing technique for temporal logic programming. Ph.D. thesis, University of Newcastle upon Tyne (1996)
12. Duan, Z.: Temporal Logic and Temporal Logic Programming. Science Press, Beijing (2006)
13. Duan, Z., Tian, C.: A practical decision procedure for propositional projection temporal logic with infinite models. Theor. Comput. Sci. **554**, 169–190 (2014)
14. Duan, Z., Tian, C., Zhang, L.: A decision procedure for propositional projection temporal logic with infinite models. Acta Informatica **45**(1), 43–78 (2008)
15. Emerson, E.A., Clarke, E.M.: Characterizing correctness properties of parallel programs using fixpoints. In: de Bakker, J.W., van Leeuwen, J. (eds.) ICALP 1980. LNCS, vol. 85, pp. 169–181. Springer, Heidelberg (1980)
16. Fischer, M.J., Ladner, R.E.: Propositional dynamic logic of regular programs. J. Comput. Syst. Sci. **18**(2), 194–211 (1979)
17. Kaivola, R.: A simple decision method for the linear time mu-calculus. In: Desel, J. (ed.) Proceedings of the International Workshop on Structures in Concurrency Theory, pp. 190–204. Springer, London (1995)
18. Kozen, D.: Results on the propositional μ-calculus. Theor. Comput. Sci. **27**(3), 333–354 (1983)
19. Liu, Y., Duan, Z., Tian, C.: A decision procedure for a fragment of linear time mu-calculus. In: Kambhampati, S. (ed.) IJCAI 2016. AAAI Press, Palo Alto (2016)
20. Patrizi, F., Lipoveztky, N., De Giacomo, G., Geffner, H.: Computing infinite plans for LTL goals using a classical planner. In: Walsh, T. (ed.) IJCAI 2011, pp. 2003–2008. AAAI Press, Palo Alto (2011)
21. Pnueli, A.: The temporal logic of programs. In: FOCS 1977, pp. 46–57. IEEE Press, New York (1977)
22. Stirling, C., Walker, D.: CCS, liveness, and local model checking in the linear time mu-calculus. In: Sifakis, J. (ed.) Automatic Verification Methods for Finite State Systems. LNCS, vol. 407, pp. 166–178. Springer, Heidelberg (1990)
23. Streett, R.S., Emerson, E.A.: The propositional mu-calculus is elementary. In: Paredaens, J. (ed.) ICALP 1984. LNCS, vol. 172, pp. 465–472. Springer, Heidelberg (1984)
24. Torres, J., Baier, J.A.: Polynomial-time reformulations of LTL temporally extended goals into final-state goals. In: Yang, Q., Wooldridge, M. (eds.) IJCAI 2015, pp. 1696–1703. AAAI Press, Palo Alto (2015)
25. Vardi, M.Y.: A temporal fixpoint calculus. In: Ferrante, J., Mager, P. (eds.) POPL 1988, pp. 250–259. ACM Press, New York (1988)

On the Complexity of Insertion Propagation with Functional Dependency Constraints

Dongjing Miao[1]([⊠]), Zhipeng Cai[2], Xianmin Liu[1,2], and Jianzhong Li[1,2]

[1] Harbin Institute of Technology, Harbin 150001, Heilongjiang, China
miaodongjing@hit.edu.cn
http://db.hit.edu.cn/~djmiao/
[2] Department of Computer Science, Georgia State University,
Atlanta, GA 30303, USA
zcai@gsu.edu

Abstract. Insertion propagation problem is a class of view update problem in relational databases [1]. Given a source database D, a monotone relational algebraic query Q, and the view V generated by the query $Q(D)$, insertion propagation problem is to find a set ΔD of tuples whose insertion into D will add the given tuples ΔV into the view V via Q without producing side-effect on view. In this paper, we consider the *fd*-restricted version insertion propagation problem '*fd*-vsef-ip', in which we aim to find the ΔD not only view side-effect free but also without introducing inconsistency with respect to the predefined functional dependencies. We study both data and combined complexity of *fd*-vsef-ip under both single and group insertion. Interestingly, the problem ranges from PTime to Σ_2^P-complete, for queries in different classes in either complexity aspect.We show that the *fd*-restricted version will be harder to get the optimal solution, contrary to its counterpart under deletion. Our study of this *fd*-restricted version insertion propagation problem generalize the computational issues involved in data lineage – the process by which databases updated through view insertion under *fd*.

Keywords: Insertion propagation · View update · Database · Complexity

1 Introduction

1.1 *fd*-restricted Insertion Propagation

In the study of view update in database [2–6], propagation analysis [1] of view update has been studied for more than a decade. Propagation analysis mainly focus on minimizing side-effect over source database or view which caused by the asymmetry between update on view and the source database. View side effect problem fundamental to propagation analysis, is identified in [1] stated as follows: given a source database D, an monotone relational query Q, the view $V = Q(D)$, and update on view (a set of tuples) ΔV, the view side-effect problem is to find a smallest ΔD such that $Q(f(D, \Delta D)) = f(V, \Delta)$,

© Springer International Publishing Switzerland 2016
T.N. Dinh and M.T. Thai (Eds.): COCOON 2016, LNCS 9797, pp. 623–632, 2016.
DOI: 10.1007/978-3-319-42634-1_50

$$f(a,b) = \begin{cases} a \setminus b, & \text{for deletion update,} \\ a \cup b, & \text{for insertion update} \end{cases} \qquad \begin{array}{c} (1) \\ (2) \end{array}$$

i.e., side effect free whenever such ΔD exists.

Example 1. Let's visit an example of the insertion update. Consider an archive management database of a company including two relations, Group(group, user) records groups each user belongs to, and Access(group, file) records files that each group has the authority to access. There is also a view defined as a conjunctive query (Selection-Projection-Join) "show the files and users has authority to access it" as follow,

Group :

Group	User
g1	usr1
g2	usr2

Access :

group	file
g1	f1
g2	f2
g2	f3

$\pi_{\text{user,file}}$ Group $\bowtie$ Access :

user	file
usr1	f1
usr2	f2
usr2	f3

now given an view insertion $\Delta V = (\text{usr1}, \text{f2})$, the task is to find a side-effect free insertion, here are some possible ways to update the source database,

(a) insert (g2, usr1) into Group,
(b) insert (g1, f2) into Access,
(c) insert $(x, \text{usr1}), (x, \text{f2})$ into Group, Access respectively, where x is a value taken from the domain of attribute group different from g1 and g2,

we see that method(a) must produce the a side effect '(usr1, f3)'on the view besides ΔV. Fortunately, the other two methods is side effect free solution that we want. However, in practice, the domain of 'group' is often finite, therefore, if no new value can be taken from the domain, then method(c) is never valid.

There have been some complexity results on the view side effect problem for insertion [7,8] and deletion propagation [1,7–12], moreover, for deletion propagation, Kimfield et al. [9] showed the dichotomy '*head domination*' for every conjunctive query without self joins, deletion propagation is either APX-hard or solvable (in polynomial time) by the unidimensional algorithm, they also showed the dichotomy '*functional head domination*' [10] for *fd*-restricted version, For multiple deletion [11], they especially showed the trichotomy for group deletion a more general case. However, there are no results on insertion propagation at the present of functional dependency (*fd*), note that database with *fd* is a very general case. It gets radically different from the case without *fd*.

Example 2. Generally, there are often some functional dependencies defined in a database to guarantee the semantic correctness. Continue the example above, possibly a functional dependency user $\rightarrow$ group is defined on Group(group, user) in order to guarantee that *each user can take part in only one group*. In such case, different from the example above, method(c) is not a valid solution any more due to the functional dependency.

In this paper, we investigate such *fd*-restricted insertion propagation, mainly on the complexity aspect of this problem.

1.2   Formal Statement

A *schema* is a finite sequence $\mathbf{R} = \langle R_1, \ldots, R_m \rangle$ of distinct relation symbols, where each R_i has an arity $r_i > 0$ and includes several attributes, denoted by $R_i = \{A_1, \ldots, A_{r_i}\}$. Each attribute A_j has a corresponding set $dom(A_j)$ which is the domain of values appearing in A_j. An database instance D (over $\mathbf{R}$) is a sequence $\langle R_1^D, \ldots, R_m^D \rangle$, such that each R_i^D is a finite set of tuples $\{t_1, \ldots, t_N\}$, each tuple t_k belongs to the set $dom(A_1) \times \cdots \times dom(A_{r_i})$. We use $R.A_i$ to indicate the attribute A_i of relation R, and also we denote R^D as R without loss of clarity.

An *fd* (functional dependency [13]) φ over a relation R can be represented by $\varphi : (X \to A)$, where both X and A are a set of attributes from R. Such dependency means the values of any two tuples' attributes A should be same if they have same value in attributes X. Given a database instance I and an *fd* φ, if there is no tuple pair violate the *fd* rule, we denote that $D \models \varphi$. Usually, we use Σ to denote the set of *fd*s. Given a database D and an *fd* set Σ, if for every *fd* $\varphi \in \Sigma$, $D \models \varphi$, we say $D \models \Sigma$.

Definition 1 (*fd*-restricted view side-effect-free insertion propagation, *fd*-vsef-ip). *Given a database D, fd set Σ, a query Q, its view V and a set of tuples inserted ΔV, it is to decide whether there is a tuple set ΔD such that*

(1) $D \cup \Delta D \models \Sigma$,
(2) $Q(D \cup \Delta D) = V \cup \Delta V$.

In this paper, we study the complexity of *fd*-vsef-ip in two cases *single* insertion when $|\Delta V| = 1$, and *group* insertion when $|\Delta V| > 1$. The query Q is written by operations in relational algebra including S (selection), P (projection), J (join), U (union).

We examine the impact of different combinations of these factors on both data and combined complexity of these problems. *Data complexity* is the complexity expressed in terms of the size of the database only, while *combined complexity* is the complexity expressed in terms of both the size of the database and the query expression [14]. The complexity measure follows the work [1] where the complexity results of propagation problem were first established and the studies [3,4,15] where the complexity of view update problems was studied. We provide a complete picture of the complexity on these problems for views defined in various fragments of SPJU queries, identifying all those cases that are intractable. For the sake of clarity, we list notations again as follows.

$D, \Delta D$	Source database, and its update
$V, \Delta V$	View, update on view
Σ, Q	Set of functional dependency, relational algebraic query
S,P,J,U	Selection, Projection, Join, Union
SPJ	Selection-Projection-Join
SPU	Selection-Projection-Union

2   Complexity Results

In this section, the complete picture we map out as the follow table,

Query	Data Comp.		Combined Comp.	
	Single	Group	Single	Group
SP	PTime	PTime	PTime	PTime 1
U	PTime	NP-complete 2	PTime	NP-complete
SJ	PTime	PTime	coNP-complete	coNP-complete 3
SPJ	PTime 4 (no self-join)	NP-complete 4 (finite dom., even without fd)	coNP-hard	Σ_2^P-complete 5 (finite dom.)
SJU	PTime	NP-complete	coNP-complete	Σ_2^P-complete 6

In the following, we show specifically the results in domination positions including both data and combined complexity of group and single insertion case.

Theorem 1. fd-vsef-ip *is* PTime *for SP query under group insertion on combined complexity.*

One can refer proof to the fd set consistency checking [13], here omitted.

Theorem 2. fd-vsef-ip *for* U *query is* NP-complete *under group insertion on data complexity.*

Proof. We construct a PTime reduction from *3-colorability* problem. The main idea is to use ΔV to encode graph, and then to decide whether ΔV can be divided into Red, Green, Blue tables under fd-restriction. Specifically,

Base Instance D. Let D include three relations Red, $Green$ and $Blue$ with the same schema $\{A_1, B_1, \cdots, A_d, B_d\}$, where d is the maximum degree of the input graph G. Initially, they are all empty.

fd Set Σ. For Red, $Green$ and $Blue$, set d fds $A_i \rightarrow B_i$ for each one.

Query Q. Let query $Q := Red \cup Green \cup Blue$.

View V. Initially, V is empty.

Insertion ΔV. Let it have n tuples, and each tuple t_i simulates a vertex v_i in the input graph G such that '$(B, \cdots, B)$' initially. Then, for the each edge $e(v_i, v_j)$ of G, find an k, such that $t_i[A_k, B_k]$ and $t_j[A_k, B_k]$ are both '(B, B)'. Note that, we can always find such $k(< d)$ since d is the maximum degree. Let (1) $t_i[A_k] = t_j[A_k] = a_{ij}$; (2) $t_i[B_k] = b_i$; (3) $t_j[B_k] = b_j$.

One can easily verify that there is an fd-restricted side-effect-free insertion Δ if and only if G can be colored by 3 colors.

Fact. To continue the analysis, we will use the term *fact*. Given a database instance D and a SPJ query Q in a form of $\pi_{\mathbf{A}}(\sigma_{con}(R_1 \times \cdots \times R_n))$. A *fact* μ of D is a tuple sequence $(t_1, t_2, \ldots, t_n) \in R_1 \times \cdots \times R_n$, where $t_i \in R_i$ for each $1 \leq i \leq n$. If $(t_1, t_2, \ldots, t_n)$ satisfies the selection condition con, then we denote it as $Q(\mu) \in Q(D)$.

Theorem 3. *fd*-vsef-ip *for an* SJ *query is* coNP-complete *under group insertion on combined complexity.*

Proof. For the *upper bound*, we should first briefly introduce some definition. Given any SJ query Q with an equivalent standard form $\sigma_c(R_1 \times \cdots \times R_m)$. Here, each R_i $(1 \leq i \leq m)$ is a table included in D, and any two tables in $R_1, \cdots, R_m$ may be the same table in the database, if Q includes self-join. For view insertion ΔV, we define its *inverse* on D with respect to Q. For each i, let $\Delta V_{R_i}^-$ be the projection on R_i of ΔV. In any query, if a self-joined table R_i has h occurrences, say $R_{i_1}, \cdots, R_{i_h}$, let $\Delta V_{R_i}^{-\star}$ be the union $\Delta V_{R_{i_1}}^- \cup \cdots \cup \Delta V_{R_{i_h}}^-$. Given any SJ query and database D, for view insertion ΔV, we define its *inverse* that $\Delta V^{-1} = \{\Delta V_{R_1}^{-\star}, ..., \Delta V_{R_m}^{-\star}\}$.

Now, we show an NP algorithm to decide if there is no *fd*-restricted side-effect-free insertion as follows, (*a*) first compute the ΔV^{-1}; (*b*) check whether $D \cup \Delta V^{-1} \models \Sigma$, if *no*, then return *yes*; (*c*) guess a fact μ of $Q(D \cup \Delta V^{-1})$, if $Q(\mu) \notin V \cup \Delta V$, then return *yes*.

Correctness. We claim that ΔV^{-1} is the minimum necessary insertion into D in order to produce the result ΔV.

For the *lower bound*, as the insertion propagation problem for J query is coNP-hard [7].

Theorem 4. *fd*-vsef-ip *for an* SPJ *query is* NP-complete *under group insertion with finite domain on data complexity, even without functional dependency.* PTime *for* SPJ *query without self-join under single insertion on data complexity.*

Proof (*Proof Sketch*). The former part is actually a dual case of the singleton deletion propagation. To prove the former part, the main idea is to modify the reduction from *monotone-3-sat* of Theorem 2.1 in [1], we build the source database D as the same, but initially only tuples with form like (a_i, x_j) of R_1 and (x_i, c_j) of R_2, therefore the view only includes tuples like (a_i, c_j). Then build the insertion ΔV including tuples like (a_i, c) and (a, c_j). Refer to [1], one can verify the correctness.

For the later part, we show an algorithm for this problem without functional dependency, *w.l.o.g*, let query is defined as a join on k relations having no more than n tuples,

Definition 2 ((semi-)free attribute). *A semi-free attribute is an attribute restricted by at least a selection condition but projected out by the query. A free attribute is a semi-free attribute but not restricted in any selection condition.*

Proposition. If the query is *monotone* without self-join, there are at most one tuple need to be inserted into each table in the source database D by the side-effect free insertion, where the value of projected attribute is the same as the insertion. This proposition guarantees that there are at most 2^k possible insertion choices need to check for a single insertion, whether it has view side-effect free solution.

A. For a possible insertion choice, we check all the value invention as follow,
(a) fix the projected attribute of the intended inserted tuples;
(b) find the scope values taken from for each semi-free attributes;
(c) try all the possible values of each semi-free attribute and check whether it produces side-effect, here we take all the absent values of a semi-free attribute as a variable x different all the active value of this attribute, therefore, this step can be finished in $O((n+1)^c)$ ('c' is the number of selection condition);
(d) take the side-effect free candidate choice of the step above, check whether it has a consistent fill for the rest free attributes and all the variables, then check whether this step produces Δ after fixing the variable, if *yes*, we can claim that *fd*-vsef-ip has a solution.
B. If all the choices are not valid or consistent, we can decide that *fd*-vsef-ip has no view side-effect free insertion.

We claim that in step (d), there is no side-effect will be produced by fixing the variable, except the new tuple producing by all the newly inserted tuples, since there will no value of other corresponding semi-free attribute equals to it. Then one can check the correctness of this algorithm, we omit the proof details due to the lack of space.

Compare with the result above, we next show SPJ query is even harder on combined complexity under finite domain unless P=NP.

Theorem 5. *fd*-vsef-ip *for an* SPJ *query is* Σ_2^P-*complete under group insertion with finite domain on combined complexity.*

Proof. i. We first show a Σ_2^P algorithm for SPJ query as follows.
    (a) guess a ΔD base on ΔV by filling all the attributes projected out.
    (b) if $D \cup \Delta D \models \Sigma$, then decide whether $Q(D \cup \Delta D) = V \cup \Delta V$, i.e., there is no side-effect on view. As SPJ query is monotone, therefore we only need to decide whether $Q(D \cup \Delta D) \subseteq V \cup \Delta V$ and this decision problem is in coNP, because we can decide whether there is side-effect on view by guessing a fact μ of D and checking whether $Q(\mu) \notin V \cup \Delta V$.

ii. We prove the lower bound of *fd*-vsef-ip for SPJ query is Σ_2^P-hard by a reduction from 3-CNF-SAT$_2$ problem. An instance of 3-CNF-SAT$_2$ problem includes two variable sets $X_1 = \{x_1, ..., x_n\}$ and $X_2 = \{x_{n+1}, ..., x_{n'}\}$, and a 3-CNF boolean expression ϕ with m clauses $\{C_1, ..., C_m\}$, the task is to determine whether there is an assignment τ for X_1 such that ϕ is satisfied by all assignments for X_2. We show the reduction as follows. (An example of the reduction for a 3-CNF-SAT$_2$ instance $\phi = \exists x_1 x_2 \forall x_3 x_4 (x_1 + x_2 + x_3)(x_1 + \overline{x}_2 + x_3)(\overline{x}_1 + \overline{x}_3 + x_4)$ is shown in Fig. 1.)

Base Instance D. Let D has $n+m+1$ relations $S_1, \cdots, S_n, R_1, \cdots, R_m$ and T, where S_i simulates existential variable x_i, R_j simulates clause C_j and universal variables in it, and T is an auxiliary relational table. Concretely,

S_i $\boxed{\begin{array}{c} A_1\ A_2 \\ \hline B \end{array}}$

R_1

A_1	A_2	A_3	A_4	A_5
0	0	1	+	1
0	1	0	+	2
0	1	1	+	3
1	0	0	+	4
1	0	1	+	5
1	1	0	+	6
1	1	1	+	7
0	0	0	−	8

R_2

A_1	A_2	A_3	A_4	A_5
0	0	0	+	1
0	0	1	+	2
0	1	1	+	3
1	0	0	+	4
1	0	1	+	5
1	1	0	+	6
1	1	1	+	7
0	1	0	−	8

R_3

A_1	A_2	A_3	A_4	A_5
0	0	0	+	1
0	0	1	+	2
0	1	0	+	3
0	1	1	+	4
1	0	0	+	5
1	0	1	+	6
1	1	1	+	7
1	1	0	−	8

Fig. 1. Example for the reduction of Theorem 5

(1) For each existential variable x_i ($1 \leq i \leq n$), construct relation $S_i = \{A_1, A_2\}$, where A_1's domain of $dom(A_1)$ is $\{B\}$ and $dom(A_2)$ is $\{0,1\}$. Initially, all S_i is *empty*.

(2) Let the auxiliary one is a unary relation $T(A)$, where $dom(A)$ is also $\{0,1\}$. Initially, T is also *empty*.

(3) For each clause C_i ($1 \leq i < m$), build a quintuple relation $R_i(A_1, A_2, A_3, A_4, A_5)$. We add 8 tuples into table R_i, whose values of A_1, A_2, A_3 refer to the 8 value assignments of the 3 variables, values of A_4 are the result of the corresponding assignments, i.e., seven '+'s or one '−', and the values of A_5 are the ids of these tuples from '1' to '8'.

fd Set Σ. Set fd: $A_2 \to A_1$ for each relation S_i, and set fd: $A_1A_2A_3 \to A_5$ for each relation R_i.

Query Q. We first prepare all the join conditions used in query as follows.
Existential variable, for each existential variable x_i ($i \in [1,n]$), build a join condition

$$con_\exists := S_1.A_1 = \cdots = S_n.A_1. \tag{3}$$

For the occurrences in clauses, we build a join condition with conjunctive form such that

$$con_i := q_1 \wedge q_2 \wedge \cdots \wedge q_k, \tag{4}$$

where each q is an equation $S_i.A_2 = R_j.A_p$, if x_i occurs in the p-th position of clauses C_j.

Universal variable, for each universal variable $x_i (i \in [n+1, n'])$, we build a join condition con_i also as a conjunctive form such that each q is also an equation $R_l.A_p = R_{l'}.A_{p'} = T.A$, if x_i occurs in the p-th and p'-th positions of clauses C_l and $C_{l'}$.

Let the query $Q = Q_1 \times Q_2 \times Q_3$ such that,

$$Q_1 := \pi_{R_1.A_4, \ldots, R_m.A_4}(\sigma_{con_1 \wedge \cdots \wedge con_{n+n'}} \mathsf{S} \times \mathsf{R} \times T), \tag{5}$$

$$Q_2 := \pi_{S_1.A_1, \ldots, S_n.A_1}(\sigma_{con_\exists}(\mathsf{S})) \tag{6}$$

$$Q_3 := \pi_A(T) \tag{7}$$

where R is $R_1 \times \cdots \times R_m$, and S is $S_1 \times \cdots \times S_n$.

View V. Initially, $V = \emptyset$ since T is empty.

Insertion ΔV. Let $\Delta V = t \times \{0,1\}$, where $t_1 = (\underbrace{+, \ldots, +}_{m}, \underbrace{B, \ldots, B}_{n})$.

One can verify the correctness of this reduction.

Surprisingly, we show that on combined complexity, for SJU query, fd-vsef-ip is in Σ_2^P-complete, not coNP-complete as expectation,

Theorem 6. fd-vsef-ip *for a* SJU *query is* Σ_2^P-complete *under group insertion with finite domain on combined complexity.*

Proof. i. We first show a Σ_2^P algorithm for SJU query as follows. Given a normal form of SJU query $Q := Q_1 \cup \cdots \cup Q_k$,
   (a) guess a grouping of ΔV into k groups $\Delta V_1, \cdots, \Delta V_k$ (empty permitted);
   (b) Compute $D \cup \Delta V_1^{-1} \cup \cdots \cup \Delta V_k^{-1}$ denoted as $D^\star$;
   (c) If $D^\star$ satisfies the fd set, then use the NP-oracle decide that 'each SJ query $Q_i(D^\star)$ has no side-effect on view $V \cup \Delta V$' and 'each tuple of ΔV included in $Q(D^\star)$'. If so, return yes.
   Its correctness is guaranteed by the upper bound proof in Theorem 5.

   ii. We prove the lower bound of fd-vsef-ip for SJU query is Σ_2^P-hard by a reduction from 3-DNF-SAT$_2$ problem. Similar to 3-CNF-SAT$_2$, its instance also includes a set of existential variables $X_1 = \{x_1, ..., x_n\}$ and a set of universal variables $X_2 = \{x_{n+1}, ..., x_{n'}\}$, and a 3-DNF boolean expression ϕ with m clauses $\{C_1, \ldots, C_m\}$, the task is to decide whether there is an assignment τ for X_1 such that ϕ is satisfied by all assignments for X_2. We show the reduction as follows. (An example of the reduction for a 3-DNF-SAT$_2$ instance $\phi = \exists x_1 x_2 \forall x_3 x_4 (x_1 \wedge x_2 \wedge x_3)(x_1 \wedge \overline{x}_2 \wedge x_3)(\overline{x}_1 \wedge \overline{x}_3 \wedge x_4)$ is shown in Fig. 2.)

Base Instance D. Let D has $m + 4$ relations $S_+, S_-, R_1, \cdots, R_m$ and G, T, where S_+ and S_- simulates existential variable x_i, R_j simulates clause C_j and universal variables in it, G and T both auxiliary tables. Concretely,
   (1) For the existential variables in X_1, let unary relation $S_+ = \{A\}$ and $S_- = \{A\}$. Initially, both are *empty*.

$$S_+ : \emptyset \to A \quad \boxed{\begin{array}{c} A \\ \\ \end{array}} \quad S_- : \emptyset \to A \quad \boxed{\begin{array}{c} A \\ \\ \end{array}} \quad G_3 \; \boxed{\begin{array}{cccc} A_1 & A_2 & \ldots & A_{4m} \\ \hline B & B & \ldots & B \end{array}} \quad T \; \boxed{\begin{array}{c} A \\ \\ \end{array}}$$

	$A_1 A_2 A_3 A_4$		$A_1 A_2 A_3 A_4$		$A_1 A_2 A_3 A_4$
	$\overline{x}_1\ \overline{x}_2\ 0\ -$		$\overline{x}_1\ \overline{x}_2\ 0\ -$		$\overline{x}_1\ 0\ 0\ -$
	$\overline{x}_1\ \overline{x}_2\ 1\ -$		$\overline{x}_1\ \overline{x}_2\ 1\ -$		$\overline{x}_1\ 1\ 0\ -$
R_1	$\overline{x}_1\ x_2\ 0\ -$	R_2	$\overline{x}_1\ x_2\ 0\ -$	R_3	$\overline{x}_1\ 1\ 1\ -$
	$x_1\ x_2\ 1\ -$		$\overline{x}_1\ x_2\ 1\ -$		$x_1\ 0\ 0\ -$
	$x_1\ \overline{x}_2\ 0\ -$		$x_1\ \overline{x}_2\ 0\ -$		$x_1\ 0\ 1\ -$
	$x_1\ \overline{x}_2\ 1\ -$		$x_1\ x_2\ 0\ -$		$x_1\ 1\ 0\ -$
	$x_1\ x_2\ 0\ -$		$x_1\ x_2\ 1\ -$		$x_1\ 1\ 1\ -$

Fig. 2. Example for the reduction of Theorem 6

(2) Let the auxiliary T is a unary relation $T\{A\}$. Initially, T is also *empty*. Let G is a $4m$-ary relation $G\{A_1, \cdots, A_{4m}\}$. Initially, has only one tuple $(B, \cdots, B)$.

(3) For each clause C_i, let quaternary relation $R_i\{A_1, A_2, A_3, A_4\}$. Initially, it is filled with 7 tuples into table R_i, whose values of A_1, A_2, A_3 refer to the 7 value assignments of the 3 variables making the clause C_i *false*, and the values of A_4 are all '$-$'. At last, if there exists existential variable, say x_i, in the p-th position clause C_i, then we substitute the value '1' of A_p with x_i and the value '0' with $\overline{x}_i$

fd Set Σ. Set fd: $\emptyset \rightarrow A$ for S_+ and S_-.

Query Q. We first prepare all the join conditions used in query as follows. *Existential variable*, for the occurrences in clauses, we build a join condition with conjunctive form such that

$$con_\exists := q_1 \wedge q_2 \wedge \cdots \wedge q_k, \tag{8}$$

where each q is an equation $S_+.A = R_j.A_p$, if x_i occurs in the p-th position of clauses C_j.

Universal variable, the same as the proof for SPJ query, for each universal variable $x_i (i \in [n+1, n'])$, we build a join condition con_i also as a conjunctive form such that each q is also an equation $R_l.A_p = R_{l'}.A_{p'} = T.A$, if x_i occurs in the p-th and p'-th positions of clauses C_l and $C_{l'}$.

Let the query $Q = Q_1 \cup Q_2$ such that,

$$Q_1 := \sigma_{con_\exists \wedge con_{n+1} \wedge \cdots \wedge con_{n+n'}}(S_+ \times R_1 \times \cdots \times R_m \times T), \tag{9}$$

$$Q_2 := (S_+ \cup S_-) \times G \times T \tag{10}$$

View V. Initially, $V = \emptyset$ since T is empty.

Insertion ΔV. Let $\Delta V = t \times \{x_1, \overline{x}_1, \cdots, x_n, \overline{x}_n\} \times \{0, 1\}$, where $t = (\underbrace{B, \ldots, B}_{4m})$.

Note that, $|\Delta V| = 4n$, i.e., it's *poly(n)*.

One can verify the correctness of this reduction.

3   Conclusion

We study the complexity of *fd*-vsef-ip problem a general case of insertion propagation, map out the complete picture of it. We can summary our result on it as the table of the complete picture of complexity classes. while due to the lack of space limitation, some proofs of data complexity results are omitted.

We are currently finding the tractable condition and approximation algorithms for intractable cases, also the study another objective of this problem – side effect on source database. We plan to study the cases at the present of other types of dependency constraints on database, such as independent dependencies in the future work.

References

1. Buneman, P., Khanna, S., Tan, W.C.: On propagation of deletions and annotations through views. In: Proceedings of the Twenty-First ACM SIGMOD-SIGACT-SIGART Symposium on Principles of Database Systems, PODS 2002, pp. 150–158. ACM, New York (2002)
2. Dayal, U., Bernstein, P.A.: On the correct translation of update operations on relational views. ACM Trans. Database Syst. **7**(3), 381–416 (1982)
3. Bancilhon, F., Spyratos, N.: Update semantics of relational views. ACM Trans. Database Syst. **6**(4), 557–575 (1981)
4. Cosmadakis, S.S., Papadimitriou, C.H.: Updates of relational views. J. ACM **31**(4), 742–760 (1984)
5. Bohannon, A., Pierce, B.C., Vaughan, J.A.: Relational lenses: a language for updatable views. In: Proceedings of the Twenty-Fifth ACM SIGMOD-SIGACT-SIGART Symposium on Principles of Database Systems, PODS 2006, pp. 338–347. ACM, New York (2006)
6. Keller, A.M.: Algorithms for translating view updates to database updates for views involving selections, projections, and joins. In: Proceedings of the Fourth ACM SIGACT-SIGMOD Symposium on Principles of Database Systems, PODS 1985, pp. 154–163. ACM, New York (1985)
7. Cong, G., Fan, W., Geerts, F., Li, J., Luo, J.: On the complexity of view update analysis and its application to annotation propagation. IEEE Trans. Knowl. Data Eng. **24**(3), 506–519 (2012)
8. Cong, G., Fan, W., Geerts, F.: Annotation propagation revisited for key preserving views. In: Proceedings of the 15th ACM International Conference on Information and Knowledge Management, CIKM 2006, pp. 632–641. ACM, New York (2006)
9. Kimelfeld, B., Vondrák, J., Williams, R.: Maximizing conjunctive views in deletion propagation. ACM Trans. Database Syst. **37**(4), 24:1–24:37 (2012)
10. Kimelfeld, B.: A dichotomy in the complexity of deletion propagation with functional dependencies. In: Proceedings of the 31st Symposium on Principles of Database Systems, PODS 2012, pp. 191–202. ACM, New York (2012)
11. Kimelfeld, B., Vondrák, J., Woodruff, D.P.: Multi-tuple deletion propagation: approximations and complexity. Proc. VLDB Endow. **6**(13), 1558–1569 (2013)
12. Miao, D., Liu, X., Li, J.: On the complexity of sampling query feedback restricted database repair of functional dependency violations. Theoret. Comput. Sci. **609**, 594–605 (2016)
13. Abiteboul, S., Hull, R., Vianu, V.: Foundations of Databases. Addison-Wesley, Reading (1995)
14. Vardi, M.Y.: The complexity of relational query languages (extended abstract). In: Proceedings of the Fourteenth Annual ACM Symposium on Theory of Computing, STOC 1982, pp. 137–146. ACM, New York (1982)
15. Lechtenbörger, J., Vossen, G.: On the computation of relational view complements. ACM Trans. Database Syst. **28**(2), 175–208 (2003)

Author Index

Printed in the United States
By Bookmasters